The Main Functional Groups *(continued)*

Structure	*Class of Compound*	*Specific Example*	*Name*	*Use*
C. Containing nitrogen				
$-NH_2$	primary amine	$CH_3CH_2NH_2$	ethylamine	intermediate for dyes, medicinals
$-NHR$	secondary amine	$(CH_3CH_2)_2NH$	diethylamine	pharmaceuticals
$-NR_2$	tertiary amine	$(CH_3)_3N$	trimethylamine	insect attractant
$-C\equiv N$	nitrile (cyanide)	$CH_2=CH-C\equiv N$	acrylonitrile	orlon manufacture
D. Containing oxygen and nitrogen				
$-\overset{+}{N}(=O)-O^-$	nitro compounds	CH_3NO_2	nitromethane	rocket fuel
$-\overset{O}{\overset{\Vert}{C}}-NH_2$	primary amide	$H\overset{O}{\overset{\Vert}{C}}NH_2$	formamide	softener for paper
E. Containing halogen				
$-X$	al[illegible] h[illegible]	CH_3Cl	methyl chloride	refrigerant, local anesthetic
$-\overset{O}{\overset{\Vert}{C}}-X$	acid (acyl) halide	$CH_3\overset{O}{\overset{\Vert}{C}}Cl$	acetyl chloride	acetylating agent
F. Containing sulfur				
$-SH$	thiol	CH_3CH_2SH	ethanethiol	odorant to detect gas leaks
$-S-$	thioether	$(CH_2=CHCH_2)_2S$	allyl sulfide	odor of garlic
$-\overset{O}{\overset{\Vert}{S}}(=O)-OH$	sulfonic acid	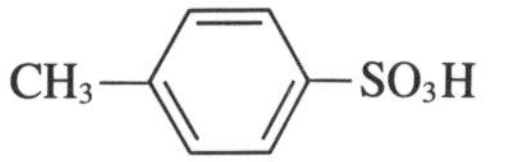 $CH_3-C_6H_4-SO_3H$	*para*-toluenesulfonic acid	strong organic acid

Organic Chemistry
A Short Course

Organic Chemistry
A Short Course

NINTH EDITION

Harold Hart
Michigan State University

David J. Hart
The Ohio State University

Leslie E. Craine
College of the Holy Cross

Houghton Mifflin Company ***Boston*** ***Toronto***
Geneva, Illinois Palo Alto Princeton, New Jersey

Senior Associate Editor: Richard Stratton
Editorial Assistant: Sunna Chung
Senior Project Editor: Cathy Labresh Brooks
Senior Production/Design Coordinator: Jill Haber
Senior Manufacturing Coordinator: Priscilla Bailey
Marketing Manager: Michael Ginley

Cover

Cover Design: Harold Burch, Harold Burch Design, New York City
Cover Illustration: Sandra Ray

Interior design

George McLean

Credits

Page 66, Wide World Photos; page 101, Union Carbide Agricultural Products, Inc.; page 107, Wide World Photos; page 137, American Cancer Society; page 203, Yoram Lehmann/Peter Arnold, Inc.; page 243, Stacy Pick/Stock Boston; page 246, U.S. Department of Agriculture; page 248, AAA Photo/Phototake; page 276, Courtesy of Bausch and Lomb, Incorporated; page 316, Tennessee Valley Authority; page 370, Paul Shambroom/Photo Researchers, Inc.; page 423, Courtesy of Prof. Dr. Dieter Sübach, Dr. H-M Müller, Laboratorium für Org. Chemie ETH-Zentrum, Zürich; page 424, Courtesy of Second Chance Body Armor, Inc.; page 488, Wide World Photos; page 506, David Leah/Science Photo Library; page 515, Courtesy of Linus Pauling; page 547, David Parker/Science Photo Library.

Printed in the U.S.A.

Library of Congress Catalog Number: 94-76506

ISBN: 0-395-70838-9

123456789-DH-98 97 96 95 94

Contents

2 *Alkanes and Cycloalkanes; Conformational and Geometric Isomerism* **42**

3 *Alkenes and Alkynes* ***70***

Preface

Purpose

Several decades have passed since the first edition of this text was published. Although the content continues to change, our purpose in writing it remains much the same—to present as clearly as possible a brief introduction to modern organic chemistry.

This book was written for students who, for the most part, will not major in chemistry, but whose main interest—agriculture, biology, human or veterinary medicine, pharmacy, nursing, medical technology, health sciences, engineering, home economics, forestry, or whatever—requires some knowledge of organic chemistry. To maintain the interest of these students, we have made a special effort to illustrate the practical applications of organic chemistry to everyday life and to biological processes. The success of this approach is demonstrated by the widespread use of this text by hundreds of thousands of students here in the U.S. and worldwide, via numerous translations.

The text is designed for a one-semester introductory course, but it is easily adapted to other formats. It is often used in either a one- or two-quarter course. In some countries (France and Japan, for example) it is an introductory text for chemistry majors, followed by a longer and more detailed full-year text. And in a number of high schools, it is used as the text for a second-year course, following the usual introductory general chemistry.

Organization

The organization is fairly classical, with some exceptions. After an introductory chapter on bonding, isomerism, and an overview of the subject (Chapter 1), the next three chapters treat in sequence saturated, unsaturated, and aromatic hydrocarbons. The concept of reaction mechanism is presented early, and examples are included in virtually all subsequent chapters. Stereoisomerism is also introduced early, briefly in Chapters 2 and 3, and then given separate attention in a full chapter (Chapter 5). Halogen compounds are used in Chapter 6 as a vehicle for introducing aliphatic substitution and elimination mechanisms and dynamic stereochemistry.

Chapters 7 through 10 take up oxygen functionality in order of increasing oxidation state of carbon (alcohols and phenols, ethers, aldehydes and ketones, acids and their derivatives). Brief mention of sulfur analogs is made in these chapters. Chapter 11 deals with amines.

Chapters 2 through 11 treat all of the main functional groups and constitute the heart of the course. Chapter 12 then takes up spectroscopy, with an emphasis on NMR and applications to structure determination. It handles the student's question—how do you know that those molecules really have the structures you say they have?

Next come two chapters on topics not always treated in introductory texts but especially important in practical organic chemistry—Chapter 13 on heterocyclic compounds and Chapter 14 on polymers. The book ends with four chapters on biologically important substances—lipids, carbohydrates, amino acids and proteins, and nucleic acids.

"A Word About" Essays

Although relevant applications of organic chemistry are stressed throughout the text, short sections under the general rubric *A Word About* emphasize applications to other branches of science and to human life. These sections, which have been a popular feature, appear at appropriate places within the text rather than as isolated essays. Numbered and printed in special type, they stand out from the text so that instructors can easily require these sections or not, as desired. There are thirty-five of these essays, three new in this edition.

Examples and Problems

Problem solving is essential to learning organic chemistry. **Examples** (worked-out problems) appear at appropriate places within each chapter to help students develop these skills. These examples and their solutions are clearly marked. Unsolved **problems** that provide immediate learning reinforcement are included within each chapter and are supplemented with an abundance of end-of-chapter problems. The combined number of examples and problems is 925, or an average of more than 51 per chapter.

New in the Ninth Edition

The entire text was carefully revised to sharpen the writing and clarify difficult sections. In addition to many small changes, users of the previous edition will notice the following more substantial changes: (1) The footnote on arrow formalism in Chapter 1 has been upgraded to a section. (2) In response to reviewers' interest, two sections introducing the thermodynamics and kinetics of organic reactions have been added to Chapter 3. Reaction energy diagrams are introduced here and are used again in Chapter 6. (3) In Chapter 5 and throughout subsequent chapters, the terms "chiral center" and "asymmetric carbon atom" have been replaced with the currently accepted "stereogenic center" and "stereogenic carbon atom." (4) A *Reaction Summary* section, located before the end-of-chapter problems, is included in each chapter where new reactions are introduced. After each reaction type, a reference to the appropriate section of the text is provided. This summary collects new reactions in one easy-to-find location and will help students organize their study of new materials effectively.

Three new *A Word About* sections have been added in this edition, and three former sections of this type have been deleted. We hope that students and teachers alike will enjoy the following timely topics: C_{60}, an Aromatic Sphere: The Fullerenes; Degradable Polymers; and Bacterial Cell Walls, Enzyme Inhibitors, and Antibiotics. Please write to us with your comments.

We are very conscious of the need to keep the book to a manageable size for the one-semester course. Wherever possible, some old material has been deleted to make room for the new material that has been added. Users will find that this edition is nearly identical in length to the previous one.

Ancillaries

Two accompanying books are available to help the student in this course learn organic chemistry.

The **Study Guide and Solutions Book** contains answers to all text problems, a guide on how to reason out the answers, a summary of each chapter, a summary of the new reactions in each chapter, a list of learning objectives for each chapter, a summary of important reaction mechanisms, and sample test questions.

The **Laboratory Manual** contains experiments that have been time-tested with thousands of students. A substantial number of the preparative experiments contain procedures on both the **macro-** and the **microscale**, thus adding considerable flexibility for the instructor and the opportunity for both types of laboratory experience for the student. We have been careful to avoid hazardous chemicals on the OSHA list and to minimize contact with solvents, and so forth. The student and instructor are clearly warned whenever caution or special care is required, and thorough waste disposal instructions are consistently specified. The manual has tear-out, perforated report sheets convenient for student and instructor. It is also a convenient size for the nonmajor lab. Most experiments can be completed in the relatively short two- or three-hour lab period for nonmajors. The manual contains appendices giving atomic weights, other properties of common reagents, instructions for the teacher on how to make or obtain special reagents, and a list of chemicals and equipment required for each experiment that will simplify ordering and stocking the labs. Experiments are a good mix of techniques, preparations, tests, and applications.

An **Instructor's Resource Manual** and a set of transparencies and black line transparency masters are also available.

Acknowledgments

For their frankness and diligence in reviewing the proposed revisions and later, the completed manuscript, we would like to thank the following professors:

Vasu Dev, California State Polytechnic University; Ihsan Erden, San Francisco State University; Marc M. Greenberg, Colorado State College; Lonnie Haynes, Tennessee State University; Dwight Klaassen, University of Wisconsin—Platteville; Brien Love, Auburn University; Deb Mlsna, Clemson University; Roger K. Murray, University of Delaware; Daniel O'Brien, Texas A & M University; Philip D. Roskos, Lakeland Community College; Kathleen M. Trahanovsky, Iowa State University; Mel Usselman, University of Western Ontario; George H. Wahl, North Carolina State University.

We have incorporated many of their recommendations, and the book is much improved as a consequence.

One pleasure of authorship is receiving letters from students who have benefited from the book, and from their teachers. We thank here all who have written to us, from all parts of the world, since the last edition; we have incorporated many of their suggestions in this revision. We will be happy to hear from users and nonusers, faculty and students, with suggestions for further improvements.

HAROLD HART
EMERITUS PROFESSOR OF CHEMISTRY
MICHIGAN STATE UNIVERSITY
EAST LANSING, MI 48824

DAVID J. HART
DEPARTMENT OF CHEMISTRY
THE OHIO STATE UNIVERSITY
120 WEST 18TH AVENUE
COLUMBUS, OH 43210

LESLIE E. CRAINE
DEPARTMENT OF CHEMISTRY
COLLEGE OF THE HOLY CROSS
WORCESTER, MA 01610

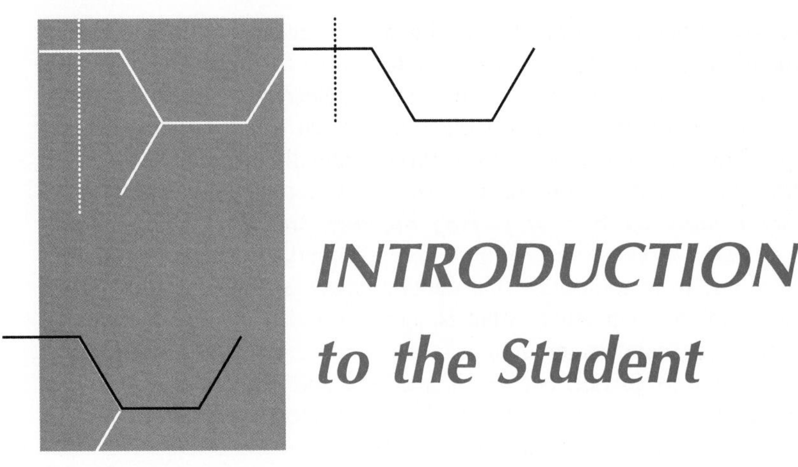

INTRODUCTION to the Student

In this introduction we will tell you briefly about organic chemistry and why it is important in a technological society. We will also explain how this course is organized and give you a few hints that may help you to study more effectively.

What is Organic Chemistry About?

The term *organic* suggests that this branch of chemistry has something to do with *organisms,* or living things. Originally, organic chemistry did deal only with substances obtained from living matter. Years ago, chemists spent much of their time extracting, purifying, and analyzing substances from animals and plants. They were motivated by a natural curiosity about living matter and also by the desire to obtain from nature ingredients for medicines, dyes, and other useful products.

It gradually became clear that most compounds in plants and animals differ in several respects from those that occur in nonliving matter, such as minerals. In particular, most compounds in living matter are made up of the same few elements: **carbon, hydrogen, oxygen, nitrogen,** and sometimes sulfur, phosphorus, and a few others. Carbon is virtually always present. This fact led to our present definition: ***Organic chemistry is the chemistry of carbon compounds.*** This definition broadens the scope of the subject to include not only compounds from nature but also synthetic compounds—compounds invented by organic chemists and prepared in their laboratories.

Synthetic Organic Compounds

Scientists used to think that compounds that occurred in living matter were different from other substances and that they contained some sort of intangible **vital force** that imbued them with life. This idea discouraged chemists from trying to make organic compounds in the laboratory. But in 1828 the German chemist Friedrich Wöhler, then 28 years old, accidentally prepared **urea,** a well-known constituent of urine, by heating the inorganic (or mineral) substance ammonium cyanate. He was quite excited about this result, and in a letter to his former teacher, the Swedish chemist J. J. Berzelius, he wrote, "I can make urea without the necessity of a kidney, or even of an animal, whether man or dog." This experiment and others like it gradually discredited the vital-force theory and opened the way for modern synthetic organic chemistry.

Synthesis usually consists of piecing together small, relatively simple molecules to make larger, more complex ones. To make a molecule that contains many atoms from molecules that contain fewer atoms, one must know how to link atoms to each other—that is, how to make and break chemical bonds. Wöhler's preparation of urea was accidental, but synthesis is much more effective if it is carried out in a controlled and rational way, so that, when all the atoms are assembled, they will be connected to one another in the correct manner to give the desired product.

Chemical bonds are made or broken during chemical reactions. In this course, you will learn about quite a few reactions that can be used to make new bonds and that are therefore useful in synthesis.

Why Synthesis?

At present, the number of organic compounds that have been synthesized in research laboratories is far greater than the number isolated from nature. Why is it important to know how to synthesize molecules? There are several reasons. For one, it might be important to synthesize a natural product in the laboratory in order to make the substance more widely available at lower cost than it would be if the compound had to be extracted from its natural source. Some examples of compounds first isolated from nature but now produced synthetically for commercial use are vitamins, amino acids, the dye indigo, the moth-repellent camphor, and the antibiotic penicillin. Although the term *synthetic* is sometimes frowned on as implying something artificial or unnatural, these synthetic natural products are in fact identical to the same compounds extracted from nature.

Another reason for synthesis is to create new substances that may have new and useful properties. Synthetic fibers such as nylon and Orlon, for example, have properties that make them superior for some uses to natural fibers such as silk, cotton, and hemp. Most drugs used in medicine are synthetic (including aspirin, ether, Novocain, and barbiturates). The list of synthetic products that we take for granted is long indeed—plastics, detergents, insecticides, and oral contraceptives are just a few. All of these are compounds of carbon; all are organic compounds.

Finally, organic chemists sometimes synthesize new compounds to test chemical theories or sometimes just for the fun of it. Certain geometric structures, for example, are aesthetically pleasing, and it can be a challenge to make a molecule in which the carbon atoms are arranged in some regular way. One example is the hydrocarbon cubane, C_8H_8. First synthesized in 1964, its molecules have eight carbons at the corners of a cube, each carbon with one hydrogen and three other carbons connected to it. Cubane is more than just aesthetically pleasing. The bond angles in cubane are distorted from normal because of its geometry. Studying the chemistry of cubane therefore gives chemists information about how the distortion of carbon–carbon and carbon–hydrogen bonds affects their chemical behavior.

cubane, C_8H_8
mp 130–131°C
P. E. Eaton (U. of Chicago), 1964

Here are a few other examples of organic chemists' fascination with geometry and the challenge of synthesis.

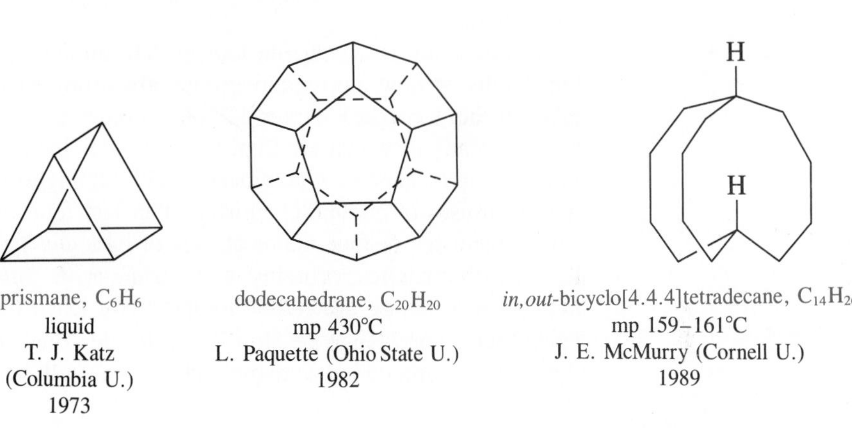

prismane, C_6H_6
liquid
T. J. Katz
(Columbia U.)
1973

dodecahedrane, $C_{20}H_{20}$
mp 430°C
L. Paquette (Ohio State U.)
1982

in,out-bicyclo[4.4.4]tetradecane, $C_{14}H_{26}$
mp 159–161°C
J. E. McMurry (Cornell U.)
1989

Organic Chemistry in Everyday Life

Organic chemistry touches our daily lives. We are made of and surrounded by organic compounds. Almost all the reactions in living matter involve organic compounds, and it is impossible to understand life, at least from the physical point of view, without knowing some organic chemistry. The major constituents of living matter—proteins, carbohydrates, lipids (fats), nucleic acids (DNA, RNA), cell membranes, enzymes, hormones—are organic, and later in the book, we will describe their chemical structures. These structures are quite complex. To understand them, we will first have to discuss simpler molecules.

Other organic substances include the gasoline, oil, and tires for our cars, the clothing we wear, the wood of our furniture and the paper of our books, the medicines we take, plastic containers, camera film, perfume, carpeting, and fabrics. Name it, and the chances are good that it is organic. Daily, in the paper or on television, we encounter references to polyethylene, epoxys, Styrofoam, nicotine, polyunsaturated fats, and cholesterol. All of these terms refer to organic substances; we will study them and many more in this book.

In short, organic chemistry is more than just a branch of science for the professional chemist or for the student preparing to become a physician, dentist, veterinarian, pharmacist, nurse, or agriculturist. It is part of our technological culture.

Organization

Organic chemistry is a vast subject. Some molecules and reactions are simple; others are quite complex. We will proceed from the simple to the complex and begin with a chapter on bonding, with special emphasis on bonds to carbon. Next, we have three chapters on organic compounds that contain only two elements, carbon and hydrogen (called hydrocarbons). The second of these chapters (Chapter 3) contains an introduction to organic reaction mechanisms and a discussion of reaction equilibria and rates. These are followed by a chapter that deals with the three dimensionality of organic compounds. Next we add other elements to the carbon and hydrogen framework, halogens in Chapter 6, oxygen and sulfur in Chapters 7 through 10, and nitrogen in Chapter 11. At that point, we will have completed an introduction to all the main classes of organic compounds.

Spectroscopy is a valuable tool in determining organic structures—that is, the details of how atoms and groups are arranged in organic molecules. We take up this topic in Chapter 12. Next comes a chapter on heterocyclic compounds, many of which are important in medicine and in natural products. It is followed by a chapter on polymers, which highlights one of the most important industrial uses of organic chemistry. The last four chapters deal with the organic chemistry of four major classes of biologically important molecules: the lipids, carbohydrates, proteins, and nucleic acids. Since the structures of these molecules of nature are rather complex, we leave them for last. But with the background knowledge of simpler molecules that you will have acquired by then, these compounds and their chemistry will be clearer and more understandable.

To help you organize and review new material, we have placed a *Reaction Summary* at the end of each chapter in which new reactions are introduced.

A Word About

In each chapter after the first, you will find special sections under the general heading "A Word About." These are short, self-contained articles that expand on the main subject of the chapter. They may deal with intellectual curiosities (the first one, on impossible organic structures); industrial applications (petroleum, gasoline, and octane number in Chapter 3 or epoxy resins in Chapter 8); organic chemistry in biology or medicine (polycyclic aromatic hydrocarbons and cancer in Chapter 4 or morphine and other painkillers in Chapter 11); or just fun topics (sweetness and sweeteners in Chapter 16). They provide a convenient break at various points in each chapter, and we hope that you will enjoy them.

The Importance of Problem Solving

One key to success in studying organic chemistry is problem solving.

Each chapter in this book contains a large number of facts that must be digested. Also, the subject matter builds continuously, so that to understand each new topic, it is essential to have the preceding information clear in your mind and available for recall. To learn all this material, careful study of the text is necessary, but it is *not sufficient*. Practical knowledge of how to use the facts is required, and such skill can be obtained only through the solving of problems.

This book contains several types of problems. Some, called ***Examples,*** contain a ***Solution,*** so that you can see how to work such problems. Throughout a chapter, examples are usually followed by similar ***Problems,*** designed to reinforce your learning immediately by allowing you to be sure that you understand the new material just presented. At the end of each chapter, ***Additional Problems*** enable you to practice your problem-solving skills. Problems that simply test your knowledge begin the list and then gradually become more challenging.

Try to work as many problems as you can. If you have trouble, seek help from your instructor or from the study guide that accompanies this text. The study guide provides answers to all the problems and explains how to solve them. Problem solving is time consuming, but it will pay off in an understanding of the subject and in a good grade.

And now let us begin.

1 Bonding and Isomerism

The ways in which atoms form bonds with one another to make molecules are important to understand because they help to explain the structures of molecules and why particular molecules react as they do. Perhaps you have already studied some of these ideas in a beginning chemistry course. Browse through each section of this chapter to see whether it is familiar, and try to work the problems. If you can work the problems, you can safely skip that section. But if you have difficulty with any of the problems within or at the end of this chapter, study the entire chapter carefully because we will use the ideas developed here throughout the rest of the book.

1.1 *How Electrons Are Arranged in Atoms*

Atoms contain a small, dense **nucleus** surrounded by **electrons.** The nucleus is positively charged and contains most of the mass of the atom. The nucleus consists of **protons,** which are positively charged, and **neutrons,** which are neutral. (The only exception is hydrogen, whose nucleus consists of only a single proton.) In a neutral atom the positive charge of the nucleus is exactly balanced by the negative charge of the electrons that surround it. The **atomic number** of an element is equal to the number of protons in the nucleus (and to the number of electrons around the nucleus in a neutral atom). The **atomic weight** is approximately equal to the sum of the number of protons and the number of neutrons in the nucleus; the electrons are not counted because they are very light by comparison. The periodic table on the inside back cover of this book shows all the elements with their atomic numbers and weights.

We are concerned here mainly with the atom's electrons because their number and arrangement provide the key to how a particular atom reacts with other atoms to form molecules. Also, we will deal only with electron arrangements in the lighter elements because these elements are the most important in organic molecules.

Electrons are concentrated in certain regions of space around the nucleus called **orbitals.** *Each orbital can contain a maximum of two electrons.* The orbitals, which differ in shape, are designated by the letters *s*, *p*, and *d*. In addition, orbitals are grouped in **shells** designated by the numbers 1, 2, 3, and so on. Each shell contains different types and numbers of orbitals, corresponding to the shell number. For example, shell 1 contains only one type of orbital, designated the 1*s* orbital. Shell 2 contains two types of orbitals, 2*s* and 2*p*, and shell 3 contains three types, 3*s*, 3*p*, and 3*d*. Within a particular shell, the number of *s*, *p*, and *d* orbitals is 1, 3, and 5, respectively (Table 1.1).

These rules permit us to count how many electrons each shell will contain when it is filled (last column in Table 1.1). Table 1.2 shows how the electrons of the first 18 elements are arranged.

The first shell is filled for helium (He) and all elements beyond, and the second shell is filled for neon (Ne) and all elements beyond. Filled shells play almost no role in chemical bonding. Rather, the outer shells, or **valence shells,** are mainly involved in chemical bonding, and we will focus our attention on them.

Table 1.3 shows the **valence electrons,** the electrons in the outermost shell, for the first 18 elements. The element's symbol stands for the **kernel** of the element (the nucleus plus the filled electron shells), and the dots represent the valence electrons. The elements are arranged in groups according to the periodic table, and (except for helium) these group numbers correspond to the number of valence electrons.

Armed with this information about atomic structure, we are now ready to tackle the problem of how elements combine to form chemical bonds.

TABLE 1.1 Numbers of orbitals and electrons in the first three shells

	Number of orbitals of each type			
Shell number	*s*	*p*	*d*	*Total number of electrons when shell is filled*
1	1	0	0	2
2	1	3	0	8
3	1	3	5	18

TABLE 1.2 Electron arrangements of the first 18 elements

Atomic number	Element	Number of electrons in each orbital				
		1s	*2s*	*2p*	*3s*	*3p*
1	H	1				
2	He	2				
3	Li	2	1			
4	Be	2	2			
5	B	2	2	1		
6	C	2	2	2		
7	N	2	2	3		
8	O	2	2	4		
9	F	2	2	5		
10	Ne	2	2	6		
11	Na	2	2	6	1	
12	Mg	2	2	6	2	
13	Al	2	2	6	2	1
14	Si	2	2	6	2	2
15	P	2	2	6	2	3
16	S	2	2	6	2	4
17	Cl	2	2	6	2	5
18	Ar	2	2	6	2	6

TABLE 1.3 Valence electrons of the first 18 elements

Group	I	II	III	IV	V	VI	VII	VIII
	H·							He:
	Li·	Ḃe·	·Ḃ·	·Ċ· (with one dot below)	·N̈: (with one dot below)	·Ö:	:F̈: (with two dots below)	:N̈e: (with two dots below)
	Na·	Ṁg·	·Ȧl·	·Ṡi· (with one dot below)	·Ṗ: (with one dot below)	·S̈:	:C̈l: (with two dots below)	:Är: (with two dots below)

1.2 *Ionic and Covalent Bonding*

An early, but still useful, theory of chemical bonding was proposed in 1916 by Gilbert Newton Lewis, then a professor at the University of California in Berkeley. Lewis noticed that the **inert gas** helium had only two electrons surrounding its nucleus and that the next inert gas neon had ten such electrons (2 + 8; see Table 1.2). He concluded that atoms of these gases must have very stable electron arrangements because these elements do not combine with other atoms. He further suggested that other atoms might react in such a way as to achieve these stable arrangements. This stability could be achieved in one of two ways: by complete transfer of electrons from one atom to another or by sharing of electrons between atoms.

1.2a Ionic Compounds ***Ionic bonds are formed by the transfer of one or more valence electrons from one atom to another.*** Because electrons are negatively charged, the atom that gives up the electron(s) becomes positively charged, a **cation.** The atom that receives the electron(s) becomes negatively charged, an **anion.** The reaction between sodium and chlorine atoms to form sodium chloride (ordinary table salt) is a typical electron-transfer reaction.*

$$\underset{\text{sodium atom}}{Na\cdot} + \underset{\text{chlorine atom}}{\cdot\ddot{\underset{..}{Cl}}:} \longrightarrow \underset{\text{sodium cation}}{Na^{+}} + \underset{\text{chloride anion}}{:\ddot{\underset{..}{Cl}}:^{-}} \quad (1.1)$$

The sodium atom has only one valence electron (it is in the third shell; see Table 1.2). By giving up that electron it achieves the electron arrangement of neon. At the same time, it becomes positively charged, a sodium cation. The chlorine atom has seven valence electrons. By accepting an additional electron, it achieves the electron arrangement of argon and becomes negatively charged, a chloride anion. *Atoms,* such as sodium, *that tend to give up electrons are said to be* **electropositive.** *Atoms,* such as chlorine, *that tend to accept electrons are said to be* **electronegative.**

EXAMPLE 1.1 Write an equation for the reaction of magnesium (Mg) with fluorine atoms (F).

Solution

$$\dot{Mg}\cdot + \cdot\ddot{\underset{..}{F}}: + \cdot\ddot{\underset{..}{F}}: \longrightarrow Mg^{2+} + 2:\ddot{\underset{..}{F}}:^{-}$$

Magnesium has two valence electrons. Since each fluorine atom can accept only one electron (from the magnesium) to complete its valence shell, two fluorine atoms are needed to react with one magnesium atom.

The product of eq. 1.1 is sodium chloride, an ionic compound made up of equal numbers of sodium and chloride ions. In general, ionic compounds form when strongly electropositive atoms and strongly electronegative atoms interact. The ions in a crystal of an ionic substance are held together by the attractive force between their opposite charges, as shown in Figure 1.1 for a sodium chloride crystal.

In a sense, the ionic bond is not really a bond at all. Being oppositely charged, the ions attract one another like the opposite poles of a magnet. In the crystal, the ions are packed in a definite arrangement, but we cannot say that

*The curved arrow in eq. 1.1 shows the movement of one electron from the valence shell of the sodium atom to the valence shell of the chlorine atom. The use of curved arrows to show the movement of electrons is explained in greater detail in Sec. 1.13.

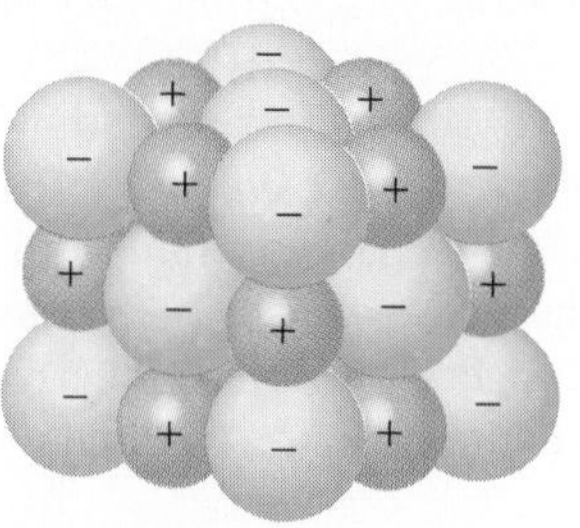

FIGURE 1.1 Sodium chloride, Na^+Cl^-, is an ionic crystal. The colored spheres represent sodium ions, Na^+, and the gray spheres are chloride ions, Cl^-. Each ion is surrounded by six oppositely charged ions, except for those ions that are at the surface of the crystal.

any particular ion is bonded or connected to any other particular ion. And, of course, when the substance is dissolved, the ions separate and are able to move about in solution relatively freely.

EXAMPLE 1.2 What charge will a beryllium ion carry?

Solution As seen in Table 1.3, beryllium (Be) has two valence electrons. To achieve the filled-shell electron arrangement of helium, it must lose both its valence electrons. Thus, the beryllium cation will carry two positive charges and is represented by Be^{2+}.

PROBLEM 1.1 Using Table 1.3, tell what charge the ion will carry when each of the following elements reacts to form an ionic compound: Al, Li, S, H.

In general, within a given horizontal row in the periodic table, the more electropositive elements are those farthest to the left, and the more electronegative elements are those farthest to the right. Within a given vertical column, the more electropositive elements are those toward the bottom, and the more electronegative elements are those toward the top.

EXAMPLE 1.3 Which atom in the second row of the periodic table is more electropositive, lithium or beryllium?

Solution The lithium nucleus has less positive charge (+3) to attract electrons than the beryllium nucleus (+4). It takes less energy, therefore, to remove an electron from lithium than it does to remove one from beryllium. Since lithium loses an electron more easily than beryllium, lithium is the more electropositive atom.

PROBLEM 1.2 Using Table 1.3, tell which is the more electropositive element: sodium or aluminum, boron or carbon, boron or aluminum.

PROBLEM 1.3 Using Table 1.3, tell which is the more electronegative element: oxygen or fluorine, oxygen or nitrogen, fluorine or chlorine.

PROBLEM 1.4 Judging from its position in Table 1.3, do you expect carbon to be electropositive or electronegative?

1.2b The Covalent Bond Elements that are neither strongly electronegative nor strongly electropositive, or that have similar electronegativities, tend to form bonds by sharing electron pairs instead of completely transferring electrons. ***A covalent bond** involves the mutual sharing of one or more electron pairs between atoms.* Two (or more) atoms joined together by covalent bonds constitute a **molecule.** When the two atoms are identical or have equal electronegativities, the electron pairs are shared equally. The hydrogen molecule is an example.

$$\underset{\text{hydrogen atoms}}{\text{H}\cdot + \text{H}\cdot} \longrightarrow \underset{\text{hydrogen molecule}}{\text{H}{:}\text{H}} + \text{heat} \qquad (1.2)$$

Each hydrogen atom can be considered to have filled its first electron shell by the sharing process. That is, each atom is considered to "own" all the electrons it shares with the other atom, as shown by the loops in these structures.

(H:) H   H(:H)

EXAMPLE 1.4 Write an equation similar to eq. 1.2 for the formation of a chlorine molecule from two chlorine atoms.

Solution

$$:\ddot{\underset{\cdot\cdot}{\text{Cl}}}\cdot + \cdot\ddot{\underset{\cdot\cdot}{\text{Cl}}}: \longrightarrow :\ddot{\underset{\cdot\cdot}{\text{Cl}}}:\ddot{\underset{\cdot\cdot}{\text{Cl}}}: + \text{heat}$$

One electron pair is shared by the two chlorine atoms. In that way, each chlorine completes its valence shell with eight electrons (three unshared pairs and one shared pair).

When two hydrogen atoms combine to form a molecule, heat is liberated. Conversely, this same amount of heat (energy) has to be supplied to a hydrogen molecule to break it apart into atoms. To break apart 1 mol of hydrogen molecules (1 gram molecular weight, in this case 2 g) into atoms requires 104 kcal (or 435 kJ*) of heat, quite a lot of energy. This energy is called the **bond energy,** or BE, and is different for bonds between different atoms. (See Appendix A.)

The H—H bond is a very strong bond. The main reason for this is that the shared electron pair is attracted to *both* hydrogen nuclei, whereas in a hydrogen atom, the valence electron is associated with only one nucleus. But other forces

*Although most organic chemists use the kilocalorie as the unit of heat, the currently used international unit is the kilojoule; 1 kcal = 4.184 kJ.

in the hydrogen molecule tend to counterbalance the attraction between the electron pair and the nuclei. These forces are the repulsion between the two like-charged nuclei and the repulsion between the two like-charged electrons. A balance is struck between the attractive and the repulsive forces. The hydrogen atoms neither fly apart nor fuse together. Instead, they remain connected, or bonded, and vibrate about some equilibrium distance, which we call the **bond length.** For a hydrogen molecule, the bond length (that is, the average distance between the two hydrogen nuclei) is 0.74 Å.* The length of a covalent bond depends on the atoms that are bonded and the number of electron pairs shared between the atoms. Bond lengths for some typical covalent bonds are given in Appendix B.

1.3 Carbon and the Covalent Bond

Now let us look at carbon and its bonding. We represent atomic carbon by the symbol ·Ċ· where the letter C stands for the kernel (the nucleus plus the two 1*s* electrons) and the dots represent the valence electrons.

With four valence electrons, the valence shell of carbon is half filled (or half empty). Carbon atoms have neither a strong tendency to lose all their electrons (and become C^{4+}) nor a strong tendency to gain four electrons (and become C^{4-}). Being in the middle of the periodic table, carbon is neither strongly electropositive nor strongly electronegative. Instead, it usually forms covalent bonds with other atoms by sharing electrons. For example, carbon combines with four hydrogen atoms (each of which supplies one valence electron) by sharing four electron pairs.** The substance formed is known as **methane.** Carbon can also share electron pairs with four chlorine atoms, forming **tetrachloromethane.**†

```
  H                  H
  •x                 |
H x C x H    or   H—C—H
  x•                 |
  H                  H
```

methane

```
    ..                      Cl
   :Cl:                     |
 ..  •x  ..           Cl—C—Cl
:Clx C xCl:    or           |
 ..  x•  ..                 Cl
   :Cl:
    ..
```

tetrachloromethane
(carbon tetrachloride)

---

* 1 Å, or angstrom unit, is 10^{-8} cm, so the H—H bond length is 0.74×10^{-8} cm.

** To designate electrons from different atoms, the symbols • and x are often used. But the electrons are, of course, identical.

† Tetrachloromethane is the systematic name, but carbon tetrachloride is the common name. We discuss how to name organic compounds later.

By sharing electron pairs, the atoms complete their valence shells. In both examples, carbon has eight valence electrons around it. In methane, each hydrogen atom completes its valence shell with two electrons, and in tetrachloromethane each chlorine atom fills its valence shell with eight electrons. In this way, all valence shells are filled and the compounds are quite stable.

The shared electron pair is called a **covalent bond** because it bonds or links the atoms together (by its attraction to both nuclei). The single bond is usually represented by a dash, or single line, as shown in the formulas above for methane and tetrachloromethane.

EXAMPLE 1.5 Draw the formula for chloromethane (also called methyl chloride), CH_3Cl.

Solution

```
                          H
   H  ..                  |
 H:C:Cl:      or     H—C—Cl
   H  ..                  |
                          H
```

PROBLEM 1.5 Draw the formulas for dichloromethane (also called methylene chloride), CH_2Cl_2, and trichloromethane (chloroform), $CHCl_3$.

1.4 Carbon–Carbon Single Bonds

The unique property of carbon atoms—that is, the property that makes it possible for millions of organic compounds to exist—is their ability to share electrons not only with different elements but also with other carbon atoms. For example, two carbon atoms may be bonded to one another, and each of these carbon atoms may be linked to other atoms. In **ethane** and **hexachloroethane,** each carbon is connected to the other carbon *and* to three hydrogen atoms or three chlorine atoms. Although they have two carbon atoms instead of one, these compounds have chemical properties similar to those of methane and tetrachloromethane, respectively.

```
                 H   H                                     Cl  Cl
  H H            |   |              :Cl: :Cl:              |   |
H:C:C:H   or   H—C—C—H           :Cl:C : C:Cl:   or   Cl—C—C—Cl
  H H            |   |              :Cl: :Cl:              |   |
                 H   H                                     Cl  Cl
```

ethane hexachloroethane

The carbon–carbon bond in ethane, like the hydrogen–hydrogen bond in a hydrogen molecule, is a purely covalent bond, with the electrons shared *equally* between the two identical carbon atoms. As with the hydrogen molecule, heat is required to break the carbon–carbon bond of ethane to give two CH_3 fragments (called **methyl radicals**). ***A radical is a molecular fragment with an odd number of unshared electrons.***

$$\underset{\text{ethane}}{\begin{array}{c} \text{H}\;\;\text{H} \\ |\;\;\;| \\ \text{H—C:C—H} \\ |\;\;\;| \\ \text{H}\;\;\text{H} \end{array}} \xrightarrow{\text{heat}} \underset{\text{two methyl radicals}}{\begin{array}{c} \text{H}\qquad\text{H} \\ |\qquad\;\;| \\ \text{H—C}\cdot + \cdot\text{C—H} \\ |\qquad\;\;| \\ \text{H}\qquad\text{H} \end{array}} \qquad (1.3)$$

However, less heat is required to break the carbon–carbon bond in ethane than is required to break the hydrogen–hydrogen bond in a hydrogen molecule. The actual amount is 88 kcal (or 368 kJ) per mole of ethane. The carbon–carbon bond in ethane is longer (1.54 Å) than the hydrogen–hydrogen bond (0.74 Å) and also somewhat weaker. Breaking carbon–carbon bonds by heat, as represented in eq. 1.3, is the first step in the *cracking* of petroleum, an important process in the manufacture of gasoline (see "A Word About Petroleum" on page 107).

EXAMPLE 1.6 What do you expect the bond length of a C—H bond (as in methane or ethane) to be?

Solution It should measure somewhere between the H—H bond length in a hydrogen molecule (0.74 Å) and the C—C bond length in ethane (1.54 Å). The actual value is about 1.09 Å, close to the average of the H—H and C—C bond lengths.

PROBLEM 1.6 The Cl—Cl bond length is 1.98 Å. Which bond will be longer, the C—C bond in ethane or the C—Cl bond in chloromethane?

There is almost no limit to the number of carbon atoms that can be linked together, and some molecules contain as many as 100 or more carbon–carbon bonds in a row. This ability of an element to form chains as a result of bonding between the same atoms is called **catenation.**

PROBLEM 1.7 Using the structure of ethane as a guide, draw the structure for propane, C_3H_8.

1.5 Polar Covalent Bonds

As we have seen, covalent bonds can be formed not only between identical atoms (H—H, C—C) but also between different atoms (C—H, C—Cl), provided that the atoms do not differ too greatly in electronegativity. However, *if the atoms are different from one another, the electron pair may not be shared equally between them.* Such a bond is sometimes called a **polar covalent bond** because the atoms that are linked carry a partial negative and a partial positive charge.

The hydrogen chloride molecule provides an example of a polar covalent bond. Chlorine atoms are more electronegative than hydrogen atoms, but even so, the bond that they form is covalent rather than ionic. However, the shared electron pair is attracted more toward the chlorine, which therefore is slightly negative with respect to the hydrogen. This bond polarization is indicated by an arrow whose head is negative and whose tail is marked with a plus sign. Alternatively, a partial charge, written as $\delta+$ or $\delta-$ (read as "delta plus" or "delta minus"), may be shown:

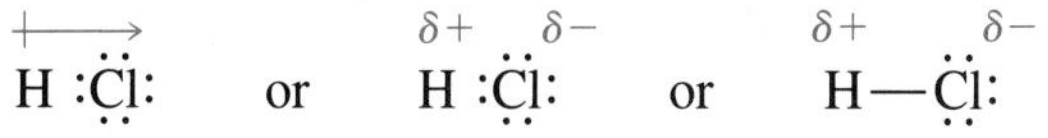

$$\overset{+\!\!\longrightarrow}{\mathrm{H}\,:\ddot{\underset{..}{\mathrm{Cl}}}:} \quad \text{or} \quad \overset{\delta+}{\mathrm{H}}\,:\overset{\delta-}{\ddot{\underset{..}{\mathrm{Cl}}}}: \quad \text{or} \quad \overset{\delta+}{\mathrm{H}}-\overset{\delta-}{\ddot{\underset{..}{\mathrm{Cl}}}}:$$

The bonding electron pair, which is shared *unequally,* is displaced toward the chlorine.

You can usually rely on the periodic table to determine which end of a polar covalent bond is more negative and which end is more positive. As we proceed from left to right across the table within a given period, the elements become *more* electronegative, owing to increasing atomic number or charge on the nucleus. The increasing nuclear charge attracts valence electrons more strongly. As we proceed from the top to the bottom of the table within a given group (down a column), the elements become *less* electronegative because the valence electrons are shielded from the nucleus by an increasing number of inner-shell electrons. From these generalizations, we can safely predict that the atom on the right in each of the following bonds will be negative with respect to the atom on the left:

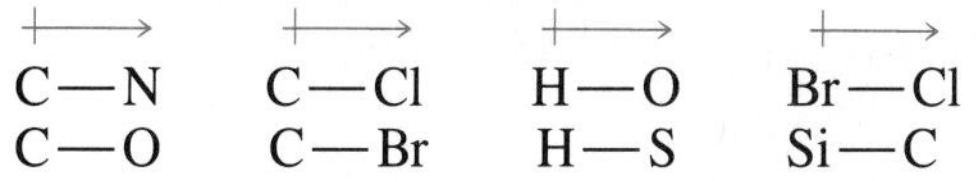

$$\begin{array}{cccc} +\!\!\longrightarrow & +\!\!\longrightarrow & +\!\!\longrightarrow & +\!\!\longrightarrow \\ \mathrm{C-N} & \mathrm{C-Cl} & \mathrm{H-O} & \mathrm{Br-Cl} \\ \mathrm{C-O} & \mathrm{C-Br} & \mathrm{H-S} & \mathrm{Si-C} \end{array}$$

The carbon–hydrogen bond, which is so common in organic compounds, requires special mention. Carbon and hydrogen have nearly identical electronegativities, so the C—H bond is almost purely covalent. The electronegativities of some common elements are listed in Table 1.4.

EXAMPLE 1.7 Indicate any bond polarization in the structure of tetrachloromethane.

Solution

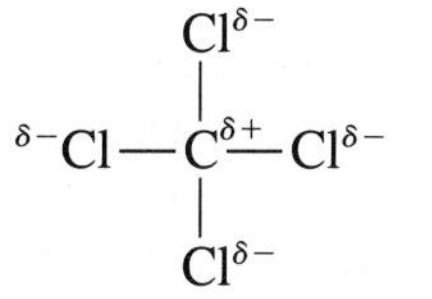

$$\begin{array}{c} \mathrm{Cl}^{\delta-} \\ | \\ {}^{\delta-}\mathrm{Cl}-\mathrm{C}^{\delta+}-\mathrm{Cl}^{\delta-} \\ | \\ \mathrm{Cl}^{\delta-} \end{array}$$

Chlorine is more electronegative than carbon. The electrons in each C—Cl bond are therefore displaced toward the chlorine.

PROBLEM 1.8 Predict the polarity of the N—Cl bond and of the S—O bond.

TABLE 1.4 Electronegativities of some common elements

Group

I	*II*		*III*	*IV*	*V*	*VI*	*VII*
H 2.2							
Li 1.0	Be 1.6		B 2.0	C 2.5	N 3.0	O 3.4	F 4.0
Na 0.9	Mg 1.3		Al 1.6	Si 1.9	P 2.2	S 2.6	Cl 3.2
K 0.8	Ca 1.0						Br 3.0
							I 2.7

PROBLEM 1.9 Draw the structure of the refrigerant dichlorodifluoromethane, CCl_2F_2 (CFC-12), and indicate the polarity of the bonds.

PROBLEM 1.10 Draw the formula for methanol, CH_3OH, and (where appropriate) indicate bond polarity with an arrow, $+\!\!\longrightarrow$.

1.6 Multiple Covalent Bonds

To complete their valence shells, atoms may sometimes share more than one electron pair. Carbon dioxide, CO_2, is an example. The carbon atom has four valence electrons, and each oxygen has six valence electrons. A formula that allows each atom to complete its valence shell with eight electrons is

$$\underset{\text{A}}{:\!\ddot{O}\!::\!C\!::\!\ddot{O}\!:} \quad \text{or} \quad \underset{\text{B}}{\ddot{\underset{\cdot\cdot}{O}}\!=\!C\!=\!\ddot{\underset{\cdot\cdot}{O}}} \quad \text{or} \quad \underset{\text{C}}{O\!=\!C\!=\!O}$$

In formula A, the dots represent the electrons from carbon, and the x's are the electrons from the oxygens. Formula B shows the bonds and oxygen's unshared electrons, and formula C shows only the covalent bonds. Two electron pairs are shared between carbon and oxygen. Consequently, the bond is called a **double bond.** Each oxygen atom also has two pairs of **nonbonding electrons,** or **unshared electron pairs.** The loops in the following formulas show that each atom in carbon dioxide has a complete valence shell of eight electrons:

$$:\!\ddot{O}\!::\!C\!::\!\ddot{O}\!: \qquad :\!\ddot{O}\!::\!C\!::\!\ddot{O}\!: \qquad :\!\ddot{O}\!::\!C\!::\!\ddot{O}\!:$$

Hydrogen cyanide, HCN, is an example of a simple compound with a **triple bond,** a bond in which three electron pairs are shared.

H:C:::N: or H—C≡N: or H—C≡N

hydrogen cyanide

PROBLEM 1.11 Show with loops how each atom in hydrogen cyanide completes its valence shell.

EXAMPLE 1.8 Tell what, if anything, is wrong with the following electron arrangement for carbon dioxide:

:O:::C::Ö:

Solution The formula contains the correct total number of valence electrons (16), and each oxygen is surrounded by 8 valence electrons, which is correct. What is wrong is that the carbon atom has 10 valence electrons, two more than is allowable.

PROBLEM 1.12 Show what is wrong with each of the following electron arrangements for carbon dioxide:

a. :O:::C:::O: b. :Ö:C̈:Ö: c. :Ö:C:::O:

PROBLEM 1.13 Methanal (formaldehyde) has the formula H_2CO. Draw a formula that shows how the valence electrons are arranged.

PROBLEM 1.14 Draw an electron-dot formula for carbon monoxide, CO.

Carbon atoms can be connected to one another by double bonds or triple bonds, as well as by single bonds. Thus there are three **hydrocarbons** (compounds with just carbon and hydrogen atoms) that have two carbon atoms per molecule: ethane, ethene, and ethyne.

```
    H  H
    |  |            H       H
H — C — C — H        \     /         H—C≡C—H
    |  |              C = C
    H  H             /     \
                    H       H
  ethane          ethene            ethyne
                (ethylene)        (acetylene)
```

They differ in that the carbon–carbon bond is single, double, or triple, respectively. They also differ in number of hydrogens. As we will see later, these compounds have different chemical reactivities because of the different types of bonds between the carbon atoms.

EXAMPLE 1.9 Draw the formula for C_3H_6 having one carbon–carbon double bond.

Solution First, draw the three carbons with one double bond.

C=C—C

Then add the hydrogens in such a way that each carbon has eight electrons around it (or in such a way that each carbon has four bonds).

```
   H  H  H
   |  |  |
H—C=C—C—H
         |
         H
```

PROBLEM 1.15 Draw three different structures that have the formula C_4H_8 and have one carbon–carbon double bond.

1.7 Valence

The valence of an element is simply the number of bonds that an atom of the element can form. The number is usually equal to the *number of electrons needed to fill the valence shell.* Table 1.5 gives the common valences of several elements.

Notice the difference between the number of valence electrons and the valence. Oxygen, for example, has six valence electrons but a valence of only 2. The *sum* of the two numbers is equal to the number of electrons in the filled shell.

The valences in Table 1.5 apply whether the bonds are single, double, or triple. For example, carbon has four bonds in each of the formulas we have written so far: methane, tetrachloromethane, ethane, ethene, ethyne, carbon dioxide, and so on. These common valences are worth remembering, because they will help you to write correct formulas.

EXAMPLE 1.10 Using dashes for bonds, draw a formula for C_3H_4 that has the proper valence of 1 for each hydrogen and 4 for each carbon.

Solution There are three possibilities:

```
   H
   |              H         H      H         H
H—C—C≡C—H          \       /        \       /
   |                C=C=C            C = C
   H               /       \          \ /
                  H         H          C
                                      / \
                                     H   H
```

A compound that corresponds to each of these three different arrangements of the atoms is known.

TABLE 1.5 Valences of common elements

Element	H·	·C·	·N:	·O:	:F:	:Cl:
Valence	1	4	3	2	1	1

PROBLEM 1.16 Use dashes for bonds, and use the valences given in Table 1.5 to write a structure for each of the following:

a. CH_5N b. CH_4O

PROBLEM 1.17 Does C_2H_5 represent a stable molecule?

In Example 1.10, we saw that three carbon atoms and four hydrogen atoms can be connected to one another in three different ways, each of which satisfies the valences of both kinds of atoms. Let us take a closer look at this phenomenon.

1.8 Isomerism

The **molecular formula** of a substance tells us the numbers of different atoms present, but a **structural formula** tells us how those atoms are arranged. For example, H_2O is the molecular formula for water. It tells us that each water molecule contains two hydrogen atoms and one oxygen atom. But the structural formula H—O—H tells us more than that. It tells us that the hydrogens are connected to the oxygen (and not to each other).

It is sometimes possible to arrange the same atoms in more than one way and still satisfy their valences. Molecules that have the same kinds and numbers of atoms but different arrangements are called **isomers,** a term that comes from the Greek (*isos,* equal, and *meros,* part). **Structural (or constitutional) isomers** ***are compounds that have the same molecular formula but different structural formulas.*** Let us look at a particular pair of isomers.

Two very different chemical substances are known, each with the molecular formula C_2H_6O. One of these substances is a colorless liquid that boils at 78.5°C, whereas the other is a colorless gas at ordinary temperatures (bp −23.6°C). The only possible explanation is that the atoms must be arranged differently in the molecules of each substance and that these arrangements are somehow responsible for the fact that one substance is a liquid and the other, a gas.

For the molecular formula C_2H_6O, two (and only two) structural formulas are possible that satisfy the valence requirement of 4 for carbon, 2 for oxygen, and 1 for hydrogen. They are

```
    H  H                      H         H
    |  |                      |         |
H—C—C—O—H      and      H—C—O—C—H
    |  |                      |         |
    H  H                      H         H
```

ethanol
(ethyl alcohol)
bp 78.5°C

methoxymethane
(dimethyl ether)
bp −23.6°C

In one formula, the two carbons are connected to one another by a single covalent bond; in the other formula, each carbon is connected to the oxygen. When we complete the valences by adding hydrogens, each arrangement requires six hydrogens. Many kinds of experimental evidence verify these structural assignments. We leave for later (in Chapters 7 and 8) an explanation of why these arrangements of atoms produce substances that are so different from one another.

Ethanol and methoxymethane are **structural isomers.** They have the same molecular formula but different structural formulas. Ethanol and methoxymethane differ in physical and chemical properties as a consequence of their different molecular structures. *In general, structural isomers are different compounds. They differ in physical and chemical properties as a consequence of their different molecular structures.*

PROBLEM 1.18 Draw structural formulas for the three possible isomers of C_3H_8O.

1.9 Writing Structural Formulas

You will be writing structural formulas throughout this course. Perhaps a few hints about how to do so will be helpful. Let's look at another case of isomerism. Suppose we want to write out all possible structural formulas that correspond to the molecular formula C_5H_{12}. We begin by writing all five carbons in a **continuous chain.**

C—C—C—C—C

a continuous chain

This chain uses up one valence for each of the end carbons and two valences for the carbons in the middle of the chain. Each end carbon therefore has three valences left for bonds to hydrogens. Each middle carbon has only two valences for bonds to hydrogens. As a consequence, the structural formula in this case is written

```
    H   H   H   H   H
    |   |   |   |   |
H—C—C—C—C—C—H
    |   |   |   |   |
    H   H   H   H   H
```

pentane, bp 36°C

To find structural formulas for the other isomers, we must consider **branched chains.** For example, we can reduce the longest chain to only four carbons and connect the fifth carbon to one of the middle carbons, as in the following structural formula:

```
C—C—C—C
  |
  C
```

a branched chain

If we add the remaining bonds so that each carbon has a valence of 4, we see that three of the carbons have three hydrogens attached, but the other carbons have only one or two hydrogens. The molecular formula, however, is still C_5H_{12}.

```
     H   H   H   H
     |   |   |   |
 H—C—C—C—C—H
     |   |   |   |
     H   |   H   H
         |
     H—C—H
         |
         H
```

2-methylbutane, bp 28°C
(isopentane)

Suppose we keep the chain of four carbons and try to connect the fifth carbon somewhere else. Consider the following chains:

```
C—C—C—C     C—C—C—C     C—C—C—C
|                 |           |
C                 C           C
```

Do we have anything new here? *No!* The first two structures have five-carbon chains, exactly as in the formula for pentane, and the third structure is identical to the branched chain we have already drawn for 2-methylbutane—a four-carbon chain with a one-carbon branch attached to the second carbon in the chain (counting now from the right instead of from the left).

But there is a third isomer of C_5H_{12}. We can find it by reducing the longest chain to only three carbons and connecting two one-carbon branches to the middle carbon.

```
    C
    |
C—C—C
    |
    C
```

If we fill in the hydrogens, we see that the middle carbon has no hydrogens attached to it.

```
          H
          |
        H—C—H
          |
   H      |      H
   |      |      |
 H—C——————C——————C—H
   |      |      |
   H      |      H
          |
        H—C—H
          |
          H
```

2,2-dimethylpropane, bp 10°C
(neopentane)

So we can draw three (and only three) different structural formulas that correspond to the molecular formula C_5H_{12}, and in fact we find that only three different chemical substances with this formula exist. They are commonly called *n*-pentane (*n* for normal, with an unbranched carbon chain), isopentane, and neopentane.

PROBLEM 1.19 To which isomer of C_5H_{12} does each of the following structural formulas correspond?

```
   H          H                   H                  H     H     H
   |          |                   |                  |     |     |
H—C—H      H—C—H              H—C—H             H—C—————C—————C—H
   |     H     |           H      |      H           |     |     |
   |     |     |           |      |      |           H     |     H
H—C—————C—————C—H       H—C——————C——————C—H               |     H
   |     |     |           |      |      |                 |     |
   H     H     H           H      H      |              H—C—————C—H
                                      H—C—H                |     |
                                         |                 H     H
                                         H
```

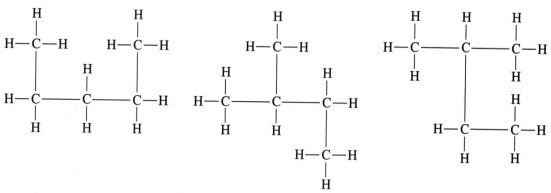

1.10 Abbreviated Structural Formulas

Structural formulas like the ones we have written so far are useful, but they are also somewhat cumbersome. They take up a lot of space and are tiresome to write out. Consequently, we often take some shortcuts that still convey the meaning of structural formulas. For example, we may abbreviate the structural formula of ethanol (ethyl alcohol) from

```
   H  H
   |  |
H—C—C—O—H     to    CH3—CH2—OH     or     CH3CH2OH
   |  |
   H  H
```

Each formula clearly represents ethanol rather than methoxymethane (dimethyl ether), which can be represented by any of the following structures:

```
   H     H
   |     |
H—C—O—C—H     to    CH3—O—CH3     or     CH3OCH3
   |     |
   H     H
```

The structural formulas for the three pentanes can be abbreviated in a similar fashion.

```
                                                 CH3
                                                  |
CH3CH2CH2CH2CH3      CH3CHCH2CH3          CH3—C—CH3
                         |                        |
                        CH3                      CH3
   n-pentane          isopentane             neopentane
```

Sometimes these formulas are abbreviated even further. For example, they can be printed on a single line in the following ways:

$CH_3(CH_2)_3CH_3$	$(CH_3)_2CHCH_2CH_3$	$(CH_3)_4C$
n-pentane	isopentane	neopentane

EXAMPLE 1.11 Write a structural formula that shows all bonds for each of the following:

a. $CH_3CCl_2CH_3$ b. $(CH_3)_2C(CH_2CH_3)_2$

Solution

```
                                                     H
                                                     |
                                                   H—C—H
                                                     |
   H  Cl  H                 H   H                    |            H   H
   |   |  |                 |   |                    |            |   |
 H—C—C—C—H               H—C—C————————C————————C—C—H
   |   |  |                 |   |                    |            |   |
   H  Cl  H                 H   H                    |            H   H
                                                     |
                                                   H—C—H
                                                     |
                                                     H
```

This is the carbon atom to which two —CH_3 and two —CH_2CH_3 groups are attached.

PROBLEM 1.20 Write a structural formula that shows all bonds for each of the following:

a. $(CH_3)_2CHCH_2OH$ b. $Cl_2C{=}CCl_2$

Perhaps the ultimate abbreviation of structures is the use of lines to represent the carbon framework:

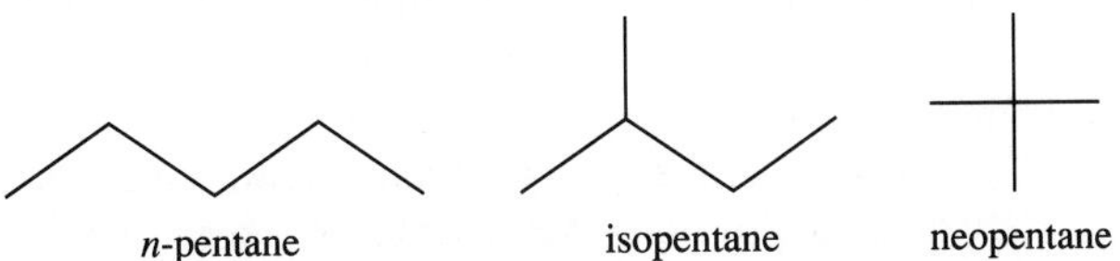

n-pentane   isopentane   neopentane

In these formulas, *each line segment is understood to have a carbon atom at each end.* The hydrogens are omitted, but we can quickly find the number of hydrogens on each carbon by subtracting from four the number of line segments that emanate from any point. Multiple bonds are represented by multiple line segments. For example, the hydrocarbon with a chain of five carbon atoms and a double bond between the second and third carbon atoms (that is, $CH_3CH{=}CHCH_2CH_3$) is represented as follows:

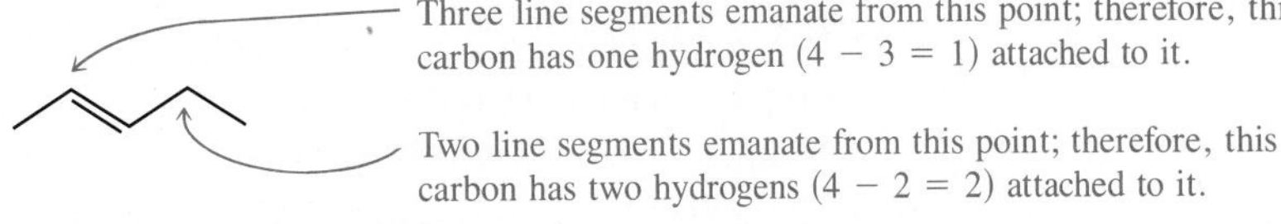

Three line segments emanate from this point; therefore, this carbon has one hydrogen (4 − 3 = 1) attached to it.
Two line segments emanate from this point; therefore, this carbon has two hydrogens (4 − 2 = 2) attached to it.

EXAMPLE 1.12 Write a more detailed structural formula for [line-segment formula].

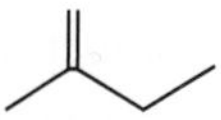

Solution

$$CH_3-\overset{\overset{\displaystyle CH_2}{\|}}{C}-CH_2-CH_3 \quad \text{or} \quad H-\overset{\overset{\displaystyle H}{|}}{\underset{\underset{\displaystyle H}{|}}{C}}-\overset{\overset{\displaystyle H_2C}{\|}}{C}-\overset{\overset{\displaystyle H}{|}}{\underset{\underset{\displaystyle H}{|}}{C}}-\overset{\overset{\displaystyle H}{|}}{\underset{\underset{\displaystyle H}{|}}{C}}-H$$

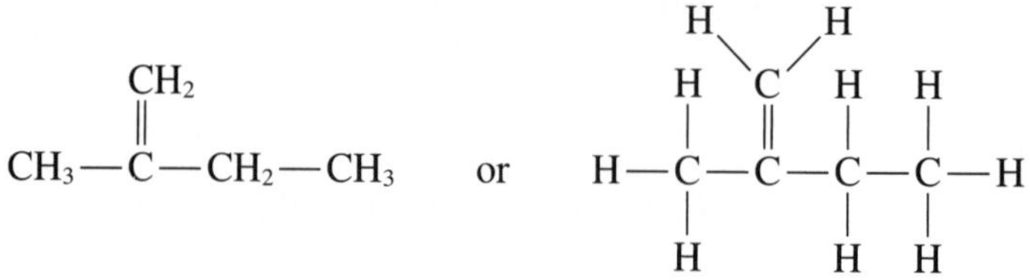

PROBLEM 1.21 Write a line-segment formula for $(CH_3)_2CHCH(CH_3)_2$.

1.11 *Formal Charge*

So far we have considered only molecules whose atoms are neutral. But in some compounds one or more atoms may be charged, either positive or negative. Because such charges usually affect the chemical reactions of such molecules, it is important to know how to tell where the charge is located.

Consider the formula for hydronium ion, H_3O^+, the product of the reaction of a water molecule with a proton.

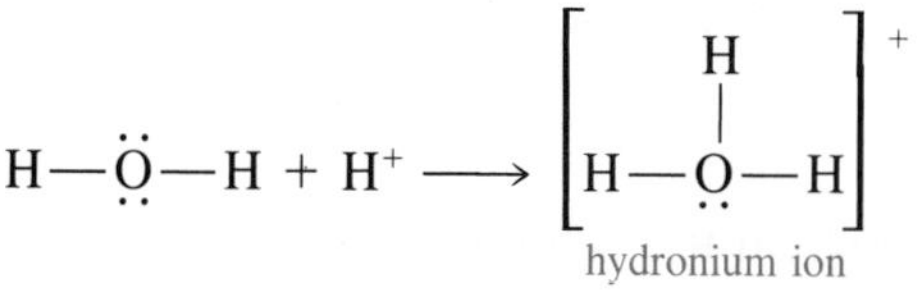

$$H-\ddot{\underset{\cdot\cdot}{O}}-H + H^+ \longrightarrow \left[H-\overset{\overset{\displaystyle H}{|}}{\underset{\cdot\cdot}{O}}-H \right]^+ \quad (1.4)$$

hydronium ion

The structure has eight electrons around the oxygen and two electrons around each hydrogen, so that all valence shells are complete. Note that there are eight valence electrons altogether. Oxygen contributes six, and each hydrogen contributes one, for a total of nine, but the ion has a single positive charge, so one electron must have been given away, leaving eight. Six of these eight electrons are used to form three O—H single bonds, leaving one unshared electron pair on the oxygen.

Although the entire hydronium ion carries a positive charge, we can ask, "Which atom, in a formal sense, bears the charge?" To determine **formal charge,** we consider that each atom "owns" *all* of its unshared electrons plus only *half* of its shared electrons (one electron from each covalent bond). We then subtract this total from the number of valence electrons in the neutral atom to get the formal charge. This definition can be expressed in equation form as follows:

$$\textbf{Formal charge} = \text{number of valence electrons in the neutral atom} - \left(\text{unshared electrons} + \text{half the shared electrons}\right) \quad (1.5)$$

or, in a simplified form,

$$\textbf{Formal charge} = \text{number of valence electrons in the neutral atom} - (\text{dots} + \text{bonds})$$

Let us apply this definition to the hydronium ion.

For each hydrogen atom:
Number of valence electrons in the neutral atom = 1
Number of unshared electrons = 0
Half the number of the shared electrons = 1
Therefore, the formal charge = $1 - (0 + 1) = 0$

For the oxygen atom:
Number of valence electrons in the neutral atom = 6
Number of unshared electrons = 2
Half the number of the shared electrons = 3
Therefore, the formal charge = $6 - (2 + 3) = +1$

It is the oxygen atom that formally carries the +1 charge in the hydronium ion.

EXAMPLE 1.13 On which atom is the formal charge in the hydroxide ion, OH^-?

Solution The electron-dot formula is

$$[:\ddot{\underset{..}{O}}:H]^-$$

Oxygen contributes six electrons, hydrogen contributes one, and there is one more for the negative charge, for a total of eight electrons. The formal charge on oxygen is $6 - (6 + 1) = -1$, so the oxygen carries the negative charge. The hydrogen is neutral.

PROBLEM 1.22 Calculate the formal charge on the nitrogen atom in ammonia, NH_3; in the ammonium ion, NH_4^+; and in the amide ion, NH_2^-.

Now let us look at a slightly more complex situation involving electron-dot formulas and formal charge.

1.12 *Resonance*

Sometimes an electron pair is involved with more than two atoms in the process of forming bonds. As an example, consider the structure of the carbonate ion, CO_3^{2-}.

The total number of valence electrons in the carbonate ion is 24 (4 from the carbon, $3 \times 6 = 18$ from the three oxygens, *plus* 2 more electrons that give the ion its negative charge; these 2 electrons presumably have been donated by some metal, perhaps one each from two sodium atoms). An electron-dot formula that completes the valence shell of eight electrons around the carbon and each oxygen is

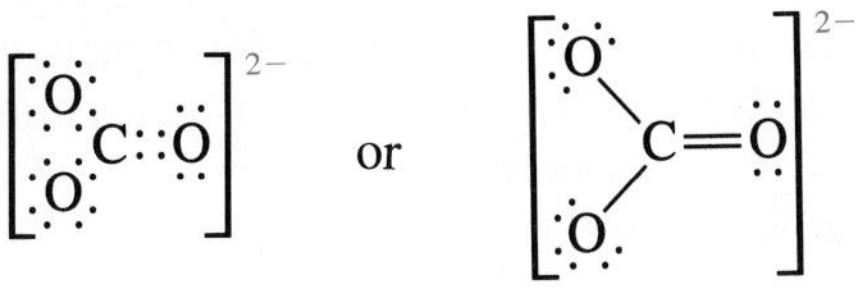

[electron-dot structure: $\left[\begin{array}{l}:\ddot{O}: \\ \quad C::\ddot{O} \\ :\ddot{O}:\end{array}\right]^{2-}$] or [$\left[\begin{array}{l}:\ddot{O}: \\ \quad \diagdown \\ \quad\quad C{=}\ddot{O} \\ \quad \diagup \\ :\ddot{O}:\end{array}\right]^{2-}$]

carbonate ion, CO_3^{2-}

The structure contains two carbon–oxygen *single* bonds and one carbon–

oxygen *double* bond. Application of the definition for formal charge shows that the carbon is formally neutral, each singly bonded oxygen has a formal charge of −1, and the doubly bonded oxygen is formally neutral.

PROBLEM 1.23 Show that the last sentence of the preceding paragraph is correct.

When we wrote the electron-dot formula for the carbonate ion, our choice of which oxygen atom would be doubly bonded to the carbon atom was purely arbitrary. There are in fact *three exactly equivalent* structures that we might write.

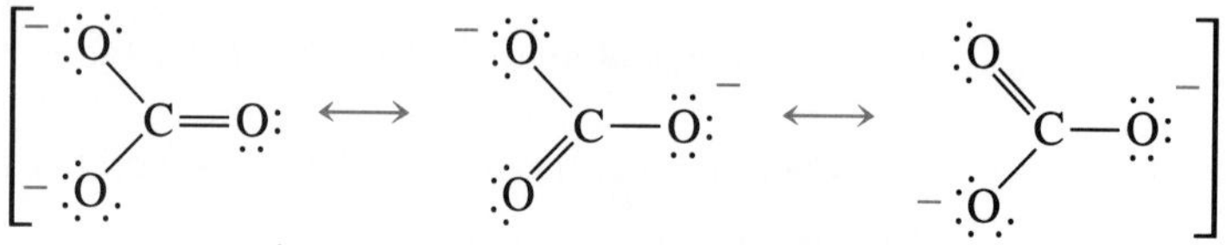

three equivalent structures for the carbonate ion

In each structure there is one C=O bond and there are two C—O bonds. These structures have the same arrangement of the atoms. They differ from one another *only* in the arrangement of the electrons.

The three structures for the carbonate ion are redrawn below, with curved arrows to show how electron pairs can be moved to interrelate the structures:

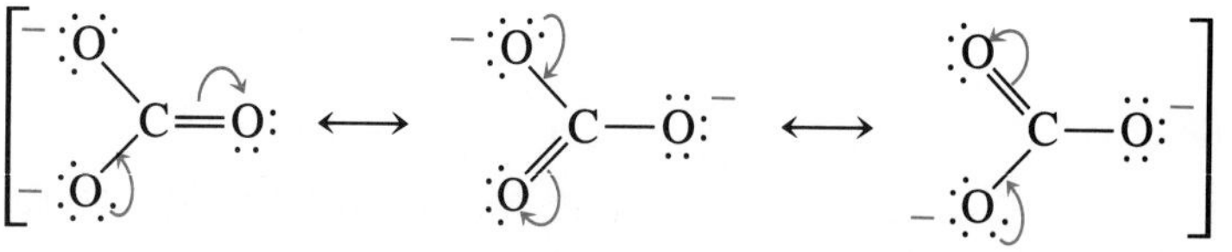

Chemists use curved arrows to keep track of a change in the location of electrons. A detailed explanation of the use of curved arrows is given in Sec. 1.13.

Physical measurements tell us that none of the foregoing structures accurately describes the real carbonate ion. For example, although each structure shows two different types of bonds between carbon and oxygen, we find experimentally that ***all three carbon–oxygen bond lengths are identical: 1.31 Å.*** This distance is intermediate between the normal C=O (1.20 Å) and C—O (1.41 Å) bond lengths. To explain this fact, we usually say that the real carbonate ion has a structure that is a **resonance hybrid** of the three contributing resonance structures. It is as if we could take an average of the three structures. In the real carbonate ion, the two formal negative charges are spread equally over the three oxygen atoms, so that each oxygen atom carries two-thirds of a negative charge. It is important to note that the carbonate ion does not physically alternate among three resonance structures but has in fact one structure—a ***hybrid*** of the three resonance structures.

Resonance *arises whenever we can write two or more structures for a molecule with different arrangements of the electrons but identical arrangements of the atoms.* Resonance is very different from isomerism, for which the atoms themselves are arranged differently. *When resonance is possible, the substance is said to have a structure that is a* **resonance hybrid** *of the vari-*

ous* contributing structures.** ***We use a double-headed arrow (↔) between contributing structures to distinguish resonance from an equilibrium, for which we use ⇌.

Each carbon–oxygen bond in the carbonate ion is neither single nor double, but something in between—perhaps a one-and-one-third bond (any particular carbon–oxygen bond is single in two contributing structures and double in one). Sometimes we represent a resonance hybrid with one formula by writing a solid line for each full bond and a dotted line for each partial bond (in carbonate ion, the dots represent one-third of a bond).

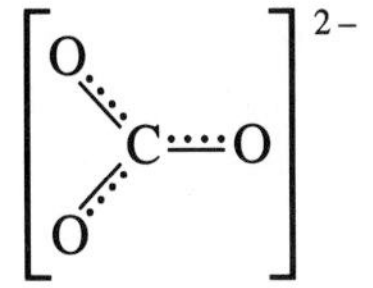

$$\left[\begin{array}{l} O \\ \quad \text{C}\cdots\text{O} \\ O \end{array}\right]^{2-}$$

carbonate ion
resonance hybrid

PROBLEM 1.24 Draw the three equivalent contributing resonance structures for the nitrate ion, NO_3^-. What is the formal charge on the nitrogen atom and on each oxygen atom in the individual structures? What is the charge on the oxygens and on the nitrogen in the resonance hybrid structure? Show with curved arrows how the structures can be interconverted.

1.13 Arrow Formalism

Arrows in chemical drawings have specific meanings. For example, in Sec. 1.12 we used curved arrows to move electrons to show the relatedness of the three resonance structures of the carbonate ion. Just as it is important to learn the structural representations and names of molecules, it is important to learn the language of arrow formalism in organic chemistry.

1. ***Curved arrows*** *are used to show the* ***movement of electrons*** *in resonance structures and in reactions.* Therefore, curved arrows always start at the initial position of electrons and end at their final position. In the example given below, the arrow that points from the C=O bond to the oxygen atom in the structure on the left indicates that the two electrons in one of the covalent bonds between carbon and oxygen moves onto the oxygen atom:

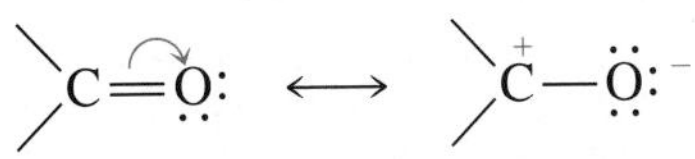

$$\text{C}=\ddot{\text{O}}\text{:} \longleftrightarrow \overset{+}{\text{C}}-\ddot{\underset{\cdot\cdot}{\text{O}}}\text{:}^{-}$$

Note that the carbon atom in the structure on the right now has a formal positive charge, and the oxygen has a formal negative charge. Notice also that when a pair of electrons in a polar covalent bond is moved to one of the bonded atoms, *it is moved to the more electronegative atom,* in this case oxygen. In the following example, the arrow that points from the unshared pair of electrons on the oxygen atom to a point between the carbon and oxygen atoms in the structure on the left indicates that the unshared pair of electrons on the oxygen atom moves between the oxygen and carbon atoms to form a covalent bond:

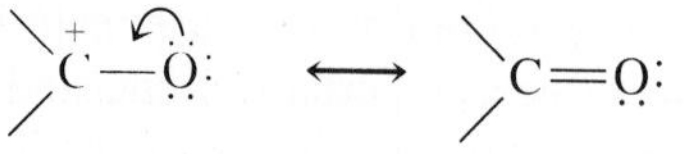

$$\overset{+}{C}-\ddot{\underset{..}{O}}: \longleftrightarrow C=\ddot{\underset{..}{O}}:$$

Note that both carbon and oxygen have formal charges of 0 in the structure on the right.

A curved arrow with half a head is called a ***fishhook.*** *This kind of arrow is used to indicate the movement of a* ***single electron.*** In eq. 1.6, two fishhooks are used to show the movement of each of the two electrons in the C—C bond of ethane to a carbon atom, forming two methyl radicals (see eq. 1.3):

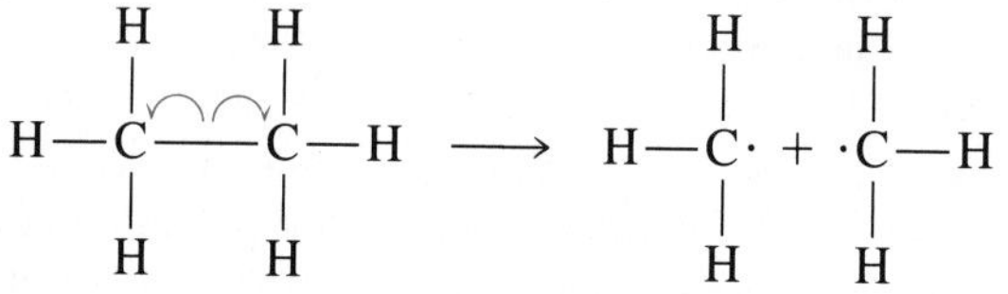

$$H_3C-CH_3 \longrightarrow H_3C\cdot + \cdot CH_3 \qquad (1.6)$$

2. ***Straight arrows*** *point* ***from reactants to products*** *in chemical reaction equations.* An example is the straight arrow pointing from ethane to the two methyl radicals in eq. 1.6. Straight arrows with half-heads are commonly used in pairs to indicate that a reaction is *reversible.*

$$A + B \rightleftharpoons C + D$$

A ***double-headed straight arrow*** *between two structures indicates that they are* ***resonance structures.*** Such an arrow does not indicate the occurrence of a chemical reaction. The double-headed arrows between the resonance structures shown above are examples.

EXAMPLE 1.14 Using correct arrow formalism, write the contributors to the resonance hybrid structure of the acetate ion, $CH_3CO_2^-$. Indicate any formal charges.

Solution There are two equivalent resonance structures for the acetate ion. Each one has a formal negative charge on one of the oxygen atoms.

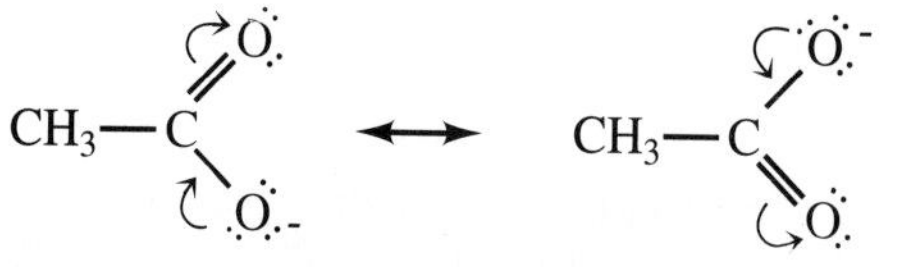

$$CH_3-C(=\ddot{O}:)-\ddot{O}:^- \longleftrightarrow CH_3-C(-\ddot{O}:^-)=\ddot{O}:$$

Notice that when one pair of electrons from oxygen is moved to form a covalent bond with carbon, a pair of electrons in a covalent bond between carbon and the other oxygen atom is moved to oxygen. This is necessary to ensure that the carbon atom does not exceed its valence of 4.

PROBLEM 1.25 Using correct arrow formalism, write the contributors to the resonance hybrid of azide ion, a linear ion with three connected nitrogens, N_3^-. Indicate the formal charge on each nitrogen atom.

We will use curved arrows throughout this text as a way of keeping track of electron movement. Several curved-arrow problems are included at the end of this chapter to help you get used to drawing them.

1.14 The Orbital View of Bonding; the Sigma Bond

Although electron-dot formulas are often useful, they have some limitations. The Lewis theory of bonding itself has some limitations, especially in explaining the three-dimensional geometries of molecules. For this purpose in particular, we will discuss how another theory of bonding, involving orbitals, is more useful.

The atomic orbitals named in Sec. 1.1 have definite shapes. The *s* orbitals are spherical. The electrons that fill an *s* orbital confine their movement to a spherical region of space around the nucleus. The three *p* orbitals are dumbbell-shaped and mutually perpendicular, oriented along the three coordinate axes, *x*, *y*, and *z*. Figure 1.2 shows the shapes of these orbitals.

In the orbital view of bonding, atoms approach each other in such a way that their atomic orbitals can *overlap* to form a bond. For example, if two hydrogen atoms form a hydrogen molecule, their two spherical 1*s* orbitals combine to form a new orbital that encompasses both of the atoms (see Figure 1.3). This orbital contains both valence electrons (one from each hydrogen). Like atomic orbitals, each **molecular orbital** can contain no more than two electrons. In

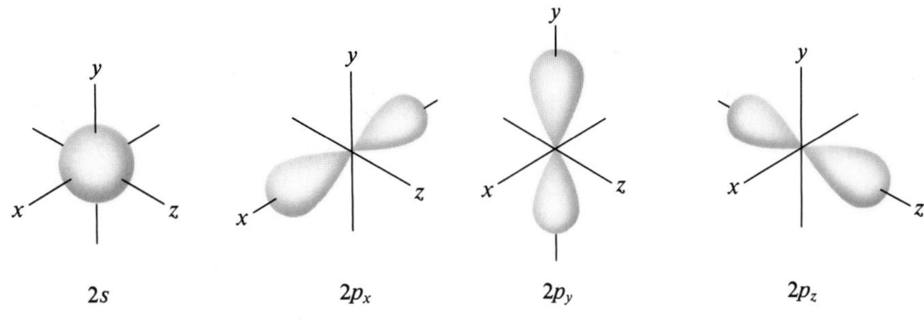

FIGURE 1.2 The shapes of the *s* and *p* orbitals used by the valence electrons of carbon. The nucleus is at the origin of the three coordinate axes.

H + H ⟶ H—H

1*s* atomic orbitals

s-*s* molecular orbital

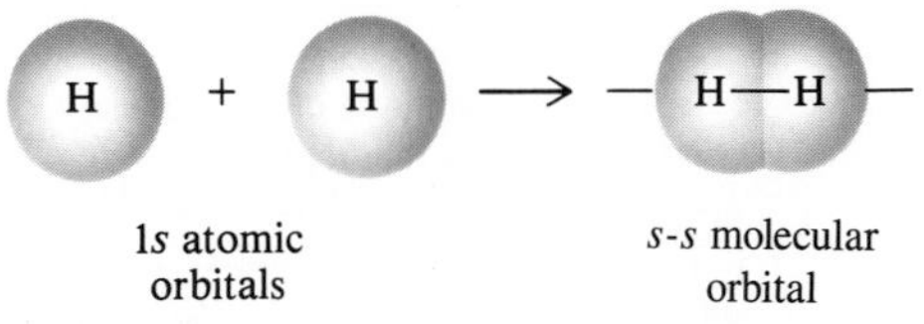

FIGURE 1.3 The molecular orbital representation of covalent bond formation between two hydrogen atoms.

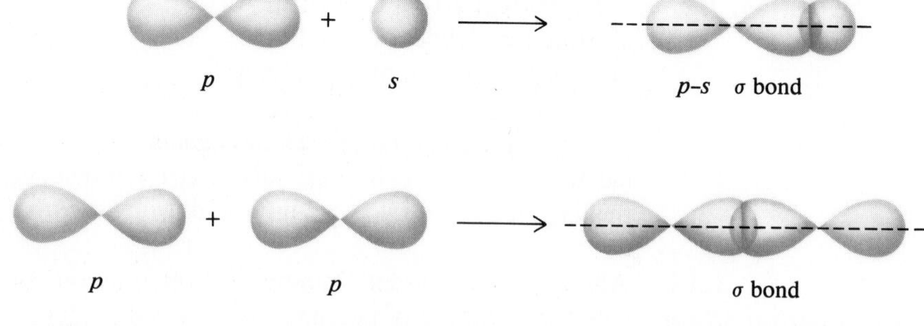

FIGURE 1.4 Orbital overlap to form σ bonds.

the hydrogen molecule these electrons mainly occupy the space between the two nuclei.

The orbital in the hydrogen molecule is cylindrically symmetric along the H—H internuclear axis. Such orbitals are called **sigma (σ) orbitals,** and the bond is referred to as a **sigma bond.** Sigma bonds may also be formed by the overlap of an *s* and a *p* orbital or of two *p* orbitals, as shown in Figure 1.4.*

Let us see how these ideas apply to bonding in carbon compounds.

1.15 Carbon sp^3 Hybrid Orbitals

In a carbon atom, the six electrons are arranged as shown in Figure 1.5 (compare with carbon in Table 1.2). The 1*s* shell is filled, and the four valence electrons are in the 2*s* orbital and two different 2*p* orbitals. There are a few things to notice about Figure 1.5. The energy scale at the left represents the energy of electrons in the various orbitals. The farther the electron is from the nucleus, the greater its potential energy, because it takes energy to keep the electron (negatively charged) and the nucleus (positively charged) apart. The 2*s* orbital has a slightly lower energy than the three 2*p* orbitals, which have equal energies (they differ from one another only in orientation around the nucleus, as shown in Figure 1.2). The two highest energy electrons are placed in different 2*p* orbitals rather than in the same orbital, because this keeps them farther

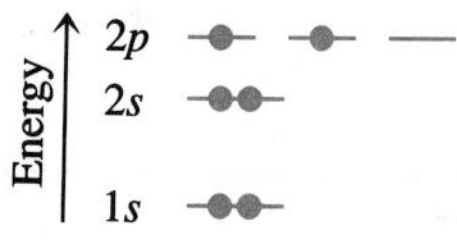

FIGURE 1.5 Distribution of the six electrons in a carbon atom. Each dot stands for an electron.

---

*Two properly aligned *p* orbitals can also overlap to form another type of bond, called a π (pi) bond. We discuss this type of bond in Chapter 3.

p + p → π bond

apart and thus reduces the repulsion between these like-charged particles. One p orbital is vacant.

We might get a misleading idea about the bonding of carbon from Figure 1.5. For example, we might think that carbon should form only two bonds (to complete the partially filled $2p$ orbitals) or perhaps three bonds (if some atom donated two electrons to the empty $2p$ orbital). But we know from experience that this picture is wrong. Carbon usually forms *four* single bonds, and often these bonds are all equivalent, as in CH_4 or CCl_4. How can this discrepancy between theory and fact be resolved?

One solution, illustrated in Figure 1.6, is to mix or combine the four atomic orbitals of the valence shell to form four identical hybrid orbitals, each containing one valence electron. In this model, the hybrid orbitals are called ***sp*³ *hybrid orbitals*** because each one has one part s character and three parts p character. As shown in Figure 1.6, each sp^3 orbital has the same energy: less than that of the $2p$ orbitals but greater than that of the $2s$ orbital. The shape of sp^3 orbitals resembles the shape of p orbitals, except that the dumbbell is lopsided, and the electrons are more likely to be found in the lobe that extends out the greater distance from the nucleus, as shown in Figure 1.7. The four sp^3 hy-

FIGURE 1.6
Unhybridized vs sp^3 hybridized orbitals on carbon. The dots stand for electrons. (Only the electrons in the valence shell are shown; the electrons in the 1s orbital are omitted because they are not involved in bonding.)

Energy
$2p$
$2s$
sp^3
Atomic orbitals of carbon
Four equivalent sp^3 hybrid orbitals

FIGURE 1.7
An sp^3 orbital extends mainly in one direction from the nucleus and forms bonds with other atoms in that direction. The four sp^3 orbitals of any particular carbon atom are directed toward the corners of a regular tetrahedron, as shown in the right-hand part of the figure (in this part of the drawing, the small "back" lobes of the orbitals have been omitted for simplification).

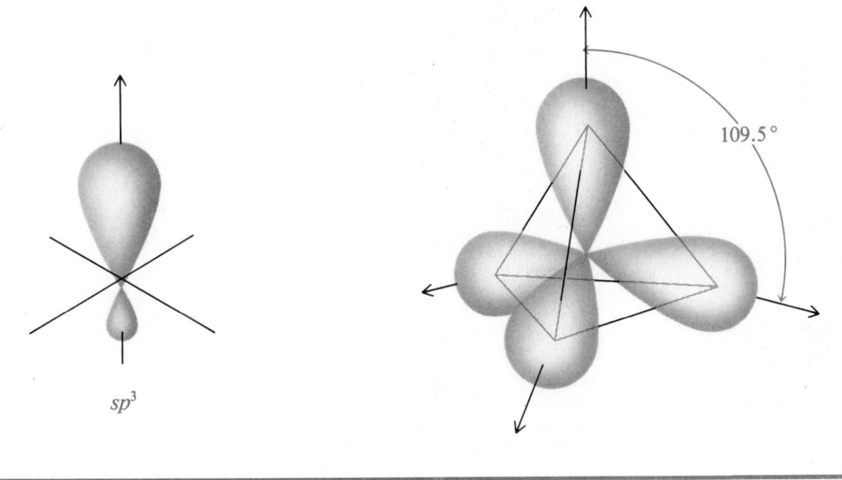

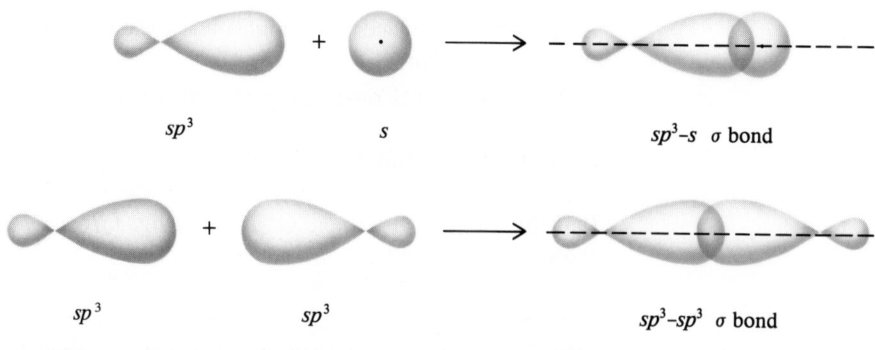

FIGURE 1.8 Examples of sigma (σ) bonds formed from sp^3 hybrid orbitals.

brid orbitals of a single carbon atom are directed toward the corners of a regular tetrahedron, also shown in Figure 1.7. This particular geometry puts each orbital as far from the other three orbitals as it can be and thus minimizes repulsion when the orbitals are filled with electron pairs. The angle between any two of the four bonds formed from sp^3 orbitals is approximately 109.5°, the angle made by lines drawn from the center to the corners of a regular tetrahedron.

Hybrid orbitals can form sigma bonds by overlap with other hybrid orbitals or with non-hybridized atomic orbitals. Figure 1.8 shows some examples.

1.16 Tetrahedral Carbon; the Bonding in Methane

We can now describe the way a carbon atom combines with four hydrogen atoms to form methane. This process is pictured in Figure 1.9. The carbon atom is joined to each hydrogen atom by a sigma bond, which is formed by the overlap of a carbon sp^3 orbital with a hydrogen 1*s* orbital. The four sigma bonds are directed from the carbon nucleus to the corners of a regular tetrahedron. In this way, the electron pair in any one bond experiences minimum repulsion from the electrons in the other bonds. Each H—C—H **bond angle** is

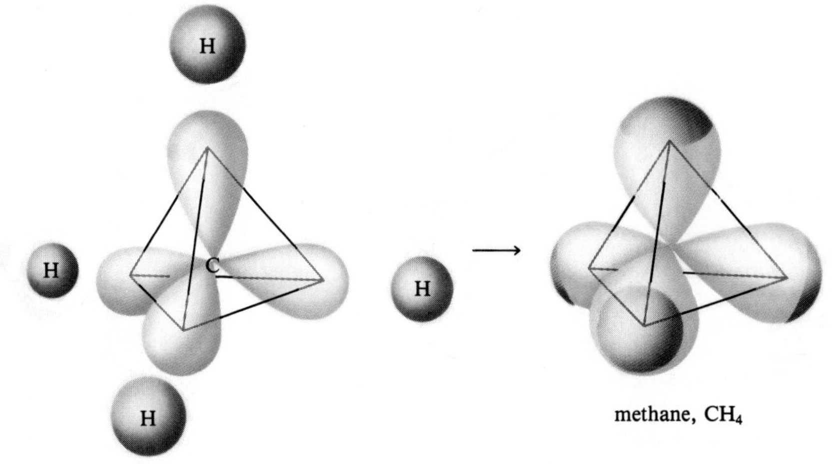

FIGURE 1.9 A molecule of methane, CH_4, formed by the overlap of the four sp^3 carbon orbitals with the 1*s* orbitals of four hydrogen atoms. The resulting molecule has the geometry of a regular tetrahedron and contains four sigma bonds of the sp^3–s type.

the same, 109.5°. To summarize, in methane there are four sp^3–s C—H sigma bonds, each directed from the carbon atom to one of the four corners of a regular tetrahedron.

PROBLEM 1.26 Considering the repulsion that exists between electrons in different bonds, give a reason why a planar geometry for methane would be less stable than the tetrahedral geometry.

Because the tetrahedral geometry of carbon plays such an important role in organic chemistry, it is a good idea to become familiar with the features of a regular tetrahedron. One feature is that *the center and any two corners of a tetrahedron form a plane that is the perpendicular bisector of a similar plane formed by the center and the other two corners.* In methane, for example, any two hydrogens and the carbon form a plane that perpendicularly bisects the plane formed by the carbon and the other two hydrogens. These planes are illustrated in Figure 1.10.

The geometry of carbon with four single bonds, as in methane, may be represented as

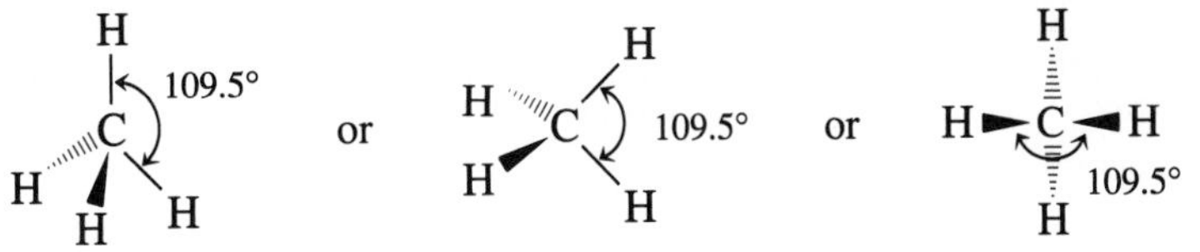

H, C, H, H, H, 109.5° or H, C, H, H, H, 109.5° or H, C, H, H, H, 109.5°

where the solid lines lie in the plane of the page, the dashed wedge goes behind the plane of the paper, and the solid wedge extends out of the plane of the paper toward you. Formulas drawn in this way are sometimes called 3D (that is, three-dimensional) formulas.

Now that we have described single covalent bonds and their geometry, we are ready to tackle, in the next chapter, the structure and chemistry of saturated hydrocarbons. But before we do that, we present a brief overview of organic chemistry, so that you can see how the subject will be organized for study.

Because carbon atoms can be linked to one another or to other atoms in so many different ways, the number of possible organic compounds is almost limitless. Literally millions of organic compounds have been characterized, and the number grows daily. How can we hope to study this vast subject systemati-

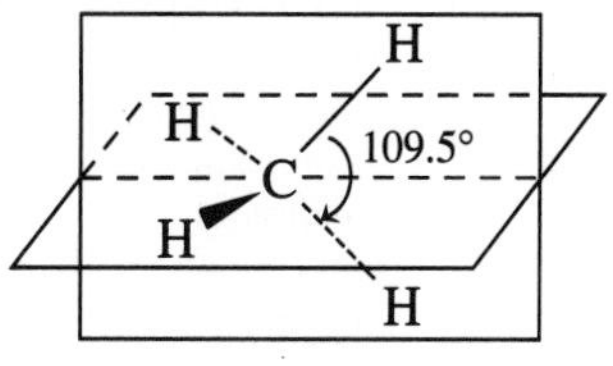

H, H, C, 109.5°, H, H

FIGURE 1.10 The carbon and two of the hydrogens in methane form a plane that perpendicularly bisects the plane formed by the carbon and the other two hydrogens.

cally? Fortunately, organic compounds can be classified according to their structures into a relatively small number of groups. Structures can be classified both according to the carbon framework (sometimes called the carbon *skeleton*) and according to the groups that are attached to that framework.

1.17 Classification According to Molecular Framework

There are three main classes of molecular frameworks for organic structures.

1.17a Acyclic Compounds By *acyclic* we mean *not cyclic*. Acyclic organic molecules have chains of carbon atoms but no rings. As we have seen, the chains may be unbranched or branched.

unbranched chain of eight carbon atoms

branched chain of eight carbon atoms

Pentane is an example of an acyclic compound with an unbranched carbon chain, whereas isopentane and neopentane are also acyclic but have branched carbon frameworks (Sec. 1.9). Figure 1.11 shows the structures of a few acyclic compounds that occur in nature.

1.17b Carbocyclic Compounds Carbocyclic compounds contain rings of carbon atoms. The smallest possible carbocyclic ring has three carbon atoms, but carbon rings come in many sizes and shapes. The rings may have chains of carbon atoms attached to them and may contain multiple bonds. Many compounds with more than one carbocyclic ring are known. Figure 1.12 shows the structures of a few carbocyclic compounds that occur in nature. Five- and six-membered rings are most common, but smaller and larger rings are also found.

1.17c Heterocyclic Compounds Heterocyclic compounds make up the third and largest class of molecular frameworks for organic compounds. In het-

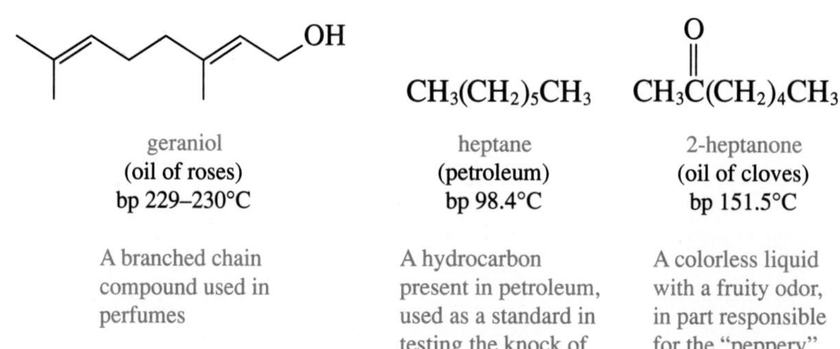
OH
$CH_3(CH_2)_5CH_3$
O
$CH_3C(CH_2)_4CH_3$
geraniol
(oil of roses)
bp 229–230°C
A branched chain compound used in perfumes
heptane
(petroleum)
bp 98.4°C
A hydrocarbon present in petroleum, used as a standard in testing the knock of gasoline engines
2-heptanone
(oil of cloves)
bp 151.5°C
A colorless liquid with a fruity odor, in part responsible for the "peppery" odor of blue cheese

FIGURE 1.11 Examples of natural acyclic compounds, their sources (in parentheses), and selected characteristics.

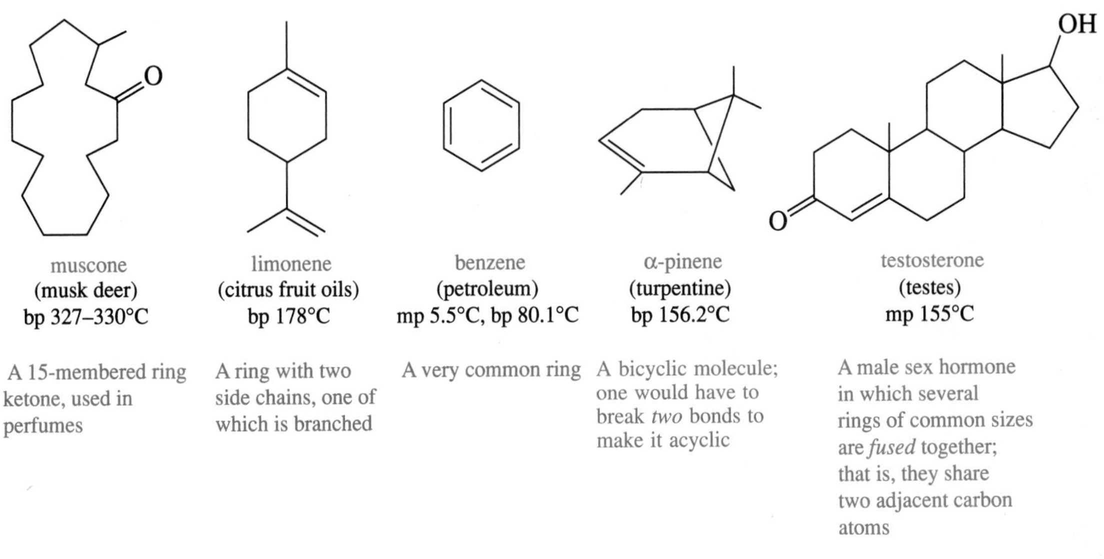
O
OH
O
muscone
(musk deer)
bp 327–330°C
limonene
(citrus fruit oils)
bp 178°C
benzene
(petroleum)
mp 5.5°C, bp 80.1°C
α-pinene
(turpentine)
bp 156.2°C
testosterone
(testes)
mp 155°C
A 15-membered ring ketone, used in perfumes
A ring with two side chains, one of which is branched
A very common ring
A bicyclic molecule; one would have to break *two* bonds to make it acyclic
A male sex hormone in which several rings of common sizes are *fused* together; that is, they share two adjacent carbon atoms

FIGURE 1.12 Examples of natural carbocyclic compounds with rings of various sizes and shapes. The source and special features of each structure are indicated below it.

erocyclic compounds, at least one atom in the ring must be a heteroatom, an atom that is *not* carbon. The most common heteroatoms are **oxygen, nitrogen,** and **sulfur,** but heterocyclics with other elements are also known. More than one heteroatom may be present and, if so, the heteroatoms may be alike or different. Heterocyclic rings come in many sizes, may contain multiple bonds, may have carbon chains or rings attached to them, and in short may exhibit a great variety of structures. Figure 1.13 shows the structures of a few natural products that contain heterocyclic rings. In these abbreviated structural formulas, the symbols for the heteroatoms are shown, but the carbons are indicated using lines only.

The formulas in Figures 1.11 through 1.13 show not only the molecular frameworks, but also various groups of atoms that may be part of or attached to the frameworks. Fortunately, these groups can also be classified in a way that helps simplify the study of organic chemistry.

1.18 *Classification According to Functional Group*

Certain groups of atoms have chemical properties that depend only moderately on the molecular framework to which they are attached. These groups of atoms are called **functional groups.** The **hydroxyl group, —OH,** is an example of a functional group, and compounds with this group attached to a carbon framework are called **alcohols.** In most organic reactions, some chemical change occurs at the functional group, but the rest of the molecule keeps its original structure. This maintenance of most of the structural formula throughout a chemical reaction greatly simplifies our study of organic chemistry. It allows us to focus attention on the chemistry of the various functional groups. We can

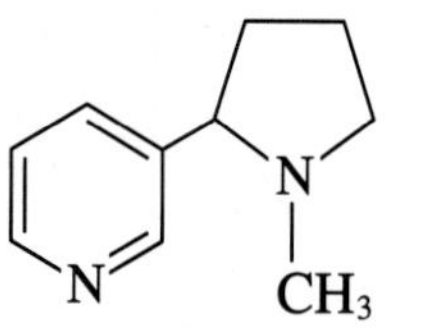

nicotine
bp 246°C

Present in tobacco, nicotine has two heterocyclic rings of different sizes, each containing one nitrogen.

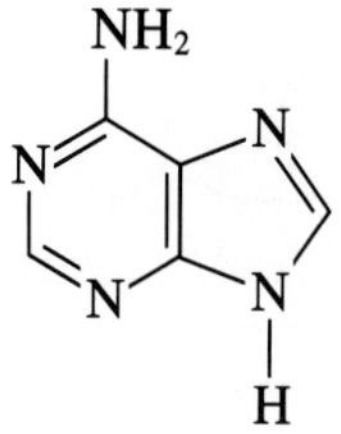

adenine
mp 360–365°C
(decomposes)

One of the four heterocyclic bases of DNA, adenine contains two fused heterocyclic rings, each of which contains two heteroatoms (nitrogen).

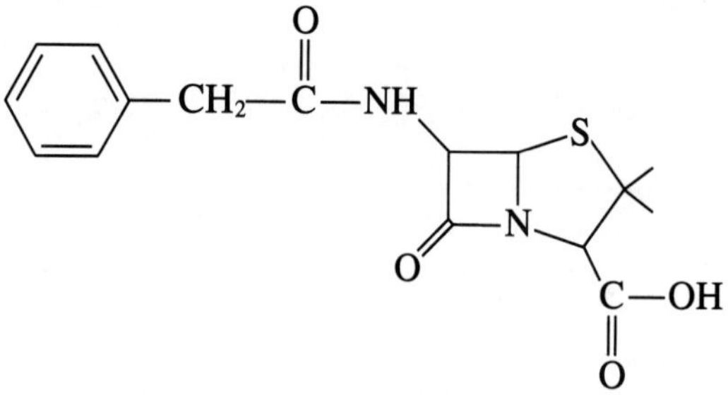

penicillin-G
(amorphous solid)

One of the most widely used antibiotics, penicillin has two heterocyclic rings, the smaller of which is crucial to biological activity.

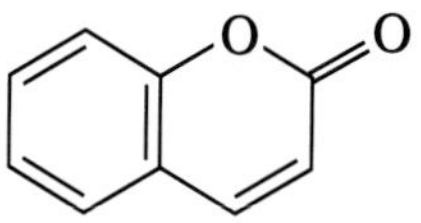

coumarin
mp 71°C

Found in clover and grasses, coumarin produces the pleasant odor of new-mown hay.

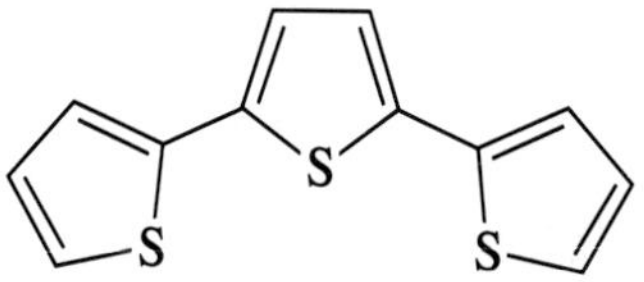

α-terthienyl
mp 92–93°C

This compound, with three linked sulfur-containing rings, is present in certain marigold species.

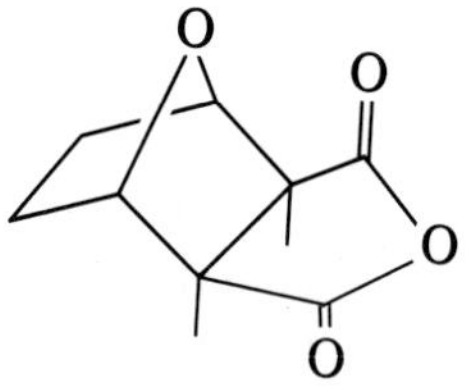

cantharidin
mp 218°C

This compound, an oxygen heterocycle, is the active principle in cantharis (also known as Spanish fly), a material isolated from certain dried beetles of the species *Cantharis vesicatoria* and incorrectly thought by some to increase sexual desire.

FIGURE 1.13 Examples of natural heterocyclic compounds having a variety of heteroatoms and ring sizes.

study classes of compounds instead of having to learn the chemistry of each individual compound.

Some of the main functional groups that we will study are listed in Table 1.6, together with a typical compound of each type. Although we will describe these classes of compounds in greater detail in later chapters, it would be a

TABLE 1.6 The main functional groups

	Structure	Class of compound	Specific example	Common name of the specific example
A. Functional groups that are a part of the molecular framework	$-\overset{\vert}{\underset{\vert}{C}}-\overset{\vert}{\underset{\vert}{C}}-$	alkane	CH_3-CH_3	ethane, a component of natural gas
	$>C=C<$	alkene	$CH_2=CH_2$	ethylene, used to make polyethylene
	$-C\equiv C-$	alkyne	$HC\equiv CH$	acetylene, used in welding
B. Functional groups containing oxygen *1. With carbon–oxygen single bonds*	$-\overset{\vert}{\underset{\vert}{C}}-OH$	alcohol	CH_3CH_2OH	ethyl alcohol, found in beer, wines, and liquors
	$-\overset{\vert}{\underset{\vert}{C}}-O-\overset{\vert}{\underset{\vert}{C}}-$	ether	$CH_3CH_2OCH_2CH_3$	diethyl ether, the common anesthetic
*2. With carbon–oxygen double bonds**	$-\overset{O}{\overset{\Vert}{C}}-H$	aldehyde	$CH_2=O$	formaldehyde, used to preserve biological specimens
	$-\overset{\vert}{\underset{\vert}{C}}-\overset{O}{\overset{\Vert}{C}}-\overset{\vert}{\underset{\vert}{C}}-$	ketone	$CH_3\overset{O}{\overset{\Vert}{C}}CH_3$	acetone, a solvent for varnish and rubber cement
3. With single and double carbon–oxygen bonds	$-\overset{O}{\overset{\Vert}{C}}-OH$	carboxylic acid	$CH_3\overset{O}{\overset{\Vert}{C}}-OH$	acetic acid, a component of vinegar
	$-\overset{O}{\overset{\Vert}{C}}-O-\overset{\vert}{\underset{\vert}{C}}-$	ester	$CH_3\overset{O}{\overset{\Vert}{C}}-OCH_2CH_3$	ethyl acetate, a solvent for nail polish and model airplane glue
*C. Functional groups containing nitrogen***	$-\overset{\vert}{\underset{\vert}{C}}-NH_2$	primary amine	$CH_3CH_2NH_2$	ethylamine, smells like ammonia
	$-C\equiv N$	cyanide or nitrile	$CH_2=CH-C\equiv N$	acrylonitrile, raw material for making Orlon

* The $>C=O$ group, present in several functional groups, is called a **carbonyl group.** The $-\overset{O}{\overset{\Vert}{C}}-OH$ group of acids is called a **carboxyl group** (a contraction of *carb*onyl and hydr*oxyl*).
** The $-NH_2$ group is called an **amino group.**

TABLE 1.6 Continued

	Structure	Class of compound	Specific example	Common name of the specific example
D. Functional group with oxygen and nitrogen	$-\overset{\overset{\displaystyle O}{\parallel}}{C}-NH_2$	primary amide	$H-\overset{\overset{\displaystyle O}{\parallel}}{C}-NH_2$	formamide, a softener for paper
*E. Functional groups containing sulfur**	$-\overset{\vert}{\underset{\vert}{C}}-SH$	thiol (also called mercaptan)	CH_3SH	methanethiol, has the odor of rotten cabbage
	$-\overset{\vert}{\underset{\vert}{C}}-S-\overset{\vert}{\underset{\vert}{C}}-$	thioether (also called sulfide)	$(CH_2{=}CHCH_2)_2S$	diallyl sulfide, has the odor of garlic

* Thiols and thioethers are the sulfur analogs of alcohols and ethers.

good idea for you to become familiar with their names and structures now. If a particular functional group is mentioned before its chemistry is discussed in detail, and you forget what it is, you can refer to Table 1.6 or to the inside front cover of this book.

PROBLEM 1.27 What functional groups can you find in the following natural products? (Their formulas are given in Figures 1.11 and 1.12.)

a. geraniol b. muscone c. limonene d. testosterone

ADDITIONAL PROBLEMS

1.28 Show the number of valence electrons in each of the following atoms. Let the element's symbol represent its kernel, and use dots for the valence electrons.

a. carbon **b.** bromine **c.** silicon
d. boron **e.** sulfur **f.** phosphorus

1.29. When a solution of salt (sodium chloride) in water is treated with a silver nitrate solution, a white precipitate forms immediately. When tetrachloromethane is shaken with aqueous silver nitrate, no such precipitate is produced. Explain these facts in terms of the types of bonds present in the two chlorides.

1.30. Use the relative positions of the elements in the periodic table (Table 1.3 or inside back cover) to classify the following substances as ionic or covalent:

a. NaF **b.** F_2 **c.** $MgCl_2$ **d.** P_2O_5
e. S_2Cl_2 **f.** LiCl **g.** ClF **h.** $SiCl_4$

1.31. For each of the following elements, tell (1) how many valence electrons it has and (2) what its common valence is:

a. oxygen **b.** hydrogen **c.** fluorine
d. carbon **e.** nitrogen **f.** sulfur

1.32. Write a structural formula for each of the following compounds, using a line to represent each single bond and dots for any unshared electron pairs:

a. CH_3F **b.** C_3H_8 **c.** C_2H_5Cl
d. CH_3NH_2 **e.** CH_3CH_2SH **f.** CH_2O

1.33. Draw a structural formula for each of the following covalent molecules. Which bonds are polar? Indicate the polarity by proper placement of the symbols $\delta+$ and $\delta-$.

a. Cl_2 **b.** CH_3Cl **c.** CO_2 **d.** HF
e. SF_6 **f.** CH_4 **g.** SO_2 **h.** CH_3OH

1.34. Considering bond polarity, which hydrogen in acetic acid, $CH_3\overset{\overset{\large O}{\|}}{C}—OH$, do you expect to be most acidic? Write an equation for the reaction between acetic acid and metallic sodium.

1.35. Draw structural formulas for all possible isomers having the following molecular formulas:

a. C_3H_8 **b.** C_3H_7Cl **c.** $C_2H_4F_2$ **d.** $C_3H_6Cl_2$
e. C_4H_9I **f.** $C_2H_2Br_2$ **g.** C_3H_6 **h.** $C_4H_{10}O$

1.36. Draw structural formulas for the five isomers of C_6H_{14}. As you write them out, try to be systematic.

1.37. For each of the following abbreviated structural formulas, write a structural formula that shows all the bonds:

a. $CH_3(CH_2)_3CH_3$ **b.** $(CH_3)_3CCH_2CH_3$ **c.** $(CH_3)_2CHOH$
d. $(CH_3CH_2)_2S$ **e.** CH_2ClCH_2OH **f.** $CH_3N(CH_2CH_3)_2$

1.38. Write structural formulas that correspond to the following abbreviated structures, and show the correct number of hydrogens on each carbon:

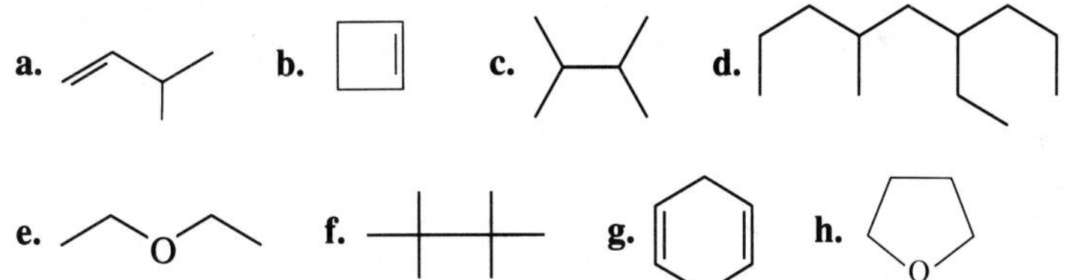

1.39. An abbreviated formula of geraniol is shown in Figure 1.11.

a. How many carbons does geraniol have?
b. What is its molecular formula?
c. Write a more detailed structural formula for it.

1.40. What is the *molecular formula* for each of the following compounds? Consult Figures 1.12 and 1.13 for the abbreviated structural formulas.

a. muscone **b.** benzene **c.** testosterone
d. nicotine **e.** coumarin **f.** adenine

1.41. Write electron-dot formulas for the following species. Show where the formal charges, if any, are located.

a. nitrous acid, HONO **b.** nitric acid, $HONO_2$
c. formaldehyde, H_2CO **d.** ammonium ion, NH_4^+
e. cyanide ion, CN^- **f.** carbon monoxide, CO
g. sulfur trioxide, SO_3 **h.** boron trifluoride, BF_3
i. hydrogen peroxide, H_2O_2 **j.** bicarbonate ion, HCO_3^-

1.42. Consider each of the following highly reactive carbon species. What is the formal charge on carbon in each species?

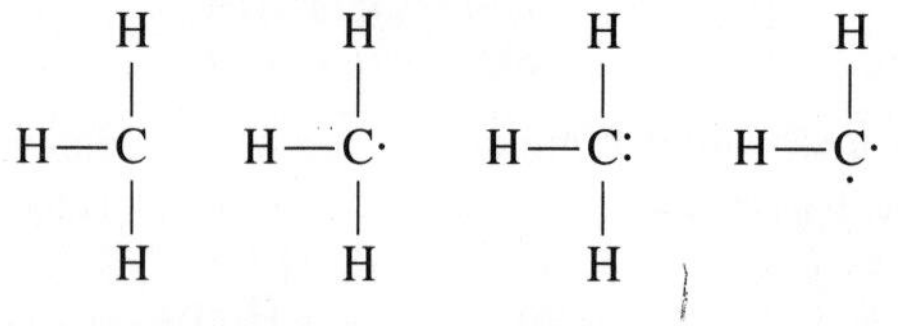

```
   H        H        H        H
   |        |        |        |
H—C     H—C·     H—C:     H—C·
   |        |        |        ·
   H        H        H
```

1.43. Draw electron-dot formulas for the two contributors to the resonance hybrid structure of the nitrite ion, NO_2^-. (Each oxygen is connected to the nitrogen.) What is the charge on each oxygen in each contributor and in the hybrid structure? Show by curved arrows how the electron pairs can relocate to interconvert the two structures.

1.44. Write the structure obtained when electrons move as indicated by the curved arrows in the following structure:

```
        :Ö:
         ||
CH3—C—N̈H2
```

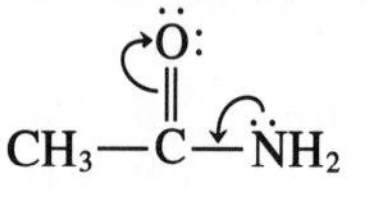

Does each atom in the resulting structure have a complete valence shell of electrons? Locate any formal charges in each structure.

1.45. Add curved arrows to the following structures to show how electron pairs must be moved to interconvert the structures, and locate any formal charges.

[C6H5—Ö—H ⟷ (:C6H5)=Ö—H]

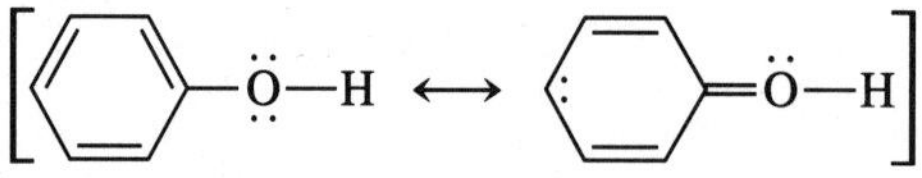

1.46. Add curved arrows to show how electrons must move to form the product from the reactants in the following equation, and locate any formal charges.

```
                         :Ö                    :Ö:
                          ||                     |
CH3—N̈H2 + CH3—C—OCH3  ⟶  CH3—C—OCH3
                                                 |
                                              H2N—CH3
```

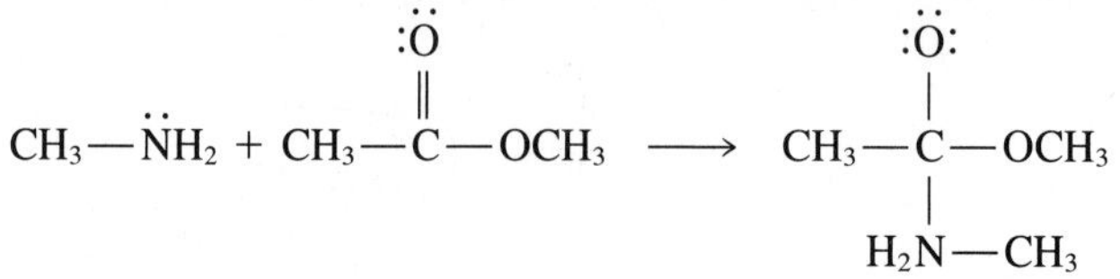

1.47. Each of the following substances contains ionic and covalent bonds. Draw their electron-dot formulas.

a. CH_3ONa **b.** NH_4Cl

1.48. Fill in any unshared electron pairs that are missing from the following formulas:

a. CH_3CH_2SH **b.** $CH_3C(=O)—OH$ **c.** $(CH_3)_2NH$ **d.** $CH_3OCH_2CH_2OH$

1.49. Make a drawing (similar to the right-hand part of Figure 1.6) of the electron distribution that will be expected in nitrogen atoms if the *s* and *p* orbitals are hybridized to sp^3. Based on this model, predict the geometry of the ammonia molecule, NH_3.

1.50. The ammonium ion, NH_4^+, has a tetrahedral geometry analogous to that of methane. Explain this structure in terms of atomic and molecular orbitals.

1.51. Silicon is just below carbon in the periodic table. Predict the geometry of silane, SiH_4.

1.52. Use lines, dashed wedges, and solid wedges to show the geometry of CCl_4 and CH_3OH.

1.53. Write a structural formula that corresponds to the molecular formula C_4H_8O and is

a. acyclic **b.** carbocyclic **c.** heterocyclic

1.54. Divide the following compounds into groups that might be expected to exhibit similar chemical behavior:

a. CH_3OH **b.** CH_3OCH_3 **c.** $CH_2(OH)CH(OH)CH_2(OH)$
d. C_5H_{12} **e.** C_4H_9OH **f.** C_8H_{18}
g. C_3H_7OH **h.** C_6H_{14} **i.** $CH_3OCH_2CH_2OCH_3$

1.55. Using Table 1.6, write a structural formula for each of the following:

a. an alcohol, $C_4H_{10}O$ **b.** an ether, C_3H_8O
c. an aldehyde, C_3H_6O **d.** a ketone, C_4H_8O
e. a carboxylic acid, $C_3H_6O_2$ **f.** an ester, $C_5H_{10}O_2$
g. an alcohol that is an isomer of the one in part a **h.** an amine, C_3H_9N

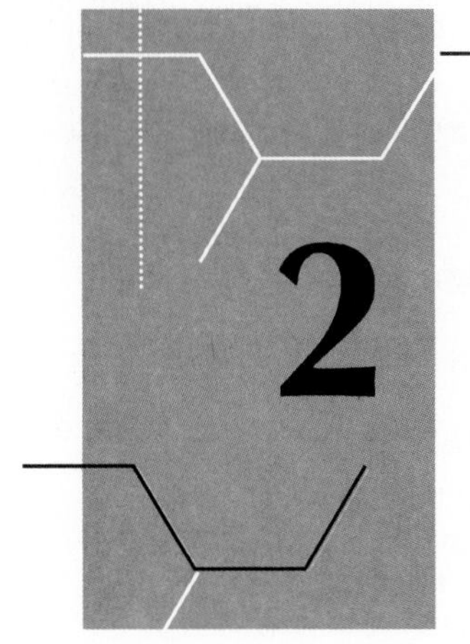

2 Alkanes and Cycloalkanes; Conformational and Geometric Isomerism

2.1 Introduction

The main components of petroleum and natural gas, resources that now supply most of our fuel for energy, are **hydrocarbons, compounds that contain only carbon and hydrogen.** There are three main classes of hydrocarbons, based on the types of carbon–carbon bonds present. **Saturated hydrocarbons** contain only carbon–carbon *single* bonds. **Unsaturated hydrocarbons** contain carbon–carbon *multiple* bonds—double bonds, triple bonds, or both. **Aromatic hydrocarbons** are a special class of cyclic compounds related in structure to benzene.*

Saturated hydrocarbons are known as **alkanes** if they are acyclic, or as **cycloalkanes** if they are cyclic. Let us look at the structure and properties of alkanes.

2.2 The Structures of Alkanes

The simplest alkane is methane. Its tetrahedral three-dimensional structure was described in the previous chapter (see Figure 1.9). Additional alkanes are constructed by lengthening the carbon chain and adding an appropriate number of hydrogens to complete the carbon valences (for examples, see Figure 2.1** and Table 2.1).

All alkanes fit the general molecular formula C_nH_{2n+2}, where n is the number of carbon atoms. Alkanes with carbon chains that are unbranched (Table

---

* Unsaturated and aromatic hydrocarbons are discussed in Chapters 3 and 4, respectively.

** Molecular models can help you visualize organic structures in three dimensions. They will be extremely useful to you throughout this course, especially when we consider various types of isomerism. Relatively inexpensive sets are usually available at stores that sell textbooks, and your instructor can suggest which kind to buy. If you cannot locate or afford a set, you can create models that are adequate for some purposes from toothpicks (for bonds) and marshmallows, gum drops, or jelly beans (for atoms).

FIGURE 2.1
Three-dimensional models of ethane, propane, and butane. The stick-and-ball models at the left show the way in which the atoms are connected and depict the correct bond angles. The space-filling models at the right are constructed to scale and give a better idea of the molecular shape.

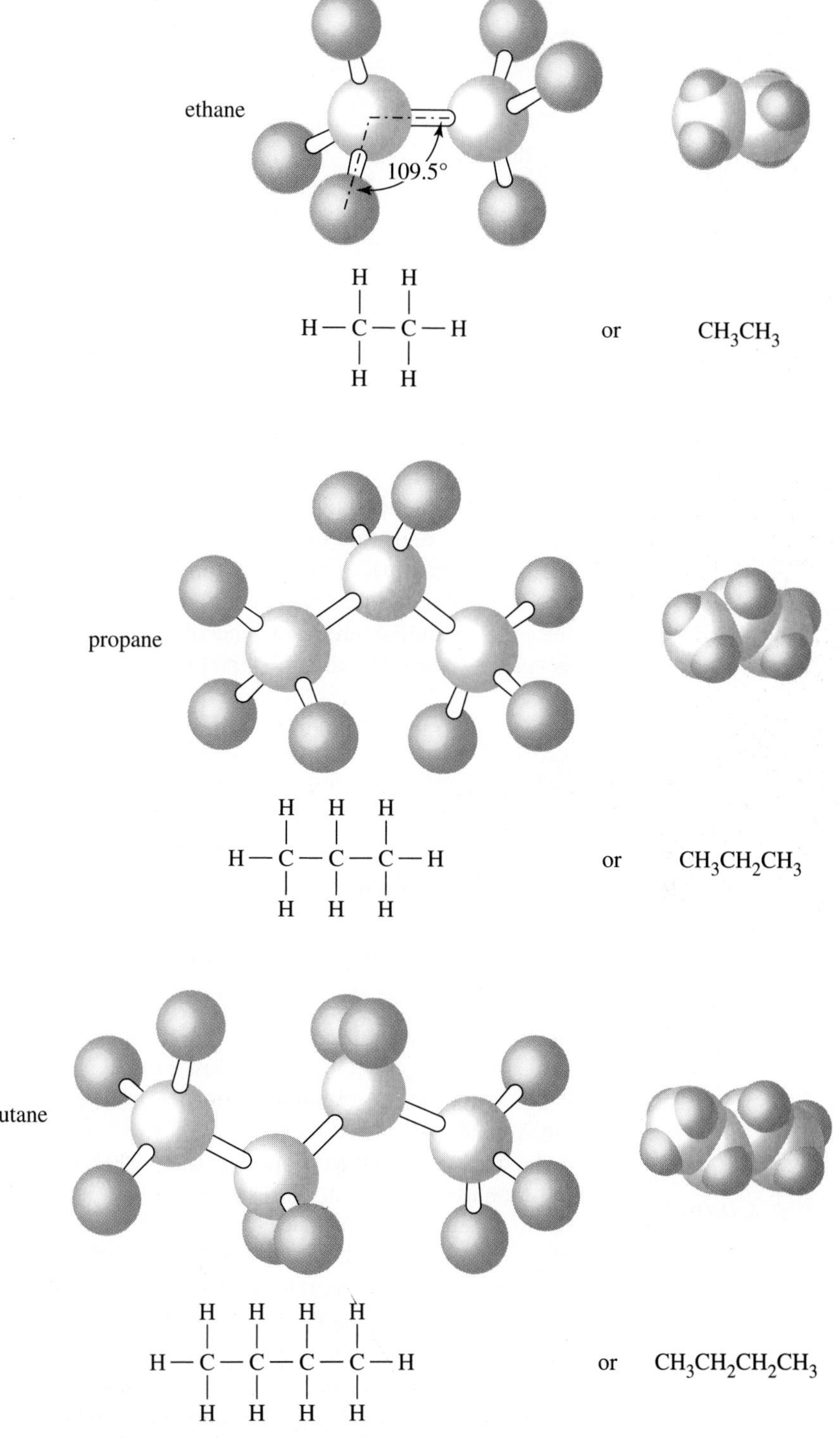

TABLE 2.1 Names and formulas of the first ten unbranched alkanes

Name	*Number of carbons*	*Molecular formula*	*Structural formula*	*Number of structural isomers*
methane	1	CH_4	CH_4	1
ethane	2	C_2H_6	CH_3CH_3	1
propane	3	C_3H_8	$CH_3CH_2CH_3$	1
butane	4	C_4H_{10}	$CH_3CH_2CH_2CH_3$	2
pentane	5	C_5H_{12}	$CH_3(CH_2)_3CH_3$	3
hexane	6	C_6H_{14}	$CH_3(CH_2)_4CH_3$	5
heptane	7	C_7H_{16}	$CH_3(CH_2)_5CH_3$	9
octane	8	C_8H_{18}	$CH_3(CH_2)_6CH_3$	18
nonane	9	C_9H_{20}	$CH_3(CH_2)_7CH_3$	35
decane	10	$C_{10}H_{22}$	$CH_3(CH_2)_8CH_3$	75

2.1) are called **normal alkanes.** Each member of this series differs from the next higher and the next lower member by a $—CH_2—$ group (called a **methylene group).** A series of compounds in which the members are built up in a regular, repetitive way is called a **homologous series.** Members of such a series have similar chemical and physical properties, which change gradually as carbon atoms are added to the chain.

EXAMPLE 2.1 What is the molecular formula of an alkane with 6 carbon atoms?

Solution If $n = 6$, then $2n + 2 = 14$. The formula is C_6H_{14}.

PROBLEM 2.1 What is the molecular formula of an alkane with 16 carbon atoms?

PROBLEM 2.2 Which of the following are alkanes?

a. C_8H_{16} b. C_7H_{16} c. C_7H_{18} d. $C_{27}H_{56}$

2.3 *Nomenclature of Organic Compounds*

In the early days of organic chemistry, each new compound was given a name that was usually based on its source or use. Examples (Figures 1.12 and 1.13) include limonene (from lemons), α-pinene (from pine trees), coumarin (from the tonka bean, known to South American natives as *cumaru*), and penicillin (from the mold that produces it, *Pencillium notatum*). Even today, this method of naming may be used to give a short, simple name to a molecule with a complex structure. For example, cubane was named after its shape.

It became clear many years ago, however, that one could not rely only on common or trivial names and that a systematic method for naming compounds was needed. Ideally, the rules of the system should result in a unique name for each compound. Knowing the rules and seeing a structure, one should be able to write the systematic name. Seeing the systematic name, one should be able to write the correct structure.

Eventually, an internationally recognized system of nomenclature was devised by a commission of the International Union of Pure and Applied Chemistry; it is known as the IUPAC (pronounced "eye-you-pack") system. In this book, we will use mainly IUPAC names. However, in some cases, the common name is so widely used that we will ask you to learn it (for example, formaldehyde [common] is used in preference to methanal [systematic], and cubane is much easier to remember than its systematic name pentacyclo$[4.2.0.0^{2,5}.0^{3,8}.0^{4,7}]$octane).

2.4 IUPAC Rules for Naming Alkanes

1. The general name for acyclic saturated hydrocarbons is **alkanes.** The *-ane* ending is used for all saturated hydrocarbons. This is important to remember because later other endings will be used for other functional groups.
2. Alkanes without branches are named according to the *number of carbon atoms*. These names, up to 10 carbons, are given in the first column of Table 2.1.
3. For alkanes with branches, ***the root name is that of the longest continuous chain of carbon atoms.*** For example, in the structure

$$\begin{array}{c} \quad CH_3 \quad CH_3 \\ \quad | \qquad | \\ CH_3{-}CH{-}CH{-}CH_2{-}CH_3 \end{array} \quad \text{or} \quad \begin{array}{c} \quad CH_3 \quad CH_3 \\ \quad | \qquad | \\ CH_3{-}CH{-}CH{-}CH_2{-}CH_3 \end{array}$$

   the longest continuous chain (in color) has five carbon atoms. The compound is therefore named as a substituted *pent*ane, even though there are seven carbon atoms altogether.
4. Groups attached to the main chain are called **substituents.** Saturated substituents that contain only carbon and hydrogen are called **alkyl groups.** *An alkyl group is named by taking the name of the alkane with the same number of carbon atoms and changing the* -ane *ending to* -yl.

   In the example above, each substituent has only one carbon. Derived from methane by removing one of the hydrogens, a one-carbon substituent is called a **methyl group.**

$$\underset{\text{methane}}{\begin{array}{c} H \\ | \\ H{-}C{-}H \\ | \\ H \end{array}} \qquad \begin{array}{c} H \\ | \\ H{-}C{-} \\ | \\ H \end{array} \quad \text{or} \quad CH_3{-} \quad \text{or} \quad Me{-}$$

methyl group

   The names of substituents with more than one carbon atom will be described in Section 2.5.
5. *The main chain is numbered in such a way that the first substituent encountered along the chain receives the lowest possible number*. Each substituent is then located by its name and by the number of the carbon atom to which it is attached. When two or more identical groups are attached to the main chain, prefixes such as *di-*, *tri-*, and *tetra-* are used. *Every substituent must be named and numbered,* even if two identical substituents are attached to

the same carbon of the main chain. The compound

$$\overset{1}{CH_3}-\overset{2}{\underset{}{CH}}(CH_3)-\overset{3}{CH}(CH_3)-\overset{4}{CH_2}-\overset{5}{CH_3}$$

is correctly named **2,3-dimethylpentane.** The name tells us that there are two methyl substituents, one attached to carbon-2 and one attached to carbon-3 of a five-carbon saturated chain.

6. If two or more different types of substituents are present, they are listed alphabetically, except that prefixes such as *di-* and *tri-* are not considered when alphabetizing.
7. *Punctuation is important when writing IUPAC names.* IUPAC names for hydrocarbons are written as one word. Numbers are separated from each other by commas and are separated from letters by hyphens. There is no space between the last named substituent and the name of the parent alkane that follows it.

To summarize and amplify these rules, we take the following steps to find an acceptable IUPAC name for an alkane:

1. Locate the longest continuous carbon chain. This gives the name of the parent hydrocarbon. For example,

$$C-C-\underset{}{C}(-C-C)-C \quad \textit{not} \quad C-C-C(-C-C)-C$$

2. Number the longest chain beginning at the end nearest the first branch point. For example,

$$\underset{6}{C}-\underset{5}{C}-\underset{4}{C}(C)-\underset{3}{C}-\underset{2}{C}(C)-\underset{1}{C} \quad \textit{not} \quad \underset{1}{C}-\underset{2}{C}-\underset{3}{C}(C)-\underset{4}{C}-\underset{5}{C}(C)-\underset{6}{C}$$

If there are two equally long continuous chains, select the one with the most branches. For example,

$$\overset{1}{C}-\overset{2}{C}(C)-\overset{3}{C}(C-C)-\overset{4}{C}-\overset{5}{C}-\overset{6}{C} \quad \textit{not} \quad C-C(C)-\overset{3}{C}(\underset{2}{C}-\underset{1}{C})-\overset{4}{C}-\overset{5}{C}-\overset{6}{C}$$

two branches — one branch

If there is a branch equidistant from each end of the longest chain, begin numbering nearest to a third branch:

$$\underset{1}{C}-\underset{2}{C}(C)-\underset{3}{C}(C)-\underset{4}{C}-\underset{5}{C}-\underset{6}{C}(C)-\underset{7}{C} \quad \textit{not} \quad \underset{7}{C}-\underset{6}{C}(C)-\underset{5}{C}(C)-\underset{4}{C}-\underset{3}{C}-\underset{2}{C}(C)-\underset{1}{C}$$

2,3,6-trimethylheptane — 2,5,6-trimethylheptane

If there is no third branch, begin numbering nearest the substituent whose name has alphabetic priority:

```
         C—C   C                       C—C   C
           |   |                         |   |
C—C—C—C—C—C—C      not      C—C—C—C—C—C—C
1 2 3 4 5 6 7               7 6 5 4 3 2 1
3-ethyl-5-methylheptane     5-ethyl-3-methylheptane
```

3. Write the name as one word, placing substituents in alphabetic order and using proper punctuation.

EXAMPLE 2.2 Give an IUPAC name for $CH_3—\overset{\displaystyle CH_3}{\underset{\displaystyle CH_3}{\overset{|}{\underset{|}{C}}}}—CH_2CH_2CH_3$.

Solution

$$\overset{1}{CH_3}—\overset{\displaystyle CH_3}{\underset{\displaystyle CH_3}{\overset{2|}{\underset{|}{C}}}}—\overset{3}{CH_2}\overset{4}{CH_2}\overset{5}{CH_3}$$ 2,2-dimethylpentane

PROBLEM 2.3 Give an IUPAC name for the following compounds:

a. $CH_3\underset{\displaystyle CH_3}{\underset{|}{C}}HCH_2CH_3$ b. $CH_3CH_2\underset{\displaystyle CH_3}{\underset{|}{C}}HCH_3$ c. $CH_3—\overset{\displaystyle CH_3}{\underset{\displaystyle CH_3}{\overset{|}{\underset{|}{C}}}}—CH_3$

2.5 *Alkyl and Halogen Substituents*

As illustrated for the methyl group, alkyl substituents are named by changing the *-ane* ending of alkanes to *-yl*. Thus the two-carbon alkyl group is called the **ethyl group,** from ethane.

CH_3CH_3	$CH_3CH_2—$	or	$C_2H_5—$	or	Et—
ethane			ethyl group		

When we come to propane, there are two possible alkyl groups, depending on which type of hydrogen is removed. If a *terminal* hydrogen is removed, the group is called a **propyl group.**

```
   H  H  H            H  H  H
   |  |  |            |  |  |
H—C—C—C—H        H—C—C—C—     or   CH3CH2CH2—   or   Pr—
   |  |  |            |  |  |
   H  H  H            H  H  H
   propane                      propyl group
```

But if a hydrogen is removed from the *central* carbon atom, we get a different isomeric propyl group, called the **isopropyl** (or 1-methylethyl) **group.**

$$\begin{array}{ccccccc} & H & & H & & H & \\ & | & & | & & | & \\ H- & C & - & C & - & C & -H \\ & | & & | & & | & \\ & H & & H & & H & \end{array} \qquad \begin{array}{ccccccc} & H & & H & & H & \\ & | & & | & & | & \\ H- & C & - & C & - & C & -H \\ & | & & | & & | & \\ & H & & & & H & \end{array} \quad \text{or} \quad \begin{array}{c} CH_3CHCH_3 \\ | \end{array} \quad \text{or} \quad i\text{-Pr}-$$

propane — isopropyl or 1-methylethyl* group

There are four different butyl groups:

$$CH_3CH_2CH_2CH_2- \qquad \begin{array}{c} CH_3CHCH_2CH_3 \\ | \end{array} \qquad \begin{array}{l} CH_3 \\ \quad \diagdown \\ \quad\; CH-CH_2- \\ \quad \diagup \\ CH_3 \end{array} \qquad \begin{array}{c} CH_3 \\ | \\ CH_3-C- \\ | \\ CH_3 \end{array}$$

butyl — *sec*-butyl (or 1-methylpropyl) — isobutyl (or 2-methylpropyl) — *tert*-butyl (or 1,1-dimethylethyl)

These names for the alkyl groups with up to four carbon atoms are very commonly used, so you should memorize them.

The letter **R** is used as a general symbol for an alkyl group. The formula R—H therefore represents any alkane, and the formula R—Cl stands for any alkyl chloride (methyl chloride, ethyl chloride, and so on).

Halogen substituents are named by changing the *-ine* ending of the element to *-o*.

F—	Cl—	Br—	I—
fluoro-	chloro-	bromo-	iodo-

EXAMPLE 2.3 Give the common and IUPAC names for $CH_3CH_2CH_2Br$.

Solution The common name is propyl bromide (the common name of the alkyl group is followed by the name of the halide). The IUPAC name is 1-bromopropane, the halogen being named as a substituent on the three-carbon chain.

PROBLEM 2.4 Give an IUPAC name for CH_2BrCl.

---

* The name 1-methylethyl for this group comes about by regarding it as a substituted ethyl group.

$$\overset{2}{C}H_3\overset{1}{C}H_2- \qquad \begin{array}{c} \overset{2}{C}H_3\overset{1}{C}H- \\ | \\ CH_3 \end{array}$$

ethyl — 1-methylethyl

PROBLEM 2.5 Write the formula for each of the following compounds:

a. propyl iodide
b. isopropyl chloride
c. 2-chloropropane
d. *tert*-butyl iodide
e. isobutyl bromide
f. all alkyl fluorides

2.6 Use of the IUPAC Rules

The examples given in Table 2.2 illustrate how the IUPAC rules are applied for particular structures. Study each example to see how a correct name is obtained and how to avoid certain pitfalls.

It is important not only to be able to write a correct IUPAC name for a given structure, but also to do the converse: write the structure given the IUPAC name. In this case, first write the longest carbon chain and number it, then add the substituents to the correct carbon atoms, and finally fill in the formula with the correct number of hydrogens at each carbon. For example, to

Table **2.2** Examples of use of the IUPAC rules

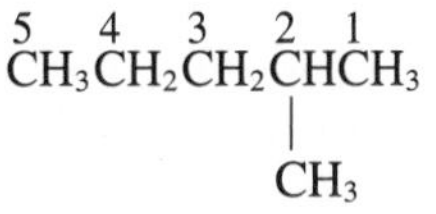 $\overset{5}{C}H_3\overset{4}{C}H_2\overset{3}{C}H_2\overset{2}{C}H\overset{1}{C}H_3$ with CH_3 on C-2 2-methylpentane (*not* 4-methylpentane)	The ending *-ane* tells us that all the carbon–carbon bonds are single; *pent-* indicates five carbons in the longest chain. We number them from right to left, starting closest to the branch point.
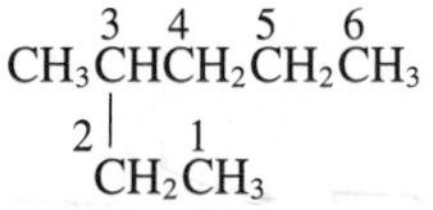 $CH_3\overset{3}{C}H\overset{4}{C}H_2\overset{5}{C}H_2\overset{6}{C}H_3$ with $\overset{2}{C}H_2\overset{1}{C}H_3$ on C-3 3-methylhexane (*not* 2-ethylpentane or 4-methylhexane)	A six-carbon saturated chain with a methyl group on the third carbon. We would usually write the structure as $CH_3CH_2CHCH_2CH_2CH_3$
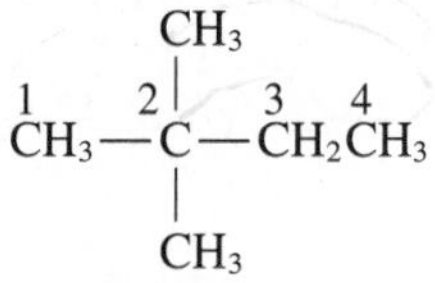 $\overset{1}{C}H_3{-}\overset{2}{C}(CH_3)_2{-}\overset{3}{C}H_2\overset{4}{C}H_3$ 2,2-dimethylbutane (*not* 2,2-methylbutane or 2-dimethylbutane)	There must be a number for each substituent, and the prefix *di-* says that there are two methyl substituents.
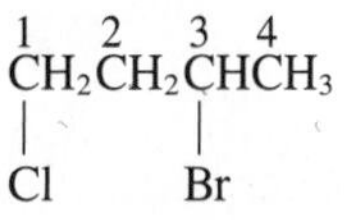 $\overset{1}{C}H_2\overset{2}{C}H_2\overset{3}{C}H\overset{4}{C}H_3$ with Cl on C-1 and Br on C-3 3-bromo-1-chlorobutane (*not* 1-chloro-3-bromobutane or 2-bromo-4-chlorobutane)	First, we number the butane chain from the end closest to the first substituent. Then we name the substituents in alphabetical order, regardless of position number.

write the formula for **2,2,4-trimethylpentane,** we go through the following steps:

C—C—C—C—C (Write down the pentane chain.) —Add the numbers.→ C1—C2—C3—C4—C5 —Add the three methyl substituents.→ C—C(CH$_3$)$_2$—C—C(CH$_3$)—C (methyls on carbons 2, 2 and 4) —Fill in the hydrogens.→

$$CH_3—C(CH_3)_2—CH_2—CH(CH_3)—CH_3$$

2,2,4-trimethylpentane

PROBLEM 2.6 Name the following compounds by the IUPAC system:

a. CH_3CHFCH_3 b. $(CH_3)_3CCH_2CHClCH_3$

PROBLEM 2.7 Write the structure for 3,3-dimethylpentane.

PROBLEM 2.8 Explain why 1,3-dichlorobutane is a correct IUPAC name, but 1,3-dimethylbutane is *not* a correct IUPAC name.

A Word About . . .

1. Isomers, Possible and Impossible

Table 2.1 shows that there are 75 structural isomers of the alkane $C_{10}H_{22}$. How many such isomers do you think there might be if we double the number of carbons ($C_{20}H_{42}$)? The answer is 366,319! And if we double the number of carbons again ($C_{40}H_{82}$)? Exactly 62,481,801,147,341. Of course, no one sits down with pencil and paper or molecular models and determines these numbers by constructing all the possibilities; it could take a lifetime. Complex mathematical formulas have been developed to compute these numbers.

Although we can write some isomers' formulas on paper, they are structurally impossible and cannot be synthesized. Consider, for example, the series of alkanes obtained by replacing the hydrogens of methane with methyl groups and then repeating that process on the product indefinitely. You can see from the drawings on the next page, even though they are only two dimensional, that in this way we build up molecules with a central core of carbon atoms and a surface of hydrogen atoms.

$$CH_4 \longrightarrow C(CH_3)_4 \longrightarrow C[C(CH_3)_3]_4 \longrightarrow C\{C[C(CH_3)_3]_3\}_4$$

$$CH_4 \longrightarrow C_5H_{12} \longrightarrow C_{17}H_{36} \longrightarrow C_{53}H_{108}$$

In three dimensions, the molecules are nearly spherical. Of these compounds, only the first two are known (methane and 2,2-dimethylpropane). The $C_{17}H_{36}$ hydrocarbon (tetra-*t*-butylmethane or, more accurately, 3,3-di-*t*-butyl-2,2,4,4-tetramethylpentane) has not yet been synthesized, and if it ever is, it will be an exceptionally strained molecule. The reason is simply that there is not enough room for all the methyl groups on the surface of the molecule. Calculations from space-filling models show that the inner five carbons form a sphere with a surface area of about 85 Å^2 whereas the 12 methyl groups require a surface area of about 107 Å^2. So the synthesis of this $C_{17}H_{36}$ isomer may be barely possible, but if prepared, its bond angles and bond lengths are likely to be severely distorted from the normal. Synthesizing this hydrocarbon, then, presents an interesting challenge for organic chemists—one that has not yet been successfully met.

There is almost no possibility of synthesizing the $C_{53}H_{108}$ isomer in this series; its structure is too strained. Like these hydrocarbons, the growth of trees, sponges, and other biological forms is similarly limited by the ratio of surface area to volume. (For more on this subject, see the article by R. E. Davies and P. J. Freyd, *J. Chem. Educ.* **1989,** *66,* 278–81.)

2.7 Sources of Alkanes

The two most important natural sources of alkanes are **petroleum** and **natural gas.** *Petroleum* is a complex liquid mixture of organic compounds, many of which are alkanes or cycloalkanes. For more details about how petroleum is refined to obtain gasoline, fuel oil, and other useful substances, read "A Word About Petroleum, Gasoline, and Octane Number" on page 107.

Natural gas, often found associated with petroleum deposits, consists mainly of methane (about 80%) and ethane (5 to 10%), with lesser amounts of some higher alkanes. Propane is the major constituent of liquefied petroleum gas (LPG), a domestic fuel used mainly in rural areas and mobile homes. Butane is the gas of choice in some areas. Natural gas is becoming an energy source that can compete with and possibly surpass oil. In the United States, there are more than 250,000 miles of natural gas pipelines distributing this energy source to all parts of the country. Natural gas is also distributed worldwide via huge tankers. To conserve space, the gas is liquefied (−160°C), because 1 cubic meter (m^3) of liquefied gas is equivalent to about 600 m^3 of gas at atmospheric pressure. Large tankers can carry more than 100,000 m^3 of liquefied gas. In the future, less-developed countries are likely to obtain energy more cheaply by developing local resources of natural gas rather than by importing oil.

2.8 Physical Properties of Alkanes

Alkanes are insoluble in water. The reason is that water molecules are polar and attract one another, whereas alkanes are nonpolar (all the C—C and C—H bonds are nearly purely covalent). To intersperse alkane and water molecules, we would have to break up the attractive force between the water molecules, which would require considerable energy.

The mutual insolubility of alkanes and water is used to advantage by many plants. Alkanes often constitute part of the protective coating on leaves and fruits. If you have ever polished an apple, you know that the skin, or cuticle, contains waxes. Among them are the normal alkanes $C_{27}H_{56}$ and $C_{29}H_{60}$. The leaf wax of cabbage and broccoli is mainly n-$C_{29}H_{60}$, and the main alkane of tobacco leaves is n-$C_{31}H_{64}$. Similar hydrocarbons are found in beeswax. The major function of plant waxes is to prevent water loss from the leaves or fruit.

Alkanes have lower boiling points for a given molecular weight than most other organic compounds. This is because the attractive forces between nonpolar molecules are weak, and the process of separating molecules from one another (which is what we do when we convert a liquid to a gas) requires relatively little energy. Figure 2.2 shows the boiling points of some alkanes. The greater the molecular surface area, the greater the attractive forces between molecules. Therefore, boiling points rise as the chain length increases and fall as the chains become branched and more nearly spherical in shape.

2.9 Conformations of Alkanes

The shapes of molecules often affect their properties. A simple molecule like ethane, for example, can have an infinite number of shapes as a consequence of rotating one carbon atom (and its attached hydrogens) with respect to the other carbon atom. These arrangements are called **conformations** or **conformers.** Two possibilities for ethane are shown in Figure 2.3.

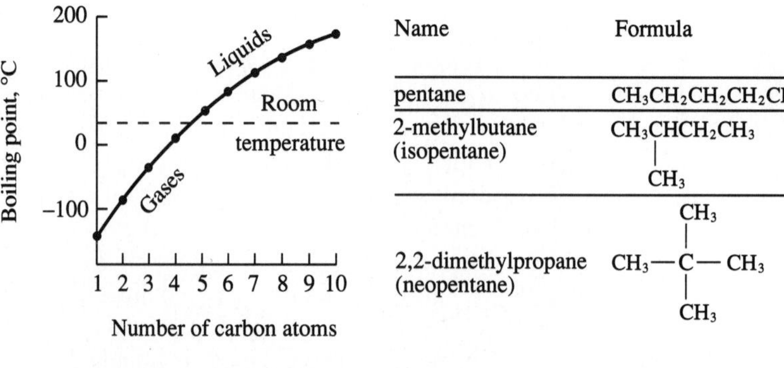

Name	Formula	Boiling point, °C
pentane	$CH_3CH_2CH_2CH_2CH_3$	36
2-methylbutane (isopentane)	$CH_3CH(CH_3)CH_2CH_3$	28
2,2-dimethylpropane (neopentane)	$CH_3—C(CH_3)_2—CH_3$	10

FIGURE 2.2
As shown by the curve, the boiling points of the normal alkanes rise smoothly as the length of the carbon chain increases. Note from the table, however, that chain branching causes a decrease in boiling point (each compound in the table has the same number of carbons and hydrogens, C_5H_{12}).

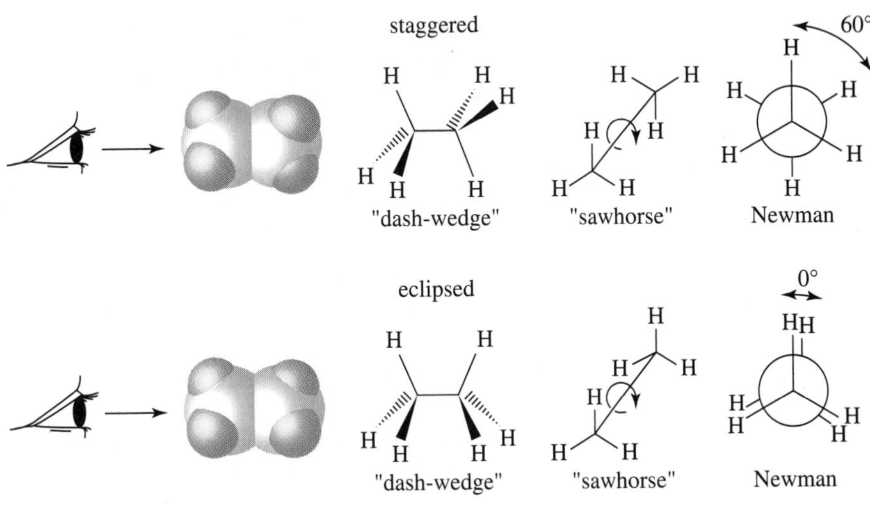

FIGURE 2.3 Two of the possible conformations of ethane: staggered and eclipsed. Interconversion is easy via a 60° rotation about the C—C bond, as shown by the curved arrows. The structures at the left are space-filling models. In each case, the next structure is a "dash-wedge" structure which, if viewed as shown by the eyes, converts to the "sawhorse" drawing, or the Newman projection at the right, an end-on view down the C—C axis. In the Newman projection, the circle represents two eclipsed carbon atoms. Bonds on the "front" carbon go to the center of the circle, and bonds on the "rear" carbon go only to the edge of the circle.

In the **staggered conformation** of ethane, each C—H bond on one carbon bisects an H—C—H angle on the other carbon. In the **eclipsed conformation,** C—H bonds on the front and back carbons are aligned. By rotating one carbon 60° with respect to the other, we can interconvert staggered and eclipsed conformations. Between these two extremes, there is an infinite number of intermediate conformations of ethane.

The staggered and eclipsed conformations of ethane can be regarded as **rotamers** because each is convertible to the other by rotation about the carbon–carbon bond. Such rotation about a single bond occurs easily because the amount of overlap of the sp^3 orbitals on the two carbon atoms is unaffected by rotation about the sigma bond (see Figure 1.8). Indeed, there is enough energy available at room temperature for the staggered and eclipsed conformers of ethane to interconvert rapidly. Consequently, the conformers cannot be separated from one another. We know from various types of physical evidence, however, that both forms are not equally stable. The staggered conformation is the most stable (has the lowest potential energy) of all ethane conformations, and at room temperature more than 99% of ethane molecules have the staggered arrangement. In contrast, the eclipsed conformation is the least stable (highest potential energy) conformation. The reasons why the staggered conformation is preferred are quite complex, but one of them is probably that the bonding electrons on adjacent carbons are farthest apart in the staggered conformer and therefore experience less mutual repulsion in this arrangement.

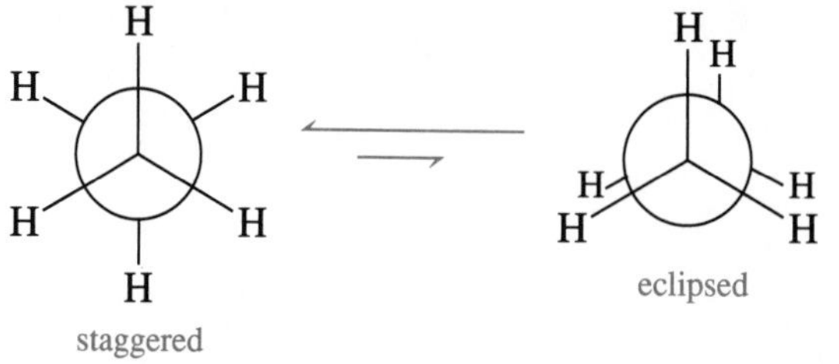

staggered ⇌ eclipsed (2.1)

EXAMPLE 2.4 Draw the Newman projections for the staggered and eclipsed conformations of propane.

Solution

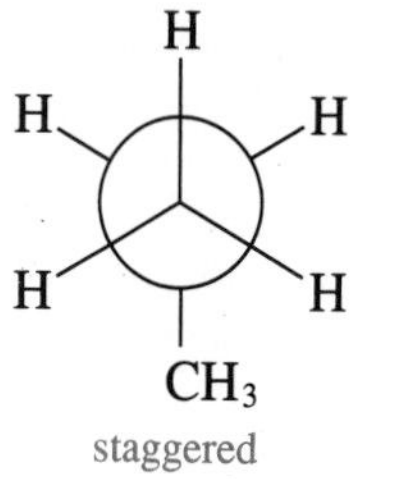

staggered

The projection formula is similar to that of ethane, except for the replacement of one hydrogen with methyl.

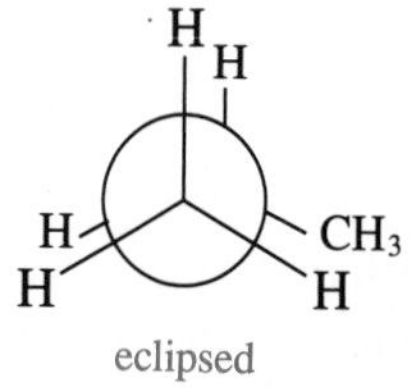

eclipsed

Rotation of the "rear" carbon of the staggered conformation by 60° gives the eclipsed conformation shown.

We are looking down the $C_1—C_2$ bond.

PROBLEM 2.9 Draw Newman projections for two different *staggered* conformations of butane (looking end-on at the bond between carbon-2 and carbon-3), and predict which of the two conformations is more stable.

The most important thing to remember about conformers is that they are just different forms of a single molecule that can be interconverted by rotational motions about single (sigma) bonds. More often than not, there is sufficient thermal energy for this rotation at room temperature. Consequently, at room temperature it is usually not possible to separate conformers from one another.

Now let us look at the structures of cycloalkanes and their conformations.

2.10 Cycloalkane Nomenclature and Conformation

Cycloalkanes are saturated hydrocarbons that have at least one ring of carbon atoms. A common example is cyclohexane.

Structural and abbreviated structural formulas for cyclohexane

Cycloalkanes are named by placing the prefix *cyclo* before the alkane name that corresponds to the number of carbon atoms in the ring. The structures and names of the first six unsubstituted cycloalkanes are as follows:

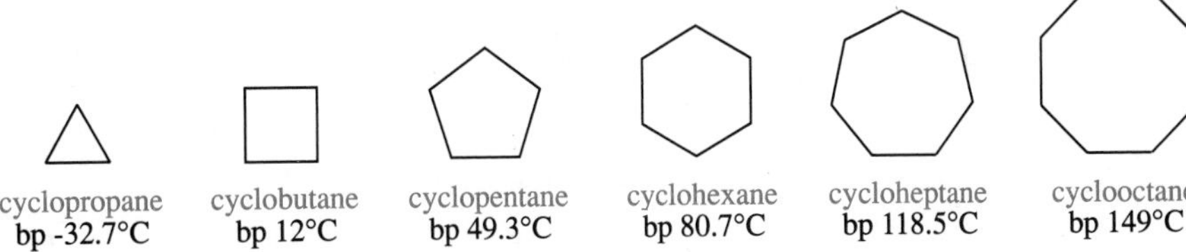

Alkyl or halogen substituents attached to the rings are named in the usual way. If only one substituent is present, no number is needed to locate it. If there are several substituents, numbers are required. One substituent is always located at ring carbon number 1, and the remaining ring carbons are then numbered consecutively in a way that gives the other substituents the lowest possible numbers. With different substituents, the one with highest alphabetic priority is located at carbon 1. The following examples illustrate the system:

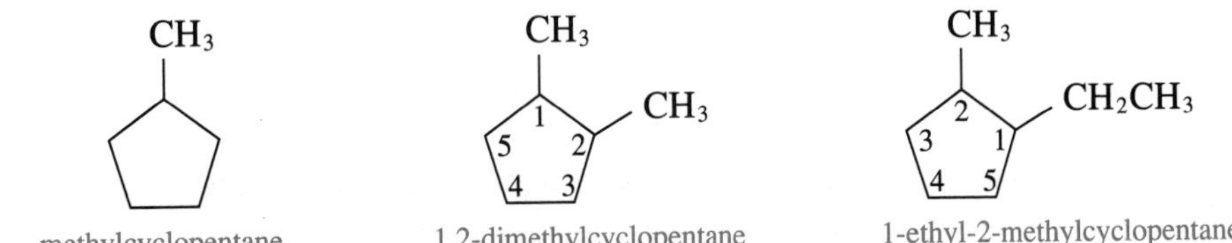

methylcyclopentane (*not* 1-methylcyclopentane) | 1,2-dimethylcyclopentane (*not* 1,5-dimethylcyclopentane) | 1-ethyl-2-methylcyclopentane (not 2-ethyl-1-methylcyclopentane)

PROBLEM 2.10 The general formula for an alkane is C_nH_{2n+2}. What is the corresponding formula for a cycloalkane with one ring?

PROBLEM 2.11 Draw the structural formulas for

a. 1,3-dimethylcyclohexane.
b. 1,2,3-trichlorocyclopropane.

PROBLEM 2.12 Give IUPAC names for

a. CH_2CH_3 b. Cl Cl c. Cl CH_3

What are the conformations of cycloalkanes? Cyclopropane, with only three carbon atoms, is necessarily planar (because three points determine a plane). The C—C—C angle is only 60° (the carbons form an equilateral triangle), much less than the usual tetrahedral angle of 109.5°. The hydrogens lie above and below the carbon plane, and hydrogens on adjacent carbons are eclipsed.

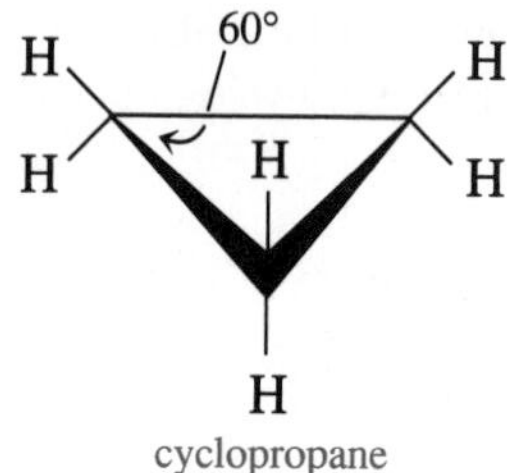

cyclopropane

EXAMPLE 2.5 Explain why the hydrogens in cyclopropane lie above and below the carbon plane.

Solution Refer to Figure 1.10. The carbons in cyclopropane have a geometry similar to that shown there, except that the C—C—C angle is "squeezed" and is smaller than tetrahedral. In compensation, the H—C—H angle is expanded and is larger than tetrahedral, approximately 120°.

H 120° H C 60° C C

The H—C—H plane perpendicularly bisects the C—C—C plane, which, as drawn here, lies in the plane of the paper.

Cycloalkanes with more than three carbon atoms are nonplanar and have "puckered" conformations. In cyclobutane and cyclopentane, puckering allows the molecule to adopt the most stable conformation (with the least strain energy). Puckering introduces strain by making the C—C—C angles a little smaller than they would be if the molecules were planar; however, less eclipsing of the adjacent hydrogens compensates for this.

	cyclobutane	cyclopentane
C—C—C angle for planar molecule	90°	108°
observed experimentally	88°	105°

Six-membered rings are rather special and have been studied in great detail because they are so common in nature. If cyclohexane were planar, the internal C—C—C angles would be those of a regular hexagon, 120°—quite a bit larger than the normal tetrahedral angle (109.5°). The strain resulting from such angles prevents cyclohexane from being planar (flat). The most favored conformation of cyclohexane is the **chair conformation,** an arrangement in which all the C—C—C angles are 109.5° and all the hydrogens on adjacent carbon atoms are perfectly staggered. Figure 2.4 shows models of the cyclohexane chair conformation.* (If a set of molecular models is available, it would be a good idea for you to construct a cyclohexane model to better visualize the concepts discussed in this and the next two sections.)

PROBLEM 2.13 How are the H—C—H and C—C—C planes at any one carbon atom in cyclohexane related? (Refer, if necessary, to Example 2.5.)

In the chair conformation, the hydrogens in cyclohexane fall into two sets, called **axial** and **equatorial.** Three axial hydrogens lie above and three lie below the average plane of the carbon atoms; the six equatorial hydrogens lie approximately in that plane. By a motion in which alternate ring carbons (say, 1, 3, and 5) move in one direction (down) and the other three ring carbons move in the opposite direction (up), one chair conformation can be converted into

---

* **Diamond** is one naturally occurring form of carbon. In the diamond crystal, the carbon atoms are connected to one another in a structure similar to the chair form of cyclohexane, except that all of the hydrogens are replaced by carbon atoms, resulting in a continuous network of carbon atoms. The hydrocarbons **adamantane** and **diamantane** show the beginnings of the diamond structure in their fusing of chair cyclohexanes. For a fascinating article on diamond structure, see "Diamond Cleavage" by M. F. Ansell in *Chemistry in Britain,* **1984,** 1017–1021.

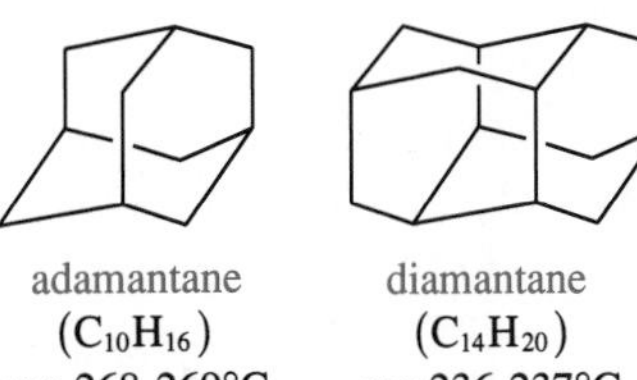

adamantane ($C_{10}H_{16}$) mp 268-269°C   diamantane ($C_{14}H_{20}$) mp 236-237°C

FIGURE 2.4
The chair conformation of cyclohexane, shown in ball-and-stick and space-filling models. The axial hydrogens lie above or below the mean plane of the carbons, and the six equatorial hydrogens lie approximately in that mean plane. The origin of the chair terminology is illustrated at the right.

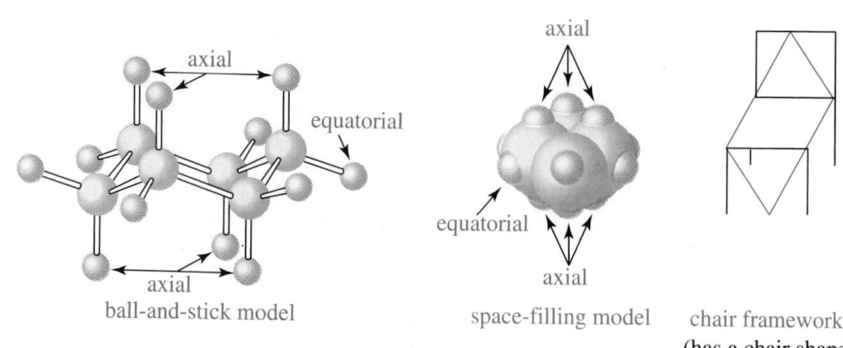

another chair conformation in which all axial hydrogens have become equatorial, and all equatorial hydrogens have become axial.

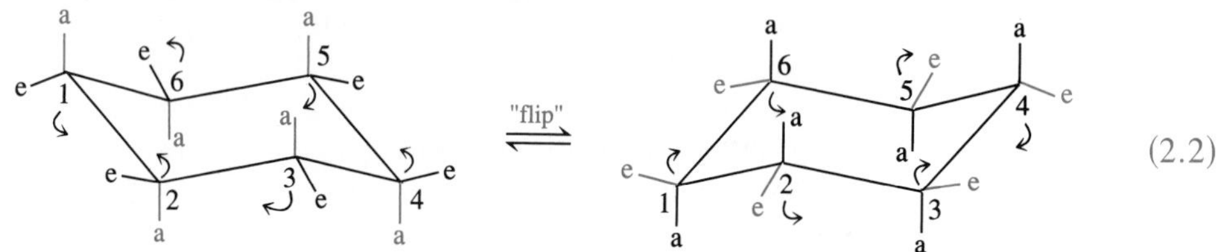

(2.2)

Axial bonds (in color) in the left structure become equatorial bonds (in color) in the right structure when the ring "flips."

At room temperature this flipping process is rapid, but at low temperatures (say −90°C), the flipping process slows down enough that the two different types of hydrogens can actually be detected by proton nuclear magnetic resonance (NMR) spectroscopy (see Chapter 12). Cyclohexane conformations have another important feature. If you look carefully at the space-filling model of cyclohexane (Figure 2.4), you will notice that *the three axial hydrogens on the same face of the ring nearly touch each other*. If an axial hydrogen is replaced by a larger substituent (such as a methyl group), the axial crowding is even worse. Therefore, the preferred conformation is the one in which the larger substituent, in this case the methyl group, is equatorial.

steric repulsion

methyl axial
5%

"flip"

methyl equatorial
95%

(2.3)

PROBLEM 2.14 Another puckered conformation for cyclohexane, one in which all C—C—C angles are the normal 109.5°, is the boat conformation.

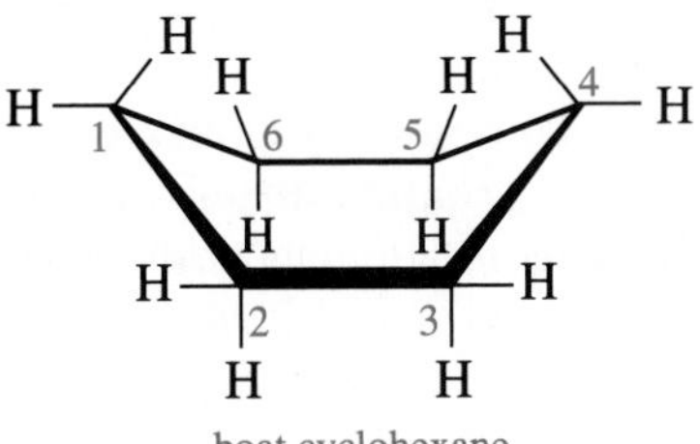

boat cyclohexane

Explain why this conformation is very much less stable than the chair conformation. (*Hint:* Note the arrangement of hydrogens as you sight along the bond between carbon-2 and carbon-3; a molecular model will help you answer this problem.)

PROBLEM 2.15 For *tert*-butylcyclohexane only one conformation, with the *tert*-butyl group equatorial, is detected. Explain why this conformational preference is greater than that for methylcyclohexane.

Before we proceed to reactions of alkanes and cycloalkanes, we need to consider a type of isomerism that may arise when two or more carbon atoms in a cycloalkane have substituents.

2.11 Cis–trans *Isomerism in Cycloalkanes*

Stereoisomerism deals with molecules that have the same order of attachment of the atoms, but different arrangements of the atoms in space. *Cis–trans* isomerism (sometimes called **geometric isomerism)** is one kind of stereoisomerism, and it is most easily understood with a specific case. Consider, for example, the possible structures of 1,2-dimethylcyclopentane. For simplicity, let us neglect the slight puckering of the ring and draw it as if it were planar. The two methyl groups may be on the same side of the ring plane or they may be on opposite sides.

do not interconvert

cis-1,2-dimethylcyclopentane
bp 99°C

trans-1,2-dimethylcyclopentane
bp 92°C

The methyl groups are said to be ***cis*** (Latin, on the same side) or ***trans*** (Latin, across) to each other.

***Cis–trans* isomers differ from one another only in the way the atoms or groups are positioned in space.** Yet this difference is sufficient to give them different physical and chemical properties. (Note, for example, the boiling points under the 1,2-dimethylcyclopentane structures.) ***Cis–trans* isomers are separate and unique compounds. Unlike conformers, they cannot be interconverted by rotation around carbon–carbon bonds.** In this example, the cyclic structure prevents rotation about the ring bonds. *Cis–trans* isomers can be separated from each other and kept separate, usually without interconversion at room temperature. *Cis–trans* isomerism can be important in determining the biological properties of molecules. For example, a molecule in which two reactive groups are *cis* will interact differently with an enzyme or biological receptor site than will its isomer in which the same two groups are *trans*.

PROBLEM 2.16 Draw the structure for the *cis* and *trans* isomers of

a. 1,2-dichlorocyclopropane.
b. 1-bromo-3-chlorocyclobutane.

2.12 Summary of Isomerism

At this point, it may be useful to summarize the relationships of the several types of isomers we have discussed so far. These relationships are outlined in Figure 2.5.

The first thing to look at in a pair of isomers is their bonding patterns (or atom connectivities). If the bonding patterns are *different,* the compounds are **structural (or constitutional) isomers.** But if the bonding patterns are the *same,* the compounds are **stereoisomers.** Examples of structural isomers are ethanol and methoxymethane (page 19) or the three isomeric pentanes (page 21). Examples of stereoisomers are the staggered and eclipsed forms of ethane (page 53) or the *cis* and *trans* isomers of 1,2-dimethylcyclopentane (page 59).

If compounds are stereoisomers, we can make a further distinction as to isomer type. If *bond rotation* easily interconverts the two stereoisomers (as with staggered and eclipsed ethane), we call them **conformers.** If the two

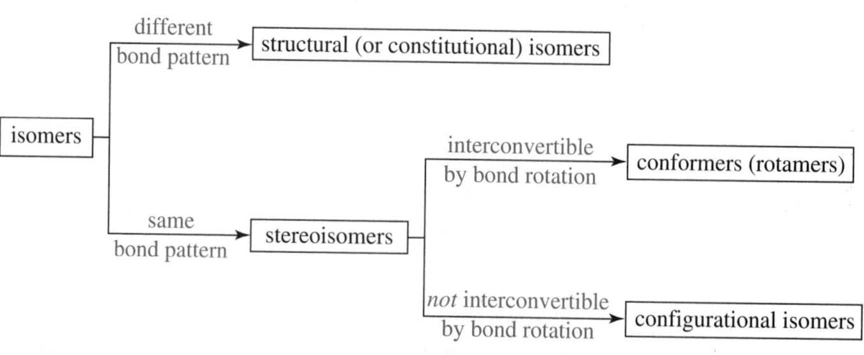

FIGURE 2.5 The relationships of the various types of isomers.

stereoisomers can be interconverted only by breaking and remaking bonds (as with *cis*- and *trans*-1,2-dimethylcyclopentane), we call them **configurational isomers.***

PROBLEM 2.17 Classify each of the following isomer pairs according to the scheme in Figure 2.5:

a. 1-bromopropane and 2-bromopropane
b. *cis*- and *trans*-1,2-dimethylcyclohexane
c. chair and boat forms of cyclohexane

2.13 Reactions of Alkanes

All the bonds in alkanes are single, covalent, and nonpolar. Hence alkanes are relatively inert. Alkanes ordinarily do not react with most common acids, bases, or oxidizing and reducing agents. Because of this inertness, alkanes can be used as solvents for extraction or crystallization or for carrying out chemical reactions with other substances. However, alkanes do react with some reagents, such as oxygen and the halogens. We will discuss those reactions here.

2.13a Oxidation and Combustion; Alkanes as Fuels The most important use of alkanes is as fuel. With excess oxygen, alkanes burn to form carbon dioxide and water. Most important, the reactions evolve considerable heat (that is, the reactions are **exothermic**).

$$\underset{\text{methane}}{CH_4} + 2\,O_2 \longrightarrow CO_2 + 2\,H_2O + \text{heat (212.8 kcal/mol)} \qquad (2.4)$$

$$\underset{\text{butane}}{C_4H_{10}} + \tfrac{13}{2}\,O_2 \longrightarrow 4\,CO_2 + 5\,H_2O + \text{heat (688.0 kcal/mol)} \qquad (2.5)$$

These combustion reactions are the basis for the use of hydrocarbons for heat (natural gas and heating oil) and for power (gasoline). An initiation step is required—usually ignition by a spark or flame. Once initiated, the reaction proceeds spontaneously and exothermically.

In methane, all four bonds to the carbon atom are C—H bonds. In carbon dioxide, its combustion product, all four bonds to the carbon are C—O bonds. *Combustion is an oxidation reaction, the replacement of C—H bonds by C—O bonds.* In methane, carbon is in its most reduced form. In carbon dioxide, carbon is in its most oxidized form. Intermediate oxidation states of carbon are also known, in which only one, two, or three of the C—H bonds are converted to C—O bonds. It is not surprising, then, that if insufficient oxygen is available for complete combustion of a hydrocarbon, *partial* oxidation may occur, as illustrated in eqs. 2.6 through 2.9.

*Remember that conformers are different conformations of the same molecule, whereas configurational isomers are different molecules.

$$2\ CH_4 + 3\ O_2 \longrightarrow \underset{\text{carbon monoxide}}{2\ CO} + 4\ H_2O \quad (2.6)$$

$$CH_4 + O_2 \longrightarrow \underset{\text{carbon}}{C} + 2\ H_2O \quad (2.7)$$

$$CH_4 + O_2 \longrightarrow \underset{\text{formaldehyde}}{CH_2O} + H_2O \quad (2.8)$$

$$2\ C_2H_6 + 3\ O_2 \longrightarrow \underset{\text{acetic acid}}{2\ CH_3CO_2H} + 2\ H_2O \quad (2.9)$$

Toxic carbon monoxide in exhaust fumes (eq. 2.6), soot emitted copiously from trucks with diesel engines (eq. 2.7), smog resulting in part from aldehydes (eq. 2.8), and acid build-up in lubricating oils (eq. 2.9) are all prices we pay for being a motorized society. However, incomplete hydrocarbon combustion is occasionally useful, as in the manufacture of carbon blacks (eq. 2.7) used for automobile tires, and lampblack, a pigment used in ink.

EXAMPLE 2.6 In which compound is carbon more oxidized, formaldehyde (CH_2O) or formic acid (HCO_2H)?

Solution Draw the structures:

```
H                      O
 \                   //
  C=O          H—C
 /                   \
H                      OH
formaldehyde     formic acid
```

Formic acid is the more oxidized form (3 C—O and 1 C—H bond, compared to 2 C—O and 2 C—H bonds in formaldehyde).

PROBLEM 2.18 Which of the following represents the more oxidized form of carbon?:
a. methanol (CH_3OH) or formaldehyde
b. methanol or dimethyl ether (CH_3OCH_3)

2.13b Halogenation of Alkanes When a mixture of an alkane and chlorine gas is stored at low temperatures in the dark, no reaction occurs. In sunlight or at high temperatures, however, an exothermic reaction occurs. One or more hydrogen atoms of the alkane are replaced by chlorine atoms. This reaction can be represented by the general equation

$$R—H + Cl—Cl \xrightarrow[\text{heat}]{\text{light or}} R—Cl + H—Cl \quad (2.10)$$

or, specifically for methane:

$$\underset{\text{methane}}{CH_4} + Cl—Cl \xrightarrow[\text{or heat}]{\text{sunlight}} \underset{\substack{\text{chloromethane} \\ \text{(methyl chloride)} \\ \text{bp } -24.2°\text{C}}}{CH_3Cl} + HCl \quad (2.11)$$

The reaction is called **chlorination.** It is a **substitution reaction;** a chlorine is substituted for a hydrogen.

An analogous reaction, called **bromination,** occurs when the halogen is bromine.

$$R{-}H + Br{-}Br \xrightarrow[\text{heat}]{\text{light or}} R{-}Br + HBr \qquad (2.12)$$

If excess halogen is present, the reaction can continue further to give polyhalogenated products. Thus, methane and excess chlorine can give products with two, three, or four chlorines.*

$$CH_3Cl \xrightarrow{Cl_2} CH_2Cl_2 \xrightarrow{Cl_2} CHCl_3 \xrightarrow{Cl_2} CCl_4 \qquad (2.13)$$

dichloromethane (methylene chloride) bp 40°C; trichloromethane (chloroform) bp 61.7°C; tetrachloromethane (carbon tetrachloride) bp 76.5°C

By controlling the reaction conditions and the ratio of chlorine to methane, we can favor formation of one or another of the possible products.

PROBLEM 2.19 Write the names and structures of all possible products of bromination of methane.

With longer chain alkanes, mixtures of products may be obtained even at the first step.** For example, with propane,

$$CH_3CH_2CH_3 + Cl_2 \xrightarrow[\text{or heat}]{\text{light}} CH_3CH_2CH_2Cl + CH_3\underset{\displaystyle Cl}{\underset{|}{C}}HCH_3 + HCl \qquad (2.14)$$

propane; 1-chloropropane (n-propyl chloride); 2-chloropropane (isopropyl chloride)

When larger alkanes are halogenated, the mixture of products becomes even more complex; individual isomers become difficult to separate and obtain pure, so halogenation tends not to be a useful way to synthesize specific alkyl halides. With unsubstituted *cycloalkanes,* however, where all the hydrogens are equivalent, a single pure organic product can be obtained:

$$\text{cyclopentane} + Br_2 \xrightarrow{\text{light}} \text{bromocyclopentane (H, Br on one ring carbon)} + HBr \qquad (2.15)$$

cyclopentane; bromocyclopentane (cyclopentyl bromide)

---

* Note that we sometimes write the formula of one of the reactants (in this case Cl_2) over the arrow for convenience, as in eq. 2.13. We also sometimes omit obvious inorganic products (in this case HCl).

** Note that we often do not write a balanced equation, especially when more than one product is formed from a single organic reactant. Instead, we show on the right side of the equation the structures of *all* the important organic products, as in eq. 2.14.

PROBLEM 2.20 Write the structures of all possible products of *mono*bromination of pentane. Note the complexity of the product mixture, compared to that from the corresponding reaction with cyclopentane (eq. 2.15).

PROBLEM 2.21 How many organic products can be obtained from the monochlorination of octane? of cyclooctane?

PROBLEM 2.22 Do you think that the chlorination of 2,2-dimethylpropane might be synthetically useful?

2.14 The Free Radical Chain Mechanism of Halogenation

One may well ask how halogenation occurs. Why is light or heat necessary? Equations 2.10 and 2.11 express the *overall* reaction for halogenation. They describe the structures of the reactants and the products, and they show necessary reaction conditions or catalysts over the arrow. But they do *not* tell us exactly how the products are formed from the reactants.

A reaction mechanism is a step-by-step description of the bond-breaking and bond-making processes that occur when reagents react to form products. In the case of halogenation, various experiments show that this reaction occurs in several steps, not in one. Halogenation occurs via a **free-radical chain** of reactions.

The **chain-initiating step** is the breaking of the halogen molecule into two halogen atoms.

$$\text{initiation} \quad :\ddot{\underset{..}{Cl}}:\ddot{\underset{..}{Cl}}: \xrightarrow[\text{heat}]{\text{light or}} :\ddot{\underset{..}{Cl}}\cdot + :\ddot{\underset{..}{Cl}}\cdot \tag{2.16*}$$

chlorine molecule — chlorine atoms

The halogen bond is weaker than either the C—H bond or the C—C bond, and is therefore the easiest bond to break. When light is the energy source, chlorine absorbs visible light but alkanes do not, so again it is the Cl—Cl bond that breaks.

The **chain-propagating steps** are

propagation

$$R{-}H + \cdot\ddot{\underset{..}{Cl}}: \longrightarrow R\cdot + H{-}Cl \tag{2.17}$$

alkyl radical

$$R\cdot + Cl{-}Cl \longrightarrow R{-}Cl + \cdot\ddot{\underset{..}{Cl}}: \tag{2.18}$$

alkyl chloride

*Note (Sec. 1.13) that we use a "fishhook," or half-headed arrow, ⌒, to show the movement of only *one* electron whereas we use a complete arrow, ⌒, to describe the movement of an electron *pair*.

Chlorine atoms are very reactive, because they have an incomplete valence shell (7 electrons instead of the required 8). They may either recombine to form chlorine molecules (the reverse of eq. 2.16) or, if they collide with an alkane molecule, abstract a hydrogen atom to form hydrogen chloride and an alkyl radical R·. Recall from Sec. 1.4 that a radical is a fragment with an odd number of unshared electrons. The space-filling models in Figure 2.1 show that alkanes seem to have an exposed surface of hydrogens covering the carbon skeleton. So it is most likely that, if a halogen atom collides with an alkane molecule, it will hit the hydrogen end of a C—H bond.

Like a chlorine atom, the alkyl radical formed in the first step of the chain (eq. 2.17) is very reactive (incomplete octet). If it were to collide with a chlorine molecule, it could form an alkyl chloride molecule and a chlorine atom (eq. 2.18). The chlorine atom formed in this step can then react to repeat the sequence. When you add eq. 2.17 and eq. 2.18, you get the overall equation for chlorination (eq. 2.10). *In each chain-propagating step, a radical (or atom) is consumed, but another radical (or atom) is formed and can continue the chain. Almost all of the reactants are consumed, and almost all of the products are formed in these steps.*

Were it not for **chain-terminating steps,** all of the reactants could, in principle, be consumed by initiating a single chain. However, because many chlorine molecules react to form chlorine atoms in the chain-initiating step, many chains are started simultaneously. Quite a few radicals are present as the reaction proceeds. If any two radicals combine, the chain will be terminated. Three possible chain-terminating steps are

termination

$$2\,:\ddot{\underset{\cdot\cdot}{Cl}}\cdot \longrightarrow Cl{-}Cl \qquad (2.19)$$

$$2\,R\cdot \longrightarrow R{-}R \qquad (2.20)$$

$$R\cdot + :\ddot{\underset{\cdot\cdot}{Cl}}\cdot \longrightarrow R{-}Cl \qquad (2.21)$$

No new radicals are formed in these reactions, so the chain is broken or, as we say, terminated.

PROBLEM 2.23 Show that when eq. 2.17 and eq. 2.18 are added, the overall equation for chlorination (eq. 2.10) results.

PROBLEM 2.24 Write equations for all the steps (initiation, propagation, termination) in the free-radical chlorination of methane to form methyl chloride.

PROBLEM 2.25 Account for the experimental observation that small amounts of ethane and chloroethane are produced during the monochlorination of methane. (*Hint:* Consider the possible chain-terminating steps.)

A Word About . . .

2. Methane, Marsh Gas, and Miller's Experiment

Methane is commonly found in nature wherever bacteria decompose organic matter in the absence of oxygen, as in marshes, swamps, or the muddy sediment of lakes—hence, its common name, *marsh gas*. In China, methane has been collected from the mud at the bottom of swamps for use in domestic cooking and lighting. Methane is similarly formed from bacteria in the digestive tracts of certain ruminant animals, such as cows.

The scale of methane production by bacteria is considerable. The earth's atmosphere contains an average of 1 part per million of methane. Because our planet is small and because methane is light compared to most other air constituents (O_2, N_2), one would expect most of the methane to have escaped from our atmosphere, and it has been calculated that the equilibrium concentration should be very much less than is observed. The reason, then, for the relatively high observed concentration is that at the same time methane escapes from the atmosphere, it is constantly being produced by bacterial decay of plant matter.

In cities, the amount of methane in the atmosphere reaches much higher levels, up to several parts per million. The peak concentrations come in the early morning and late afternoon, directly correlated with the peaks of automobile traffic. Fortunately, methane, which constitutes about 50% of urban atmospheric hydrocarbon pollutants, seems to have no direct harmful effect on human health.

Methane can accumulate in coal mines, where it is a hazard because, when mixed with 5 to 14% of air, it is explosive. Also, miners can be asphyxiated by it (due to lack of sufficient oxygen). Dangerous concentrations of methane can be detected readily by a variety of safety devices.

Hydrogen is the most common element in the solar system (it constitutes about 87% of the sun's mass). It therefore seemed reasonable to think that, when the planets were formed, other elements should have been present in reduced (not oxidized) forms: carbon as methane, nitrogen as ammonia, and oxygen as water. Indeed, some of the outer planets (Saturn and Jupiter) still have atmospheres that are rich in methane and ammonia.

A now-famous experiment carried out in 1953 by Stanley L. Miller (working in the laboratory of H. C. Urey at Columbia University) supports the idea that life could have arisen in a reducing atmosphere. Miller found that when mixtures of methane, ammonia, water, and hydrogen were subjected to electric discharges (to simulate lightning), some organic compounds were formed (amino acids, for example) that are important to biology and necessary for life. Similar results have since been obtained using heat or ultraviolet light in place of electric discharges (it seems likely that the earth's early atmosphere was subjected to much more ultraviolet radiation than it is now). When oxygen was added to these simulated primeval atmospheres, no amino acids were produced—strong evidence that the earth's original atmosphere did *not* contain free oxygen. Miller's experiment provided the model for much work in the branch of science now called **chemical evolution,** the study of chemical events that may have taken place on earth or elsewhere in the universe leading to the appearance of the first living cell.

In the years since Miller's experiment, ideas about the chemistry of life's origin have become more precise as a consequence of much experimentation and of exploration in outer space. We now know that the earth's primary atmosphere was formed mainly by degassing the molten interior rather than by accretion from the solar nebula. It now seems likely that the main carbon sources in the earth's early atmosphere were CO_2 and CO, *not* methane as assumed by Miller, and that nitrogen was present mainly as N_2 rather than ammonia. Repetition of Miller-type experiments with these assumed primordial atmospheres again gave biomolecules.

For additional reading, see *Chemical Evolution* by Stephen F. Mason, Clarendon Press, Oxford, 1991; especially Ch. 13 on Prebiotic Chemistry, pp. 233–259, and the extensive bibliography, pp. 285–302.

REACTION SUMMARY

Reactions of Alkanes and Cycloalkanes

1. Combustion (Sec. 2.13a)

$$C_nH_{2n+2} + \left(\frac{3n+1}{2}\right)O_2 \longrightarrow nCO_2 + (n+1)H_2O$$

2. Halogenation (Sec. 2.13b)

$$R\text{—}H + X_2 \xrightarrow[\text{or light}]{\text{heat}} R\text{—}X + H\text{—}X \quad (X = Cl, Br)$$

ADDITIONAL PROBLEMS

2.26. Write structural formulas for the following compounds:

a. 2-methylpentane **b.** 2,3-dimethylbutane
c. 4-ethyl-2,2-dimethylhexane **d.** 3-bromo-2-methylpentane
e. 1,1-dichlorocyclopropane **f.** 2-iodopropane
g. 1,1,3-trimethylcyclohexane **h.** 1,1,3,3-tetrachloropropane

2.27. Write expanded formulas for the following compounds, and name them using the IUPAC system:

a. $CH_3(CH_2)_2CH_3$ **b.** $CH_3CH_2CH(CH_3)CH_2CH_3$
c. $(CH_3)_3CCH_2CH_2CH_3$ **d.** $(CH_2)_3$
e. $CH_3CH_2CHFCH_3$ **f.** $CH_3CCl_2CBr_3$
g. *i*-PrCl **h.** MeBr
i. CH_2ClCH_2Cl **j.** $(CH_3CH_2)_4C$

2.28. Give both common and IUPAC names for the following compounds:

a. CH_3F **b.** CH_3CH_2Br **c.** CH_2Cl_2 **d.** CHI_3
e. $(CH_3)_2CHBr$ **f.** $CH_3CH_2CH_2I$ **g.** $(CH_3)_3CCl$ **h.** $\begin{array}{l}CH_2\text{—}CH\text{—}Br\\ |\qquad\quad |\\ CH_2\text{—}CH_2\end{array}$
i. $CH_3CHFCH_2CH_3$

2.29. Write a structure for each of the compounds listed. Explain why the name given here is incorrect, and give a correct name in each case.

a. 1-methylbutane **b.** 2-ethylpentane
c. 2,3-dichloropropane **d.** 1,3-dimethylcyclopropane
e. 4-bromo-3-methylbutane **f.** 1,1,3-trimethylpentane

2.30. Chemical substances used for communication in nature are called *pheromones*. The pheromone used by the female tiger moth to attract the male is the alkane 2-methylheptadecane. Write its structural formula.

2.31. Write the structural formula for all the isomers for each of the following compounds, and name each isomer by the IUPAC system (The number of isomers is indicated in parentheses.):

a. C_4H_{10} (2) **b.** C_4H_9Cl (4) **c.** C_5H_{12} (3)
d. $C_2H_2ClBr_3$ (3) **e.** $C_3H_6F_2$ (4) **f.** C_3H_6BrCl (5)

2.32. Write structural formulas and names for all possible cycloalkanes having each of the following molecular formulas. Be sure to include *cis–trans* isomers when appropriate. Name each compound by the IUPAC system.

a. C_5H_{10} **b.** C_6H_{12} (there are 16)

2.33. Without referring to tables, arrange the following five hydrocarbons in order of increasing boiling point:

a. 2-methylhexane **b.** heptane **c.** 3,3-dimethylpentane
d. hexane **e.** 2-methylpentane

2.34. In Problem 2.9 you drew two staggered conformations of butane (looking end-on down the bond between carbon-2 and carbon-3). There are also two eclipsed conformations around this bond. Draw Newman projections for them. Arrange all four conformations in order of decreasing stability.

2.35. Draw all possible staggered and eclipsed conformations of 1-bromo-2-chloroethane, using Newman projections. Underneath each, draw the corresponding "dash-wedge" and "sawhorse" structures. Rank the conformations in order of decreasing stability.

2.36. Draw the formula for the perferred conformation of

a. ethylcyclohexane. **b.** *trans*-1,4-dimethylcyclohexane.
c. *cis*-1-chloro-3 (1-methylethyl)cyclohexane. **d.** 1,1-dimethylcyclohexane.

2.37. Name the following *cis–trans* pairs:

a.

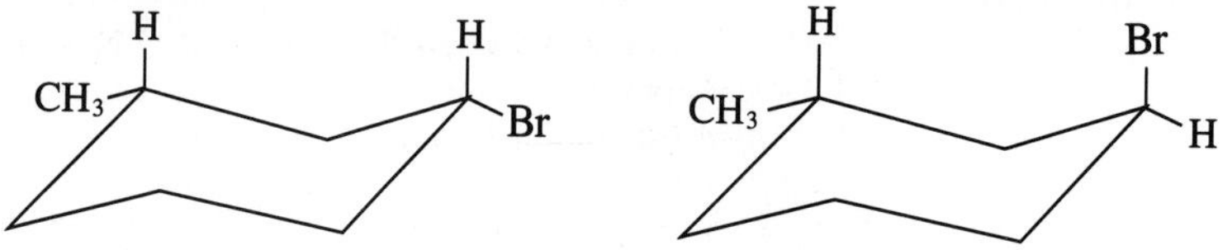

H, CH_3, H, Br H, CH_3, Br, H

b.

H, Br, Br, H Br, H, Br, H

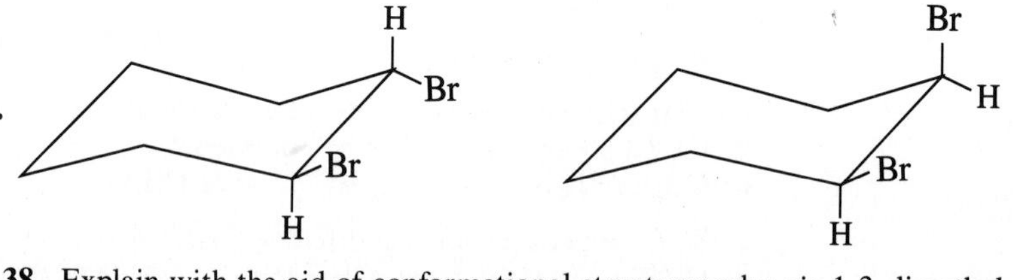

2.38. Explain with the aid of conformational structures why *cis*-1,3-dimethylcyclohexane is more stable than *trans*-1,3-dimethylcyclohexane, whereas the reverse order of stability is observed for the 1,2 and 1,4 isomers.

2.39. Which will be more stable, *cis*- or *trans*-1,4-di-*tert*-butylcyclohexane? Explain your answer by drawing conformational structures for each compound.

2.40. Classify the following pairs of structures according to the scheme in Figure 2.5:

a. the pairs of compounds in Problem 2.37

b. 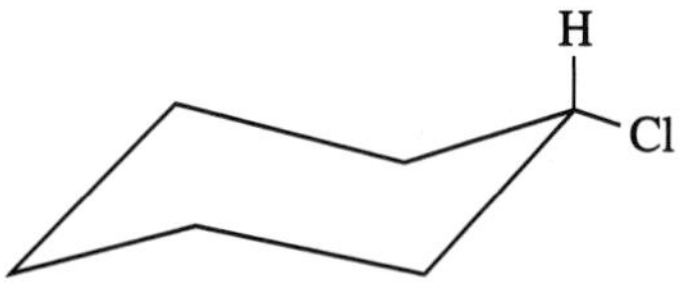 and

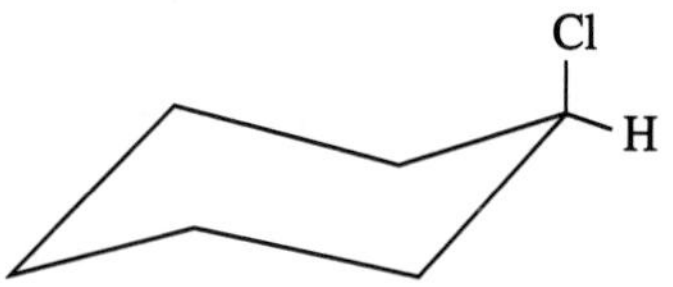

c. 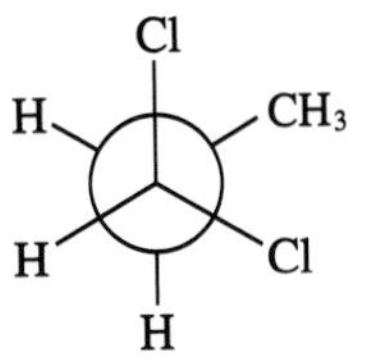 and 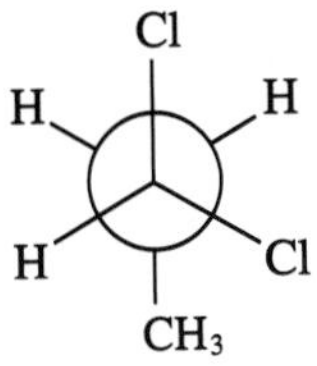 **d.** 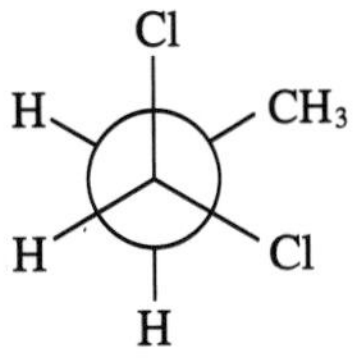 and

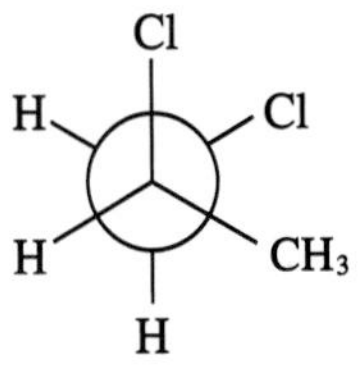

e. $CH_3CHCH_2CH_2CH_3$ (with CH_3 on C-2) and $CH_3CH_2CH_2CHCH_3$ (with CH_3 on C-4) (careful!)

2.41. Draw structural formulas for all possible dichlorocyclohexanes. Include *cis–trans* isomers.

2.42. How many monochlorination products can be obtained from each of the following polycyclic alkanes?:

2.43. Using structural formulas, write equations for the following reactions, and name each organic product.

a. the complete combustion of heptane
b. the complete combustion of cyclopentane
c. the monobromination of butane
d. the monochlorination of cyclohexane
e. the complete chlorination of propane

2.44. From the dichlorination of propane, four isomeric products with the formula $C_3H_6Cl_2$ were isolated and designated A, B, C, and D. Each was separated and further chlorinated to give one or more trichloropropanes, $C_3H_5Cl_3$. A and B gave three trichloro compounds, C gave one, and D gave two. Deduce the structures of C and D. One of the products from A was identical with the product from C. Deduce structures for A and B.

2.45. Write all the steps in the free-radical chain mechanism for the monochlorination of ethane.

$$CH_3CH_3 + Cl_2 \longrightarrow CH_3CH_2Cl + HCl$$

What trace by-products would you expect to be formed as a consequence of the chain-terminating steps?

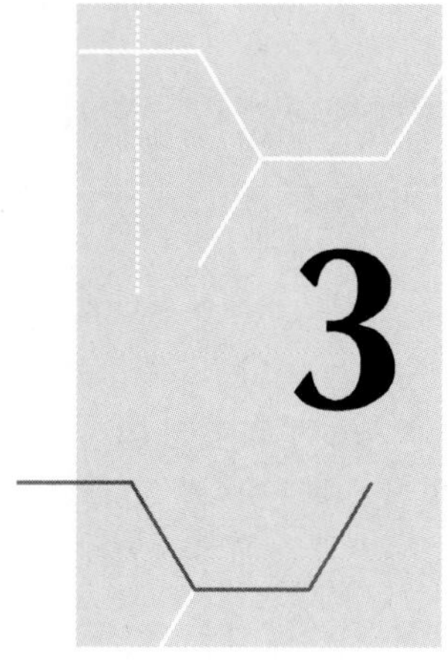

3 Alkenes and Alkynes

3.1 Definition, Classification, and Physical Properties

Hydrocarbons that contain a carbon–carbon double bond are called **alkenes;** those with a carbon–carbon triple bond are **alkynes.*** Their general formulas are

$$\underset{\text{alkenes}}{C_nH_{2n}} \qquad \underset{\text{alkynes}}{C_nH_{2n-2}}$$

Both of these classes of hydrocarbons are **unsaturated,** because they contain fewer hydrogens per carbon than alkanes (C_nH_{2n+2}). Alkanes can be obtained from alkenes or alkynes by adding one or two moles of hydrogen.

$$\left.\begin{array}{l} RCH{=}CHR \xrightarrow[\text{catalyst}]{H_2} \\ \\ RC{\equiv}CR \xrightarrow[\text{catalyst}]{2H_2} \end{array}\right\} \longrightarrow RCH_2CH_2R \tag{3.1}$$

Compounds with more than one double or triple bond exist. If two double bonds are present, the compounds are called **alkadienes** or, more commonly, **dienes.** There are also trienes, tetraenes, and even polyenes (compounds with *many* double bonds, from the Greek *poly,* many). Compounds with more than one triple bond, or with double and triple bonds, are also known.

---

*An old but still used synonym for alkenes is *olefins,* which means oil-forming. This name was originally given to ethylene because it formed an oily liquid when treated with chlorine. Alkynes are also called *acetylenes* after the first member of the series.

EXAMPLE 3.1 What are all the structural possibilities for the compound C_3H_4?

Solution The formula C_3H_4 corresponds to the general formula C_nH_{2n-2}. The compound could have one triple bond, two double bonds, or one ring and one double bond. For their structures, see the solution to Example 1.10 on page 18.

PROBLEM 3.1 What are all the structural possibilities for C_4H_6? (Nine compounds, four acyclic and five cyclic, are known.)

When two or more multiple bonds are present in a molecule, it is useful to classify the structure further, depending on the relative positions of the multiple bonds. Double bonds are said to be **cumulated** when they are right next to one another. When multiple bonds *alternate* with single bonds, they are called **conjugated.** When more than one single bond comes between multiple bonds, the latter are **isolated** or **nonconjugated.**

cumulated	conjugated	isolated
C=C=C	C=C—C=C	C=C—C—C=C
C=C=C=C	C=C—C≡C	C≡C—C—C—C≡C

PROBLEM 3.2 Which of the following compounds have conjugated multiple bonds?

a.  b.

c. 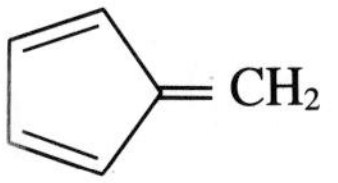 CH_2 d.

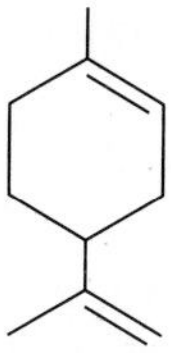

Unsaturated hydrocarbons have physical properties similar to those of alkanes. They are less dense than water and, being nonpolar, are not very soluble in it. As with the alkanes, compounds with four or fewer carbons are colorless gases, whereas higher homologs are volatile liquids.

3.2 Nomenclature

The IUPAC rules for naming alkenes and alkynes are similar to those for alkanes (Sec. 2.4), but a few rules must be added for naming and locating the multiple bonds.

1. The ending *-ene* is used to designate a carbon–carbon double bond. When more than one double bond is present, the ending is *-diene, -triene,* and so on. The ending *-yne* is used for a triple bond (*-diyne* for two triple bonds, and so on). Compounds with a double *and* a triple bond are *-enynes.*

2. Select the longest chain that includes *both* carbons of the double or triple bond. For example,

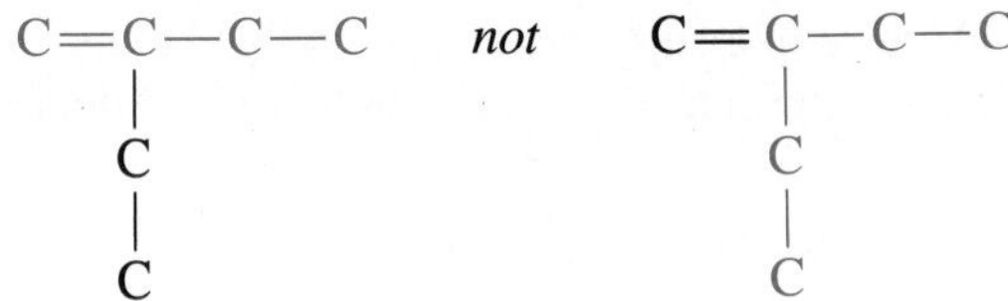

named as a butene, not as a pentene

3. Number the chain from the end nearest the multiple bond, so that the carbon atoms in that bond have the lowest possible numbers.

$$\overset{1}{C}-\overset{2}{C}=\overset{3}{C}-\overset{4}{C}-\overset{5}{C} \quad \textit{not} \quad \overset{5}{C}-\overset{4}{C}=\overset{3}{C}-\overset{2}{C}-\overset{1}{C}$$

If the multiple bond is equidistant from both ends of the chain, number from the end nearest the first branch point.

$$\overset{1}{C}-\underset{\displaystyle C}{\underset{|}{\overset{2}{C}}}=\overset{3}{C}-\overset{4}{C} \quad \textit{not} \quad \overset{4}{C}-\underset{\displaystyle C}{\underset{|}{\overset{3}{C}}}=\overset{2}{C}-\overset{1}{C}$$

4. Indicate the position of the multiple bond using the *lower-numbered carbon atom* of that bond. For example,

$$\overset{1}{C}H_2=\overset{2}{C}H\overset{3}{C}H_2\overset{4}{C}H_3$$ 1-butene, *not* 2-butene

5. If more than one multiple bond is present, number from the end nearest the first multiple bond.

$$\overset{1}{C}=\overset{2}{C}-\overset{3}{C}=\overset{4}{C}-\overset{5}{C} \quad \textit{not} \quad \overset{5}{C}=\overset{4}{C}-\overset{3}{C}=\overset{2}{C}-\overset{1}{C}$$

If a double and a triple bond are equidistant from the end of the chain, the *double* bond receives the lowest numbers. For example,

$$\overset{1}{C}=\overset{2}{C}-\overset{3}{C}\equiv\overset{4}{C} \quad \textit{not} \quad \overset{4}{C}=\overset{3}{C}-\overset{2}{C}\equiv\overset{1}{C}$$

Let us see how these rules are applied. The first two members of each series are

CH_3CH_3	$CH_2=CH_2$	$HC\equiv CH$
ethane	ethene	ethyne
$CH_3CH_2CH_3$	$CH_2=CHCH_3$	$HC\equiv CCH_3$
propane	propene	propyne

The root of the name (*eth-* or *prop-*) tells us the number of carbons, and the ending (*-ane, -ene,* or *-yne*) tells us whether the bonds are single, double, or triple. No number is necessary in these cases, because in each instance, only one structure is possible.

With four carbons, a number is necessary to locate the double or triple bond.

$\overset{1}{C}H_2{=}\overset{2}{C}H\overset{3}{C}H_2\overset{4}{C}H_3$	$\overset{1}{C}H_3\overset{2}{C}H{=}\overset{3}{C}H\overset{4}{C}H_3$	$H\overset{1}{C}{\equiv}\overset{2}{C}\overset{3}{C}H_2\overset{4}{C}H_3$	$\overset{1}{C}H_3\overset{2}{C}{\equiv}\overset{3}{C}\overset{4}{C}H_3$
1-butene	2-butene	1-butyne	2-butyne

Branches are named in the usual way.

$\overset{1}{C}H_2{=}\overset{2}{C}(CH_3)—\overset{3}{C}H_3$	$\overset{1}{C}H_2{=}\overset{2}{C}(CH_3)—\overset{3}{C}H_2\overset{4}{C}H_3$	$\overset{1}{C}H_3—\overset{2}{C}(CH_3){=}\overset{3}{C}H\overset{4}{C}H_3$	$\overset{1}{C}H_2{=}\overset{2}{C}(CH_3)—\overset{3}{C}H{=}\overset{4}{C}H_2$
methylpropene (isobutylene)	2-methyl-1-butene	2-methyl-2-butene	2-methyl-1,3-butadiene (isoprene)

Note how the rules are applied in the following examples:

$\overset{1}{C}H_3—\overset{2}{C}H{=}\overset{3}{C}H—\overset{4}{C}H(CH_3)—\overset{5}{C}H_3$

4-methyl-2-pentene
(Not 2-methyl-3-pentene; the chain is numbered so that the double bond gets the lower number.)

$\overset{1}{C}H_2{=}\overset{2}{C}(CH_2CH_3)—\overset{3}{C}H_2\overset{4}{C}H_3$

2-ethyl-1-butene
(Named this way, even though there is a five-carbon chain present, because that chain does not include both carbons of the double bond.)

$\overset{1}{C}H_2{=}\overset{2}{C}H—\overset{3}{C}H{=}\overset{4}{C}H_2$

1,3-butadiene
(Note the *a* inserted in the name, to help in pronunciation.)

With cyclic hydrocarbons, we start numbering the ring with the carbons of the multiple bond.

cyclopentene
(No number is necessary, because there is only one possible structure.)

CH_3 3 2 4 1 5

3-methylcyclopentene
(Start numbering at, and number through, the double bond; 5-methycyclopentene and 1-methyl-2-cyclopentene are incorrect names.)

1 2 6 3 5 4

1,3-cyclohexadiene

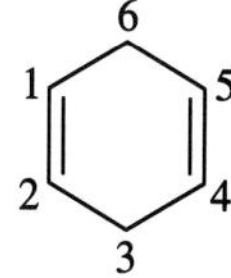

1,4-cyclohexadiene

PROBLEM 3.3 Name each of the following structures by the IUPAC system:

a. $CH_2{=}C(Cl)CH_3$
b. $(CH_3)_2C{=}C(CH_3)_2$
c. $BrCH{=}CHCH_3$
d. CH_3 (on cyclohexene ring)
e. $CH_2{=}C(CH_3)CH{=}CH_2$
f. $CH_3(CH_2)_3C{\equiv}CH$

EXAMPLE 3.2 Write the structural formula for 3-methyl-2-pentene.

Solution To get the structural formula from the IUPAC name, first write the longest chain or ring, number it, and locate the multiple bond. In this case, note that the chain has five carbons and that the double bond is located between carbon-2 and carbon-3:

$$\overset{1}{C}-\overset{2}{C}=\overset{3}{C}-\overset{4}{C}-\overset{5}{C}$$

Next, add the substituent:

$$\overset{1}{C}-\overset{2}{C}=\underset{\displaystyle CH_3}{\underset{|}{\overset{3}{C}}}-\overset{4}{C}-\overset{5}{C}$$

Finally, fill in the hydrogens:

$$CH_3-CH=\underset{\displaystyle CH_3}{\underset{|}{C}}-CH_2-CH_3$$

PROBLEM 3.4 Write the structural formula for:

a. 2,3-dimethyl-2-pentene.
b. 3-hexyne.
c. 1,2-dichlorocyclobutene.
d. 2-bromo-1,3-butadiene.

In addition to the IUPAC rules, it is important to learn a few common names. For example, the simplest members of the alkene and alkyne series are frequently referred to by their older common names, **ethylene, acetylene,** and **propylene.**

$CH_2=CH_2$	$HC\equiv CH$	$CH_3CH=CH_2$
ethylene (ethene)	acetylene (ethyne)	propylene (propene)

Three important groups also have common names. They are the **vinyl, allyl,** and **propargyl** groups (their IUPAC names are in parentheses).

$CH_2=CH-$	$\overset{1}{C}H_2=\overset{2}{C}H-\overset{3}{C}H_2-$	$\overset{1}{H}C\equiv\overset{2}{C}-\overset{3}{C}H_2-$
vinyl (ethenyl)	allyl (3-propenyl)	propargyl (3-propynyl)

These groups are used in common names.

$CH_2=CHCl$	$CH_2=CH-CH_2Cl$	$HC\equiv C-CH_2Br$
vinyl chloride (chloroethene)	allyl chloride (3-chloropropene)	propargyl bromide (3-bromopropyne)

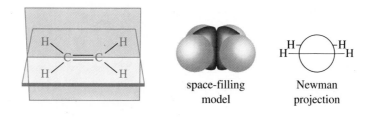

FIGURE 3.1 Three models of ethylene, each showing that the four atoms attached to a carbon–carbon double bond lie in a single plane.

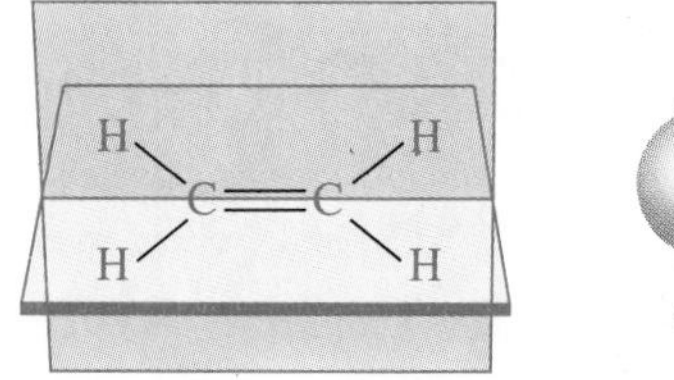

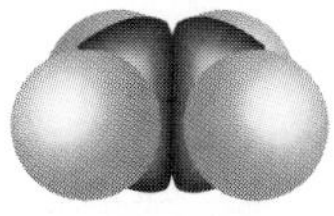

space-filling model

Newman projection

PROBLEM 3.5 Write the structural formula for

a. vinylcyclohexane.
b. allylcyclopropane.
c. propargyl iodide.

3.3 Some Facts About Double Bonds

Carbon–carbon double bonds have some special features that are different from those of single bonds. For example, each carbon atom of a double bond is connected to only *three* other atoms (instead of four atoms, as with tetrahedral carbon). We speak of such a carbon as being **trigonal.** Furthermore, the two carbon atoms of a double bond and the four atoms that are attached to them lie in a single plane. This planarity is shown in Figure 3.1 for ethylene. The H—C—H and H—C=C angles in ethylene are approximately 120°. Although rotation occurs freely around single bonds, *rotation around double bonds is restricted.* Ethylene does not adopt any other conformation except the planar one. The doubly bonded carbons with two attached hydrogens do not rotate with respect to each other. Finally, carbon–carbon double bonds are shorter than carbon–carbon single bonds.

These differences between single and double bonds are summarized in Table 3.1. Let us see how the orbital model for bonding can explain the structure and properties of double bonds.

TABLE 3.1 Comparison of C—C and C=C bonds

Property	*C—C*	*C=C*
1. Number of atoms attached to a carbon	4 (tetrahedral)	3 (trigonal)
2. Rotation	relatively free	restricted
3. Geometry	many conformations are possible; staggered is preferred	planar
4. Bond angle	109.5°	120°
5. Bond length	1.54 Å	1.34 Å

FIGURE 3.2 Unhybridized vs. sp^2-hybridized orbitals on carbon.

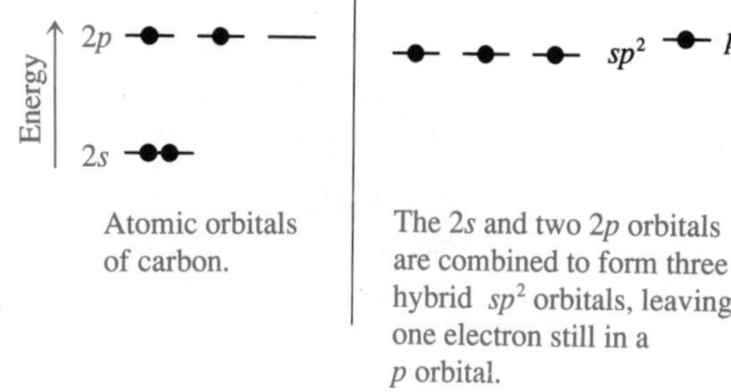

3.4 The Orbital Model of a Double Bond; the Pi Bond

Figure 3.2 shows what must happen with the atomic orbitals of carbon to accommodate trigonal bonding, bonding to only three other atoms. The first part of this figure is exactly the same as Figure 1.6. But now we combine only *three* of the orbitals, to make *three equivalent* sp^2*-hybridized orbitals* (called sp^2 because they are formed by combining one s and two p orbitals). These orbitals lie in a plane and are directed to the corners of an equilateral triangle. The angle between them is 120°. This angle is preferred because repulsion between electrons in each orbital is minimized. Three valence electrons are placed in the three sp^2 orbitals. The fourth valence electron is placed in the remaining $2p$ orbital, whose axis is perpendicular to the plane formed by the three sp^2 hybrid orbitals (see Figure 3.3).

Now let us see what happens when two sp^2-hybridized carbons are brought together to form a double bond. The process can be imagined as occurring stepwise (Figure 3.4). One of the two bonds, formed by end-on overlap of two sp^2 orbitals, is a sigma (σ) bond. The second bond of the double bond is

FIGURE 3.3 A trigonal carbon showing three sp^2 hybrid orbitals in a plane with a 120° angle between them. The remaining p orbital is perpendicular to the sp^2 orbitals. There is a small back lobe to each sp^2 orbital, which has been omitted for ease of representation.

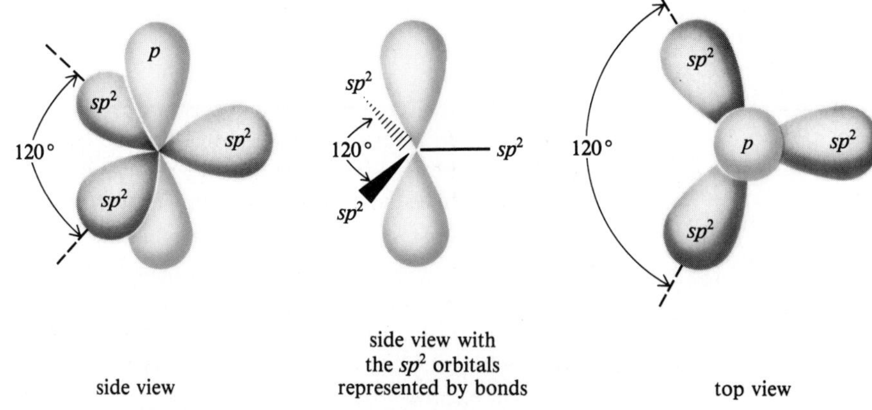

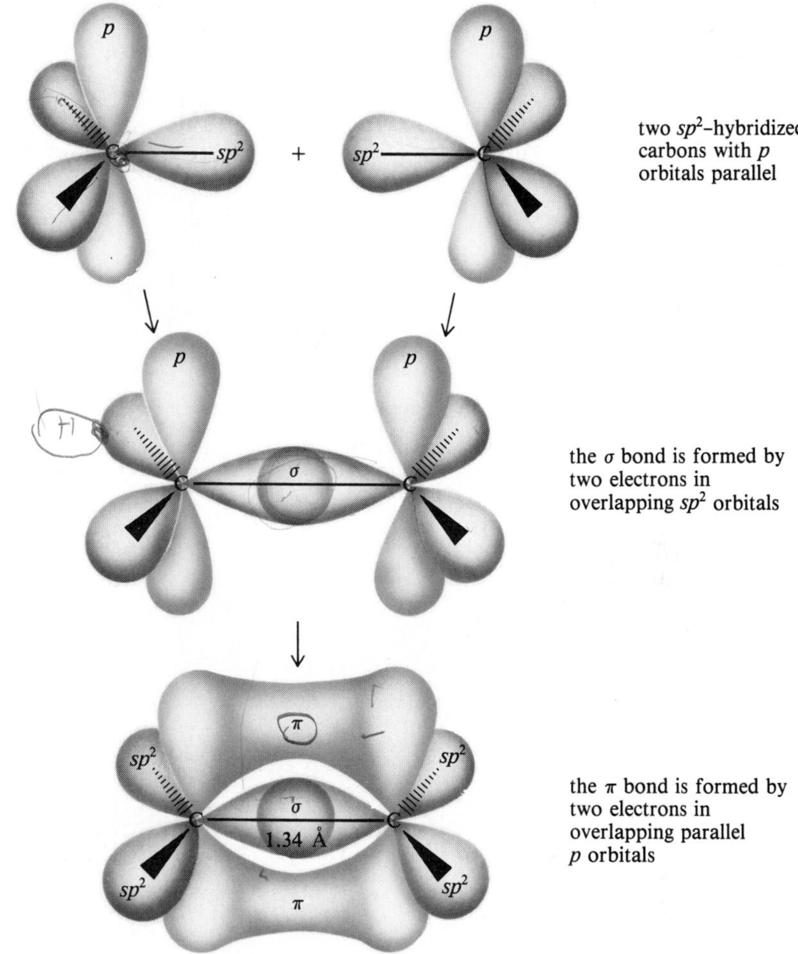
p
p
sp²
+
sp²
two sp²-hybridized carbons with p orbitals parallel
p
p
σ
the σ bond is formed by two electrons in overlapping sp² orbitals
π
sp²
sp²
σ
1.34 Å
sp²
sp²
π
the π bond is formed by two electrons in overlapping parallel p orbitals

FIGURE 3.4
Schematic formation of a carbon–carbon double bond. Two sp^2 carbons form a sigma (σ) bond (end-on overlap of two sp^2 orbitals) and a pi (π) bond (lateral overlap of two properly aligned *p* orbitals).

formed differently. If the two carbons are aligned with the *p* orbitals on each carbon parallel, lateral overlap can occur, as shown at the bottom of Figure 3.4. The bond formed by lateral *p*-orbital overlap is called a **pi (π) bond.** The bonding in ethylene is summarized in Figure 3.5.

The orbital model explains the facts about double bonds listed in Table 3.1. Rotation about a double bond is restricted because, for rotation to occur, we would have to "break" the pi bond, as seen in Figure 3.6. For ethylene, it takes about 62 kcal/mol (259 kJ/mol) to break the pi bond, much more thermal energy than is available at room temperature. With the pi bond intact, the sp^2 orbitals on each carbon lie in a single plane. The 120° angle between those orbitals minimizes repulsion between the electrons in them. Finally, the carbon–

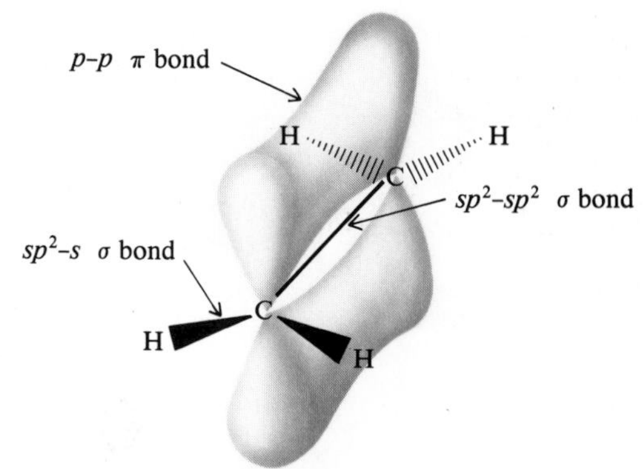

FIGURE 3.5 The bonding in ethylene consists of one sp^2–sp^2 carbon–carbon σ bond, four sp^2–s carbon–hydrogen σ bonds, and one p–p π bond.

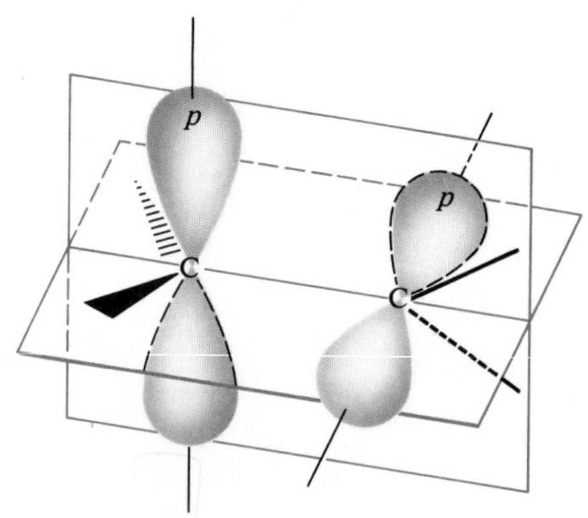

FIGURE 3.6 Rotation of one sp^2 carbon 90° with respect to another orients the p orbitals perpendicular to one another so that no overlap (and therefore no π bond) is possible.

carbon double bond is shorter than the carbon–carbon single bond because the two shared electron pairs draw the nuclei closer together than a single pair does.

To recap, according to the orbital model, the carbon–carbon double bond consists of one sigma bond and one pi bond. The two electrons in the sigma bond lie along the internuclear axis; the two electrons in the pi bond lie in a region of space above and below the plane formed by the two carbons and the four atoms attached to them. The pi electrons are more exposed than the σ electrons and, as we will see, can be attacked by various electron-seeking reagents.

But before we consider reactions at the double bond, let us examine an important result of the restricted rotation around double bonds.

3.5 *Cis–trans* Isomerism in Alkenes

Because rotation at carbon–carbon double bonds is restricted, *cis–trans* isomerism is possible in appropriately substituted alkenes. For example, 1,2-dichloroethene exists in two different forms:

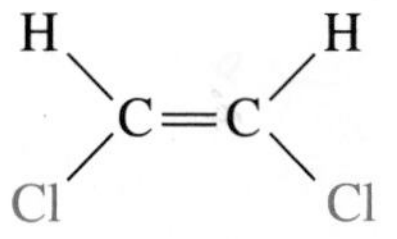

cis-1,2-dichloroethene (H and H on top, Cl and Cl on bottom)
bp 60°C, mp −80°C

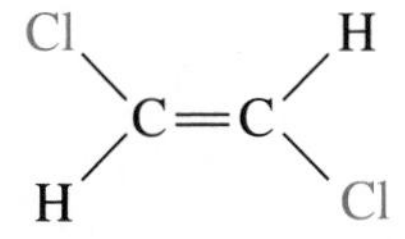

trans-1,2-dichloroethene (Cl and H on top, H and Cl on bottom)
bp 47°C, mp −50°C

These stereoisomers are *not* readily interconverted by rotation around the double bond at room temperature. They are configurational stereoisomers and can be separated from one another by distillation, taking advantage of the difference in their boiling points.

EXAMPLE 3.3 Are *cis–trans* isomers possible for 1-butene and 2-butene?

Solution Only 2-butene has *cis–trans* isomers.

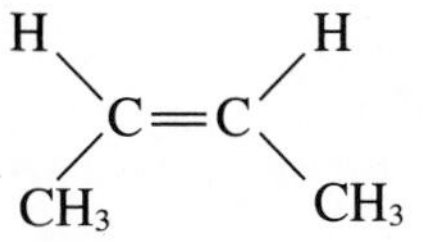

cis-2-butene (H, H on top; CH_3, CH_3 on bottom)
bp 3.7°C, mp −139°C

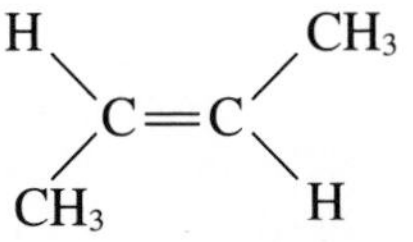

trans-2-butene (H, CH_3 on top; CH_3, H on bottom)
bp 0.3°C, mp −106°C

For 1-butene, carbon-1 has two identical hydrogen atoms attached to it; therefore, only one structure is possible.

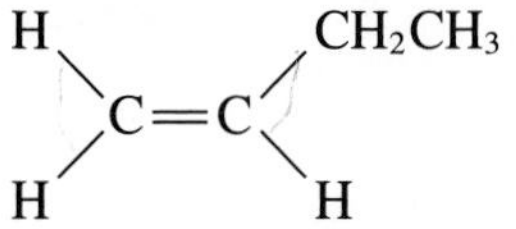

H(H)C=C(H)CH_2CH_3 (with CH_2CH_3 on top right, H bottom right) is identical to H(H)C=C(H)CH_2CH_3 (with H top right, CH_2CH_3 bottom right)

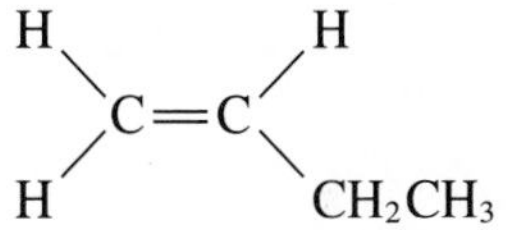

For *cis–trans* isomerism in alkenes, *each* carbon of the double bond must have two different atoms or groups attached to it.

PROBLEM 3.6 Which of the following compounds can exist as *cis–trans* isomers? Draw their structures.

a. propene b. 3-hexene c. 2-hexene d. 2-methyl-2-butene

Geometric isomers of alkenes can be interconverted if sufficient energy is supplied to break the pi bond and allow rotation about the remaining, somewhat stronger, sigma bond (eq. 3.2). The required energy may take the form of light or heat.

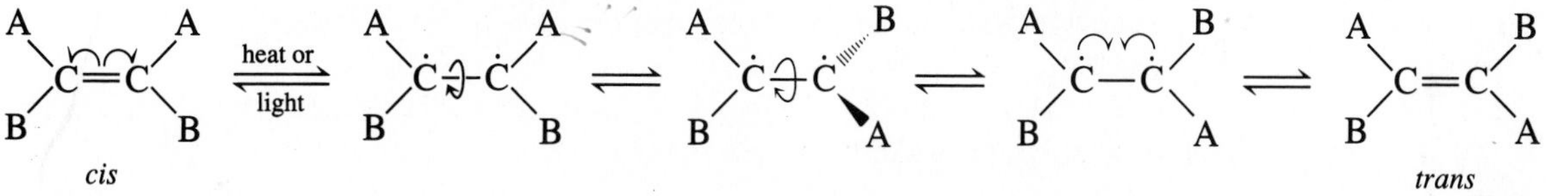

$$\underset{cis}{A(B)C{=}C(A)B} \xrightleftharpoons[\text{light}]{\text{heat or}} A(B)\dot{C}{-}\dot{C}(A)B \rightleftharpoons A(B)\dot{C}{-}\dot{C}(B)A \rightleftharpoons A(B)\dot{C}{-}\dot{C}(B)A \rightleftharpoons \underset{trans}{A(B)C{=}C(B)A} \tag{3.2}$$

A Word About . . .

3. The Chemistry of Vision

Color in organic molecules is usually associated with extended conjugated systems of double bonds. A good example is **β-carotene,** a yellow-orange pigment found in carrots and many other plants. This $C_{40}H_{56}$ hydrocarbon has 11 carbon–carbon double bonds in conjugation. It is the biological precursor of the C_{20} unsaturated alcohol **vitamin A** (also called retinol), which in turn leads to the key substance involved in vision, **11-*cis*-retinal.** Notice in Figure 3.7 that the conversion of vitamin A to 11-*cis*-retinal involves not only oxidation of the alcohol group ($—CH_2OH$) to an aldehyde (—CH═O), but also *trans* → *cis* isomerism at the C_{11}–C_{12} double bond.

Cis–trans isomerism plays a key role in the process of vision. The rod cells in the retina of the eye contain a red, light-sensitive pigment called **rhodopsin.** This pigment consists of the protein **opsin** combined at its active site with 11-*cis*-retinal. When visible light with the appropriate energy is absorbed by rhodopsin, the complexed *cis*-retinal is isomerized to the *trans isomer*. This process is fantastically fast, occurring in only picoseconds (10^{-12} seconds). As you can see from their structures, the shapes of the *cis* and *trans* isomers are very different.

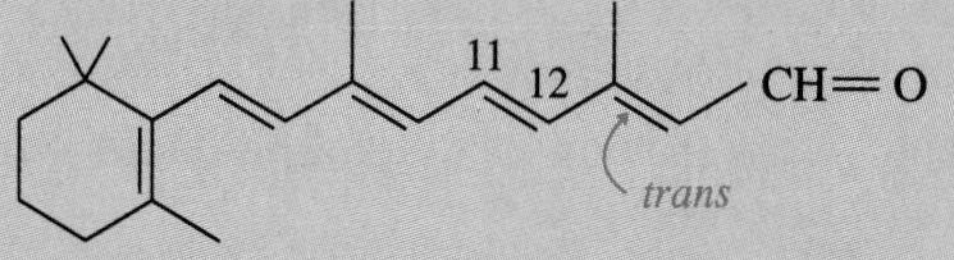

trans-retinal

The *trans*-retinal complex with opsin (called metarhodopsin-II) is less stable than the *cis*-retinal complex, and it dissociates into opsin and *trans*-retinal. This change in geometry triggers a response in the rod nerve cells that is transmitted to the brain and perceived as vision.

If this were all that happened, we would be able to see for only a few moments, because all of the 11-*cis*-retinal present in the rod cells would be quickly consumed. Fortunately, the enzyme *retinal isomerase,* in the presence of light, converts the *trans*-retinal back to the 11-*cis* isomer, so that the cycle can be repeated. Calcium ions in the cell and its membrane control how fast the visual system recovers after exposure to light. They also mediate the way in which cells adapt to various light levels. The following sequence summarizes the visual cycle.

$$\begin{array}{ccc} \text{rhodopsin} & \xrightarrow[\text{energy}]{\text{light}} & \text{metarhodopsin-II (+ nerve impulse)} \\ \Updownarrow & & \Updownarrow \\ \text{opsin + 11-}cis\text{-retinal} & \xleftarrow[\text{+ light}]{\text{retinal isomerase}} & trans\text{-retinal + opsin} \end{array}$$

This representation is simplified because there are actually several additional intermediates between rhodopsin and the fully dissociated *trans*-retinal and opsin.

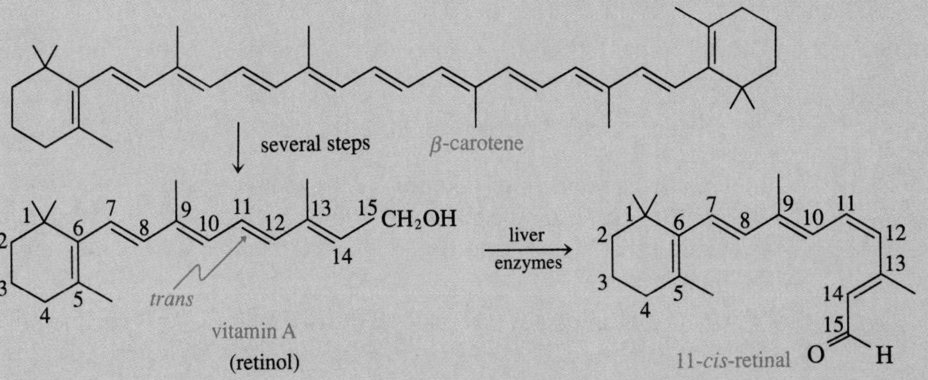

FIGURE 3.7 In the liver, β-carotene is converted into vitamin A first and then into 11-*cis*-retinal.

3.6 Addition and Substitution Reactions Compared

We saw in Chapter 2 that, aside from combustion,the most common reaction of alkanes is **substitution** (for example, halogenation). This reaction type can be expressed by a general equation.

$$\text{R—H} + \text{A—B} \longrightarrow \text{R—A} + \text{H—B} \tag{3.3}$$

where R—H stands for an alkane and A—B stands for the halogen molecule.

With alkenes, on the other hand, the most common reaction is **addition:**

$$\rangle\text{C}{=}\text{C}\langle + \text{A—B} \longrightarrow \text{—}\underset{\text{A}}{\overset{|}{\underset{|}{\text{C}}}}\text{—}\underset{\text{B}}{\overset{|}{\underset{|}{\text{C}}}}\text{—} \tag{3.4}$$

In an addition reaction, group A of the reagent A—B becomes attached to one carbon atom of the double bond, group B becomes attached to the other carbon atom, and the product has only a single bond between the two carbon atoms.

What bond changes take place in an addition reaction? The pi bond of the alkene is broken and the sigma bond of the reagent is also broken. Two new sigma bonds are formed. In other words, we break a pi and a sigma bond, and we make two sigma bonds. Because sigma bonds are usually stronger than pi bonds, the net reaction is favorable.

PROBLEM 3.7 Why, in general, is a sigma bond between two atoms stronger than a pi bond between the same two atoms?

3.7 Polar Addition Reactions

Several reagents add to double bonds by a two-step polar process. In this section we will describe examples of this reaction type, after which we will consider details of the reaction mechanism.

3.7a Addition of Halogens

Alkenes readily add chlorine or bromine.

$$CH_3CH{=}CHCH_3 + Cl_2 \longrightarrow CH_3\underset{Cl}{\underset{|}{C}}H\text{—}\underset{Cl}{\underset{|}{C}}HCH_3 \tag{3.5}$$

2-butene, bp 1–4°C → 2,3-dichlorobutane, bp 117–119°C

$$CH_2{=}CH\text{—}CH_2\text{—}CH{=}CH_2 + 2\,Br_2 \longrightarrow \underset{Br}{\underset{|}{CH_2}}\text{—}\underset{Br}{\underset{|}{CH}}\text{—}CH_2\text{—}\underset{Br}{\underset{|}{CH}}\text{—}\underset{Br}{\underset{|}{CH_2}} \tag{3.6}$$

1,4-pentadiene, bp 26.0°C → 1,2,4,5-tetrabromopentane, mp 85–86°C

Usually the halogen is dissolved in some inert solvent such as tri- or tetrachloromethane, and then this solution is added dropwise to the alkene. Reaction is nearly instantaneous, even at room temperature or below. No light or heat is required, as in the case of substitution reactions.

PROBLEM 3.8 Write an equation for the reaction of bromine at room temperature with

a. 1-butene. b. cyclopentene.

The addition of bromine can be used as a **chemical test** for the presence of unsaturation in an organic compound. Bromine solutions in tetrachloromethane are dark reddish-brown, and the unsaturated compound and its bromine adduct are usually both colorless. As the bromine solution is added to the unsaturated compound, the bromine color disappears. If the compound being tested is saturated, it will not react with bromine under these conditions, and the color will persist.

3.7b Addition of Water (Hydration) If an acid catalyst is present, water adds to alkenes. It adds as H—OH, and the products are alcohols.

$$CH_2{=}CH_2 + H{-}OH \xrightarrow{H^+} \underset{\substack{|\\H}}{CH_2}{-}\underset{\substack{|\\OH}}{CH_2} \quad (\text{or } CH_3CH_2OH) \tag{3.7}$$

ethanol

$$\text{cyclohexene} + H{-}OH \xrightarrow{H^+} \text{cyclohexanol} \tag{3.8}$$

cyclohexene
bp 83.0°

cyclohexanol
bp 161.1°

An acid catalyst is required to serve as an electrophile in this case, because the neutral water molecule is not acidic enough to provide protons to start the reaction. The stepwise mechanism for this reaction is given later in eq. 3.20. Hydration is used industrially and occasionally in the laboratory to synthesize alcohols from alkenes.

PROBLEM 3.9 Write an equation for the acid-catalyzed addition of water to

a. 2-butene.
b. 3-hexene.

3.7c Addition of Acids A variety of acids add to the double bond of alkenes. The hydrogen ion (or proton) adds to one carbon of the double bond, and the remainder of the acid becomes connected to the other carbon.

$$\gt C{=}C\lt + \overset{\delta+}{H}{-}\overset{\delta-}{A} \longrightarrow -\underset{\substack{|\\H}}{\overset{|}{C}}-\underset{\substack{|\\A}}{\overset{|}{C}}- \tag{3.9}$$

Acids that add in this way are the hydrogen halides (HF, HCl, HBr, HI), sulfuric acid ($H{-}OSO_3H$), and organic carboxylic acids $\left(H{-}O\overset{\overset{\displaystyle O}{\|}}{C}R\right)$. Here are two typical examples:

$$\underset{\text{ethene}}{CH_2{=}CH_2} + \underset{\text{hydrogen chloride}}{H{-}Cl} \longrightarrow \underset{\substack{\text{chloroethane}\\\text{(ethyl chloride)}}}{\underset{\substack{|\\H}}{CH_2}{-}\underset{\substack{|\\Cl}}{CH_2}} \quad (\text{or } CH_3CH_2Cl) \qquad (3.10)$$

$$\underset{\text{cyclopentene}}{\text{(cyclopentene, H H)}} + \underset{\text{sulfuric acid}}{H{-}OSO_3H} \longrightarrow \underset{\text{cyclopentyl hydrogen sulfate}}{\text{(cyclopentane, H, H, H, } OSO_3H)} \qquad (3.11)$$

PROBLEM 3.10 Write an equation for each of the following reactions:

a. 2-butene + HI
b. cyclohexene + HBr

Before we discuss the mechanism of addition reactions, we must introduce a complication that we have carefully avoided in all the examples given so far.

3.8 Addition of Unsymmetric Reagents to Unsymmetric Alkenes; Markovnikov's Rule

Reagents and alkenes can be classified as either *symmetric* or *unsymmetric* with respect to addition reactions. Table 3.2 illustrates what this means. If a reagent and/or an alkene is symmetric, only one addition product is possible. If you check back through all the equations and problems for addition reactions up to now, you will see that either the alkene or the reagent (or both) were symmetric. But if *both* the reagent *and* the alkene are *unsymmetric,* two products are, in principle, possible.

$$\underset{\text{unsymmetric alkene}}{\overset{R\qquad H}{\gt C{=}C \lt}} + \underset{\text{unsymmetric reagent}}{X{-}Y} \longrightarrow -\underset{\substack{|\\X}}{\overset{\substack{R\\|}}{C}}-\underset{\substack{|\\Y}}{\overset{\substack{H\\|}}{C}}- \text{ and/or } -\underset{\substack{|\\Y}}{\overset{\substack{R\\|}}{C}}-\underset{\substack{|\\X}}{\overset{\substack{H\\|}}{C}}- \qquad (3.12)$$

The products of eq. 3.12 are sometimes called **regioisomers.** If a reaction of this type gives *only one* of the two possible regioisomers, it is said to be **regiospecific.** If it gives *mainly one* product, it is said to be **regioselective.**

TABLE 3.2 Classification of reagents and alkenes by symmetry with regard to addition reactions

	Symmetric	*Unsymmetric*
Reagents	Br—Br	H—Br
	Cl—Cl	H—OH
	H—H	H—OSO_3H
Alkenes	$CH_2{=}CH_2$	$CH_3CH{=}CH_2$
	$CH_3CH{=}CHCH_3$	$CH_3CH_2CH{=}CHCH_3$
	cyclopentene	1-methylcyclopentene (CH_3)

Let us consider as a specific example the acid-catalyzed addition of water to propene. In principle, two products could be formed: 1-propanol or 2-propanol.

$$\underset{\text{propene}}{\overset{3\quad 2\qquad 1}{CH_3CH{=}CH_2}} \begin{cases} \xrightarrow[H^+]{H-OH} \underset{\text{2-propanol}}{CH_3\underset{\displaystyle OH}{\underset{|}{C}}HCH_3} \\ \xrightarrow[H^+]{H-OH} \underset{\text{1-propanol}}{CH_3CH_2CH_2{-}OH} \end{cases} \tag{3.13}$$

That is, the hydrogen of the water could add to C-1 and the hydroxyl group to C-2 of propene, or vice versa. When the experiment is carried out, *only one product is observed. The addition is regiospecific, and the only product is 2-propanol.*

Most addition reactions of alkenes show a similar preference for the formation of only (or mainly) one of the two possible addition products. Here are some examples.

$$\mathrm{CH_3CH{=}CH_2 + \overset{\delta^+}{H}{-}\overset{\delta^-}{Cl} \longrightarrow CH_3\underset{\displaystyle Cl}{\underset{|}{C}}HCH_3 \qquad \underset{\text{not observed}}{(CH_3CH_2CH_2Cl)}} \tag{3.14}$$

$$\mathrm{CH_3\underset{\displaystyle CH_3}{\underset{|}{C}}{=}CH_2 + \overset{\delta^+}{H}{-}\overset{\delta^-}{OH} \xrightarrow{H^+} CH_3\overset{\displaystyle OH}{\overset{|}{\underset{\displaystyle CH_3}{\underset{|}{C}}}}CH_3 \qquad \underset{\text{not observed}}{(CH_3\underset{\displaystyle CH_3}{\underset{|}{C}}HCH_2OH)}} \tag{3.15}$$

$$\text{1-methylcyclopentene} + \mathrm{H{-}\overset{\delta-}{I}} \longrightarrow \text{1-iodo-1-methylcyclopentane} \qquad \underset{\text{not observed}}{(\text{1-iodo-2-methylcyclopentane})} \tag{3.16}$$

Notice that the reagents are all polar, with a positive and a negative end. After studying a number of such addition reactions, the Russian chemist Vladimir Markovnikov formulated the following rule more than 100 years ago: ***When an unsymmetric reagent adds to an unsymmetric alkene, the electropositive part of the reagent bonds to the carbon of the double bond that has the greater number of hydrogen atoms attached to it.****

PROBLEM 3.11 Use Markovnikov's rule to predict which regioisomer predominates in each of the following reactions:

a. 1-butene + HCl b. 2-methyl-2-butene + H_2O (H^+ catalyst)

PROBLEM 3.12 What two products are *possible* from the addition of HCl to 2-pentene? Would you expect the reaction to be regiospecific?

Let us now develop a rational explanation for Markovnikov's rule in terms of modern chemical theory.

---

*Actually, Markovnikov stated the rule a little differently. The form given here is easier to remember and apply. For an interesting historical article on what he actually said, when he said it, and how his name is spelled, see J. Tierney, *J. Chem. Educ.* **1988,** *65,* 1053–54.

3.9 *Mechanism of Electrophilic Addition to Alkenes*

The pi electrons of a double bond are more exposed to an attacking reagent than are the sigma electrons. The π bond is also weaker than the σ bond. It is the pi electrons, then, that are involved in additions to alkenes. The double bond can act as a supplier of pi electrons to an electron-seeking reagent.

Polar reactants can be classified as either **electrophiles** or **nucleophiles.** **Electrophiles (literally, electron lovers) are electron poor reagents; in reactions with some other molecule, they seek electrons.** They are often positive ions (cations) or otherwise electron-deficient species. **Nucleophiles (literally, nucleus lovers), on the other hand, are electron rich; they form bonds by donating electrons to an electrophile.**

$$\underset{\text{electrophile}}{E^+} + \underset{\text{nucleophile}}{:Nu^-} \longrightarrow E:Nu \qquad (3.17)$$

Let us now consider the mechanism of a polar addition to a carbon–carbon double bond, specifically the addition of acids to alkenes. The carbon–carbon double bond, because of its pi electrons, is a nucleophile. The proton (H^+) is the attacking electrophile. As the proton approaches the π bond, the two pi electrons are used to form a σ bond between the proton and one of the two carbon atoms. Because this bond uses *both* pi electrons, the other carbon acquires a positive charge, producing a **carbocation.**

$$H^+ + \text{C}{=}\text{C} \longrightarrow \underset{\text{carbocation}}{\text{H}{-}\text{C}{-}\text{C}^+} \qquad (3.18)$$

The resulting carbocations are, however, extremely reactive because there are only six electrons (instead of the usual eight) around the positive carbon. The carbocation rapidly combines with some species that can supply it with two electrons, a nucleophile.

$$\text{H}{-}\text{C}{-}\text{C}^+ + \underset{\text{nucleophile}}{\text{Nu:}^-} \longrightarrow \underset{\text{product of addition of H—Nu to an alkene}}{\text{H}{-}\text{C}{-}\text{C}{-}\text{Nu}} \qquad (3.19)$$

Examples include the addition of HCl, $HOSO_3H$, and HOH to alkenes:

$$H^+ + \mathrm{C{=}C} \rightleftharpoons \mathrm{H{-}C{-}C^+} \begin{cases} \xrightarrow{:\ddot{Cl}:^-} \mathrm{H{-}C{-}C{-}\ddot{Cl}:} \\ \xrightarrow{^-:\ddot{O}SO_3H} \mathrm{H{-}C{-}C{-}\ddot{O}SO_3H} \\ \xrightarrow{H{-}\ddot{O}{-}H} \mathrm{H{-}C{-}C{-}O^+H_2} \rightleftharpoons \mathrm{H{-}C{-}C{-}\ddot{O}H} + H^+ \end{cases} \quad (3.20)$$

In these reactions, the electrophile H^+ first adds to the alkene to give a carbocation. Then the carbocation combines with a nucleophile, in these examples, a chloride ion, a bisulfate ion, or a water molecule.

With most alkenes, the first step in this process—the formation of the carbocation—is the slower of the two steps. The resulting carbocation is usually so reactive that combination with the nucleophile is extremely rapid. **Since the initiating step in these additions is attack by the electrophile, the whole process is called an electrophilic addition.**

EXAMPLE 3.4 Since carbocations are involved in the electrophilic addition reactions of alkenes, it is important to understand the bonding in these chemical intermediates. Describe the bonding in carbocations in orbital terms.

Solution The carbon atom is positively charged and therefore has only three valence electrons to use in bonding. Each of these electrons is in an sp^2 orbital. The three sp^2 orbitals lie in one plane with 120° angles between them, an arrangement that minimizes repulsion between the electrons in the three bonds. The remaining p orbital is perpendicular to that plane and vacant.

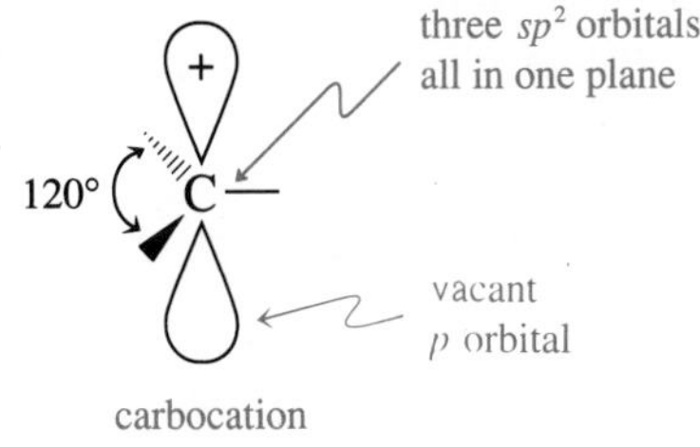

carbocation

3.10 Markovnikov's Rule Explained

To explain Markovnikov's rule, let us consider a specific example, the addition of H—Cl to propene. The first step is addition of a proton to the double bond. This can occur in two ways, to give either an isopropyl cation or a propyl cation.

$$\underset{\text{propene}}{\overset{3}{C}H_3-\overset{2}{C}H=\overset{1}{C}H_2} \xrightarrow{H^+} \begin{cases} \xrightarrow{\text{adds to C-1}} \underset{\text{isopropyl cation}}{CH_3\overset{+}{C}HCH_3} \\ \xrightarrow[\not\ ]{\text{adds to C-2}} \underset{\text{propyl cation}}{CH_3CH_2CH_2^+} \end{cases} \tag{3.21}$$

At this stage of the reaction, the structure of the product is already determined; when combining with chloride ion, the isopropyl cation can give only 2-chloropropane, and the propyl cation can give only 1-chloropropane. The only observed product is 2-chloropropane, so we must conclude that the *proton adds to C-1 to form only the isopropyl cation*. Why?

Carbocations are classified as tertiary, secondary, or primary, depending on whether the positive carbon atom has attached to it three organic groups, two groups, or only one group. From many studies, it has been established that the stability of carbocations decreases in the following order:

$$\underset{\substack{\text{tertiary (3°)}\\ \text{most stable}}}{R_3C^+} > \underset{\text{secondary (2°)}}{R_2\overset{+}{C}H} \gg \underset{\text{primary (1°)}}{R-\overset{+}{C}H_2} > \underset{\substack{\text{methyl (unique)}\\ \text{least stable}}}{\overset{+}{C}H_3}$$

One reason for this order is the following: A carbocation will be more stable when the positive charge can be spread out, or delocalized, over several atoms in the ion, instead of being concentrated on a single carbon atom. In alkyl cations, this delocalization occurs by drift of electron density to the positive carbon from the other bonds in the ion. The more bonds, the more the charge is delocalized. If the positive carbon is surrounded by other carbon atoms (alkyl groups), instead of by hydrogen atoms, there will be more bonding electrons to help delocalize the charge. This is the main reason for the observed stability order of carbocations.

Markovnikov's rule can now be restated in modern and more generally useful terms: ***The electrophilic addition of an unsymmetric reagent to an unsymmetric double bond proceeds in such a way as to involve the most stable carbocation.***

PROBLEM 3.13 Classify each of the following carbocations as primary, secondary, or tertiary:

a. $CH_3CH_2\overset{+}{C}HCH_3$ b. c. $(CH_3)_2CH\overset{+}{C}H_2$

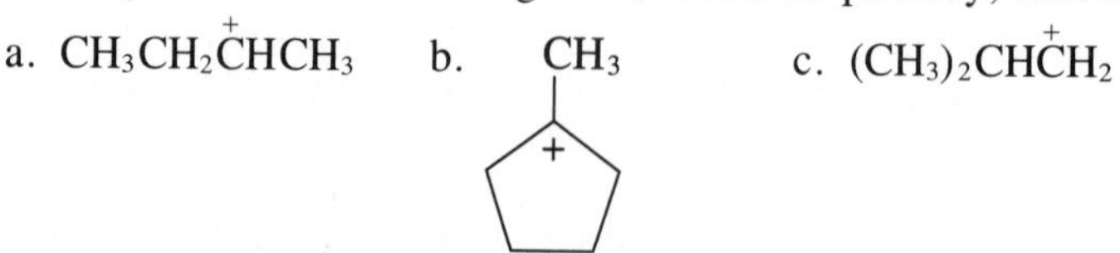

PROBLEM 3.14 Which carbocation in Problem 3.13 is most stable? least stable?

PROBLEM 3.15 Write the steps in the electrophilic additions in eqs. 3.15 and 3.16, and in each case, show that reaction occurs via the more stable carbocation.

Discussion of Markovnikov's rule raises two important general questions about chemical reactions: (1) Under what conditions is a reaction likely to proceed? (2) How rapidly will a reaction occur? We will consider these questions briefly in the next two sections before continuing our survey of the reactions of alkenes.

3.11 Reaction Equilibrium: What Makes a Reaction Go?

A chemical reaction can proceed in two directions. Reactant molecules can form product molecules, and product molecules can react to reform the reactant molecules. For the reaction*

$$\text{aA} + \text{bB} \rightleftharpoons \text{cC} + \text{dD} \qquad (3.22)$$

we describe the chemical equilibrium for the forward and backward reactions by the following equation:

$$K_{eq} = \frac{[\text{C}]^c[\text{D}]^d}{[\text{A}]^a[\text{B}]^b} \quad \left(\frac{\text{product of product concentrations}}{\text{product of reactant concentrations}}\right) \qquad (3.23)$$

In this equation, K_{eq}, the equilibrium constant, is equal to the product of the concentrations of the products divided by the product of the concentrations of the reactants. (The small letters a, b, c, and d are the numbers of molecules of reactants and products in the balanced reaction equation.)

The equilibrium constant tells us the direction that is favored for the reaction. If K_{eq} is greater than 1, when the reactants are added together, the formation of products C and D will be favored over the formation of reactants A and B. The preferred direction for the reaction is from left to right. Conversely, if K_{eq} is less than 1, the preferred direction for the reaction is from right to left.

What determines whether a reaction will proceed to the right, toward products? A reaction will occur when the products are lower in energy (more stable) than the reactants. A reaction in which products are higher in energy than reactants will proceed to the left, toward reactants. When products are lower in energy than reactants, heat is given off in the course of the reaction. For example, heat is given off when an acid such as HBr is added to ethylene (eq. 3.24). Such a reaction is **exothermic**.

$$H_2C{=}CH_2 + HBr \rightleftharpoons CH_3CH_2Br \qquad (3.24)$$

On the other hand, heat must be added to ethane to produce two methyl radicals (eq. 1.3). This reaction is **endothermic** (takes in heat). The term used by chemists for heat energy is **enthalpy** and is designated by the symbol H. The

*The double arrow indicates that this reaction goes both ways and reaches chemical equilibrium.

difference in enthalpy between products and reactants is designated by the symbol ΔH (pronounced "delta H").

For the addition of HBr to ethylene, the product (bromoethane) is more stable than the reactants (ethylene and HBr), and the reaction proceeds to the right. For this reaction ΔH is negative (heat is given off), and K_{eq} is much greater than 1 (Figure 3.8a). For the formation of two methyl radicals from

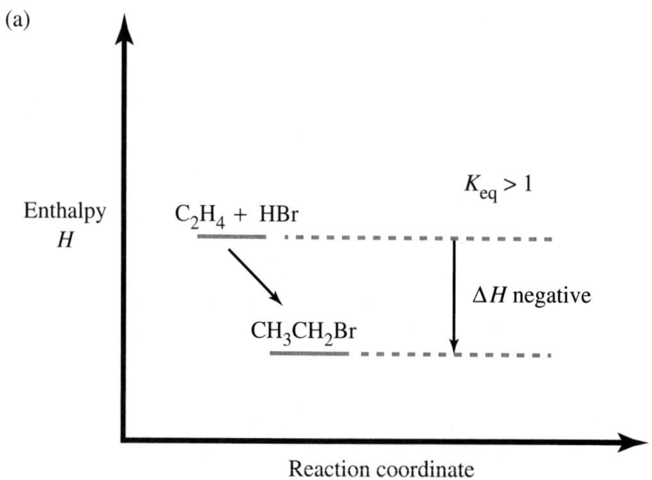

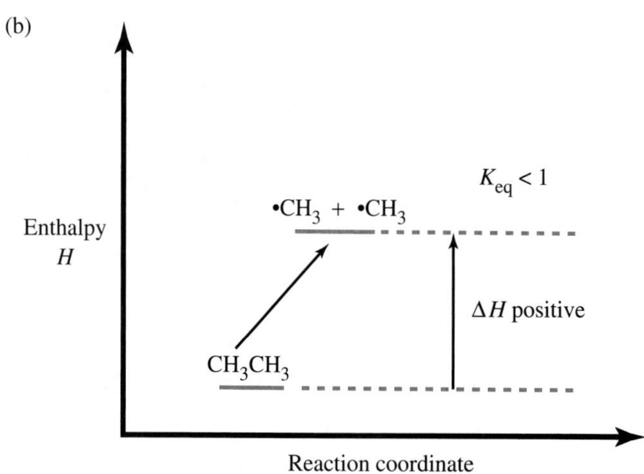

FIGURE 3.8 (a) The addition of HBr to ethylene; the reaction equilibrium lies to the right. (b) The formation of methyl radicals from ethane; the reaction equilibrium lies to the left.

ethane, ΔH is positive (heat is absorbed), and K_{eq} is much less than 1 (Figure 3.8b).*

3.12 Reaction Rates: How Fast Does a Reaction Go?

The equilibrium constant for a reaction tells us whether or not products are more stable than reactants in a reaction. However, *the equilibrium constant does not tell us anything about the rate of a reaction.* For example, the equilibrium constant for the reaction of gasoline with oxygen is very large, but gasoline can be safely handled in air because the reaction is very slow unless heat is used to initiate it. The rate of addition of HBr to ethylene is also very slow, although the reaction is exothermic.

In order to react, molecules must collide with each other with enough energy and with the right orientation so that the breaking and making of bonds can occur. The energy required for this process is a barrier to reaction. The higher the barrier, the slower the reaction.

Chemists use **reaction energy diagrams** to show the changes in energy that occur in the course of a reaction. Figure 3.9 shows the reaction energy diagram for the polar addition of the acid HBr to ethylene (eq. 3.24). This reaction occurs in two steps. In the first step, the π bond of the alkene is broken and a C—H σ bond is formed, giving a carbocation intermediate product. The reactants start with the energy shown in the diagram. As the π bond begins to break and the new σ bond begins to form, the structure formed by the reactants reaches a maximum energy. This structure with maximum energy is called the **transition state** for the first step. This structure cannot be isolated and continues to change until the carbocation product of the first step is fully formed.

The difference in energy between the transition state and the reactants is called the **activation energy,** E_a. It is this energy that determines the rate of the reaction. If E_a is great, the reaction will be slow. A small E_a means that the reaction will proceed rapidly.

In the second step of the reaction, a new carbon–bromine σ bond is formed. Again, the approach of the bromide ion to the positively charged carbon of the carbocation intermediate causes a rise in energy to a maximum. The structure at this energy maximum is the transition state for the second step. The difference in energy between the carbocation and this transition state is the activation energy E_a for this step. This structure cannot be isolated and continues to change until the σ bond is fully formed, completing the formation of the product.

Notice on Figure 3.9 that although the final product of the reaction is lower in energy (ΔH) than the reactants, the reactants must surmount two energy

*Actually, enthalpy is not the only factor that contributes to the energy difference between products and reactants. A factor called **entropy**, *S*, also contributes to the total energy difference, which is known as the **Gibbs free-energy difference**, ΔG, in the equation $\Delta G = \Delta H - T\Delta S$. For most organic reactions, however, the entropy contribution is very small compared to the enthalpy contribution.

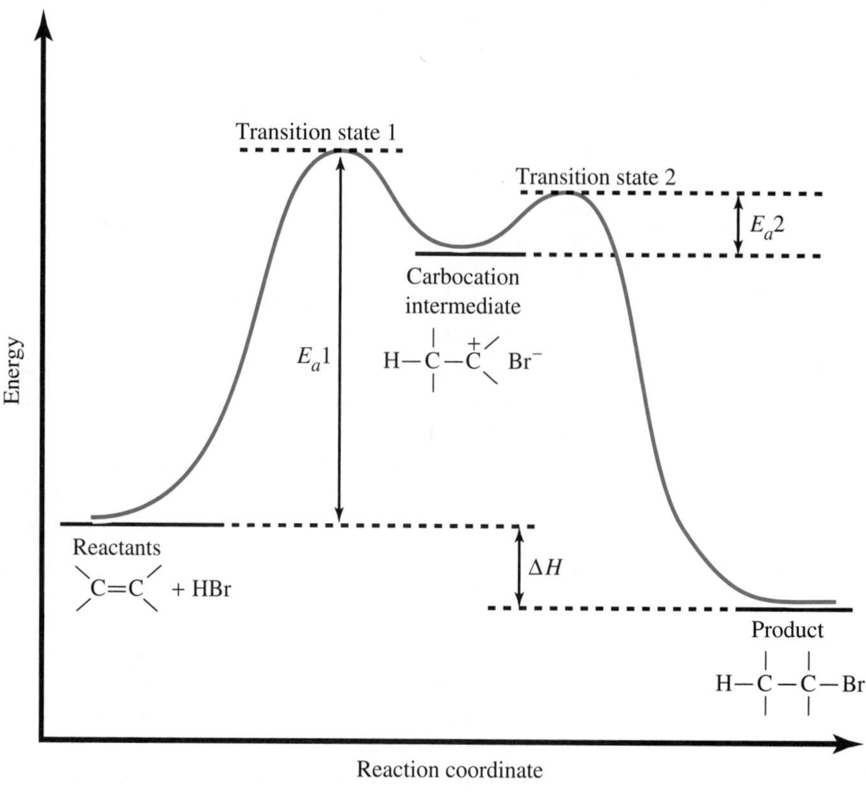

FIGURE 3.9 Reaction energy diagram for the addition of HBr to ethylene.

barriers (E_a1 and E_a2), one for each step of the reaction. Between the two transition states, the carbocation intermediate is at an energy minimum that is higher than reactants or products. The first step of the reaction is endothermic, because the carbocation intermediate product is higher in energy than the reactants. The second step is exothermic, because the product is lower in energy than the carbocation. The overall reaction is exothermic, because the product is lower in energy than the reactants. However, the **rate** of the reaction is determined by the highest energy barrier, E_a1. The second activation energy, E_a2, is very low compared to the activation energy for the first step. Therefore, as described in Sec. 3.9, the first step is the slower of the two steps, and the rate of the reaction is determined by the rate of this first step.

EXAMPLE 3.5 Sketch a reaction energy diagram for a one-step reaction that is very slow and slightly exothermic.

Solution A very slow reaction has a large E_a, and a slightly exothermic reaction has a small negative ΔH. Therefore, the diagram will look like this:

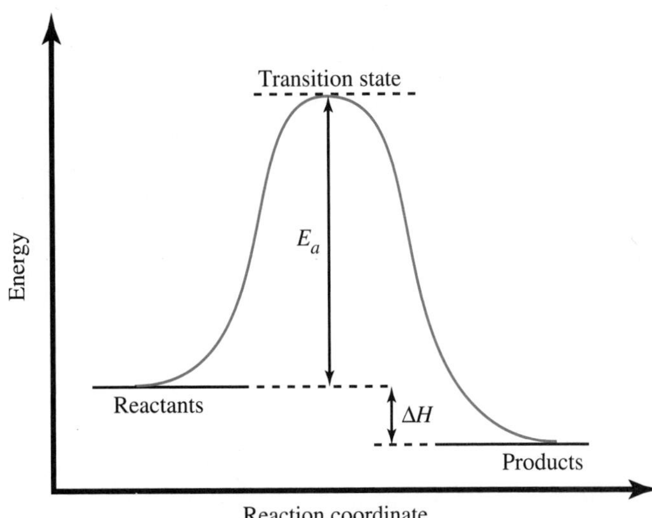

PROBLEM 3.16 Draw a reaction energy diagram for a one-step reaction that is very fast and very exothermic.

PROBLEM 3.17 Draw a reaction energy diagram for a one-step reaction that is very slow and slightly endothermic.

PROBLEM 3.18 Draw a reaction energy diagram for a two-step reaction that has an endothermic first step and an exothermic second step. Label the reactants, transition states, reaction intermediate, activation energies, and enthalpy differences.

Let us see how reaction rates are related to Markovnikov's rule. In electrophilic addition reactions, more stable carbocations are formed more rapidly than less stable carbocations. This is because more stable carbocations are lower in energy than less stable carbocations and it follows that the activation energy E_a for the formation of more stable carbocations is also lower. The regioselectivity of electrophilic additions is thus the result of competing first steps, in which the more stable carbocation is formed at a faster rate.

Other factors that affect reaction rates are **temperature** and **catalysts**. Heating a reaction generally increases the rate at which the reaction occurs by providing the reactant molecules with more energy to surmount activation energy barriers. Catalysts are substances that increase reaction rates but are not changed by the reaction. Catalysts speed up a reaction by providing an alternative pathway or mechanism for the reaction in which the activation energy is lower. Enzymes play this role in biochemical reactions.

In the next five sections, we will continue our survey of the reactions of alkenes.

3.13 Hydroboration of Alkenes

Hydroboration was discovered by Professor Herbert C. Brown (Purdue University). This reaction is so useful in synthesis that Brown's work earned him a Nobel Prize in 1979. We will describe here only one practical example of hydroboration, a two-step alcohol synthesis from alkenes.

Hydroboration involves addition of a hydrogen–boron bond to an alkene. The $H{-}B\langle$ bond is polarized with the hydrogen $\delta-$ and the boron $\delta+$. Addition occurs so that the boron (the electrophile) adds to the less substituted carbon.

$$R{-}CH{=}CH_2 + \overset{\delta-}{H}{-}\overset{\delta+}{B}\langle \longrightarrow R{-}CH_2{-}CH_2{-}B\langle \qquad (3.25)$$

Thus, it resembles a normal electrophilic addition to an alkene, following Markovnikov's rule, even though the addition is concerted (that is, all bond-breaking and bond-making occurs in one step).

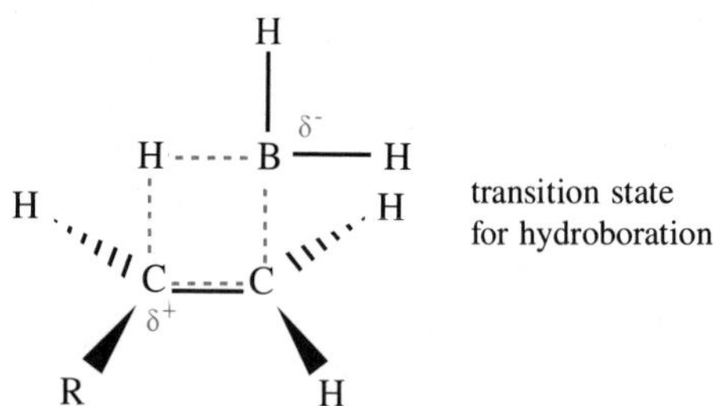

transition state for hydroboration

Because it has three B—H bonds, one molecule of borane, BH_3, can react with three molecules of an alkene. For example, propene gives tri-*n*-propylborane.

$$3\,CH_3CH{=}CH_2 + BH_3 \longrightarrow CH_3CH_2CH_2{-}B(CH_2CH_2CH_3)_2 \qquad (3.26)$$

propene   borane   tri-*n*-propylborane

The trialkylboranes made in this way are usually not isolated but are treated with some other reagent to obtain the desired final product. For example, trialkylboranes can be oxidized by hydrogen peroxide and base to give alcohols.

$$(CH_3CH_2CH_2)_3B + 3\,H_2O_2 + 3\,NaOH \longrightarrow 3\,CH_3CH_2CH_2OH + Na_3BO_3 + 3\,H_2O \qquad (3.27)$$

tri-*n*-propylborane; *n*-propyl alcohol; sodium borate

One great advantage of this hydroboration-oxidation sequence is that it provides a route to alcohols that *cannot* be obtained by the acid-catalyzed hydration of alkenes (review eq. 3.13).

$$R{-}CH{=}CH_2 \begin{cases} \xrightarrow[H^+]{H{-}OH} R{-}\underset{\displaystyle OH}{\underset{|}{C}H}{-}CH_3 \quad \text{Markovnikov product} \\ \xrightarrow[2.\ H_2O_2,\ OH^-]{1.\ BH_3} R{-}CH_2{-}CH_2OH \quad \text{anti-Markovnikov product} \end{cases} \tag{3.28}$$

The overall result of the two-step hydroboration sequence ***appears*** *to be the addition of water to the carbon–carbon double bond in the reverse of the usual Markovnikov sense.*

EXAMPLE 3.6 What alcohol is obtained from this sequence?

$$CH_3{-}\overset{\displaystyle CH_3}{\overset{|}{C}}{=}CH_2 \xrightarrow{BH_3} \xrightarrow[OH^-]{H_2O_2}$$

Solution The boron adds to the less substituted carbon; oxidation gives the corresponding alcohol. Compare this result with that of eq. 3.15.

$$3\,CH_3{-}\overset{\displaystyle CH_3}{\overset{|}{C}}{=}CH_2 \xrightarrow{BH_3} (CH_3{-}\overset{\displaystyle CH_3}{\overset{|}{C}H}{-}CH_2)_3{-}B \xrightarrow[OH^-]{H_2O_2} 3\,CH_3{-}\overset{\displaystyle CH_3}{\overset{|}{C}H}{-}CH_2OH$$

PROBLEM 3.19 What alcohol is obtained by applying the hydroboration-oxidation sequence to 2-methyl-2-butene?

PROBLEM 3.20 What alkene is needed to obtain (cyclohexyl)$-CH_2CH_2OH$ via the hydroboration-oxidation sequence?

3.14 Addition of Hydrogen

Hydrogen adds to alkenes in the presence of an appropriate catalyst. The process is called **hydrogenation.**

$$\gt C{=}C\lt + H_2 \xrightarrow{\text{catalyst}} -\underset{\displaystyle H}{\underset{|}{\overset{|}{C}}}-\underset{\displaystyle H}{\underset{|}{\overset{|}{C}}}- \tag{3.29}$$

The catalyst is usually a finely divided metal, such as nickel, platinum, or palladium. These metals adsorb hydrogen gas on their surfaces and activate the hydrogen–hydrogen bond. Both hydrogen atoms usually add from the catalyst surface to the same face of the double bond. For example, 1,2-dimethylcyclopentene gives mainly *cis*-1,2-dimethylcyclopentane.

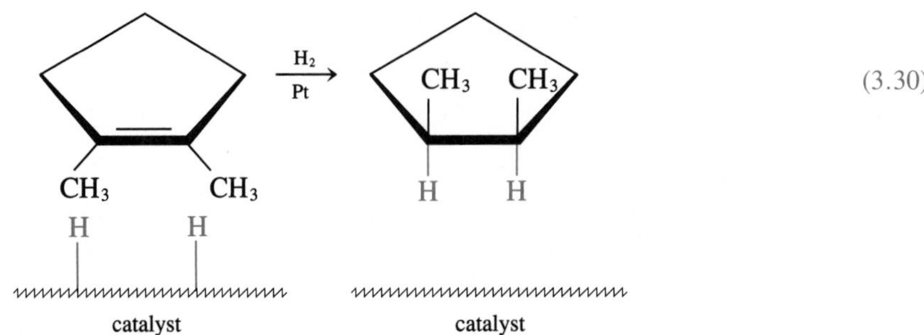

(3.30)

Catalytic hydrogenation of double bonds is used commercially to convert vegetable oils to margarine and other cooking fats (Sec. 15.3).

PROBLEM 3.21 Write an equation for the catalytic hydrogenation of

a. methylpropene.
b. 1,2-dimethylcyclobutene.

3.15 Electrophilic Additions to Conjugated Systems

Alternate double and single bonds of conjugated systems have special consequences for their addition reactions. When 1 mole of hydrogen bromide adds to 1 mole of 1,3-butadiene, a rather surprising result is obtained. Two products are isolated.

$$\overset{1}{C}H_2{=}\overset{2}{C}H{-}\overset{3}{C}H{=}\overset{4}{C}H_2 \xrightarrow{HBr} \begin{cases} \underset{\text{H}}{\underset{|}{CH_2}}{-}\underset{\text{Br}}{\underset{|}{CH}}{-}CH{=}CH_2 & \text{(1,2-addition)} \quad \text{3-bromo-1-butene} \\ \underset{\text{H}}{\underset{|}{CH_2}}{-}CH{=}CH{-}\underset{\text{Br}}{\underset{|}{CH_2}} & \text{(1,4-addition)} \quad \text{1-bromo-2-butene} \end{cases} \tag{3.31}$$

1,3-butadiene

In one of these products, HBr has added to one of the two double bonds, and the other double bond is still present in its original position. We call this the product of **1,2-addition.** The other product may at first seem unexpected. The hydrogen and bromine have added to carbon-1 and carbon-4 of the original diene, and a new double bond has appeared between carbon-2 and carbon-3. This process, called **1,4-addition,** is quite a general reaction for electrophilic additions to conjugated systems. How can we explain it?

In the first step, the proton adds to the terminal carbon atom, according to Markovnikov's rule.

$$H^{+} + CH_2{=}CH{-}CH{=}CH_2 \longrightarrow CH_3{-}\overset{+}{C}H{-}CH{=}CH_2 \qquad (3.32)$$

The resulting carbocation can be stabilized by resonance; in fact, it is a hybrid of two contributing resonance structures.

$$[CH_3{-}\overset{+}{C}H{-}CH{=}CH_2 \longleftrightarrow CH_3{-}CH{=}CH{-}\overset{+}{C}H_2]$$

The positive charge is delocalized over carbon-2 and carbon-4. When, in the next step, the carbocation reacts with bromide ion (the nucleophile), it can react either at carbon-2 to give the product of 1,2-addition, or at carbon-4 to give the product of 1,4-addition.

$$\left.\begin{matrix} \underset{1}{CH_3}{-}\underset{2}{\overset{+}{C}H}{-}\underset{3}{CH}{=}\underset{4}{CH_2} \\ \updownarrow \\ \underset{1}{CH_3}{-}\underset{2}{CH}{=}\underset{3}{CH}{-}\underset{4}{\overset{+}{C}H_2} \end{matrix}\right\} \xrightarrow{Br^-} \begin{matrix} CH_3{-}\underset{\substack{|\\ Br}}{CH}{-}CH{=}CH_2 \\ + \\ CH_3{-}CH{=}CH{-}\underset{\substack{|\\ Br}}{CH_2} \end{matrix} \qquad (3.33)$$

PROBLEM 3.22 Explain why, in the first step in the addition of HBr to 1,3-butadiene, the proton adds to C-1 (eq. 3.32) and not to C-2.

The carbocation intermediate in these reactions is a single species, a resonance hybrid. ***This type of carbocation, with a carbon–carbon double bond adjacent to the positive carbon, is called an allylic cation.*** The parent allyl cation, shown below as a resonance hybrid, is a primary carbocation, but it is more stable than simple primary ions (such as propyl), because its positive charge is delocalized over the two end carbon atoms.

$$CH_2{=}CH{-}\overset{+}{C}H_2 \longleftrightarrow \overset{+}{C}H_2{-}CH{=}CH_2$$

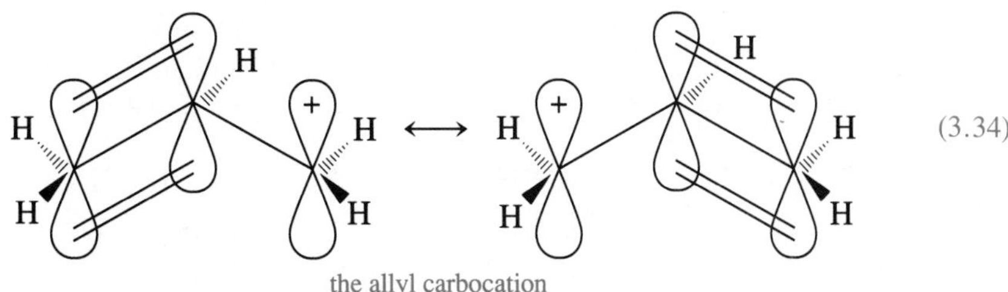

the allyl carbocation (3.34)

PROBLEM 3.23 Draw the contributors to the resonance hybrid structure of the 3-cyclopentenyl cation

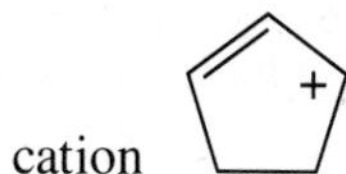

PROBLEM 3.24 Write an equation for the expected products of 1,2-addition and 1,4-addition of bromine to 1,3-butadiene.

3.16 Free-Radical Additions; Polyethylene

Some reagents add to alkenes by a free-radical mechanism instead of by an ionic mechanism. From a commercial standpoint, the most important of these free-radical additions are those that lead to polymers.

A **polymer** is a large molecule, usually with a high molecular weight, built up from small repeating units. The simple molecule from which these repeating units are derived is called a **monomer,** and the process of converting a monomer to a polymer is called **polymerization.**

The free-radical polymerization of ethylene gives **polyethylene,** a material that is produced on a very large scale (more than 10 billion pounds annually in the United States alone). The reaction is carried out by heating ethylene under pressure with a catalyst (eq. 3.35). How does this reaction occur?

$$\underset{\text{ethylene}}{CH_2{=}CH_2} \xrightarrow[\text{1000 atm, >100°C}]{\text{ROOR}} \underset{\substack{\text{polyethylene} \\ (n\,=\,\text{several thousand})}}{\left(CH_2{-}CH_2\right)_n} \qquad (3.35)$$

One common type of catalyst for polymerization is an organic peroxide. The O—O single bond is weak, and on heating this bond breaks, with one electron going to each of the oxygens.

$$\underset{\text{organic peroxide}}{R{-}O{-}O{-}R} \xrightarrow{\text{heat}} \underset{\text{two radicals}}{2\,R{-}O\cdot} \qquad (3.36)$$

A catalyst radical then adds to the carbon–carbon double bond:

$$\underset{\substack{\text{catalyst}\\\text{radical}}}{RO\cdot} \quad CH_2{=}CH_2 \longrightarrow \underset{\substack{\text{a carbon}\\\text{free radical}}}{RO{-}CH_2{-}CH_2\cdot} \qquad (3.37)$$

The result of this addition is a carbon free radical, which may add to another ethylene molecule, and another, and another, and so on.

$$ROCH_2CH_2\cdot \xrightarrow{CH_2{=}CH_2} ROCH_2CH_2CH_2CH_2\cdot \xrightarrow{CH_2{=}CH_2} ROCH_2CH_2CH_2CH_2CH_2CH_2\cdot \text{ and so on} \qquad (3.38)$$

The carbon chain continues to grow in length until some chain-termination reaction occurs (perhaps a combination of two radicals).

We might think that only a single long chain of carbons will be formed in this way, but this is not always the case. A "growing" polymer chain may abstract a hydrogen atom from its back, so to speak, to cause **chain branching.**

$$\cdot CH_2 \text{~~~~H} \longrightarrow H{-}CH_2 \text{~~~~} \xrightarrow{CH_2{=}CH_2} CH_3 \text{~~~~} CH_2{-}CH_2\cdot \quad \text{and so on} \qquad (3.39)$$

A giant molecule with long and short branches is thus formed:

branched polyethylene

The degree of chain branching and other features of the polymer structure can often be controlled by the choice of catalyst and reaction conditions.

A polyethylene molecule is mainly saturated despite its name (polyethyl*ene*) and consists mostly of linked CH_2 groups, but with CH groups at the branch points and CH_3 groups at the ends of the branches. It also contains an OR group from the catalyst at one end, but since the molecular weight is very large, this OR group constitutes a minor and, as far as properties go, relatively insignificant fraction of the molecule.

Polyethylene made in this way is transparent and used in packaging and film (for example, for freezer and sandwich bags).

In Chapter 14, we will describe many other polymers, some made by the process just described for polyethylene and some made by other methods.

3.17 Oxidation of Alkenes

In general, alkenes are more easily oxidized than alkanes by chemical oxidizing agents. These reagents attack the pi electrons of the double bond. The reactions may be useful as chemical tests for the presence of a double bond or for synthesis.

3.17a Oxidation with Permanganate; a Chemical Test

Alkenes react with alkaline potassium permanganate to form **glycols** (compounds with two adjacent hydroxyl groups).

$$3\ \underset{\text{alkene}}{>C{=}C<} + \underset{\substack{\text{potassium}\\\text{permanganate}\\\text{(purple)}}}{2\ K^+MnO_4^-} + 4\ H_2O \longrightarrow 3{-}\underset{\substack{\text{OH OH}\\\text{a glycol}}}{\overset{|}{\underset{|}{C}}{-}\overset{|}{\underset{|}{C}}{-}} + \underset{\substack{\text{manganese}\\\text{dioxide}\\\text{(brown-black)}}}{2\ MnO_2} + 2\ K^+OH^- \qquad (3.40)$$

As the reaction occurs, the purple color of the permanganate ion is replaced by the brown precipitate of manganese dioxide. Because of this color change, the reaction can be used as a chemical test to distinguish alkenes from alkanes, which normally do not react.

PROBLEM 3.25 Write an equation for the reaction of 2-butene with potassium permanganate.

3.17b Ozonolysis of Alkenes Alkenes react rapidly and quantitatively with ozone, O_3. Ozone is generated by passing oxygen over a high-voltage electric discharge. The resulting gas stream is then bubbled at low temperature into a solution of the alkene in an inert solvent, such as dichloromethane. The first product, a **molozonide,** rearranges rapidly to an **ozonide.** Since these products may be explosive if isolated, they are usually treated directly with a reducing agent, commonly zinc and aqueous acid, to give carbonyl compounds as the isolated products.

$$\text{>C=C<} \xrightarrow{O_3} [\text{molozonide}] \longrightarrow \text{ozonide} \xrightarrow[H_3O^+]{Zn} \text{>C=O} + \text{O=C<} \quad (3.41)$$

alkene — a molozonide — an ozonide — two carbonyl groups

The net result of this reaction is to break the double bond of the alkene and to form two carbon–oxygen double bonds (carbonyl groups), one at each carbon of the original double bond. The overall process is called ozonolysis.

Ozonolysis can be used to locate the position of a double bond. For example, ozonolysis of 1-butene gives two different aldehydes, whereas 2-butene gives a single aldehyde.

$$\underset{\text{1-butene}}{CH_2{=}CHCH_2CH_3} \xrightarrow[\text{2. Zn, H}^+]{\text{1. O}_3} \underset{\text{formaldehyde}}{CH_2{=}O} + \underset{\text{propanal}}{O{=}CHCH_2CH_3} \quad (3.42)$$

$$\underset{\text{2-butene}}{CH_3CH{=}CHCH_3} \xrightarrow[\text{2. Zn, H}^+]{\text{1. O}_3} \underset{\text{ethanal}}{2\ CH_3CH{=}O} \quad (3.43)$$

Using ozonolysis, one can easily tell which butene isomer is which. By working backward from the structures of ozonolysis products, one can deduce the structure of an unknown alkene.

EXAMPLE 3.7 Ozonolysis of an alkene produces equal amounts of acetone and formaldehyde, $(CH_3)_2C{=}O$ and $CH_2{=}O$, respectively. Deduce the alkene structure.

Solution Connect to each other by a double bond the carbons that are bound to oxygen in the ozonolysis products. The alkene is $(CH_3)_2C{=}CH_2$.

PROBLEM 3.26 Which alkene will give only acetone, $(CH_3)_2C{=}O$, as the ozonolysis product?

3.17c Other Alkene Oxidations Various reagents can convert alkenes to epoxides (eq. 3.44).

$$\text{alkene} \longrightarrow \text{epoxide} \tag{3.44}$$

This reaction and the chemistry of epoxides are discussed in Chapter 8.

Like alkanes (and all other hydrocarbons), alkenes can be used as fuels. Complete combustion gives carbon dioxide and water.

$$C_nH_{2n} + \tfrac{3n}{2}O_2 \longrightarrow nCO_2 + nH_2O \tag{3.45}$$

Before we turn to alkynes and their chemistry, you might want to read "A Word About" the importance of ethylene in our economy.

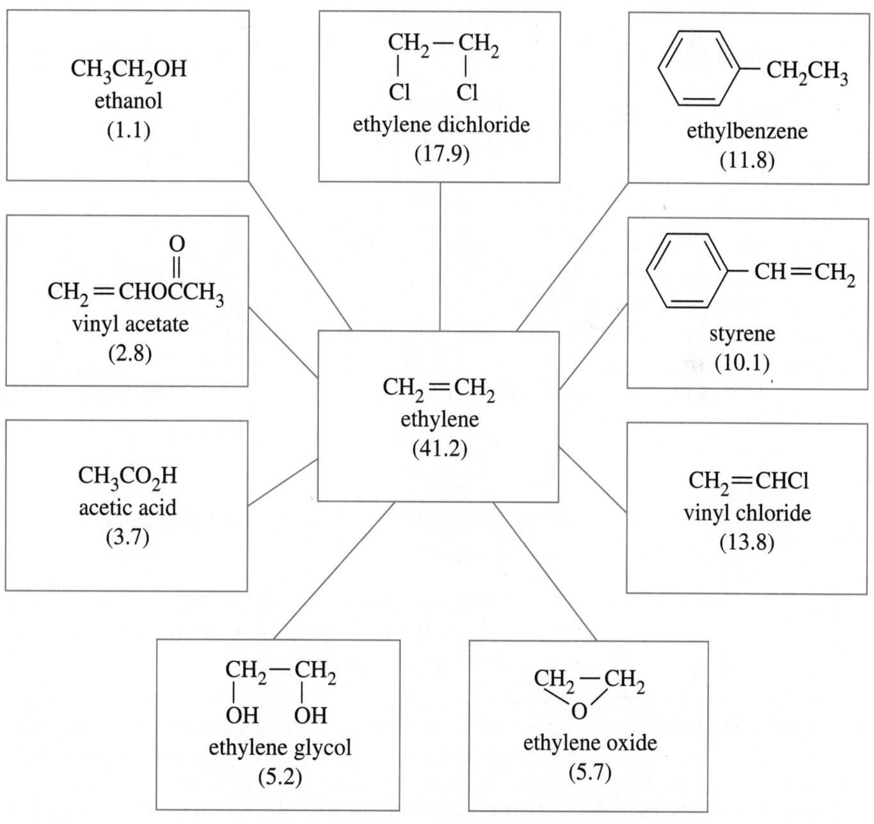

FIGURE 3.10 Ethylene is central to the manufacture of many industrial organic chemicals. The numbers in parentheses give the recent U.S. production of these chemicals in billions of pounds.

A Word About . . .

4. Ethylene and Acetylene

Ethylene, the simplest alkene, ranks first among organic chemicals in industrial production. Current U.S. annual production of ethylene is well over 40 billion pounds. Propene comes in second with about half that amount.

How is all this ethylene produced, and what is it used for? Most hydrocarbons can be "cracked" to give ethylene. (See "A Word About Petroleum, Gasoline, and Octane Number" on page 107.) In the United States, the major raw material for this purpose is ethane.

$$CH_3{-}CH_3 \xrightarrow{700-900°C} CH_2{=}CH_2 + H_2$$

A substantial fraction of industrial ethylene is, of course, converted to polyethylene, as described in Sec. 3.16; but ethylene is also a key raw material for the manufacture of other industrial organic chemicals, because of the reactivity of the carbon–carbon double bond. Shown in Figure 3.10 are 9 of the top 50 organic chemicals; each is produced from ethylene.

Ethylene is not only the most important industrial source of organic chemicals, it also has some biochemical properties that are crucial to agriculture. Ethylene is a **plant hormone** that can cause seeds to sprout, flowers to bloom, fruit to ripen and fall, and leaves and petals to shrivel and turn brown. It is produced naturally by plants from the amino acid *methionine* via an unusual cyclic amino acid, *1-aminocyclopropane-1-carboxylic acid (ACC),* which is then, in several steps, converted to ethylene.

$$\underset{\text{methionine}}{CH_3{-}S{-}CH_2CH_2{-}\underset{\displaystyle NH_3^+}{\underset{|}{CH}}{-}CO_2^-} \xrightarrow[\text{steps}]{\text{several}}$$

$$\underset{\text{ACC}}{\text{(cyclopropane } CH_2{-}C(CO_2^-)(NH_3^+){-}CH_2)} \xrightarrow[\text{steps}]{\text{several}} \underset{\text{ethylene}}{CH_2{=}CH_2} + CO_2 + HCN$$

The mode by which ethylene functions biologically is still being studied.

Chemists have prepared synthetic compounds that can release ethylene in plants in a controlled manner. One such example is 2-chloroethylphosphonic acid, $ClCH_2CH_2PO(OH)_2$. Sold by Union Carbide under the trade name *ethrel,* it is water soluble and is taken up by plants, where it breaks down to ethylene, chloride, and phosphate. It has been used commercially to induce fruits, such as pineapples and tomatoes, to ripen uniformly so that an entire field can be harvested efficiently, as shown in the photo above. Ethylene has also been used to regulate the growth of other crops, such as wheat, apples, cherries, and cotton. Only a small amount need be used, since plants are very sensitive to ethylene and respond to concentrations lower than 0.1 part per million of the gas.

Acetylene, like ethylene, is made by pyrolysis, but from methane instead of ethane. A much higher temperature and a very short reaction time are required.

$$2CH_4 \xrightarrow[<0.1\ s]{1500°C} HC{\equiv}CH + 3H_2$$

Acetylene is considerably more expensive than ethylene. Most of it is used directly, in arc welding, rather than as a raw material for industrial chemicals.

3.18 Some Facts About Triple Bonds

In the final sections of this chapter, we will describe some of the special features of triple bonds and alkynes.

A carbon atom that is part of a triple bond is directly attached to only *two* other atoms, and the bond angle is 180°. Thus, acetylene is linear, as shown in Figure 3.11. The carbon–carbon triple bond distance is about 1.21 Å, appreciably shorter than that of an ordinary double bond (1.34 Å) or single bond (1.54 Å). Apparently, three electron pairs between two carbons draw them even closer together than do two pairs. Because of their linear geometry, no *cis–trans* isomerism is possible for alkynes.

Now let us see how the orbital theory of bonding can be adapted to explain these facts.

3.19 The Orbital Model of a Triple Bond

The carbon atom of an acetylene is connected to only *two* other atoms. Therefore, we combine the 2*s* orbital with only one 2*p* orbital to make two *sp*-hybrid orbitals (Figure 3.12). These orbitals extend in opposite directions from the carbon atom. The angle between the two hybrid orbitals is 180° so as to minimize repulsion between any electrons placed in them. One valence electron is placed in each *sp*-hybrid orbital. The remaining two valence electrons occupy two different *p* orbitals that are both mutually perpendicular and also perpendicular to the hybrid *sp* orbitals.

The formulation of a triple bond from two *sp*-hybridized carbons is shown in Figure 3.13. *End-on overlap of two* sp *orbitals forms a sigma bond between the two carbons, and lateral overlap of the properly aligned p orbitals forms two pi bonds (designated π_1 and π_2 in the figure).* This model nicely explains the linearity of acetylenes.

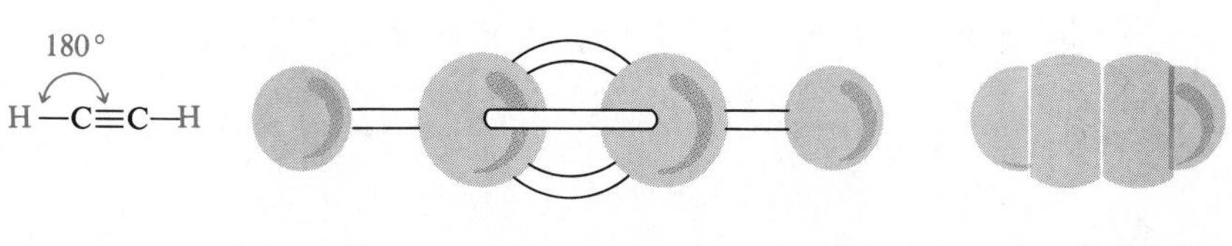

180°

H—C≡C—H

FIGURE 3.11
Models of acetylene, showing its linearity.

FIGURE 3.12
Unhybridized versus *sp*-hybridized orbitals on carbon.

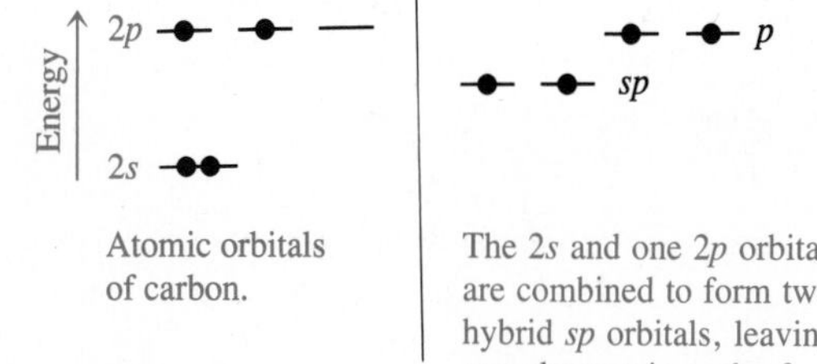

FIGURE 3.13
A triple bond consists of the end-on overlap of two *sp* hybrid orbitals to form a σ bond and the lateral overlap of two sets of parallel-oriented *p* orbitals to form two mutually perpendicular π bonds.

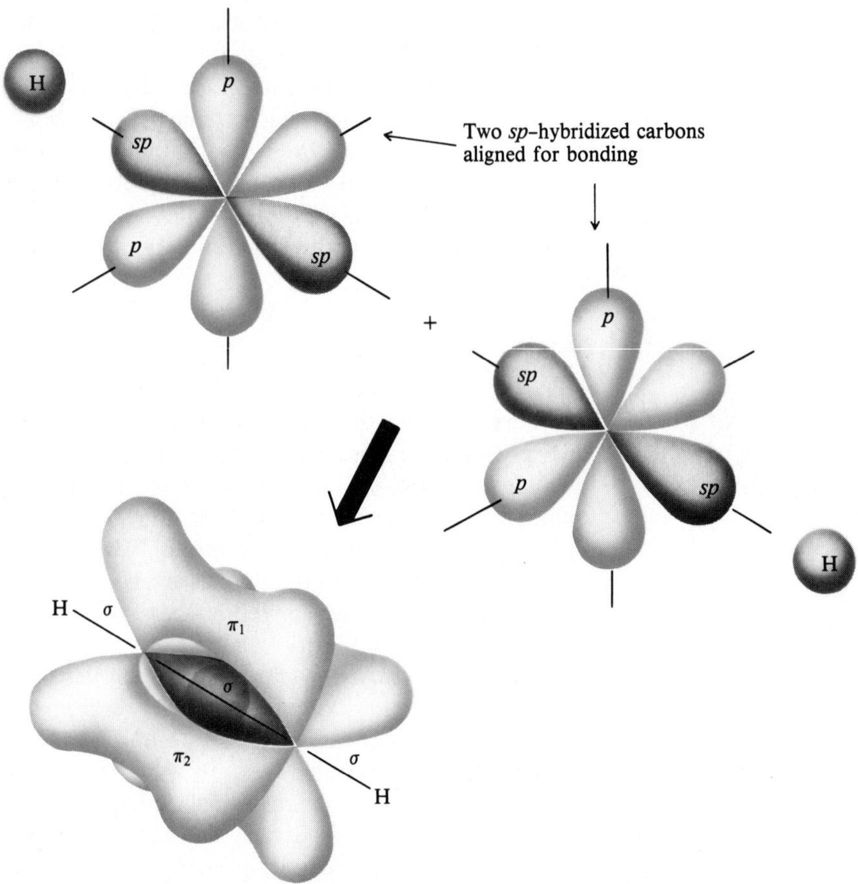

3.20 Addition Reactions of Alkynes

Many addition reactions described for alkenes also occur, though usually more slowly, with alkynes. For example, bromine adds as follows:

$$\underset{\text{ethyne}}{H-C\equiv C-H} \xrightarrow{Br_2} \underset{trans\text{-1,2-dibromoethene}}{(H)(Br)C=C(Br)(H)} \xrightarrow{Br_2} \underset{\text{1,1,2,2-tetrabromoethane}}{H-CBr_2-CBr_2-H} \quad (3.46)$$

In the first step, the addition occurs mainly in a *trans* manner.

With an ordinary nickel or platinum catalyst, alkynes are hydrogenated all the way to alkanes (eq. 3.1). However, a special palladium catalyst (called **Lindlar's catalyst**) can control hydrogen addition so that only *one* mole of hydrogen adds. In this case, the product is a *cis* alkene, because both hydrogens add to the same face of the triple bond from the catalyst surface.

$$\underset{\substack{\text{2-butyne}\\ \text{bp 27°C}}}{CH_3-C\equiv C-CH_3} \xrightarrow[\text{Pd (Lindlar catalyst)}]{H-H} \underset{\substack{cis\text{-2-butene}\\ \text{bp 3.7°C}}}{(CH_3)(H)C=C(CH_3)(H)} \quad (3.47)$$

With unsymmetric triple bonds and unsymmetric reagents, Markovnikov's rule is followed in each step, as shown in the following example:

$$CH_3C\equiv CH + H-Br \longrightarrow CH_3\overset{+}{C}=CH_2 + Br^- \longrightarrow \underset{\text{2-bromopropene}}{CH_3C(Br)=CH_2}$$

$$CH_3C(Br)=CH_2 + H-Br \longrightarrow CH_3\overset{+}{C}(Br)-CH_3 + Br^- \longrightarrow \underset{\text{2,2-dibromopropane}}{CH_3-CBr_2-CH_3} \quad (3.48)$$

Addition of water to alkynes requires not only an acid catalyst but mercuric ion as well. The mercuric ion forms a complex with the triple bond and activates it for addition. Although the reaction is similar to that of alkenes, the initial product—a **vinyl alcohol,** or **enol**—rearranges to a carbonyl compound.

$$R-C\equiv CH + H-OH \xrightarrow[HgSO_4]{H^+} \underset{\text{a vinyl alcohol, or enol}}{\left[R-C(OH)=C(H)-H\right]} \longrightarrow R-\overset{O}{\overset{\|}{C}}-CH_3 \quad (3.49)$$

The product is a methyl ketone or, in the case of acetylene itself (R = H), acetaldehyde. We will discuss the chemistry of enols and the mechanism of the second step of eq. 3.49 in Chapter 9.

PROBLEM 3.27 Write equations for the following reactions:

a. $CH_3C{\equiv}CH + Br_2$ (1 mol) b. $CH_3C{\equiv}CH + Cl_2$ (2 mol)
c. 1-pentyne + HBr (1 and 2 mol) d. 1-butyne + H_2O (Hg^{2+}, H^+)

3.21 Acidity of Alkynes

A hydrogen atom on a triply bonded carbon is weakly acidic and can be removed by a very strong base. Sodium amide, for example, converts acetylenes to acetylides.

$$R{-}C{\equiv}C{-}H + Na^+NH_2^- \xrightarrow{\text{liquid } NH_3} R{-}C{\equiv}C{:}^-Na^+ + NH_3 \quad (3.50)$$

sodium amide; a sodium acetylide

this hydrogen is weakly acidic

This type of reaction occurs easily with a hydrogen adjacent to a triple bond, but with increasing difficulty when the hydrogens are adjacent to a double or single bond. Why? Consider the hybridization of the carbon atom in C—H bonds.

$-CH_3$	$=CH_2$	$\equiv C{-}H$
sp^3	sp^2	sp
25% *s*, 75% *p*	33⅓% *s*, 66⅔% *p*	50% *s*, 50% *p*

increasing acidity →

As the hybridization at carbon becomes more s-like and less p-like, the acidity of the attached hydrogen increases. Recall that *s* orbitals are closer to the nucleus than are *p* orbitals. Consequently, the bonding electrons are closest to the carbon atom in the $\equiv$C—H bond, making it easiest for a base to remove that type of proton. Sodium amide is a sufficiently strong base for this purpose.

PROBLEM 3.28 Write an equation for the reaction of 1-butyne with sodium amide in liquid ammonia.

Although 1-alkynes are acidic, they are much less acidic than water. Internal alkynes, in which the triple bond is not at the end of the chain, have no acidic hydrogens.

PROBLEM 3.29 Write an equation for the reaction of a sodium acetylide with water.

PROBLEM 3.30 Will 2-butyne react with sodium amide? Explain.

TABLE 3.3 Common petroleum fractions

Boiling range, °C	*Name*	*Range of carbon atoms per molecule*	*Use*
below 20	gases	C_1 to C_4	heating, cooking, petrochemical raw material
20–200	naphtha; straight-run gasoline	C_5 to C_{12}	fuel; lighter fractions (such as petroleum ether, bp 30–60°C) also used as laboratory solvents
200–300	kerosene	C_{12} to C_{15}	fuel
300–400	fuel oil	C_{15} to C_{18}	heating homes, diesel fuel
more than 400		over C_{18}	lubricating oil, greases, paraffin waxes, asphalt

A Word About . . .

5. Petroleum, Gasoline, and Octane Number

Petroleum is at present our most important fossil fuel. The need for petroleum to keep our industrial society going sometimes seems second only to our need for food, air, water, and shelter. What is black gold, as petroleum has been called, and how do we use it?

Petroleum is a complex mixture of hydrocarbons formed over eons of time through the gradual decay of buried animal and vegetable matter. **Crude oil** is a viscous black liquid that collects in vast underground pockets in sedimentary rock (the word *petroleum* literally means rock oil, from the Latin *petra,* rock, and *oleum,* oil). It must be brought to the surface via drilling and pumping. To be most useful, the crude oil must be refined.

The first step in petroleum refining is usually **distillation.** The crude oil is heated to about 400°C, and the vapors rise through a tall fractionating column. The lower boiling fractions rise faster and higher in the column before condensing to liquids; higher boiling fractions do not rise as high. By drawing off liquid at various column levels, technicians separate crude oil roughly into the fractions shown in Table 3.3.

The gasoline fraction comprises only about 25% of crude oil. It is the most valuable fraction, however, both as a fuel and as a source material for the petrochemical industry, the industry that furnishes our synthetic fibers, plastics, and many other useful materials. For this reason, many processes have been developed for converting the other fractions into gasoline.

Higher-boiling fractions can be "cracked" by heat and catalysts (mainly silica and alumina), to give products with shorter carbon chains and therefore lower boiling points. The carbon chain can break at many points.

$$C_{10}H_{22}\ (\text{alkane}) \longrightarrow \begin{cases} C_5H_{12}\ (\text{alkane}) + C_5H_{10}\ (\text{alkene}) \\ C_8H_{18} + C_2H_4 \\ C_2H_6 + C_8H_{16} \\ C_4H_{10} + (C_4H_8 + C_2H_4) \end{cases}$$

To balance the number of hydrogens, any particular alkane must give at least one alkane and one alkene as products. Thus, catalytic cracking converts larger alkanes into a mixture of smaller alkanes and alkenes and increases the yield of gasoline from petroleum.

During cracking, large amounts of the lower gaseous hydrocarbons—ethene, propene, butanes, and butenes—are formed. Some of these, especially ethene, are used as petrochemical raw materials. To obtain more gasoline, scientists sought methods to convert these low-molecular-weight hydrocarbons to somewhat larger hydrocarbons that boil in the gasoline range. One such process is **alkylation,** the combination of an alkane with an alkene to form a higher boiling alkane.

$$C_2H_6 + C_4H_8 \xrightarrow{\text{catalyst}} C_6H_{14}$$

$$C_4H_{10} + C_4H_8 \xrightarrow{\text{catalyst}} C_8H_{18}$$

These processes, which were developed in the 1930s, were important for producing aviation fuel during World War II and are still used to make high-octane gasoline.

This brings us to **octane number** and why it is important. Some hydrocarbons, especially those with highly branched structures, burn smoothly in an engine and drive the piston forward evenly. Other hydrocarbons, especially those with unbranched carbon chains, tend to explode in the cylinder and drive the piston forward violently. These undesirable explosions produce audible knocks. A scale was set up many years ago to evaluate this important knock property of gasolines. **Isooctane** (2,2,4-trimethylpentane), an excellent fuel with a highly branched structure, was arbitrarily given a rating of 100, and **heptane,** a very poor automotive fuel, was given a rating of 0. A "regular" gasoline with an octane number of 87 has the same "knock" properties as a mixture that is 87% isooctane and 13% heptane.

The addition of small amounts of **tetraethyllead,** $(CH_3CH_2)_4Pb$, to gasoline improves its octane rating but is undesirable for environmental reasons. For example, there is some evidence that, where automobile fumes are intense, young children in particular accumulate high levels of lead in their blood. However, unleaded gasoline must contain a very high percentage of hydrocarbons with a high octane rating. It became important, then, to develop methods for converting straight-chain hydrocarbons to branched-chain hydrocarbons, which have higher octane ratings, or to develop additives less toxic than tetraethyllead.

Certain catalysts can produce branched-chain alkanes from straight-chain alkanes. This process, called **isomerization,** is carried out on a large scale commercially.

$$\underset{n\text{-butane}}{CH_3CH_2CH_2CH_3} \xrightarrow[\text{alumina}]{AlCl_3,\ HCl} \underset{\text{isobutane}}{CH_3\overset{\overset{CH_3}{|}}{C}HCH_3}$$

Aromatic hydrocarbons, such as benzene and toluene, also have a high octane rating. A platinum catalyst used in a process called **platforming** cyclizes and dehydrogenates alkanes to cycloalkanes and to aromatic hydrocarbons. Of course, large amounts of hydrogen gas are also formed during platforming. Millions of gallons of aromatic hydrocarbons are produced daily by such processes, not only to add to unleaded gasoline to improve its octane rating, but also to supply raw materials for many other petrochemically based products, as we will see in the next chapter.

For further reading on the history of the petrochemical industry, see the book by P. H. Spitz, *Petrochemicals: The Rise of an Industry,* Wiley, New York, 1988.

$$CH_3(CH_2)_5CH_3 \xrightarrow[\text{catalyst}]{Pt} \text{methylcyclohexane} \xrightarrow[\text{catalyst}]{Pt} \text{toluene}$$

CH₃ (on cyclohexane ring)

methylcyclohexane

CH₃ (on benzene ring)

toluene
(an aromatic hydrocarbon)

REACTION SUMMARY

1. Reactions of Alkenes

a. Addition of Halogens (Sec. 3.7)

$$\mathrm{>C{=}C<} + X_2 \longrightarrow \mathrm{-\overset{|}{\underset{X}{C}}-\overset{|}{\underset{X}{C}}-} \quad (X = Cl, Br)$$

b. Addition of Polar Reagents (Sec. 3.7)

$$\mathrm{>C{=}C<} + H{-}OH \longrightarrow \mathrm{-\overset{|}{\underset{H}{C}}-\overset{|}{\underset{OH}{C}}-}$$

$$\mathrm{>C{=}C<} + H{-}X \longrightarrow \mathrm{-\overset{|}{\underset{H}{C}}-\overset{|}{\underset{X}{C}}-} \quad \left(X = F, Cl, Br, I, -OSO_3H, -O-\underset{\underset{O}{\|}}{C}-R\right)$$

c. Hydroboration-Oxidation (Sec. 3.13)

$$RCH{=}CH_2 \xrightarrow{BH_3} (RCH_2CH_2)_3B \xrightarrow[HO^-]{H_2O_2} RCH_2CH_2OH$$

d. Addition of Hydrogen (Sec. 3.14)

$$\mathrm{>C{=}C<} + H_2 \xrightarrow{\text{Pd, Pt, or Ni}} \mathrm{-\overset{|}{\underset{H}{C}}-\overset{|}{\underset{H}{C}}-}$$

e. Addition of X_2 and HX to Conjugated Dienes (Sec. 3.15)

$$C{=}C{-}C{=}C + X_2 \longrightarrow \underset{\text{1,2-addition}}{\underset{X}{C}-\underset{X}{C}-C{=}C} + \underset{\text{1,4-addition}}{\underset{X}{C}-C{=}C-\underset{X}{C}}$$

$$C{=}C{-}C{=}C + H{-}X \longrightarrow \underset{\text{1,2-addition}}{\underset{H}{C}-\underset{X}{C}-C{=}C} + \underset{\text{1,4-addition}}{\underset{H}{C}\quad C{=}C-\underset{X}{C}}$$

(X = Cl, Br)

f. Polymerization of Ethylene (Sec. 3.16)

$$n\ H_2C{=}CH_2 \xrightarrow{\text{catayst}} \left(CH_2-CH_2\right)_n$$

g. Oxidation to Diols or Carbonyl-Containing Compounds (Sec. 3.17)

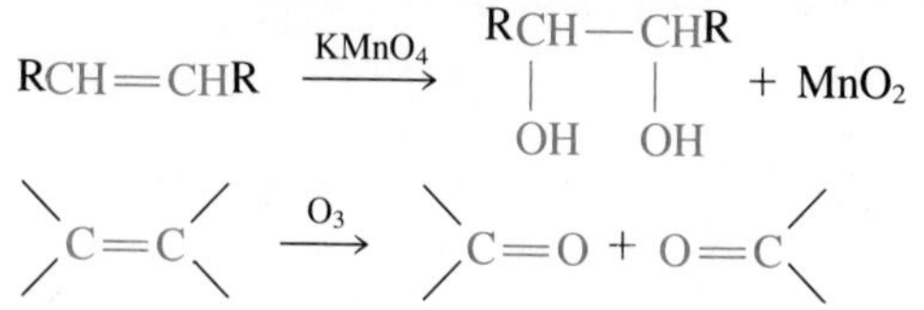

2. Reactions of Alkynes

a. Additions to the Triple Bond (Sec. 3.20)

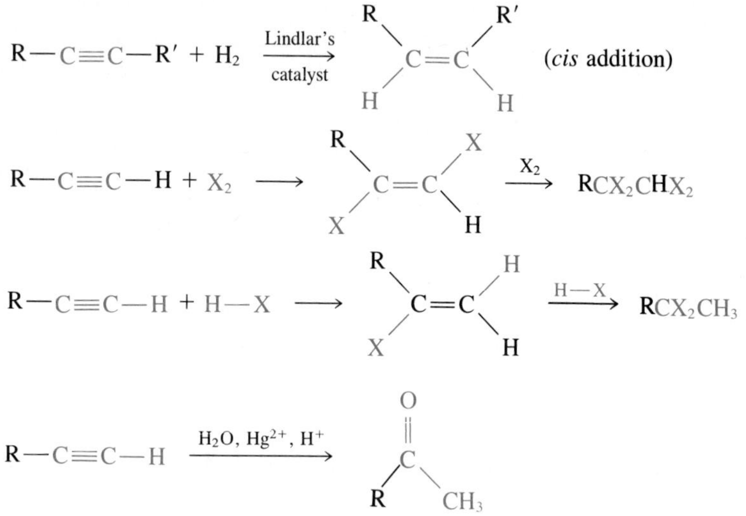

b. Formation of Acetylide Anions (Sec. 3.21)

$$R-C\equiv C-H + NaNH_2 \xrightarrow{NH_3} R-C\equiv C:^{\ominus}\,Na^{\oplus} + NH_3$$

ADDITIONAL PROBLEMS

3.31. For the following compounds, write structural formulas and IUPAC names for all possible isomers having the indicated number of multiple bonds:

a. C_4H_8(one double bond) **b.** C_5H_8(two double bonds)
c. C_5H_8(one triple bond)

3.32. Name the following compounds by the IUPAC system:

a. $CH_3CH=C(CH_3)_2$
b. $CH_3CH=CHCH_2CH_3$
c. 
d. $CH_3CH_2C\equiv CCH_2CH_3$
e. $CH_2=CH-CBr=CH_2$
f. $CH_2=CH-C\equiv C-CH_3$
g. 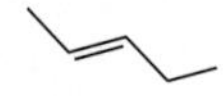
h. 

3.33. Write a structural formula for each of the following compounds:

a. 2-hexene
b. cyclopropene
c. 1,3-dichloro-2-butene
d. 4-methyl-1-pentyne
e. 1,4-cyclohexadiene
f. vinyl chloride
g. allyl bromide
h. vinylcyclobutane
i. 4-methylcyclopentene
j. 2,3-dibromo-1,3-cyclopentadiene

3.34. Explain why the following names are incorrect, and give a correct name in each case:

a. 3-pentene
b. 3-butyne
c. 2-ethyl-1-propene
d. 2-methylcyclohexene
e. 3-methyl-1,3-butadiene
f. 1-methyl-2-butene
g. 3-buten-1-yne
h. 3-pentyne-1-ene

3.35.

a. What are the usual lengths for the single (sp^3–sp^3), double (sp^2–sp^2), and triple (sp–sp) carbon–carbon bonds?

b. The *single* bond in each of the following compounds has the length shown. Suggest a possible explanation for the observed shortening.

$$CH_2{=}CH\underset{\uparrow \atop 1.47\ \text{Å}}{-}CH{=}CH_2 \qquad CH_2{=}CH\underset{\uparrow \atop 1.43\ \text{Å}}{-}C{\equiv}CH \qquad HC{\equiv}C\underset{\uparrow \atop 1.37\ \text{Å}}{-}C{\equiv}CH$$

3.36. Which of the following compounds can exist as *cis–trans* isomers? If such isomerism is possible, draw the structures in a way that clearly illustrates the geometry.

a. 1-pentene
b. 2-hexene
c. 1-bromopropene
d. 3-chloropropene
e. 1,3,5-hexatriene
f. 1,2-dichlorocyclodecene

3.37. The mold metabolite and antibiotic *mycomycin* has the formula

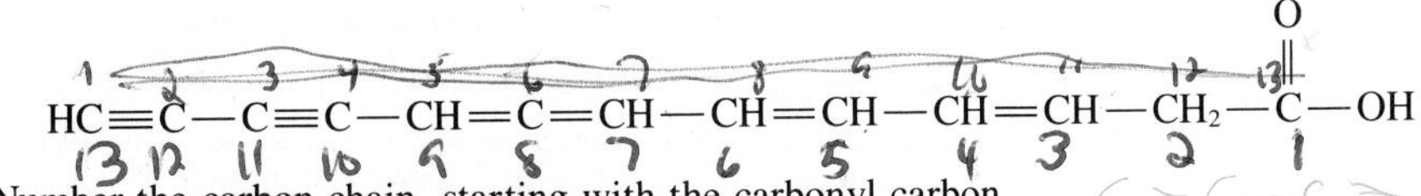

Number the carbon chain, starting with the carbonyl carbon.

a. Which multiple bonds are conjugated?
b. Which multiple bonds are cumulative?
c. Which multiple bonds are isolated?

3.38. Draw a reaction energy diagram for the reaction in eq. 3.14. Show both reactions, including the one for which no product is observed. (Hint: Consider eq. 3.21 in constructing your diagram.)

3.39. Write the structural formula and name of the product when each of the following reacts with 1 equivalent of bromine:

a. 2-pentene
b. vinyl chloride
c. 1,4-cyclohexadiene
d. 1,3-cyclohexadiene
e. 2,3-dimethyl-1-butene

3.40. Write an equation for the reaction of 1-butene with each of the following reagents:

a. bromine
b. hydrogen bromide
c. hydrogen (Pt catalyst)
d. ozone, followed by Zn, H^+
e. H_2O, H^+
f. BH_3 followed by H_2O_2, OH^-
g. $KMnO_4$, OH^-
h. oxygen (combustion)

3.41. What reagent will react by addition to what unsaturated hydrocarbon to form each of the following compounds?

a. $CH_3CHClCHClCH_3$ **b.** $(CH_3)_2CHOSO_3H$ **c.** $(CH_3)_3COH$

d. 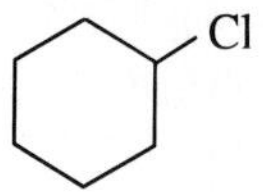 (cyclohexane)–Cl **e.** $CH_3CH{=}CHCH_2Br$ **f.** $CH_3CBr_2CBr_2CH_3$

g. 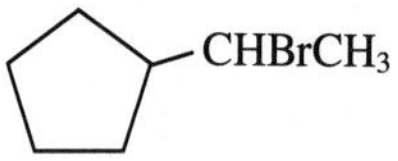 (cyclopentane)–$CHBrCH_3$

3.42. *Caryophyllene* is an unsaturated hydrocarbon mainly responsible for the odor of oil of cloves. It has the molecular formula $C_{15}H_{24}$. Hydrogenation of caryophyllene gives a saturated hydrocarbon $C_{15}H_{28}$. Does caryophyllene contain any rings? How many?

3.43. Which of the following reagents are electrophiles? Which are nucleophiles?

a. HBr **b.** H_3O^+ **c.** Br^- **d.** $AlCl_3$ **e.** OH^-

3.44. Water can act as an electrophile or as a nucleophile. Explain.

3.45. The acid-catalyzed hydration of 1-methylcyclohexene gives 1-methylcyclohexanol.

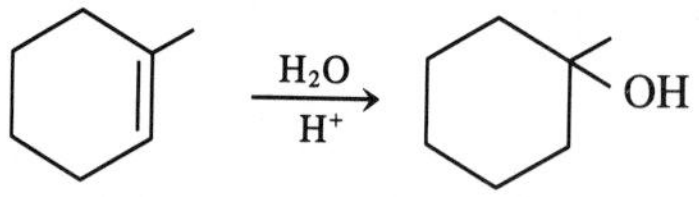

$$\xrightarrow[H^+]{H_2O}$$ OH

Write every step in the mechanism of this reaction.

3.46. Predict the structures of the two possible monohydration products of limonene (Figure 1.12). These alcohols are called *terpineols*. Predict the structure of the diol (dialcohol) obtained by hydrating both double bonds in limonene. These alcohols are used in the cough medicine "elixir of terpin hydrate."

3.47. Draw the resonance contributors to the carbocation

$$(CH_3)_2CH\overset{+}{C}HCH{=}CHCH(CH_3)_2.$$

Does the ion have a symmetric structure?

3.48. Adding 1 equivalent of hydrogen bromide to 1,3-hexadiene gives two products. Give their structures, and write all the steps in a reaction mechanism that explains how each product is formed.

3.49. Given the information that free-radical stability follows the same order as carbocation stability (3° > 2° > 1°), predict the structure of polypropylene produced by the free-radical polymerization of propene. It should help to write out each step in the mechanism, as in eqs. 3.37 and 3.38.

3.50. Write an equation that clearly shows the structure of the alcohol obtained from the sequential hydroboration and H_2O_2/OH^- oxidation of

a. 2,3-dimethyl-1-butene **b.** 1-methylcyclopentene.

3.51. Write equations to show how (cyclohexane)$=CH_2$ could be converted to

a. (cyclohexane with CH_3 and OH on the same carbon) **b.** (cyclohexane)$-CH_2OH$

3.52. Describe two simple chemical tests that could be used to distinguish cyclopentane from cyclopentene.

3.53. Give the formulas of the alkenes that, on ozonolysis, give

a. only $CH_3CH_2CH_2CH{=}O$. **b.** $(CH_3)_2C{=}O$ and $CH_3CH_2CH{=}O$.
c. $CH_2{=}O$ and $(CH_3)_2CHCH{=}O$. **d.** $O{=}CHCH_2CH_2CH_2CH{=}O$.

3.54. Write equations for the following reactions:
a. 2-pentyne + H_2 (1 mol, Lindlar's catalyst)
b. 3-hexyne + Cl_2 (2 mol)
c. propyne + sodium amide in liquid ammonia
d. 1-butyne + H_2O (H^+, $HgSO_4$ catalyst)

3.55. Determine what alkyne and what reagent will give
a. 2,2-dichlorobutane. **b.** 2,2,3,3-tetrabromobutane.

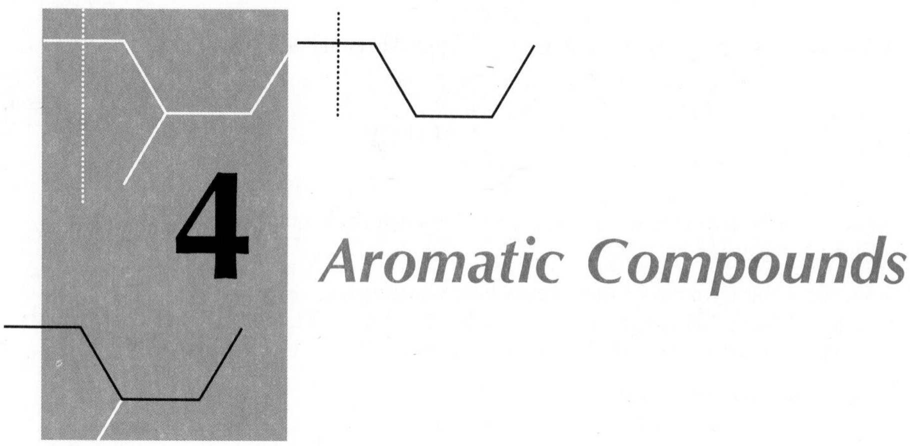

4 Aromatic Compounds

4.1 Historical Introduction

Spices and herbs have long played a romantic role in the course of history. They bring to mind frankincense and myrrh and the great explorers of past centuries—Vasco da Gama, Christopher Columbus, Ferdinand Magellan, Sir Francis Drake—whose quest for spices helped to open the Western world. Trade in spices was immensely profitable. It was natural, therefore, that spices and herbs were among the first natural products studied by organic chemists. If one could isolate from plants the pure compounds with these desirable fragrances and flavors and determine their structures, perhaps one could synthesize them in large quantity and at low cost.

It turned out that many of these aromatic substances have relatively simple structures. Many contain a six-carbon unit that passes unscathed through various chemical reactions that alter only the rest of the structure. This group, C_6H_5—, is common to many substances, including **benzaldehyde** (isolated from the oil of bitter almonds), **benzyl alcohol** (isolated from gum benzoin, a balsam resin obtained from certain Southeast Asian trees), and **toluene** (a hydrocarbon isolated from tolu balsam). When any of these three compounds is oxidized, the C_6H_5 group remains intact. The product in each case is **benzoic acid** (another constituent of gum benzoin). The calcium salt of this acid, when heated, yields the parent hydrocarbon C_6H_6 (eq. 4.1).

$$
\begin{array}{l}
\underset{\text{benzaldehyde}}{C_6H_5CH{=}O} \xrightarrow{\text{oxidize}} \\
\underset{\text{benzyl alcohol}}{C_6H_5CH_2OH} \xrightarrow{\text{oxidize}} \underset{\text{benzoic acid}}{C_6H_5CO_2H} \xrightarrow[\text{2. heat}]{\text{1. CaO}} \underset{\text{benzene}}{C_6H_6} \\
\underset{\text{toluene}}{C_6H_5CH_3} \xrightarrow{\text{oxidize}}
\end{array}
\qquad (4.1)
$$

This same hydrocarbon, first isolated from compressed illuminating gas by Michael Faraday in 1825, is now called **benzene.** It is the parent hydrocarbon of a class of substances that we now call **aromatic compounds,** *not because of their aroma,* but because of their special chemical properties, in particular their stability.*

4.2 Some Facts About Benzene

The carbon-to-hydrogen ratio in benzene, C_6H_6, suggests a highly unsaturated structure. Compare the number of hydrogens, for example, with that in hexane, C_6H_{14}, or in cyclohexane, C_6H_{12}, both of which also have six carbons but are saturated.

PROBLEM 4.1 Draw at least five isomeric structures that have the molecular formula C_6H_6. Note that all are highly unsaturated or contain small, strained rings.

Despite its molecular formula, benzene for the most part does not behave as if it were unsaturated. For instance, it does not decolorize bromine solutions the way alkenes and alkynes do, nor is it easily oxidized by potassium permanganate. It does not undergo the typical addition reactions of alkenes or alkynes. Instead, *benzene reacts mainly by substitution.* For example, when treated with bromine in the presence of ferric bromide as a catalyst, benzene gives bromobenzene and hydrogen bromide.

$$\underset{\text{benzene}}{C_6H_6} + Br_2 \xrightarrow[\text{catalyst}]{FeBr_3} \underset{\text{bromobenzene}}{C_6H_5Br} + HBr \qquad (4.2)$$

Chlorine, with a ferric chloride catalyst, reacts similarly.

$$\underset{\text{benzene}}{C_6H_6} + Cl_2 \xrightarrow[\text{catalyst}]{FeCl_3} \underset{\text{chlorobenzene}}{C_6H_5Cl} + HCl \qquad (4.3)$$

Only *one* monobromobenzene or monochlorobenzene has ever been isolated; that is, no isomers are obtained in either of these reactions. This result implies that *all six hydrogens in benzene are chemically equivalent.* It does not matter which hydrogen is replaced by bromine; we get the same monobromobenzene. This fact has to be accounted for in any structure proposed for benzene.

When bromobenzene is treated with a second equivalent of bromine and the same type of catalyst, *three di*bromobenzenes are obtained.

$$C_6H_5Br + Br_2 \xrightarrow[\text{catalyst}]{FeBr_3} \underset{\substack{\text{dibromobenzenes} \\ \text{(three isomers)}}}{C_6H_4Br_2} + HBr \qquad (4.4)$$

---

* Today, benzene is one of the most important commercial organic chemicals. Approximately 12 billion pounds are produced annually in the United States alone. Benzene is obtained mostly from petroleum by catalytic reforming of alkanes and cycloalkanes or by cracking certain gasoline fractions. It is used to make styrene, phenol, acetone, cyclohexane, and other industrial chemicals.

The isomers are not formed in equal amounts. Two of them predominate, and only a small amount of the third isomer is formed. The important point is that there are three isomers—no more and no less. Similar results are obtained when chlorobenzene is further chlorinated to give dichlorobenzenes. These facts also have to be explained by any structure proposed for benzene.

The problem of benzene's structure does not sound overwhelming, yet it took many decades to solve. Let us examine the main ideas that led to our modern view of its structure.

4.3 The Kekulé Structure of Benzene

In 1865 Kekulé proposed the first reasonable structure for benzene.* He suggested that the six carbon atoms are located at the corners of a regular hexagon, with one hydrogen atom attached to each carbon atom. To give each carbon atom a valence of 4, he suggested that single and double bonds alternate around the ring (what we now call a *conjugated* system of double bonds). But this structure is highly unsaturated. To explain benzene's negative tests for unsaturation (that is, its failure to decolorize bromine or to give a permanganate test), Kekulé suggested that the single and double bonds exchange positions around the ring *so rapidly* that the typical reactions of alkenes cannot take place.

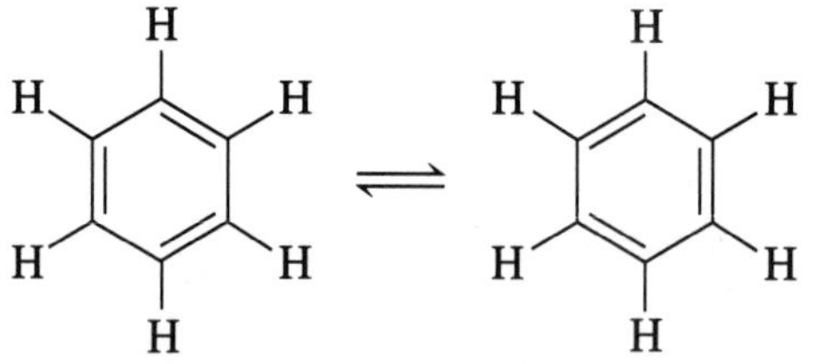

the Kekulé structures for benzene

PROBLEM 4.2 Write out eqs. 4.2 and 4.4 using a Kekulé structure for benzene. Does this model explain the existence of only one monobromobenzene? only three dibromobenzenes?

---

*Friedrich August Kekulé (1829–1896) was a pioneer in the development of structural formulas in organic chemistry. He was among the first to recognize the tetracovalence of carbon and the importance of carbon chains in organic structures. He is best known for his proposal regarding the structure of benzene and other aromatic compounds. It is interesting that Kekulé first studied architecture, and only later switched to chemistry. Judging from his contributions, he apparently viewed chemistry as molecular architecture. Kekulé is supposed to have arrived at his structure for benzene while daydreaming before a fireplace; the flames reminded him of a snake swallowing its own tail. The truth or myth of this bit of chemical folklore is discussed in an interesting article by J. H. Wotiz and S. Rudofsky in *Chemistry in Britain*, **1984,** *20,* 720–723. This controversial article stimulated a strong response from A. J. Rocke and O. B. Ramsay, *Chemistry in Britain*, **1984,** *20,* 1093. The subject, which has implications for the history and philosophy of science, is still being debated (see *Chemical and Engineering News*, Nov. 4, 1985, pp. 22–23). Further controversy about the history of the development of the structural formula of benzene has recently been raised by C. R. Noe and A. Bader, *Chemistry in Britain*, **1993,** *29,* 126–128. They claim that the Austrian physicist/chemist Joseph Loschmidt was the first person to publish a circular structure for benzene in 1861. Who deserves most credit? This question is still being debated.

PROBLEM 4.3 How might Kekulé explain the fact that there is only one dibromobenzene with the bromines on adjacent carbon atoms, even though we can draw two different structures, with either a double or a single bond between the bromine-bearing carbons?

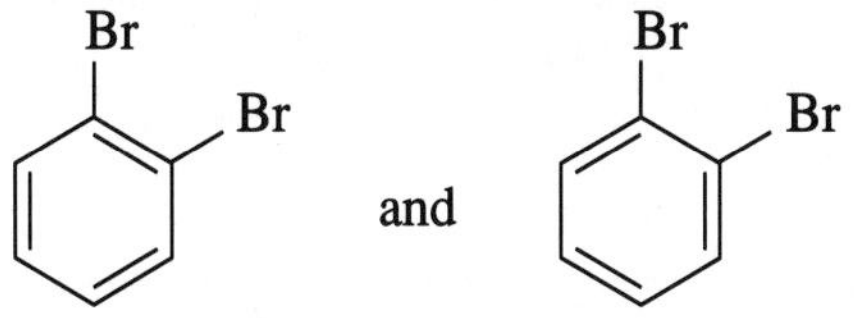

Br Br and Br Br

4.4 The Resonance Model for Benzene

Kekulé's model for the structure of benzene is nearly, but not entirely, correct. *Kekulé's two structures for benzene differ only in the arrangement of the electrons;* all the atoms occupy the same positions in both structures. *This is precisely the requirement for resonance* (review Sec. 1.12). Kekulé's formulas represent two identical contributing structures to a *single* resonance hybrid structure of benzene. Instead of writing an equilibrium symbol between them, as Kekulé did, we now write the double-headed arrow used to indicate a resonance hybrid:

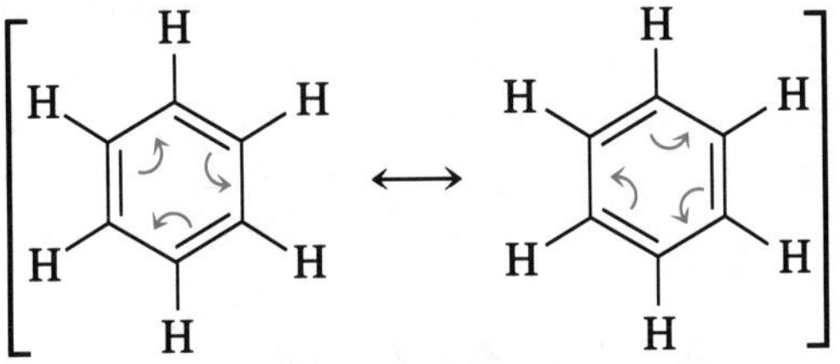

Benzene is a resonance hybrid of these two contributing structures.

To express this model another way, all benzene molecules are identical, and their structure is not adequately represented by either of Kekulé's contributing structures. Being a resonance hybrid, benzene is more stable than its contributing Kekulé structures. There are no single or double bonds in benzene—only one type of carbon–carbon bond, which is of some intermediate type. Consequently, it is not surprising that benzene does not react chemically exactly like alkenes.

Modern physical measurements support this model for the benzene structure. *Benzene is planar, and each carbon atom is at the corner of a regular hexagon. All the carbon–carbon bond lengths are identical,* and this bond length is 1.39 Å, intermediate between typical single (1.54 Å) and double (1.34 Å) carbon–carbon bond lengths. Figure 4.1 shows a space-filling model of the benzene molecule.*

---

*Notice the difference in the shapes of benzene and cyclohexane (Figure 2.4).

FIGURE 4.1 Space-filling model of benzene.

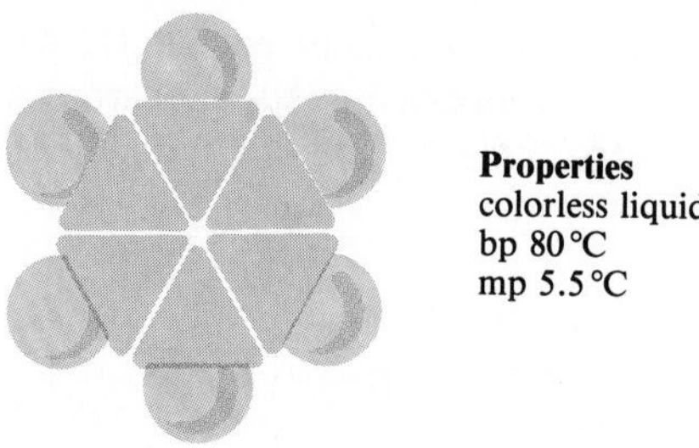

Properties
colorless liquid
bp 80 °C
mp 5.5 °C

4.5 Orbital Model for Benzene

Orbital theory, which is so useful in rationalizing the geometries of alkanes, alkenes, and alkynes, is also useful in explaining the structure of benzene. Each carbon atom in benzene is connected to only *three* other atoms (two carbons and a hydrogen). Each carbon is therefore sp^2-hybridized, as in ethylene. Two sp^2 orbitals of each carbon atom overlap with similar orbitals of adjacent carbon atoms to form the sigma bonds of the hexagonal ring. The third sp^2 orbital of each carbon overlaps with a hydrogen 1*s* orbital to form the C—H sigma bonds. Perpendicular to the plane of the three sp^2 orbitals at each carbon is a *p* orbital containing one electron, the fourth valence electron. The *p* orbitals on all six carbon atoms can overlap laterally to form pi orbitals that create a cloud of electrons above and below the plane of the ring. The construction of a benzene ring from six sp^2-hybridized carbons is shown schematically in Figure 4.2. This model nicely explains the planarity of benzene. It also explains its hexagonal shape, with H—C—C and C—C—C angles of 120°.

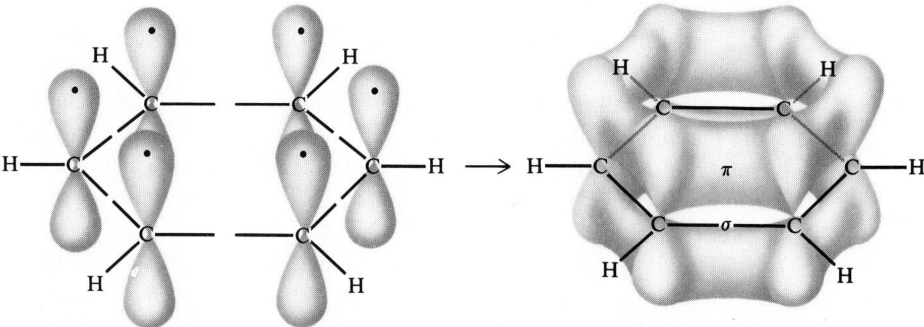

FIGURE 4.2 An orbital representation of the bonding in benzene. Sigma bonds are formed by the end-on overlap of sp^2 orbitals. In addition, each carbon contributes one electron to the pi system by lateral overlap of its *p* orbital with the *p* orbitals of its two neighbors.

4.6 Symbols for Benzene

Two symbols are used to represent benzene. One is the Kekulé structure, and the other is a hexagon with an inscribed circle, to represent the idea of a delocalized pi electron cloud.

Kekulé

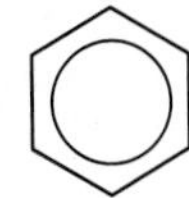
delocalized pi cloud

Regardless of which symbol is used, the hydrogens are usually not written explicitly, but we must remember that one hydrogen atom is attached to the carbon at each corner of the hexagon.

The symbol with the inscribed circle emphasizes the fact that the electrons are distributed evenly around the ring, and in this sense, it is perhaps the more accurate of the two. The Kekulé symbol, however, reminds us very clearly that there are six pi electrons in benzene. For this reason, it is particularly useful in allowing us to keep track of the valence electrons during chemical reactions of benzene. In this book, we will use the Kekulé symbol. However, we must keep in mind that the "double bonds" are not fixed in the positions shown, nor are they really double bonds at all.

EXAMPLE 4.1 Write the structural formula for benzaldehyde (eq. 4.1).

Solution One hydrogen in the formula for benzene is replaced by the aldehyde group.

O
‖
C—H

PROBLEM 4.4 Write the formulas for benzyl alcohol, toluene, and benzoic acid (eq. 4.1).

4.7 Nomenclature of Aromatic Compounds

Because aromatic chemistry developed in a haphazard fashion many years before systematic methods of nomenclature were developed, common names have acquired historic respectability and are accepted by IUPAC. Examples include

benzene	toluene	cumene	styrene	phenol
	CH_3	CH_3CHCH_3	$CH{=}CH_2$	OH

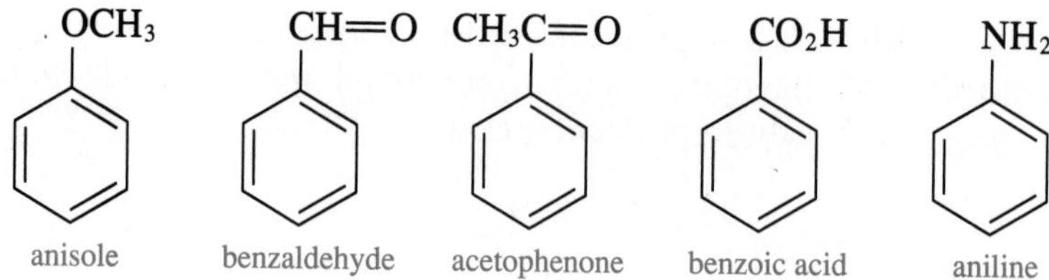

anisole benzaldehyde acetophenone benzoic acid aniline

Monosubstituted benzenes are named as derivatives of benzene.

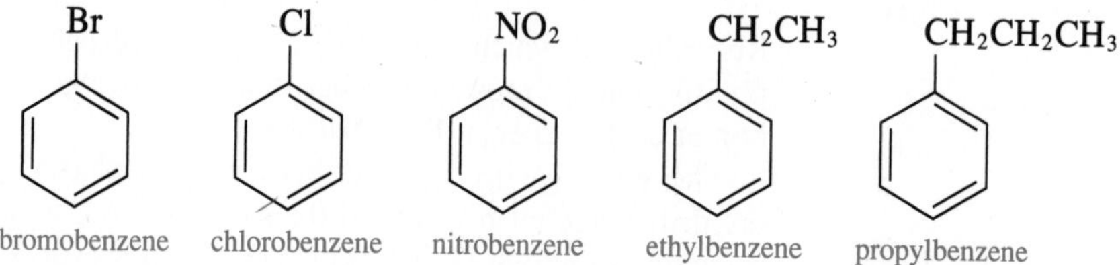

bromobenzene chlorobenzene nitrobenzene ethylbenzene propylbenzene

When two substituents are present, three isomeric structures are possible. They are designated by the prefixes *ortho*-, *meta*-, and *para*-, which are usually abbreviated as *o*-, *m*-, and *p*-, respectively. If substituent X is attached (by convention) to carbon 1, then *o*-groups are on carbons 2 and 6, *m*-groups are on carbons 3 and 5, and *p*-groups are on carbon 4.*

X
o *o*
m *m*
p

Specific examples are

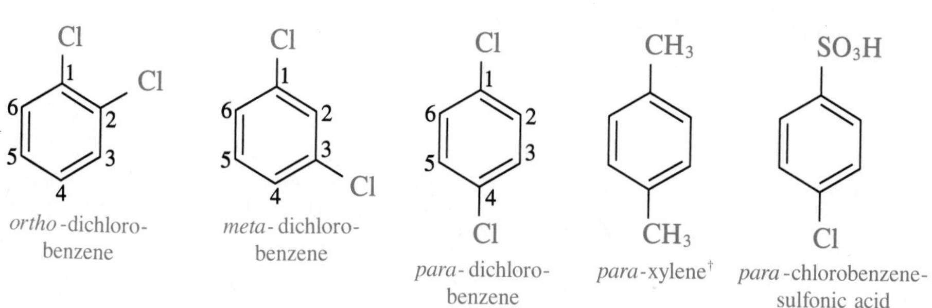

ortho-dichlorobenzene *meta*-dichlorobenzene *para*-dichlorobenzene *para*-xylene[†] *para*-chlorobenzenesulfonic acid

---

* Note that X can be on any carbon of the ring. It is the ***position of the second substituent relative to that of X*** that is important.

† The common and IUPAC name is xylene, *not p*-methyltoluene.

PROBLEM 4.5 Draw the structures for *ortho*-xylene and *meta*-xylene.

The prefixes *ortho*-, *meta*-, and *para*- are used even when the two substituents are not identical.

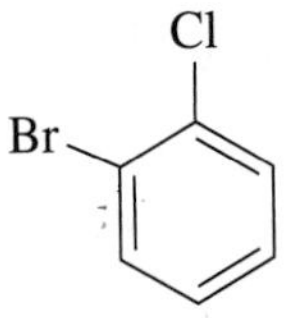

o-bromochlorobenzene (note alphabetical order) 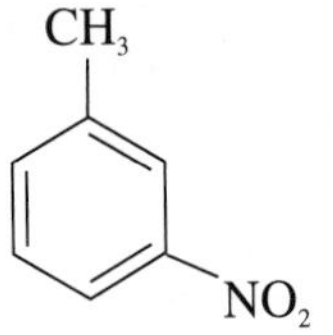 *m*-nitrotoluene 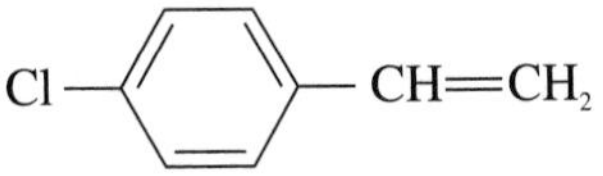 *p*-chlorostyrene 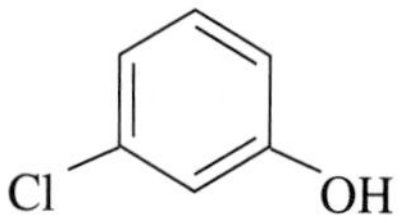 *m*-chlorophenol 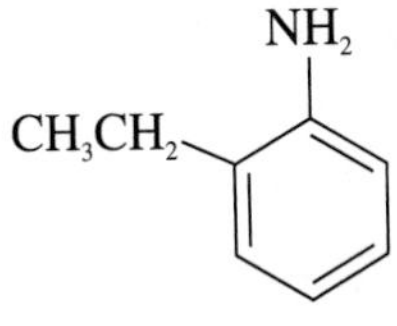 *o*-ethylaniline

When more than two substituents are present, their positions are designated by numbering the ring.

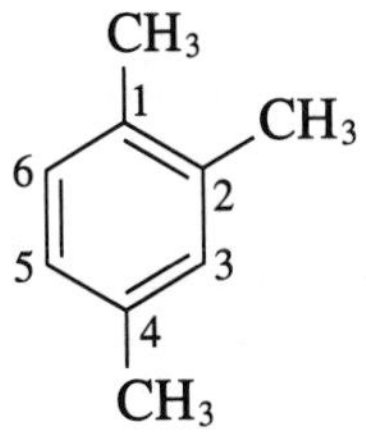

1,2,4-trimethylbenzene 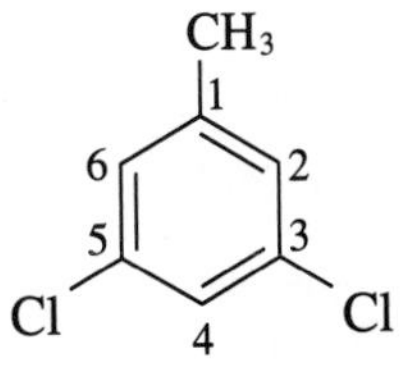 3,5-dichlorotoluene 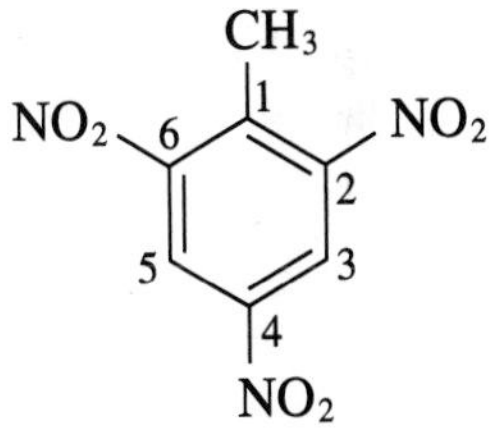 2,4,6-trinitrotoluene (TNT)

PROBLEM 4.6 Draw the structure of

a. *p*-nitrophenol.
b. *o*-bromotoluene.
c. *m*-dinitrobenzene.
d. *p*-divinylbenzene.

PROBLEM 4.7 Draw the structure of

a. 1,3,5-trimethylbenzene.
b. 4-bromo-2,6-dichlorotoluene.

Aromatic hydrocarbons, as a class, are called **arenes.** The symbol **Ar** is used for an **aryl group,** just as the symbol R is used for an alkyl group. The formula Ar-R would therefore represent any arylalkane.

Two groups with special names occur frequently in aromatic compounds. They are the **phenyl group** and the **benzyl group.**

$C_6H_5—$ or

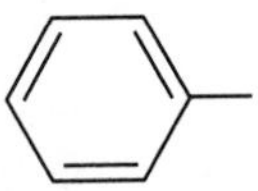

phenyl group

$C_6H_5CH_2—$ or $—CH_2—$

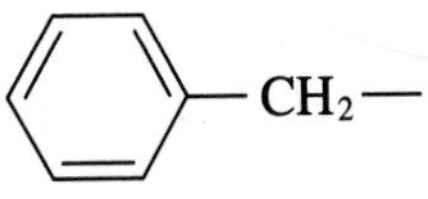

benzyl group

The symbol Ph is sometimes used as an abbreviation for the phenyl group. The use of these group names is illustrated in the following examples:

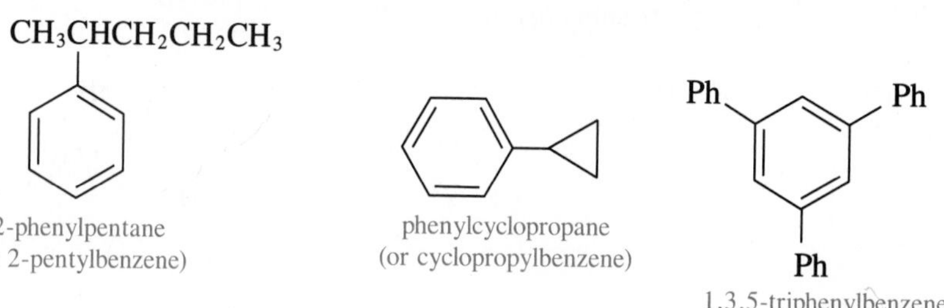

2-phenylpentane (or 2-pentylbenzene)   phenylcyclopropane (or cyclopropylbenzene)   1,3,5-triphenylbenzene

CH_2Cl O_2N CH_2OH

biphenyl   benzyl chloride   *m*-nitrobenzyl alcohol

PROBLEM 4.8 Draw the structure of

a. cyclohexylbenzene.
b. benzyl bromide.
c. *m*-phenylstyrene.
d. dibenzyl.

PROBLEM 4.9 Name the following structures:

a.   b. OH CH_2

4.8 The Resonance Energy of Benzene

We have asserted that a resonance hybrid is always more stable than any of its contributing structures. Fortunately, in the case of benzene, this assertion can be proved experimentally, and we can even measure how much more stable benzene is than the hypothetical molecule 1,3,5-cyclohexatriene (the IUPAC name for one Kekulé structure).

Hydrogenation of a carbon–carbon double bond is an exothermic reaction. The amount of energy (heat) released is about 26 to 30 kcal/mol for each double bond.

$$\gt C{=}C\lt \; + \; H{-}H \longrightarrow -\underset{H}{\underset{|}{\overset{|}{C}}}-\underset{H}{\underset{|}{\overset{|}{C}}}- \; + \text{ heat (26–30 kcal/mol)} \qquad (4.5)$$

(The exact value depends on the substituents attached to the double bond.) When two double bonds in a molecule are hydrogenated, twice as much heat is evolved, and so on.

Hydrogenation of cyclohexene releases 28.6 kcal/mol.

$$\text{cyclohexene} + \text{H—H} \longrightarrow \text{cyclohexane} + \text{heat (28.6 kcal/mol)} \qquad (4.6)$$

We expect that the complete hydrogenation of 1,3-cyclohexadiene should release twice that amount of heat, or $2 \times 28.6 = 57.2$ kcal/mol; experimentally the value is close to what we expect.

$$\text{1,3-cyclohexadiene} + 2\text{H—H} \longrightarrow \text{cyclohexane} + \text{heat (55.4 kcal/mol)} \qquad (4.7)$$

It seems reasonable, therefore, to expect that the heat of hydrogenation of a Kekulé structure (the *hypothetical* triene 1,3,5-cyclohexatriene) should correspond to that for *three* double bonds, or about 84 to 86 kcal/mol. However, we find experimentally that benzene is more difficult to hydrogenate than simple alkenes, and the heat evolved when benzene is hydrogenated to cyclohexane is *much lower* than expected: only 49.8 kcal/mol.

$$\text{benzene} + 3\text{H—H} \longrightarrow \text{cyclohexane} + \text{heat (49.8 kcal/mol)} \qquad (4.8)$$

We conclude that *real benzene molecules are more stable than the contributing resonance structures* (the hypothetical molecule 1,3,5-cyclohexatriene) *by about 36 kcal/mol* $(86 - 50 = 36)$.

We define the **stabilization energy,** or **resonance energy,** of a substance as the difference between the actual energy of the real molecule (the resonance hybrid) and the calculated energy of the most stable contributing structure. For benzene this value is about 36 kcal/mol. This is a substantial amount of energy. Consequently, as we will see, *benzene and other aromatic compounds usually react in such a way as to preserve their aromatic structure and therefore retain their resonance energy.*

4.9 Electrophilic Substitution

The most common reactions of aromatic compounds involve substitution of other atoms or groups for a ring hydrogen. Here are some typical substitution reactions of benzene.

$$C_6H_6 + Cl_2 \xrightarrow{FeCl_3} C_6H_5Cl + HCl \qquad \text{chlorination} \tag{4.9}$$

$$C_6H_6 + Br_2 \xrightarrow{FeBr_3} C_6H_5Br + HBr \qquad \text{bromination} \tag{4.10}$$

$$C_6H_6 + \underset{(HONO_2)}{HNO_3} \xrightarrow{H_2SO_4} C_6H_5NO_2 + H_2O \qquad \text{nitration} \tag{4.11}$$

$$C_6H_6 + \underset{(HOSO_3H)}{H_2SO_4} \longrightarrow C_6H_5SO_3H + H_2O \qquad \text{sulfonation} \tag{4.12}$$

$$C_6H_6 + RCl \xrightarrow{AlCl_3} C_6H_5R + HCl \qquad \text{alkylation} \tag{4.13}$$

(R = an alkyl group such as $CH_3—$, $CH_3CH_2—$)

$$C_6H_6 + CH_2{=}CH_2 \xrightarrow{H_2SO_4} C_6H_5CH_2CH_3 \qquad \text{alkylation} \tag{4.14}$$

$$C_6H_6 + R{-}\overset{O}{\overset{\|}{C}}Cl \xrightarrow{AlCl_3} C_6H_5\overset{O}{\overset{\|}{C}}{-}R + HCl \qquad \text{acylation} \tag{4.15}$$

Most of these reactions are carried out at temperatures between about 0° and 50°C, but these conditions may have to be milder or more severe if other substituents are already present on the benzene ring. Also, the conditions usually can be adjusted to introduce more than one substituent, if desired.

How do these reactions take place? And why do we observe substitution instead of addition? In the next sections, we will try to answer these questions.

4.10 The Mechanism of Electrophilic Aromatic Substitution

Much evidence indicates that all of the substitution reactions listed in the previous section involve initial attack on the benzene ring by an electrophile. Consider chlorination (eq. 4.9) as a specific example. The reaction of benzene with chlorine is exceedingly slow without a catalyst, but it occurs quite briskly with one. What does the catalyst do? It acts as a Lewis acid and converts chlorine to a strong electrophile by polarizing the Cl—Cl bond.

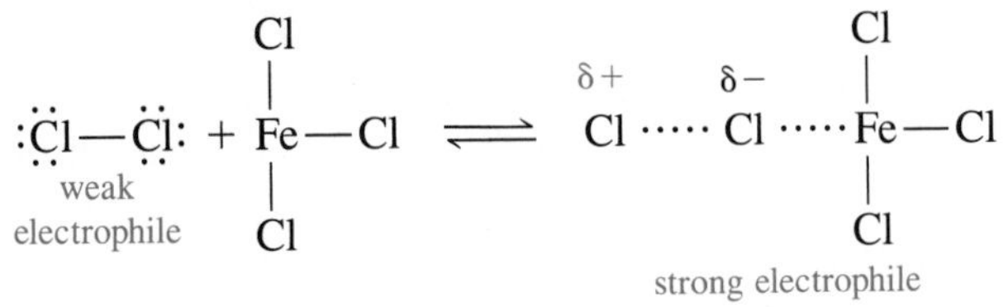

$$:\ddot{C}l-\ddot{C}l: + FeCl_3 \rightleftharpoons \overset{\delta+}{Cl}\cdots\overset{\delta-}{Cl}\cdots FeCl_3 \quad (4.16)$$

weak electrophile — strong electrophile

The reason why a *strong* electrophile is required will become apparent shortly.

The electrophile bonds to one carbon atom of the benzene ring, using two of the pi electrons from the pi cloud to form a sigma bond with a ring carbon atom. This carbon atom becomes sp^3-hybridized. The benzene ring acts as a pi-electron donor, or nucleophile, toward the electrophilic reagent.

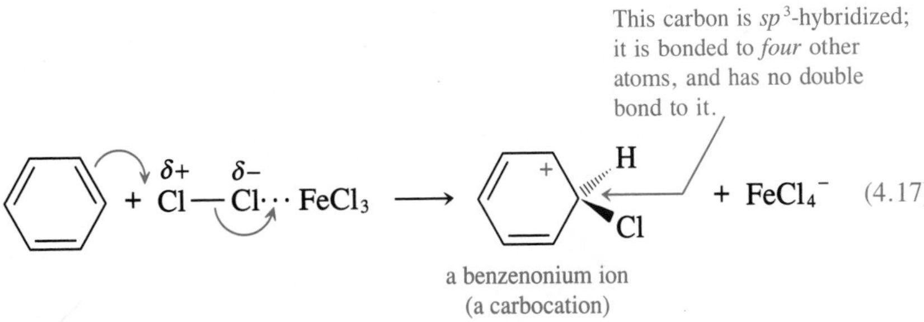

$$C_6H_6 + \overset{\delta+}{Cl}-\overset{\delta-}{Cl}\cdots FeCl_3 \longrightarrow C_6H_6Cl^+ + FeCl_4^- \quad (4.17)$$

This carbon is sp^3-hybridized; it is bonded to *four* other atoms, and has no double bond to it.

a benzenonium ion (a carbocation)

The resulting carbocation is a **benzenonium ion,** in which the positive charge is delocalized by resonance to the carbon atoms *ortho* and *para* to the carbon to which the chlorine atom became attached; that is, *ortho* and *para* to the sp^3 carbon atom.

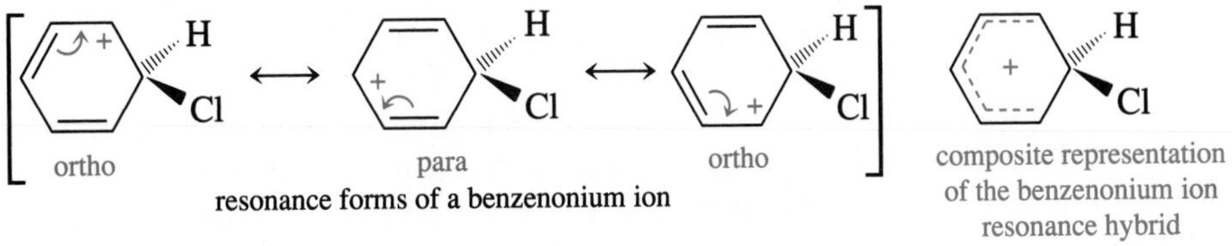

ortho ⟷ para ⟷ ortho

resonance forms of a benzenonium ion

composite representation of the benzenonium ion resonance hybrid

A benzenonium ion is similar to an allylic carbocation (sec. 3.15), but the positive charge is delocalized over five carbon atoms instead of only three. Although stabilized by resonance compared with other carbocations, its resonance energy is much less than that of the starting benzene ring.

Substitution is completed by loss of a proton from the sp^3 carbon atom, the same atom to which the electrophile became attached.

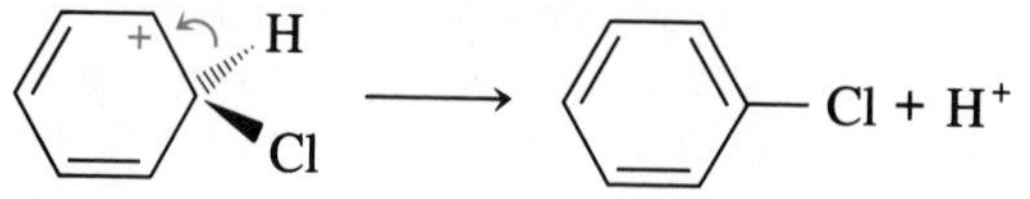

$$\longrightarrow \quad \text{C}_6\text{H}_5\text{—Cl} + H^+ \tag{4.18}$$

We can generalize this two-step mechanism for all the electrophilic aromatic substitutions in Sec. 4.9 with the following equation:

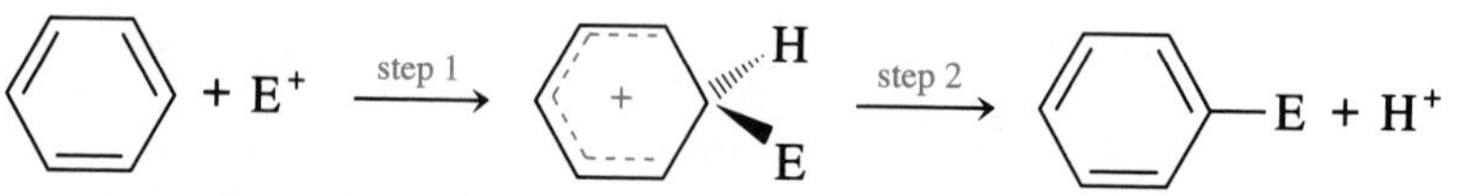

$$+ E^+ \xrightarrow{\text{step 1}} \xrightarrow{\text{step 2}} \text{—E} + H^+ \tag{4.19}$$

The reason why a strong electrophile is important, and why we observe substitution instead of addition, now becomes clear. In step 1, the stabilization energy (resonance energy) of the aromatic ring is lost, due to disruption of the aromatic pi system. This disruption, caused by addition of the electrophile to one of the ring carbons, requires energy and a strong electrophile. In step 2, the aromatic resonance energy is regained by loss of a proton. This would not be the case if the intermediate carbocation added a nucleophile (as in electrophilic *additions* to double bonds, Sec. 3.9).

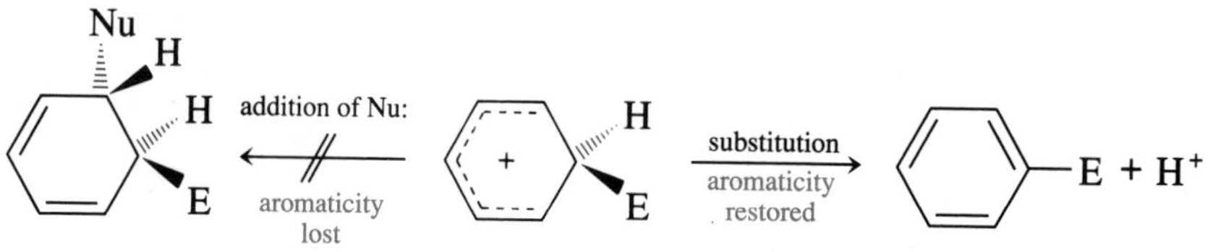

$$\xleftarrow[\text{aromaticity lost}]{\text{addition of Nu:}} \qquad \xrightarrow[\text{aromaticity restored}]{\text{substitution}} \text{—E} + H^+ \tag{4.20}$$

The first step in eq. 4.19 is usually slow or rate-determining because it disrupts the aromatic system. The second step is usually fast because it regenerates the aromatic system.

Now let us briefly consider separately each of the various types of electrophilic aromatic substitutions listed in Sec. 4.9.

4.10a Halogenation Chlorine or bromine is readily introduced into aromatic rings by using the halogen together with the corresponding iron halide as a catalyst (that is, $Cl_2 + FeCl_3$ or $Br_2 + FeBr_3$). Usually the reaction is carried out by adding the halogen slowly to a mixture of the aromatic compound and iron filings. The iron reacts with the halogen to form the iron halide, which then catalyzes the halogenation.

Direct fluorination or iodination of aromatic rings is also possible but requires special methods.

4.10b Nitration In aromatic nitrations (eq. 4.11), the sulfuric acid catalyst* protonates the nitric acid, which then loses water to generate the **nitronium ion**, which contains a positively charged nitrogen atom.

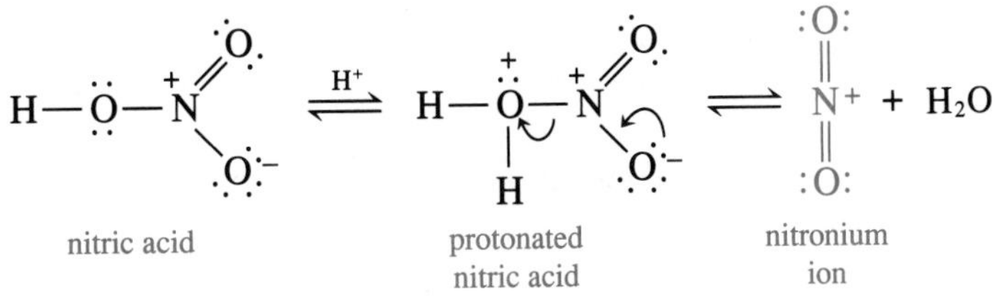

$$\mathrm{H{-}\ddot{O}{-}\overset{+}{N}(=O)O^-} \underset{}{\overset{H^+}{\rightleftharpoons}} \mathrm{H{-}\overset{+}{O}(H){-}\overset{+}{N}(=O)O^-} \rightleftharpoons \mathrm{O{=}N^+{=}O} + H_2O \qquad (4.21)$$

nitric acid — protonated nitric acid — nitronium ion

This is the electrophile that then attacks the aromatic ring.

EXAMPLE 4.2 Write out the steps in the mechanism for the nitration of benzene.

Solution The first step, formation of the electrophile NO_2^+, is shown in eq. 4.21. Then

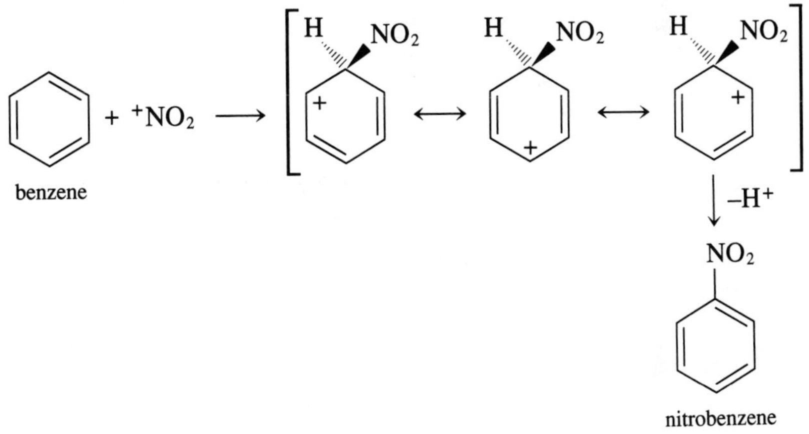

benzene + $^+NO_2$ ⟶ [three resonance structures of the H, NO_2 cation] $\xrightarrow{-H^+}$ nitrobenzene

4.10c Sulfonation In sulfonation (eq. 4.12), we use either concentrated or fuming sulfuric acid, and the electrophile may be sulfur trioxide (SO_3) or protonated sulfur trioxide, $^+SO_3H$. The following resonance structures demonstrate that SO_3 is a strong electrophile at sulfur.

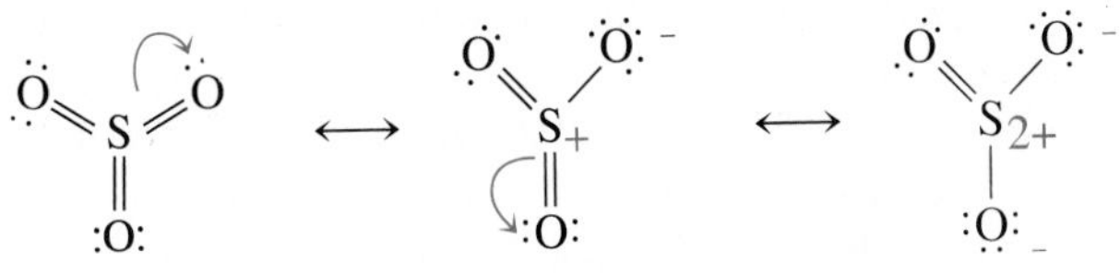

$$\mathrm{O{=}S(=O){=}O} \longleftrightarrow \mathrm{O{=}S^+(O^-){=}O} \longleftrightarrow \mathrm{O{=}S^{2+}(O^-){-}O^-}$$

---

*Sulfuric acid provides a proton catalyst as shown in the following equilibrium: $\mathrm{H{-}OSO_3H \rightleftharpoons H^+ + {}^-OSO_3H}$.

The products, sulfonic acids, are strong organic acids. Also, they can be converted to phenols by reaction with base at high temperatures.

$$C_6H_6 \xrightarrow[H_2SO_4]{SO_3} C_6H_5SO_3H \xrightarrow[200°C]{NaOH} C_6H_5OH \quad (4.22)$$

benzenesulfonic acid phenol

PROBLEM 4.10 Write out the steps in the mechanism for the sulfonation of benzene.

4.10d Alkylation and Acylation Alkylation of aromatic compounds (eqs. 4.13 and 4.14) is referred to as the **Friedel-Crafts reaction,** after Charles Friedel (French) and James Mason Crafts (American), who first discovered the reaction in 1877. The electrophile is a carbocation, which may be formed either by removing halide ion from an alkyl halide with a Lewis acid catalyst (for example, $AlCl_3$) or by adding a proton to an alkene. For example, the synthesis of ethylbenzene may be carried out as follows:

$$AlCl_3 + ClCH_2CH_3 \rightleftharpoons Cl{-}Al^{-}Cl_3 + {}^{+}CH_2CH_3 \xleftarrow{H^+} CH_2{=}CH_2 \quad (4.23)$$

ethyl cation

$$C_6H_6 + {}^{+}CH_2CH_3 \longrightarrow \left[\text{resonance structures of } C_6H_6(H)(CH_2CH_3)^{+} \right] \xrightarrow{-H^+} C_6H_5CH_2CH_3 \quad (4.24)$$

benzene ethylbenzene

PROBLEM 4.11 Which product would you expect if propene were used in place of ethene in eq. 4.14, propylbenzene or isopropylbenzene? Explain.

The Friedel-Crafts alkylation reaction has some limitations. It cannot be applied to an aromatic ring that already has on it a nitro or sulfonic acid group, because these groups form complexes with and deactivate the aluminum chloride catalyst.

Friedel-Crafts **acylations** (eq. 4.15) occur similarly. The electrophile is an acyl cation generated from an acid derivative, usually an acyl halide. The reaction provides a useful general route to aromatic ketones.

$$CH_3\overset{O}{\overset{\|}{C}}Cl + AlCl_3 \rightleftharpoons CH_3\overset{+}{C}{=}O + AlCl_4^{-} \quad (4.25)$$

acetyl chloride acetyl cation

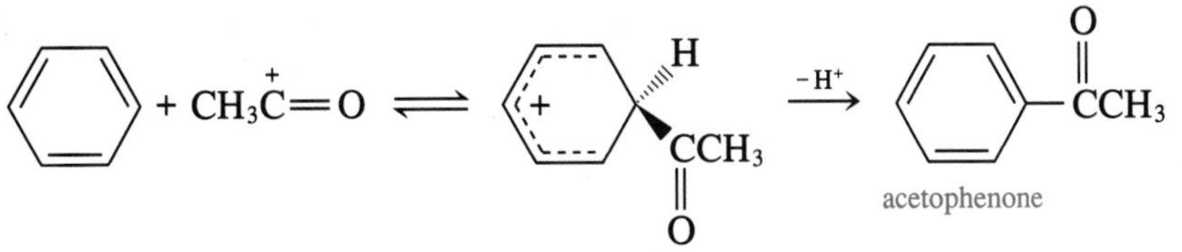

$$\text{C}_6\text{H}_6 + CH_3\overset{+}{C}{=}O \rightleftharpoons [\text{arenium ion: H, }C(=O)CH_3] \xrightarrow{-H^+} C_6H_5-\overset{O}{\overset{\|}{C}}CH_3 \quad (4.26)$$

acetophenone

4.11 *Ring-Activating and Ring-Deactivating Substituents*

In this section and the next, we will present experimental evidence that supports the electrophilic aromatic substitution mechanism just described. We will do this by examining how substituents already present on an aromatic ring affect further substitution reactions.

For example, consider the relative nitration rates of the following compounds, all under the same reaction conditions:

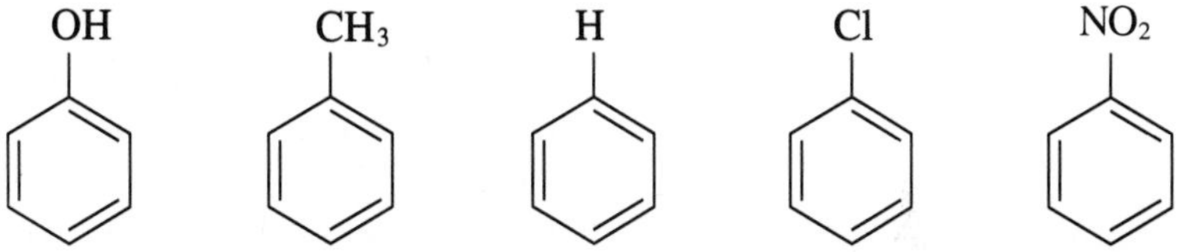

	OH	CH_3	H	Cl	NO_2
nitration rate (relative)	1000	24.5	1.0	0.033	0.0000001

decreasing rate →

Taking benzene as the standard, we see that some substituents (for example, OH and CH_3) speed up the reaction, and other substituents (Cl and NO_2) retard the reaction. We know from other evidence that hydroxyl and methyl groups are more electron donating than hydrogen, whereas chloro and nitro groups are more electron withdrawing than hydrogen.

These observations support the electrophilic mechanism for substitution. If the reaction rate depends on electrophilic (that is, electron-seeking) attack on the aromatic ring, then substituents that donate electrons to the ring will speed up the reaction; substituents that withdraw electrons from the ring will decrease electron density in the ring and therefore slow down the reaction. This reactivity pattern is exactly what is observed, not only with nitration, but with all electrophilic aromatic substitution reactions.

4.12 Ortho, Para-*Directing and* Meta-*Directing Groups*

Substituents already present on an aromatic ring determine the position taken by a new substituent. For example, nitration of toluene gives mainly a mixture of *o*- and *p*-nitrotoluene.

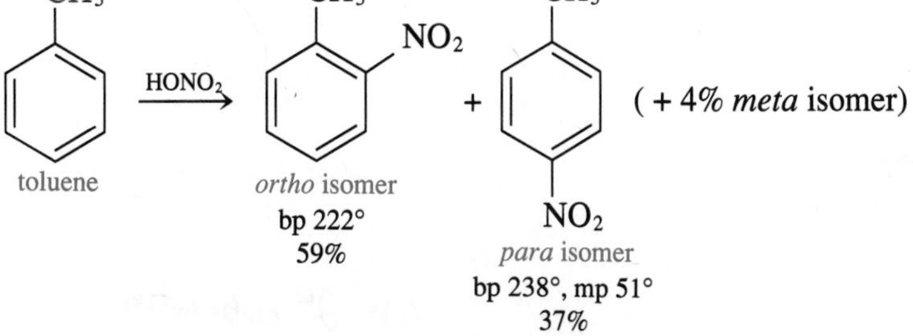

$$\text{toluene} \xrightarrow{HONO_2} \textit{ortho}\text{ isomer} + \textit{para}\text{ isomer} \quad (+\,4\%\ \textit{meta}\text{ isomer}) \quad (4.27)$$

toluene

ortho isomer
bp 222°
59%

para isomer
bp 238°, mp 51°
37%

On the other hand, nitration of nitrobenzene under similar conditions gives mainly the *meta* isomer.

$$\text{nitrobenzene} \xrightarrow{HONO_2} \textit{meta}\text{ isomer (mp 89°C, 93\%)} \quad (+\,7\%\ \textit{ortho}\ \text{isomer}) \tag{4.28}$$

This pattern is also followed for other electrophilic aromatic substitutions—chlorination, bromination, sulfonation, and so on. Toluene undergoes mainly *ortho,para* substitution, whereas nitrobenzene undergoes *meta* substitution.

In general, groups fall into one of two categories. Certain groups are ***ortho,para*-directing,** and others are ***meta*-directing.** Table 4.1 lists some of the common groups in each category. Let us see how the electrophilic substitution mechanism accounts for the behavior of these two classes of substituents.

4.12a *Ortho,Para*-Directing Groups Consider the nitration of toluene. In the first step, the nitronium ion may attack a ring carbon that is *ortho, meta,* or *para* to the methyl group.

Ortho, para attack

toluene —*ortho*→ [resonance structures with CH_3, H, NO_2] ... 3° carbocation

toluene —*para*→ [resonance structures with CH_3, H, NO_2] ... 3° carbocation (4.29)

Meta attack

toluene —*meta*→ [resonance structures with CH_3, H, NO_2] (4.30)

all 2° carbocations

TABLE 4.1 Directing and activating effects of common functional groups (groups are listed in decreasing order of activation)

Directing	Substituent group	Name of group	Effect
Ortho,Para-Directing	$-\ddot{N}H_2$, $-\ddot{N}HR$, $-\ddot{N}R_2$	amino	ACTIVATING
Ortho,Para-Directing	$-\ddot{O}H$, $-\ddot{O}CH_3$, $-\ddot{O}R$	hydroxy, alkoxy	ACTIVATING
Ortho,Para-Directing	$-\ddot{N}HC(=O)-R$	acylamino	ACTIVATING
Ortho,Para-Directing	$-CH_3$, $-CH_2CH_3$, $-R$	alkyl	ACTIVATING
Ortho,Para-Directing	$-\ddot{F}:$, $-\ddot{Cl}:$, $-\ddot{Br}:$, $-\ddot{I}:$	halo	DEACTIVATING
Meta-Directing	$-C(=\ddot{O})-R$, $-C(=\ddot{O})-\ddot{O}H$	acyl, carboxy	DEACTIVATING
Meta-Directing	$-C(=\ddot{O})-\ddot{N}H_2$, $-C(=\ddot{O})-\ddot{O}R$	carboxamido, carboalkoxy	DEACTIVATING
Meta-Directing	$-S(=\ddot{O})_2-\ddot{O}H$	sulfonic acid	DEACTIVATING
Meta-Directing	$-C\equiv N:$	cyano	DEACTIVATING
Meta-Directing	$-\overset{+}{N}(=\ddot{O}:)-\ddot{O}:^-$	nitro	DEACTIVATING

In one of the three resonance contributors to the benzenonium ion intermediate for *ortho* or *para* substitution (shown in dashed boxes), the positive charge is on the methyl-bearing carbon. That contributor is a *tertiary* carbocation and more stable than the other contributors, which are secondary carbocations. However, with *meta* attack, *all* the contributors are secondary carbocations; the positive charge in the intermediate benzenonium ion is never adjacent to the methyl substituent. Therefore the methyl group is *ortho,para*-directing, so that the reaction can proceed via the most stable carbocation intermediate.

Similarly, all other alkyl groups are *ortho, para* directing.

Consider now the other *ortho,para*-directing groups listed in Table 4.1. ***In each of them, the atom attached to the aromatic ring has an unshared electron pair.***

—$\ddot{F}$: —$\ddot{O}$H —$\ddot{N}H_2$

This unshared electron pair can stabilize an adjacent positive charge. Let's consider, as an example, the bromination of phenol.

preferred
Ortho,para attack

stabilizes (+) charge

ortho

OH

phenol

:ÖH H + Br ⟷ +ÖH H Br ⟷ :ÖH H Br + ⟷ :ÖH H + Br

para

:ÖH + H Br ⟷ :ÖH + H Br ⟷ +ÖH H Br ⟷ :ÖH + H Br (4.31)

Meta attack

:ÖH phenol —meta→ :ÖH + H Br ⟷ :ÖH + H Br ⟷ :ÖH + H Br (4.32)

In the case of *ortho* or *para* attack, one of the contributors to the intermediate benzenonium ion places the positive charge on the hydroxyl-bearing carbon. *Shift of an unshared electron pair from the oxygen to the positive carbon allows the positive charge to be delocalized onto the oxygen* (see the structures in the dashed boxes). No such structures are possible for *meta* attack. Therefore, the hydroxyl group is *ortho,para*-directing.

We can generalize this observation. ***All groups with unshared electrons on the atom attached to the ring are ortho,para-directing.***

PROBLEM 4.12 Draw the important resonance contributors for the intermediate in the bromination of aniline, and explain why *ortho,para* substitution predominates.

$\ddot{N}H_2$

aniline

4.12b *Meta*-Directing Groups Now let us examine the nitration of nitrobenzene in the same way, to see if we can explain the *meta*-directing effect of the nitro group. In nitrobenzene, the nitrogen has a formal charge of +1, as shown on the structures. The equations for forming the intermediate benzenonium ion are

Ortho, para attack

nitrobenzene — *ortho* / *para*

(4.33)

Meta attack

nitrobenzene — *meta*

(4.34)

In eq. 4.33, one of the contributors to the resonance hybrid intermediate for *ortho* or *para* substitution (shown in the boxes) has *two adjacent positive charges,* a highly *undesirable* arrangement, because like charges repel each other. No such intermediate is present for *meta* substitution (eq. 4.34). For this reason, *meta* substitution is preferred.

Can we generalize this explanation to the other *meta*-directing groups in Table 4.1? Notice that each *meta*-directing group is connected to the aromatic ring by an atom that is part of a double or triple bond, at the other end of which is an atom more electronegative than carbon (for example, an oxygen or nitrogen atom). In such cases, ***the atom directly attached to the benzene ring*** *will carry a partial positive charge* (like the nitrogen in the nitro group). This is because of resonance contributors, such as

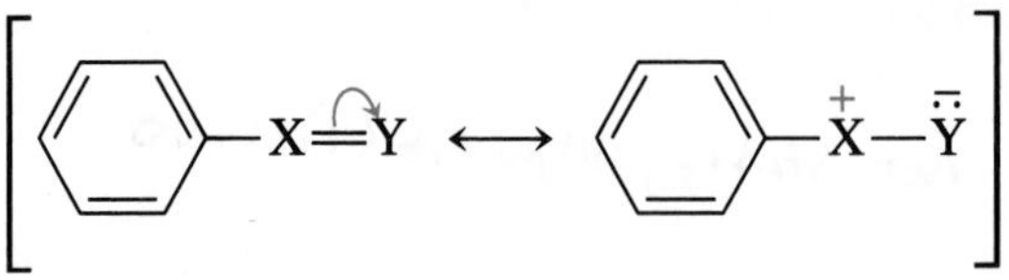

$$\left[\text{C}_6\text{H}_5-\text{X}{=}\text{Y} \longleftrightarrow \text{C}_6\text{H}_5-\overset{+}{\text{X}}-\overset{..}{\underset{..}{\text{Y}}}{}^{-} \right]$$

Y is an electron-withdrawing atom such as oxygen or nitrogen; atom X carries a positive charge in one of the resonance contributors.

All such groups will be *meta*-directing for the same reason that the nitro group is *meta*-directing: to avoid having two adjacent positive charges in the intermediate benzenonium ion. We can generalize. **All groups in which the atom directly attached to the aromatic ring is positively charged or is part of a multiple bond to a more electronegative element will be *meta*-directing.**

PROBLEM 4.13 Compare the intermediate benzenonium ions for *ortho,meta* and *para* bromination of benzoic acid, and explain why the main product is *m*-bromobenzoic acid.

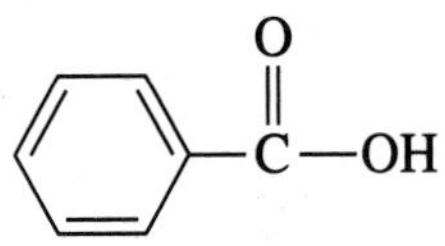

$$\text{C}_6\text{H}_5-\overset{\overset{\displaystyle\text{O}}{\|}}{\text{C}}-\text{OH}$$

benzoic acid

4.12c Substituent Effects on Reactivity Substituents not only affect the position of substitution, they also affect the *rate* of substitution, whether it will occur slower or faster than for benzene. A substituent is considered to be **activating** if the rate is faster and **deactivating** if the rate is slower (see Table 4.1) than for benzene. Is this rate effect related to the orientation effect?

In all *meta*-directing groups, the atom connected to the ring carries a full or partial positive charge and will therefore withdraw electrons from the ring. *All* meta-*directing groups are therefore ring-deactivating groups.* On the other hand, ortho,para-*directing groups in general supply electrons to the ring and are therefore ring-activating.* With the halogens (F, Cl, Br, and I), two opposing effects bring about the only important exception to these rules. *Because they are strongly electron withdrawing, the halogens are ring-deactivating; but because they have unshared electron pairs, they are* ortho,para-*directing.*

4.13 The Importance of Directing Effects in Synthesis

When designing a multistep synthesis involving electrophilic aromatic substitution, we must keep in mind the directing and activating effects of the groups involved. Consider, for example, the bromination and nitration of benzene to make bromonitrobenzene. If we brominate first and then nitrate, we will get a mixture of the *ortho* and *para* isomers.

$$\text{benzene} \xrightarrow[\text{FeBr}_3]{\text{Br}_2} \text{bromobenzene} \xrightarrow[\text{H}_2\text{SO}_4]{\text{HONO}_2} \underset{\text{ortho}}{o\text{-BrC}_6\text{H}_4\text{NO}_2} + \underset{\text{para}}{p\text{-BrC}_6\text{H}_4\text{NO}_2} \qquad (4.35)$$

This is because the bromine atom in bromobenzene is *ortho,para*-directing. On the other hand, if we nitrate first and then brominate, we will get mainly the *meta isomer* because the nitro group is *meta*-directing.

$$\text{benzene} \xrightarrow[\text{H}_2\text{SO}_4]{\text{HONO}_2} \text{nitrobenzene} \xrightarrow[\text{FeBr}_3]{\text{Br}_2} \underset{\text{meta}}{m\text{-BrC}_6\text{H}_4\text{NO}_2} \qquad (4.36)$$

The sequence in which we carry out the reactions of bromination and nitration is therefore very important. It determines which type of product is formed.

PROBLEM 4.14 Devise a synthesis for each of the following, starting with benzene:

a. *m*-chlorobenzenesulfonic acid
b. *p*-nitrotoluene

PROBLEM 4.15 Explain why it is *not* possible to prepare *m*-bromochlorobenzene or *p*-nitrobenzenesulfonic acid by carrying out two successive electrophilic aromatic substitutions.

4.14 Polycyclic Aromatic Hydrocarbons

The concept of **aromaticity**—*the unusual stability of certain fully conjugated* ***cyclic*** *systems*—can be extended well beyond benzene itself or simple substituted benzenes.

Coke, required in huge quantities for the manufacture of steel, is obtained by heating coal in the absence of air. A by-product of this conversion of coal to coke is a distillate called **coal tar,** a complex mixture containing many aromatic hydrocarbons (including benzene, toluene, and xylenes). **Naphthalene,** $C_{10}H_8$, was the *first* pure compound to be obtained from the higher-boiling fractions of coal tar. It was easily isolated because it sublimes from the tar as a beautiful colorless crystalline solid, mp 80°C. *Naphthalene is a planar molecule with two fused benzene rings.* The two rings share two carbon atoms.

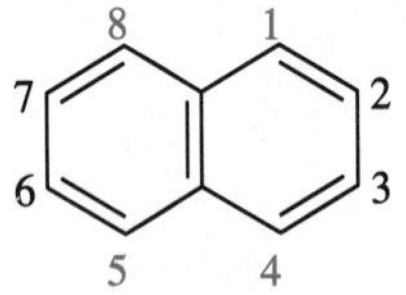

naphthalene
mp 80°C

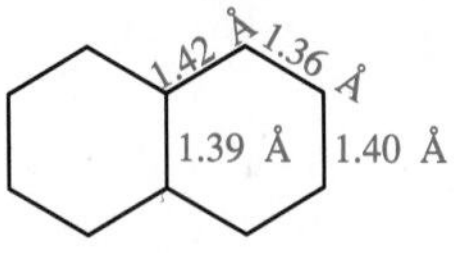

bond lengths in naphthalene

The bond lengths in naphthalene are not all identical, but they all approximate the bond length in benzene (1.39 Å). Although it has two six-membered rings, naphthalene has a resonance energy somewhat less than twice that of benzene, about 60 kcal/mol. Because of its symmetry, naphthalene has two sets of equivalent carbon atoms: C-1, C-4, C-5, and C-8; and C-2, C-3, C-6, and C-7. Like benzene, naphthalene undergoes electrophilic substitution reactions (halogenation, nitration, and so on), usually under somewhat milder conditions than benzene. Although two monosubstitution products are possible, substitution at C-1 usually predominates.

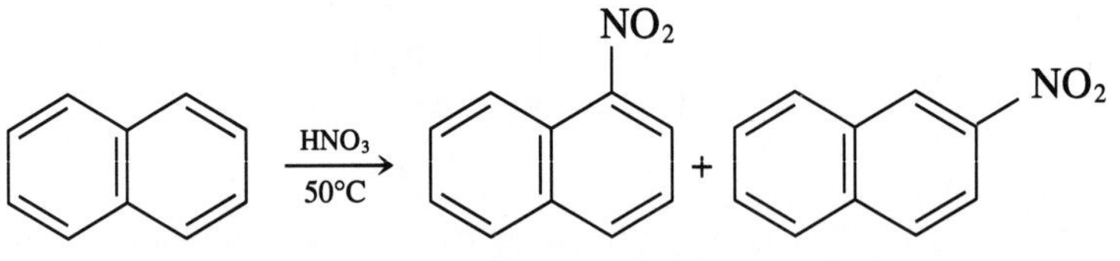

$\xrightarrow[50°C]{HNO_3}$ 1-nitronaphthalene + 2-nitronaphthalene (4.37)

(ratio 10:1)

EXAMPLE 4.3 Draw the resonance contributors for the carbocation intermediate in nitration of naphthalene at C-1; include only structures that retain benzenoid aromaticity in the unsubstituted ring.

Solution Four such contributors are possible.

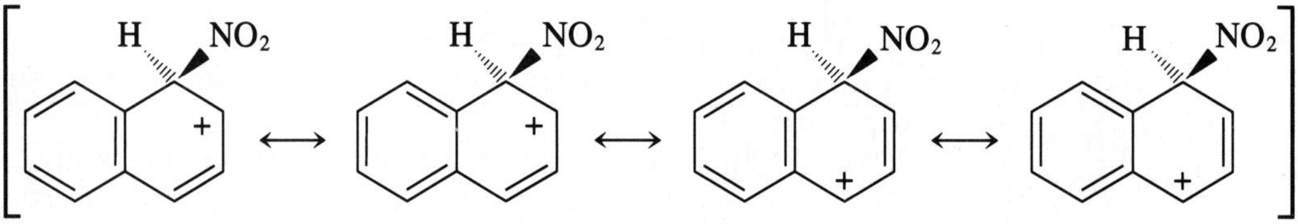

PROBLEM 4.16 Repeat Example 4.3 for nitration at C-2. Can you suggest why substitution at C-1 is preferred?

Naphthalene is the parent compound of a series of **fused polycyclic hydrocarbons,** a few other examples of which are

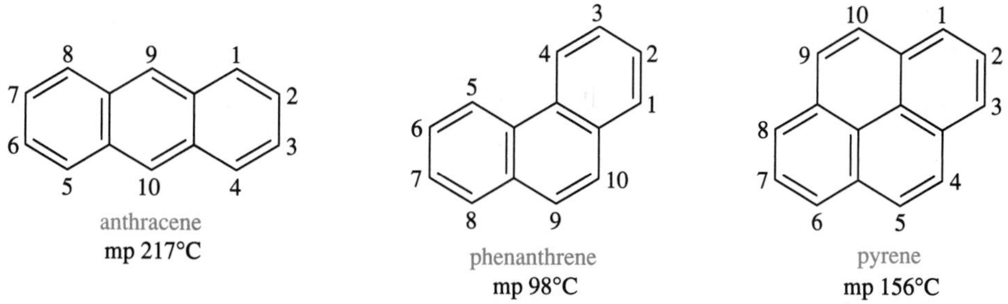

anthracene
mp 217°C

phenanthrene
mp 98°C

pyrene
mp 156°C

Infinite extension of such rings leads to sheets of hexagonally arranged carbons, the structure of graphite (a form of elemental carbon).

PROBLEM 4.17 Calculate the ratio of carbons to hydrogens in benzene, naphthalene, anthracene, and pyrene. Notice that as the number of fused rings increases, the proportion of carbon also increases, so that extrapolation to a very large number of such rings leads to graphite (see "A Word About" on page 138).

A Word About . . .

6. Polycyclic Aromatic Hydrocarbons and Cancer

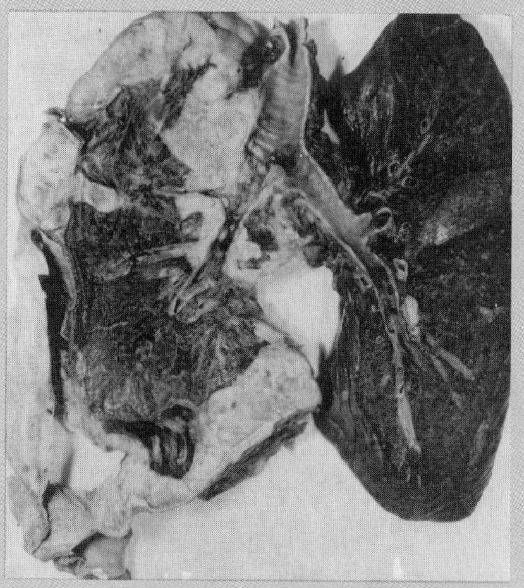

Certain polycyclic aromatic hydrocarbons are carcinogenic (that is, they produce cancers). They can produce a tumor on mice in a short time when only trace amounts are painted on the skin. These carcinogenic hydrocarbons are present not only in coal tar but also in soot and tobacco smoke and can be formed in barbecuing meat. Their biological effect was noted as long ago as 1775 when soot was identified as the cause of the high incidence of scrotal cancer in chimney sweeps. A similar occurrence of lung and lip cancer is common in habitual smokers.

The way these carcinogens produce cancer is now fairly well understood. To eliminate hydrocarbons, the body usually oxidizes them to render them more water soluble, so that they can be excreted. The metabolic oxidation products seem to be the real culprits in causing cancer. For example, one of the most potent carcinogens of this type is benzo[a]pyrene. Enzymatic oxidation converts it to the diol-epoxide shown.

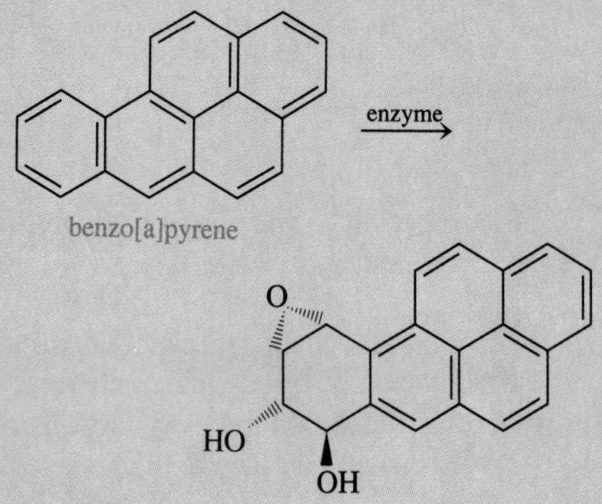

enzyme →

benzo[a]pyrene

O HO OH

a diol-epoxide

The diol-epoxide reacts with cellular DNA, causing mutations that eventually prevent the cells from reproducing normally.

Benzene itself is quite toxic to humans and can cause severe liver damage, but toluene is much less toxic. How can this different behavior of two very similar compounds be possible? To eliminate benzene from the body, the aromatic ring must be oxidized, and intermediates in this oxidation are damaging. However, the *methyl side chain* of toluene can be oxidized to give benzoic acid, which can be excreted. None of the intermediates in this process causes problems.

Although some chemicals may cause cancer, others can help prevent or cure it. Many substances inhibit cancer growth, and the study of cancer chemotherapy has contributed substantially to human health.

A Word About . . .

7. C_{60}, an Aromatic Sphere: The Fullerenes

No recent chemical discovery has been as dramatic nor stimulated such an explosive burst of research activity as that of C_{60}. Until the mid-1980s only two major allotropic forms of elemental carbon were known, *diamond* (page 57) and *graphite*. But in September 1985 scientists interested in small carbon fragments blown out of giant red stars in interstellar space were trying to produce these same particles here on earth for closer study. To do so, they subjected graphite to a high-energy pulsed laser beam, and passed the vaporized fragments into a mass spectrometer for analysis. Although they did find some of the small fragments they sought, they also saw to their great surprise an exceptionally intense mass spectral peak at the very high mass of 720, corresponding to C_{60} ($C = 12 \times 60 = 720$). Certain other less intense high-mass peaks were also observed (for example, C_{70} at mass 840). They proposed a unique structure for C_{60} to account for its remarkable stability; the proposed structure turned out to be correct and five years later methods were developed for producing this new form of carbon in quantities sufficient to study its chemistry. What is C_{60} and why is it (and not C_{59} or C_{61}) so stable?

First let us consider how the carbon atoms are arranged in graphite, C_{60}'s precursor. Graphite consists of layers of planar hexagonal carbon rings. Each carbon atom (except for the few at the outer edges) is connected to three other carbon atoms in the same layer, by bonds that are approximately the same length (1.42 Å) as the carbon–carbon bonds in benzene (1.39 Å).

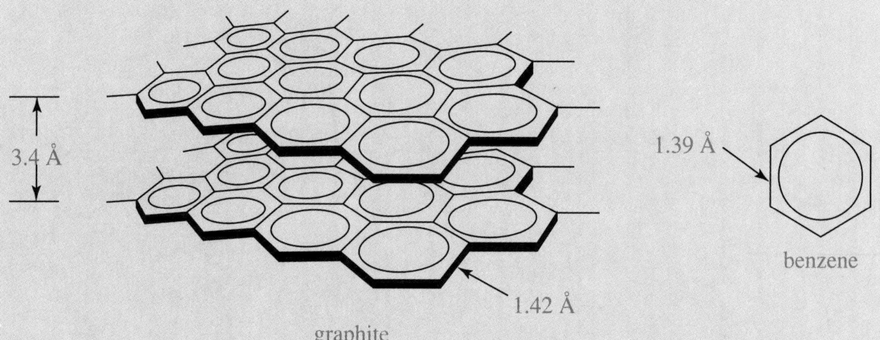

In a very real sense, then, the sheets in graphite are like an infinite number of fused aromatic rings. There are no covalent bonds between carbon atoms in different layers, which are 3.4 Å apart. Only weak forces hold the layers together. The lubricant properties of graphite are thought to result from sliding of the layers with respect to each other.

Consider, now, what might happen if we blast a graphite layer with sufficient energy (as in the laser beam experiment) to expel a carbon atom from one of the rings, and reduce one ring from a hexagon to a pentagon. *The structure will no longer be flat, but will curve!* The known aromatic hydrocarbon **corannulene,** for example,

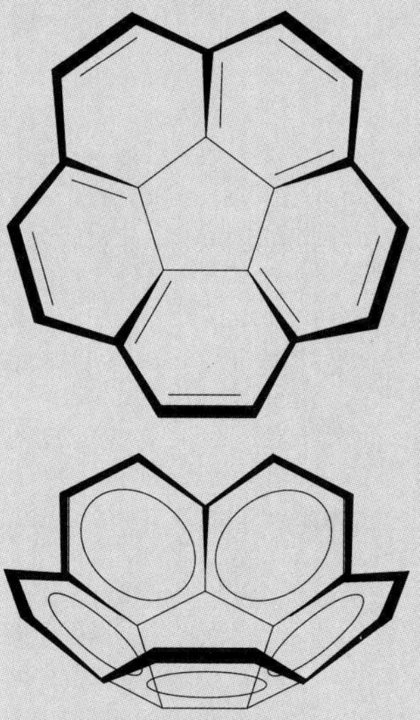

corannulene, a saucer-shaped aromatic molecule

with five benzene rings around a pentagon, is saucer-shaped, not flat. Ejection of additional carbon atoms from the graphite layers leads to increasing curvature and eventual closure to spherelike structures, although details of how this happens are still being investigated.

The structure originally guessed at and later verified for C_{60} is the same as for a soccer ball, and is shown in Figure 4.3. It is a polygon (technically, a truncated icosahedron) with 60 vertices, one carbon at each. There are 32 faces, 12 of which are *pentagons* (12 pentagons × 5 carbons each = 60 carbons) and 20 of which are hexagons (20 hexagons × 6 carbons each = 120 carbons ÷ 2 = 60 carbons; we must divide by 2 because, as you can see in Figure 4.3, each carbon atom is shared by *two* hexagons). Each pentagon is surrounded by five hexagons—no two pentagons are adjacent. However, the six bonds of each hexagon are fused alternately to three pentagons and three hexagons.

C_{60} was trivially named "buckminsterfullerene" after R. Buckminster Fuller, the engineer, architect, and philosopher who used similar shapes to construct geodesic domes. Due to their spherical shape, C_{60} molecules are sometimes colloquially called "buckyballs." Compounds derived from C_{60} and related carbon clusters that enclose space (such as C_{70}, which is egg-shaped, C_{76}, C_{84}, and others) are called "fullerenes."

Why is C_{60} so stable? As with benzene and graphite, each carbon atom in C_{60} is attached to three other atoms, therefore is sp^2-hybridized. The fourth valence electron of each carbon lies in a *p* orbital that is perpendicular to the spherical surface. These orbitals overlap to form a pi cloud

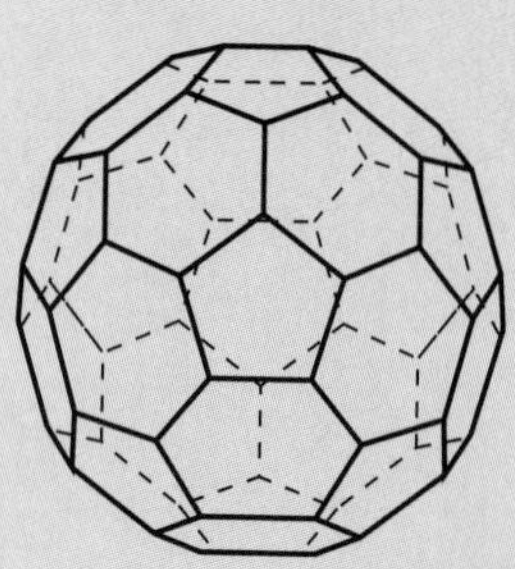

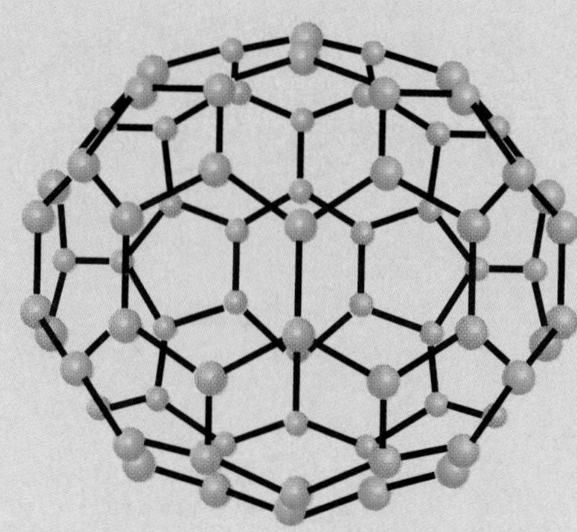

FIGURE 4.3 C_{60} or "buckminsterfullerene" (left) and C_{70} (right).

outside and *inside* the sphere, like the pi cloud above and below the plane of a benzene ring. Hence the structure is in a sense aromatic and exceptionally stable. Indeed, C_{60} was recently detected in certain black, lustrous rocks where it may have been trapped more than half a billion years ago!

C_{60} can now be made in gram quantities. It is extracted by organic solvents (in which it is sparingly soluble) from a specially prepared soot, and is separated chromatographically from higher fullerenes that are simultaneously formed. Its solutions are beautifully colored, for example magenta in hexane.

What about its reactions? Since substitution is impossible (there are no hydrogens in C_{60}), it reacts mainly by addition. Although all 60 carbons are equivalent (just like the six carbons in benzene), the bonds are not. C_{60} has two types of bonds, the 6-6 bonds shared by adjacent hexagons, and the 5-6 bonds shared by a pentagon and a hexagon. The 6-6 bonds are a bit shorter (1.39 Å) than the 5-6 bonds (1.43 Å) and are more like double bonds. Most additions to C_{60} therefore occur across a 6-6 bond. Multiple additions are possible, but lead to complex mixtures of isomers that are difficult to separate, and the orientation rules (analogous to those for benzene) have yet to be worked out.

$$C_{60} + A{-}B \longrightarrow$$

We have lessons to learn from the serendipitous and entirely unanticipated discovery of C_{60}. This advance, which opened up a whole new field of chemistry, was the unexpected result of studies in fundamental science. Yet new types of polymers, superconductors, structures with metals or other atoms trapped inside the carbon clusters, new catalysts, and pharmaceuticals—all these and other commercial possibilities not yet imagined—will most assuredly follow from this exciting discovery. The C_{60} story illustrates once more why it is so important, in a technological world, to support research in the fundamental sciences. Where the research will lead cannot be predicted with certainty, but experience shows that the eventual practical benefits that follow, even from only a small fraction of fundamental discoveries, compensate many times over for the initial investment.

REACTION SUMMARY

Electrophilic Aromatic Substitution (Sec. 4.9)

1. Halogenation

$$C_6H_6 + X_2 \xrightarrow{FeX_3} C_6H_5X + HX$$

X = Cl, Br

2. Nitration

$$C_6H_6 + HNO_3 \xrightarrow{H_2SO_4} C_6H_5NO_2 + H_2O$$

3. Sulfonation

$$C_6H_6 + HOSO_3H \xrightarrow{heat} C_6H_5SO_3H + H_2O$$

4. Alkylation (Friedel-Crafts)

$$C_6H_6 + RCl \xrightarrow{AlCl_3} C_6H_5R + HCl$$

R = alkyl group

5. Alkylation

$$C_6H_6 + CH_2{=}CH_2 \xrightarrow{H^+} C_6H_5CH_2CH_3$$

6. Acylation (Friedel-Crafts)

$$C_6H_6 + R{-}\overset{O}{\overset{\|}{C}}{-}Cl \xrightarrow{AlCl_3} C_6H_5{-}\overset{O}{\overset{\|}{C}}{-}R + HCl$$

ADDITIONAL PROBLEMS

4.18. Write structural formulas for the following compounds:

a. 1,2,4-tribromobenzene
b. *o*-chlorotoluene
c. *m*-diethylbenzene
d. isopropylbenzene
e. *p*-chlorophenol
f. benzyl bromide
g. 2,3-diphenylpentane
h. *p*-chlorostyrene
i. 2-bromo-4-ethyl-3,5-dinitrotoluene
j. *p*-chlorobenzenesulfonic acid
k. *m*-bromobenzoic acid
l. *o*-fluoroacetophenone
m. *p*-nitroanisole
n. 2,4,6-trimethylaniline

4.19. Name the following compounds:

a. $CH_2CH_2CH_3$

b. CH=O, Br

c. Cl Cl

d. Cl, CH_3, Cl

e. $(CH_3)_2CH$— —OH

f.

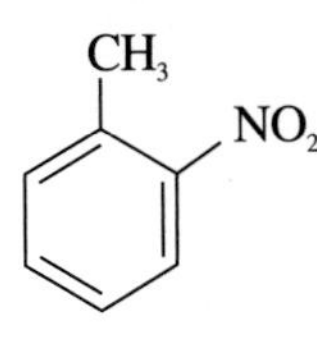

g. F, F, F, F, F, F

h. $CH{=}CH_2$, Br, Br

i. CH_3CH_2

4.20. Give the structures and names for all possible
a. trimethylbenzenes. **b.** dichloronitrobenzenes.

4.21. There are three dibromobenzenes (*o*-, *m*-, and *p*-). Suppose we have samples of each in separate bottles, but we don't know which is which. Let us call them A, B, and C. On nitration, compound A (mp 87°C) gives only *one* nitrodibromobenzene. What is the structure of A? B and C are both liquids. On nitration, B gives *two* nitrodibromobenzenes, and C gives *three* nitrodibromobenzenes (of course, not in equal amounts). What are the structures of B and C? of their mononitration products? (This method, known as Körner's method, was used years ago to assign structures to isomeric benzene derivatives.)

4.22. Give the structure and name of each of the following aromatic hydrocarbons:
a. C_8H_{10}; has three possible ring-substituted monobromo derivatives
b. C_9H_{12}; can give only one mononitro product on nitration
c. C_9H_{12}; can give four mononitro derivatives on nitration

4.23. The observed amount of heat evolved when 1,3,5,7-cyclooctatetraene is hydrogenated is 110 kcal/mol. What does this tell you about the possible resonance energy of this compound?

4.24. The structure of the nitro group —NO_2 is usually shown as

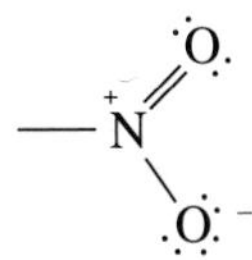

yet experiments show that the two nitrogen–oxygen bonds have the same length of 1.21 Å. This length is intermediate between 1.36 Å for the N—O single bond and 1.18 Å for the N═O double bond. Draw structural formulas that explain this observation.

4.25. Draw all reasonable electron-dot formulas for the nitronium ion, $(NO_2)^+$, the electrophile in aromatic nitrations. Show any formal charges. Which structure is favored and why?

4.26. Write out all steps in the mechanism for the reaction of
a. *p*-dibromobenzene + nitric acid (H_2SO_4 catalyst).
b. toluene + *t*-butyl chloride + $AlCl_3$.

4.27. Draw all possible contributing structures to the carbocation intermediate in the chlorination of chlorobenzene. Explain why the major products are *o*- and *p*-dichlorobenzene. (*Note: p*-dichlorobenzene is produced commercially this way, for use against clothes moths.)

4.28. Repeat Problem 4.27 for the chlorination of benzoic acid, and explain why the product is *m*-chlorobenzoic acid.

4.29. Indicate the main *mono*substitution products in each of the following reactions. Keep in mind that certain substituents are *meta*-directing and others are *ortho,para*-directing.
a. anisole + chlorine (Fe catalyst)
b. nitrobenzene + concentrated sulfuric acid (heat)
c. bromobenzene + chlorine (Fe catalyst)
d. toluene + bromine (Fe catalyst)
e. benzenesulfonic acid + concentrated nitric acid (heat)
f. iodobenzene + chlorine (Fe catalyst)
g. ethylbenzene + concentrated nitric acid (H_2SO_4 catalyst)
h. toluene + acetyl chloride ($AlCl_3$ catalyst)

4.30. Suggest a reason why $FeCl_3$ is used as a catalyst for aromatic chlorinations and $FeBr_3$ for brominations (that is, why the iron halide used has the same halogen as the halogenating agent).

4.31. Using benzene or toluene as the only aromatic organic starting material, devise a synthesis for each of the following:
a. *m*-chloronitrobenzene
b. *p*-toluenesulfonic acid
c. *p*-nitroethylbenzene
d. methylcyclohexane
e. 2,6-dichloro-4-nitrotoluene
f. *p*-bromonitrobenzene
g. 2-bromo-4-nitrotoluene
h. *m*-bromoacetophenone

4.32. When benzene is treated with excess D_2SO_4 at room temperature, the hydrogens on the benzene ring are gradually replaced by deuterium. Write a mechanism that explains this observation.

4.33. Predict whether the following substituents on the benzene ring are likely to be *ortho,para*-directing or *meta*-directing and whether they are likely to be ring-activating or ring-deactivating:

a. $-NHC(=O)CH_3$ **b.** $-C(=O)-OCH_3$ **c.** $-SCH_3$ **d.** $-\overset{+}{N}H(CH_3)_2$

4.34. The explosive TNT (2,4,6-trinitrotoluene) can be made by nitrating toluene with a mixture of nitric and sulfuric acids, but the reaction conditions must gradually be made more severe as the nitration proceeds. Explain why.

4.35. Which compound is more reactive toward electrophilic substitution (for example, nitration)?

a. phenol or benzoic acid **b.** chlorobenzene or ethylbenzene

4.36. For a one-step synthesis of 3-bromo-5-nitrobenzoic acid, which is the better starting material, 3-bromobenzoic acid or 3-nitrobenzoic acid? Why?

4.37. Show how pure 3,5-dinitrochlorobenzene can be prepared, starting from a disubstituted benzene.

4.38. How many possible monosubstitution products are there for each of the following?

a. anthracene **b.** phenanthrene

4.39. Bromination of anthracene gives mainly 9-bromoanthracene. Write out the steps in the mechanism of this reaction.

4.40. Draw a molecular orbital picture for the resonance hybrid benzenonium ion shown in eq. 4.19, and describe the hybridization of each ring carbon atom.

5 Stereoisomerism

5.1 Introduction

Stereoisomers *have the same atom connectivities, or order of attachment of the atoms, but different arrangements of the atoms in space*. We have already seen that stereoisomers may be characterized according to the ease with which they can be interconverted. That is, they may be **conformers**, which can be interconverted by rotation about a single bond, or they may be **configurational isomers**, which can be interconverted only by breaking and remaking covalent bonds.

Here we will consider other useful ways to categorize stereoisomers, ways that are particularly helpful in describing their properties.

5.2 Chirality and Enantiomers

Consider the difference between a pair of gloves and a pair of socks. A sock, like its partner, can be worn on either the left or the right foot. But a left-hand glove, unlike its partner, cannot be worn on the right hand. Like a pair of gloves, certain molecules possess this property of "handedness," which affects their chemical behavior. Let us examine the idea of molecular handedness.

A molecule (or object) is either **chiral** or **achiral.** The word *chiral,* pronounced "kai-ral" to rhyme with spiral, comes from the Greek $\chi\epsilon\iota\rho$ (*cheir,* hand.) ***A chiral molecule (or object) is one that exhibits the property of handedness.*** An **achiral** molecule does not have this property.

What test can we apply to tell whether a molecule (or object) is chiral or achiral? We examine the molecule (or object) *and its mirror image*. ***The mirror image of a chiral molecule cannot be superimposed on the molecule itself. The mirror image of an achiral molecule, however, is identical with or superimposable on the molecule itself.***

Let us apply this test to some specific examples. Figure 5.1 shows one of the more obvious examples. The mirror image of a left hand is not another left hand, but a right hand. A hand and its mirror image are not superimposable. A hand is chiral. But the mirror image of a ball (sphere) is also a ball (sphere), so a ball (sphere) is achiral.

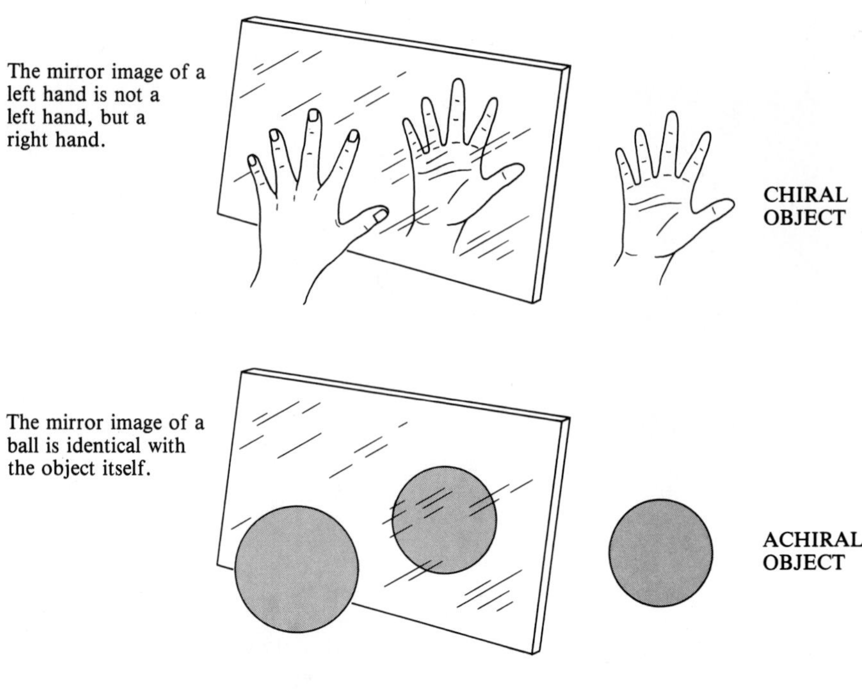
The mirror image of a left hand is not a left hand, but a right hand.
CHIRAL OBJECT
The mirror image of a ball is identical with the object itself.
ACHIRAL OBJECT

FIGURE 5.1 The mirror-image relationships of chiral and achiral objects.

PROBLEM 5.1 Which of the following objects are chiral and which are achiral?

a. golf club b. teacup c. football d. corkscrew
e. tennis racket f. shoe g. portrait h. pencil

Now let us look at two molecules, 2-chloropropane and 2-chlorobutane, and their mirror images.

Figure 5.2 shows that 2-chloropropane is achiral. Its mirror image is superimposable on the molecule itself. Therefore 2-chloropropane has only one possible structure.

On the other hand, as Figure 5.3 shows, 2-chlorobutane has two possible structures, related to one another as nonsuperimposable mirror images. **We call a pair of molecules that are related as nonsuperimposable mirror images *enantiomers*.** Every molecule, of course, has a mirror image. Only those that are *nonsuperimposable* are called enantiomers.

5.3 *Stereogenic Centers; the Stereogenic Carbon Atom*

What is it about their structures that leads to chirality in 2-chlorobutane but not in 2-chloropropane? Notice that, in 2-chlorobutane, carbon atom 2, the one marked with an asterisk, has four different groups attached to it (Cl, H, CH_3, and CH_3CH_2). A carbon atom with four different groups attached to it is called a **stereogenic carbon atom.** This type of carbon is also called a **stereogenic center** because it gives rise to stereoisomers.

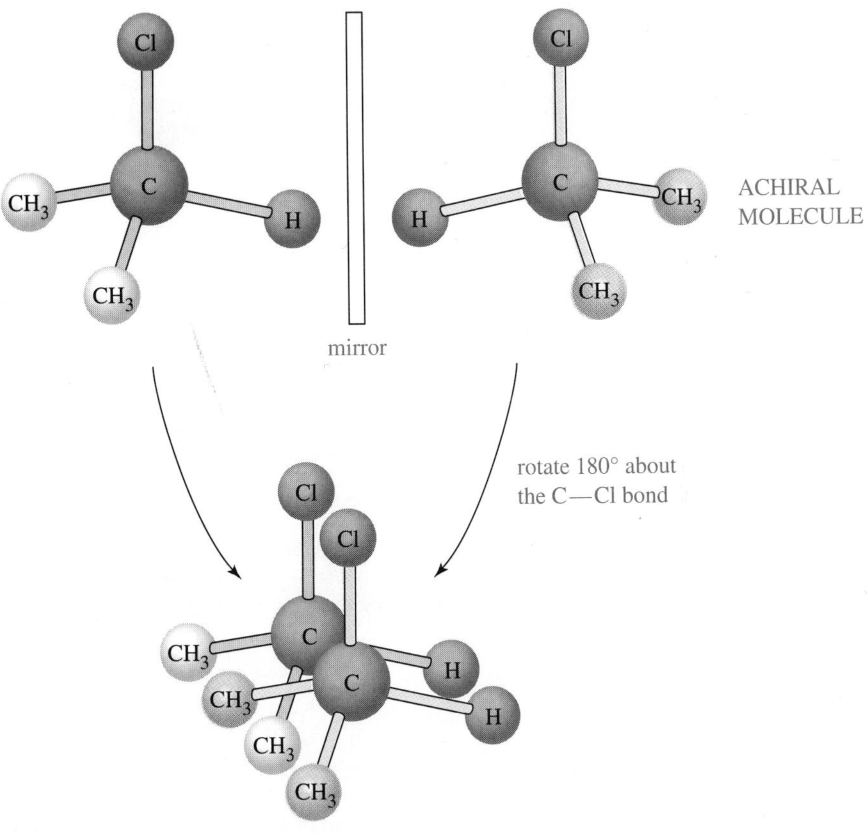

FIGURE 5.2 Model of 2-chloropropane and its mirror image. The mirror image is superimposable on the original molecule.

$$\begin{array}{c} \mathrm{Cl} \\ | \\ \mathrm{CH_3}-\overset{*}{\mathrm{C}}-\mathrm{CH_2CH_3} \\ | \\ \mathrm{H} \end{array}$$

Let us examine the more general case of a carbon atom with any four different groups attached; let us call the groups A, B, D, and E. Figure 5.4 shows such a molecule and its mirror image. That the molecules on each side of the mirror in Figure 5.4 are nonsuperimposable mirror images (enantiomers) becomes clear by examining Figure 5.5. (We strongly urge you to use molecular models when studying this chapter. It is sometimes difficult to visualize three-dimensional structures when they are drawn on a two-dimensional surface [this page or a blackboard], though with experience, your ability to do so will improve.)

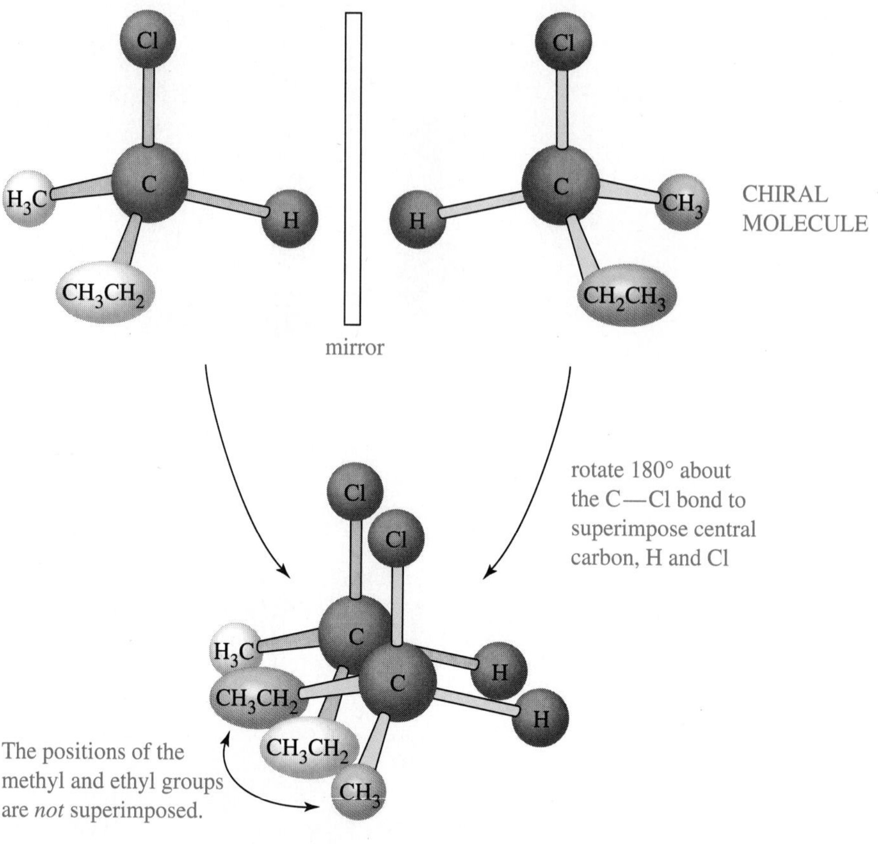

FIGURE 5.3 Model of 2-chlorobutane and its mirror image. The mirror image is *not* superimposable on the original molecule. The two forms of 2-chlorobutane are enantiomers.

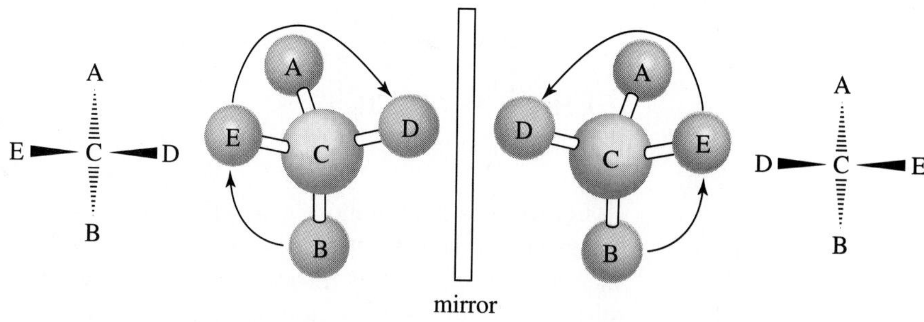

FIGURE 5.4 The chirality of enantiomers. Looking down the C—A bond, we have to read clockwise to spell BED for the model on the left, but we must read counterclockwise for its mirror image.

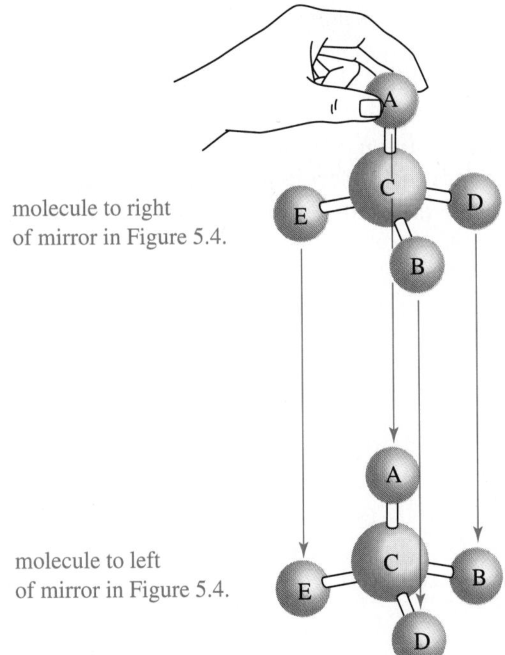

FIGURE 5.5 When the four different groups attached to a stereogenic carbon atom are arranged to form mirror images, the molecules are not superimposable. The models may be twisted or turned in any direction, but as long as no bonds are broken, only two of the four attached groups can be made to coincide.

The handedness of these molecules is also illustrated in Figure 5.4, where the clockwise or counterclockwise arrangement of the groups (we might call them right- or left-handed arrangements) is apparent.

What happens when all four of the groups attached to the central carbon atom are *not* different from one another? Suppose two of the groups are identical—say, A, A, B, and D. Figure 5.6 describes this situation. The molecule

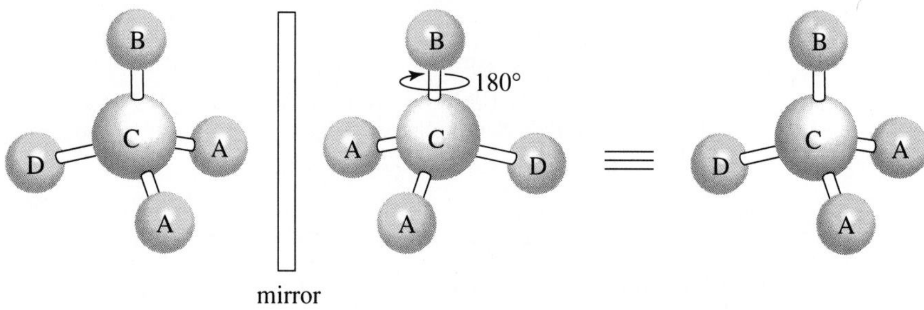

FIGURE 5.6 The tetrahedral model at the left has two corners occupied by identical groups (A). It has a plane of symmetry that passes through atoms B, C, and D and bisects angle ACA. Its mirror image is identical to itself, seen by a 180° rotation of the mirror image about the C—B bond. Hence the model is achiral.

and its mirror image are now *identical,* and the molecule is achiral. This is exactly the situation with 2-chloropropane, where two of the four groups attached to carbon 2 are identical (CH_3, CH_3, H, and Cl).

Notice that the molecule in Figure 5.6 has a plane of symmetry. This plane passes through atoms B, C, and D and bisects the ACA angle. On the other hand, the molecule in Figure 5.4 does *not* have a symmetry plane.

A **plane of symmetry** (sometimes called a mirror plane) is a plane that passes through a molecule (or object) in such a way that what is on one side of

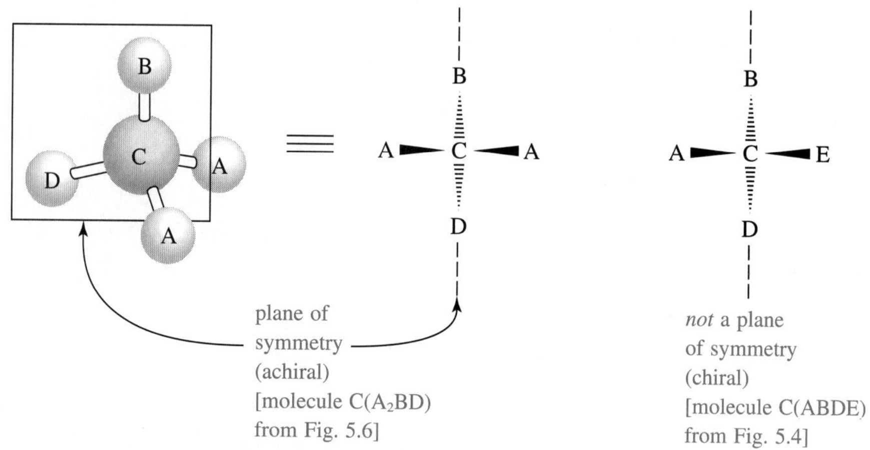

the plane is the exact reflection of what is on the other side. *Any molecule with a plane of symmetry is achiral. Chiral molecules do not have a plane of symmetry.* Seeking a plane of symmetry is usually one quick way to tell whether a molecule is chiral or achiral.

To summarize, a molecule with a stereogenic center (in our examples, the stereogenic center is a carbon atom with four different groups attached to it) can exist in two stereoisomeric forms, that is, as a pair of enantiomers. Such a molecule does not have a symmetry plane. Compounds with a symmetry plane are achiral.

EXAMPLE 5.1 Locate the stereogenic center in 3-methylhexane.

Solution Draw the structure, and look for a carbon atom with four different groups attached.

$$\overset{1}{CH_3}\overset{2}{CH_2}\overset{3}{\underset{\displaystyle CH_3}{\underset{|}{CH}}}\overset{4}{CH_2}\overset{5}{CH_2}\overset{6}{CH_3}$$

All of the carbons except carbon 3 have at least two hydrogens (two identical groups) and therefore cannot be stereogenic centers. But carbon 3 has four different groups attached (H, CH_3—, CH_3CH_2—, and $CH_3CH_2CH_2$—) and is therefore a stereogenic center. By convention, we sometimes mark such centers with an asterisk.

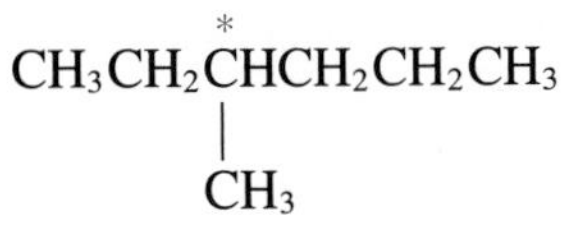

$$CH_3CH_2\overset{*}{C}HCH_2CH_2CH_3$$
$$\qquad\quad |$$
$$\qquad\quad CH_3$$

EXAMPLE 5.2 Draw the two enantiomers of 3-methylhexane.

Solution There are many ways to do this. Here are two of them. First draw carbon 3 with four tetrahedral bonds.

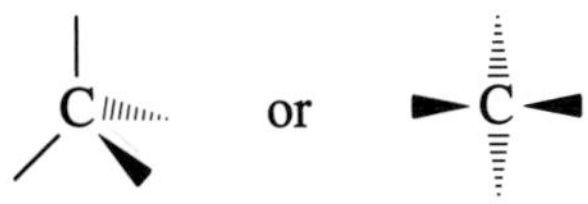

C or C

Then attach the four different groups, in any order.

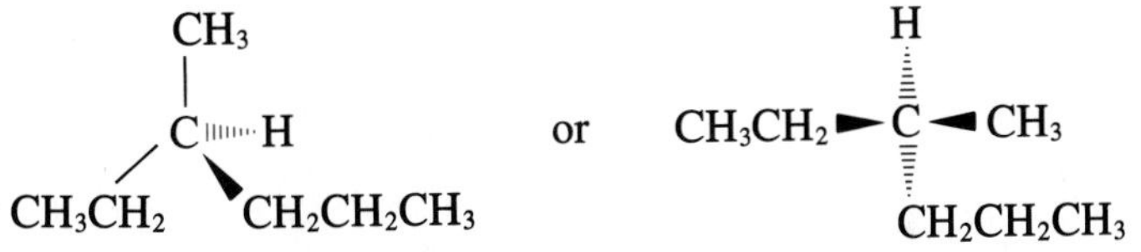

CH₃ above C, C···H, CH₃CH₂ and CH₂CH₂CH₃ below — or — H above, CH₃CH₂►C◄CH₃, CH₂CH₂CH₃ below

Now draw the mirror image, or interchange the positions of any two groups.

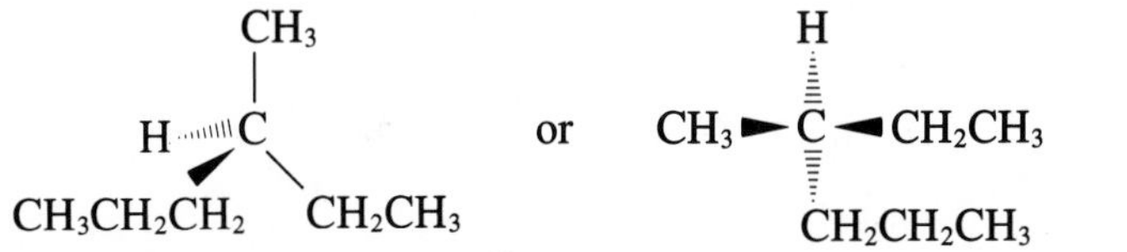

CH₃ above, H···C, CH₃CH₂CH₂ and CH₂CH₃ below — or — H above, CH₃►C◄CH₂CH₃, CH₂CH₂CH₃ below

To convince yourself that the ***interchange of any two groups at a stereogenic center produces the enantiomer,*** work with molecular models.

PROBLEM 5.2 Find the stereogenic centers in

a. 3-iodohexane. b. 2,3-dibromobutane.
c. 3-chlorocyclohexene. d. 1-bromo-1-chloroethane.

PROBLEM 5.3 Which of the following compounds is chiral?

a. 1-bromo-1-phenylethane b. 1-bromo-2-phenylethane

PROBLEM 5.4 Draw three-dimensional structures for the two enantiomers of the chiral compound in Problem 5.3.

PROBLEM 5.5 Locate the planes of symmetry in the eclipsed conformation of ethane. In this conformation, is ethane chiral or achiral?

PROBLEM 5.6 Does the staggered conformation of ethane have planes of symmetry? In this conformation, is ethane chiral or achiral? *(Careful!)*

PROBLEM 5.7 Locate the planes of symmetry in *cis*- and *trans-1,2*- dichloroethene. Are these molecules chiral or achiral? *(Careful!)*

5.4 Configuration and the R-S Convention

Enantiomers differ in the arrangement of the groups attached to the chiral center. This arrangement of groups is called the **configuration** of the chiral center. *Enantiomers are configurational isomers; they are said to have opposite configurations.*

When referring to a particular enantiomer, we would like to be able to specify which configuration we mean without having to draw the structure. A convention for doing this is known as the *R-S* or Cahn-Ingold-Prelog (CIP)* system. Here is how it works.

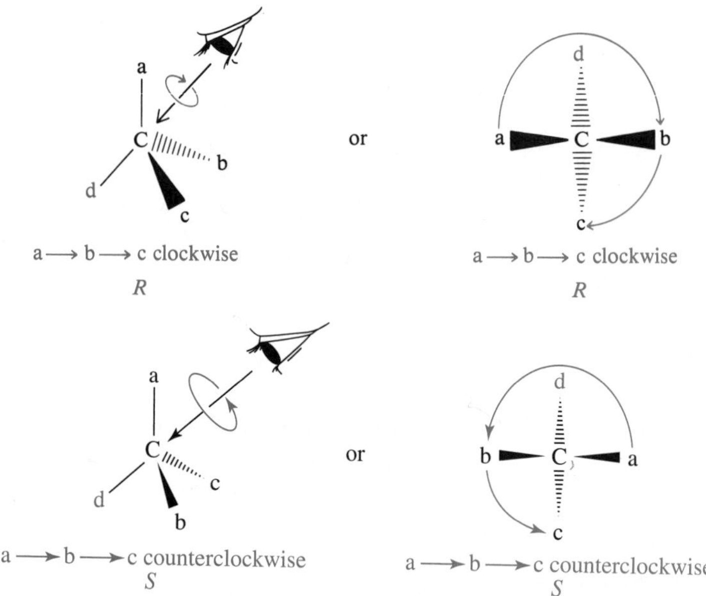

---

*After R. S. Cahn and C. K. Ingold, both British organic chemists, and V. Prelog, a Swiss chemist and Nobel Prize winner.

The four groups attached to the stereogenic center are placed in a priority order (by a system we will describe next), $a \rightarrow b \rightarrow c \rightarrow d$. The stereogenic center is then observed *from the side opposite the lowest priority group, d*. If the remaining three groups ($a \rightarrow b \rightarrow c$) form a clockwise array, the configuration is designated *R* (from the Latin *rectus*, right).* If they form a *counterclockwise* array, the configuration is designated as *S* (from the Latin *sinister*, left).

The priority order of the four groups is set in the following way:

Rule 1. The atoms directly attached to the stereogenic center are ranked according to *atomic number*: the higher the atomic number, the higher the priority.

$$Cl > O > C > H$$

high priority ⟶ low priority

Rule 2. If a decision cannot be reached with rule 1 (that is, if two or more of the directly attached atoms are the same), work outward from the stereogenic center until a decision is reached. For example, the ethyl group has a higher priority than the methyl group, because at the first point of difference, working outward from the stereogenic center, we come to a *carbon* (higher priority) in the ethyl group and a *hydrogen* (lower priority) in the methyl group.

stereogenic center $-CH_2-CH_2-H$ (ethyl) > stereogenic center $-CH_2-H$ (methyl)

ethyl　　methyl

EXAMPLE 5.3 Assign a priority order to the following groups: H, Br, $-CH_2CH_3$, and $-CH_2OCH_3$.

Solution $Br > -CH_2OCH_3 > -CH_2CH_3 > H$

The atomic numbers of the directly attached atoms are ordered Br > C > H.

---

*More precisely, *rectus* means "right" in the sense of "correct, or proper," and not in the sense of direction (which is *dexter* = right, opposite to left). It may not be entirely coincidental that the initials of one of the inventors of this system are R. S.

To prioritize the two carbon groups, we must go out until a point of difference is reached.

$$-CH_2OCH_3 > -CH_2CH_3 \qquad (O > C)$$

PROBLEM 5.8 Assign a priority order to each of the following sets of groups:

a. $—CH_3$, $—CH(CH_3)_2$, $—H$, $—OH$
b. $—OH$, $—F$, $—CH_3$, $—CH_2OH$
c. $—OCH_3$, $—NHCH_3$, $—CH_2NH_2$, $—OH$
d. $—CH_2CH_3$, $—CH_2CH_2CH_3$, $—C(CH_3)_3$, $—CH(CH_3)_2$

For stereogenic centers in cyclic compounds, the same rule for assigning priorities is followed. For example, in 1,1,3-trimethylcyclohexane, the four groups attached to carbon 3 in order of priority are $—CH_2C(CH_3)_2CH_2 >$ $—CH_2CH_2 > —CH_3 > —H$.

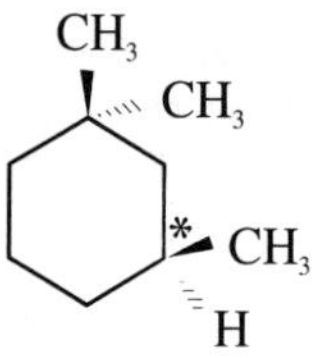

CH3 CH3 * CH3 H

1,1,3-trimethylcyclohexane

A third, somewhat more complicated, rule is required to handle double or triple bonds and aromatic rings (which are written in the Kekulé fashion).

Rule 3. Multiple bonds are treated as if they were an equal number of single bonds. For example, the vinyl group $—CH{=}CH_2$ is counted as

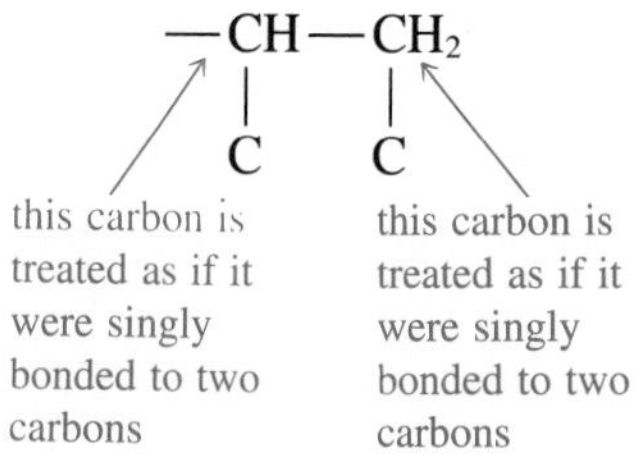

$$—\underset{\underset{\displaystyle C}{|}}{CH}—\underset{\underset{\displaystyle C}{|}}{CH_2}$$

this carbon is treated as if it were singly bonded to two carbons

this carbon is treated as if it were singly bonded to two carbons

Similarly,

$$—C{\equiv}CH \quad \text{is treated as} \quad —\overset{\overset{\displaystyle C}{|}}{\underset{\underset{\displaystyle C}{|}}{C}}—\overset{\overset{\displaystyle C}{|}}{\underset{\underset{\displaystyle C}{|}}{C}}—H$$

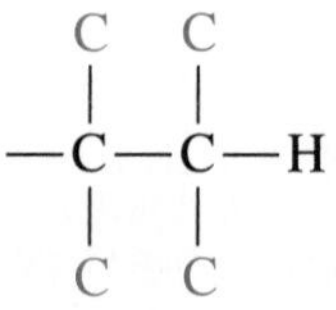

and

$$-\mathrm{CH}{=}\mathrm{O} \quad \text{is treated as} \quad \begin{array}{c} \mathrm{H} \\ | \\ -\mathrm{C}-\mathrm{O} \\ |\quad\;\; | \\ \mathrm{O}\quad\; \mathrm{C} \end{array}$$

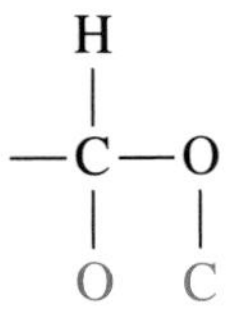

EXAMPLE 5.4 Which group has the higher priority, isopropyl or vinyl?

Solution The vinyl group has the higher priority. We go out until we reach a difference, shown in color.

$$\underset{\text{vinyl}}{-\mathrm{CH}{=}\mathrm{CH_2}} \equiv \begin{array}{c} -\mathrm{CH}-\mathrm{CH_2} \\ |\qquad\;\; | \\ \mathrm{C}\qquad\;\; \mathrm{C} \end{array}$$

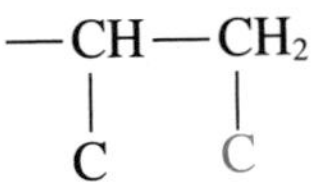

$$\underset{\text{isopropyl}}{-\mathrm{CH(CH_3)_2}} \equiv \begin{array}{c} -\mathrm{CH}-\mathrm{CH_2} \\ |\qquad\;\; | \\ \mathrm{CH_3}\quad\;\; \mathrm{H} \end{array}$$

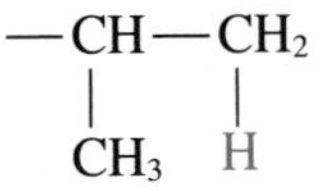

PROBLEM 5.9 Assign a priority order to

a. $-\mathrm{C}{\equiv}\mathrm{CH}$ and $-\mathrm{CH}{=}\mathrm{CH_2}$ b. $-\mathrm{CH}{=}\mathrm{CH_2}$ and $-\mathrm{C_6H_5}$ (phenyl)

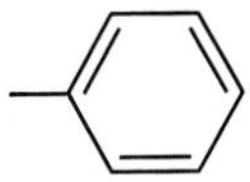

c. $-\mathrm{CH}{=}\mathrm{O}$, $-\mathrm{CH}{=}\mathrm{CH_2}$, $-\mathrm{CH_2CH_3}$, and $-\mathrm{CH_2OH}$

Now let us see how these rules are applied.

EXAMPLE 5.5 Assign the configuration (*R* or *S*) to the following enantiomer of 3-methylhexane (see Example 5.2).

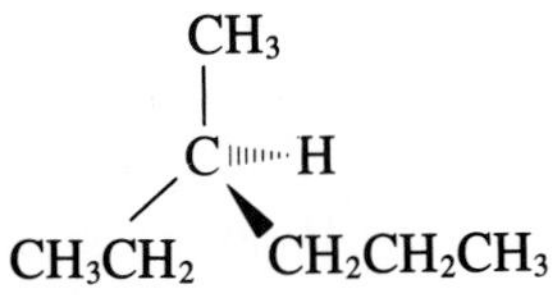

$$\begin{array}{c} \mathrm{CH_3} \\ | \\ \mathrm{C}\cdots\mathrm{H} \\ \swarrow\quad\searrow \\ \mathrm{CH_3CH_2}\quad \mathrm{CH_2CH_2CH_3} \end{array}$$

Solution First assign the priority order to the four different groups attached to the stereogenic center.

$$-\mathrm{CH_2CH_2CH_3} > -\mathrm{CH_2CH_3} > -\mathrm{CH_3} > -\mathrm{H}$$

Now view the molecule *from the side opposite the lowest priority group* (−H) and determine whether the remaining three groups, from high to low priority, form a clockwise (*R*) or counterclockwise (*S*) array.

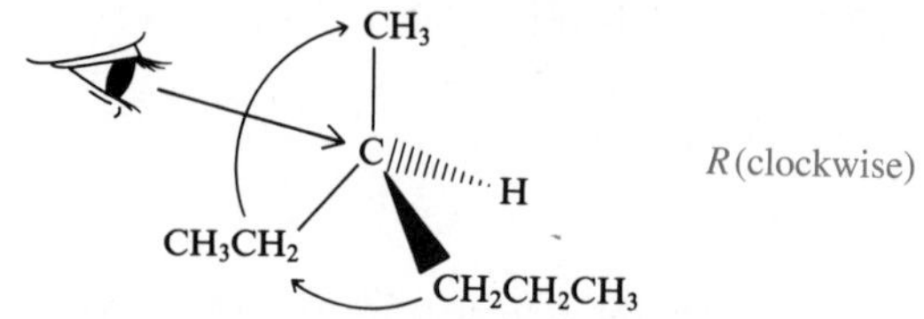

R (clockwise)

We write the name (*R*)-3-methylhexane.
If we view the other representation of this molecule shown in Example 5.2, we come to the same conclusion.

H

CH_3CH_2—C—CH_3

$CH_2CH_2CH_3$

view down the C—H bond; the configuration is *R*

PROBLEM 5.10 Determine the configuration (*R* or *S*) at the stereogenic center in

a. CH=O, H, CH_3, OH

b. H_2N—C—(phenyl), H, CH_3

EXAMPLE 5.6 Draw the structure of (*R*)-2-bromobutane.

Solution First, write out the structure and prioritize the groups attached to the stereogenic center.

$CH_3\overset{*}{C}HCH_2CH_3$
|
Br

Br > CH_3CH_2— > CH_3— > H

Now make the drawing with the H (lowest priority group) "away" from you, and place the three remaining groups (Br → CH_3CH_2 → CH_3) in a clockwise (*R*) array.

C—H → Br, C—H, CH_3, CH_2CH_3 or —C— (H) → CH_3—C—Br (H, CH_2CH_3)

Of course, we could have started with the top-priority group at either of the other two bonds to give the following structures, which are equivalent to those above:

CH_3 | C(H)(Br)(CH_3CH_2) or CH_2CH_3 | C(H)(CH_3)(Br) and CH_3CH_2►C◄CH_3 (H above, Br below) or Br►C◄CH_2CH_3 (H above, CH_3 below)

PROBLEM 5.11 Draw the structure of

a. (*S*)-2-phenylbutane.
b. (*R*)-3-methyl-1-pentene.
c. (*R*)-3-methylcyclopentene.

5.5 The E-Z Convention for Cis-Trans *Isomers*

Before we continue with other aspects of chirality, let us digress briefly to describe a useful extension of the Cahn-Ingold-Prelog system of nomenclature to *cis–trans* isomers. Sometimes *cis–trans* nomenclature is ambiguous, as in the following examples:

F(Cl)C=C(Br)(I) — *cis* or *trans?*

CH_3CH_2(CH_3)C=C(Cl)(Br) — *cis* or *trans?*

The system we have just discussed for stereogenic centers has been extended to double-bond isomers. We use exactly the same priority rules. *The two groups attached to each carbon of the double bond are assigned priorities.* If the two higher priority groups are on *opposite* sides of the double bond, the prefix *E* (from the German *entgegen,* opposite) is used. If the two higher priority groups are on the *same* side of the double bond, the prefix is *Z* (from the German *zusammen,* together). The higher priority groups for the above examples are shown here in color, and the correct names are given below the structures.

F(Cl)C=C(Br)(I) — (*Z*)-1-bromo-2-chloro-2-fluoro-1-iodoethene

CH_3CH_2(CH_3)C=C(Cl)(Br) — (*E*)-1-bromo-1-chloro-2-methyl-1-butene

PROBLEM 5.12 Name each compound by the *E-Z* system.

a. CH_3(H)C=C(H)(CH_2CH_3)

b. F(Br)C=C(H)(Cl)

PROBLEM 5.13 Write the structure for

a. (Z)-2-pentene. b. (E)-1,3-pentadiene.

5.6 Polarized Light and Optical Activity

The concept of molecular chirality follows logically from the tetrahedral geometry of carbon, as developed in Secs. 5.2 and 5.3. Historically, however, these concepts were developed in the reverse order; how this happened is one of the most elegant and logically beautiful stories in the history of science. The story began in the early eighteenth century with the discovery of polarized light and with studies on how molecules placed in the path of such a light beam affect it.

An ordinary light beam consists of waves vibrating in all possible planes perpendicular to its path. However, if this light beam is passed through certain types of substances, the transmitted beam will have all of its waves vibrating in parallel planes. Such a light beam, said to be **plane-polarized,** is illustrated in Figure 5.7. One convenient way to polarize light is to pass it through a device composed of Iceland spar (crystalline calcium carbonate) called a **Nicol prism** (invented in 1828 by the British physicist William Nicol). A more recently developed polarizing material is **Polaroid,** which was invented by E. H. Land, an American. It contains a crystalline organic compound properly oriented and embedded in a transparent plastic. Sunglasses, for example, are often made from Polaroid.

A light beam will pass through *two* samples of polarizing material only if their polarizing axes are aligned. If the axes are perpendicular, no light will pass through. This result, illustrated in Figure 5.8, is the basis of an instrument used to study the effect of various substances on plane-polarized light.

A **polarimeter** is shown schematically in Figure 5.9. Here is how it works. With the light on and the sample tube empty, the analyzer prism is rotated so that the light beam that has been polarized by the polarizing prism is com-

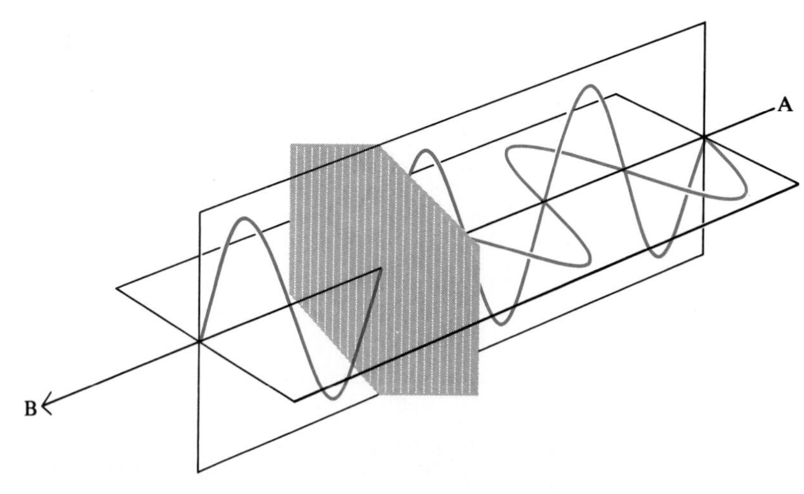

FIGURE 5.7 A beam of light, AB, initially vibrating in all directions, passes through a polarizing substance that "strains" the light so that only the vertical component emerges.

FIGURE 5.8 The two sheets of polarizing material shown have their axes aligned perpendicularly. Although each disk alone is almost transparent, the area where they overlap is opaque. You can duplicate this effect using two pairs of Polaroid sunglasses. (Courtesy of the Polaroid Corporation.)

pletely blocked and the field of view is dark. At this point, the prism axes of the polarizer prism and the analyzer prism are perpendicular to one another. Now the sample is placed in the sample tube. If the substance is **optically inactive,** nothing changes. The field of view remains dark. But if an **optically active** substance is placed in the tube, it rotates the plane of polarization, and some light passes through the analyzer to the observer. By turning the analyzer prism clockwise or counterclockwise, the observer can again block the light beam and restore the dark field.

The angle through which the analyzer prism must be rotated in this experiment is called α, the **observed rotation.** It is equal to the number of degrees that the optically active substance rotated the beam of plane-polarized light. If the analyzer must be rotated to the *right* (clockwise), the optically active substance is said to be **dextrorotatory** (+); if rotated to the *left* (counterclockwise), the substance is **levorotatory** (−).*

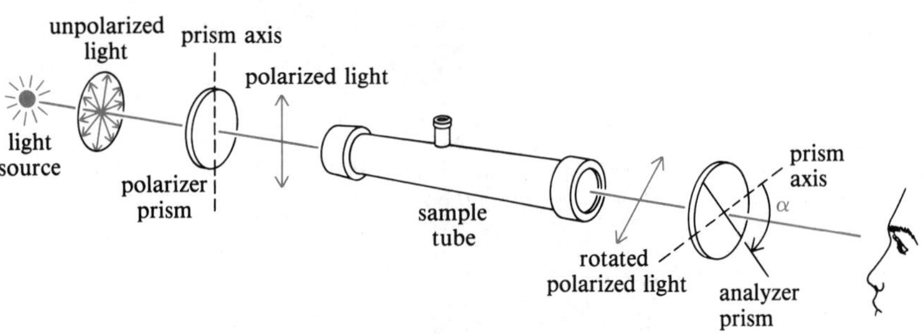

FIGURE 5.9 Diagram of a polarimeter.

* It is not possible to tell from a single measurement whether a rotation is + or −. For example, is a reading +10° or −350°? We can distinguish between these alternatives by, for example, increasing the sample concentration by 10%. Then a + 10° reading would change to +11°, and a −350° reading would change to −385° (that is, −35°).

The observed rotation, α, of a sample of an optically active substance depends on its molecular structure and also on the number of molecules in the sample tube, the length of the tube, the wavelength of the polarized light, and the temperature. All of these have to be standardized if we want to compare the optical activity of different substances. The **specific rotation** $[\alpha]$, of an optically active substance is defined as follows:

$$\text{Specific rotation} = [\alpha]_{\lambda}{}^{t} = \frac{\alpha}{l \times c}(\text{solvent})$$

where l is the length of the sample tube in *decimeters, c* is the concentration in *grams per milliliter, t* is the temperature of the solution, and λ is the wavelength of light. The solvent used is indicated in parentheses. Measurements are usually made at room temperature, and the most common light source is the D-line of a sodium vapor lamp (λ = 589.3 nm), although modern instruments called **spectropolarimeters** allow the light wavelength to be varied at will. The specific rotation of an optically active substance at a particular wavelength is as definite a property of the substance as its melting point, boiling point, or density.

PROBLEM 5.14 Camphor is optically active. A camphor sample (1.5 g) dissolved in ethanol (optically inactive) to a total volume of 50 mL, placed in a 5-cm polarimeter sample tube, gives an observed rotation of +0.66° at 20°C (using the sodium D-line). Calculate and express the specific rotation of camphor.

In the early nineteenth century, the French physicist Jean Baptiste Biot (1774–1862) studied the behavior of a great many substances in a polarimeter. Some, such as turpentine, lemon oil, solutions of camphor in alcohol, and solutions of cane sugar in water, were optically active. Others, such as water, alcohol, and solutions of salt in water, were optically inactive. Later, many natural products (carbohydrates, proteins, and steroids, to name just a few) were added to the list of optically active compounds. What is it about the structure of molecules that causes some to be optically active and others inactive?

When plane-polarized light passes through a single molecule, the light and the electrons in the molecule interact. This interaction causes the plane of polarization to rotate slightly.* But when we place a substance in a polarimeter, *we do not place a single molecule there, we place a large collection of molecules there* (recall that even as little as a thousandth of a mole contains 6×10^{20} molecules).

Now, if the substance is achiral, then for every single molecule in one orientation that rotates the plane of polarization in one direction, there is apt to be another molecule with the mirror-image orientation that will rotate the plane

---

*This is because the electric and magnetic fields that result from electronic motions in the molecule affect the electric and magnetic fields of the light.

of polarization an equal amount in the opposite direction. The net result is such that the light beam passes through a sample of achiral molecules without any net change in the plane of polarization. *Achiral molecules are optically inactive.*

But for chiral molecules the situation is different. Consider a sample containing one enantiomer (say, *R*) of a chiral molecule. For any molecule with a given orientation in the sample, *there can be no mirror-image orientation* (because the mirror image gives a different molecule, the *S* enantiomer). Therefore, the rotation in the polarization plane caused by one molecule is *not* cancelled by any other molecule, and the light beam passes through the sample with a net change in the plane of polarization. *Chiral molecules are optically active.*

A Word About . . .

8. Pasteur's Experiments and the van't Hoff-LeBel Explanation

The great French scientist Louis Pasteur (1822–1895) was the first to recognize that optical activity is related to what we now call chirality. He realized that similar molecules that rotate plane-polarized light through equal angles but in opposite directions must be related to one another as an object and its nonsuperimposable mirror image (that is, as a pair of enantiomers). Here is how he came to that conclusion.

Working in the mid-nineteenth century in a country famous for its wine industry, Pasteur was aware of two *isomeric* acids that deposit in wine casks during fermentation. One of these, called **tartaric acid,** was optically active and dextrorotatory. The other, at the time called **racemic acid,** was optically inactive.

Pasteur prepared various salts of these acids. He noticed that *crystals* of the sodium ammonium salt of *tartaric acid* were *not* symmetric (that is, they were not identical to their mirror images). In other words, they exhibited the property of handedness (chirality). Let us say that the crystals were all left-handed.

When Pasteur next examined crystals of the same salt of *racemic acid,* he found that they, too, were chiral but that some of the crystals were left-handed and others were right-handed. The crystals were related to one another as an object and its nonsuperimposable mirror image, and they were present in equal amounts. With a magnifying lens and a pair of tweezers, Pasteur carefully separated these crystals into two piles: the left-handed ones and the right-handed ones.

Then Pasteur made a crucial observation. When he *dissolved* the two types of crystals *separately* in water and placed the solutions in a polarimeter, he found that each solution was optically active (remember, he obtained these crystals from racemic acid, which was optically inactive). One solution had a specific rotation identical to that of the sodium ammonium salt of tartaric acid! The other had an equal but *opposite* specific rotation. This meant that it must be the mirror image, or levorotatory tartaric acid. Pasteur correctly concluded that racemic acid was not really a single substance, but a 50:50 mixture of (+) and (−)-tartaric acids. Racemic acid was optically inactive because it contained equal amounts of two enantiomers. *We define a* ***racemic mixture*** *as a 50:50 mixture of enantiomers,* and of course, such a mixture is optically inactive because the rotations of the two enantiomers cancel out.

Pasteur recognized that optical activity must be due to some property of tartaric acid molecules themselves, not to some property of the crystals, because the crystalline shape was lost when the crystals were dissolved in water in order to measure the specific rotation. However, the precise explanation in terms of molecular structure eluded Pasteur and was not to come for another 25 years.

Pasteur's experiments were performed at about the same time that Kekulé in Germany was developing his theories about organic structures. Kekulé recognized that carbon is tetravalent, and there is even a hint in some of his writings (about 1867) and also in the writings of the Russian chemist A. M. Butlerov (1862) and the Italian E. Paterno (1869) that carbon might be tetrahedral. But it was not until 1874 that the Dutch physical chemist J. H. van't Hoff (1852–1911) and the Frenchman J. A. LeBel (1874–1930) simultaneously but quite independently made a bold hypothesis about carbon that would explain the optical activity of some organic molecules and the optical inactivity of others.

These scientists knew their solid geometry. They knew that four different objects can be arranged in two different ways at the corners of a tetrahedron, and that these two ways are related to one another as an object and its nonsuperimposable mirror image. They also knew that this arrangement resulted in right- and left-handedness, as shown in the drawing.

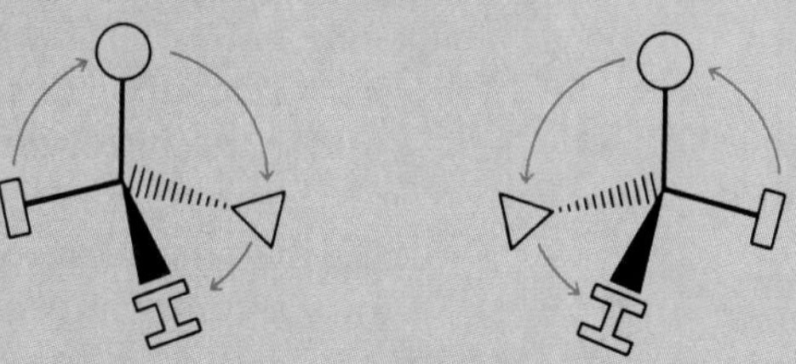

They made the bold hypothesis that the four valences of carbon were directed toward the corners of a tetrahedron, and that optically active molecules would contain at least one carbon atom with four different groups attached to it. This idea would explain why Pasteur's (+) and (−)-tartaric acids rotated plane-polarized light to equal extents but in opposite (right- and left-handed) directions. Optically inactive organic substances would either contain no asymmetric carbon atom or be a 50:50 mixture of enantiomers.

This, then, is how the tetrahedral geometry of carbon was first recognized.* The logic that led to the proposal is admirable, especially when one realizes that neither the electron nor the nucleus of an atom had yet been discovered and that almost nothing was then known about the physical nature of chemical bonds. At the time they made their proposal, van't Hoff and LeBel were relatively unknown chemists. Their hypothesis was ridiculed by at least one establishment chemist at the time, but it was soon generally accepted, has survived many tests, and can now be regarded as fact.

5.7 Properties of Enantiomers

In what properties do enantiomers differ from one another? Enantiomers differ only with respect to chirality. In all other respects they are identical. For this reason, they differ from one another *only in properties that are also chiral.* Let us illustrate this idea first with familiar objects.

A left-handed baseball player (chiral) can use the same ball (achiral) as can a right-handed player. But of course a left-handed player (chiral) can use only a

*Actually, LeBel developed his ideas based on symmetry considerations, whereas van't Hoff based his (at least at first) on the idea of an asymmetric carbon atom. For a stimulating article on these differences, see R. B. Grossman, *J. Chem. Educ.* **1988,** *66,* 30–33.

left-handed baseball glove (chiral). A bolt with a right-handed thread (chiral) can use the same washer (achiral) as a bolt with a left-handed thread, but it can only fit into a nut (chiral) with a right-handed thread. To generalize, *the chirality of an object is most significant when the object interacts with another chiral object.*

Enantiomers have identical achiral properties, such as melting point, boiling point, density, and various types of spectra. Their solubilities in an ordinary, achiral solvent are also identical. However, *enantiomers have different chiral properties,* one of which is the *direction* in which they rotate plane-polarized light (clockwise or counterclockwise). Although enantiomers rotate plane-polarized light in opposite directions, they have specific rotations of the same magnitude (but with opposite signs), because the *number of degrees* is not a chiral property. Only the *direction* of rotation is a chiral property. Here is a specific example.

Lactic acid is an optically active hydroxyacid that is important in several biological processes. It has one stereogenic center. Its structure and some of its properties are shown in Figure 5.10. Note that both enantiomers have identical melting points and, *except for sign,* identical specific rotations.

A 50 : 50 mixture of enantiomers, called a **racemic mixture,** is optically inactive because the rotations of the two enantiomers cancel out.

There is no obvious relationship between configuration (R or S) and sign of rotation (+ or −). For example, (*R*)-lactic acid is levorotatory. When (*R*)-lactic acid is converted to its methyl ester (eq. 5.1), the configuration is unchanged because none of the bonds to the chiral carbon is involved in the reaction. Yet the sign of rotation of the product, a physical property, changes from − to +.

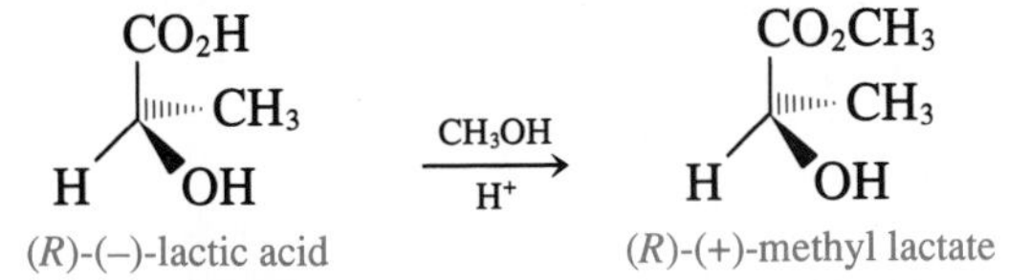

CO_2H / CH_3 / H / OH — (*R*)-(−)-lactic acid $\xrightarrow[H^+]{CH_3OH}$ CO_2CH_3 / CH_3 / H / OH — (*R*)-(+)-methyl lactate (5.1)

Enantiomers often have different biological properties because the biological property usually involves a reaction with another chiral molecule. For example, the enzyme *lactic acid dehydrogenase* will oxidize (+)-lactic acid to pyruvic acid, but it will *not* oxidize (−)-lactic acid.

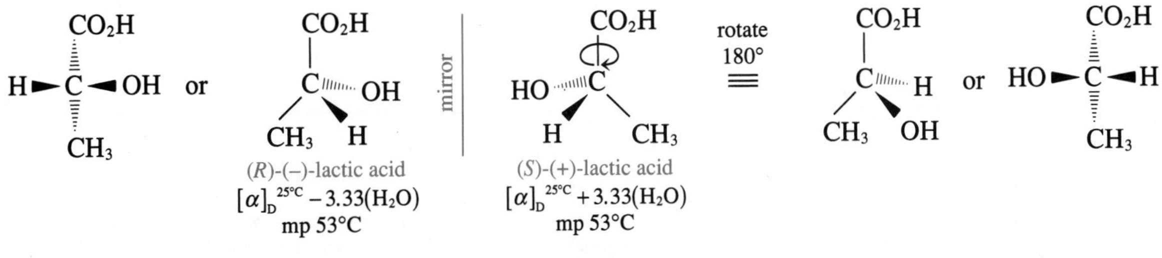

CO_2H, H, C, OH, CH_3 or CO_2H, C, OH, CH_3, H — (*R*)-(−)-lactic acid, $[\alpha]_D^{25°C}$ −3.33(H_2O), mp 53°C

mirror

CO_2H, HO, C, H, CH_3 — (*S*)-(+)-lactic acid, $[\alpha]_D^{25°C}$ +3.33(H_2O), mp 53°C

rotate 180° ≡ CO_2H, C, H, CH_3, OH or CO_2H, HO, C, H, CH_3

FIGURE 5.10 The structures and properties of the lactic acid enantiomers.

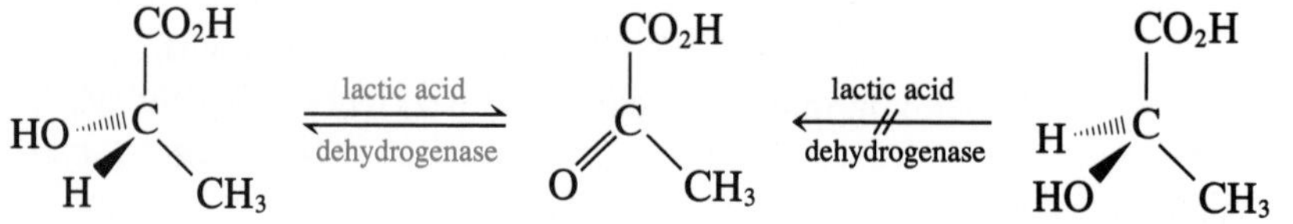

$$\text{(+)-lactic acid} \underset{\text{dehydrogenase}}{\overset{\text{lactic acid}}{\rightleftharpoons}} \text{pyruvic acid} \underset{\text{dehydrogenase}}{\overset{\text{lactic acid}}{\not\leftarrow}} \text{(–)-lactic acid} \qquad (5.2)$$

Why? The enzyme itself is chiral and can distinguish between right- and left-handed lactic acid molecules.

Enantiomers differ in many types of biological activity. One enantiomer may be a drug, whereas the other enantiomer may be ineffective. For example, only (−)-adrenalin is a cardiac stimulant; (+)-adrenalin is ineffective. One enantiomer may be toxic, another harmless. One may be an antibiotic, the other useless. One may be an insect sex attractant, the other without effect or actually a repellant. Chirality is of paramount importance in the biological world.

5.8 Fischer Projection Formulas

Instead of using dashed and solid wedges to show the three-dimensional arrangements of groups in a chiral molecule, it is convenient to have a two-dimensional way of doing so. A useful way to do this was devised many years ago by Emil Fischer*; the formulas are called **Fischer projections.**

Consider the formula for (*R*)-lactic acid, to the left of the mirror in Figure 5.10. If we project that three-dimensional formula onto a plane, as illustrated in Figure 5.11, we obtain the flattened Fischer projection formula.

There are two important things to notice about Fischer projection formulas. First, the C for the stereogenic carbon atom is omitted and is represented simply as the crossing point of the horizontal and vertical lines. Second, horizon-

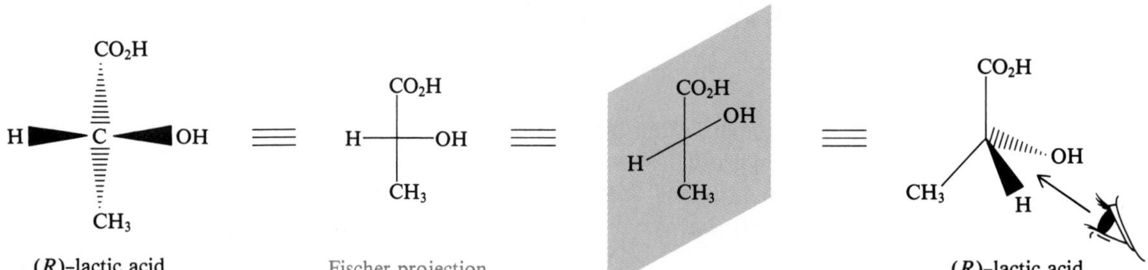

(*R*)-lactic acid

Fischer projection formula of (*R*)-lactic acid

(*R*)-lactic acid

FIGURE 5.11 Projecting the model at the right onto a plane gives the Fischer projection formula.

* Emil Fischer (1852–1919), who devised these formulas, was one of the early giants in the field of organic chemistry. He did much to elucidate the structures of carbohydrates, proteins, and other natural products and received the 1902 Nobel Prize in chemistry.

tal lines connect the stereogenic center to groups that project *above* the plane of the page, *toward* the viewer; vertical lines lead to groups that project *below* the plane of the page, *away* from the viewer.

PROBLEM 5.15 Draw a Fischer projection formula for (*S*)-lactic acid.

Because of the manner in which they are defined, Fischer projections can only be manipulated in certain allowed ways and *still maintain their stereochemical integrity* (that is, *not* be changed to the enantiomer).

1. Fischer projections may be *turned 180° in the plane of the paper* (but *not* 90°).

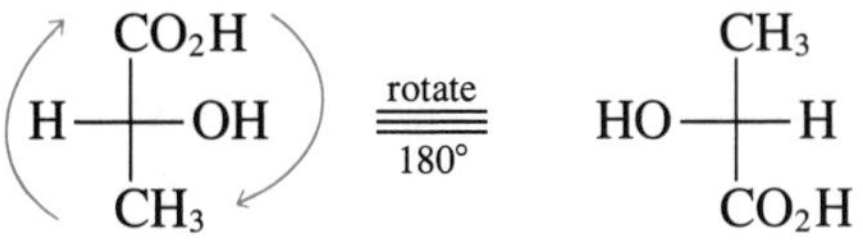

CO_2H / H—+—OH / CH_3 (*R*)-lactic acid $\xrightarrow{\text{rotate } 180°}$ CH_3 / HO—+—H / CO_2H (*R*)-lactic acid

Since the direction in which horizontal and vertical lines project is strictly defined, rotation by 90° in the plane of the paper gives a projection that represents the enantiomer.

CO_2H / H—+—OH / CH_3 (*R*)-lactic acid $\xrightarrow{\text{rotate 90° to the right}}$ H / CH_3—+—CO_2H / OH (*S*)-lactic acid

mirror images

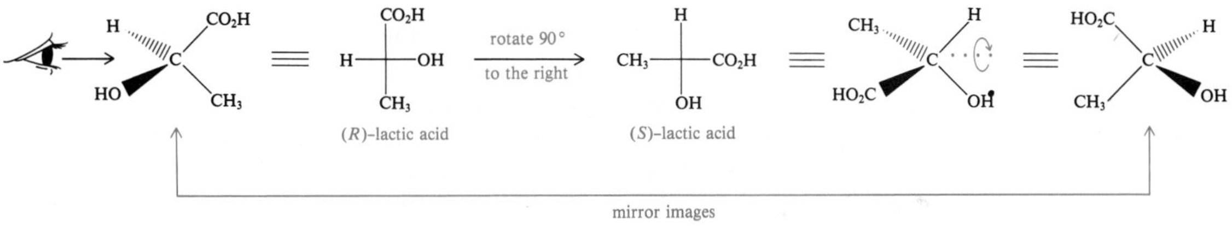

For the same reason, lifting a Fischer projection out of the plane of the paper and turning it over also gives a projection that represents the enantiomer:

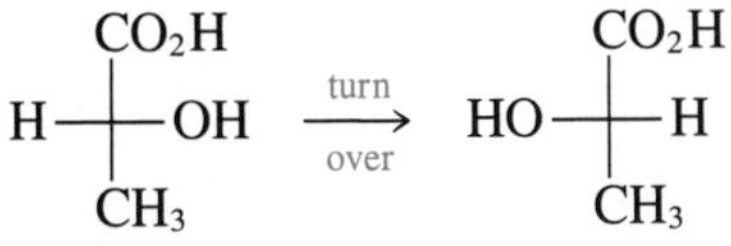

CO_2H / H—+—OH / CH_3 (*R*)-lactic acid $\xrightarrow{\text{turn over}}$ CO_2H / HO—+—H / CH_3 (*S*)-lactic acid

2. If any one group is held fixed, the remaining three groups to a stereogenic center in a Fischer projection can be rotated either clockwise or counterclockwise without changing the configuration. For example,

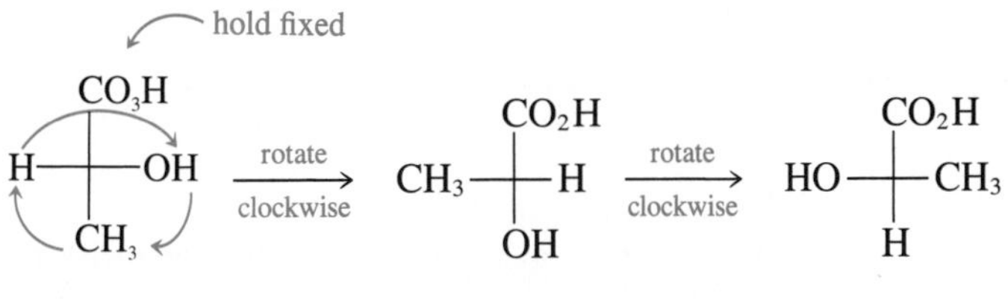

hold fixed

CO_2H / H—+—OH / CH_3 (*R*)-lactic acid $\xrightarrow{\text{rotate clockwise}}$ CO_2H / CH_3—+—H / OH (*R*)-lactic acid $\xrightarrow{\text{rotate clockwise}}$ CO_2H / HO—+—CH_3 / H (*R*)-lactic acid

This process is simply equivalent to rotating the stereogenic center around the C—CO_2H single bond.

$$\underset{CH_3}{\overset{CO_2H}{H \blacktriangleright C \blacktriangleleft OH}} \underset{\text{counterclockwise}}{\overset{\text{clockwise}}{\rightleftharpoons}} \underset{OH}{\overset{CO_2H}{CH_3 \blacktriangleright C \blacktriangleleft H}} \underset{\text{counterclockwise}}{\overset{\text{clockwise}}{\rightleftharpoons}} \underset{H}{\overset{CO_2H}{HO \blacktriangleright C \blacktriangleleft CH_3}}$$

PROBLEM 5.16 Show that the Fischer projections

$$\underset{CH_3}{\overset{H}{HO \!-\!\!\!+\!\!\!-\! CO_2H}} \quad \text{and} \quad \underset{CO_2H}{\overset{CH_3}{HO \!-\!\!\!+\!\!\!-\! H}}$$

can be derived from $\underset{CH_3}{\overset{CO_2H}{H \!-\!\!\!+\!\!\!-\! OH}}$ by allowed motions and that all represent (*R*)-lactic acid.

It is easy to convert from the Fischer to the *R-S* convention. *First,* assign the four groups attached to the stereogenic center a priority order in the usual way. *Next,* by allowed motions for Fischer projection formulas, locate the *lowest* priority group either at the *top* or at the *bottom* of the Fischer formula. *Finally,* determine the clockwise (*R*) or counterclockwise (*S*) direction of the remaining groups as you proceed from priority $a \rightarrow b \rightarrow c$.

EXAMPLE 5.7 What is the configuration of the following, *R* or *S?*

$$\underset{CH_2CH_3}{\overset{CH_3}{H \!-\!\!\!+\!\!\!-\! OH}}$$

Solution First, assign priority orders

$$\underset{(b)\,CH_2CH_3}{\overset{(c)\,CH_3}{(d)\,H \!-\!\!\!+\!\!\!-\! OH\,(a)}}$$

Next, holding the ethyl group fixed, rotate the remaining three groups one turn clockwise, so as to bring the lowest priority group (H) to the top.

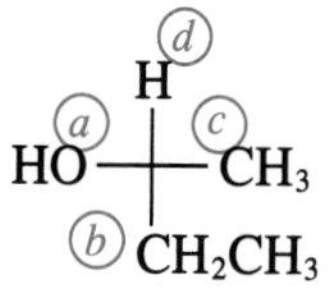

$$\begin{array}{c} \textcircled{d} \\ H \\ \textcircled{a}\,HO-\!\!\!+\!\!\!-CH_3\,\textcircled{c} \\ \textcircled{b}\,CH_2CH_3 \end{array}$$

Now observe that the three top-priority groups are arranged counterclockwise, so the configuration is *S*.

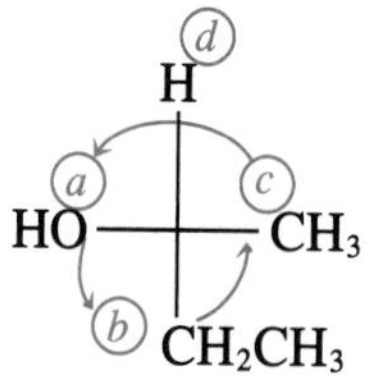

$$\begin{array}{c} \textcircled{d} \\ H \\ \textcircled{a}\,HO-\!\!\!+\!\!\!-CH_3\,\textcircled{c} \\ \textcircled{b}\,CH_2CH_3 \end{array}$$

PROBLEM 5.17 Show that you get the same result as in Example 5.7 by holding the methyl group fixed and rotating the three remaining groups so that the hydrogen is at the bottom of the Fischer formula.*

Fischer projection formulas are especially useful in dealing with compounds containing more than one stereogenic center, as we will now see.

5.9 Compounds with More Than One Stereogenic Center; Diastereomers

Compounds may have more than one stereogenic center, so it is important to be able to determine how many isomers exist in such cases and how they are related to one another. Consider the molecule 2-bromo-3-chlorobutane.

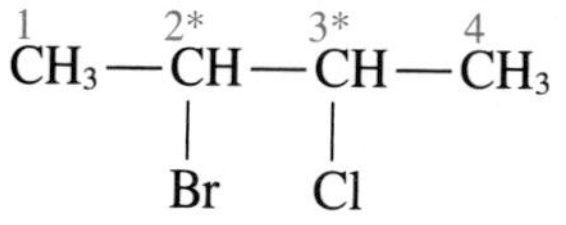

$$\overset{1}{CH_3}-\underset{Br}{\overset{2*}{CH}}-\underset{Cl}{\overset{3*}{CH}}-\overset{4}{CH_3}$$

2-bromo-3-chlorobutane

As indicated by the asterisks, the molecule has two stereogenic centers. Each of these could have the configuration *R* or *S*. Thus, four isomers in all are possible: (2*R*,3*R*), (2*S*,3*S*), (2*R*,3*S*) and (2*S*,3*R*). We can draw these four isomers as shown in Figure 5.12. Note that there are two pairs of enantiomers. The (2*R*,3*R*) and (2*S*,3*S*) forms are nonsuperimposable mirror images, and the (2*R*,3*S*) and (2*S*,3*R*) forms are another such pair.

Let us see how to use Fischer projection formulas for these molecules. Consider the (2*R*,3*R*) isomer, the one at the left in Figure 5.12. The solid-dashed

---

* There is an even faster way to get to the R-S assignment. First assign the priorities. If the lowest priority group is in a *vertical* position (top or bottom), then a clockwise arrangement of the remaining groups in priority order $a \rightarrow b \rightarrow c$ signifies *R* (and counterclockwise, *S*). *But*, if the lowest priority group is in a *horizontal* position (left or right), then the reverse directions apply; a clockwise arrangement of the remaining groups $a \rightarrow b \rightarrow c$ signifies *S* (and counterclockwise, *R*).

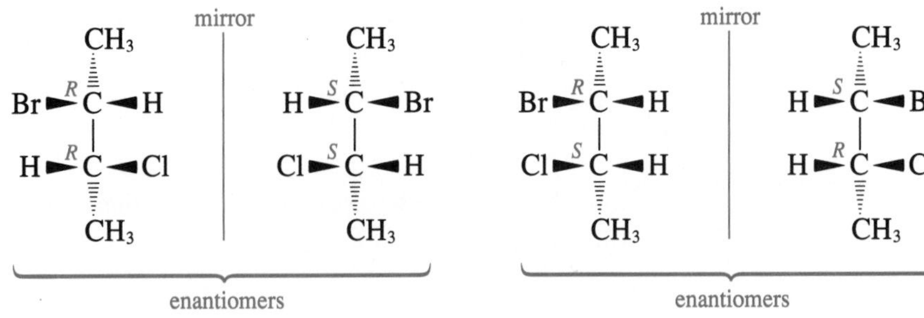

FIGURE 5.12 The four stereoisomers of 2-bromo-3-chlorobutane, a compound with two stereogenic centers.

wedge drawing has horizontal groups projecting out of the plane of the paper toward us and vertical groups going away from us, behind the paper. These facts are expressed in the equivalent Fischer projection formula as shown.*

solid-dashed wedge formula ≡ Fischer projection formula

PROBLEM 5.18 Using the method described in Sec. 5.8, show that the above Fischer projection has the *R* configuration at carbon 2 and at carbon 3.

PROBLEM 5.19 Draw the Fischer projection formulas for the remaining stereoisomers of 2-bromo-3-chlorobutane shown in Figure 5.12, and check the *R-S* assignments.

---

* Notice that these structures are derived from an eclipsed conformation of the molecule, viewed from above so that horizontal groups project toward the viewer. The actual molecule is probably an equilibrium mixture of several staggered conformations, one of which is shown. Fischer formulas are used to represent the correct *configurations,* but not necessarily the lowest energy *conformations* of a molecule.

eclipsed ⇌ staggered

Now we come to an extremely important new idea. Consider the relationship between, for example, the (2*R*,3*R*) and (2*R*,3*S*) forms of the isomers in Figure 5.12. These forms are *not* mirror images because they have the *same* configuration at carbon 2, though they have opposite configurations at carbon 3. They are certainly stereoisomers, but they are not enantiomers. For such pairs of stereoisomers we use the term **diastereomers.** ***Diastereomers are stereoisomers that are not mirror images of one another.***

There is an important, fundamental difference between enantiomers and diastereomers. Because they are mirror images, enantiomers differ *only* in mirror-image (chiral) properties. They have the same achiral properties, such as melting point, boiling point, and solubility in ordinary solvents. Enantiomers cannot be separated from one another by methods that depend on achiral properties, such as recrystallization or distillation. On the other hand, diastereomers are *not* mirror images. They may differ in *all* properties, whether chiral or achiral. As a consequence, diastereomers may differ in melting point, boiling point, solubility, and not only direction but also the number of degrees that they rotate plane-polarized light—in short, they behave as two different chemical substances.

PROBLEM 5.20 How do you expect the specific rotations of the (2*R*,3*R*) and (2*S*,3*S*) forms of 2-bromo-3-chlorobutane to be related? Answer the same question for the (2*R*,3*R*) and (2*S*,3*R*) forms.

Can we generalize about the number of stereoisomers possible when a larger number of stereogenic centers is present? Suppose, for example, that we add a third stereogenic center to the compounds shown in Figure 5.12 (say, 2-bromo-3-chloro-4-iodopentane). The new stereogenic center added to each of the four structures can once again have either an *R* or an *S* configuration, so that with three different stereogenic centers, eight stereoisomers are possible. The situation is summed up in a single rule: *If a molecule has* n *different stereogenic centers, it may exist in a maximum of* 2^n *stereoisomeric forms. There will be a maximum of* $2^n/2$ *pairs of enantiomers.*

PROBLEM 5.21 The Fischer projection formula for glucose (blood sugar) is

CH=O
H—|—OH
HO—|—H
H—|—OH
H—|—OH
CH_2OH
glucose

Altogether, how many stereoisomers of this sugar are possible?

Actually, the number of isomers predicted by this rule is the *maximum* number possible. Sometimes certain structural features reduce the actual number of isomers. In the next section, we examine a case of this type.

5.10 Meso Compounds; the Stereoisomers of Tartaric Acid

Consider the stereoisomers of 2,3-dichlorobutane. There are two stereogenic centers.

$$\overset{1}{CH_3}-\overset{2*}{\underset{\underset{Cl}{|}}{CH}}-\overset{3*}{\underset{\underset{Cl}{|}}{CH}}-\overset{4}{CH_3}$$

2,3-dichlorobutane

We can write out the stereoisomers just as we did in Figure 5.12; they are shown in Figure 5.13. Once again, the (*R*,*R*) and (*S*,*S*) isomers constitute a pair of nonsuperimposable mirror images, or enantiomers. *However, the other "two" structures, (R,S) and (S,R), in fact, now represent a single compound.* You can easily see this by performing one of the allowed manipulations of the Fischer projection formulas at the right. A 180° rotation in the plane of the paper interconverts these structures; therefore, they are identical. So altogether there are only *three* stereoisomers of 2,3-dichlorobutane, not four. The reason is that each stereogenic center has the *same* four groups attached.

Now let us look more closely at the structures to the right of Figure 5.13. Notice that they have a plane of symmetry that is perpendicular to the plane of the paper and bisects the central C—C bond. The structures are identical, superimposable mirror images and therefore *achiral*. We call such a structure a ***meso* compound.** *A* meso *compound is an achiral diastereomer of a compound with stereogenic centers.* Its stereogenic centers have opposite configurations. *Meso* compounds become possible whenever the stereogenic centers have identical groups attached. Being achiral, *meso* compounds are optically inactive.

Now let us take a look at tartaric acid, the compound whose optical activity was so carefully studied by Louis Pasteur (see "A Word About" on page 161). It has two identical stereogenic centers.

$$HO-\overset{\overset{O}{\|}}{C}-\overset{*}{\underset{\underset{OH}{|}}{CH}}-\overset{*}{\underset{\underset{OH}{|}}{CH}}-\overset{\overset{O}{\|}}{C}-OH$$

tartaric acid

The structures of these three stereoisomers and two of their properties are shown in Figure 5.14. Note that the enantiomers have identical properties ex-

CH₃	CH₃	CH₃		CH₃
Cl —(R)— H	H —(S)— Cl	Cl —(R)— H	≡	H —(S)— Cl
H —(R)— Cl	Cl —(S)— H	Cl —(S)— H		H —(R)— Cl
CH₃	CH₃	CH₃		CH₃

enantiomers, chiral (left pair); identical, achiral, a *meso* form (right pair); plane of symmetry (dashed line between the two stereocenters of the right pair)

FIGURE 5.13 Fischer projections of the stereoisomers of 2,3-dichlorobutane.

FIGURE 5.14 The stereoisomers of tartaric acid.

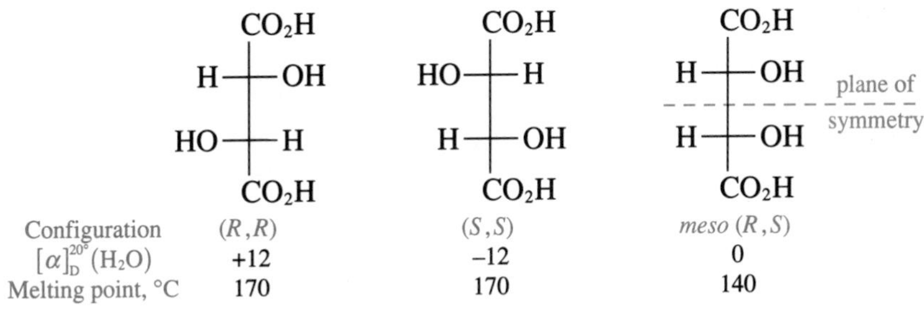

cept for the *sign* of the specific rotation, whereas the *meso* form, being a diastereomer of each enantiomer, differs from them in both properties.

For about 100 years after Pasteur's research, it was still not possible to determine the configuration associated with a particular enantiomer of tartaric acid. For example, it was not known whether (+)-tartaric acid had the (*R*,*R*) or the (*S*,*S*) configuration. It was known that (+)-tartaric acid had to have one of these two configurations and that (−)-tartaric acid had to have the opposite configuration, but which isomer had which?

In 1951, the Dutch scientist J. M. Bijvoet developed a special x-ray technique that solved the problem. Using this technique on crystals of the sodium rubidium salt of (+)-tartaric acid, Bijvoet showed that it had the (*R*,*R*) configuration. So this was the tartaric acid studied by Pasteur, and racemic acid was a 50 : 50 mixture of the (*R*,*R*) and (*S*,*S*) isomers. The *meso* form was not studied until later.

Since tartaric acid had been converted chemically into other chiral compounds and these in turn into still others, it became possible as a result of Bijvoet's work to assign **absolute configurations** (that is, the correct *R* or *S* configuration for each stereocenter) to many pairs of enantiomers.

PROBLEM 5.22 Show that *trans*-1,2-dimethylcyclopentane can exist in chiral, enantiomeric forms.

PROBLEM 5.23 Is *cis*-1,2-dimethylcyclopentane chiral or achiral? What stereochemical term can we give to it?

5.11 Stereochemistry; a Recap of Definitions

We have seen here and in Sec. 2.12 that *stereoisomers* can be classified in three different ways. They may be either *conformers* or *configurational isomers;* they may be *chiral* or *achiral;* and they may be *enantiomers* or *diastereomers*.

A
- *Conformers:* interconvertible by rotation about single bonds
- *Configurational Isomers:* not interconvertible by rotation, only by breaking and making bonds

B
- *Chiral:* mirror image not superimposable on itself
- *Achiral:* molecule and mirror image are identical

C
- *Enantiomers:* mirror images; have opposite configurations at all stereogenic centers
- *Diastereomers:* stereoisomers but not mirror images; have same configuration at one or more centers, but differ at the remaining stereogenic centers

Various combinations of these three sets of terms can be applied to any pair of stereoisomers. Here are a few examples:

1. *Cis*- and *trans*-2-butene.

CH_3(H)C=C(CH_3)(H) and CH_3(H)C=C(H)(CH_3)

These isomers are *configurational* (not interconverted by rotation about single bonds), *achiral* (the mirror image of each is superimposable on the original), and *diastereomers* (although they are stereoisomers, they are *not* mirror images of one another; hence they must be diastereomers).

2. Staggered and eclipsed ethane.

(sawhorse formulas) and (sawhorse formulas)

These are *achiral conformers*. They are *diastereomeric conformers* (but without stereogenic centers) because they are not mirror images.

3. (*R*)- and (*S*)-lactic acid.

CH_3, H, HO_2C, OH and CO_2H, H, CH_3, OH

These isomers are *configurational*, each is *chiral*, and they constitute a pair of *enantiomers*.

4. *Meso*- and (*R*,*R*)-tartaric acids.

CO_2H	CO_2H
H—OH	H—OH
H—OH	HO—H
CO_2H	CO_2H
meso	(*R*,*R*)

These isomers are *configurational* and *diastereomers*. One is *achiral*, and the other is *chiral*.

Enantiomers, such as (*R*)- and (*S*)-lactic acid, differ only in chiral properties and therefore cannot be separated by ordinary achiral methods such as distillation or recrystallization. Diastereomers differ in all properties, chiral or achiral. *If* they are also configurational isomers (such as *cis*- and *trans*-2-butene, or *meso*- and *R*,*R*-tartaric acid), they can be separated by ordinary achiral methods, such as distillation or recrystallization. *If,* on the other hand, they are conformers (such as staggered and eclipsed ethane), they may interconvert so readily by bond rotation as to not be separable.

PROBLEM 5.24 Draw the two stereoisomers of 1,3-dimethylcyclobutane, and classify the pair according to the categories listed in A, B, and C above.

5.12 Stereochemistry and Chemical Reactions

How important is stereochemistry in chemical reactions? The answer depends on the nature of the reactants. First, consider the formation of a chiral product from achiral reactants; for example, the addition of hydrogen bromide to 1-butene to give 2-bromobutane in accord with Markovnikov's rule.

$$\underset{\text{1-butene}}{CH_3CH_2CH{=}CH_2} + HBr \longrightarrow \underset{\text{2-bromobutane}}{CH_3CH_2\overset{*}{C}HCH_3 \atop \quad | \atop \quad Br} \tag{5.3}$$

The product has one stereogenic center, marked with an asterisk, but both enantiomers are formed in exactly equal amounts. The product is a racemic mixture. Why? Although this result will be obtained *regardless* of the reaction mechanism, let us consider the generally accepted mechanism.

$$CH_3CH_2CH{=}CH_2 + H^+ \longrightarrow \underset{\text{2-butyl cation}}{CH_3CH_2\overset{+}{C}HCH_3} \xrightarrow{Br^-} CH_3CH_2CHCH_3 \atop \quad\quad | \atop \quad\quad Br \tag{5.4}$$

The intermediate 2-butyl cation obtained by adding a proton to the end carbon is planar, and bromide ion can combine with it from the "top" or "bottom" side with exactly equal probability.

$$CH_3CH_2{-}\overset{+}{C}(H)(CH_3) \;\; \text{(2-butyl cation)} \;\; \xrightarrow{:\ddot{Br}:^-} \begin{cases} \text{top} \rightarrow & (S)\text{-2-bromobutane: } Br\text{ above C, with } CH_3CH_2,\ H,\ CH_3 \\ \text{bottom} \rightarrow & (R)\text{-2-bromobutane: } Br\text{ below C, with } CH_3CH_2,\ H,\ CH_3 \end{cases} \tag{5.5}$$

The product is therefore a racemic mixture, an optically inactive 50 : 50 mixture of the two enantiomers.

We can generalize this result. *When chiral products are obtained from achiral reactants, both enantiomers are formed at the same rates, in equal amounts.*

PROBLEM 5.25 Show that, if the mechanism of addition of HBr to 1-butene involved *no* intermediates, but *simultaneous one-step* addition (in the Markovnikov sense), the product would still be racemic 2-bromobutane.

PROBLEM 5.26 Show that the chlorination of butane at carbon-2 will give a 50 : 50 mixture of enantiomers.

Now consider a different situation, the reaction of a *chiral* molecule with an achiral reagent to create a second stereogenic center. Consider, for example, the addition of HBr to 3-chloro-1-butene.

$$\underset{\text{3-chloro-1-butene}}{CH_3\overset{*}{C}H(Cl)CH{=}CH_2} + HBr \longrightarrow \underset{\text{2-bromo-3-chlorobutane}}{CH_3\overset{*}{C}H(Cl){-}\overset{*}{C}H(Br)CH_3} \quad (5.6)$$

Suppose we start with one pure enantiomer of 3-chloro-1-butene, say, the *R* isomer. What can we say about the stereochemistry of the products? One way to see the answer quickly is to draw Fischer projections.

$$\underset{(R)\text{-3-chloro-1-butene}}{\begin{array}{c} CH_3 \\ Cl \overset{R}{-\!\!+\!\!-} H \\ CH{=}CH_2 \end{array}} \xrightarrow{HBr} \underset{(2R,3R)\text{-2-bromo-3-chlorobutane}}{\begin{array}{c} CH_3 \\ Cl \overset{R}{-\!\!+\!\!-} H \\ H \overset{R}{-\!\!+\!\!-} Br \\ CH_3 \end{array}} + \underset{(2S,3R)\text{-2-bromo-3-chlorobutane}}{\begin{array}{c} CH_3 \\ Cl \overset{R}{-\!\!+\!\!-} H \\ Br \overset{S}{-\!\!+\!\!-} H \\ CH_3 \end{array}} \quad (5.7)$$

The configuration where the chloro substituent is located remains unchanged and *R*, but the new stereogenic center can be either *R* or *S*. Therefore, the products are *diastereomers*. Are they formed in equal amounts? No. Looking at the starting material in eq. 5.7, we can see that it has no plane of symmetry. Approach of the bromine to the double bond from the H side or from the Cl side of the stereogenic center should not occur with equal ease.

We can generalize this result. *Reaction of a chiral reagent with an achiral reagent, when it creates a new stereogenic center, leads to diastereomeric products at different rates and in unequal amounts.*

PROBLEM 5.27 Let us say that the (2*R*,3*R*) and (2*S*,3*R*) products in eq. 5.7 are formed in a 60 : 40 ratio. What products would be formed and in what ratio by adding HBr to pure (*S*)-3-chloro-1-butene? by adding HBr to a racemic mixture of (*R*)- and (*S*)-3-chloro-1-butene?

5.13 Resolution of a Racemic Mixture

We have just seen (eq. 5.5) that, when reaction between two achiral reagents leads to a chiral product, it always gives a racemic (50:50) mixture of enantiomers. Suppose we want to obtain each enantiomer pure and free of the other. *The process of separating a racemic mixture into its enantiomers is called* ***resolution.*** Since enantiomers have identical achiral properties, how can we resolve a racemic mixture into its components? The answer is to convert them to diastereomers, separate the *diastereomers,* and then reconvert the now-separated diastereomers back to enantiomers.

To separate two enantiomers, we first let them react with a chiral reagent. The product will be a pair of *diastereomers.* These, as we have seen, differ in all types of properties and can be separated by ordinary methods. This principle is illustrated in the following equation:

$$\left\{\begin{matrix} R \\ S \end{matrix}\right\} + R \longrightarrow \left\{\begin{matrix} R\text{—}R \\ S\text{—}R \end{matrix}\right\} \tag{5.8}$$

pair of enantiomers (not separable)	chiral reagent	diastereomeric products (separable)

After the diastereomers are separated, we then carry out reactions that regenerate the chiral reagent and the separated enantiomers.

$$R\text{—}R \longrightarrow R + R$$

and (5.9)

$$S\text{—}R \longrightarrow S + R$$

Louis Pasteur was the first to resolve a racemic mixture when he separated the sodium ammonium salts of (+)- and (−)-tartaric acid. In a sense, he was a chiral reagent, since he could distinguish between the right- and left-handed crystals. In Chapter 11, we will see a specific example of how this is done chemically.

The principle behind the resolution of racemic mixtures is the same as the principle involved in the specificity of many biological reactions. That is, a chiral reagent (in cells, usually an enzyme) can discriminate between enantiomers because the two possible products of the reaction are diastereomers.

ADDITIONAL PROBLEMS

5.28. Define or describe the following terms.

a. stereogenic center **b.** chiral molecule
c. enantiomers **d.** plane-polarized light
e. specific rotation **f.** diastereomers
g. plane of symmetry **h.** *meso* form
i. racemic mixture **j.** resolution

5.29. Which of the following substances can exist in optically active forms?

a. 2,2-dichloropropane **b.** 1,2-dibromopropane
c. 2-methylpentane **d.** 2,3-dichlorohexane
e. 1-deuterioethanol (CH_3CHDOH) **f.** methylcyclobutane

5.30. Locate with an asterisk the stereogenic centers (if any) in the following structures.

a. $C_6H_5CH(OH)CO_2H$ **b.** $CH_2(OH)CH(OH)CH(OH)CHO$

c. $CH_3CHClCF_3$ **d.** ▷—$CH(OH)CH_3$

e. (cyclopentenyl)—$CH(OH)CH_3$ **f.** CH_3—(cyclohexane)—OH

5.31. What would happen to the observed and to the *specific* rotation if, in measuring the optical activity of a solution of sugar in water, we
a. doubled the concentration of the solution?
b. doubled the length of the sample tube?

5.32. The observed rotation for 100 mL of an aqueous solution containing 1 g of sucrose (ordinary sugar), placed in a 2-decimeter sample tube, is +1.33° at 25°C (using a sodium lamp). Calculate and express the specific rotation of sucrose.

5.33. Tell whether the following structures are identical or enantiomers.

a. CH_3 above C; C bonded to H (dashed), Cl (wedge), CH_3CH_2 — and — Cl above C; C bonded to CH_3 (dashed), CH_2CH_3 (wedge), H

b. Cl above C; C bonded to H (dashed), CH_3 (wedge), CH_3CH_2 — and — Cl above C; C bonded to CH_2CH_3 (dashed), H (wedge), CH_3

5.34. Draw a structural formula for an optically active compound with the molecular formula
a. $C_4H_{10}O$ **b.** $C_5H_{11}Cl$ **c.** $C_4H_8(OH)_2$ **d.** C_6H_{12}

5.35. Draw the formula of an unsaturated bromide, C_5H_9Br, that can show
a. neither *cis–trans* isomerism nor optical activity.
b. *cis–trans* isomerism but no optical activity.
c. no *cis–trans* isomerism but optical activity.
d. *cis–trans* isomerism and optical activity.

5.36. Place the members of the following groups in order of decreasing priority according to the *R-S* convention:
a. CH_3—, H—, HS—, CH_3CH_2—
b. H—, CH_3—, C_6H_5—, I—
c. CH_3—, HO—, —CH_2Br, —CH_2OH
d. CH_3CH_2—, $CH_3CH_2CH_2$—, CH_2=CH—, —CH=O

5.37. Assume that the four groups in each part of Problem 5.36 are attached to one carbon atom. Draw a three-dimensional formula for the *R* configuration of the molecule.

5.38. Tell whether the stereogenic centers marked with an asterisk in the following structures have the *R* or the *S* configuration:

a. 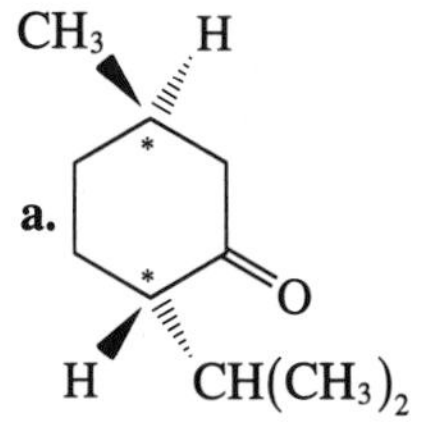 (–)-menthone (found in peppermint)

b. 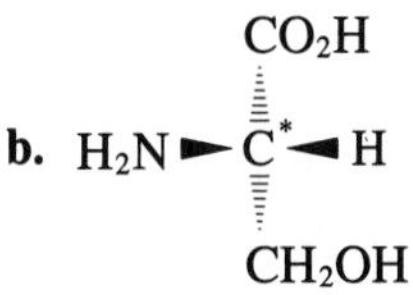 (–)-serine (an amino acid found in proteins)

c. 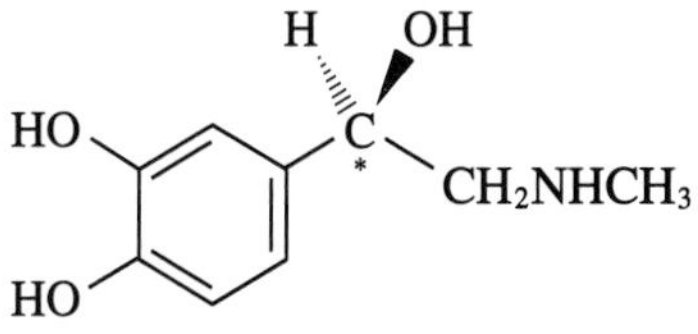 (–)-epinephrine (also called adrenalin)

5.39. Determine the configuration, *R* or *S*, of (+)-carvone, the compound responsible for the odor of caraway seeds.

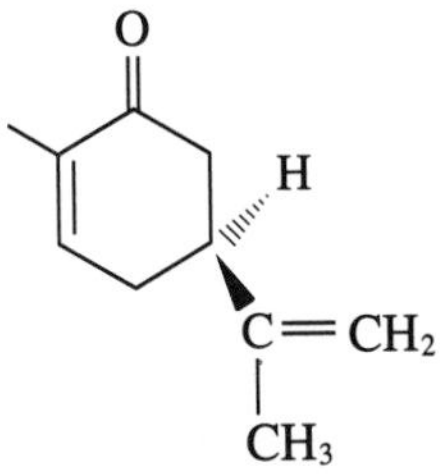

(+)-carvone

5.40. Name the following compounds, using *E-Z* notation:

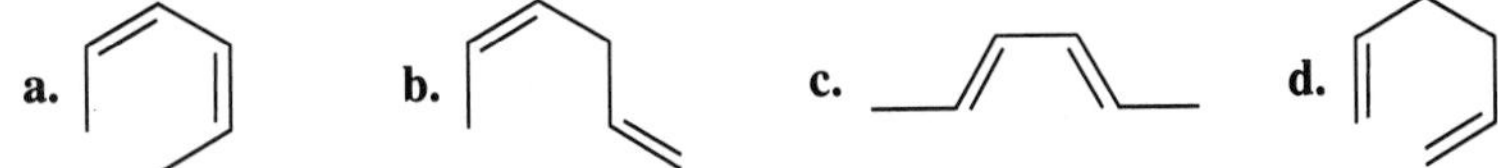

5.41. 4-Bromo-2-pentene has a double bond that can have either the *E* or the *Z* configuration and a stereogenic center that can have either the *R* or the *S* configuration. How many stereoisomers are possible altogether? Draw the structure of each, and group the pairs of enantiomers.

5.42. How many stereoisomers are possible for each of the following structures? Draw them, and name each by the *R-S* and *E-Z* conventions.

a. 3-methyl-1,4-pentadiene **b.** 3-methyl-1,4-hexadiene
c. 2-bromo-5-chloro-3-hexene **d.** 2,5-dibromo-3-hexene

5.43. Which of the following Fischer projection formulas have the same configuration as

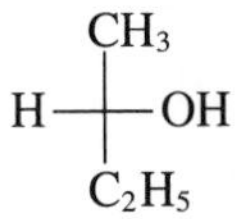

(A): top CH_3, left H, right OH, bottom C_2H_5

and which are its enantiomer?

a. 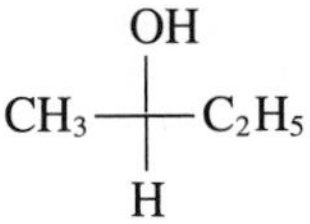 top OH, left CH_3, right C_2H_5, bottom H
b. top H, left C_2H_5, right CH_3, bottom OH
c. top CH_3, left H, right OH, bottom C_2H_5
d. 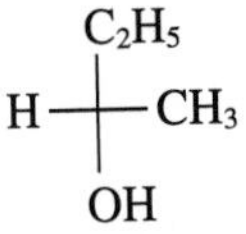 top C_2H_5, left H, right CH_3, bottom OH

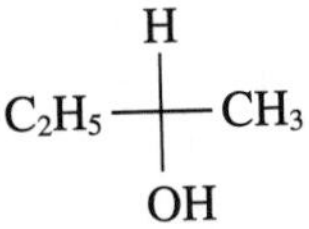

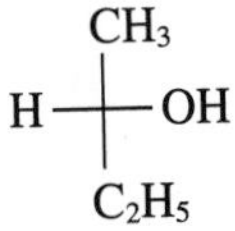

5.44. What is the configuration, *R* or *S*, of

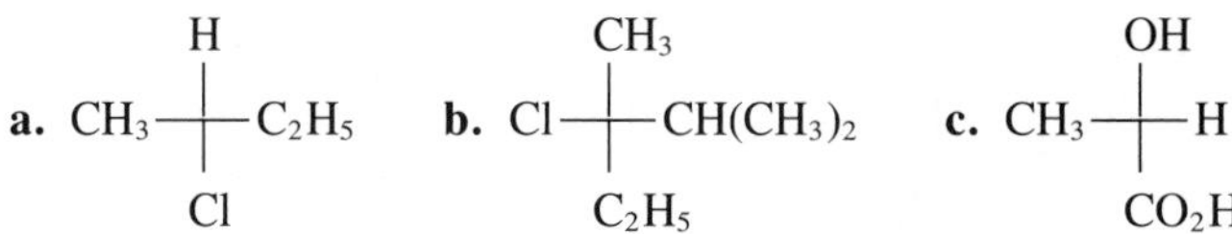

5.45. What is the configuration, *R* or *S*, at each stereogenic center in

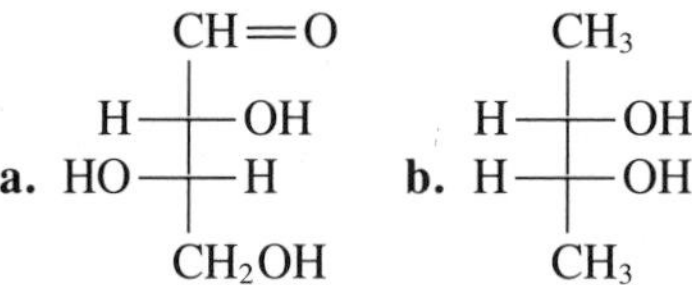

5.46. What is the configuration, *R* or *S*, at each stereogenic center in glucose (Problem 5.21, p. 169)?

5.47. Two possible configurations for a molecule with three different stereogenic centers are (*R,R,R*) and its mirror image (*S,S,S*). What are all the remaining possibilities? Repeat for a compound with four different stereogenic centers.

5.48. When racemic 2-chlorobutane is chlorinated, we obtain some 2,3-dichlorobutane. It consists of 71% *meso* isomer and 29% racemic isomers. Explain why the mixture need not be 50 : 50 *meso* and racemic 2,3-dichlorobutane. (*Hint:* It will help if you draw three-dimensional structures or Fischer projections.)

5.49. Below are Newman projections for the three tartaric acids (*R,R*), (*S,S*), and *meso*. Which is which?

5.50. Convert the sawhorse formula below for one isomer of tartaric acid to a Fischer projection formula. Which isomer of tartaric acid is it?

5.51. Two possible isomeric structures of 1,2-dichloroethane are

and

Classify them fully, according to the discussion in Sec. 5.11.

5.52. Two other possible isomeric structures of 1,2-dichloroethane are

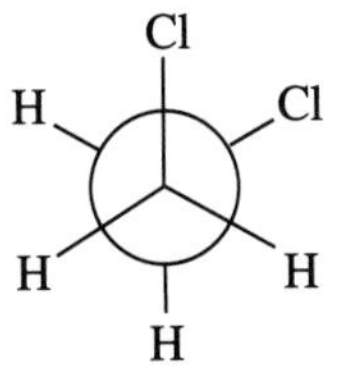

(Newman projections: front Cl at top, H and Cl; back H, H, H — and — front Cl at top, Cl and H; back H, H, H)

Cl, H, Cl, H, H, H and Cl, Cl, H, H, H, H

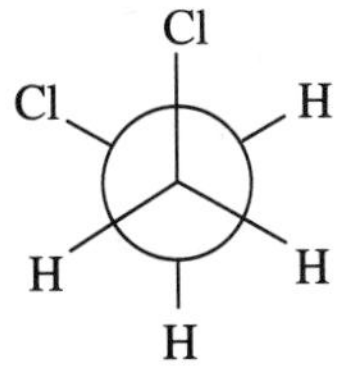

Classify them fully, as in Problem 5.51.

5.53. The formula for muscarine, a toxic constituent of poisonous mushrooms, is

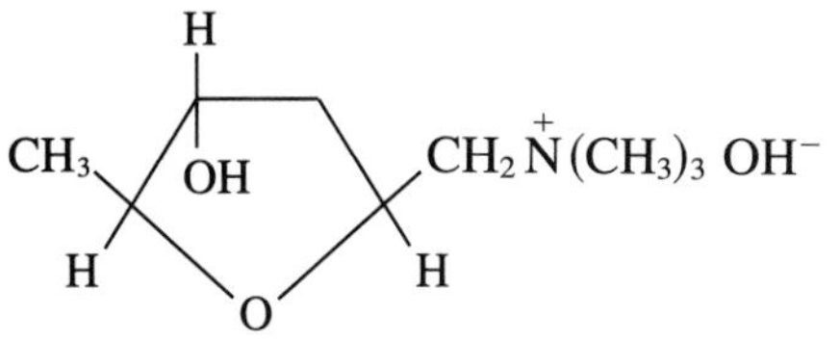

H; CH_3; OH; $CH_2\overset{+}{N}(CH_3)_3$ OH^-; H; O; H

Is it chiral? How many stereoisomers of this structure are possible? An interesting murder mystery, which you might enjoy reading and which depends for its solution on the distinction between optically active and racemic forms of this poison, is Dorothy L. Sayers's *The Documents in the Case,* published in paperback by Avon Books. (See an article by H. Hart, "Accident, Suicide, or Murder? A Question of Stereochemistry," *J. Chem. Educ.*, **1975,** *52,* 444.)

5.54. Chloramphenicol is an antibiotic that is particularly effective against typhoid fever. Its structure is

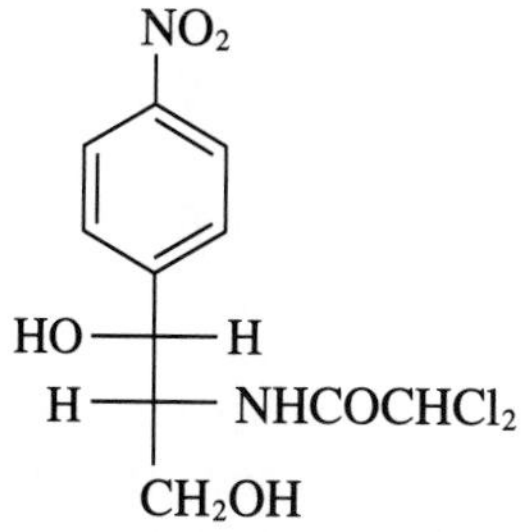

NO_2 (para-substituted benzene ring)

HO—C—H

H—C—$NHCOCHCl_2$

CH_2OH

What is the configuration (*R*,*S*) at each stereogenic center?

5.55. What can you say about the stereochemistry of the products in the following reactions?

a. 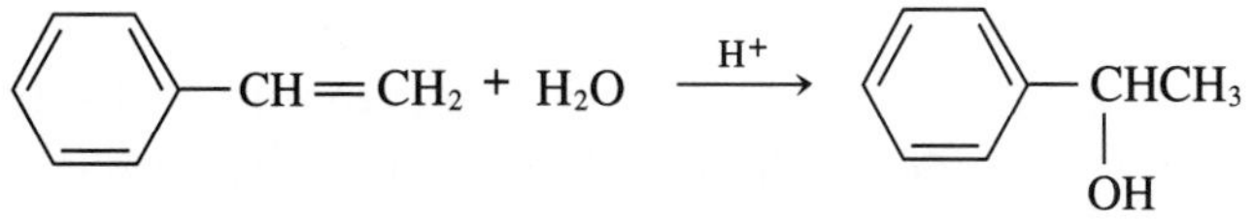 C_6H_5—$CH{=}CH_2$ + H_2O $\xrightarrow{H^+}$ C_6H_5—$\underset{\large OH}{CHCH_3}$

b. $\underset{\large OH}{CH_3CHCH{=}CH_2}$ + H_2O $\xrightarrow{H^+}$ $CH_3\underset{\large OH}{CH}{-}\underset{\large OH}{CH}CH_3$

(*R*-enantiomer)

5.56. (+)- and (−)-Carvone (see Problem 5.39 for the structure) are enantiomers that have very different odors and are responsible for the odors of caraway seeds and spearmint, respectively. Suggest a possible explanation.

6 Organic Halogen Compounds; Substitution and Elimination Reactions

6.1 Introduction

Chlorine- and bromine-containing natural products have been isolated from various species that live in the sea—sponges, mollusks, and other ocean creatures that have adapted to their environment by metabolizing inorganic chlorides and bromides that are prevalent there. With these exceptions, most organic halogen compounds are creatures of the laboratory. We have already seen that they can be made by the direct halogenation of alkanes and aromatic compounds and by the addition of hydrogen halides to alkenes and alkynes.* And in Chapter 7, we will learn how alkyl halides can be prepared from alcohols.

Halogen compounds are important for several reasons. Simple alkyl and aryl halides, especially chlorides and bromides, are versatile reagents in syntheses. Through *substitution reactions*, which we will discuss in this chapter, halogens can be replaced by many other functional groups. Organic halides can be converted to unsaturated compounds through dehydrohalogenation. Also, some halogen compounds, especially those that contain two or more halogen atoms per molecule, have practical uses—as solvents, insecticides, herbicides, fire retardants, cleaning fluids, and refrigerants, for example. In this chapter, we will discuss all these aspects of halogen compounds.

6.2 Nucleophilic Substitution

Let us look at a typical **nucleophilic substitution reaction.** Ethyl bromide reacts with hydroxide ion to give ethyl alcohol and bromide ion.**

$$HO^- + \underset{\text{ethyl bromide}}{CH_3CH_2{-}Br} \xrightarrow{H_2O} \underset{\text{ethyl alcohol}}{CH_3CH_2{-}OH} + Br^- \qquad (6.1)$$

* If you do not remember these reactions, review Secs. 2.13b, 3.7a, 3.11, 3.15, 3.20, and 4.9. Also review the definitions of electrophiles and nucleophiles in Sec. 3.9.

** The nomenclature of alkyl halides is discussed in Sec. 2.5.

Hydroxide ion is the **nucleophile.** It reacts with the **substrate** (ethyl bromide) and displaces bromide ion. The bromide ion is called the **leaving group.**

In reactions of this type, one covalent bond is broken, and a new covalent bond is formed. In this particular example, the carbon–bromine bond is broken and the carbon–oxygen bond is formed. The leaving group (bromide) takes with it *both* of the electrons from the carbon–bromine bond, and the nucleophile (hydroxide ion) supplies *both* electrons for the new carbon–oxygen bond.

These ideas are generalized in the following equations for a nucleophilic substitution reaction:

$$\underset{\text{nucleophile (neutral)}}{\text{Nu:}} + \underset{\text{substrate}}{\text{R:L}} \longrightarrow \underset{\text{product}}{\text{R:}\overset{+}{\text{Nu}}} + \underset{\text{leaving group}}{\text{:L}^-} \qquad (6.2)$$

$$\underset{\text{nucleophile (anion)}}{\text{Nu:}^-} + \underset{\text{substrate}}{\text{R:L}} \longrightarrow \underset{\text{product}}{\text{R:Nu}} + \underset{\text{leaving group}}{\text{:L}^-} \qquad (6.3)$$

If the nucleophile and substrate are neutral, the product will be positively charged (eq. 6.2). If the nucleophile is a negative ion and the substrate is neutral, the product will also be neutral (eq. 6.3). In either case, an unshared electron pair on the nucleophile supplies the electrons for the new covalent bond.

In principle, of course, these reactions may be reversible because the leaving group also has an unshared electron pair that can be used to form a covalent bond. However, we can use various methods to force the reactions to go in the forward direction. For example, we can choose the nucleophile so it is a *stronger* nucleophile than the leaving group. Or we can shift the equilibrium by using a large excess of one reagent or by removing one of the products as it is formed.

6.3 *Examples of Nucleophilic Substitutions*

Nucleophiles can be classified according to the kind of atom that forms a new covalent bond. For example, the hydroxide ion in eq. 6.1 is an *oxygen* nucleophile. In the product, a new carbon–*oxygen* bond is formed. **The most common nucleophiles are *oxygen, nitrogen, sulfur, halogen,* and *carbon* nucleophiles.** Table 6.1 shows some examples of nucleophiles and the products that they form when they react with an alkyl halide.

Let us consider a few specific examples of these reactions, to see how they may be used in synthesis.

EXAMPLE 6.1 Use Table 6.1 to write an equation for the reaction of sodium ethoxide with bromoethane.

Solution

$$\underset{\text{sodium ethoxide}}{CH_3CH_2\ddot{\underset{\cdot\cdot}{O}}:^-\ Na^+} + \underset{\text{bromoethane}}{CH_3CH_2Br} \longrightarrow \underset{\text{diethyl ether}}{CH_3CH_2OCH_2CH_3} + Na^+Br^-$$

Ethoxide is the nucleophile, bromoethane is the substrate, and bromide ion is the leaving group (item 2 in Table 6.1). The product is diethyl ether, which is used as an anesthetic. Notice that the counterion of the nucleophile, Na^+, is merely a spectator during the reaction. It is present at the beginning and end of the reaction.

EXAMPLE 6.2 Devise a synthesis for propyl cyanide using a nucleophilic substitution reaction.

Solution First, write the structure of the desired product.

$$\underset{\text{propyl cyanide}}{CH_3CH_2CH_2{-}CN}$$

If we use cyanide ion as the nucleophile (item 14 in Table 6.1), the alkyl halide must have the halogen (Cl, Br, or I) attached to a propyl group. The equation is

$$CN^- + CH_3CH_2CH_2Br \longrightarrow CH_3CH_2CH_2CN + Br^-$$

Some salt, such as sodium or potassium cyanide, can be used to supply the nucleophile.

EXAMPLE 6.3 Show how 1-butyne could be converted to 3-hexyne using a nucleophilic substitution reaction.

Solution Compare the starting material with the product.

$$\underset{\text{1-butyne}}{CH_3CH_2C{\equiv}CH} \qquad \underset{\text{3-hexyne}}{CH_3CH_2C{\equiv}CCH_2CH_3}$$

From Table 6.1, line 15, we see that acetylides react with alkyl halides to give acetylenes. We therefore need to convert 1-butyne to an acetylide (review eq. 3.53), then treat it with a 2-carbon alkyl halide.

$$CH_3CH_2C{\equiv}CH + NaNH_2 \xrightarrow{NH_3} CH_3CH_2C{\equiv}C{:}^-Na^+$$

$$CH_3CH_2C{\equiv}C{:}^-Na^+ + CH_3CH_2Br \longrightarrow CH_3CH_2C{\equiv}CCH_2CH_3 + Na^+Br^-$$

EXAMPLE 6.4 Complete the following equation:

$$NH_3 + CH_3CH_2CH_2Br \longrightarrow$$

Solution Ammonia is a nitrogen nucleophile (Table 6.1, line 6). Since both reactants are neutral, the product has a positive charge (the formal +1 charge is on the nitrogen—check it out!)

$$NH_3 + CH_3CH_2CH_2Br \longrightarrow CH_3CH_2CH_2\overset{+}{N}H_3 + Br^-$$

Table 6.1 Reactions of common nucleophiles with alkyl halides* (eqs. 6.2 and 6.3)

Nu		R—Nu		Comments	
Formula	*Name*	*Formula*	*Name*		
Oxygen nucleophiles					
1. $H\ddot{O}:^-$	hydroxide	R—OH	alcohol		
2. $R\ddot{O}:^-$	alkoxide	R—OR	ether		
3. $H\ddot{O}H$	water	$R-\overset{+}{O}H_2$	alkyloxonium ion	These ions lose a proton and the products are alcohols and ethers.	$\xrightarrow{-H^+}$ ROH (alcohol)
4. $R\ddot{O}H$	alcohol	$R-\overset{+}{O}(R)H$	dialkyloxonium ion		$\xrightarrow{-H^+}$ ROR (ether)
5. $R-C(=O)\ddot{O}:^-$	carboxylate	$R-OC(=O)-R$	ester		
Nitrogen nucleophiles					
6. $\ddot{N}H_3$	ammonia	$R-\overset{+}{N}H_3$	alkylammonium ion	With a base, these ions readily lose a proton to give amines.	$\xrightarrow{-H^+}$ RNH_2
7. $R\ddot{N}H_2$	primary amine	$R-\overset{+}{N}H_2R$	dialkylammonium ion		$\xrightarrow{-H^+}$ R_2NH
8. $R_2\ddot{N}H$	secondary amine	$R-\overset{+}{N}HR_2$	trialkylammonium ion		$\xrightarrow{-H^+}$ R_3N
9. $R_3\ddot{N}$	tertiary amine	$R-\overset{+}{N}R_3$	tetraalkylammonium ion		
Sulfur nucleophiles					
10. $H\ddot{S}:^-$	hydrosulfide ion	R—SH	thiol		
11. $R\ddot{S}:^-$	mercaptide ion	R—SR	thioether (sulfide)		
12. $R_2\ddot{S}:$	thioether	$R-\overset{+}{\ddot{S}}R_2$	trialkylsulfonium ion		

*Aryl halides and vinyl halides normally do *not* undergo this type of nucleophilic substitution reaction.

TABLE 6.1 (continued)

Nu		R—Nu		
Formula	*Name*	*Formula*	*Name*	*Comments*
Halogen nucleophiles				
13. $:\ddot{\underset{..}{I}}:^-$	iodide	R—I	alkyl iodide	The usual solvent is acetone. Sodium iodide is soluble in acetone, but sodium bromide and sodium chloride are not.
Carbon nucleophiles				
14. $^-:C{\equiv}N:$	cyanide	R—CN	alkyl cyanide (nitrile)	Sometimes the isonitrile, RNC, is formed.
15. $^-:C{\equiv}CR$	acetylide	R—C≡CR	acetylene	

PROBLEM 6.1 Using Table 6.1, write complete equations for the following nucleophilic substitution reactions:

a. $NaOH + CH_3CH_2CH_2Br$
b. $(CH_3CH_2)_3N + CH_3CH_2Br$
c. NaSH + C_6H_5—CH_2Br (benzene ring)

PROBLEM 6.2 Write an equation for the preparation of each of the following compounds, using a nucleophilic substitution reaction. In each case, label the nucleophile, the substrate, and the leaving group.

a. $CH_3CH_2CH_2CH_2OH$ b. $(CH_3)_2CHCH_2C{\equiv}N$
c. $(CH_3CH_2CH_2)_3N$ d. $(CH_3CH_2)_3S^+Br^-$
e. $CH_2{=}CHCH_2I$ f. C_6H_5—OCH_3 (benzene ring) *(Careful!* See footnote to Table 6.1.)

The substitution reactions in Table 6.1 have some serious limitations, particularly with respect to the structure of the *R* group in the alkyl halide. These limitations occur most often when the nucleophile is either an anion or a base or both. For example,

$$\underset{\text{anion}}{CN^-} + \underset{\text{primary halide*}}{CH_3CH_2CH_2CH_2Br} \longrightarrow CH_3CH_2CH_2CH_2CN + Br^- \quad (6.4)$$

---

* For the definition of primary, secondary, and tertiary alkyl groups, review Sec. 3.13.

but

$$\underset{\text{anion}}{CN^-} + \underset{\text{tertiary halide}}{CH_3-\overset{\overset{\displaystyle CH_3}{|}}{\underset{\underset{\displaystyle Br}{|}}{C}}-CH_3} \longrightarrow \underset{\text{methylpropene}}{CH_3-\overset{\overset{\displaystyle CH_2}{\|}}{C}-CH_3} + HCN + Br^- \qquad (6.5)$$

Another example is

$$\underset{\text{neutral, not very basic}}{H_2O} + \underset{\text{tertiary halide}}{CH_3-\overset{\overset{\displaystyle CH_3}{|}}{\underset{\underset{\displaystyle Br}{|}}{C}}-CH_3} \longrightarrow \underset{\text{(about 80\%; some methylpropene is also formed)}}{CH_3-\overset{\overset{\displaystyle CH_3}{|}}{\underset{\underset{\displaystyle OH}{|}}{C}}-CH_3} + H^+ + Br^- \qquad (6.6)$$

but

$$\underset{\text{strong base}}{OH^-} + \underset{\text{tertiary halide}}{CH_3-\overset{\overset{\displaystyle CH_3}{|}}{\underset{\underset{\displaystyle Br}{|}}{C}}-CH_3} \longrightarrow \underset{\text{methylpropene}}{CH_3-\overset{\overset{\displaystyle CH_2}{\|}}{C}-CH_3} + H_2O + Br^- \qquad (6.7)$$

To understand these differences, we must consider the mechanisms by which the substitutions in Table 6.1 take place.

6.4 Nucleophilic Substitution Mechanisms

As a result of experiments that began more than 60 years ago, we now understand the mechanisms of nucleophilic substitution reactions rather well. We use the plural because such ***nucleophilic substitutions occur by more than one mechanism.*** The mechanism observed in a particular case depends on the structures of the nucleophile and the alkyl halide, the solvent, the reaction temperature, and other factors.

There are two main nucleophilic substitution mechanisms. These are described by the symbols $\mathbf{S_N2}$ and $\mathbf{S_N1}$, respectively. The S_N part of each symbol stands for "substitution, nucleophilic." The meaning of the numbers 2 and 1 will become clear as we discuss each mechanism.

6.5 The S_N2 Mechanism

The S_N2 mechanism is a one-step process, represented by the following equation:

$$\underset{\text{nucleophile}}{Nu:^-} + \underset{\text{substrate}}{C-L} \longrightarrow \underset{\text{transition state}}{\left[\overset{\delta^-}{Nu}\cdots C\cdots\overset{\delta^-}{L}\right]^-} \longrightarrow \underset{\text{product}}{\overset{+}{Nu}-C} + \underset{\text{leaving group}}{:L^-} \qquad (6.8)$$

The nucleophile attacks from the *back* side of the C—L bond. At some stage (the transition state) the nucleophile *and* the leaving group are *both* partly bonded to the carbon at which substitution occurs. As the leaving group departs *with its electron pair*, the nucleophile supplies another electron pair to the carbon atom.

The number 2 is used in describing this mechanism because the reaction is *bi*molecular. That is, two molecules—the nucleophile and the substrate—are involved in the key step (the *only* step) in the reaction mechanism. The reaction shown in eq. 6.1 occurs by an S_N2 mechanism. A reaction energy diagram is shown in Figure 6.1.

PROBLEM 6.3 Draw a reaction energy diagram for the reaction between $CH_3CH_2CH_2Br$ and sodium cyanide (NaCN). Label the energy of activation (E_a) and ΔH for the reaction. (Refer to Sec. 3.12 if you need help.)

How can we recognize when a particular nucleophile and substrate react by the S_N2 mechanism? There are several tell-tale signs.

1. ***The rate of the reaction depends on both the nucleophile and the substrate concentrations.*** The reaction of hydroxide ion with ethyl bromide (eq. 6.1) is an example of an S_N2 reaction. If we double the base concentration (OH^-), the reaction goes twice as fast. The same thing happens if we

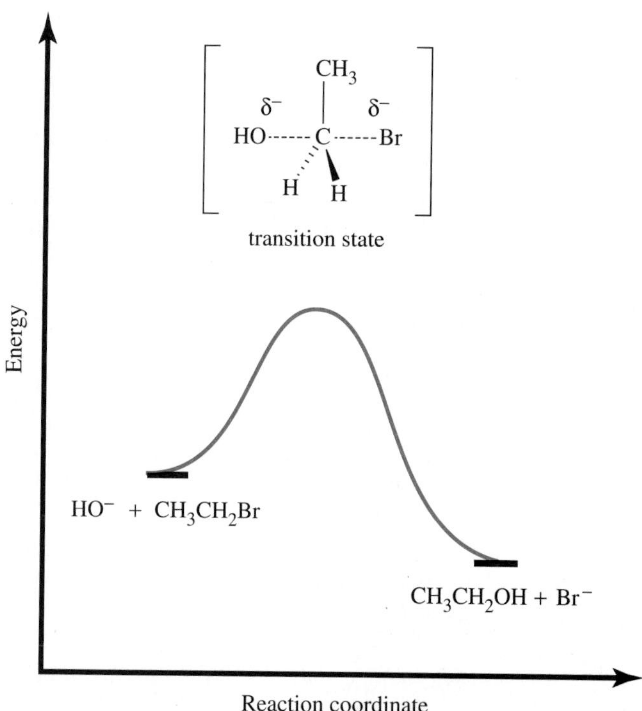

FIGURE 6.1 Reaction energy diagram for an S_N2 reaction.

double the ethyl bromide concentration. We will see shortly that this rate behavior is *not* observed in the S_N1 mechanism.

2. ***Every S_N2 displacement occurs with inversion of configuration.*** For example, if we treat (*R*)-2-bromobutane with sodium hydroxide, we obtain (*S*)-2-butanol.

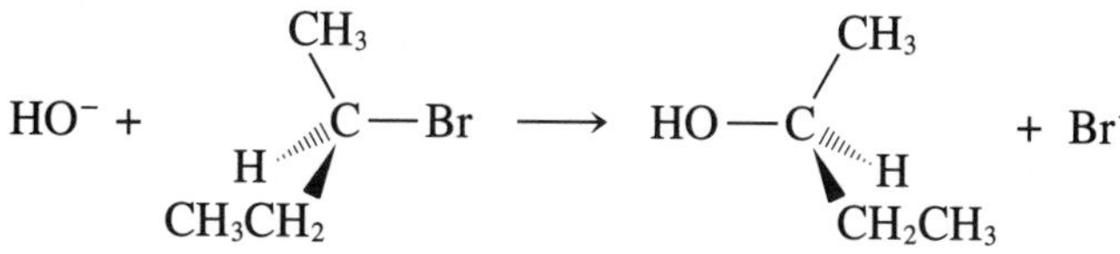

$$HO^- + \underset{(R)\text{-2-bromobutane}}{(CH_3)(H)(CH_3CH_2)C{-}Br} \longrightarrow \underset{(S)\text{-2-butanol}}{HO{-}C(CH_3)(H)(CH_2CH_3)} + Br^- \quad (6.9)$$

This experimental result, which at first came as a surprise to chemists, meant that the OH group did *not* take the exact position occupied by the Br. If it had, the configuration would have been retained; (*R*)-bromide would have given (*R*)-alcohol. What is the only reasonable explanation? The hydroxide ion must attack the C—Br bond from the rear. As substitution occurs, the three groups attached to the sp^3 carbon *invert,* somewhat like an umbrella caught in a strong wind.

3. ***The reaction is fastest when the alkyl group of the substrate is methyl or primary and slowest when it is tertiary.*** Secondary alkyl halides react at an intermediate rate. The reason for this reactivity order is fairly obvious if we think about the S_N2 mechanism. The rear side of the carbon, where displacement occurs, is more crowded if more alkyl groups are attached to it, thus slowing down the reaction rate.

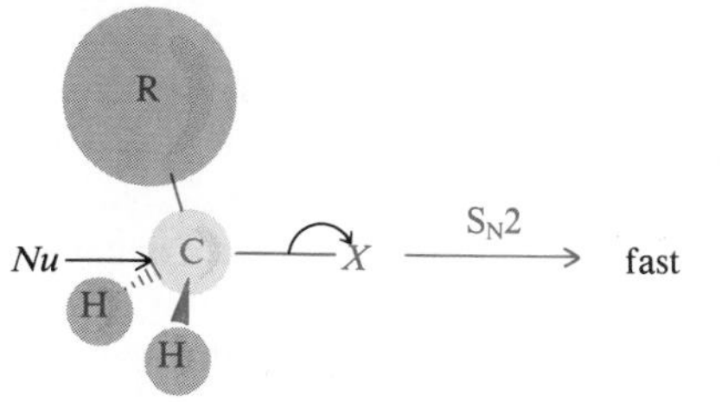

(6.10)

primary halide
(rear side not crowded)

(6.11)

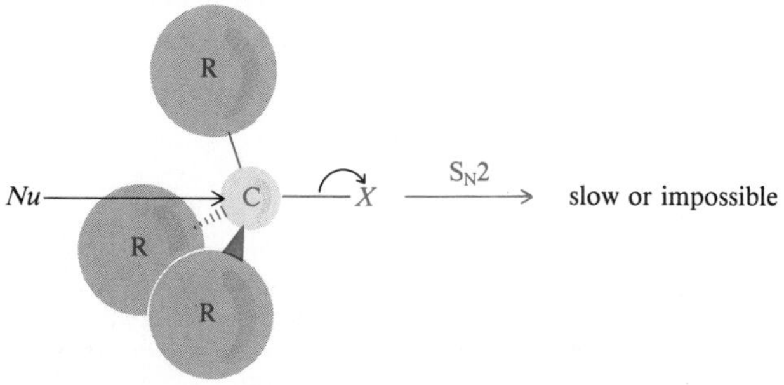

tertiary halide
(rear side crowded)

EXAMPLE 6.5 Predict the product from the S_N2 reaction of *cis*-4-methylcyclohexane with cyanide ion.

Solution

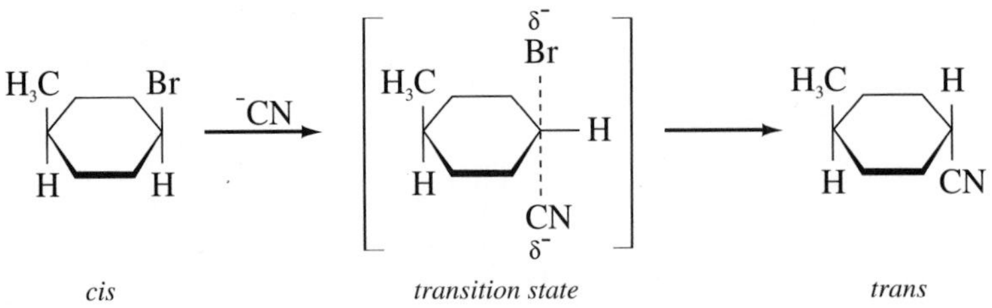

cis — *transition state* — *trans*

Cyanide ion attacks the C—Br bond from the rear and therefore the cyano group ends up *trans* to the methyl group.

PROBLEM 6.4 Draw a Fischer projection formula for the product of this S_N2 reaction:

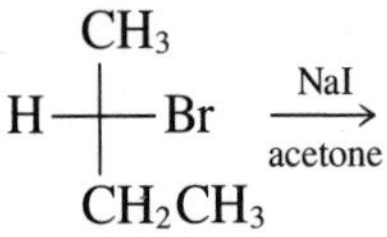

$$\begin{array}{c} CH_3 \\ H-\!\!\!+\!\!\!-Br \\ CH_2CH_3 \end{array} \xrightarrow[\text{acetone}]{\text{NaI}}$$

PROBLEM 6.5 Arrange the following compounds in order of *decreasing* S_N2 reactivity toward sodium ethoxide:

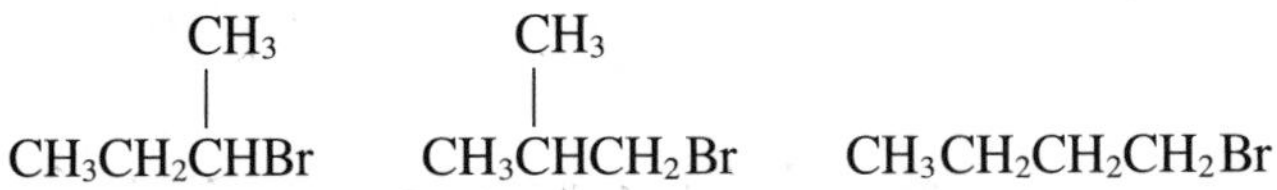

$CH_3CH_2CH(CH_3)Br$ $\quad$ $CH_3CH(CH_3)CH_2Br$ $\quad$ $CH_3CH_2CH_2CH_2Br$

To summarize, the S_N2 mechanism is a one-step process favored for methyl and primary halides. It occurs more slowly with secondary halides and usually not at all with tertiary halides. An S_N2 reaction occurs with inversion of configuration, and its rate depends on the concentration of *both* the nucleophile and the substrate (the alkyl halide).

Now let us see how these features differ for the S_N1 mechanism.

6.6 The S_N1 Mechanism

The S_N1 mechanism is a two-step process. In the first step, which is slow, the bond between the carbon and the leaving group breaks as the substrate dissociates (ionizes).

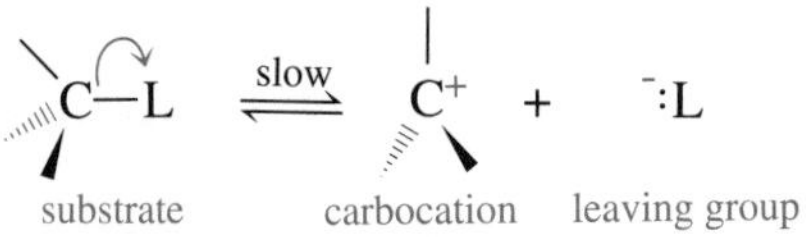

$$\text{C—L (substrate)} \underset{}{\overset{\text{slow}}{\rightleftharpoons}} \text{C}^+ \text{ (carbocation)} + {}^-\!:\!\text{L (leaving group)} \qquad (6.12)$$

The electrons of the C—L bond go with the leaving group, and a carbocation is formed.

In the second step, which is fast, the carbocation combines with the nucleophile to give the product.

$$\text{C}^+ + \text{:Nu} \xrightarrow{\text{fast}} \text{C}-\overset{+}{\text{Nu}} \quad \text{and} \quad \overset{+}{\text{Nu}}-\text{C} \tag{6.13}$$

carbocation nucleophile

When the nucleophile is a neutral molecule, such as water or an alcohol, loss of a proton from the nucleophilic oxygen, in a third step, gives the final product.

The number 1 is used to designate this mechanism because the slow, or rate-determining, step involves *only one* of the two reactants: the substrate (eq. 6.12). It does *not* involve the nucleophile at all. That is, the first step is *unimolecular*. The reaction shown in eq. 6.6 occurs by an S_N1 mechanism and a reaction energy diagram for that reaction is shown in Figure 6.2. Notice that the energy diagram for this reaction, and all S_N1 reactions, resembles that of an electrophilic addition to an alkene (Figure 3.9), another reaction that has a carbocation intermediate. Also notice that the energy of activation for the first step (the rate-determining step) is much greater than for subsequent steps.

PROBLEM 6.6 What are the products expected from the reaction of $(CH_3)_3C—CL$ with $CH_3—OH$? Draw a reaction energy diagram for the reaction.

How can we recognize when a particular nucleophile and substrate react by the S_N1 mechanism? Here are the signs:

1. ***The rate of the reaction does not depend on the concentration of the nucleophile.*** The first step is rate determining, and the nucleophile is not involved in this step. The bottleneck in the reaction rate is therefore the rate of formation of the carbocation, not its rate of reaction with the nucleophile, which is nearly instantaneous.

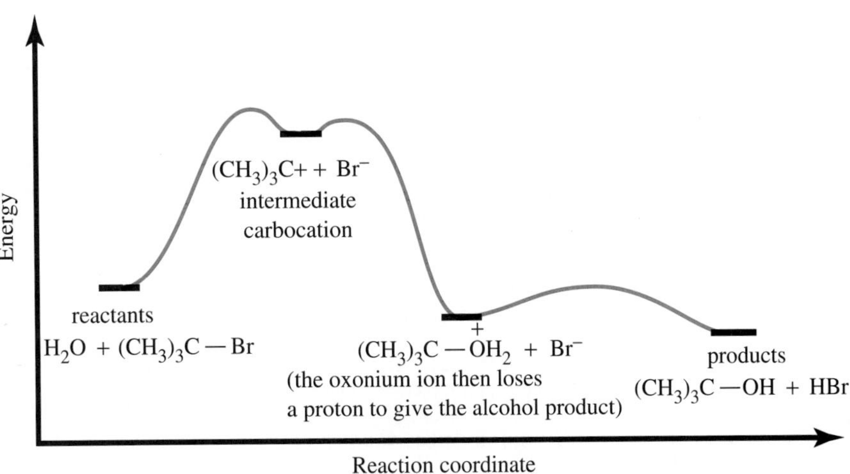

FIGURE 6.2 Reaction energy diagram for an S_N1 reaction.

2. ***If the carbon bearing the leaving group is stereogenic, the reaction occurs mainly with loss of optical activity (that is, with racemization).*** In carbocations, only three groups are attached to the positive carbon. Therefore, the positive carbon is sp^2-hybridized and planar. As shown in eq. 6.13, the nucleophile can react at either "face" of the carbocation to give a 50 : 50 mixture of two enantiomers, a racemic mixture. For example, the reaction of (*R*)-3-bromo-3-methylhexane with water gives the racemic alcohol.

Br
CH_3CH_2 C CH_3
$CH_3CH_2CH_2$
(R)-3-bromo-3-methylhexaneaqueous

$\xrightarrow[\text{acetone}]{\text{aqueous}}$

CH_3CH_2 C — CH_3
$CH_3CH_2CH_2$ (+)
intermediate carbocation

H_2O | $-H^+$ (see item 3 in Table 6.1)

(6.14)

CH_3CH_2 CH_3
$CH_3CH_2CH_2$ C
OH
50% *S*
(product from attack on the bottom face of the carbocation)

+

OH
CH_3CH_2 C CH_3
$CH_3CH_2CH_2$
50% *R*
(product from attack on the top face of the carbocation)

The intermediate carbocation is planar and achiral. Combination with H_2O from the "top" or "bottom" is equally probable, giving the *R* and *S* alcohols, respectively, in equal amounts.

3. ***The reaction is fastest when the alkyl group of the substrate is tertiary and slowest when it is primary.*** The reason is that S_N1 reactions proceed via carbocations, so the reactivity order corresponds to that of carbocation stability (3° > 2° > 1°). That is, the easier it is to form the carbocation, the faster the reaction will proceed. For this reason, S_N1 reactivity is also favored for resonance-stabilized carbocations, such as allylic carbocations (see Sec. 3.15).

PROBLEM 6.7 Which of the following bromides will react faster with methanol (S_N1)? What are the reaction products in each case?

a. $CH_3CH_2CH(CH_3)Br$ or $CH_3CH_2C(CH_3)_2Br$
b. $CH_3CH_2CH_2Br$ or $H_2C{=}CHCH_2Br$

To summarize, the S_N1 mechanism is a two-step process favored when the alkyl halide is tertiary. Primary halides normally do not react by this mechanism. The S_N1 process occurs with racemization, and its rate is independent of the nucleophile's concentration.

6.7 The S_N1 and S_N2 Mechanisms Compared

How can we tell whether a particular nucleophilic substitution reaction will proceed by an S_N2 or an S_N1 mechanism? And why do we care? First of all, we care for several reasons. When we perform a reaction in the laboratory, we want to be sure that the reaction will proceed at a rate such that we obtain the product in a reasonable time. If the reaction has stereochemical consequences, we want to know in advance what that outcome will be: inversion or racemization.

Table 6.2 should be helpful in answering our first question. It summarizes what we have said so far about the two substitution mechanisms, and it compares them with respect to two other variables, solvent and nucleophile structure, which we will discuss here.

Notice that primary halides almost always react by the S_N2 mechanism, whereas tertiary halides react by the S_N1 mechanism. Only with secondary halides are we likely to encounter both possibilities.

One experimental variable that we can use to help control mechanism is the solvent polarity. Water and alcohols are **polar protic solvents** (protic because of the proton-donating ability of the hydroxyl groups). How will such solvents affect S_N1 and S_N2 reactions?

The first step of the S_N1 mechanism involves the formation of ions. Since polar solvents can solvate ions, the reaction rate of S_N1 processes is enhanced by polar solvents. On the other hand, solvation of nucleophiles ties up their unshared electron pairs. Therefore, S_N2 reactions, whose rates depend on nucleophile effectiveness, are usually retarded by polar protic solvents.*

TABLE **6.2** Comparison of S_N2 and S_N1 substitutions

Halide structure	*S_N2*	*S_N1*
Primary or CH_3	Common	Rarely**
Secondary	Sometimes	Sometimes
Tertiary	Rarely	Common
Stereochemistry	Inversion	Racemization
Solvent	Rate is retarded by polar protic solvents	Because the intermediates are ions, the rate is highly favored with polar solvents
Nucleophile	Rate depends on nucleophile concentration; mechanism is favored when the nucleophile is an anion	Rate is independent of nucleophile concentration; mechanism is more likely with neutral nucleophiles

**Allyl and benzyl substrates are the common exceptions (see Problem 6.7).

* Polar but *aprotic* solvents [examples are dimethyl sulfoxide, $(CH_3)_2S{=}O$, or dimethylformamide, $(CH_3)_2NCHO$] solvate cations instead of anions. These solvents *accelerate* S_N2 reactions because, by solvating the cation (say, K^+ in K^+CN^-), they leave the anion more "naked" or unsolvated and improve its nucleophilicity.

Let us apply these ideas to a *secondary halide*, which might react by either mechanism. We may choose the mechanism simply by adjusting the solvent polarity. For example, we can change the mechanism by which a secondary halide reacts with water (to form an alcohol) from S_N2 to S_N1 by changing the solvent from 95% acetone–5% water (relatively nonpolar) to 50% acetone–50% water (more polar and a better ionizing solvent).

EXAMPLE 6.6 The reaction of (*R*)-2-iodobutane with 95% acetone–5% water gives mainly (*S*)-2-butanol. With 30% acetone–70% water, the product has a much lower optical activity, namely being about 60% (*S*)-2-butanol and 40% (*R*)-2-butanol. Explain.

Solution In 95% acetone–5% water (a relatively nonpolar solvent), substitution occurs by the S_N2 mechanism with inversion.

(*R*)-2-iodobutane $\xrightarrow[-I^-]{S_N2}$ oxonium ion $\underset{}{\overset{-H^+}{\rightleftharpoons}}$ (*S*)-2-butanol

When the percentage of water in the solvent is increased, the solvent becomes more polar, and some reaction occurs by the S_N1 process.

(*R*)-2-iodobutane $\underset{\text{slow}}{\overset{S_N1}{\rightleftharpoons}}$ $CH_3C^+HCH_2CH_3$ $\xrightarrow[\text{fast}]{H-\ddot{O}-H,\ -H^+}$ (*R*)-2-butanol + (*S*)-2-butanol

There is a slight preference for the (*S*) isomer (60 : 40) for one of two reasons. Either some of the reaction still occurs by the S_N2 mechanism, or else the departing iodide ion in the S_N1 mechanism is still close enough to the face of the carbocation from which it departed to partially block that face from nucleophilic attack.

Now let us consider the other variable in Table 6.2—the nucleophile. As we have seen, the rate of an S_N2 reaction (but *not* an S_N1 reaction) depends on the nucleophile. If the nucleophile is *strong*, the S_N2 mechanism will be favored. How can we tell whether a nucleophile is strong or weak, or whether one nucleophile is stronger than another? Here are a few useful generalizations.

1. ***Negative ions are more nucleophilic, or better electron suppliers, than the corresponding neutral molecules.*** Thus,

$$HO^- > HOH \qquad RS^- > RSH$$

$$RO^- > ROH \qquad R{-}\underset{\underset{O}{\|}}{C}{-}O^- > R{-}\underset{\underset{O}{\|}}{C}{-}OH$$

2. ***Elements low in the periodic table tend to be more nucleophilic than elements above them in the same column.*** Thus,

$$HS^- > HO^- \qquad I^- > Br^- > Cl^- > F^- \quad \text{(in protic solvents)}$$

$$R\ddot{\underset{..}{S}}H > R\ddot{\underset{..}{O}}H \qquad (CH_3)_3P\!: > (CH_3)_3N\!:$$

3. ***Elements in the same row in the periodic table tend to be less nucleophilic, the more electronegative the element (that is, the more tightly it holds electrons to itself).*** Thus,

$$R_3C^- > R_2N^- > R{-}O^- > F^- \quad \text{and} \quad H_3N\!: > H_2\ddot{O}\!: > H\ddot{\underset{..}{F}}\!:$$

Can we juggle all these factors to make some predictions about particular substitution reactions? Here are some examples.

EXAMPLE 6.7 Which mechanism, S_N1 or S_N2, would you predict for this reaction?

$$(CH_3)_3CBr + CH_3OH \longrightarrow (CH_3)_3COCH_3 + HBr$$

Solution S_N1, because the substrate is a tertiary alkyl halide. Also, methanol is a weak, neutral nucleophile and, if used as the reaction solvent, rather polar. Thus, it favors ionization.

EXAMPLE 6.8 Which mechanism, S_N1 or S_N2, would you predict for this reaction?

$$CH_3CH_2I + NaCN \longrightarrow CH_3CH_2CN + NaI$$

Solution S_N2, because the substrate is a primary halide, and cyanide, an anion, is a rather strong nucleophile.

PROBLEM 6.8 Which mechanism, S_N1 or S_N2, would you predict for each of the following reactions?

a. $$CH_3\underset{\underset{Br}{|}}{C}HCH_2CH_2CH_3 + Na^+SH^- \longrightarrow CH_3\underset{\underset{SH}{|}}{C}HCH_2CH_2CH_3 + NaBr$$

b. $CH_3\underset{\displaystyle Br}{\underset{|}{C}}HCH_2CH_2CH_3 + CH_3OH \longrightarrow CH_3\underset{\displaystyle OCH_3}{\underset{|}{C}}HCH_2CH_2CH_3 + HBr$

c. $CH_3\underset{\displaystyle Br}{\underset{|}{C}}HCH_2CH_2CH_3 + CH_3O^-Na^+ \longrightarrow CH_3\underset{\displaystyle OCH_3}{\underset{|}{C}}HCH_2CH_2CH_3 + NaBr$

Now let us turn to reactions like those in eqs. 6.5 and 6.7 to try to explain why a reaction other than substitution occurred.

6.8 Dehydrohalogenation, an Elimination Reaction; the E2 and E1 Mechanisms

We have now seen several examples of reactions in which two reactants give, not a single product, but mixtures. Examples include halogenation of alkanes (eq. 2.14), addition to double bonds (eqs. 3.14 and 3.29), and electrophilic aromatic substitutions (Sec. 4.12), where more than one isomer may be formed from the same two reactants. Even in nucleophilic substitution, more than one substitution product may form. For example, hydrolysis of a single alkyl bromide gives a mixture of two alcohols in eq. 6.14. But sometimes we find **two entirely different reaction types occurring at the same time between the same two reactants,** to give two (or more) entirely different types of products. Let us consider one example.

When an alkyl halide with a hydrogen attached to the carbon *adjacent* to the halogen-bearing carbon reacts with a nucleophile, two competing reaction paths are possible: **substitution** or **elimination.**

$$-\overset{\overset{\displaystyle H}{|_2}}{\underset{|}{C}}-\overset{|_1}{\underset{|}{C}}-X + Nu:^- \xrightarrow{\text{substitution } (S)} -\overset{\overset{\displaystyle H}{|}}{\underset{|}{C}}-\overset{|}{\underset{|}{C}}-Nu + X^- \quad (6.15)$$

$$\xrightarrow{\text{elimination } (E)} \;\; \diagdown C=C \diagup \;\; + \; Nu-H + X^- \quad (6.16)$$

In the substitution reaction, the nucleophile replaces the halogen X. In the elimination reaction, the nucleophile acts as a base and removes a proton from carbon 2, the carbon next to the one that bears the halogen X. The halogen X and the hydrogen from the *adjacent* carbon atom are *eliminated,* and a new bond (a pi bond) is formed between carbons 1 and 2.* The symbol E is used to designate an elimination process. Since, in this case, a hydrogen halide is eliminated, the reaction is called **dehydrohalogenation.** Elimination reactions provide a useful way to prepare compounds with double or triple bonds.

Often substitution and elimination reactions occur simultaneously with the same set of reactants—a nucleophile and a substrate. One reaction type or the other may predominate, depending on the structure of the nucleophile, the structure of the substrate, and other reaction conditions. As with substitution

---

* For a discussion of pi bonds and bonding in alkenes, review Sec. 3.4.

reactions, **there are two main mechanisms for elimination reactions, designated E2 and E1.** To learn how to control these reactions, we must first understand their mechanisms.

Like the S_N2 mechanism, **the E2 mechanism is a one-step process.** *The nucleophile, acting as a base, removes the proton (hydrogen) on a carbon atom adjacent to the one that bears the leaving group. At the same time, the leaving group departs and a double bond is formed.* The bond-breaking and bond-making that occurs during an E2 reaction is shown by the curved arrows:

$$\text{Nu}^{-}: + \text{H}-\text{C}-\text{C}-\text{L} \xrightarrow{\text{E2}} \text{C}=\text{C} + \text{Nu}-\text{H} + :\text{L}^{-} \quad (6.17)$$

The preferred conformation for the substrate in an E2 reaction is also shown in eq. 6.17. The H—C—C—L atoms lie in a single plane, with H and L in an *anti*-arrangement. The reason for this preference is that the C—H and C—L bonds are parallel in this conformation. This alignment is needed to make the new pi bond as the C—H and C—L bonds break.

The E1 mechanism is a two-step process and has the same first step as the S_N1 mechanism, the slow and rate-determining ionization of the substrate to give a carbocation (compare with eq. 6.12).

$$\underset{\text{substrate}}{\text{H}-\text{C}-\text{C}-\text{L}} \underset{}{\overset{\text{slow}}{\rightleftharpoons}} \underset{\text{carbocation}}{\text{H}-\text{C}-\text{C}^{+}} + :\text{L}^{-} \quad (6.18)$$

Two reactions are then possible for the carbocation. It may either combine with a nucleophile (the S_N1 process) or lose a proton from a carbon atom adjacent to the positive carbon, as shown by the curved arrow, to give an alkene (the E1 process).

$$\underset{\text{carbocation}}{\text{H}-\text{C}-\text{C}^{+}} \begin{cases} \xrightarrow{\text{Nu:}} \text{H}-\text{C}-\text{C}-\text{Nu} & S_N1 \\ \xrightarrow{-\text{H}^{+}} \text{C}=\text{C} + \text{H}^{+} & \text{E1} \end{cases} \quad (6.19)$$

6.9 Substitution and Elimination in Competition

Now we can consider specific examples of how substitution and elimination reactions compete with one another. Let us consider the options for each class of alkyl halide.

6.9a. Tertiary Halides Substitution can only occur by the S_N1 mechanism, but elimination can occur by either the E1 or E2 mechanism. With weak nucleophiles and polar solvents, the S_N1 and E1 mechanisms compete. For example,

$$(CH_3)_3CBr \underset{}{\overset{H_2O}{\rightleftharpoons}} (CH_3)_3C^+ + Br^- \begin{cases} \xrightarrow{H_2O,\ S_N1} (CH_3)_3COH \text{ (about 80\%)} \\ \xrightarrow{E1} (CH_3)_2C{=}CH_2 + H^+ \text{ (about 20\%)} \end{cases} \tag{6.20}$$

t-butyl bromide

Can the ratio of the S_N1/E1 products be altered under these conditions? No, because both products arise from a common intermediate, the carbocation.

If we use a strong nucleophile instead of a weak one, and if we use a less polar solvent, we favor elimination by the E2 mechanism. Thus, with OH^- or CN^- as nucleophiles, only elimination occurs (eqs. 6.5 and 6.7), and the exclusive product is the alkene.

$$HO^- + H_3C{-}C(CH_3)_2Br \xrightarrow{E2} H_2C{=}C(CH_3)_2 + H_2O + Br^- \tag{6.21}$$

t-butyl bromide → methylpropene (100%)

Remember that the tertiary carbon is too hindered sterically for S_N2 attack, so under these conditions, substitution does not compete with elimination.

6.9b. Primary Halides Only the S_N2 and E2 mechanisms are possible, because ionization to a primary carbocation, the first step that would be required for the S_N1 or E1 mechanisms, does not occur.

In general, with most nucleophiles, *primary halides give mainly substitution products* (S_N2). Only with very bulky, strongly basic nucleophiles do we see the E2 process favored. For example,

$$CH_3CH_2CH_2CH_2Br \xrightarrow[\text{in ethanol}]{CH_3CH_2O^-\,Na^+} CH_3CH_2CH_2CH_2OCH_2CH_3 + CH_3CH_2CH{=}CH_2$$

1-bromobutane → butyl ethyl ether (S_N2; 90%) + 1-butene (E2, 10%)

$$CH_3CH_2CH_2CH_2Br \xrightarrow[\text{in } t\text{-butyl alcohol}]{(CH_3)_3C{-}O^-\,K^+} CH_3CH_2CH_2CH_2OC(CH_3)_3 + CH_3CH_2CH{=}CH_2 \tag{6.22}$$

1-bromobutane → butyl t-butyl ether (S_N2; 15%) + 1-butene (E2, 85%)

Potassium *t*-butoxide is a bulky base. Hence substitution is retarded, and the main reaction is elimination.

6.9c. Secondary Halides All four mechanisms, S_N2 and E2, S_N1 and E1, are possible. The product composition is sensitive to the nucleophile (its strength as a nucleophile and as a base) and to the reaction conditions (solvent, temperature). In general, substitution is favored with good nucleophiles that are not strong bases (S_N2) or by weaker nucleophiles such as polar solvents (S_N1), but elimination is favored by strong bases (E2).

$$CH_3CH(Br)CH_3 \text{ (2-bromopropane)} \xrightarrow[\text{strong nucleophile}]{CH_3CH_2S^- Na^+} CH_3CH(SCH_2CH_3)CH_3 \; (S_N2)$$

$$CH_3CH(Br)CH_3 \xrightarrow[\text{weak nucleophile}]{CH_3CH_2OH} \underset{(S_N1;\ \text{major})}{CH_3CH(OCH_2CH_3)CH_3} + \underset{(E1;\ \text{minor})}{CH_3CH{=}CH_2} \tag{6.23}$$

$$CH_3CH(Br)CH_3 \xrightarrow[\text{strong base}]{CH_3CH_2O^- Na^+} \underset{(S_N2;\ \text{minor})}{CH_3CH(OCH_2CH_3)CH_3} + \underset{(E2;\ \text{major})}{CH_3CH{=}CH_2}$$

EXAMPLE 6.9 Predict the product of the reaction of 1-bromo-1-methylcyclohexane with

a. sodium ethoxide in ethanol.
b. refluxing ethanol.

Solution The alkyl bromide is tertiary

[cyclohexane ring bearing CH_3 and Br on the same carbon]

a. The first set of conditions favors the E2 process, because sodium ethoxide is a strong base. Two elimination products are possible, depending on whether the base attacks a hydrogen on an adjacent CH_2 or CH_3 group.

[1-methylcyclohexene, ring—CH_3] and [methylenecyclohexane, ring$=CH_2$]

b. This set of conditions favors ionization, because the ethanol is neutral (hence a weak nucleophile) and, as a solvent, fairly polar. The S_N1 process predominates, and the main product is the ether.

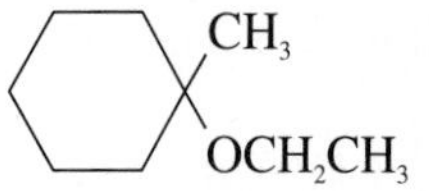

(1-ethoxy-1-methylcyclohexane structure: cyclohexane ring bearing CH_3 and OCH_2CH_3)

Some of the above alkenes will also be formed by the E1 mechanism.

EXAMPLE 6.10 Draw structures for *all* possible E2 products when 2-bromobutane reacts with concentrated sodium hydroxide.

Solution

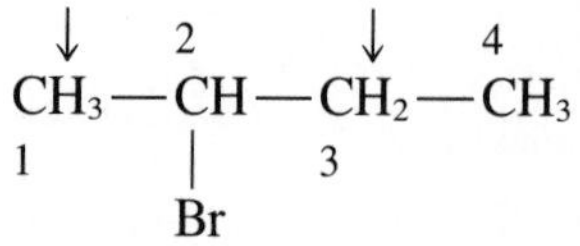

$$\overset{\downarrow}{\underset{1}{CH_3}}-\overset{2}{\underset{|\atop Br}{CH}}-\overset{\downarrow}{\underset{3}{CH_2}}-\overset{4}{CH_3}$$

The base can attack the hydrogens on the adjacent CH_3 or CH_2 group, as indicated by the arrows. Attack at the methyl group puts the double bond between carbon-1 and carbon-2.

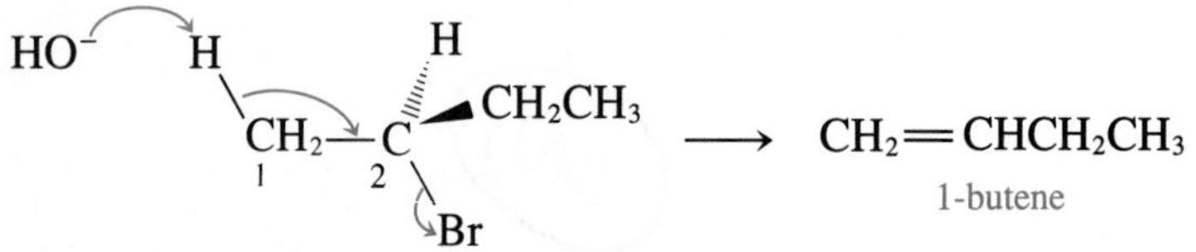

$$HO^- \; H-\underset{1}{CH_2}-\underset{2}{C}(H)(CH_2CH_3)Br \longrightarrow \underset{\text{1-butene}}{CH_2{=}CHCH_2CH_3}$$

Attack at the methylene group puts the double bond between carbon-2 and carbon-3, giving 2-butene. Either *cis*- or *trans*-2-butene can be obtained, so altogether, *three* alkenes are formed.

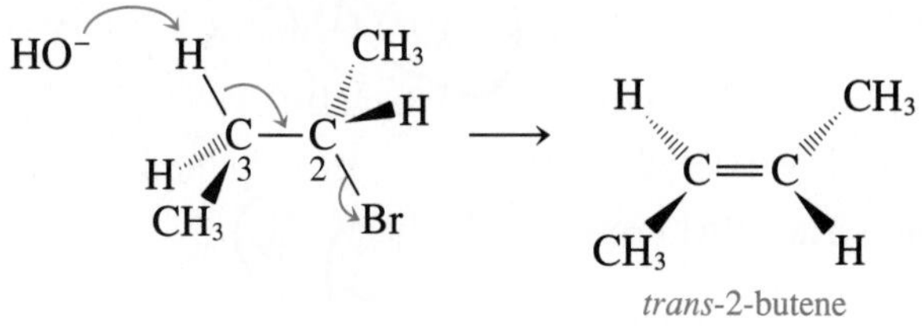

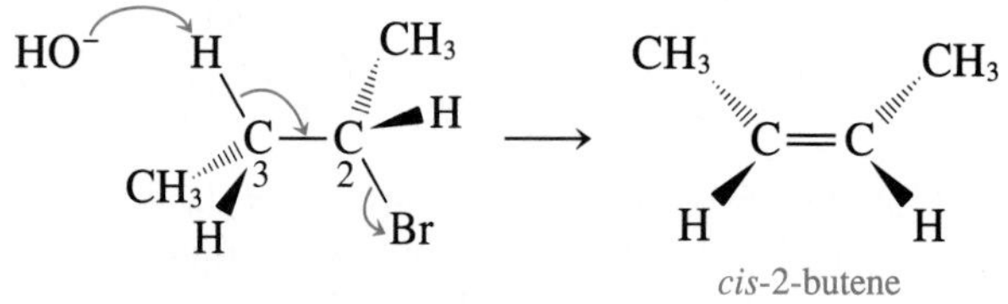

$$HO^- \; H-\underset{3}{C}(H)(CH_3)-\underset{2}{C}(CH_3)(H)Br \longrightarrow (H)(CH_3)C{=}C(CH_3)(H)$$

trans-2-butene

$$HO^- \; H-\underset{3}{C}(CH_3)(H)-\underset{2}{C}(CH_3)(H)Br \longrightarrow (CH_3)(H)C{=}C(CH_3)(H)$$

cis-2-butene

PROBLEM 6.9 Draw structures for *all* possible elimination products obtainable from 3-bromo-3-methylhexane.

PROBLEM 6.10 Treatment of the alkyl halide in Problem 6.9 with KOH in methanol gives mainly a mixture of the alkenes whose structures you drew. But treatment with only methanol gives a different product. What is it, and by what mechanism is it formed?

A Word About . . .

9. S_N2 Reactions in the Cell: Biochemical Methylations

Substitution and elimination reactions are so useful that it is not surprising that they occur in living matter. However, alkyl halides are not compatible with cytoplasm, being hydrocarbon-like and therefore water insoluble. In the cell, **alkyl phosphates** play the role that alkyl halides do in the laboratory. Adenosine triphosphate (ATP) is an example of a biological equivalent of an alkyl halide. We will abbreviate its structure here as Ad—O—Ⓟ—Ⓟ—Ⓟ (the full structure is given in Sec. 18.13). Ad- can be considered as a primary alkyl group, and the triphosphate group, —O—Ⓟ—Ⓟ—Ⓟ, acts as a leaving group, just like a halogen.

There are many compounds in nature with a methyl group attached to an oxygen or nitrogen atom. Examples include *mescaline* (a hallucinogen from the peyote cactus), which has three —OCH_3 groups; *morphine* (the pain-relieving drug from opium), which has an >N—CH_3 group; and *codeine* (a close relative of morphine, used as an anticough agent), which has an >N—CH_3 group and an —OCH_3 group.

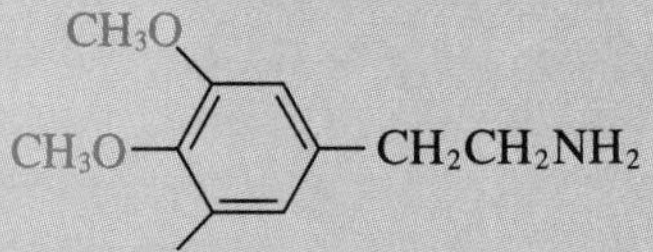

mescaline

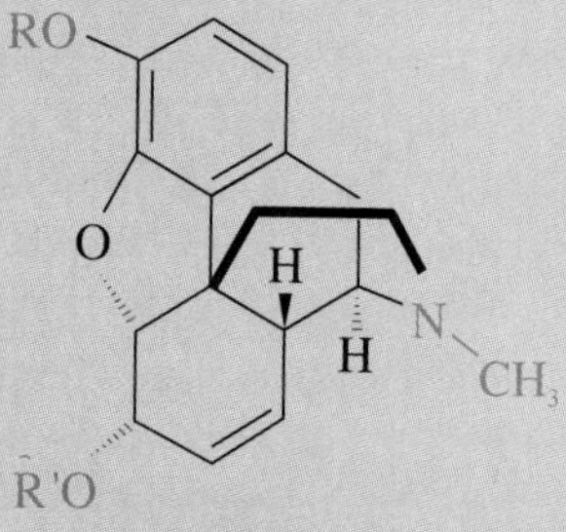

morphine (R = R' = H)
codeine (R = CH_3, R' = H)

How do the methyl groups get there? Two steps are involved, both of them nucleophilic substitutions.

The methyl carrier in most biochemical methylations is a sulfur-containing amino acid called *methionine*. In the first step, methionine is alkylated by ATP to form S-adenosylmethionine (shown in Figure 6.3). This reaction is a biological example of reaction 12 in Table 6.1. The methionine acts as a sulfur nucleophile in an S_N2 reaction and displaces the triphosphate ion.

In the second step, the oxygen or nitrogen atom to be methylated acts as a nucleophile. The S-adenosylmethionine acts just like a methyl halide. Indeed, it has been shown that these methylation reactions take place with inversion of configuration.

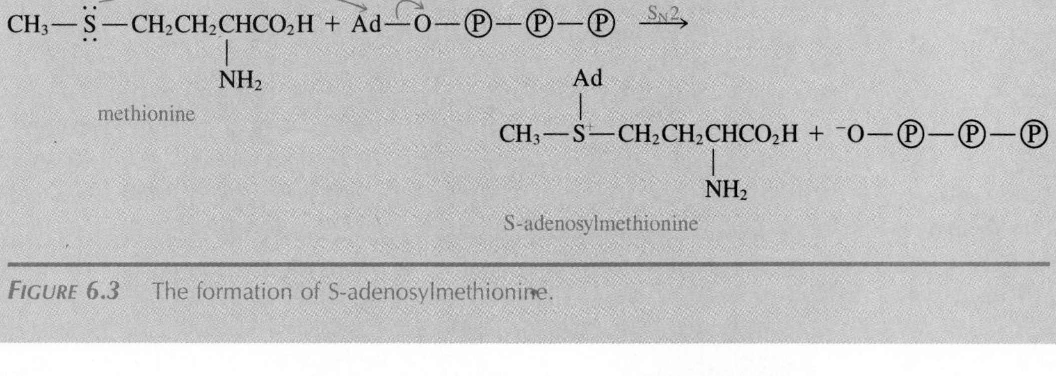

FIGURE 6.3 The formation of S-adenosylmethionine.

$$\begin{matrix} -\ddot{O}H \\ \text{or} \\ -\ddot{N}H_2 \end{matrix} + CH_3-\overset{\overset{Ad}{|}}{S^+}-CH_2CH_2\underset{\underset{NH_2}{|}}{C}HCO_2H \xrightarrow{S_N2}$$

$$\begin{matrix} -OCH_3 \\ \text{or} \\ -NHCH_3 \end{matrix} + Ad-S-CH_2CH_2\underset{\underset{NH_2}{|}}{C}HCO_2H$$

S-adenosylhomocysteine

Eventually, the S-adenosylhomocysteine formed in the second step is converted, by enzymes, back to ATP and methionine for reuse.

Biochemical methylations are just one of many examples of nucleophilic substitutions that occur in metabolic processes.

6.10 Polyhalogenated Aliphatic Compounds

Because of their useful properties, many polyhalogen compounds are produced commercially. Because they are industrial chemicals, they are usually given common names.

Chlorinated methanes* are made by the chlorination of methane (eqs. 2.11 and 2.13). **Carbon tetrachloride** (CCl_4, bp 77°C), **chloroform** ($CHCl_3$, bp 62°C), and **methylene chloride** (CH_2Cl_2, bp 40°C) are all water insoluble, but effective solvents for organic compounds. Also important for this purpose are the **tri-** and **tetrachloroethylenes,** used in dry cleaning and as degreasing agents in metal and textile processing.

$Cl_2C{=}CHCl$	$Cl_2C{=}CCl_2$
trichloroethylene	tetrachloroethylene
bp 87°C	bp 121°C

Because some of these chlorinated compounds are suspected carcinogens, adequate ventilation is essential when they are used as solvents.

Tetrafluoroethylene is the raw material for **Teflon,** a polymer related to polyethylene (Sec. 3.16) but with all of the hydrogens replaced by fluorine atoms.

$$n\ CF_2{=}CF_2 \xrightarrow[\text{catalyst}]{\text{peroxide}} \underset{\text{Teflon}}{\left(CF_2CF_2\right)_n} \quad (6.24)$$

* The analogous F, Br, and I compounds are also known, but are much more expensive and not commercially important.

Teflon has exceptional properties. It is resistant to almost all chemicals and is widely used as a nonstick coating for pots, pans, and other cooking utensils. It stands up to heat and prevents food from sticking to the utensil surface, making it easy to clean. Another use of Teflon is in Gore-Tex-like fabrics, materials with as many as nine billion pores per square inch. These pores are the right size to transmit water vapor but not liquid water. Thus, perspiration vapor can pass through the fabric, but wind, rain, and snow cannot. Gore-Tex has revolutionized cold- and wet-weather gear for both military and civilian uses. It is used in skiwear, boots, sleeping bags, tents, and other rugged outdoor gear. Teflon is also used in wire insulation, as a protective coating over glass and fabric roofs, in vascular grafts and heart patches, and recently in delicate films for certain biological experiments outside the earth's atmosphere.

Nonpolymeric perfluorochemicals (hydrocarbons, ethers, or amines in which all of the hydrogens are replaced by fluorine atoms) also have fascinating and useful properties. For example, perfluorochemicals such as perfluorotributylamine $(CF_3CF_2CF_2CF_2)_3N$ and F-decalin $(C_{10}F_{18})$ can dissolve as much as 60% oxygen by volume. By contrast, whole blood dissolves only about 20%, and blood plasma about 3%. Because of this property, these perfluorochemicals are important components of artificial blood.

Other polyhalogen compounds that contain two or three different halogens per molecule are commercially important. The best known are the **chlorofluorocarbons** (**CFCs,** formerly known as **Freons**). The two that have been produced on the largest scale are CFC-11 and CFC-12, made by fluorination of carbon tetrachloride.

$$\underset{\text{(bp 77°C)}}{CCl_4} \xrightarrow[SbF_5]{HF} \underset{\substack{\text{trichlorofluoromethane} \\ \text{(CFC-11)} \\ \text{bp 24°C}}}{CCl_3F} \xrightarrow[SbF_5]{HF} \underset{\substack{\text{dichlorodifluoromethane} \\ \text{(CFC-12)} \\ \text{bp −30°C}}}{CCl_2F_2} \quad (6.25)$$

They are used as refrigerants, as blowing agents in the manufacture of foams, as cleaning fluids, and as aerosol propellants. They are exceptionally stable. Because of this stability, they accumulate in the upper stratosphere, where they damage the earth's ozone layer. Consequently, their use for nonessential propellant purposes is now banned in the United States, and the search is on for replacements (see "A Word About" number 10).

Bromine-containing compounds of this type are now widely used to extinguish fires. Called **Halons,** the best known are

$CBrClF_2$	$CBrF_3$
bromochloro-difluoromethane (Halon-1211)	bromotrifluoro-methane (Halon-1301)

Halons are much more effective than carbon tetrachloride. They are very important in air safety because of their ability to douse fires within seconds.

A Word About . . .

10. CFCs, the Ozone Layer, and Tradeoffs

The story of CFCs has many lessons for us. Fluorine's reputation as one of the most reactive of the elements was widespread. It was quite remarkable and unexpected, therefore, that fluorocarbons and chlorofluorocarbons should be so extremely *un*reactive. (Indeed, Thomas Midgley, who discovered CFCs, demonstrated their nontoxicity, nonflammability, and noncorrosiveness to the public by inhaling them and then puffing out a candle.)

Because of these and other properties, at least *four* major commercial uses of CFCs were developed. Their low boiling points and other heat properties make them excellent *refrigerants,* far superior to ammonia, sulfur dioxide, and other rather difficult-to-handle refrigerants. Thus CFCs became widely used in freezers and refrigerators, as well as in air conditioners. CFCs make excellent *blowing agents* for rigid foams (such as those used for ice chests, fast-food take-out boxes, and other packaging materials) and for flexible foams (like those used to make pillows and furniture cushions). Their low surface tension and low viscosity give them excellent wetting properties, which led to their use as *cleaning fluids* for printed computer circuits, artificial limbs, and many other products. Finally, they were used as *propellants* in aerosol sprays. CFCs were manufactured on such a scale that they constituted a multibillion-dollar industry.

But their extreme stability led to a major world problem. CFCs are so stable that, when they are released into the atmosphere, they do not decompose in the lower atmosphere, as do most other industrial chemicals. Instead, they eventually rise to the stratosphere where, through ultraviolet radiation, the C—Cl bonds are broken and chlorine atoms are released. These chlorine atoms initiate a chain of reactions that damages the ozone layer, needed to protect life on earth from harmful ultraviolet rays.*

How to solve the problem? Ban all use of CFCs? To do so could bring about a major crisis for civilization. How, without refrigeration, could we ensure safe and adequate food supplies to urban areas, or delivery of heat-sensitive medical supplies? A return to old-fashioned refrigerants would also be fraught with environmental hazards. So the solution is not simple.

What has been done is to cut back or ban some less essential uses gradually. For example, in the United States (but *not* worldwide), all nonessential aerosol use of CFCs is banned. In 1987, twenty-four countries signed the Montreal Protocol, an agreement that called for cutting CFC use to half of the 1986 level by 1998. This has since been expanded to a complete ban on CFC production and use by 1996.

Chemists and other scientists have been seeking replacements to fill the gap left by elimination of CFC production and use. The properties of polyhalogen compounds are so uniquely useful that for the moment the best chance to find CFC substitutes is among this type of compound. For example, it was found that introducing one or more hydrogen atoms into the molecule substantially increases its decomposition rate in the lower atmosphere, where no damage to the ozone layer is possible. Compounds such as CF_3CHCl_2 (HCFC-123, bp 28°C) and CF_3CH_2F(HFC-134a, bp −26°C) are possible replacements for CFC-11 and CFC-12, respectively. These compounds are more difficult to synthesize than CFCs but continued use of the cheaper CFCs and destruction of the ozone layer is not a viable alternative.

CFCs are just one of many examples of the tradeoffs between beneficial and possible harmful effects of a new research product. During World War II, the use of CFCs as propellants for the insecticide DDT saved the lives of many troops in the Pacific zone who were suffering more casualties from malaria than from enemy action. But later, indiscriminate use of CFCs as propellants for all sorts of trivial purposes led to problems with their environmental accumulation. Chemicals (in this case, CFCs) are neither good nor evil, but we must exercise good judgment in how we use them.

* For an interesting account of the discovery and development of the CFC-ozone problem by F. Sherwood Rowland and Mario J. Molina, chemists who initiated ozone depletion research, see *Chemical and Engineering News,* **1994,** *77* (33), 8–13.

A Word About . . .

11. Insecticides and Herbicides

Many polyhalogen compounds have been used as insecticides and herbicides. Perhaps the best-known is DDT (*di*chloro*di*phenyl*tri*chloroethane). It is manufactured by the acid-catalyzed reaction of chlorobenzene with trichloroacetaldehyde, an example of an electrophilic aromatic substitution.

$$2\ C_6H_5\text{—Cl} + CCl_3CH{=}O \xrightarrow{H^+} Cl\text{—}C_6H_4\text{—}\underset{\underset{CCl_3}{|}}{CH}\text{—}C_6H_4\text{—Cl} + H_2O$$

DDT

DDT was used during World War II to control malaria by killing mosquitoes. It is also effective against flies and many agricultural pests. Unfortunately, DDT is not easily degraded biochemically, and excessive use resulted in its accumulation in the environment. It tends to accumulate in fat tissues and can cause harm, particularly to fish and birds. Currently the use of DDT is restricted, and annual world production is only about 80,000 tons.

Weeds create formidable problems for agriculture. They consume nutrients and moisture needed by crops. They also rob crops of sunlight and space, thus reducing crop yields. U.S. agricultural production is diminished by about 10% because of weeds. The annual financial loss is about \$12 billion, and an additional \$6 billion is spent annually on weed control.

One way to control weeds is through the use of herbicides. About 85% to 90% of U.S. corn, soybean, cotton, peanut, and rice acreage is sprayed with herbicides to control weeds. Some are sprayed before the crop is planted, others are applied after planting but before emergence of the crop, and still others are applied to the weed foliage itself. The rising world population makes increasing demands on food production and, ultimately, on herbicide use to meet that demand. Although most herbicides are used for this purpose, some are also employed for industrial purposes—along railway and power line rights of way, on roadsides, on rangeland, in vacant lots, and so on—as well as for lawns and gardens.

Farmers have used weed killers for many years; common salt was used for this purpose even in ancient times. Before World War II, most chemical herbicides were not very selective (they killed weeds but also damaged crops) and had to be used in large quantities per acre of land. The breakthrough came with the discovery that 2,4-dichlorophenoxyacetic acid (2,4-D) killed broadleaf weeds but allowed narrowleaf plants to grow unharmed and in greater yield. In addition, only 0.25 to 2.0 pounds were needed per acre (compared to more than 200 pounds per acre for inorganic herbicides such as sodium chlorate). 2,4-D is still the most popular herbicide used on wheat fields.

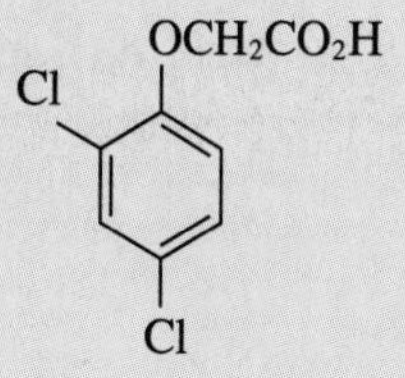

OCH_2CO_2H, Cl, Cl

2,4-dichlorophenoxyacetic acid
"2,4-D"

At least 40 herbicides are in large-scale use. Some of the most widely used are

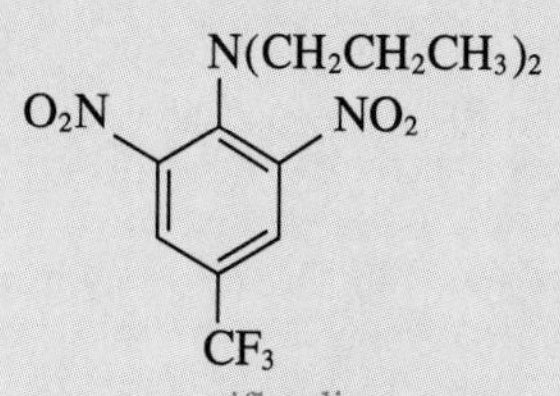

$N(CH_2CH_2CH_3)_2$, O_2N, NO_2, CF_3

trifluralin
(cotton, lima beans, cantaloupes, tomatoes, sugar beets)

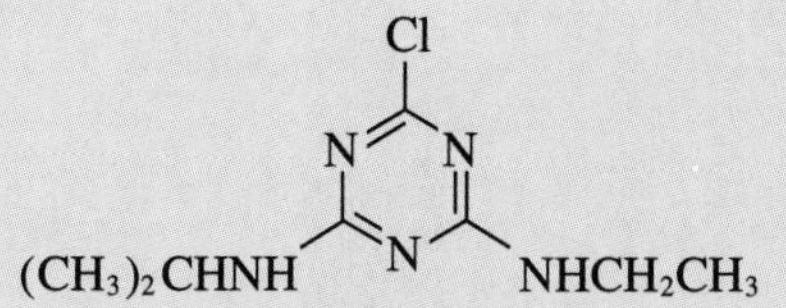

Cl, N, N, N, $(CH_3)_2CHNH$, $NHCH_2CH_3$

atrazine
(corn, sugarcane, pineapples)

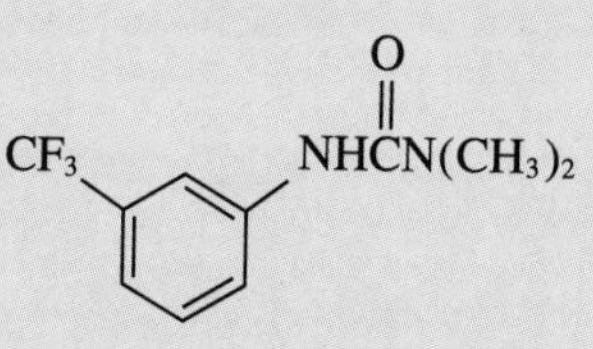

CF_3, $NHCN(CH_3)_2$ (C=O)

fluometuron
(cotton, sugarcane)

A new class of herbicides recently developed by DuPont represents a major breakthrough. One example is Glean™.

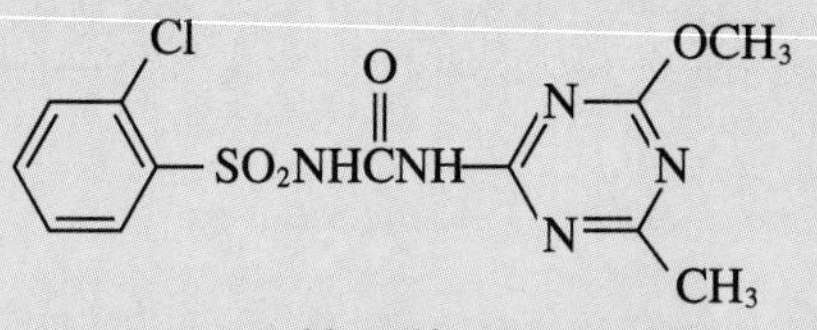

Cl, SO_2NHCNH (C=O), N, N, N, OCH_3, CH_3

chlorsulfuron
(Glean™)

It is effective against a broad spectrum of weeds in cereal grains (wheat, barley, oats) at the remarkably low application level of *less than one ounce per acre*. At present, at least a dozen pesticides of this type are marketed commercially, serving unique needs of farmers.

Because of the danger that, even in very low concentrations, pesticides may have harmful side effects and become environmental pollutants, and because agriculture without pesticides can no longer meet human needs, still other approaches to the important problem of pest control are being researched. For example, many plants have natural defense mechanisms against predators and pathogens; they produce substances that shield them from harm. These substances may kill harmful bacteria, repel insects or interfere with their reproductive cycles, prevent fungal spores from germinating, and in many other ways act as natural protection for a plant. One new approach to pesticides is to isolate these natural defense substances, learn how they work, and then either use these materials themselves (rotenone, the principal insecticidal constitutent of derris root, is a well-known example) or design synthetic pesticides based on their structures. One recent example is derived from the neem tree, native to arid regions of India, Pakistan, and Sri Lanka. It has been known for centuries that areas where it grows are virtually free of insects, nematodes, and plant diseases. Extracts of its seeds provide effective protection against more than 100 crop pests, aphids, and boll weevils; and a pesticide based on these extracts has recently been approved for limited use.

The achievements of modern agriculture and the feeding of the ever-growing world population would not be possible without the herbicides that chemists and other scientists have developed.

REACTION SUMMARY

1. **Nucleophilic Substitutions (S_N1 and S_N2)**

Alkyl halides react with a variety of nucleophiles to give alcohols, ethers, alkyl halides, alkynes, and other families of compounds. Examples are shown in Table 6.1 and Sec. 6.3.

$$\text{Nu:} + \text{R—X} \longrightarrow \text{R—Nu}^+ + \text{X}^-$$
$$\text{Nu:}^- + \text{R—X} \longrightarrow \text{R—Nu} + \text{X}^-$$

2. **Elimination Reactions (E1 and E2)**

Alkyl halides react with bases to give alkenes (Sec. 6.8).

$$\text{H}-\overset{|}{\underset{|}{\text{C}}}-\overset{|}{\underset{|}{\text{C}}}-\text{X} \xrightarrow{\text{B:}^-} \text{>C=C<} + \text{BH} + \text{X}^-$$

ADDITIONAL PROBLEMS

6.11. Using Table 6.1, write an equation for each of the following substitution reactions.

a. 1-bromobutane + sodium iodide (in acetone)
b. 2-chlorobutane + sodium ethoxide
c. *t*-butyl bromide + water
d. *p*-chlorobenzyl chloride + sodium cyanide
e. *n*-propyl iodide + sodium acetylide
f. 2-chloropropane + sodium hydrosulfide
g. allyl chloride + ammonia (2 equivalents)
h. 1,4-dibromobutane + sodium cyanide (excess)
i. 1-methyl-1-bromocyclohexane + methanol

6.12. Select an alkyl halide and a nucleophile that will give each of the following products:

a. $CH_3CH_2CH_2NH_2$
b. $CH_3CH_2OCH_2CH_3$
c. $HC{\equiv}CCH_2CH_2CH_3$
d. $(CH_3)_2CHSCH(CH_3)_2$
e. C_6H_5—CH_2CN (benzene ring—CH_2CN)
f. C_6H_5—OCH_2CH_3 (benzene ring—OCH_2CH_3)

6.13. Draw each of the following equations in a way that shows clearly the stereochemistry of the reactants and products.

a. (*S*)-2-bromobutane + sodium methoxide (in methanol) $\xrightarrow{S_N2}$ 2-methoxybutane

b. (*R*)-3-bromo-3-methylhexane + methanol $\xrightarrow{S_N1}$ 3-methoxy-3-methylhexane

c. *cis*-1-bromo-4-methylcyclohexane + NaSH $\longrightarrow$ 4-methylcyclohexanethiol

6.14. Use Fischer projection formulas to show the stereochemistry of Problem 6.13, parts a and b.

6.15. Determine the order of reactivity for $(CH_3)_2CHCH_2Br$, $(CH_3)_3CBr$, and $CH_3\underset{\displaystyle |\atop \displaystyle Br}{C}HCH_2CH_3$ in substitution reactions with

a. sodium cyanide. **b.** 50% aqueous acetone.

6.16. When treated with sodium iodide, a solution of (*R*)-2-iodooctane in acetone gradually loses all its optical activity. Explain.

6.17. Equation 6.20 shows that hydrolysis of *t*-butyl bromide gives about 80% $(CH_3)_3COH$ and 20% $(CH_3)_2C{=}CH_2$. The same ratio of alcohol to alkene is obtained when the starting halide is *t*-butyl chloride or *t*-butyl iodide. Explain.

6.18. Tell what product you expect, and by what mechanism it is formed, for each of the following reactions:

a. 1-chloro-1-methylcyclohexane + ethanol

b. 1-chloro-1-methylcyclohexane + sodium ethoxide (in ethanol)

6.19. Give the structures of all possible products when 2-chloro-2-methylbutane reacts by the E1 mechanism.

6.20. Explain the different products of the following two reactions by considering the mechanism by which each reaction proceeds:

$$CH_2{=}CH{-}\underset{\displaystyle Br}{\underset{|}{CH}}{-}CH_3 + Na^+\ {}^-OCH_3 \xrightarrow{CH_3OH} CH_2{=}CH{-}\underset{\displaystyle OCH_3}{\underset{|}{CH}}{-}CH_3$$

$$CH_2{=}CH{-}\underset{\displaystyle Br}{\underset{|}{CH}}{-}CH_3 + CH_3OH \longrightarrow$$

$$CH_2{=}CH{-}\underset{\displaystyle OCH_3}{\underset{|}{CH}}{-}CH_3 + \underset{\displaystyle OCH_3}{\underset{|}{CH_2}}CH{=}CHCH_3$$

6.21. Combine an electrophilic addition and a nucleophilic substitution to devise a two-step synthesis of

a. $CH_3\underset{\displaystyle OCH_3}{\underset{|}{C}}HCH_2CH_3$ from $CH_3CH{=}CHCH_3$.

b. $CH_3{-}\overset{\displaystyle CH_3}{\overset{|}{\underset{\displaystyle OCH_3}{\underset{|}{C}}}}{-}CH_2CH_3$ from $CH_3{-}\overset{\displaystyle CH_3}{\overset{|}{C}}{=}CHCH_3$.

c. $C_6H_5{-}\underset{\displaystyle CN}{\underset{|}{C}}HCH_3$ from $C_6H_5{-}CH{=}CH_2$.

6.22. Combine the reaction of an alcohol with sodium and a nucleophilic substitution to devise a two-step synthesis of

a. $CH_3OCH_2CH_3$ from an alkoxide and an alkyl halide.

b. $CH_3OC(CH_3)_3$ from an alcohol and an alkyl halide.

6.23. Combine the reaction in eq. 3.53 with a nucleophilic substitution to devise

a. a two-step synthesis of $CH_3C{\equiv}C{-}CH_2{-}C_6H_5$ (benzene ring) from

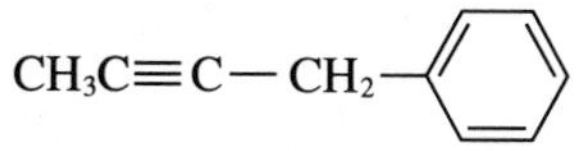

$CH_3C{\equiv}CH$ and $C_6H_5{-}CH_2Br$ (benzene ring).

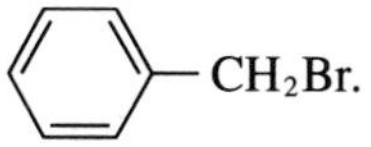

b. a four-step synthesis of $CH_3C{\equiv}CCH_2CH_3$ from acetylene and appropriate alkyl halides.

6.24. Combine an electrophilic addition with an elimination reaction to devise a two-step synthesis of

a. $(CH_3)_2C{=}CHCH_3$ from $CH_2{=}C(CH_3)CH_2CH_3$.

b. 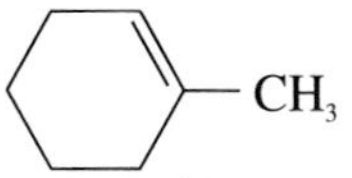 (cyclohexene ring)$-CH_3$ from (cyclohexane ring)$=CH_2$

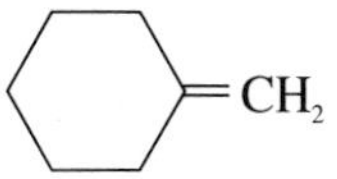

6.25. Combine a nucleophilic substitution with a catalytic hydrogenation to synthesize

a. $CH_3CH_2CH_2OH$ from $CH_2{=}CHCH_2Br$.

b. *cis*-2-butene from propyne and methyl iodide.

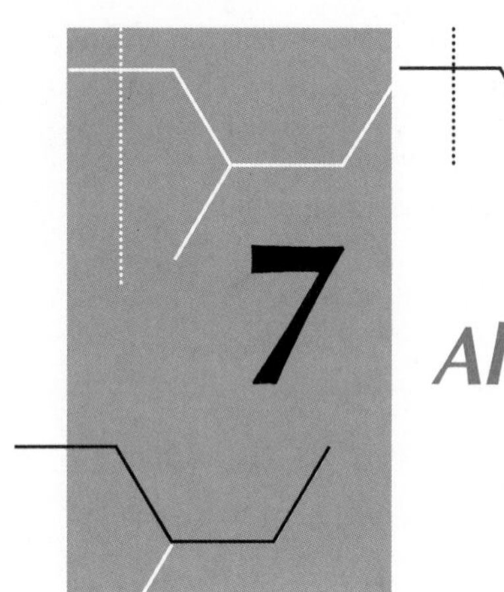

7 Alcohols, Phenols, and Thiols

7.1 Introduction

Alcohols have the general formula **R—OH,** and are structurally similar to water but with one of the hydrogens replaced by an alkyl group. Their functional group is the **hydroxyl group,** —OH. **Phenols** have the same functional group, but it is attached to an *aromatic* ring. **Thiols** are similar to alcohols and phenols, except that the oxygen is replaced by sulfur.

H—Ö—H (with two lone pairs on O)	R—Ö—H (with two lone pairs on O)	Ar—Ö—H (with two lone pairs on O)	R—S̈—H (with two lone pairs on S)	Ar—S̈—H (with two lone pairs on S)
water	an alcohol	a phenol	a thiol	a thiophenol

Alcohols, phenols, and thiols occur in nature. In this chapter, we will discuss the physical properties and main chemical reactions of these classes of compounds. We will also describe their commercial and laboratory syntheses and give examples of their biological importance.

7.2 Nomenclature of Alcohols

In the IUPAC system, the hydroxyl group in alcohols is indicated by the ending **-ol.** In common names the separate word *alcohol* is placed after the name of the alkyl group. The following examples illustrate the use of IUPAC rules, with common names given in parentheses.

CH_3OH	CH_3CH_2OH	$\overset{3}{C}H_3\overset{2}{C}H_2\overset{1}{C}H_2OH$	$\overset{1}{C}H_3\overset{2}{C}H\overset{3}{C}H_3$ (OH on C-2)
methanol (methyl alcohol)	ethanol (ethyl alcohol)	1-propanol (*n*-propyl alcohol)	2-propanol (isopropyl alcohol)

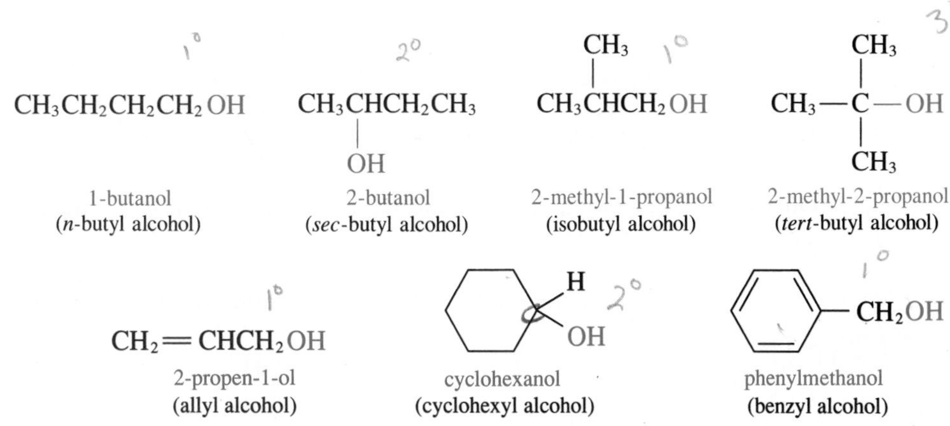

With unsaturated alcohols, two endings are needed: one for the double or triple bond and one for the hydroxyl group (see the IUPAC name for allyl alcohol). In these cases, the *-ol* suffix comes last and takes precedence in numbering.

EXAMPLE 7.1 Name the following alcohols by the IUPAC system:

a. $CLCH_2CH_2OH$ b. (cyclobutane ring bearing OH) c. $CH_3C{\equiv}CCH_2CH_2OH$

Solution a. 2-chloroethanol (number from the hydroxyl-bearing carbon)
b. cyclobutanol
c. 3-pentyne-1-ol (*not* 2-pentyne-5-ol)

PROBLEM 7.1 Name these alcohols by the IUPAC system:

a. $BrCH_2CH_2CH_2OH$ b. (cyclopentane ring bearing H and OH) c. $CH_2{=}CHCH_2CH_2OH$

PROBLEM 7.2 Write a structural formula for

a. 3-pentanol. b. 2-phenylethanol. c. 3-pentyn-2-ol.

7.3 *Classification of Alcohols*

Alcohols are classified as primary (1°), secondary (2°), or tertiary (3°), depending on whether one, two, or three organic groups are connected to the hydroxyl-bearing carbon atom.

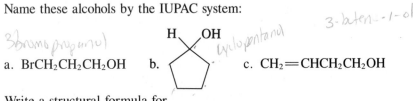

$$\underset{\text{primary (1°)}}{R—CH_2OH} \qquad \underset{\text{secondary (2°)}}{R—\overset{\displaystyle R}{\overset{|}{C}}HOH} \qquad \underset{\text{tertiary (3°)}}{R—\underset{\underset{\displaystyle R}{|}}{\overset{\overset{\displaystyle R}{|}}{C}}—OH}$$

Methyl alcohol, which is not strictly covered by this classification, is usually grouped with the primary alcohols. This classification is similar to that for carbocations (Sec. 3.10) and alkyl halides (Sec. 6.3). We will see that the chemistry of an alcohol sometimes depends on its class.

PROBLEM 7.3 Classify as 1°, 2°, or 3° the eleven alcohols listed in Sec. 7.2.

7.4 Nomenclature of Phenols

Phenols are usually named as derivatives of the parent compounds.

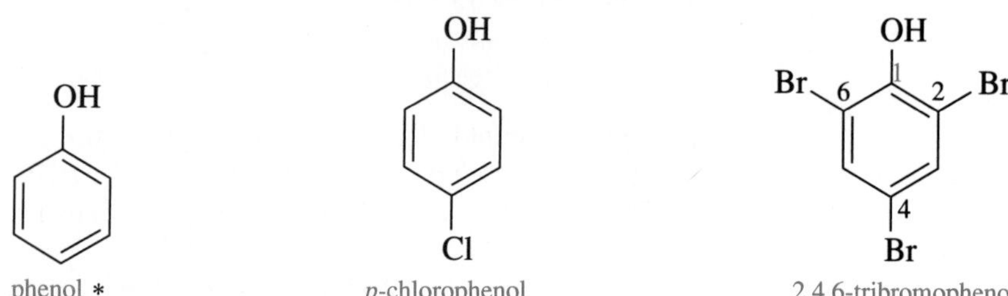

phenol *   *p*-chlorophenol   2,4,6-tribromophenol

The hydroxyl group is named as a substituent when it occurs in the same molecule with carboxylic acid, aldehyde, or ketone functionalities, which have priority in naming. Examples are

CO_2H / OH — *m*-hydroxybenzoic acid

CHO / OH — *p*-hydroxybenzaldehyde

but

OH / NO_2 — *p*-nitrophenol (*not* *p*-hydroxynitrobenzene)

PROBLEM 7.4 Write the structure for

a. *p*-ethylphenol.
b. pentachlorophenol (an insecticide for termite control, and a fungicide).
c. *o*-hydroxyacetophenone (for the structure of acetophenone see Sec. 4.7).

7.5 Hydrogen Bonding in Alcohols and Phenols

The boiling points of alcohols are much higher than those of ethers or hydrocarbons with similar molecular weights.

	CH_3CH_2OH	CH_3OCH_3	$CH_3CH_2CH_3$
mol wt	46	46	44
bp	+78.5°C	−24°C	−42°C

* The name *benzenol* has recently been introduced for phenol and its derivatives. Although this name is used by Chemical Abstracts Services, it is not in common use among organic chemists.

Why? Because their molecules form **hydrogen bonds** with one another. The O—H bond is polarized by the high electronegativity of the oxygen atom. This polarization places a partial positive charge on the hydrogen atom and a partial negative charge on the oxygen atom. Because of these charges and the small size of the hydrogen atom, it can link two electronegative atoms such as oxygen.

$$\underset{\text{two separate alcohol molecules}}{\overset{R}{\underset{\delta-}{O}}-\overset{\delta+}{H} + \overset{R}{\underset{\delta-}{O}}-\overset{\delta+}{H}} \rightleftharpoons \underbrace{\overset{R}{\underset{\delta-}{O}}-\overset{\delta+}{H}\cdots\overset{R}{\underset{\delta-}{O}}-\overset{\delta+}{H}}_{\text{a hydrogen bond}} \qquad (7.1)$$

Two or more alcohol molecules thus become loosely bonded to one another through hydrogen bonds.

Hydrogen bonds are weaker than ordinary covalent bonds.* Nevertheless, their strength is significant, about 5 to 10 kcal/mol (20 to 40 kJ/mol). Consequently, alcohols and phenols have relatively high boiling points because we must not only supply enough heat to vaporize each molecule but also supply enough heat (energy) to break the hydrogen bonds before each molecule can be vaporized.

Water, of course, is also a hydrogen-bonded liquid. The lower-molecular-weight alcohols can readily replace water molecules in the hydrogen-bonded network.

$$H-O(H)\cdots H-O(R)\cdots H-O-H$$

This accounts for the complete miscibility of the lower alcohols with water. However, as the organic chain lengthens and the alcohol becomes relatively more hydrocarbon-like, its water solubility decreases. Table 7.1 illustrates these properties.

TABLE 7.1 Boiling point and water solubility of some alcohols

Name	*Formula*	*bp, °C*	*Solubility in H_2O g/100 g at 20°C*
methanol	CH_3OH	65	completely miscible
ethanol	CH_3CH_2OH	78.5	completely miscible
1-propanol	$CH_3CH_2CH_2OH$	97	completely miscible
1-butanol	$CH_3CH_2CH_2CH_2OH$	117.7	7.9
1-pentanol	$CH_3CH_2CH_2CH_2CH_2OH$	137.9	2.7
1-hexanol	$CH_3CH_2CH_2CH_2CH_2CH_2OH$	155.8	0.59

* Covalent O—H bond strengths are about 120 kcal/mol (480 kJ/mol).

7.6 Acidity and Basicity Reviewed

The acid–base behavior of organic compounds often helps to explain their chemistry; this is certainly true of alcohols. It is a good idea, therefore, to review the fundamental concepts of acidity and basicity.

Acids and bases are defined in two ways. According to the **Brønsted-Lowry definition,** ***an acid is a proton donor, and a base is a proton acceptor.*** For example, in eq. 7.2, which represents what occurs when hydrogen chloride dissolves in water, the water accepts a proton from the hydrogen chloride.

$$\underset{\text{base}}{H-\ddot{O}(H):} + \underset{\text{acid}}{H-\ddot{Cl}:} \rightleftharpoons \underset{\text{conjugate acid of water}}{H-\overset{+}{\ddot{O}}(H)-H} + \underset{\text{conjugate base of hydrogen chloride}}{:\ddot{Cl}:^-} \quad (7.2)$$

Here water acts as a base or proton acceptor, and hydrogen chloride acts as an acid or proton donor. The products of this proton exchange are called the **conjugate acid** and the **conjugate base.**

The strength of an acid is measured quantitatively by its **acidity constant,** or **ionization constant,** $\boldsymbol{K_a}$**,** usually calculated in reference to dilute solutions of the acid in water. For example, any acid dissolved in water is in equilibrium with hydronium ions and its conjugate base.

$$HA + H_2O \rightleftharpoons H_3O^+ + A^- \quad (7.3)$$

K_a is related to the equilibrium constant for this reaction and is defined as follows:

$$K_a = \frac{[H_3O^+][A^-]}{[HA]} \quad (7.4)^*$$

The stronger the acid, the more this equilibrium is shifted to the right, thus increasing the concentration of H_3O^+ and the value of K_a.

For water, these expressions are

$$H_2O + H_2O \rightleftharpoons H_3O^+ + OH^- \quad (7.5)$$

and

$$K_a = \frac{[H_3O^+][OH^-]}{[H_2O]} = 1.8 \times 10^{-16} \quad (7.6)$$

PROBLEM 7.5 Verify from eq. 7.6 and from the molarity of water (55.5 M) that the concentrations of both H_3O^+ or OH^- in water are 10^{-7} moles per liter.

---

*The square brackets used in the expression for K_a indicate concentration, at equilibrium, of the enclosed species in moles per liter. The acidity constant K_a is related to the equilibrium constant for the reaction shown in eq. 7.4, only the concentration of water $[H_2O]$ is omitted from the denominator of the expression since it remains nearly constant at 55.5 M, very large compared to the concentrations of the other three species. For a discussion of reaction equilibria and equilibrium constants, see Sec. 3.11.

To avoid using numbers with negative exponents, such as those we have just seen for the acidity constant K_a for water, we often express acidity as **pK_a, *the negative logarithm of the acidity constant.***

$$pK_a = -\log K_a \tag{7.7}$$

The pK_a of water is

$$-\log(1.8 \times 10^{-16}) = -\log 1.8 - \log 10^{-16} = -0.26 + 16 = +15.74$$

The mathematical relationship between the values for K_a and pK_a means that ***the smaller K_a or the larger pK_a, the weaker the acid.***

It is useful to keep in mind that there is an inverse relationship between the strength of an acid and the strength of its conjugate base. In eq. 7.2, for example, hydrogen chloride is a *strong* acid since the equilibrium is shifted largely to the right. It follows, then, that the chloride ion must be a *weak* base since it has relatively little affinity for a proton. Similarly, since water is a *weak* acid, its conjugate base, hydroxide ion, must be a *strong* base.

Another way to define acids and bases was first proposed by G. N. Lewis. ***A Lewis acid is a substance that can accept an electron pair, and a Lewis base is a substance that can donate an electron pair.*** According to this definition, a proton is considered a Lewis acid because it can accept an electron pair from a donor (a Lewis base) to fill its 1*s* shell.

$$H^+ + :\ddot{O}(H)-H \rightleftharpoons \left[H-\ddot{O}(H)-H\right]^+ \tag{7.8}$$

Lewis acid (H⁺) — Lewis base (H₂O)

Any atom with an unshared electron pair can act as a Lewis base.

Compounds with an element whose valence shell is incomplete also act as Lewis acids. For example,

$$BF_3 + :\ddot{F}:^- \rightleftharpoons \left[BF_4\right]^- \tag{7.9}$$

Lewis acid (BF_3) — Lewis base (F^-)

Similarly, when $FeCl_3$ or $AlCl_3$ acts as a catalyst for electrophilic aromatic chlorination (eqs. 4.16 and 4.17) or the Friedel-Crafts reaction (eqs. 4.23 and 4.25), they are acting as Lewis acids; the metal atom accepts an electron pair from chlorine or from an alkyl or acyl chloride to complete its valence shell of electrons.

Finally, some substances can act as either an acid or a base, depending on the other reactant. For example, in eq. 7.2, water acts as a base (a proton acceptor). However, in its reaction with ammonia, water acts as an acid (a proton donor).

$$\underset{\substack{\text{water}\\\text{(acid)}}}{:\ddot{O}{-}H} \;(\text{with H on O}) + \underset{\substack{\text{ammonia}\\\text{(base)}}}{:NH_3} \rightleftharpoons \underset{\substack{\text{hydroxide}\\\text{ion}}}{H{-}\ddot{O}:^-} + \underset{\substack{\text{ammonium}\\\text{ion}}}{H{-}\overset{+}{N}H_3} \qquad (7.10)$$

Water acts as a base toward acids that are stronger than itself (HCl) and as an acid toward bases that are stronger than itself (NH_3). Substances that can act as either an acid or a base are said to be **amphoteric.**

PROBLEM 7.6 The K_a for ethanol is 1.0×10^{-16}. What is its pK_a?

PROBLEM 7.7 The pK_a's of hydrogen cyanide and acetic acid are 9.2 and 4.7, respectively. Which is the stronger acid?

PROBLEM 7.8 Which of the following are Lewis acids and which are Lewis bases?

a. $H:^-$ b. $(CH_3)_3B$ c. Mg^{2+}
d. CH_3OCH_3 e. $(CH_3)_3C^+$ f. CH_3NH_2
g. $(CH_3)_3N$ h. $(CH_3)_3C:^-$ i. Zn^{2+}

PROBLEM 7.9 In eq. 3.53, how is the amide ion, NH_2^-, functioning?

7.7 The Acidity of Alcohols and Phenols

Like water, alcohols and phenols are weak acids. The hydroxyl group can act as a proton donor, and dissociation occurs in a manner similar to that for water:

$$\underset{\text{alcohol}}{R\ddot{\underset{..}{O}}{-}H} \rightleftharpoons \underset{\substack{\text{alkoxide}\\\text{ion}}}{R\ddot{\underset{..}{O}}:^-} + H^+ \qquad (7.11)$$

The conjugate base of an alcohol is an **alkoxide ion** (for example, *meth*oxide ion from *meth*anol, *eth*oxide ion from *eth*anol, and so on).

Table 7.2 lists pK_a values for selected alcohols and phenols. Methanol and ethanol have nearly the same acid strength as water; bulky alcohols such as *t*-butyl alcohol are somewhat weaker because their bulk makes it difficult to solvate the corresponding alkoxide ion.

Phenol is a much stronger acid than ethanol. How can we explain this acidity difference between alcohols and phenols, since in both types of compounds the proton donor is a hydroxyl group?

Phenols are stronger acids than alcohols mainly because the corresponding phenoxide ions are stabilized by resonance. The negative charge of an alkoxide ion is concentrated on the oxygen atom, but the negative charge on a phenoxide ion can be delocalized to the *ortho* and *para* ring positions through resonance.

TABLE 7.2 pK_a's of selected alcohols and phenols in aqueous solution

Name	Formula	pK_a
water	HO—H	15.7
methanol	CH_3O—H	15.5
ethanol	CH_3CH_2O—H	15.9
t-butyl alcohol	$(CH_3)_3CO$—H	18
2,2,2-trifluoroethanol	CF_3CH_2O—H	12.4
phenol	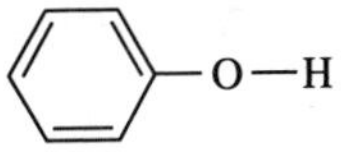 C_6H_5—O—H	10.0
p-nitrophenol	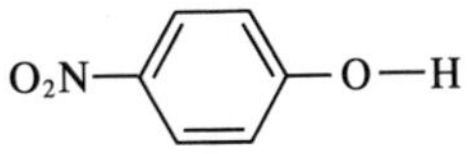 O_2N—C_6H_4—O—H	7.2
picric acid	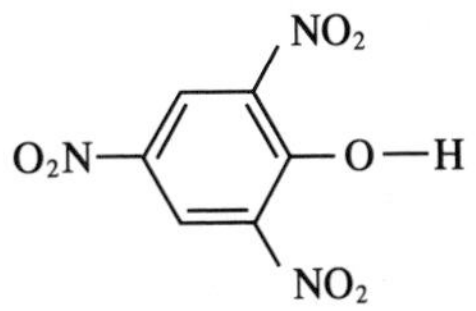 O_2N—$C_6H_2(NO_2)_2$—O—H	0.25

R—Ö:$^{(-)}$

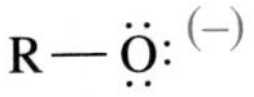

charge localized on the oxygen atom in alkoxide ions

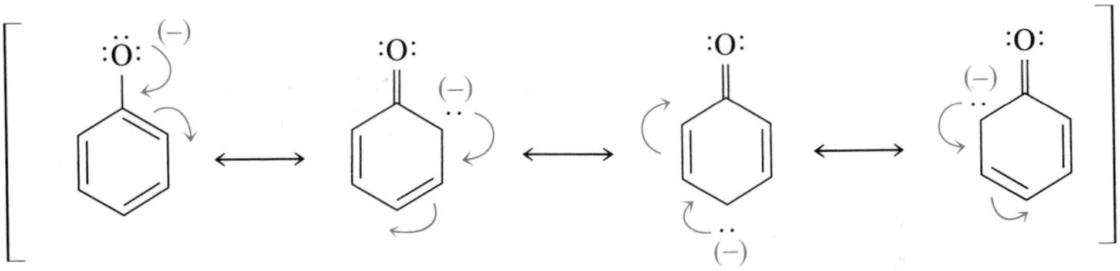

charge delocalized in phenoxide ion

Since phenoxide ions are stabilized in this way, the equilibrium for their formation is more favorable than that for alkoxide ions. Thus, phenols are stronger acids than alcohols.

We see in Table 7.2 that 2,2,2-trifluoroethanol is about 3000 times stronger an acid than is ethanol. How can we explain this effect of fluorine substitution? Again, think about the stabilities of the respective anions. Fluorine is a strongly electronegative element, so each C—F bond is polarized, with the fluorine partially negative and the carbon partially positive.

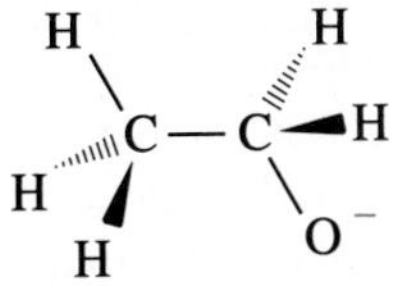

ethoxide ion

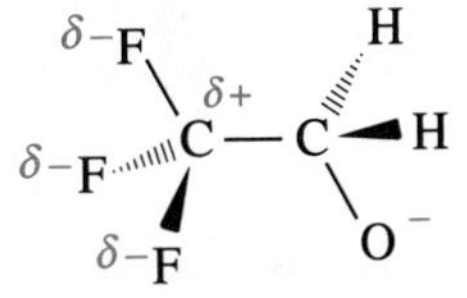

2,2,2-trifluoroethoxide ion

The positive charge on the carbon is located near the negative charge on the nearby oxygen atom, where it can partially neutralize and hence stabilize it. This **inductive effect,** as it is called, is absent in ethoxide ion.

The acidity-increasing effect of fluorine seen here is not a special case, but a general phenomenon. *All electron-withdrawing groups increase acidity* by stabilizing the conjugate base. *Electron-donating groups decrease acidity* because they destabilize the conjugate base.

Here is another example. *p*-Nitrophenol (Table 7.2) is a much stronger acid than phenol. In this case, the nitro group acts in *two* ways to stabilize the *p*-nitrophenoxide ion.

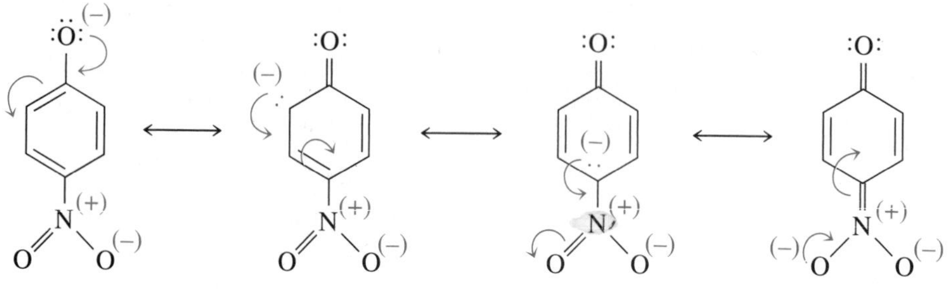

I ⟷ II ⟷ III ⟷ IV

p-nitrophenoxide ion resonance contributors

First, the nitrogen atom has a formal positive charge and is therefore strongly electron withdrawing. It therefore increases the acidity of *p*-nitrophenol through the inductive effect. Second, the negative charge on the oxygen of the hydroxyl group can be delocalized through resonance, not only to the *ortho* and *para* ring carbons, as in phenoxide itself, *but to the oxygen atoms of the nitro group as well* (structure IV). Both the inductive and the resonance effects of the nitro group are acid-strengthening.

Additional nitro groups on the benzene ring further increase phenolic acidity. Picric acid (2,4,6-trinitrophenol) is an even stronger acid than *p*-nitrophenol.

PROBLEM 7.10 Draw the resonance contributors for the 2,4,6-trinitrophenoxide (picrate) ion, and show that the negative charge can be delocalized to every oxygen atom.

PROBLEM 7.11 Rank the following five compounds in order of increasing acid strength: 2-chloroethanol, *p*-chlorophenol, *p*-methylphenol, ethanol, phenol.

Alkoxides, the conjugate bases of alcohols, are strong bases just like hydroxide ion. They are ionic compounds and are frequently used as strong bases in organic chemistry. They can be prepared by the reaction of an alcohol with sodium or potassium metal or with a metal hydride. These reactions proceed irreversibly to give the metal alkoxides that can frequently be isolated as white solids.

$$2RO{-}H + 2\,K \longrightarrow 2\,RO^{-}\,K^{+} + H_2 \quad (7.12)$$

alcohol; potassium alkoxide

$$RO{-}H + NaH \longrightarrow RO^{-}\,Na^{+} + H_2 \quad (7.13)$$

sodium hydride; sodium alkoxide

PROBLEM 7.12 Write the equation for the reaction of *t*-butyl alcohol with potassium metal. Name the product.

Ordinarily, treatment of alcohols with sodium hydroxide does not convert them to their alkoxides. This is because alkoxides are stronger bases than hydroxide ion, so the reaction goes in the reverse direction. Phenols, however, can be converted to phenoxide ions in this way.

$$ROH + Na^{+}OH^{-} \not\rightleftharpoons RO^{-}Na^{+} + H_2O \quad (7.14)$$

$$C_6H_5{-}OH + Na^{+}OH^{-} \longrightarrow C_6H_5{-}O^{-}Na^{+} + HOH \quad (7.15)$$

phenol; sodium phenoxide

PROBLEM 7.13 Write an equation for the reaction, if any, between

a. *p*-nitrophenol and aqueous potassium hydroxide.
b. cyclohexanol and aqueous potassium hydroxide.

7.8 The Basicity of Alcohols and Phenols

Alcohols (and phenols) function not only as weak acids but also as weak bases. They have unshared electron pairs on the oxygen and are therefore Lewis bases. They can be protonated by strong acids. The product, analogous to the oxonium ion H_3O^{+}, is an alkyloxonium ion.

$$R{-}\ddot{\underset{..}{O}}{-}H + H^{+} \rightleftharpoons \left[R{-}\underset{..}{\overset{H}{\overset{|}{O}}}{-}H \right]^{+} \quad (7.16)$$

alcohol acting as a base; alkyloxonium ion

This protonation is the first step in two important reactions of alcohols that are discussed in the following two sections: their dehydration of alkenes and their conversion to alkyl halides.

7.9 Dehydration of Alcohols to Alkenes

Alcohols can be dehydrated by heating them with a strong acid. For example, when ethanol is heated at 180°C with a small amount of concentrated sulfuric acid, a good yield of ethylene is obtained.

$$H{-}CH_2CH_2{-}OH \xrightarrow{H^{+},\ 180°C} CH_2{=}CH_2 + H{-}OH \quad (7.17)$$

ethanol; ethylene

This type of reaction, which can be used to prepare alkenes, is the reverse of their hydration (Sec. 3.7b). It is an *elimination reaction* and can occur by either an E1 or an E2 mechanism, depending on the class of the alcohol.

Tertiary alcohols dehydrate by the E1 mechanism. *t*-Butyl alcohol is a typical example. The first step involves reversible protonation of the hydroxyl group.

$$(CH_3)_3C{-}\ddot{\underset{\cdot\cdot}{O}}H + H^+ \rightleftharpoons (CH_3)_3C{-}\overset{+}{\ddot{O}}{-}H \;\;(\text{with } H \text{ on } O) \qquad (7.18)$$

Ionization, with water as the leaving group, occurs readily because the resulting carbocation is tertiary.

$$(CH_3)_3C{-}\overset{+}{\ddot{O}}(H){-}H \rightleftharpoons \underset{t\text{-butyl cation}}{(CH_3)_3C^+} + H_2O \qquad (7.19)$$

Proton loss from a carbon atom adjacent to the positive carbon completes the reaction.

$$H{-}CH_2{-}\overset{+}{C}(CH_3)_2 \longrightarrow CH_2{=}C(CH_3)_2 + H^+ \qquad (7.20)$$

The overall dehydration reaction is the sum of all three steps.

$$\underset{t\text{-butyl alcohol}}{H{-}CH_2{-}C(CH_3)_2{-}OH} \xrightarrow[\text{heat}]{H^+} \underset{\substack{\text{2-methylpropene}\\ \text{(isobutylene)}}}{CH_2{=}C(CH_3)_2} + H{-}OH \qquad (7.21)$$

With a primary alcohol, a primary carbocation intermediate is avoided by combining the last two steps of the mechanism. The loss of water and an adjacent proton occur simultaneously in an E2 mechanism.

$$CH_3CH_2\ddot{\underset{\cdot\cdot}{O}}H + H^+ \rightleftharpoons CH_3CH_2{-}\overset{+}{\ddot{O}}(H){-}H \qquad (7.22)$$

$$H{-}CH_2{-}CH_2{-}\overset{+}{\ddot{O}}(H){-}H \longrightarrow CH_2{=}CH_2 + H^+ + H_2O \qquad (7.23)$$

The important things to remember about alcohol dehydrations are (1) they all begin by protonation of the hydroxyl group (that is, the alcohol acts as a base) and (2) the ease of alcohol dehydration is 3° > 2° > 1° (the same as the order of carbocation stability).

Sometimes a single alcohol gives two or more alkenes because the proton lost during dehydration can come from any carbon atom that is *adjacent* to the hydroxyl-bearing carbon. For example, 2-methyl-2-butanol can give two alkenes.

$$\underset{\text{2-methyl-2-butanol}}{\overset{\text{H}}{\overset{|}{CH_2}}-\underset{\underset{CH_3}{|}}{\overset{\text{OH}}{\overset{|}{C}}}-\overset{\text{H}}{\overset{|}{CH}}-CH_3} \xrightarrow[\text{heat}\ -H_2O]{H^+} \underset{\text{2-methyl-1-butene}}{CH_2{=}\underset{\underset{CH_3}{|}}{C}-CH_2CH_3} \quad \text{and/or} \quad \underset{\text{2-methyl-2-butene}}{CH_3-\underset{\underset{CH_3}{|}}{C}{=}CHCH_3} \qquad (7.24)$$

In these cases, *the alkene with the most substituted double bond usually predominates*. By "most substituted" we mean the alkene with the greatest number of alkyl groups on the doubly bonded carbons. Thus, in the example shown, the major product is 2-methyl-2-butene.

PROBLEM 7.14 Write the structure for all the possible dehydration products of

a. 3-methyl-3-pentanol.

b. 1-methylcyclohexanol (cyclohexane ring bearing CH_3 and OH on the same carbon)

In each case, which product do you expect to predominate?

7.10 The Reaction of Alcohols with Hydrogen Halides

Alcohols react with hydrogen halides (HCl, HBr, and HI) to give alkyl halides (chlorides, bromides, and iodides).

$$\underset{\text{alcohol}}{R-OH} + H-X \longrightarrow \underset{\text{alkyl halide}}{R-X} + H-OH \qquad (7.25)$$

This substitution reaction provides a useful general route to alkyl halides. Because halide ions are good nucleophiles, we obtain mainly substitution products instead of dehydration. The reaction rate and mechanism depend on the class of alcohol (tertiary, secondary, or primary).

Tertiary alcohols react the fastest. For example, we can convert *t*-butyl alcohol to *t*-butyl chloride simply by shaking it for a few minutes at room temperature (rt) with concentrated hydrochloric acid.

$$\underset{\textit{t}\text{-butyl alcohol}}{(CH_3)_3COH} + H-Cl \xrightarrow[\text{15 min}]{\text{rt}} \underset{\textit{t}\text{-butyl chloride}}{(CH_3)_3C-Cl} + H-OH \qquad (7.26)$$

The reaction occurs by an S_N1 mechanism and involves a carbocation intermediate. The first two steps in the mechanism are identical to those shown in eq. 7.18 and 7.19. The final step involves capture of the *t*-butyl carbocation by chloride ion.

$$(CH_3)_3C^+ + Cl^- \xrightarrow{\text{fast}} (CH_3)_3CCl \tag{7.27}$$

On the other hand, 1-butanol, a primary alcohol, reacts slowly and must be heated for several hours with a mixture of concentrated hydrochloric acid and a Lewis acid catalyst such as zinc chloride to accomplish the same type of reaction.

$$\underset{\text{1-butanol}}{CH_3CH_2CH_2CH_2OH} + H{-}Cl \xrightarrow[\text{several hours}]{\text{heat, }ZnCl_2} \underset{\text{1-chlorobutane}}{CH_3CH_2CH_2CH_2{=}Cl} + H{-}OH \tag{7.28}$$

The reaction occurs by an S_N2 mechanism. In the first step, the alcohol is protonated by the acid.

$$CH_3CH_2CH_2CH_2{-}\ddot{\underset{\cdot\cdot}{O}}H + H^+ \rightleftharpoons CH_3CH_2CH_2CH_2{-}\overset{+}{\ddot{O}}(H){-}H \tag{7.29}$$

In the second step, chloride ion displaces water in a typical S_N2 process. The zinc chloride is a good Lewis acid and can serve the same role as a proton in sharing an electron pair of the hydroxyl oxygen. It also increases the chloride ion concentration, thus speeding up the S_N2 displacement.

$$Cl^- + CH_3CH_2CH_2{-}CH_2{-}\overset{+}{\ddot{O}}(H){-}H \longrightarrow CH_3CH_2CH_2CH_2Cl + H_2O \tag{7.30}$$

Secondary alcohols react at intermediate rates by both S_N1 and S_N2 mechanisms.

EXAMPLE 7.2 Explain why *t*-butyl alcohol reacts at equal rates with HCl, HBr, and HI (to form, in each case, the corresponding *t*-butyl halide).

Solution *t*-Butyl alcohol is a teriary alcohol and thus it reacts by an S_N1 mechanism. As in all S_N1 reactions, the rate-determining step involves formation of a carbocation, in this case the *t*-butyl carbocation. The rate of this step does not depend on which acid is used, so all of the reactions proceed at equal rates.

$$(CH_3)_3COH + H^+ \rightleftarrows (CH_3)_3C{-}\overset{\cdot\cdot}{\underset{+}{O}}(H){-}H \xrightarrow[\text{slow step}]{S_N1} \underset{t\text{-butyl cation}}{(CH_3)_3C^+} + H_2O$$

The reaction of the carbocation with Cl^-, Br^-, or I^- is then fast.

PROBLEM 7.15 Explain why 1-butanol reacts with hydrogen halides in the rate order HI > HBr > HCl (to form, in each case, the corresponding butyl halide).

PROBLEM 7.16 Write an equation for the reaction of 1-methylcyclohexanol with concentrated HBr.

7.11 Other Ways to Prepare Alkyl Halides from Alcohols

Since alkyl halides are extremely useful in synthesis, it is not surprising that chemists have devised several ways to prepare them from alcohols. For example, **thionyl chloride** reacts with alcohols to give alkyl chlorides.

$$ROH + Cl-\overset{\overset{\Large O}{\|}}{S}-Cl \xrightarrow{\text{heat}} RCl + HCl\uparrow + SO_2\uparrow \quad (7.31)$$

thionyl chloride

One advantage of this method is that two of the reaction products, hydrogen chloride and sulfur dioxide, are gases and evolve from the reaction mixture, leaving behind only the desired alkyl chloride. The method is not effective, however, for preparing low-boiling alkyl chlorides (in which R has only a few carbon atoms), because they easily boil out of the reaction mixture with the gaseous products.

Phosphorus halides also convert alcohols to alkyl halides.

$$3\,ROH + PX_3 \longrightarrow 3\,RX + H_3PO_3\ (X = Cl \text{ or } Br) \quad (7.32)$$

In this case, the other reaction product, phosphorous acid, has a rather high boiling point. Thus, the alkyl halide is usually the lowest-boiling component of the reaction mixture and can be isolated by distillation.

Both of these methods are used mainly with primary and secondary alcohols, whose reaction with hydrogen halides is slow.

PROBLEM 7.17 Tell how you would prepare each of the following alkyl halides from the corresponding alcohol without using HX.

a. (cyclohexyl)—CH_2Cl b. (cyclohexyl)—Br

7.12 A Comparison of Alcohols and Phenols

Because they have the same functional group, alcohols and phenols have many similar properties. But whereas it is relatively easy, with acid catalysis, to break the C—OH bond of alcohols, this bond is difficult to break in phenols. Protonation of the phenolic hydroxyl group can occur, but loss of a water molecule would give a phenyl cation.

$$C_6H_5-\overset{+}{\ddot{O}}(H)-H \not\longrightarrow C_6H_5^{+} + H_2O \quad (7.33)$$

a phenyl cation

With only two attached groups, the positive carbon in a phenyl cation should be *sp*-hybridized and linear. But this geometry is prevented by the structure of the benzene ring, so phenyl cations are exceedingly difficult to form. Consequently, phenols cannot undergo replacement of the hydroxyl group by an S_N1 mechanism. Neither can phenols undergo displacement by the S_N2 mechanism (the geometry of the ring makes the usual inversion mechanism impossible). Therefore, hydrogen halides, phosphorus halides, or thionyl halides cannot cause replacement of the hydroxyl group by halogens in phenols.

PROBLEM 7.18 Compare the reactions of cyclopentanol and phenol with

a. HBr. b. H_2SO_4, heat.

7.13 Oxidation of Alcohols to Aldehydes, Ketones, and Carboxylic Acids

Alcohols with at least one hydrogen attached to the hydroxyl-bearing carbon can be oxidized to carbonyl compounds. Primary alcohols give aldehydes, which may be further oxidized to carboxylic acids. Secondary alcohols give ketones.*

$$\underset{\text{primary alcohol}}{R-\underset{H}{\overset{OH}{C}}-H} \xrightarrow[\text{agent}]{\text{oxidizing}} \underset{\text{aldehyde}}{R-\overset{O}{\overset{\|}{C}}-H} \xrightarrow[\text{agent}]{\text{oxidizing}} \underset{\text{acid}}{R-\overset{O}{\overset{\|}{C}}-OH} \tag{7.34}$$

$$\underset{\text{secondary alcohol}}{R-\underset{H}{\overset{OH}{C}}-R'} \xrightarrow[\text{agent}]{\text{oxidizing}} \underset{\text{ketone}}{R-\overset{O}{\overset{\|}{C}}-R'} \tag{7.35}$$

Tertiary alcohols do not undergo this type of oxidation.

A common laboratory oxidizing agent for this purpose is chromic anhydride, CrO_3, dissolved in aqueous sulfuric acid and acetone **(Jones' reagent).** Typical examples are

* Notice that as an alcohol is oxidized to an aldehyde or ketone and then to a carboxylic acid, the number of bonds between the reactive carbon atom and oxygen atoms increases from one to two to three. In other words, we say that the oxidation state of that carbon increases as we go from an alcohol to an aldehyde or ketone to a carboxylic acid.

$$\text{cyclohexanol} \xrightarrow[\text{H}^+,\ \text{acetone (Jones' reagent)}]{\text{CrO}_3} \text{cyclohexanone} \qquad (7.36)$$

$$\underset{\text{1-octanol}}{CH_3(CH_2)_6CH_2OH} \xrightarrow[\text{reagent}]{\text{Jones'}} \underset{\text{octanoic acid}}{CH_3(CH_2)_6CO_2H} \qquad (7.37)$$

With primary alcohols, oxidation can be stopped at the aldehyde stage by special reagents, such as pyridinium chlorochromate (PCC).*

$$\underset{\text{1-octanol}}{CH_3(CH_2)_6CH_2OH} \xrightarrow[CH_2Cl_2,\ 25°C]{\text{PCC}} \underset{\text{octanal}}{CH_3(CH_2)_6\overset{\overset{\displaystyle O}{\|}}{C}-H} \qquad (7.38)$$

PCC is prepared by dissolving CrO_3 in hydrochloric acid and then adding pyridine:

$$CrO_3 + HCl + \underset{\text{pyridine}}{C_5H_5N:} \longrightarrow \underset{\text{pyridinium chlorochromate (PCC)}}{C_5H_5N^+{-}H\ \ CrO_3Cl^-} \qquad (7.39)$$

PROBLEM 7.19 Write an equation for the oxidation of

a. 1-pentanol with Jones' reagent.
b. 1-pentanol with PCC.
c. 4-phenyl-2-butanol with Jones' reagent.
d. 4-phenyl-2-butanol with PCC.

In the body, similar oxidations are accomplished by enzymes, together with a rather complex coenzyme called nicotinamide adenine dinucleotide, NAD^+ (for its structure, see Sec. 18.13). Oxidation occurs in the liver and is a key step in the body's attempt to rid itself of imbibed alcohol.

$$\underset{\text{ethanol}}{CH_3CH_2OH} + NAD^+ \underset{\text{dehydrogenase}}{\overset{\text{alcohol}}{\rightleftharpoons}} \underset{\text{acetaldehyde}}{CH_3\overset{\overset{\displaystyle O}{\|}}{C}-H} + NADH \qquad (7.40)$$

The resulting acetaldehyde—also toxic—is further oxidized to acetic acid and eventually to carbon dioxide and water.

* In the oxidation reactions shown in eqs. 7.36–7.38, the chromium is reduced from Cr^{6+} to Cr^{3+}. Aqueous solutions of Cr^{6+} are orange, whereas aqueous solutions of Cr^{3+} are green. This color change has been used as the basis for detecting ethanol in breathalyzer tests.

7.14 Alcohols with More Than One Hydroxyl Group

Compounds with two adjacent alcohol groups are called **glycols.** The most important example is **ethylene glycol.** Compounds with more than two hydroxyl groups are also known, and several, such as **glycerol** and **sorbitol,** are important commercial chemicals.

$$\underset{\text{OH}}{\underset{|}{CH_2}}-\underset{\text{OH}}{\underset{|}{CH_2}} \qquad \underset{\text{OH}}{\underset{|}{CH_2}}-\underset{\text{OH}}{\underset{|}{CH}}-\underset{\text{OH}}{\underset{|}{CH_2}} \qquad \underset{\text{OH}}{\underset{|}{CH_2}}-\underset{\text{OH}}{\underset{|}{CH}}-\underset{\text{OH}}{\underset{|}{CH}}-\underset{\text{OH}}{\underset{|}{CH}}-\underset{\text{OH}}{\underset{|}{CH}}-\underset{\text{OH}}{\underset{|}{CH_2}}$$

ethylene glycol (1,2-ethanediol) bp 198°C

glycerol (glycerine) (1,2,3-propanetriol) bp 290°C (decomposes)

sorbitol (1,2,3,4,5,6-hexanehexaol) mp 110–112°C

Ethylene glycol is used as the "permanent" antifreeze in automobile radiators and as a raw material in the manufacture of Dacron. Ethylene glycol is completely miscible with water. Notice that, because of its increased capacity for hydrogen bonding, ethylene glycol has an exceptionally high boiling point for its molecular weight—much higher than that of ethanol.

Glycerol is a syrupy, colorless, water-soluble, high-boiling liquid with a distinctly sweet taste. Its soothing qualities make it useful in shaving and toilet soaps and in cough drops and syrups. It is also used as a moistening agent in tobacco.

Nitration of glycerol gives **glyceryl trinitrate** (nitroglycerine), a powerful and shock-sensitive explosive.

$$\begin{array}{l} CH_2OH \\ | \\ CHOH \\ | \\ CH_2OH \end{array} + 3\ HONO_2 \xrightarrow{H_2SO_4} \begin{array}{l} CH_2ONO_2 \\ | \\ CHONO_2 \\ | \\ CH_2ONO_2 \end{array} + 3\ H_2O \qquad (7.41)$$

glycerol

glyceryl trinitrate (nitroglycerine)

Alfred Nobel, inventor of dynamite (in 1866), found that glyceryl trinitrate could be controlled by absorbing it on an inert porous material. Dynamite contains about 15% glyceryl (and glycol) nitrate. The main explosive is ammonium nitrite (55%); the other components are sodium nitrate and wood pulp (about 15% each). Dynamite is used mainly in mining and construction.

Nitroglycerine is also used in medicine as a vasodilator, to prevent heart attacks in patients who suffer from angina.

Triesters of glycerol are fats and oils, whose chemistry is discussed in Chapter 15.

Sorbitol, with its many hydroxyl groups, is water soluble. It is almost as sweet as cane sugar and is used in candy making and as a sugar substitute for diabetics.

A Word About . . .

12. Industrial Alcohols

The lower alcohols (those with up to four carbon atoms) are manufactured on a large scale. They are used as raw materials for other valuable chemicals and also have important uses in their own right.

Methanol was at one time produced from wood by distillation and is still sometimes called wood alcohol. The word *methyl* originates from the Greek (*methy*, wine, and *yle*, wood). At present, however, methanol is manufactured from carbon monoxide and hydrogen.

$$CO + 2H_2 \xrightarrow[400°C,\ 150\ atm]{ZnO—Cr_2O_3} CH_3OH$$

The world production of methanol is approximately 10 million tons per year. Most of it is used to produce formaldehyde and other chemicals, but some is used as a solvent and an antifreeze. Recently methanol has been used as the carbon source in the commercial production of single-cell proteins. Some yeasts and bacteria (single cells) can synthesize proteins from methanol and other carbon sources in the presence of aqueous nutrient salt solutions that contain certain essential sulfur, phosphorus, and nitrogen compounds. These proteins are used as an animal food supplement and may eventually also play a part in human nutrition. Methanol itself, however, is highly toxic and can cause permanent blindness and death if taken internally.

Ethanol is prepared by the fermentation of black-strap molasses, the residue that results from the purification of cane sugar.

$$\underset{\text{cane sugar}}{C_{12}H_{22}O_{11}} + H_2O \xrightarrow{\text{yeast}} \underset{\text{ethyl alcohol}}{4\ CH_3CH_2OH} + 4\ CO_2$$

The starch in grain, potatoes, and rice can be fermented similarly to produce ethanol, sometimes called grain alcohol.

Besides fermentation, ethanol is also manufactured by the acid-catalyzed hydration of ethylene (eq. 3.7). This method, using sulfuric acid or other acid catalysts, results in an annual world production of more than 1 million tons.

Commercial alcohol is a constant-boiling mixture containing 95% ethanol and 5% water and cannot be further purified by distillation. To remove the remaining water to obtain absolute alcohol, one adds quicklime (CaO), which reacts with water to form calcium hydroxide but does not react with ethanol.

Since earlier times, ethanol has been known as an ingredient in fermented beverages (beer, wine, whiskey, and so on). The term *proof*, as used in the United States in reference to alcoholic beverages, is approximately twice the volume percentage of alcohol present. For example 100-proof whiskey contains 50% ethanol.

Ethanol is used as a solvent, as a topical antiseptic, and as a starting material for the manufacture of ether and ethyl esters. It can be used as a fuel (gasohol) and as a carbon source for single-cell proteins.

2-Propanol (isopropyl alcohol) is manufactured commercially by the acid-catalyzed hydration of propene (eq. 3.13). It is the main component of rubbing alcohol. More than half the isopropyl alcohol produced (more than 1 million tons annually) is used to make acetone, by oxidation.

7.15 Aromatic Substitution in Phenols

Now we will examine some reactions that occur with phenols, but not with alcohols. Phenols undergo electrophilic aromatic substitution under very mild conditions because the hydroxyl group is strongly ring activating. For example, phenol can be nitrated with *dilute aqueous* nitric acid.

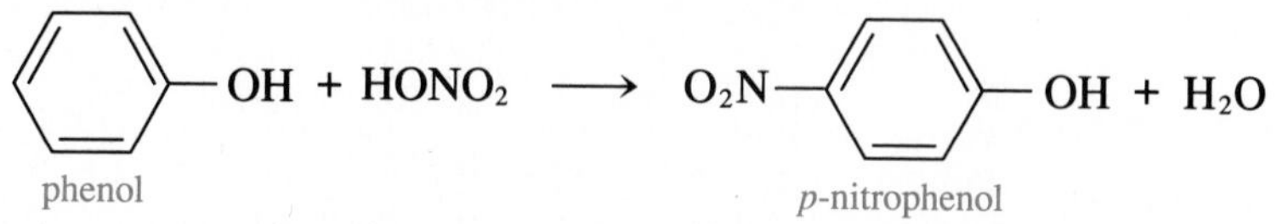

$$C_6H_5\text{—OH} + HONO_2 \longrightarrow O_2N\text{—}C_6H_4\text{—OH} + H_2O \quad (7.42)$$

phenol p-nitrophenol

Phenol is also brominated rapidly with *bromine* water, to produce 2,4,6-tribromophenol.

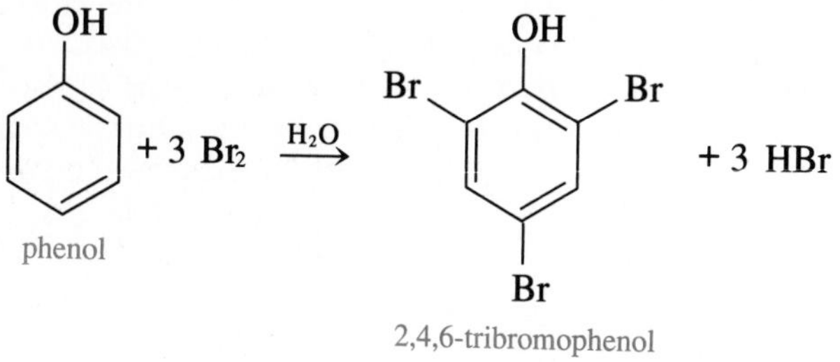

$$C_6H_5OH + 3\ Br_2 \xrightarrow{H_2O} 2,4,6\text{-}Br_3C_6H_2OH + 3\ HBr \quad (7.43)$$

phenol 2,4,6-tribromophenol

EXAMPLE 7.3 Draw the intermediate in electrophilic aromatic substitution *para* to a hydroxyl group, and show how the intermediate benzenonium ion is stabilized by the hydroxyl group.

Solution

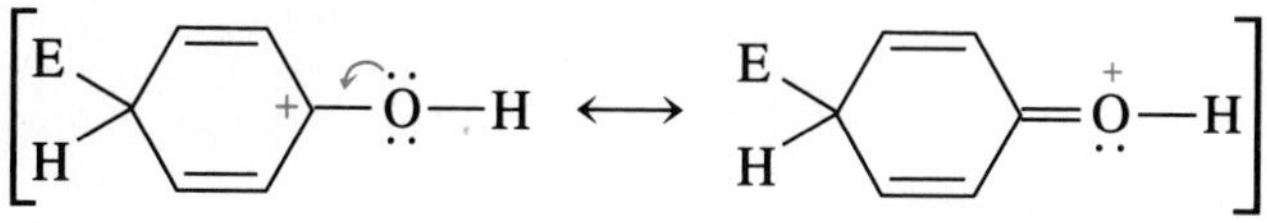

$$\left[\text{(E)(H)C}_6\text{H}_4^{+}\text{—}\ddot{\text{O}}\text{—H} \longleftrightarrow \text{(E)(H)C}_6\text{H}_4\text{=}\overset{+}{\text{O}}\text{—H} \right]$$

An unshared electron pair on the oxygen atom helps to delocalize the positive charge.

PROBLEM 7.20 Explain why phenoxide ion undergoes electrophilic aromatic substitution even more easily than phenol does.

PROBLEM 7.21 Write an equation for the reaction of

a. *p*-methylphenol + $HONO_2$ (1 mol).
b. *o*-chlorophenol + Br_2 (1 mol).

7.16 Oxidation of Phenols

Phenols are easily oxidized. Samples that stand exposed to air for some time often become highly colored due to the formation of oxidation products. With **hydroquinone** (1,4-dihydroxybenzene), the reaction is easily controlled to give **1,4-benzoquinone** (commonly called *quinone*).

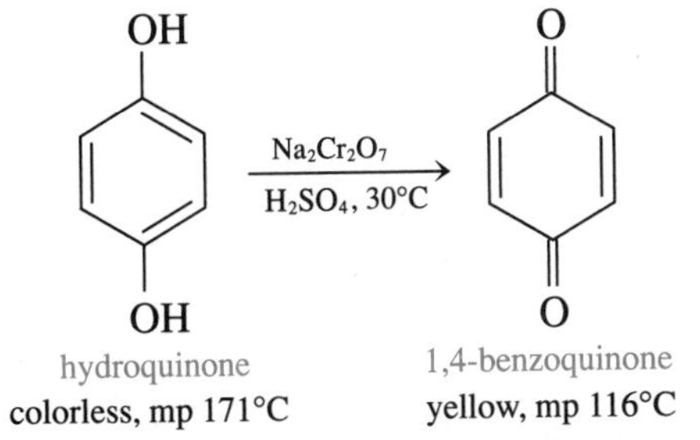

$$\text{hydroquinone (colorless, mp 171°C)} \xrightarrow[H_2SO_4,\ 30°C]{Na_2Cr_2O_7} \text{1,4-benzoquinone (yellow, mp 116°C)} \quad (7.44)$$

hydroquinone
colorless, mp 171°C

1,4-benzoquinone
yellow, mp 116°C

Hydroquinone and related compounds are used in photographic developers. They reduce silver ion that has not been exposed to light to metallic silver (and, in turn, they are oxidized to quinones). The oxidation of hydroquinones to quinones is reversible; this interconversion plays an important role in several biological oxidation–reduction reactions.

Substances that are sensitive to air oxidation, such as foods and lubricating oils, can be protected by phenolic additives. Phenols function as **antioxidants.** They react with and destroy peroxy free radicals (ROO•), which otherwise react with alkenes present in foods and oils and cause their degradation. Two commercial phenolic antioxidants are BHA (butylated hydroxy anisole) and BHT (butylated hydroxy toluene).*

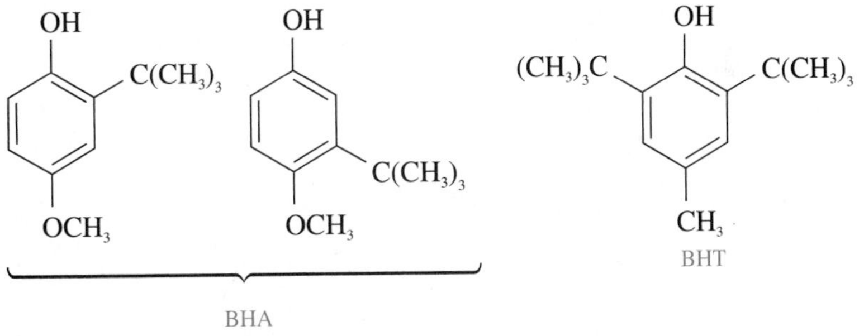

OH, $C(CH_3)_3$, OCH_3 — OH, $C(CH_3)_3$, OCH_3

BHA

OH, $(CH_3)_3C$, $C(CH_3)_3$, CH_3

BHT

BHA is used as an antioxidant in foods, especially meat products. BHT is used not only in foods, animal feeds, and vegetable oils, but also in lubricating oils, synthetic rubber, and various plastics.

* For an account of the roles peroxy free radicals and natural antioxidants may play in the aging process, see R. L. Rusting in *Scientific American,* **1992** (December), 131–141.

Vitamin E (α-tocopherol) is a common naturally occurring phenol. One of its biological functions is to act as a natural antioxidant.

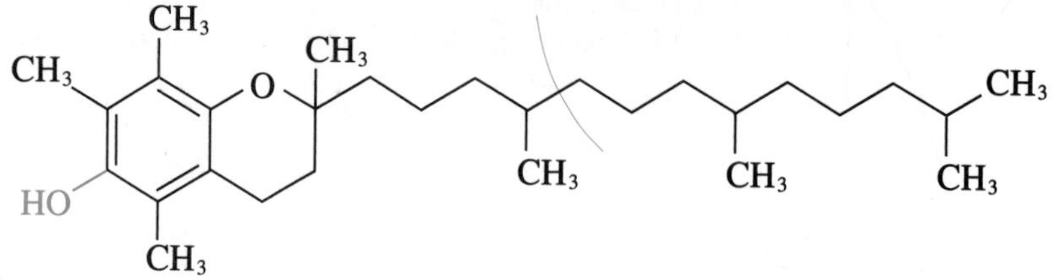

vitamin E (α-tocopherol)

A Word About . . .

13. Biologically Important Alcohols and Phenols

The hydroxyl group appears in many biologically important molecules, both as an alcohol and as a phenol.

Four metabolically important unsaturated primary alcohols are 3-methyl-2-buten-1-ol, 3-methyl-3-buten-1-ol, geraniol, and farnesol.

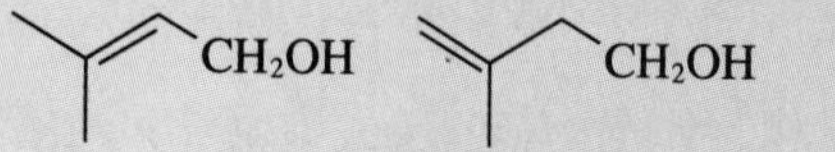

3-methyl-2-buten-1-ol 3-methyl-3-buten-1-ol

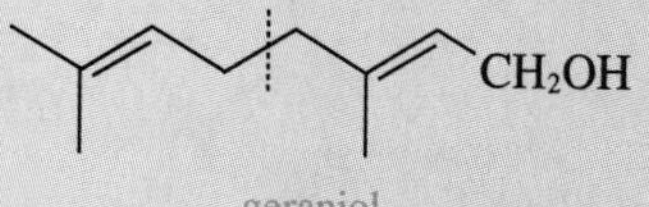

geraniol

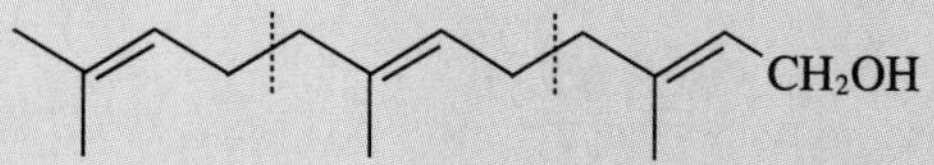

farnesol

The two smaller alcohols contain a five-carbon unit, called an isoprene unit, that is present in many natural products. This unit consists of a four-carbon chain with a one-carbon branch at carbon 2. These five-carbon alcohols can combine to give geraniol, which then can add yet another five-carbon unit to give farnesol. Note the isoprene units, marked off by dotted lines, in the structures of geraniol and farnesol.

Compounds of this type are called terpenes. Terpenes occur in the *essential oils* of many plants and flowers. They have 10, 15, 20, or more carbon atoms and are formed by linking isoprene units in various ways.

Geraniol, as its name implies, occurs in oil of geranium but also constitutes about 50% of rose oil, the extract of rose petals. Geraniol is also the biological precursor of α-pinene, a terpene that is the main component of turpentine. Farnesol, which occurs in the essential oils of rose and cyclamen, has a pleasing lily-of-the-valley odor. Both geraniol and farnesol are used in making perfumes.

Combination of two farnesol units (15 carbons each) leads to squalene, a 30-carbon hydrocarbon present in small amounts in the livers of most higher animals. Squalene is the biological precursor of steroids.

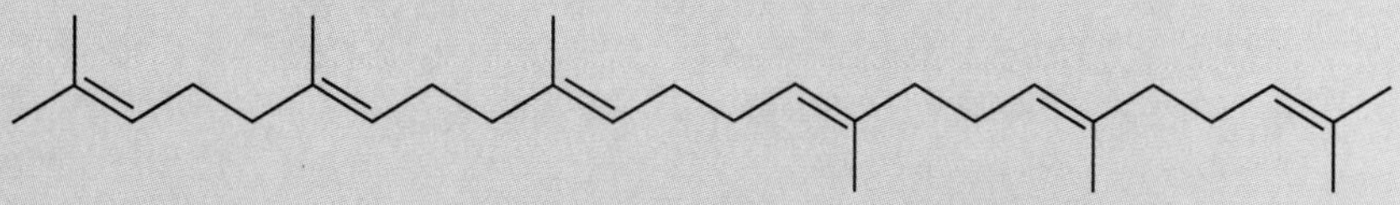

squalene

Cholesterol, a typical steroidal alcohol, has the structure

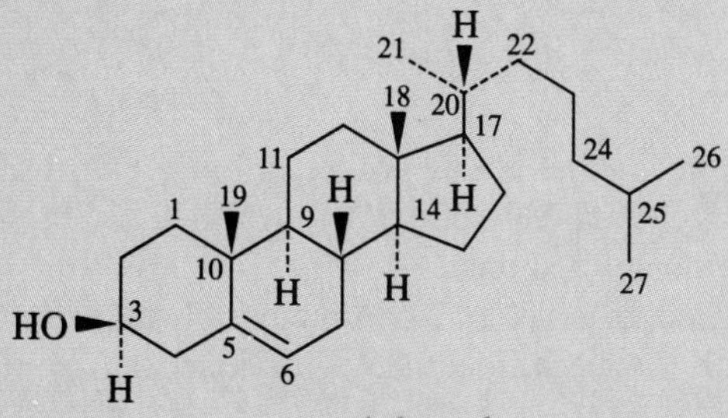

cholesterol
mp 148.5°C

Although it has 27 carbon atoms (instead of 30) and is therefore not strictly a terpene, cholesterol is synthesized in the body from the terpene squalene through a complex process that, in its final stages, involves the loss of 3 carbon atoms.

Phenols are less involved than alcohols in fundamental metabolic processes. Three phenolic alcohols do, however, form the basic building blocks of **lignins**, complex polymeric substances that, together with cellulose, form the woody parts of trees and shrubs. They have very similar structures.

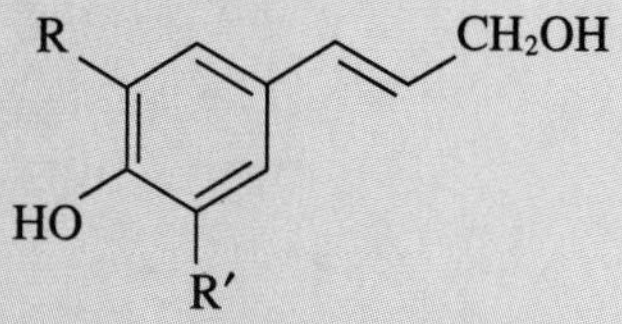

coniferyl alcohol (R = OCH_3, R′ = H)
sinapyl alcohol (R = R′ = OCH_3)
p-coumaryl alcohol (R = R′ = H)

Some phenolic natural products to be avoided are **urushiols**, the active allergenic ingredients in poison ivy and poison oak.

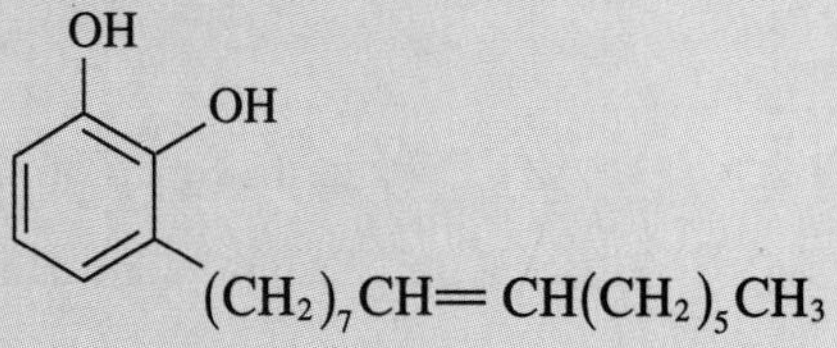

$(CH_2)_7CH{=}CH(CH_2)_5CH_3$

an urushiol

In other urushiols, the long side chain may be saturated, may have additional double bonds, or may have two more carbon atoms.

7.17 Thiols, the Sulfur Analogs of Alcohols and Phenols

Sulfur is immediately beneath oxygen in the periodic table and can often take its place in organic structures. The —SH group, called the **sulfhydryl group,** is the functional group of **thiols.** Thiols are named as follows:

CH_3SH
methanethiol
(methyl mercaptan)

$CH_3CH_2CH_2CH_2SH$
1-butanethiol
(*n*-butyl mercaptan)

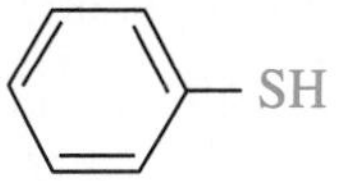

thiophenol
(phenyl mercaptan)

Thiols are sometimes called **mercaptans** because of their reaction with mercuric ion to form mercury salts called **mercaptides.**

$$2\ RSH + HgCl_2 \longrightarrow \underset{\text{a mercaptide}}{(RS)_2Hg} + 2\ HCl \qquad (7.45)$$

PROBLEM 7.22 Draw the structure for

a. 2-butanethiol. b. isopropyl mercaptan.

Alkyl thiols can be made from alkyl halides by nucleophilic displacement with sulfhydryl ion.

$$R{-}X + {}^{-}SH \longrightarrow R{-}SH + X^{-} \qquad (7.46)$$

Perhaps the most distinctive feature of thiols is their intense and disagreeable odor. The thiols $CH_3CH{=}CHCH_2SH$ and $(CH_3)_2CHCH_2CH_2SH$, for example, are responsible for the odor of a skunk.

Thiols are more acidic than alcohols. The pK_a of ethanethiol, for example, is 10.6, and that of ethanol is 15.9. Hence, thiols react with aqueous base to give **thiolates.**

$$RSH + Na^{+}OH^{-} \longrightarrow \underset{\text{a sodium thiolate}}{RS^{-}Na^{+}} + HOH \qquad (7.47)$$

PROBLEM 7.23 Write an equation for the reaction of ethanethiol with

a. KOH. b. $HgCl_2$. c. $CH_3CH_2O^{-}Na^{+}$

Thiols are easily oxidized to **disulfides** by mild oxidizing agents such as hydrogen peroxide or iodine. A naturally occurring disulfide whose smell you are probably familiar with is diallyl disulfide ($CH_2{=}CHCH_2S{-}SCH_2CH{=}CH_2$), which is responsible for the odor of fresh garlic.*

$$\underset{\text{thiol}}{2\ RS{-}H} \underset{\text{reduction}}{\overset{\text{oxidation}}{\rightleftharpoons}} \underset{\text{disulfide}}{RS{-}SR} \qquad (7.48)$$

The reaction shown in eq. 7.48 can be reversed with a variety of reducing agents. Since proteins contain disulfide links, these reversible oxidation–reduction reactions can be used to manipulate their structures.

REACTION SUMMARY

1. Alcohols

a. Conversion to Alkoxides (Sec. 7.7)

$$2\ RO{-}H + 2\ Na \longrightarrow 2\ RO^{-}Na^{+} + H_2$$
$$RO{-}H + NaH \longrightarrow RO^{-}Na^{+} + H_2$$

---

* Garlic belongs to the plant family *Allium,* from which the *allyl* group gets its name.

A Word About . . .

14. Hair, Curly or Straight

Hair consists of a fibrous protein called **keratin,** which, as proteins go, contains an unusually large percentage of the sulfur-containing amino acid **cystine.** Horse hair, for example, contains about 8% cystine:

$$\underset{\displaystyle NH_2}{HO_2CCHCH_2}S-S\underset{\displaystyle NH_2}{CH_2CHCO_2H}$$

cystine (CyS—SCy)

The disulfide link in cystine serves to cross-link the chains of amino acids that make up the protein (Figure 7.1).

The chemistry used in waving or straightening hair involves the oxidation–reduction chemistry of the disulfide bond (eq. 7.48). First, the hair is treated with a reducing agent, which breaks the S—S bonds, converting each sulfur to an —SH group. This breaks the cross-links between the long protein chains. The reduced hair can now be shaped as desired, either waved or straightened. Finally, the reduced and rearranged hair is treated with an oxidizing agent to reform the disulfide cross-links. The new disulfide bonds, no longer in their original positions, hold the hair in its new shape.

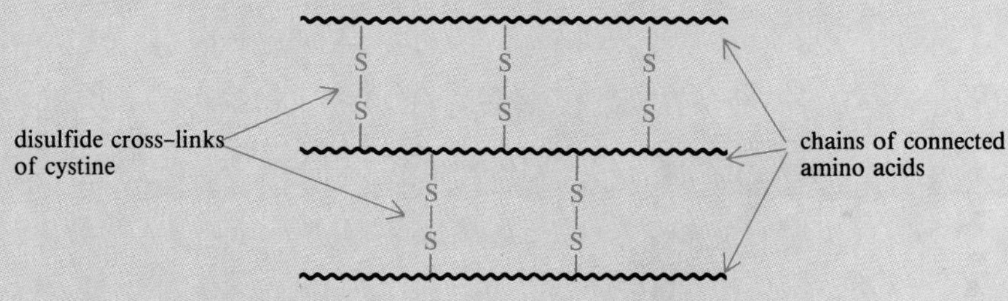

FIGURE 7.1 Schematic structure of hair

b. Dehydration to Alkenes (Sec. 7.9)

$$H-C-C-OH \xrightarrow{H^+ \text{(cat)}} C=C + H_2O$$

c. Conversion to Alkyl Halides (Secs. 7.10–7.11)

$$RO-H + HX \longrightarrow R-X + H_2O \; (X=Cl, Br, I)$$
$$RO-H + SOCl_2 \longrightarrow R-Cl + HCl + SO_2$$
$$RO-H + PX_3 \longrightarrow R-X + H_3PO_3 \; (X=Cl, Br)$$

d. Oxidation (Sec. 7.13)

$$RCH_2OH \text{ (primary)} \xrightarrow{PCC} R-\overset{O}{\overset{\|}{C}}-H \text{ (aldehyde)}$$

$$RCH_2OH \text{ (primary)} \xrightarrow[H^+]{CrO_3} R-\overset{O}{\overset{\|}{C}}-OH \text{ (carboxylic acid)}$$

$$R_2CHOH \text{ (secondary)} \xrightarrow{PCC \text{ or } CrO_3,\ H^+} R-\overset{O}{\overset{\|}{C}}-R \text{ (ketone)}$$

2. Phenols

a. Preparation of Phenoxides (Sec. 7.7)

$$ArO-H + NaOH \longrightarrow ArO^- Na^+ + H_2O$$

b. Electrophilic Aromatic Substitution (Sec. 7.15)

$$C_6H_5-OH \xrightarrow{HNO_3} O_2N-C_6H_4-OH$$

$$C_6H_5-OH \xrightarrow{Br_2} \text{2,4,6-}Br_3C_6H_2-OH$$

c. Oxidation to Quinones (Sec. 7.16)

$$HO-C_6H_4-OH \xrightarrow[H_2SO_4,\ H_2O]{Na_2Cr_2O_7} O=C_6H_4=O$$

quinone

3. Thiols

a. Conversion to Thiolates (Sec. 7.17)

$$\underset{\text{thiol}}{RS-H} + NaOH \longrightarrow \underset{\text{thiolate}}{RS^- Na^+} + H_2O$$

b. Oxidation to Disulfides (Sec. 7.17)

$$\underset{\text{thiol}}{2\ RSH} \xrightarrow{\text{oxidation}} \underset{\text{disulfide}}{RS-SR}$$

ADDITIONAL PROBLEMS

7.24. Name each of the following alcohols:

a. $CH_3CH_2CH(OH)CH_2CH_3$ **b.** $CH_3CH(Cl)CH(OH)CH_2CH_3$
c. $(CH_3)_2CHCH(OH)CH_2CH_3$ **d.** $CH_3CH(Cl)CH_2CH(OH)CH_3$

7.25. Write a structural formula for each of the following compounds:

a. 2,2-dimethyl-1-butanol **b.** *o*-bromophenol
c. 2,3-pentanediol **d.** 2-phenylethanol
e. sodium ethoxide **f.** 1-methylcyclopentanol
g. *trans*-2-methylcyclopentanol **h.** (*R*)-2-butanol
i. 2-methyl-2-propen-1-ol **j.** 2-cyclohexenol

7.26. Classify the alcohols in parts a, d, f, g, h, i, and j of Problem 7.25 as primary, secondary, or tertiary.

7.27. Name each of the following compounds:

a. $CH_3C(CH_3)_2CH(OH)CH_3$ **b.** $CH_3CHBrC(CH_3)_2OH$

c. 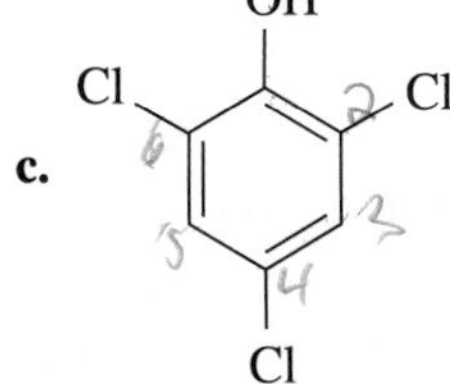

d. 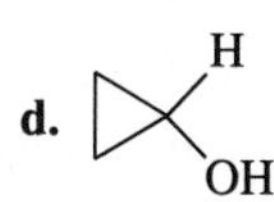

e. 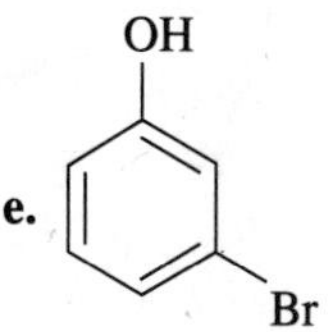

f. 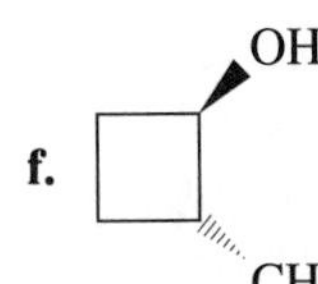

g. $CH_3CH{=}CHCH_2OH$ **h.** $CH_3CH(SH)CH_3$
i. $HOCH_2CH(OH)CH(OH)CH_2OH$ **j.** $CH_3CH_2CH_2O^-K^+$

7.28. Explain why each of the following names is unsatisfactory, and give a correct name:

a. 2,2-dimethyl-3-butanol **b.** 2-ethyl-1-propanol
c. 1-propene-3-ol **d.** 2-chloro-4-pentanol
e. 3,6-dibromophenol

7.29. Arrange the compounds in each of the following groups in order of increasing solubility in water, and briefly explain your answers.

a. ethanol; ethyl chloride; 1-hexanol
b. 1-pentanol; 1,5-pentanediol; $HOCH_2(CHOH)_3CH_2OH$

7.30. The following classes of organic compounds are Lewis bases. Write an equation that shows how each class might react with H^+.

a. ether, ROR **b.** amine, R_3N **c.** ketone, $R_2C{=}O$

7.31. Arrange the following compounds in order of increasing acidity, and explain the reasons for your choice of order: cyclohexanol, phenol, *p*-nitrophenol, 2-chlorocyclohexanol.

7.32. Which is the stronger base, potassium *t*-butoxide or potassium ethoxide? (*Hint:* Use the data in Table 7.2.)

7.33. Complete each of the following equations:

a. $CH_3CH(OH)CH_2CH_3 + K \longrightarrow$

b. $(CH_3)_2CHOH + NaH \longrightarrow$

c. $Cl-C_6H_4-OH + NaOH \longrightarrow$

d. H, OH (cyclopentane ring) $+ NaOH \longrightarrow$

e. $CH_3CH{=}CHCH_2SH + NaOH \longrightarrow$

7.34 Explain why your answers to parts c, d, and e of Problem 7.33 are consistent with the pK_a's of the starting acids and product acids (see eqs. 7.14, 7.15, and 7.47).

7.35. Show the structures of all possible acid-catalyzed dehydration products of the following. If more than one alkene is possible, predict which one will be formed in the largest amount.

a. cyclopentanol **b.** 2-butanol
c. 1-methylcyclopentanol **d.** 2-phenylethanol

7.36. Explain why the reaction shown in eq. 7.19 occurs much more easily than the reaction $(CH_3)_3C{-}OH \rightleftharpoons (CH_3)_3C^+ + OH^-$. (That is, why is it necessary to protonate the alcohol before ionization can occur?)

7.37. Write out all the steps in the mechanism for eq. 7.24, showing how each product is formed.

7.38. Although the reaction shown in eq. 7.26 occurs faster than that shown in eq. 7.28, the yield of product is lower. The yield of *t*-butyl chloride is only 80%, whereas the yield of *n*-butyl chloride is nearly 100%. What by-product is formed in eq. 7.26, and by what mechanism is it formed? Why is a similar by-product *not* formed in eq. 7.28?

7.39. Write an equation for each of the following reactions:

a. 2-methyl-2-butanol + HCl **b.** 1-pentanol + Na
c. cyclopentanol + PBr_3 **d.** 1-phenylethanol + $SOCl_2$
e. 1-methylcyclopentanol + H_2SO_4, heat **f.** ethylene glycol + $HONO_2$
g. 1-pentanol + aqueous NaOH **h.** 1-octanol + HBr + $ZnBr_2$
i. 1-pentanol + CrO_3, H^+ **j.** 2-cyclohexylethanol + PCC

7.40. Treatment of 3-buten-2-ol with concentrated hydrochloric acid gives a mixture of two products, 3-chloro-1-butene and 1-chloro-2-butene. Write a reaction mechanism that explains how both products are formed.

7.41. Which four-carbon acyclic alcohols can be manufactured commercially by acid-catalyzed hydration of alkenes? (Remember Markovnikov's rule!)

7.42. Write an equation for each of the following two-step syntheses:

a. cyclohexene to cyclohexanone
b. 1-bromobutane to butanal
c. 1-butanol to 1-butanethiol

7.43. What product do you expect from the oxidation of cholesterol with CrO_3 and H^+? (See page 229 for the formula of cholesterol.)

7.44. Draw the structure of the quinone expected from the oxidation of

a. 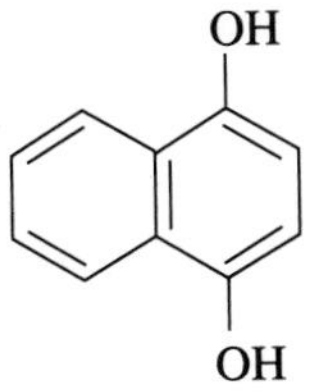 (1,4-naphthalenediol: OH, OH)

b. 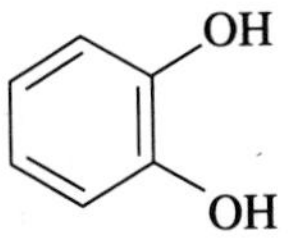 (1,2-benzenediol: OH, OH)

7.45. Dimethyl disulßde, $CH_3S{-}SCH_3$, found in the vaginal secretions of female hamsters, acts as a sexual attractant for the male hamster. Write an equation for its synthesis from methanethiol.

7.46. The disulßde $[(CH_3)_2CHCH_2CH_2S]_2$ is a component of the odorous secretion of mink. Describe a synthesis of this disulßde, starting with 3-methyl-1-butanol.

8 Ethers and Epoxides

8.1 Introduction

To most people the word *ether* is synonymous with the well-known anesthetic. That particular ether, however, is but one member of a general class of organic compounds known as **ethers.** These compounds have two organic groups connected to a single oxygen atom. The general formula for ethers is R—O—R′, where R and R′ may be identical or different, and may be alkyl or aryl groups. Specifically, in the common anesthetic, both R and R′ are ethyl groups, $CH_3CH_2—O—CH_2CH_3$.

In this chapter, we will describe the physical and chemical properties of ethers. Their excellent solvent properties are applied in the preparation of Grignard reagents, organometallic compounds with a carbon-magnesium bond. We will give special attention to **epoxides,** cyclic three-membered ethers that have important industrial utility.

8.2 Nomenclature of Ethers

Ethers are usually named by giving the name of each alkyl or aryl group, in alphabetical order, followed by the word *ether*.

$CH_3CH_2—O—CH_3$
ethyl methyl ether

$CH_3CH_2—O—CH_2CH_3$
diethyl ether (the prefix *di*- is sometimes omitted)

$C_6H_5—O—C_6H_5$
diphenyl ether

For ethers with more complex structures, it may be necessary to name the **—OR** group as an **alkoxy group.** In the IUPAC system, the smaller alkoxy group is named as a substituent.

$CH_3CHCH_2CH_2CH_3$ (with OCH_3 on C-2)
2-methoxypentane

trans-2-methoxycyclohexanol

1,3,5-trimethoxybenzene

EXAMPLE 8.1 Give a correct name for $CH_3CHCH(CH_3)_2$ (with OCH_2CH_3 on C-2)

Solution

$CH_3CHCHCH_3$ (carbons numbered 1, 2, 3, 4; OCH_2CH_3 on C-2, CH_3 on C-3)

2-ethoxy-3-methylbutane

PROBLEM 8.1 Give a correct name for

a. $(CH_3)_2CHOCH_3$ b. C_6H_5—O—$CH_2CH_2CH_3$ c. 1-methoxy-1-methylcyclohexane ring with CH_3 and OCH_3

PROBLEM 8.2 Write the structural formula for

a. dicyclopropyl ether. b. 2-ethoxyoctane.

8.3 *Physical Properties of Ethers*

Ethers are colorless compounds with characteristic, relatively pleasant odors. They have lower boiling points than alcohols with an equal number of carbon atoms. In fact, an ether has nearly the same boiling point as the corresponding hydrocarbon in which a —CH_2— group replaces the ether's oxygen. The following data illustrate these facts:

		bp	mol wt	water solubility (g/100 mL, 20°C)
1-butanol	$CH_3CH_2CH_2CH_2OH$	118°C	74	7.9
diethyl ether	CH_3CH_2—O—CH_2CH_3	35°C	74	7.5
pentane	CH_3CH_2—CH_2—CH_2CH_3	36°C	72	0.03

Because of their structures, ether molecules cannot form hydrogen bonds with one another. This is why they boil so much lower than their isomeric alcohols.

PROBLEM 8.3 Write structures for each of the following *isomers*, and arrange them in order of decreasing boiling point: 3-methoxy-1-propanol, 1,2-dimethoxyethane, 1,4-butanediol.

Although ethers cannot form hydrogen bonds with one another, they do form hydrogen bonds with alcohols:

$$\underset{\displaystyle R}{\overset{}{R-\underset{|}{O}}}\cdots\cdots H-\underset{\underset{\displaystyle R}{|}}{O}$$

For this reason, alcohols and ethers are usually mutually soluble. Low-molecular-weight ethers, such as dimethyl ether, are quite soluble in water. Likewise, the modest solubility of diethyl ether in water is similar to that of its isomer 1-butanol (see data tabulated above) because each can form a hydrogen bond to water. Ethers are less dense than water.

8.4 Ethers as Solvents

Ethers are relatively inert compounds. They do not usually react with dilute acids, with dilute bases, or with common oxidizing and reducing agents. They do not react with metallic sodium—a property that distinguishes them from alcohols. This general inertness, coupled with the fact that most organic compounds are ether-soluble, makes ethers excellent solvents in which to carry out organic reactions.

Ethers are also used frequently to extract organic compounds from their natural sources. Diethyl ether is particularly good for this purpose. Its low boiling point makes it easy to remove from an extract and easy to recover by distillation. It is highly flammable, however, and must not be used if there are any flames in the same laboratory.

Another risk is that ethers that have been in a laboratory for a long time, exposed to air, may contain organic peroxides as a result of oxidation.

$$CH_3CH_2OCH_2CH_3 + O_2 \longrightarrow CH_3CH_2O\underset{\underset{\displaystyle OOH}{|}}{C}HCH_3 \tag{8.1}$$

an ether hydroperoxide

These peroxides are extremely explosive and must be removed before the ether can be used safely. Shaking with aqueous ferrous sulfate ($FeSO_4$) destroys these peroxides by reduction.

8.5 The Grignard Reagent; an Organometallic Compound

One of the most striking examples of the solvating power of ethers is in the preparation of **Grignard reagents.** These reagents, which are exceedingly useful in organic synthesis, were discovered by the French organic chemist Victor Grignard [pronounced "greenyar(d)"]. In 1912 he received a Nobel Prize for this contribution to organic synthesis.*

---

*For a brief account of how Grignard discovered these reagents, see D. Hodson, *Chemistry in Britain* **1987,** 141–142.

Grignard found that when magnesium turnings are stirred with an ether solution of an alkyl or aryl halide, an exothermic reaction occurs. The magnesium, which is insoluble in ether, disappears as it reacts with the halide to give solutions of ether-soluble Grignard reagents.

$$\mathrm{R{-}X + Mg \xrightarrow{\text{dry ether}} \underset{\text{a Grignard reagent}}{R{-}MgX}} \qquad (8.2)$$

The carbon–halogen bond is broken and both the alkyl group and halogen become bonded to the magnesium.

Although the ether used as a solvent for this reaction is normally not shown as part of the Grignard reagent structure, it does play an important role. The unshared electron pairs on the ether oxygen help to stabilize the magnesium through coordination.

R R
Ö
R′—Mg—X
Ö
R R

Acting as a Lewis base, ether stabilizes a Grignard reagent.

The two ethers most commonly used in Grignard preparations are diethyl ether and the cyclic ether tetrahydrofuran, abbreviated THF (page 249). The Grignard reagent will not form unless the ether is scrupulously dry, free of traces of water or alcohols.

Grignard reagents are named as shown in the following equations:

$$\underset{\text{methyl iodide}}{\mathrm{CH_3{-}I}} + \mathrm{Mg} \xrightarrow{\text{ether}} \underset{\text{methylmagnesium iodide}}{\mathrm{CH_3MgI}} \qquad (8.3)$$

$$\underset{\text{bromobenzene}}{\mathrm{C_6H_5{-}Br}} + \mathrm{Mg} \xrightarrow{\text{ether}} \underset{\text{phenylmagnesium bromide}}{\mathrm{C_6H_5{-}MgBr}} \qquad (8.4)$$

Notice that there is no space between the name of the organic group and magnesium, but that there is a space before the halide name.

Grignard reagents usually react as if the alkyl or aryl group is negatively charged (a carbanion) and the magnesium atom is positively charged.

$$\overset{\delta-}{\mathrm{R}}{-}\overset{\delta+}{\mathrm{MgX}}$$

Carbanions are strong bases (they are the conjugate bases of hydrocarbons, which are very weak acids). It is not surprising, then, that Grignard reagents react vigorously with even such a weak acid as water, or with any other compound with an O—H, S—H, or N—H bond.

$$\underset{\text{stronger base}}{\overset{\delta-}{\text{R}}\text{—MgX}} + \underset{\text{stronger acid}}{\overset{\delta+}{\text{H}}\text{—OH}} \longrightarrow \underset{\text{weaker acid}}{\text{R—H}} + \underset{\text{weaker base}}{\text{Mg}^{2+}(\text{OH})^{-}\text{X}^{-}} \quad (8.5)$$

This is why the ether used as a solvent for the Grignard reagent must be scrupulously free of water or alcohol.

PROBLEM 8.4 Write an equation for the reaction between

a. methylmagnesium iodide and water.
b. phenylmagnesium bromide and methanol.

EXAMPLE 8.2 Is it possible to prepare a Grignard reagent from $HOCH_2CH_2CH_2Br$ and magnesium?

Solution No! Any Grignard reagent that forms is immediately destroyed by protons from the OH group. Grignard and hydroxyl groups in the same molecule are incompatible.

PROBLEM 8.5 Is it possible to prepare a Grignard reagent from $CH_3OCH_2CH_2CH_2Br$?

The reaction of a Grignard reagent with water can be put to useful purpose. For example, if heavy water (D_2O) is used, deuterium can be substituted for a halogen.

$$\underset{p\text{-bromotoluene}}{CH_3\text{—}C_6H_4\text{—Br}} \xrightarrow[\text{ether}]{\text{Mg}} \underset{p\text{-tolylmagnesium bromide}}{CH_3\text{—}C_6H_4\text{—MgBr}} \xrightarrow{D_2O} \underset{p\text{-deuteriotoluene}}{CH_3\text{—}C_6H_4\text{—D}} \quad (8.6)$$

This is a useful way to introduce an isotopic label into an organic compound.

EXAMPLE 8.3 Show how to prepare CH_3CHDCH_3 from $CH_2{=}CHCH_3$.

Solution

$$CH_2{=}CHCH_3 \xrightarrow{\text{HBr}} \underset{\substack{|\\ \text{Br}}}{CH_3CHCH_3} \xrightarrow[\text{ether}]{\text{Mg}} \underset{\substack{|\\ \text{MgBr}}}{CH_3CHCH_3} \xrightarrow{D_2O} CH_3CHDCH_3$$

PROBLEM 8.6 Show how to prepare CH_3CHDCH_3 from $(CH_3)_2CHOH$.

Grignard reagents are **organometallic compounds;** they contain a carbon-metal bond. Many other types of organometallic compounds are known (recall acetylides, eq. 3.50). Among the more useful in synthesis are **organolithium compounds,** which can be prepared in a manner similar to that for Grignard reagents.

$$\text{R—X} + 2\,\text{Li} \xrightarrow{\text{ether}} \underset{\text{an alkyllithium}}{\text{R—Li}} + \text{Li}^+\text{X}^- \tag{8.7}$$

PROBLEM 8.7 Write an equation for the preparation of propyllithium and for its reaction with D_2O.

Later in this chapter and elsewhere in this book, we will see examples of the synthetic utility of organometallic reagents.

8.6 Preparation of Ethers

The most important commercial ether is diethyl ether. It is prepared from ethanol and sulfuric acid.

$$\underset{\text{ethanol}}{CH_3CH_2OH + HOCH_2CH_3} \xrightarrow[140^\circ C]{H_2SO_4} \underset{\text{diethyl ether}}{CH_3CH_2OCH_2CH_3} + H_2O \tag{8.8}$$

Note that ethanol can be dehydrated by sulfuric acid to give either ethylene (eq. 7.17) or diethyl ether (eq. 8.8). Of course, the reaction conditions are different in each case. These reactions provide a good example of how important it is to control reaction conditions and to specify them in equations.

PROBLEM 8.8 The reaction in eq. 7.17 occurs by an E2 mechanism (review eqs. 7.22 and 7.23). By what mechanism does the reaction in eq. 8.8 occur?

Although it can be adapted to other ethers, the alcohol–sulfuric acid method is most commonly used to make symmetric ethers from primary alcohols.

PROBLEM 8.9 Write an equation for the synthesis of propyl ether from 1-propanol.

The commercial production of *t*-butyl methyl ether (and its ethyl analog) has become important in recent years. Used as an octane number enhancer in unleaded gasolines, it is prepared by the acid-catalyzed addition of methanol to 2-methylpropene. The reaction is related to the hydration of alkenes (Sec. 3.7b). The only difference is that an alcohol, methanol, is used as the nucleophile instead of water.

$$\underset{\text{methanol}}{CH_3OH} + \underset{\text{2-methylpropene}}{CH_2{=}C(CH_3)_2} \xrightarrow{H^+} \underset{t\text{-butyl methyl ether}}{CH_3O{-}\overset{\displaystyle CH_3}{\overset{|}{\underset{\underset{\displaystyle CH_3}{|}}{C}}}{-}CH_3} \tag{8.9}$$

PROBLEM 8.10 Write out the steps in the mechanism for eq. 8.9 (see eqs. 3.18 and 3.20).

Most important for the laboratory synthesis of unsymmetric ethers is the **Williamson synthesis,** named after the British chemist who devised it, Alexander Williamson. This method has two steps, both of which we have already discussed. In the first step, an alcohol is converted to its alkoxide by treatment with a reactive metal (sodium or potassium) or metal hydride (review eqs. 7.12 and 7.13). In the second step, an S_N2 displacement is carried out between the alkoxide and an alkyl halide (see Table 6.1, item 2). The Williamson synthesis is summarized by the general equations

$$2\,ROH + 2\,Na \longrightarrow 2\,RO^-Na^+ + H_2 \quad (8.10)$$

$$RO^-Na^+ + R'{-}X \longrightarrow ROR' + Na^+X^- \quad (8.11)$$

Since the second step is an S_N2 reaction, it works best if R′ in the alkyl halide is primary and not well at all if R′ is tertiary.

EXAMPLE 8.4 Write an equation for the synthesis of $CH_3OCH_2CH_2CH_3$ using the Williamson method.

Solution There are two possibilities, depending on which alcohol and which alkyl halide are used:

$$CH_3OCH_2CH_2CH_3 \quad \text{or} \quad CH_3OCH_2CH_2CH_3$$

$$\uparrow \qquad\qquad\qquad\qquad \uparrow$$

$$CH_3O^-\ Na^+ + XCH_2CH_2CH_3 \qquad CH_3X + Na^+\ {}^-OCH_2CH_2CH_3$$

The equations are

$$2\,CH_3OH + 2\,Na \longrightarrow 2\,CH_3O^-Na^+ + H_2$$

$$CH_3O^-Na^+ + CH_3CH_2CH_2X \longrightarrow CH_3OCH_2CH_2CH_3 + Na^+X^-$$

or

$$2\,CH_3CH_2CH_2OH + 2\,Na \longrightarrow 2\,CH_3CH_2CH_2O^-Na^+ + H_2$$

$$CH_3CH_2CH_2O^-Na^+ + CH_3X \longrightarrow CH_3CH_2CH_2OCH_3 + Na^+X^-$$

X is usually Cl, Br, or I.

PROBLEM 8.11 Write equations for the synthesis of the following ethers by the Williamson method.

a. C_6H_5—OCH_3 (benzene ring)—OCH₃   b. $(CH_3)_3COCH_3$ *(Reminder*: The second step proceeds by the S_N2 mechanism.)

A Word About . . .

15. Ether and Anesthesia

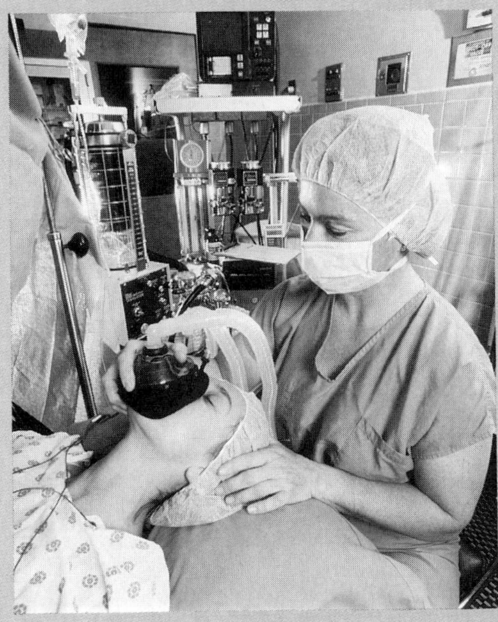

Prior to the 1840s, pain during surgery was relieved by various methods (asphyxiation, pressure on nerves, administration of narcotics or alcohol), but on the whole it was almost worse torture to undergo an operation than to endure the disease. Modern use of anesthesia during surgery has changed all that. Anesthesia stems from the work of several physicians in the mid-nineteenth century. The earliest experiments used nitrous oxide, ether, or chloroform. Perhaps the best known of these experiments was the removal of a tumor from the jaw of a patient anesthetized by ether, performed by Boston dentist William T. G. Morton in 1846.

Anesthetics fall into two major categories, general and local. *General anesthetics* are usually administered to accomplish three ends: insensitivity to pain (analgesia), loss of consciousness, and muscle relaxation. Gases such as nitrous oxide and cyclopropane and volatile liquids such as ether are administered by inhalation, but other general anesthetics such as barbiturates are injected intravenously.

The exact mechanism by which anesthetics affect the central nervous system is not completely known. Unconsciousness may result from several factors: changes in the properties of nerve cell membranes, suppression of certain enzymatic reactions, and solubility of the anesthetic in lipid membranes.

A good inhalation anesthetic should vaporize readily and have appropriate solubility in the blood and tissues. It should also be stable, inert, nonflammable, potent, and minimally toxic. It should have an acceptable odor and cause minimal side effects such as nausea or vomiting. No anesthetic that meets *all* these specifications has yet been developed. Although *diethyl ether* is perhaps the best known general anesthetic to the layperson, it fails on several counts (flammability, side effects of nausea or vomiting, and relatively slow action). It is quite potent, however, and produces good analgesia and muscle relaxation. The use of ether at present is rather limited, mainly because of its undesirable side effects. **Halothane**, $CF_3CHBrCl$, comes closest to an ideal inhalation anesthetic at present, but halogenated ethers such as **enflurane**, $CF_2H—O—CF_2CHClF$, are also used.

Local anesthetics are either applied to body surfaces or injected near nerves to desensitize a particular region of the body to pain. The best known of these anesthetics is procaine (Novocain), an aromatic amino-ester (see "A Word About" on page 402).

The discovery of anesthetics enabled physicians to perform surgery with deliberation and care, leading to many of the advances of modern medicine.

8.7 Cleavage of Ethers

Ethers have unshared electron pairs on the oxygen atom and are therefore Lewis bases. They react with strong proton acids and with Lewis acids such as the boron halides.

$$\mathrm{R{-}\ddot{\underset{\cdot\cdot}{O}}{-}R' + H^+ \rightleftharpoons R{-}\underset{\cdot\cdot}{\overset{H}{\overset{|}{O^+}}}{-}R'} \tag{8.12}$$

$$\mathrm{R{-}\ddot{\underset{\cdot\cdot}{O}}{-}R' + Br{-}\underset{\underset{Br}{|}}{B}{-}Br \rightleftharpoons R{-}\underset{Br{-}\underset{\underset{Br}{|}}{B^-}{-}Br}{\underset{|}{\ddot{O}^+}}{-}R'} \tag{8.13}$$

These reactions are similar to the reaction of alcohols with strong acids (eq. 7.18). If the alkyl groups R and/or R′ are primary or secondary, the bond to oxygen can be broken by reaction with a strong nucleophile such as I^- or Br^- (by an S_N2 process). For example,

$$\underset{\text{ethyl isopropyl ether}}{\mathrm{CH_3CH_2OCH(CH_3)_2}} + \mathrm{HI} \xrightarrow{\text{heat}} \underset{\text{ethyl iodide}}{\mathrm{CH_3CH_2I}} + \underset{\text{isopropyl alcohol}}{\mathrm{HOCH(CH_3)_2}} \tag{8.14}$$

$$\underset{\text{anisole}}{\mathrm{C_6H_5{-}OCH_3}} + \mathrm{BBr_3} \xrightarrow[\text{2. } H_2O]{\text{1. heat}} \underset{\text{phenol}}{\mathrm{C_6H_5{-}OH}} + \underset{\text{methyl bromide}}{\mathrm{CH_3Br}} \tag{8.15}$$

If R or R′ is tertiary, a strong nucleophile is not required since reaction will occur by an S_N1 (or E1) mechanism.

$$\underset{t\text{-butyl phenyl ether}}{\mathrm{C_6H_5{-}OC(CH_3)_3}} \xrightarrow[H_2O]{H^+} \underset{\text{phenol}}{\mathrm{C_6H_5{-}OH}} + \underset{\substack{t\text{-butyl alcohol} \\ \text{(and } (CH_3)_2C{=}CH_2)}}{\mathrm{(CH_3)_3COH}} \tag{8.16}$$

The net result of these reactions is **cleavage** of the ether at one of the C—O bonds. Ether cleavage is a useful reaction for determining the structure of a complex, naturally occurring ether because it allows one to break the large molecule into more easily handled, smaller fragments.

EXAMPLE 8.5 Write out the steps in the mechanism for eq. 8.14.

Solution The ether is first protonated by the acid.

$$\mathrm{CH_3CH_2\ddot{\underset{\cdot\cdot}{O}}CH(CH_3)_2 \overset{H^+}{\rightleftharpoons} CH_3CH_2\underset{\cdot\cdot}{\overset{H}{\overset{|}{O^+}}}CH(CH_3)_2}$$

The resulting oxonium ion is then cleaved by S_N2 attack of iodide ion at the primary carbon (recall that 1° > 2° in S_N2 reactions).

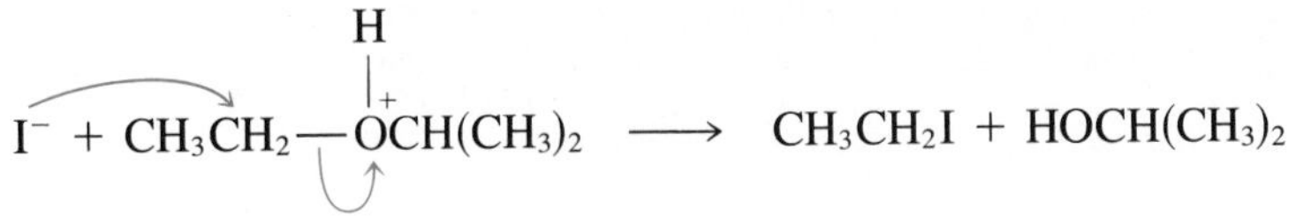

$$I^- + CH_3CH_2-\overset{\overset{H}{|}}{\overset{+}{O}}CH(CH_3)_2 \longrightarrow CH_3CH_2I + HOCH(CH_3)_2$$

PROBLEM 8.12 Write out the steps in the mechanism for formation of *t*-butyl alcohol in eq. 8.16. Which C-O bond cleaves, the one to the phenyl or the one to the *t*-butyl group?

8.8 *Epoxides (Oxiranes)*

Epoxides (or oxiranes) are cyclic ethers with a three-membered ring containing one oxygen atom.

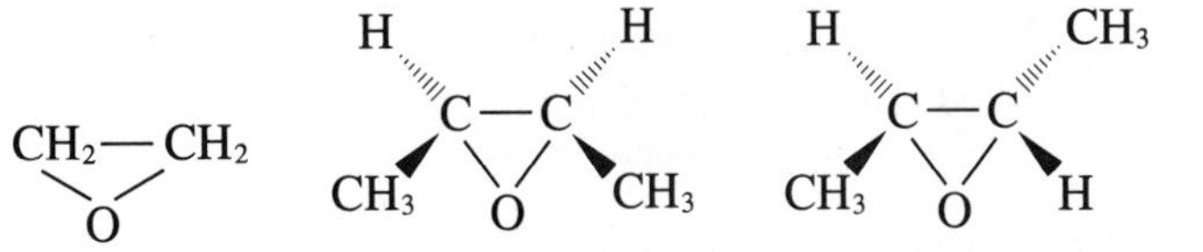

ethylene oxide (oxirane) bp 13.5°C

cis-2-butene oxide (*cis*-2,3-dimethyloxirane) bp 60°C

trans-2-butene oxide (*trans*-2,3-dimethyloxirane) bp 54°C

The most important commercial epoxide is ethylene oxide, produced by the silver-catalyzed air oxidation of ethylene.

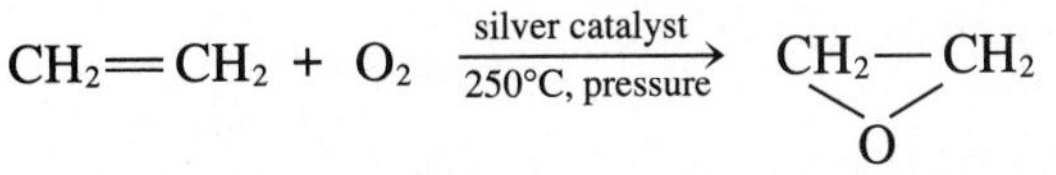

$$CH_2{=}CH_2 + O_2 \xrightarrow[250°C,\ pressure]{silver\ catalyst} \underset{\text{ethylene oxide}}{CH_2-CH_2 \ (\text{O bridge})} \quad (8.17)$$

Annual U.S. production of ethylene oxide exceeds 5.5 billion pounds. Only rather small amounts are used directly (for example, as a fumigant in grain storage). Most of the ethylene oxide constitutes a versatile raw material for the manufacture of other products, the main one being ethylene glycol.

The reaction in eq. 8.17 is suitable only for ethylene oxide. Other epoxides are usually prepared by the reaction of an alkene with an organic peroxyacid.

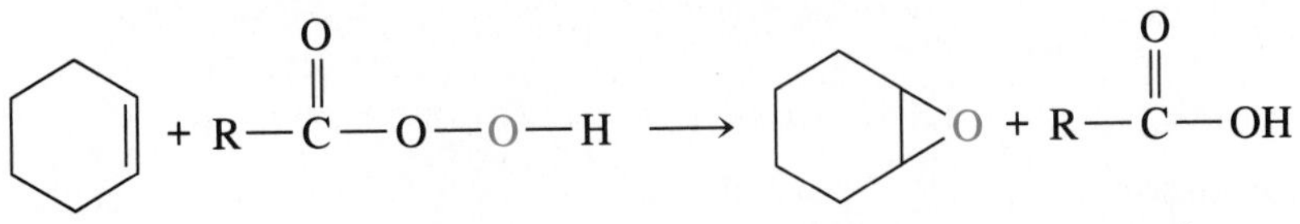

$$\text{cyclohexene} + R-\overset{\overset{O}{\|}}{C}-O-O-H \longrightarrow \text{cyclohexene oxide} + R-\overset{\overset{O}{\|}}{C}-OH \quad (8.18)$$

cyclohexene   organic peroxy acid   cyclohexene oxide   organic acid

Peroxyacids, like hydrogen peroxide H—O—O—H, to which they are structurally related, are good oxidizing agents. On a large scale, peroxyacetic acid ($R{=}CH_3$) is used, whereas in the laboratory organic peroxy acids such as *m*-chloroperoxybenzoic acid (R = *m*-chlorophenyl) are frequently used.

PROBLEM 8.13 Write an equation for the reaction of cyclopentene with *m*-chloroperoxybenzoic acid.

A Word About . . .

16. The Gypsy Moth's Epoxide

The main mode of communication among insects is via the emission and detection of specific chemical substances. These substances are called **pheromones.** The word is from the Greek (*pherein,* to carry, and *horman,* to excite). Even though they are emitted and detected in exceedingly small amounts, pheromones have profound biological effects. One of their main effects is sexual attraction and stimulation, but they are also used as alarm substances to alert members of the same species to danger, as aggregation substances to call together both sexes of a species, and as trail substances to lead members of a species to food.

Often pheromones are chemically relatively simple compounds—alcohols, esters, aldehydes, ketones, ethers, epoxides, or even hydrocarbons. Two examples are **muscalure** and **bombykol,** the sex attractants of the common housefly and the silkworm moth, respectively.

$$\underset{\text{H}}{\overset{CH_3(CH_2)_7}{}}\!\!>C=C<\!\!\underset{\text{H}}{\overset{(CH_2)_{12}CH_3}{}}$$

muscalure

$$CH_3(CH_2)_2CH=CH-CH=CH-(CH_2)_8CH_2OH$$

bombykol

Their molecular weights are low enough that the substances are volatile, yet not so low that they disperse too rapidly. Also, their molecular structures must be distinctive to make them species-specific; survival of the species would not be served by attracting another species. Often this specificity is attained through stereoisomerism (at double bonds and/or at chiral centers), but it can also be achieved by using specific ratios of two or more pheromones for a particular communication purpose.

Let us consider a specific pheromone, **disparlure,** the sex attractant of the gypsy moth (*Lymantria dispar*). The gypsy moth is a serious despoiler of forest and shade trees as well as fruit orchards. Gypsy moth larvae, which hatch each spring, are voracious eaters and can strip a tree bare of leaves in just a few weeks.

The abdominal tips (last two segments) of the virgin female moth contain the sex attractant. Extraction of 78,000 tips led to isolation of the main sex attractant, which was the following *cis*-epoxide:

$(CH_3)_2CH(CH_2)_4$ (C-7) – (C-8) $(CH_2)_9CH_3$, with H on each carbon and O bridging C-7 and C-8

(7*R*,8*S*)-(+)-7,8-epoxy-2-methyloctadecane

(disparlure)

The active isomer has the *R* configuration at carbon 7 and the *S* configuration at carbon 8. This isomer can be detected by the male gypsy moth at a concentration as low at 10^{-10}g/mL; its enantiomer is inactive in solutions a million times more concentrated.

Disparlure has been synthesized in the laboratory. The synthetic material can be used to lure the male to traps and in that way to control the insect population. This form of insect control sometimes has advantages over spraying with insecticides.

8.9 Reactions of Epoxides

Because of the strain in the three-membered ring, epoxides are much more reactive than ordinary ethers and give products in which the ring has opened. For example, with water they undergo acid-catalyzed ring opening to give glycols.

$$\underset{\text{ethylene oxide}}{\underset{\diagdown O \diagup}{CH_2—CH_2}} + H—OH \xrightarrow{H^+} \underset{\text{ethylene glycol}}{\underset{OH \quad OH}{\overset{|\qquad |}{CH_2—CH_2}}} \tag{8.19}$$

In this way, about 5 billion pounds of ethylene glycol are produced annually in the United States alone. Approximately half of it is used in automobile cooling systems as antifreeze. Most of the rest is used to prepare polyesters such as Dacron (see Sec. 14.9).

EXAMPLE 8.6 Write equations that show the mechanism for eq. 8.19.

Solution The first step is reversible protonation of the epoxide oxygen, as in eq. 8.12.

$$\underset{\diagdown \ddot{\underset{..}{O}} \diagup}{CH_2—CH_2} + H^+ \rightleftharpoons \underset{\underset{H}{|}}{\underset{\diagdown \ddot{O}^+ \diagup}{CH_2—CH_2}}$$

The second step is a nucleophilic S_N2 displacement on the primary carbon, with water as the nucleophile. Then proton loss yields the glycol (see Table 6.1, item 3).

$$H_2\ddot{O}: + \underset{\underset{H}{|}}{\underset{\diagdown \ddot{O}^+ \diagup}{CH_2—CH_2}} \longrightarrow H—\overset{+}{\underset{\underset{H}{|}}{\ddot{O}}}—CH_2—CH_2—OH \rightleftharpoons$$

$$HO—CH_2CH_2—OH + H^+$$

PROBLEM 8.14 Write an equation for the acid-catalyzed reaction of cyclohexene oxide with water. Predict the stereochemistry of the product.

Other nucleophiles add to epoxides in a similar way.

$$\underset{\diagdown O \diagup}{CH_2—CH_2} \xrightarrow{H^+} \begin{cases} \xrightarrow{CH_3OH} \underset{\text{2-methoxyethanol}}{HOCH_2CH_2OCH_3} \\ \xrightarrow{HOCH_2CH_2OH} \underset{\text{diethylene glycol}}{HOCH_2CH_2OCH_2CH_2OH} \end{cases} \tag{8.20}$$

2-Methoxyethanol is an additive for jet fuels, used to prevent water from freezing in fuel lines. Being both an alcohol and an ether, it is soluble in both water and organic solvents. *Diethylene glycol* is useful as a plasticizer (softener) in cork gaskets and tiles.

Grignard reagents and organolithium compounds are strong nucleophiles capable of opening the ethylene oxide ring. The initial product is a magnesium alkoxide or lithium alkoxide, but after hydrolysis (as in the reverse of eq. 7.14), we obtain a primary alcohol with two more carbon atoms than the organometallic reagent.

$$\overset{\delta^-}{R}—\overset{\delta^+}{MgX} + \underset{\underset{\delta^-}{O}}{H_2\overset{\delta^+}{C}—CH_2} \longrightarrow \underset{\text{a magnesium alkoxide}}{RCH_2CH_2OMgX} \xrightarrow{H—OH} RCH_2CH_2OH + Mg(OH)X \quad (8.21)$$

$$\overset{\delta^-}{R}—\overset{\delta^+}{Li} + \underset{\underset{\delta^-}{O}}{H_2\overset{\delta^+}{C}—CH_2} \longrightarrow \underset{\text{a lithium alkoxide}}{RCH_2CH_2OLi} \xrightarrow{H—OH} RCH_2CH_2OH + LiOH \quad (8.22)$$

PROBLEM 8.15 Write an equation for the reaction between ethylene oxide and

a. $CH_3CH_2CH_2MgCl$ followed by hydrolysis.
b. $H_2C{=}CHLi$ followed by hydrolysis.
c. $CH_3C{\equiv}C^-\,Na^+$ followed by hydrolysis.

A Word About . . .

17. Epoxy Resins

Most people hear the word *epoxy* in connection with *epoxy resins,* materials used as adhesives for bonding to metal, glass, and ceramics. Epoxy resins are also used in surface coatings (for example, paints) because of their exceptional inertness, hardness, and flexibility.

Two raw materials for the manufacture of epoxy resins are **epichlorhydrin** and **bisphenol-A.**

$$Cl—CH_2—\underset{\diagdown\ O\ \diagup}{CH—CH_2}$$

epichlorhydrin

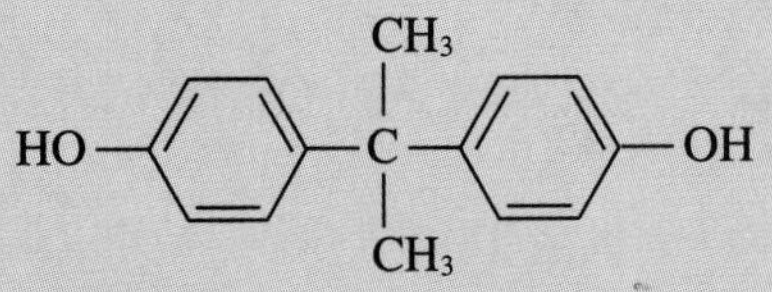

bisphenol-A

Reaction of a mixture of these two raw materials with a base gives a "linear" epoxy resin; its structure is shown in Figure 8.1.

Commercial resins of this type range from liquids (where *n* is small) to viscous adhesives to solids used in surface coatings (where *n* may be as large as 25).

It is possible to take advantage of the remaining epoxide rings and hydroxyl groups in the "linear" polymer to form cross-links between the polymer chains, thus substantially increasing the molecular weight of the polymer. This is especially important when the end use is as a surface coating.

Epoxy resins can be varied in structure. For example, the bisphenol-A can be partially or totally replaced by other di- or polyhydroxy compounds, and epoxides other than epichlorhydrin can be used. Annual world production of epoxy resins runs to about a billion pounds.

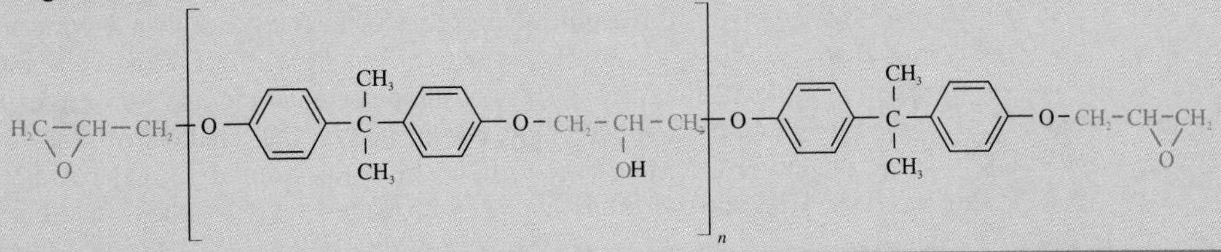

FIGURE 8.1 Structure of a "linear" epoxy resin.

8.10 Cyclic Ethers

Cyclic ethers whose rings are larger than the three-membered epoxides are known. Most commonly they have five- or six-membered rings. Some examples include

tetrahydrofuran (oxolane) bp 67°C

tetrahydropyran (oxane) bp 88°C

1,4-dioxane bp 101°C

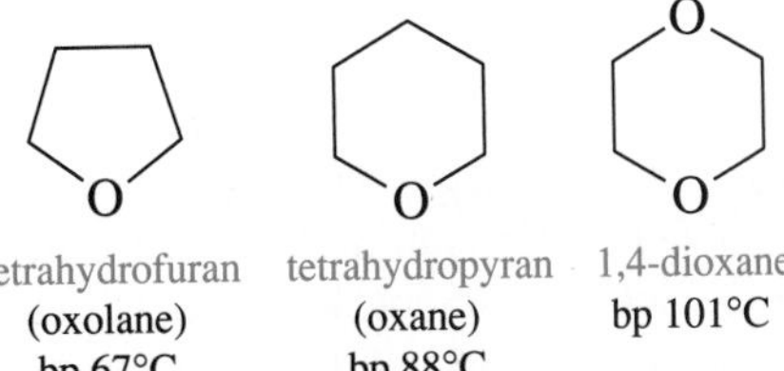

Tetrahydrofuran (THF), is a particularly useful solvent that not only dissolves many organic compounds but is miscible with water. THF is an excellent solvent—often superior to diethyl ether—in which to prepare Grignard reagents. Although it has the same number of carbon atoms as diethyl ether, they are "pinned back" in a ring. The oxygen in THF is therefore less hindered and better at coordinating with the magnesium in a Grignard reagent. **Tetrahydropyran** and **1,4-dioxane** are also soluble in both water and organic solvents.

In recent years, there has been much interest in macrocylic (large-ring) polyethers. Some examples are

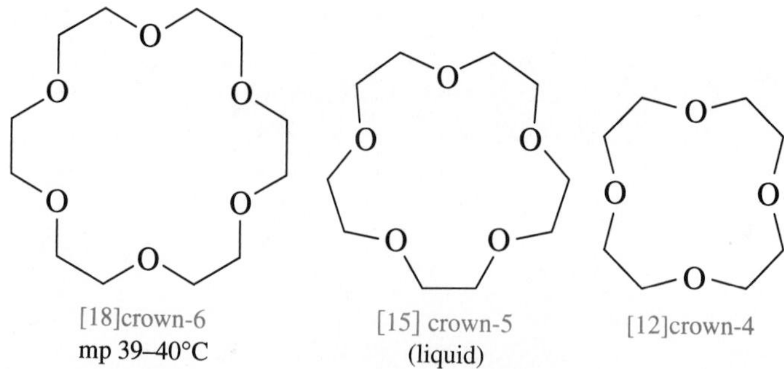

These compounds are called **crown ethers** because their molecules have a crownlike shape. The number in brackets in their common names gives the ring size, and the terminal number gives the number of oxygens. The oxygens are usually separated from one another by two carbon atoms.

Crown ethers have the unique property of forming complexes with positive ions (Na^+, K^+, and so on). The positive ions fit within the macrocyclic rings selectively, depending on the sizes of the particular ring and ion. For example, [18]crown-6 binds K^+ more tightly than it does the smaller Na^+ (too loose a fit) or the larger Cs^+ (too large to fit in the hole). Similarly, [15]crown-5 binds Na^+, and [12]crown-4 binds Li^+. The crown ethers act as hosts for their ionic guests.

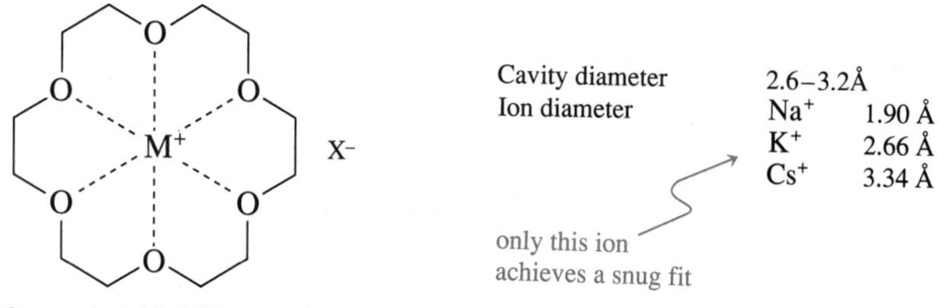

This complexing ability is so strong that ionic compounds can be dissolved in organic solvents that contain a crown ether. For example, potassium permanganate ($KMnO_4$) is soluble in water but insoluble in benzene. However, if some dicyclohexyl[18]crown-6 is dissolved in the benzene, it is possible to extract the potassium permanganate from the water into the benzene! The resulting "purple benzene," containing free, essentially unsolvated permanganate ions, is a powerful oxidizing agent.*

---

*Crown ethers were discovered by Charles J. Pedersen, working at Du Pont Company. This discovery had broad implications for a field now known as molecular recognition, or host-guest chemistry. Pedersen, Donald J. Cram (U.S.) and Jean-Marie Lehn (France) shared the 1987 Nobel Prize in chemistry for their imaginative development of this field. You might enjoy Pedersen's personal account of this discovery (*Journal of Inclusion Phenomena* **1988,** *6,* 337–350); the same journal contains the Nobel lectures by Cram and Lehn on their work.

The selective binding of metallic ions by macrocyclic compounds is important in nature. Several antibiotics, such as **nonactin,** have large rings that contain regularly spaced oxygen atoms. Nonactin (which contains four tetrahydrofuran rings joined by four ester links) selectively binds K^+ (in the presence of Na^+) in aqueous media, thus allowing the selective transport of K^+ (but not Na^+) through cell membranes.

nonactin

REACTION SUMMARY

1. Organometallic Compounds

a. Preparation of Grignard Reagents (Sec. 8.5)

$$R{-}X + Mg \xrightarrow{\text{ether or THF}} R{-}MgX$$

$X = Cl, Br, I$

b. Preparation of Organolithium Reagents (Sec. 8.5)

$$R{-}X + 2Li \longrightarrow R{-}Li + LiX$$

$X = Cl, Br, I$

c. Hydrolysis of Organometallics to Alkanes (Sec. 8.5)

$$R{-}MgX + H_2O \longrightarrow R{-}H + Mg(OH)X$$
$$R{-}Li + H_2O \longrightarrow R{-}H + LiOH$$

2. Ethers

a. Preparation by Dehydration of Alcohols (Sec. 8.6)

$$2\,R{-}OH \xrightarrow[\Delta]{H_2SO_4} R{-}O{-}R + H_2O$$

b. Preparation from Alkenes and Alcohols (Sec. 8.6).

$$>C{=}C< \xrightarrow[H^+ \text{ (catalyst)}]{ROH} H{-}\overset{|}{\underset{|}{C}}{-}\overset{|}{\underset{|}{C}}{-}OR$$

c. Preparation from Alcohols and Alkyl Halides (Sec. (8.6)

$$ROH + NaH \longrightarrow RO^- Na^+ + H_2$$
$$RO^- Na^+ + R'{-}X \longrightarrow RO{-}R' + Na^+ X^-$$

(best for R' = primary)

d. Cleavage by Hydrogen Halides (Sec. 8.7)

$$R{-}O{-}R + HX \longrightarrow R{-}X + R{-}OH \xrightarrow{HX} R{-}X + H_2O$$

e. Cleavage by Boron Tribromide (Sec. 8.7)

$$R{-}O{-}R \xrightarrow[2.\ H_2O]{1.\ BBr_3} RBr$$

3. Epoxides

a. Preparation from Alkenes (Sec. 8.8)

$$>C{=}C< + RCO_3H \longrightarrow >\overset{O}{C{-}C}< + RCO_2H$$

b. Reaction with Water and Alcohols (Sec. 8.9)

$$>\overset{O}{C{-}C}< \xrightarrow[H^+]{H{-}OH} -\overset{HO}{\underset{|}{C}}-\overset{OH}{\underset{|}{C}}-$$

$$>\overset{O}{C{-}C}< \xrightarrow[R{-}OH]{H^+} -\overset{RO}{\underset{|}{C}}-\overset{OH}{\underset{|}{C}}-$$

c. Reaction with Organometallic Reagents

$$>\overset{O}{C{-}C}< \xrightarrow{RMgX \text{ or } RLi} -\overset{R}{\underset{|}{C}}-\overset{O^-M^+}{\underset{|}{C}}- \xrightarrow{H_2O} -\overset{R}{\underset{|}{C}}-\overset{OH}{\underset{|}{C}}- + M{-}OH$$

M = MgX or Li

ADDITIONAL PROBLEMS

8.16. Write a structural formula for each of the following compounds:

a. dipropyl ether
b. *t*-butyl ethyl ether
c. 3-methoxyhexane
d. allyl propyl ether
e. *p*-bromophenyl ethyl ether
f. *cis*-2-ethoxycyclopentanol
g. ethylene glycol dimethyl ether
h. 1-methoxypropene
i. propylene oxide
j. *p*-ethoxyanisole

8.17. Name each of the following compounds:

a. $(CH_3)_2CHOCH(CH_3)_2$
b. $(CH_3)_2CHCH_2OCH_3$
c. $CH_3CH{-}CH_2$ (with O bridging the two carbons, epoxide ring)
d. Br–C_6H_4–OCH_3 (para-substituted benzene ring)
e. $CH_3CH(OCH_2CH_3)CH_2CH_2CH_3$
f. $CH_3OCH_2CH_2OH$

g. [structure: 2-methyltetrahydrofuran, ring O with CH_3]

h. $CH_3OCH_2C{\equiv}CH$

8.18. Ethers and alcohols can be isomeric. Write the structures and give the names for all possible isomers with the molecular formula $C_4H_{10}O$.

8.19. Consider four compounds that have nearly the same molecular weights: 1,2-dimethoxyethane, ethyl propyl ether, hexane, and 1-pentanol. Which would you expect to have the highest boiling point? Which would be most soluble in water? Explain the reasons for your choices.

8.20. Write equations for the reaction of each of the following with (1) Mg in ether followed by (2) addition of D_2O to the resulting solution:
a. $CH_3CH_2CH_2CH_2Br$ **b.** $CH_3OCH_2CH_2CH_2Br$

8.21. The following steps can be used to convert anisole to *o*-*t*-butylanisole. Give the reagent for each step. Explain why the overall result cannot be achieved in one step by a Friedel-Crafts alkylation.

C_6H_5–OCH_3 ⟶ Br–C_6H_4–OCH_3 ⟶

Br–C_6H_3($C(CH_3)_3$)–OCH_3 ⟶ C_6H_4($C(CH_3)_3$)–OCH_3

8.22. Write equations for the best method to prepare each of the following ethers:

a. $(CH_3CH_2CH_2CH_2)_2O$ **b.** C_6H_5–OCH_2CH_3 **c.** $CH_3CH_2OC(CH_3)_3$

8.23. Explain why the Williamson synthesis cannot be used to prepare diphenyl ether.

8.24. Ethers are soluble in cold, concentrated sulfuric acid, but alkanes are not. This difference can be used as a simple chemical test to distinguish between these two classes of compounds. What chemistry (show an equation) is the basis for this difference?

8.25. Write an equation for each of the following reactions. If no reaction occurs, say so.
a. dibutyl ether + boiling aqueous NaOH ⟶
b. methyl propyl ether + excess HBr (hot) ⟶
c. dipropyl ether + Na ⟶
d. ethyl ether + cold concentrated H_2SO_4 ⟶
e. ethyl phenyl ether + BBr_3 ⟶

8.26. When heated with excess HBr, a cyclic ether gave 1,4-dibromobutane as the only organic product. Write a structure for the ether and an equation for the reaction.

8.27. Using the peroxy acid epoxidation of an alkene and the ring opening of an epoxide, devise a two-step synthesis of 1,2-butanediol from 1-butene.

8.28. Write an equation for the reaction of ethylene oxide with
a. 1 mol of HBr. **b.** excess HBr. **c.** phenol + H^+.

8.29. $CH_3CH_2OCH_2CH_2OH$ (ethyl cellosolve) and $CH_3CH_2OCH_2CH_2OCH_2CH_2OH$ (ethyl carbitol) are solvents used in the formulation of lacquers. They are produced commercially from ethylene oxide and certain other reagents. Show with equations how this might be done.

8.30. 2-Phenylethanol, which has the aroma of oil of roses, is used in perfumes. Write equations to show how 2-phenylethanol can be synthesized from bromobenzene and ethylene oxide, using a Grignard reagent.

8.31. 1,1-Dimethyloxirane dissolved in excess methanol and treated with a little acid yields the product 2-methoxy-2-methyl-1-propanol (and no 1-methoxy-2-methyl-2-propanol). What reaction mechanism explains this result?

8.32. Write an equation for the reaction of ammonia with ethylene oxide. The product is a water-soluble organic base used to absorb and concentrate CO_2 in the manufacture of dry ice.

8.33. Design a synthesis of 3-pentyne-1-ol using propyne and ethylene oxide as the only sources of carbon atoms.

8.34. Write out the steps in the reaction mechanisms for the reactions given in eq. 8.20.

8.35. What chemical test will distinguish between the compounds in each of the following pairs? Indicate what is visually observed with each test.
a. dipropyl ether and hexane
b. ethyl phenyl ether and allyl phenyl ether
c. 1-butanol and methyl propyl ether
d. phenol and anisole

8.36. An organic compound with the molecular formula $C_4H_{10}O_3$ shows properties of both an alcohol and an ether. When treated with an excess of hydrogen bromide, it yields only one organic compound, 1,2-dibromoethane. Draw a structural formula for the original compound.

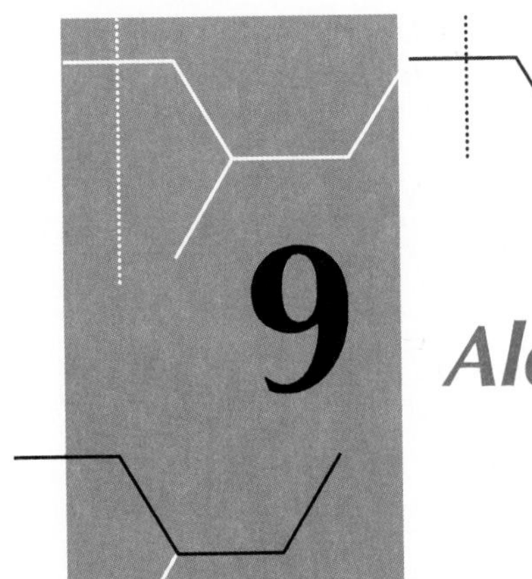

9 Aldehydes and Ketones

9.1 *Introduction*

We now come to perhaps the most important functional group in organic chemistry—the **carbonyl group,**

$$>C=O$$

This group is present in aldehydes, ketones, carboxylic acids, esters, and several other classes of compounds. Many of these compounds are important commercially and in biological processes. In this chapter, we will discuss aldehydes and ketones and, in the next chapter, carboxylic acids and related compounds.

Aldehydes have at least one hydrogen atom attached to the carbonyl group. The remaining group may be another hydrogen atom or any organic group.

$$-\overset{O}{\overset{\|}{C}}-H \text{ or } -CHO \qquad H-\overset{O}{\overset{\|}{C}}-H \qquad R-\overset{O}{\overset{\|}{C}}-H \qquad Ar-\overset{O}{\overset{\|}{C}}-H$$

aldehyde group — formaldehyde — aliphatic aldehyde — aromatic aldehyde

In **ketones,** the carbonyl carbon atom is connected to two other carbon atoms.

$$R-\overset{O}{\overset{\|}{C}}-R \qquad R-\overset{O}{\overset{\|}{C}}-Ar \qquad Ar-\overset{O}{\overset{\|}{C}}-Ar \qquad \bigcirc C=O$$

aliphatic ketone — alkyl aryl ketone — aromatic ketone — a cyclic ketone

9.2 Nomenclature of Aldehydes and Ketones

In the IUPAC system, the characteristic ending for aldehydes is *-al* (from the first syllable of aldehyde). The following examples illustrate the system:

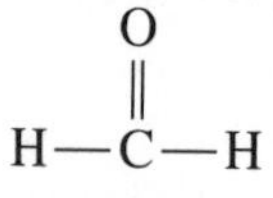

$$H{-}\overset{O}{\overset{\|}{C}}{-}H \qquad CH_3{-}\overset{O}{\overset{\|}{C}}{-}H \qquad CH_3CH_2{-}\overset{O}{\overset{\|}{C}}{-}H \qquad CH_3CH_2CH_2{-}\overset{O}{\overset{\|}{C}}{-}H$$

methanal (formaldehyde); ethanal (acetaldehyde); propanal (propionaldehyde); butanal (*n*-butyraldehyde)

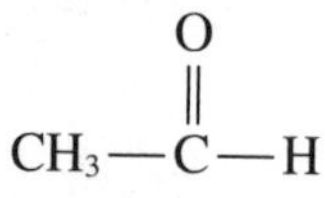

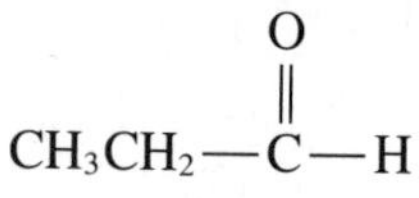

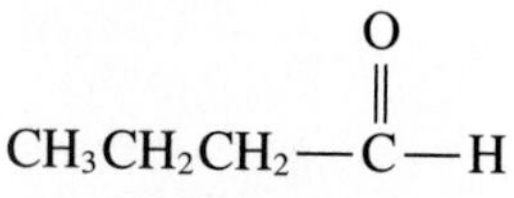

The common names shown below the IUPAC names are frequently used, so you should learn them.

For substituted aldehydes, we number the chain starting with the aldehyde carbon, as the following examples illustrate:

$$\overset{4}{C}H_3\overset{3}{C}H(CH_3)\overset{2}{C}H_2{-}\overset{1}{C}(=O){-}H \qquad \overset{4}{C}H_2{=}\overset{3}{C}H{-}\overset{2}{C}H_2{-}\overset{1}{C}(=O){-}H \qquad \overset{3}{C}H_2(OH){-}\overset{2}{C}H(OH){-}\overset{1}{C}(=O){-}H$$

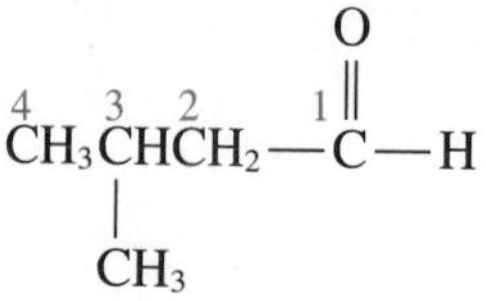

3-methylbutanal; 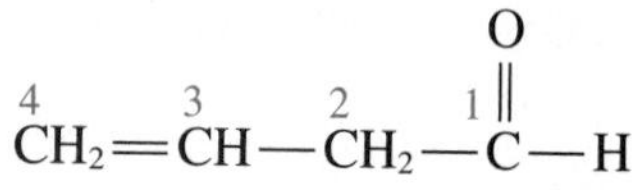 3-butenal; 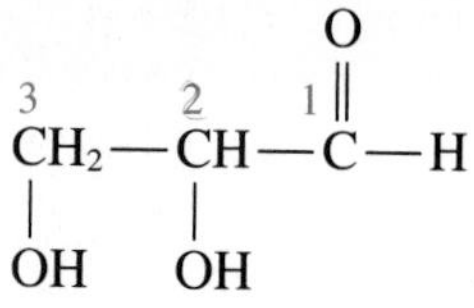 2,3-dihydroxypropanal (glyceraldehyde)

Notice from the last two examples that an aldehyde group has priority over a double bond or a hydroxyl group, not only in numbering, but also as the suffix. For cyclic aldehydes, the suffix *-carbaldehyde* is used. Aromatic aldehydes often have common names:

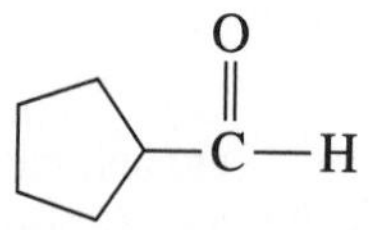

cyclopentanecarbaldehyde (formylcyclopentane); benzaldehyde (benzenecarbaldehyde); salicylaldehyde (2-hydroxybenzenecarbaldehyde)

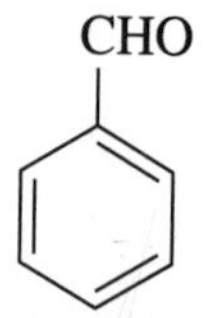

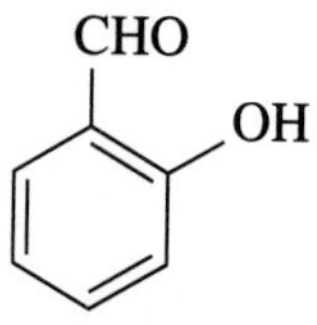

In the IUPAC system, the ending for ketones is *-one* (from the last syllable of ketone). The chain is numbered so that the carbonyl carbon has the lowest possible number. Common names of ketones are formed by adding the word *ketone* to the names of the alkyl or aryl groups attached to the carbonyl carbon. In still other cases, traditional names are used. The following examples illustrate these methods:

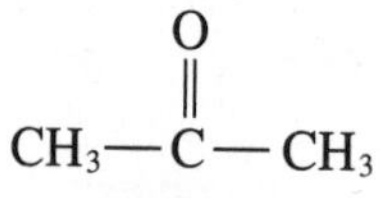

$$CH_3{-}\overset{O}{\overset{\|}{C}}{-}CH_3 \qquad \overset{1}{C}H_3{-}\overset{2}{C}(=O){-}\overset{3}{C}H_2\overset{4}{C}H_3 \qquad \overset{1}{C}H_2\overset{2}{C}H_3{-}\overset{3}{C}(=O){-}\overset{4}{C}H_2\overset{5}{C}H_3$$

propanone (acetone); 2-butanone (ethyl methyl ketone); 3-pentanone (diethyl ketone)

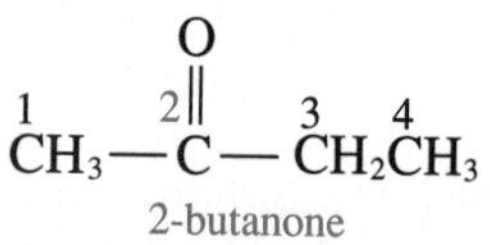

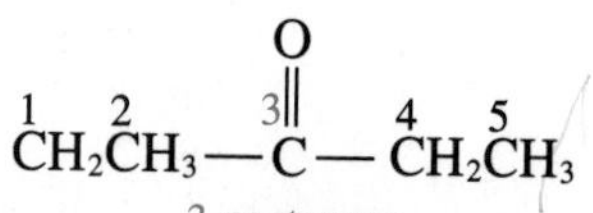

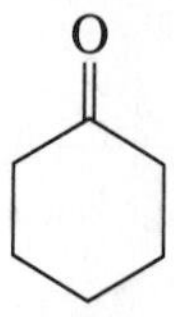

$$\overset{4}{C}H_2{=}\overset{3}{C}H{-}\overset{2}{C}(=O){-}\overset{1}{C}H_3$$

cyclohexanone; 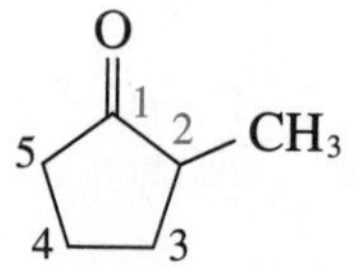 2-methylcyclopentanone; 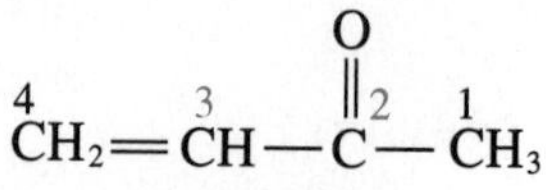 3-buten-2-one (methyl vinyl ketone)

$C_6H_5-C(=O)-CH_3$ — acetophenone (methyl phenyl ketone)

$C_6H_5-C(=O)-C_6H_5$ — benzophenone (diphenyl ketone)

$C_3H_5-C(=O)-C_3H_5$ — dicyclopropyl ketone

PROBLEM 9.1 Using the examples as a guide, write a structure for

a. pentanal.
b. *p*-bromobenzaldehyde.
c. 2-pentanone.
d. *t*-butyl methyl ketone.
e. cyclohexanecarbaldehyde.
f. 3-pentyne-2-one.

PROBLEM 9.2 Using the examples as a guide, write a correct name for

a. $(CH_3)_2CHCH_2CH{=}O$
b. $CH_3CH{=}CHCH{=}O$
c. cyclobutanone (four-membered ring with C=O)
d. $(CH_3)_2CHCH_2\overset{O}{\overset{\|}{C}}CH_3$

9.3 Some Common Aldehydes and Ketones

Formaldehyde, the simplest aldehyde, is manufactured on a very large scale by the oxidation of methanol.

$$CH_3OH \xrightarrow[600-700^\circ C]{\text{Ag catalyst}} \underset{\text{formaldehyde}}{CH_2{=}O} + H_2 \tag{9.1}$$

Annual world production is more than 8 billion pounds. Formaldehyde is a gas (bp -21°C), but it cannot be stored in a free state because it polymerizes readily.* Normally it is supplied as a 37% aqueous solution called **formalin.** In this form it is used as a disinfectant and preservative, but most formaldehyde is used in the manufacture of plastics, building insulation, particle board, and plywood.

Acetaldehyde boils close to room temperature (bp 20°C). It is manufactured mainly by the Wacker process, which involves direct selective oxidation of ethylene over a palladium-copper catalyst.

$$2\ CH_2{=}CH_2 + O_2 \xrightarrow[100-130^\circ C]{\text{Pd}-\text{Cu}} 2\ CH_3CH{=}O \tag{9.2}$$

About half the acetaldehyde produced annually is oxidized to acetic acid. The rest is used for the production of 1-butanol and other commercial chemicals.

Acetone, the simplest ketone, is also produced on a large scale—about 4 billion pounds annually. The most common methods for its commercial synthesis are the Wacker oxidation of propene (analogous to eq. 9.2), the oxida-

* The polymer derived from formaldehyde is a long chain of alternating CH_2 and oxygen units, which can be described by the structure $(CH_2O)_n$. See Sec. 3.16 and Chapter 14 for discussions of polymers.

tion of isopropyl alcohol (eq. 7.35, R = R′ = CH_3), and the oxidation of isopropylbenzene (eq. 9.3).

$$C_6H_5-CH(CH_3)_2 \xrightarrow{O_2} C_6H_5-C(OOH)(CH_3)_2 \xrightarrow[H_2O]{\text{dil } H_2SO_4} \underset{\text{phenol}}{C_6H_5OH} + \underset{\text{acetone}}{CH_3-\overset{O}{\overset{\|}{C}}-CH_3} \quad (9.3)$$

About 30% of the acetone is used directly, because it is not only completely miscible with water but is also an excellent solvent for many organic substances (resins, paints, dyes, and nail polish). The rest is used to manufacture other commercial chemicals, including bisphenol-A for epoxy resins (page 248).

$$2\ \underset{\text{phenol}}{HO-C_6H_5} + \underset{\text{acetone}}{CH_3\overset{O}{\overset{\|}{C}}CH_3} \xrightarrow[-H_2O]{H^+} \underset{\text{bisphenol-A}}{HO-C_6H_4-C(CH_3)_2-C_6H_4-OH} \quad (9.4)$$

9.4 Synthesis of Aldehydes and Ketones

We have already seen, in previous chapters, several ways to prepare aldehydes and ketones. One of the most useful is the oxidation of alcohols.

$$\gt C(H)(OH) \xrightarrow[\text{agent}]{\text{oxidizing}} \gt C{=}O \quad (9.5)$$

Oxidation of a primary alcohol gives an aldehyde, whereas oxidation of a secondary alcohol gives a ketone. Chromium reagents such as pyridinium chlorochromate (PCC) are commonly used in the laboratory for this purpose (review Sec. 7.13).

PROBLEM 9.3 Give the product expected from treatment of

a. cyclopentanol with Jones' reagent (see Sec. 7.13).
b. 2-methylhexanol with pyridinium chlorochromate (PCC).

PROBLEM 9.4 Give the structure of an alcohol that is a suitable precursor for oxidation to

a. 2-methylpropanal. b. 4-*t*-butylcyclohexanone.

Aromatic ketones can be made by Friedel-Crafts acylation of an aromatic ring (review eq. 4.15 and Sec. 4.10d). For example,

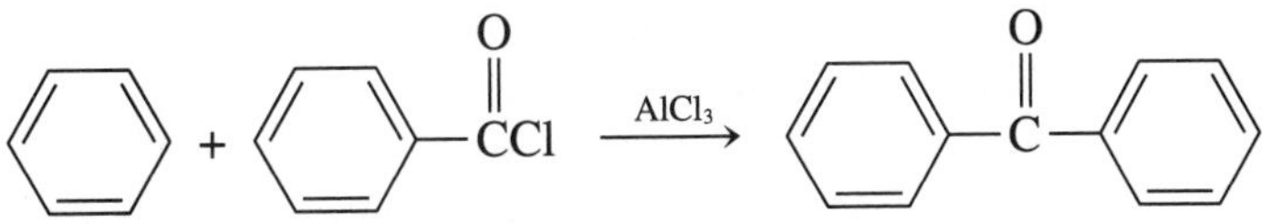

$$C_6H_6 + C_6H_5\overset{O}{\overset{\|}{C}}Cl \xrightarrow{AlCl_3} C_6H_5\overset{O}{\overset{\|}{C}}C_6H_5 + HCl \qquad (9.6)$$

benzene, benzoyl chloride, benzophenone

PROBLEM 9.5 Complete the following equation and name the product.

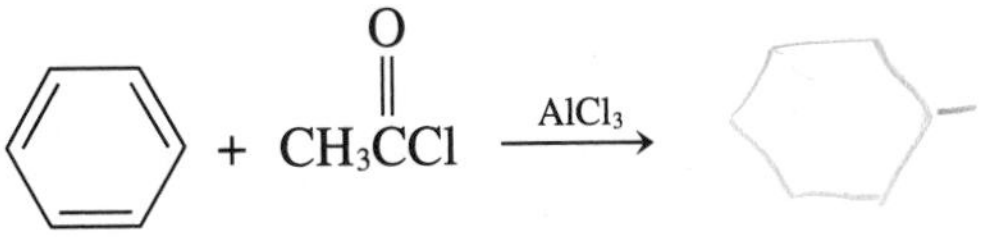

$$C_6H_6 + CH_3\overset{O}{\overset{\|}{C}}Cl \xrightarrow{AlCl_3}$$

Methyl ketones can be prepared by hydration of terminal alkynes, catalyzed by acid and mercuric ion (review eq. 3.49). For example,

$$\underset{\text{1-octyne}}{CH_3(CH_2)_5C{\equiv}CH} \xrightarrow[Hg^{2+}]{H^+,\ H_2O} \underset{\text{2-octanone}}{CH_3(CH_2)_5\overset{O}{\overset{\|}{C}}CH_3} \qquad (9.7)$$

PROBLEM 9.6 What alkyne would be useful for the synthesis of 2-heptanone (oil of cloves)?

9.5 Aldehydes and Ketones in Nature

Aldehydes and ketones occur very widely in nature. Figures 1.11 and 1.12 show three examples, and Figure 9.1 gives several more. Many aldehydes and ketones have pleasant odors and flavors and are used for these properties in

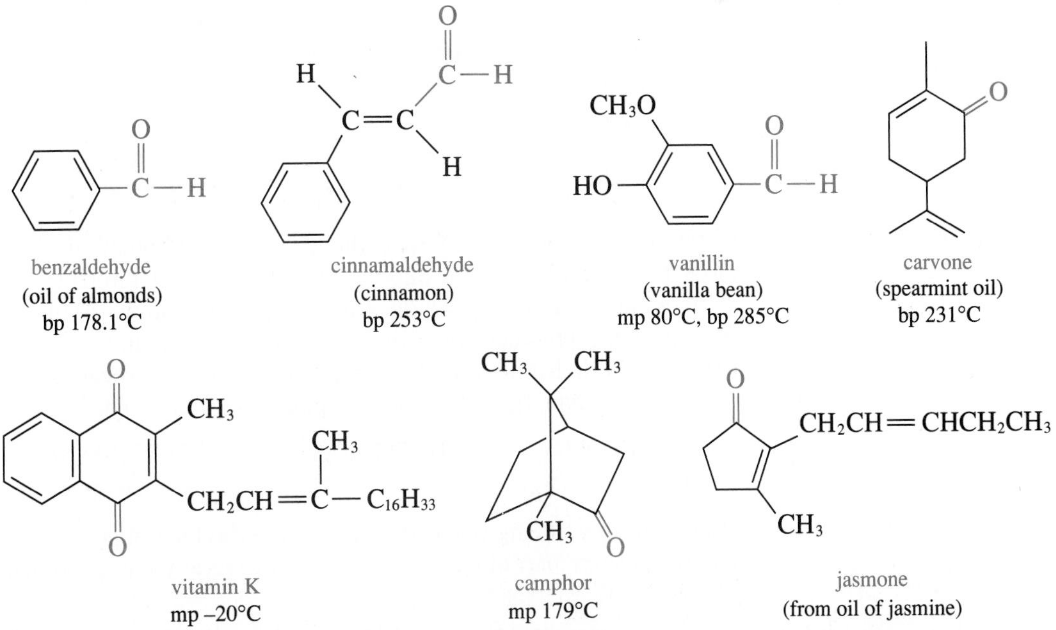

benzaldehyde (oil of almonds) bp 178.1°C

cinnamaldehyde (cinnamon) bp 253°C

vanillin (vanilla bean) mp 80°C, bp 285°C

carvone (spearmint oil) bp 231°C

vitamin K mp –20°C

camphor mp 179°C

jasmone (from oil of jasmine)

FIGURE 9.1 Some naturally occurring aldehydes and ketones.

perfumes and other consumer products (soaps, bleaches, and air fresheners, for example). The gathering and extraction of these fragrant substances from flowers, plants, and animal glands is extremely expensive, however. Chanel No. 5, introduced to the perfume market in 1921, was the first fine fragrance to use *synthetic* organic chemicals. Today most fragrances do.*

9.6 The Carbonyl Group

To best understand the reactions of aldehydes, ketones, and other carbonyl compounds, we must first appreciate the structure and properties of the carbonyl group.

The carbon-oxygen double bond consists of a sigma bond and a pi bond (Figure 9.2). The carbon atom is sp^2-hybridized. *The three atoms attached to the carbonyl carbon lie in a plane with bond angles of 120°.* The pi bond is formed by overlap of a *p* orbital on carbon with an oxygen *p* orbital. There are also two unshared electron pairs on the oxygen atom. The C=O bond distance is 1.24 Å, shorter than the C—O distance in alcohols and ethers (1.43 Å).

Oxygen is much more electronegative than carbon. Therefore the electrons in the C=O bond are attracted to the oxygen, producing a highly polarized bond. This effect is especially pronounced for the pi electrons and can be expressed in the following ways:

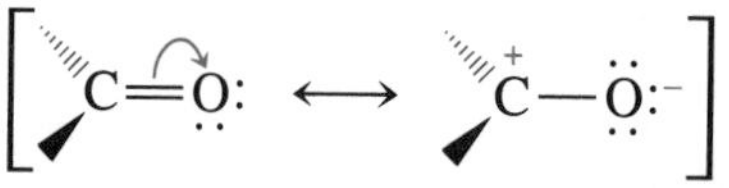

$$\left[\text{C}=\ddot{\text{O}}: \longleftrightarrow \overset{+}{\text{C}}-\ddot{\text{O}}:^- \right] \qquad \overset{\delta+}{\text{C}}=\overset{\delta-}{\text{O}} \qquad \text{C}=\text{O}\ (\text{+}\!\longrightarrow)$$

resonance contributors to the carbonyl group    polarization of the carbonyl group

As a consequence of this polarization, most carbonyl reactions involve **nucleophilic attack** at the carbonyl carbon, often accompanied by addition of a proton to the oxygen.

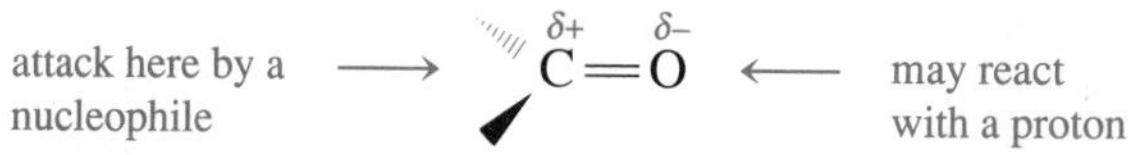

$$\text{attack here by a nucleophile} \longrightarrow \overset{\delta+}{\text{C}}=\overset{\delta-}{\text{O}} \longleftarrow \text{may react with a proton}$$

C=O bonds are quite different, then, from C=C bonds, which are not polarized and where attack at carbon is usually by an electrophile (Sec. 3.9).

In addition to its effect on reactivity, polarization of the C=O bond influences the physical properties of carbonyl compounds. For example, carbonyl compounds boil at higher temperatures than hydrocarbons, but at lower temperatures than alcohols of comparable molecular weight.

$CH_3(CH_2)_3CH_3$	$CH_3(CH_2)_3CH{=}O$	$CH_3(CH_2)_3CH_2OH$
pentane (bp 36°C)	pentanal (bp 75°C)	pentanol (bp 118°C)

Why is this so? Being polar, molecules of carbonyl compounds tend to associate; the positive part of one molecule is attracted to the negative part of another

* For a brief, interesting, and readable account of present-day perfumery, see C. S. Sell, *Chemistry in Britain* **1988**, 791–794.

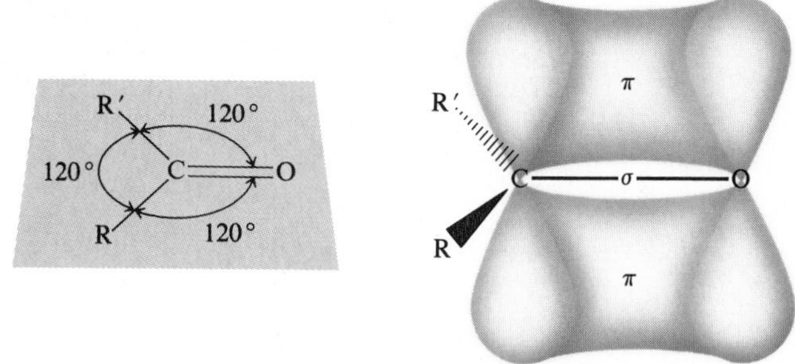

FIGURE 9.2 Bonding in the carbonyl group (see the text for a description of this bonding).

molecule. To overcome this attractive force, which is not significant in hydrocarbons, energy (heat) is required when the substance is converted from liquid to vapor. Having no O—H bonds, however, carbonyl compounds cannot form hydrogen bonds with one another, as alcohols can.

The polarity of the carbonyl group also affects the solubility properties of aldehydes and ketones. For example, carbonyl compounds with low molecular weights are soluble in water. Although they cannot form hydrogen bonds with themselves, they can form hydrogen bonds with O—H or N—H compounds.

$$\overset{\delta+}{C}=\overset{\delta-}{\ddot{O}}: \cdots \overset{\delta+}{H}-\overset{\delta-}{O}-H$$

PROBLEM 9.7 Arrange benzaldehyde (mol. wt. 106), benzyl alcohol (mol. wt. 108), and *p*-xylene (mol. wt. 106) in order of

a. increasing boiling point. b. increasing water solubility.

9.7 Nucleophilic Addition to Carbonyl Groups; an Overview

Nucleophiles attack the carbon atom of a carbon–oxygen double bond because that carbon has a partial positive charge. The pi electrons of the C=O bond move to the oxygen atom, which, because of its electronegativity, can easily accommodate the negative charge that it acquires. When these reactions are carried out in a hydroxylic solvent such as alcohol or water, the reaction is usually completed by addition of a proton to the negative oxygen. The overall reaction involves addition of a nucleophile and a proton across the pi bond of the carbonyl group.

$$\text{Nu:}^- + \underset{\text{trigonal reactant}}{C=\ddot{O}:} \rightleftharpoons \underset{\text{tetrahedral intermediate}}{\text{Nu}-C-\ddot{O}:^-} \overset{H_2O}{\rightleftharpoons} \underset{\text{tetrahedral product}}{\text{Nu}-C-\ddot{O}H} \quad (9.8)$$

The carbonyl carbon, which is trigonal and sp^2-hybridized in the starting aldehyde or ketone, becomes tetrahedral and sp^3-hybridized in the reaction product.

Because of the unshared electron pairs on the oxygen atom, carbonyl compounds are weak Lewis bases and can be protonated. ***Acids can catalyze the addition of weak nucleophiles to carbonyl compounds*** by protonating the carbonyl oxygen atom.

$$>C{=}\ddot{O}: + H^+ \longrightarrow \left[>C{=}\overset{+}{\ddot{O}}H \longleftrightarrow >\overset{+}{C}{-}\ddot{O}H \right] \xrightarrow{Nu:^-} >C(Nu){-}\ddot{O}H \qquad (9.9)$$

a resonance-stabilized carbocation

This converts the carbonyl carbon to a carbocation and enhances its susceptibility to attack by nucleophiles.

Nucleophiles can be classified as those that add reversibly to the carbonyl carbon and those that add irreversibly. Nucleophiles that add reversibly are also good leaving groups. In other words, they are the conjugate bases of relatively strong acids. Nucleophiles that add irreversibly are poor leaving groups, the conjugate bases of weak acids. This classification will be useful when we consider the mechanism of carbonyl additions.

In general, ***ketones are somewhat less reactive than aldehydes toward nucleophiles.*** There are two main reasons for this reactivity difference. *The first reason is steric*. The carbonyl carbon atom is more crowded in ketones (two organic groups) than in aldehydes (one organic group and one hydrogen atom). In nucleophilic addition, we bring these attached groups closer together because the hybridization changes from sp^2 to sp^3 and the bond angles decrease from 120° to 109.5°. Less strain is involved in additions to aldehydes than in additions to ketones because one of the groups (H) is small. *The second reason is electronic*. As we have already seen in connection with carbocation stability, alkyl groups are usually electron-donating compared to hydrogen. They therefore tend to neutralize the partial positive charge on the carbonyl carbon, decreasing its reactivity toward nucleophiles. Ketones have two such alkyl groups; aldehydes have only one. If, however, the attached groups are strongly electron-withdrawing (contain halogens, for example), they can have the opposite effect and increase carbonyl reactivity toward nucleophiles.

In the following discussion, we will classify nucleophilic additions to aldehydes and ketones according to the type of new bond formed to the carbonyl carbon. We will consider oxygen, carbon, and nitrogen nucleophiles, in that sequence.

9.8 Addition of Alcohols: Formation of Hemiacetals and Acetals

The reactions discussed in this section are extremely important because they are crucial to understanding the chemistry of carbohydrates, which we will discuss later.

Alcohols are oxygen nucleophiles. They can attack the carbonyl carbon of aldehydes or ketones, resulting in addition to the C=O bond.

$$\text{ROH} + \text{R'(H)C=O} \overset{H^+}{\rightleftharpoons} \text{RO(R')(H)C—OH} \tag{9.10}$$

alcohol aldehyde hemiacetal

Because alcohols are *weak* nucleophiles, an acid catalyst is required.* The product is a **hemiacetal,** which contains both alcohol and ether functional groups on the same carbon atom. The addition is reversible.

The mechanism of hemiacetal formation involves three steps. First, the carbonyl carbon is protonated by the acid catalyst. The alcohol oxygen then attacks the carbonyl carbon, and a proton is lost from the resulting positive oxygen. *Each step is reversible.* In terms of acid–base reactions, each step involves conversion of the starting acid into a product acid of similar strength.

$$\text{R'(H)C=\ddot{O}:} \underset{-H^+}{\overset{H^+}{\rightleftharpoons}} \text{R'(H)C=\overset{+}{\ddot{O}}H} \underset{-ROH}{\overset{R\ddot{O}H}{\rightleftharpoons}} \text{R}\overset{+}{\ddot{O}}\text{(H)—C(R')(H)—\ddot{O}H} \underset{H^+}{\overset{-H^+}{\rightleftharpoons}} \text{RO(R')(H)C—\ddot{O}H} \tag{9.11}$$

aldehyde; protonated aldehyde; protonated hemiacetal; hemiacetal

PROBLEM 9.8 Write an equation for the formation of a hemiacetal from acetaldehyde, ethanol, and H^+. Show each step in the reaction mechanism.

In the presence of *excess alcohol,* hemiacetals react further to form **acetals.**

$$\text{RO(R')(H)C—OH} + \text{ROH} \overset{H^+}{\rightleftharpoons} \text{RO(R')(H)C—OR} + \text{HOH} \tag{9.12}$$

hemiacetal acetal

The hydroxyl group of the hemiacetal is replaced by an alkoxyl group. Acetals have *two* ether functions at the same carbon atom.

The mechanism of acetal formation involves the steps shown below.

---

* Many acid catalysts can be used. Sulfuric acid and *p*-toluenesulfonic acid are commonly used in the laboratory.

$$\text{hemiacetal} \underset{-H^+}{\overset{H^+}{\rightleftharpoons}} \; RO(R')(H)C-\overset{+}{O}H_2 \underset{+H_2O}{\overset{-H_2O}{\rightleftharpoons}} \left[RO-\overset{+}{C}(R')(H) \longleftrightarrow R\overset{+}{O}=C(R')(H) \right] \tag{9.13}$$

hemiacetal — resonance-stabilized carbocation

$$\left[\text{carbocation}\right] \underset{-ROH}{\overset{ROH}{\rightleftharpoons}} RO(R')(H)C-\overset{+}{O}(H)R \underset{H^+}{\overset{-H^+}{\rightleftharpoons}} RO(R')(H)C-OR$$

acetal

Either oxygen of the hemiacetal can be protonated. When the hydroxyl oxygen is protonated, loss of water leads to a resonance-stabilized carbocation. Reaction of this carbocation with the alcohol, *which is usually the solvent and present in large excess,* gives (after proton loss) the acetal. The mechanism is like an S_N1 reaction. *Each step is reversible.*

PROBLEM 9.9 Write an equation for the reaction of the hemiacetal

$$CH_3\overset{OH}{\overset{|}{C}}HOCH_2CH_3$$

with excess ethanol and H^+. Show each step in the mechanism.

Aldehydes that have an appropriately located hydroxyl group *in the same molecule* may exist in equilibrium with a **cyclic hemiacetal,** formed by *intramolecular* nucleophilic addition. For example, 5-hydroxypentanal exists mainly in the cyclic hemiacetal form:

$$\underset{\text{5-hydroxypentanal}}{HOCH_2(5)CH_2(4)CH_2(3)CH_2(2)CH{=}O} \rightleftharpoons \text{cyclic } CH_2-CH(OH)-O-CH_2-CH_2-CH_2 \text{ or (ring with O, OH)} \tag{9.14}$$

5-hydroxypentanal — hemiacetal form of 5-hydroxypentanal (also called 2-hydroxytetrahydropyran)

The hydroxyl group is favorably located to act as a nucleophile toward the carbonyl carbon, and cyclization occurs by the following mechanism:

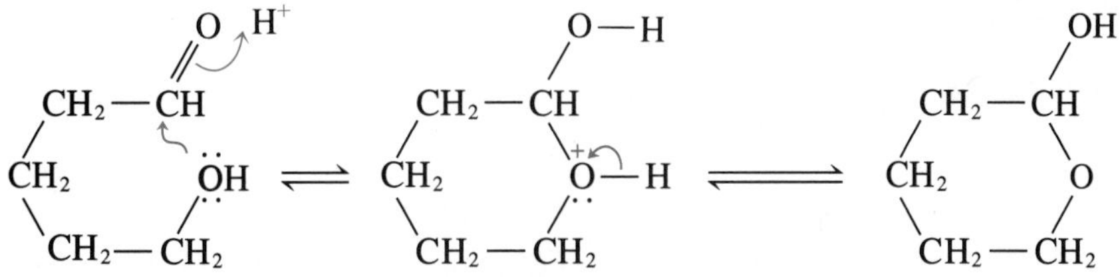

$$\text{cyclic: } CH_2{-}CH(=O\ H^+)\ \ldots\ \ddot{O}H \rightleftharpoons CH_2{-}CH(O{-}H)\ \ldots\ \overset{+}{O}{-}H \rightleftharpoons CH_2{-}CH(OH){-}O \quad (CH_2, CH_2{-}CH_2 \text{ ring}) \qquad (9.15)$$

Compounds with a hydroxyl group four or five carbons from the aldehyde group tend to form cyclic hemiacetals and acetals because the ring size (five- or six-membered) is relatively strain free. As we will see in Chapter 16, these structures are crucial to the chemistry of carbohydrates. For example, glucose is an important carbohydrate that exists as a cyclic hemiacetal.

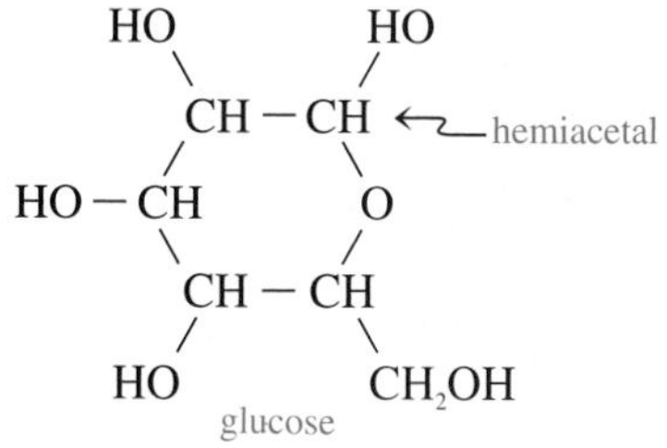

HO, HO; CH — CH ← hemiacetal; HO — CH, O; CH — CH; HO, CH_2OH

glucose

Ketones also form acetals. If, as in the following example, a glycol is used as the alcohol, the product will be cyclic.

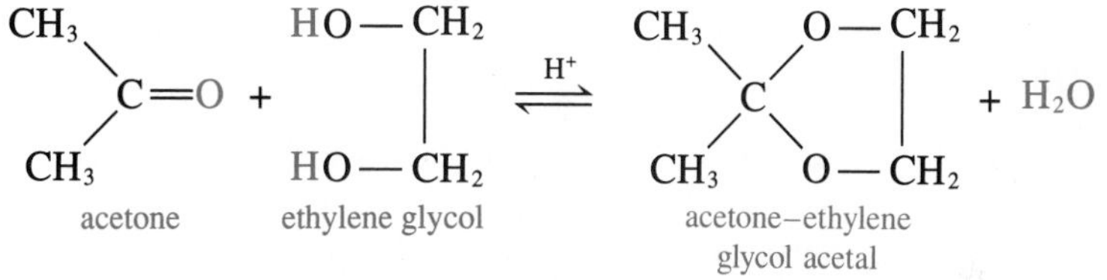

$$(CH_3)_2C{=}O + HO{-}CH_2{-}CH_2{-}OH \underset{}{\overset{H^+}{\rightleftharpoons}} (CH_3)_2C(O{-}CH_2)(O{-}CH_2) + H_2O \qquad (9.16)$$

acetone; ethylene glycol; acetone–ethylene glycol acetal

To summarize, aldehydes and ketones react with alcohols to form, first, hemiacetals and then, if excess alcohol is present, acetals.

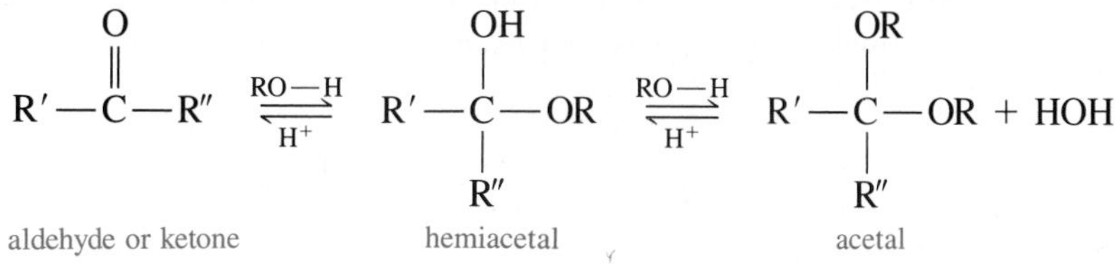

$$R'{-}\overset{O}{\overset{\|}{C}}{-}R'' \underset{H^+}{\overset{RO-H}{\rightleftharpoons}} R'{-}\overset{OH}{\underset{R''}{C}}{-}OR \underset{H^+}{\overset{RO-H}{\rightleftharpoons}} R'{-}\overset{OR}{\underset{R''}{C}}{-}OR + HOH \qquad (9.17)$$

aldehyde or ketone; hemiacetal; acetal

EXAMPLE 9.1 Write an equation for the reaction of benzaldehyde with excess methanol and an acid catalyst.

Solution

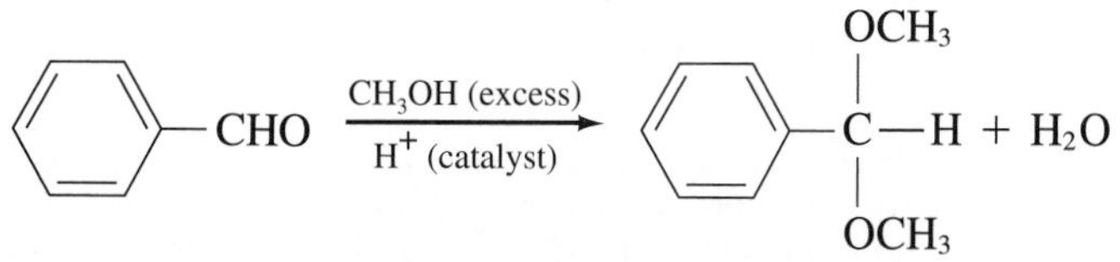

$$C_6H_5{-}CHO \xrightarrow[H^+ \text{ (catalyst)}]{CH_3OH \text{ (excess)}} C_6H_5{-}\overset{OCH_3}{\underset{OCH_3}{C}}{-}H + H_2O \qquad (9.18)$$

PROBLEM 9.10 Show the steps in the mechanism for eq. 9.18.

PROBLEM 9.11 Write an equation for the acid catalyzed reactions between cyclohexanone and

a. excess ethanol.
b. excess ethylene glycol ($HOCH_2CH_2OH$).

Notice that acetal formation is a reversible process that involves a series of equilibria (eq. 9.17). How can these reactions be driven in the forward direction? One way is to use a large excess of the alcohol. Another way is to remove water, a product of the forward reaction, as it is formed.* The reverse of acetal formation, called *acetal hydrolysis,* cannot proceed without water. On the other hand, an acetal can be hydrolyzed to its aldehyde or ketone and alcohol components by treatment with *excess water* in the presence of acid. The hemiacetal intermediate in both the forward and reverse processes usually cannot be isolated when R′ and R″ are simple alkyl or aryl groups.

EXAMPLE 9.2 Write an equation for the reaction of benzaldehyde dimethylacetal with aqueous acid

Solution

$$C_6H_5-CH(OCH_3)_2 \xrightarrow[H^+]{H_2O} C_6H_5-CH{=}O + 2\ CH_3OH \qquad (9.19)$$

PROBLEM 9.12 Show the steps in the mechanism for eq. 9.19.

The acid-catalyzed cleavage of acetals occurs much more readily than the acid-catalyzed cleavage of simple ethers (Sec. 8.7) because the intermediate carbocation is resonance-stabilized. However acetals, like ordinary ethers, are stable toward bases.

9.9 Addition of Water; Hydration of Aldehydes and Ketones

Water, like alcohols, is an oxygen nucleophile and can add reversibly to aldehydes and ketones. For example, formaldehyde in water exists mainly as its hydrate.

$$\underset{\text{formaldehyde}}{H_2C{=}O} + H{-}OH \rightleftharpoons \underset{\text{formaldehyde hydrate}}{H_2C(OH)_2} \qquad (9.20)$$

---

*In the laboratory this can be accomplished in several ways. One method involves distilling the water from the reaction mixture. Another method involves trapping the water with molecular sieves, inorganic materials with cavities of the size and shape required to hold water molecules.

With most other aldehydes or ketones, however, the hydrates cannot be isolated because they readily lose water to reform the carbonyl compound. An exception is trichloroacetaldehyde (chloral), which forms a stable crystalline hydrate, $CCl_3CH(OH)_2$. **Chloral hydrate** is used in medicine as a sedative and in veterinary medicine as a narcotic and anesthetic for horses, cattle, swine, and poultry. The potent drink known as a Mickey Finn is a combination of alcohol and chloral hydrate.

PROBLEM 9.13 Hydrolysis of $CH_3CBr_2CH_3$ with sodium hydroxide does *not* give $CH_3C(OH)_2CH_3$. Instead, it gives acetone. Explain.

9.10 Addition of Grignard Reagents and Acetylides

Grignard reagents act as carbon nucleophiles toward carbonyl compounds. The R group of the Grignard reagent adds irreversibly to the carbonyl carbon, forming a new carbon–carbon bond. In terms of acid–base reactions, the addition is favorable because the product (an alkoxide) is a much weaker base than the starting carbanion (Grignard reagent). The alkoxide can be protonated to give an alcohol.

$$\text{>C=O} + \text{RMgX} \xrightarrow{\text{ether}} \underset{\substack{\text{intermediate addition}\\ \text{product (a magnesium alkoxide)}}}{\text{R}-\text{C}-\overset{-}{\text{O}}\overset{+}{\text{MgX}}} \xrightarrow[\text{HCl}]{H_2O} \underset{\text{an alcohol}}{\text{R}-\text{C}-\text{OH}} + Mg^{2+}X^-Cl^- \qquad (9.21)$$

The reaction is normally carried out by slowly adding an ether solution of the aldehyde or ketone to an ether solution of the Grignard reagent. After all the carbonyl compound is added and the reaction is complete, the resulting magnesium alkoxide is hydrolyzed with aqueous acid.

The reaction of a Grignard reagent with a carbonyl compound is very useful. Many alcohols can be synthesized in this way by the proper choice of reagents. The type of carbonyl compound chosen determines the class of alcohol produced. *Formaldehyde gives primary alcohols.*

$$R-MgX + \underset{\text{formaldehyde}}{H-\overset{O}{\overset{\|}{C}}-H} \longrightarrow R-\underset{H}{\overset{H}{C}}-OMgX \xrightarrow[H^+]{H_2O} \underset{\text{a primary alcohol}}{R-\underset{H}{\overset{H}{C}}-OH} \qquad (9.22)$$

Other aldehydes give secondary alcohols.

$$R-MgX + \underset{\text{aldehyde}}{R'-\overset{O}{\overset{\|}{C}}-H} \longrightarrow R-\underset{H}{\overset{R'}{C}}-OMgX \xrightarrow[H^+]{H_2O} \underset{\text{a secondary alcohol}}{R-\underset{H}{\overset{R'}{C}}-OH} \qquad (9.23)$$

Ketones give tertiary alcohols.

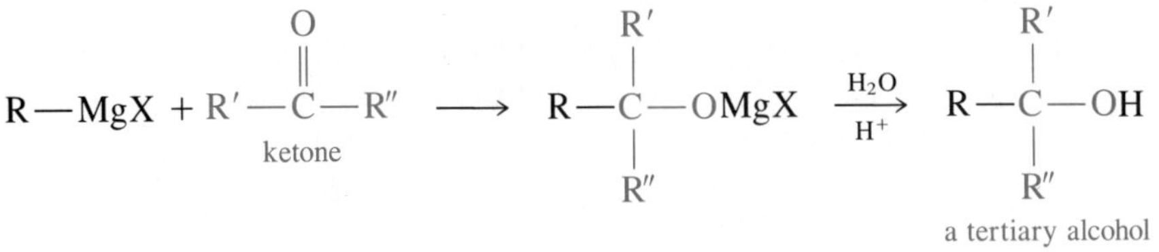

$$\text{R—MgX} + \text{R}'\text{—}\overset{\overset{\displaystyle \text{O}}{\|}}{\text{C}}\text{—R}'' \longrightarrow \text{R—}\underset{\underset{\displaystyle \text{R}''}{|}}{\overset{\overset{\displaystyle \text{R}'}{|}}{\text{C}}}\text{—OMgX} \xrightarrow[\text{H}^+]{\text{H}_2\text{O}} \text{R—}\underset{\underset{\displaystyle \text{R}''}{|}}{\overset{\overset{\displaystyle \text{R}'}{|}}{\text{C}}}\text{—OH} \quad (9.24)$$

ketone — a tertiary alcohol

Note that only *one* of the R groups (shown in black) attached to the hydroxyl-bearing carbon of the alcohol comes from the Grignard reagent. The rest of the alcohol's carbon skeleton comes from the carbonyl compound.

EXAMPLE 9.3 What is the product expected from the reaction between ethyl magnesium bromide and 3-pentanone followed by hydrolysis?

Solution 3-Pentanone is a ketone. Following eq. 9.24 as an example, the product is 2-ethyl-3 pentanol.

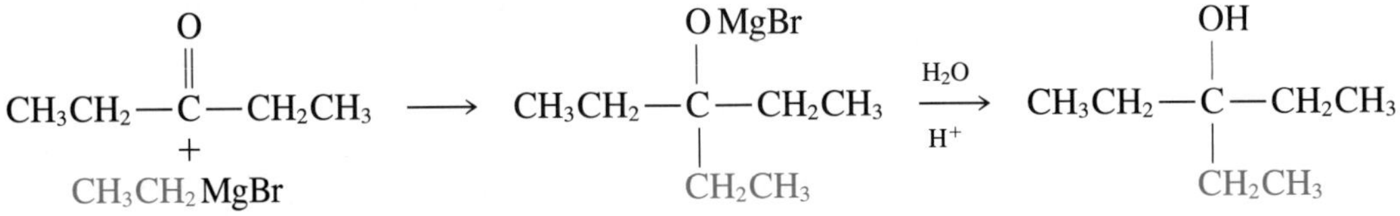

$$\underset{\displaystyle +\ \text{CH}_3\text{CH}_2\text{MgBr}}{\text{CH}_3\text{CH}_2\text{—}\overset{\overset{\displaystyle \text{O}}{\|}}{\text{C}}\text{—CH}_2\text{CH}_3} \longrightarrow \text{CH}_3\text{CH}_2\text{—}\underset{\underset{\displaystyle \text{CH}_2\text{CH}_3}{|}}{\overset{\overset{\displaystyle \text{OMgBr}}{|}}{\text{C}}}\text{—CH}_2\text{CH}_3 \xrightarrow[\text{H}^+]{\text{H}_2\text{O}} \text{CH}_3\text{CH}_2\text{—}\underset{\underset{\displaystyle \text{CH}_2\text{CH}_3}{|}}{\overset{\overset{\displaystyle \text{OH}}{|}}{\text{C}}}\text{—CH}_2\text{CH}_3$$

PROBLEM 9.14 Provide the products expected from the reaction of

a. formaldehyde with propyl magnesium bromide followed by hydrolysis.
b. pentanal with ethyl magnesium bromide followed by hydrolysis.

EXAMPLE 9.4 Show how the following alcohol can be synthesized from a Grignard reagent and a carbonyl compound.

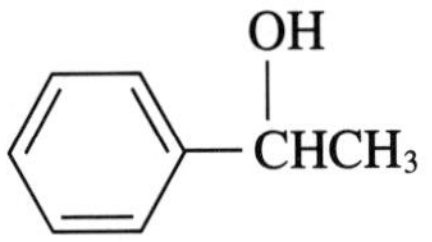

$$\text{C}_6\text{H}_5\text{—}\overset{\overset{\displaystyle \text{OH}}{|}}{\text{C}}\text{HCH}_3$$

Solution The alcohol is secondary, so the carbonyl compound must be an aldehyde. We can use either a methyl or a phenyl Grignard reagent.

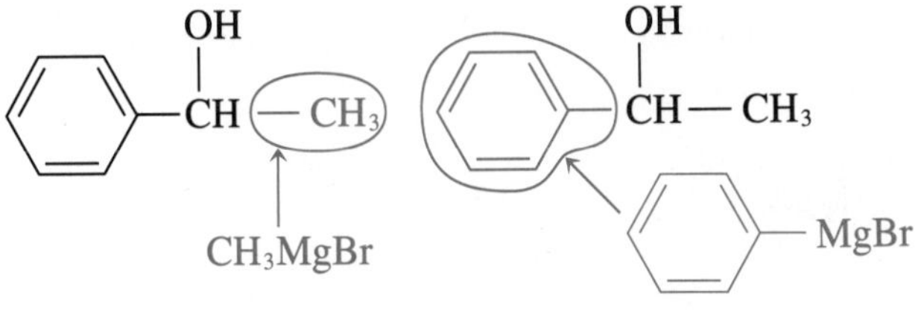

$$\text{C}_6\text{H}_5\text{—}\overset{\overset{\displaystyle \text{OH}}{|}}{\text{C}}\text{H}\text{—CH}_3 \ (\uparrow \text{CH}_3\text{MgBr}) \qquad \text{C}_6\text{H}_5\text{—}\overset{\overset{\displaystyle \text{OH}}{|}}{\text{C}}\text{H—CH}_3 \ (\nwarrow \text{C}_6\text{H}_5\text{—MgBr})$$

The equations are

$$CH_3\text{—}MgBr + C_6H_5\text{—}\overset{O}{\overset{\|}{C}}H$$

methylmagnesium bromide; benzaldehyde

$$\downarrow$$

$$CH_3\text{—}\underset{}{\overset{OMgBr}{\overset{|}{C}H}}\text{—}C_6H_5 \xrightarrow[H^+]{H_2O} CH_3\text{—}\overset{OH}{\overset{|}{C}H}\text{—}C_6H_5 \qquad (9.25)$$

alkoxide

$$\uparrow$$

$$C_6H_5\text{—}MgBr + CH_3\text{—}\overset{O}{\overset{\|}{C}}\text{—}H$$

phenylmagnesium bromide; acetaldehyde

The choice between the possible sets of reactants may be made by availability or cost, or for chemical reasons (for example, the more reactive aldehyde or ketone might be selected).

PROBLEM 9.15 Show how each of the following alcohols can be made from a Grignard reagent and a carbonyl compound.

a. $C_6H_5\text{—}CH_2OH$ b. $C_6H_5\text{—}C(CH_3)_2OH$

Other organometallic reagents, such as organolithium compounds and acetylides, react with carbonyl compounds similarly to Grignard reagents. For example,

$$\text{cyclopentanone} + Na^{+-}C{\equiv}CH \longrightarrow \text{1-}(Na^+O^-)\text{-1-}(C{\equiv}CH)\text{cyclopentane} \xrightarrow[H_2O]{H^+} \text{1-}(HO)\text{-1-}(C{\equiv}CH)\text{cyclopentane} \qquad (9.26)$$

a ketone; sodium acetylide; a tertiary acetylenic alcohol

PROBLEM 9.16 Provide the product expected from the reaction of

a. lithium acetylide with benzaldehyde followed by H_3O^+.
b. $CH_3C{\equiv}C^-Na^+$ with cyclopentanone followed by H_3O^+.

9.11 Addition of Hydrogen Cyanide; Cyanohydrins

Hydrogen cyanide adds reversibly to the carbonyl group of aldehydes and ketones to form **cyanohydrins,** compounds with a hydroxyl and a cyano group attached to the same carbon. A basic catalyst is required.

$$>C{=}O + HCN \xrightarrow{KOH} >C(CN){-}OH \qquad (9.27)$$

a cyanohydrin

Acetone, for example, reacts as follows:

$$CH_3{-}\overset{O}{\overset{\|}{C}}{-}CH_3 + HCN \xrightarrow{KOH} CH_3{-}\underset{CN}{\underset{|}{\overset{OH}{\overset{|}{C}}}}{-}CH_3 \qquad (9.28)$$

acetone — acetone cyanohydrin

Hydrogen cyanide has no unshared electron pair on its carbon, so it cannot function as a carbon nucleophile. The base converts some of the hydrogen cyanide to cyanide ion, however, which then acts as a carbon nucleophile.

$$>C{=}\ddot{O}: + {}^{-}:C{\equiv}N: \rightleftharpoons >C(CN){-}\ddot{O}:^{-} \underset{}{\overset{HCN}{\rightleftharpoons}} >C(CN){-}\ddot{O}H + {}^{-}CN \qquad (9.29)$$

cyanohydrin

PROBLEM 9.17 Write an equation for the addition of HCN to

a. acetaldehyde. b. benzaldehyde.

9.12 Addition of Nitrogen Nucleophiles

Ammonia, amines, and certain related compounds have an unshared electron pair on the nitrogen atom and act as nitrogen nucleophiles toward the carbonyl carbon atom. For example, primary amines react as follows:

$$>C{=}O + \ddot{N}H_2{-}R \rightleftharpoons \left[>C(OH){-}NHR\right] \xrightarrow{-HOH} >C{=}NR \qquad (9.30)$$

primary amine — tetrahedral addition product — imine

The tetrahedral addition product that is formed first is similar to a hemiacetal, but with an NH group in place of one of the oxygens. These addition products are normally not stable. They eliminate water to form a product with a carbon–nitrogen double bond. With primary amines, the products are called **imines.** Imines are like carbonyl compounds, except that the O is replaced by NR.

They are important intermediates in some biochemical reactions, particularly in binding carbonyl compounds to the free amino groups that are present in most enzymes.

$$\underset{\text{substrate}}{\overset{\text{enzyme}}{\text{—}NH_2}} \;\; \underset{}{>C=O} \longrightarrow \underset{\text{enzyme-substrate compound}}{\text{—}N=C<} + H_2O \qquad (9.31)$$

For example, retinal ("A Word About" on page 80) binds to the protein opsin in this way, to form rhodopsin.

PROBLEM 9.18 Write an equation for the reaction of benzaldehyde with aniline (the formula of which is $C_6H_5NH_2$).

Other ammonia derivatives containing an $\text{—}NH_2$ group react with carbonyl compounds similarly to primary amines. Table 9.1 lists some specific examples.

TABLE 9.1 Nitrogen derivatives of carbonyl compounds

Formula of ammonia derivative	*Name*	*Formula of carbonyl derivative*	*Name*
RNH_2 or $ArNH_2$	primary amine	$>C=NR$ or $>C=NAr$	imine
NH_2OH	hydroxylamine	$>C=NOH$	oxime
NH_2NH_2	hydrazine	$>C=NNH_2$	hydrazone
$NH_2NHC_6H_5$	phenylhydrazine	$>C=NNHC_6H_5$	phenylhydrazone
$NH_2NH\overset{O}{\overset{\|}{C}}NH_2$	semicarbazide	$>C=NNH\overset{O}{\overset{\|}{C}}NH_2$	semicarbazone

EXAMPLE 9.5 Using Table 9.1 as a guide, write an equation for the reaction of hydrazine with cyclohexanone.

Solution

$$\text{(cyclohexane)}{=}O + NH_2NH_2 \longrightarrow \text{(cyclohexane)}{=}NNH_2 + H_2O$$

The product is a hydrazone. Notice that water is a product of the reaction, just as in imine formation.

PROBLEM 9.19 Using Table 9.1 as a guide, write an equation for the reaction of propanal ($CH_3CH_2CH{=}O$) with

a. hydroxylamine.
b. phenylhydrazine.
c. semicarbazide.

9.13 Reduction of Carbonyl Compounds

Aldehydes and ketones are easily reduced to primary and secondary alcohols, respectively. Reduction can be accomplished in many ways, most commonly by metal hydrides.

The most common metal hydrides used to reduce carbonyl compounds are **lithium aluminum hydride** ($LiAlH_4$) and **sodium borohydride** ($NaBH_4$). The metal-hydride bond is polarized, with the metal positive and the hydrogen negative. The reaction therefore involves irreversible nucleophilic attack of the hydride (H^-) at the carbonyl carbon:

$$\overset{\delta+}{C}{=}\overset{\delta-}{\ddot{O}}: \quad \overset{\delta-}{H}{-}\overset{\delta+}{AlH_3} \; Li^+ \longrightarrow \underset{\text{aluminum alkoxide}}{H{-}C{-}O{-}\bar{A}lH_3 \; Li^+} \xrightarrow[H^+]{H_2O} \underset{\text{alcohol}}{H{-}C{-}OH} \tag{9.32}$$

The initial product is an aluminum alkoxide, which is subsequently hydrolyzed by water and acid to give the alcohol. The net result is addition of hydrogen across the carbon-oxygen double bond. A specific example is

$$\underset{\text{cyclohexanone}}{\text{(cyclohexane)}{=}O} \xrightarrow[\text{2. } H^+,\ H_2O]{\text{1. } LiAlH_4} \underset{\text{cyclohexanol}}{\text{(cyclohexane)}(H)(OH)} \tag{9.33}$$

PROBLEM 9.20 Show how each of the following alcohols can be made from lithium aluminum hydride and a carbonyl compound.

a. 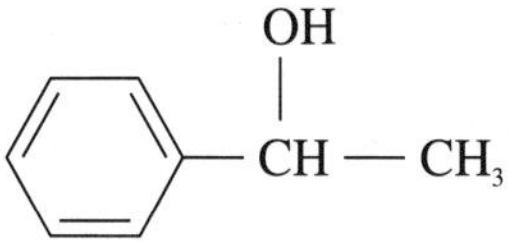 (benzene ring)–CH(OH)–CH_3 b. $CH_3CH_2CH_2CH_2CH_2OH$

Since a carbon-carbon double bond is not readily attacked by nucleophiles, metal hydrides can be used to reduce a carbon-oxygen double bond to the corresponding alcohol without reducing a carbon-carbon double bond present in the same compound.

$$\underset{\substack{\text{2-butenal}\\\text{(crotonaldehyde)}}}{CH_3{-}CH{=}CH{-}\overset{\overset{\displaystyle O}{\|}}{C}H} \xrightarrow{NaBH_4} \underset{\substack{\text{2-buten-1-ol}\\\text{(crotyl alcohol)}}}{CH_3CH{=}CH{-}CH_2OH} \tag{9.34}$$

PROBLEM 9.21 Show how (cyclohexenyl)–C(=O)–CH_3 can be reduced to (cyclohexenyl)–$CH(OH)CH_3$

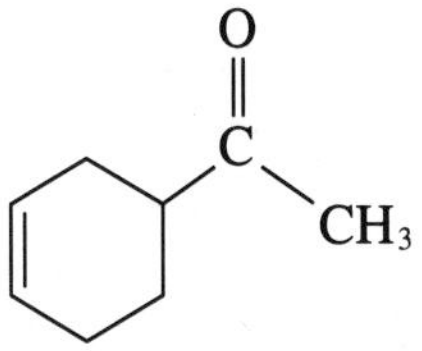

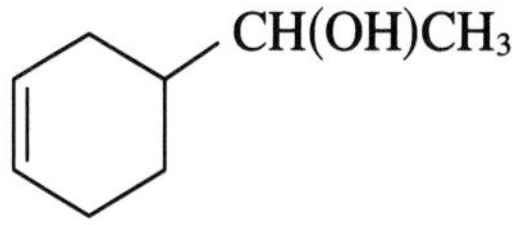

9.14 Oxidation of Carbonyl Compounds

Aldehydes are more easily oxidized than ketones. Oxidation of an aldehyde gives an acid with the same number of carbon atoms.

$$\underset{\text{aldehyde}}{R{-}\overset{\overset{\displaystyle O}{\|}}{C}{-}H} \xrightarrow{\text{oxidizing agent}} \underset{\text{acid}}{R{-}\overset{\overset{\displaystyle O}{\|}}{C}{-}OH} \tag{9.35}$$

Since the reaction occurs easily, many oxidizing agents, such as $KMnO_4$, CrO_3, Ag_2O, and peroxyacids (see eq. 8.18), will work. Specific examples are

$$CH_3(CH_2)_5CH{=}O \xrightarrow[\text{(Jones' reagent)}]{CrO_3,\ H^+} CH_3(CH_2)_5CO_2H \tag{9.36}$$

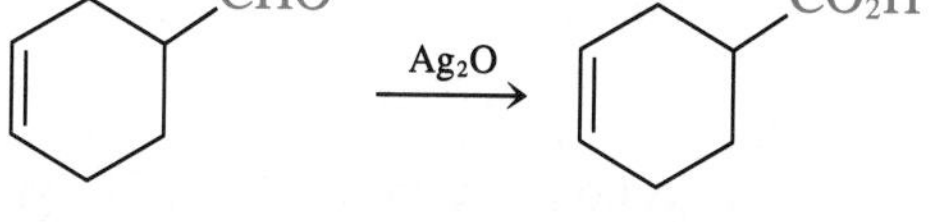

$$\text{(cyclohexenyl)}{-}CHO \xrightarrow{Ag_2O} \text{(cyclohexenyl)}{-}CO_2H \tag{9.37}$$

Silver ion as an oxidant is expensive but has the virtue that it selectively oxidizes the aldehyde group without oxidizing the double bond (eq. 9.37).

A laboratory test that distinguishes aldehydes from ketones takes advantage of their different ease of oxidation. In the **Tollens' silver mirror test,** the silver–ammonia complex ion is reduced by aldehydes (but not by ketones) to metallic silver.* The equation for the reaction may be written as follows:

* Silver hydroxide is insoluble in water, so the silver ion must be complexed with ammonia to keep it in solution in a basic medium.

$$\underset{\text{aldehyde}}{\overset{\overset{\displaystyle O}{\|}}{RCH}} + \underset{\substack{\text{silver–ammonia}\\ \text{complex ion}\\ \text{(colorless)}}}{2\,Ag(NH_3)_2^+} + 3\,OH^- \rightarrow \underset{\substack{\text{acid}\\ \text{anion}}}{\overset{\overset{\displaystyle O}{\|}}{RC}-O^-} + \underset{\substack{\text{silver}\\ \text{mirror}}}{2\,Ag\downarrow} + 4\,NH_3\uparrow + 2\,H_2O \qquad (9.38)$$

If the glass vessel in which the test is performed is thoroughly clean, the silver deposits as a mirror on the glass surface. This reaction is also employed to silver glass, using the relatively inexpensive aldehyde formaldehyde.

PROBLEM 9.22 Write an equation for the formation of a silver mirror from formaldehyde and Tollens' reagent.

Aldehydes are so easily oxidized that stored samples usually contain some of the corresponding acid. This contamination is caused by air oxidation.

$$2\,RCHO + O_2 \longrightarrow 2\,RCO_2H \qquad (9.39)$$

Ketones can be oxidized but require special oxidizing conditions. For example, cyclohexanone is oxidized commercially to **adipic acid,** an important industrial chemical used to manufacture nylon.

one of these C—C bonds is cleaved in the oxidation

$$\underset{\text{cyclohexanone}}{C_6H_{10}O} + HNO_3 \xrightarrow{V_2O_5} \underset{\text{adipic acid}}{HO-\overset{\overset{\displaystyle O}{\|}}{C}-CH_2CH_2CH_2CH_2-\overset{\overset{\displaystyle O}{\|}}{C}-OH} \qquad (9.40)$$

9.15 Keto–Enol Tautomerism

Aldehydes and ketones may exist as an equilibrium mixture of two forms, called the **keto form** and the **enol form.** The two forms differ in the location of a proton and a double bond.

$$\underset{\text{keto form}}{-\overset{\overset{\displaystyle H}{|}}{\underset{|}{C}}-\overset{\overset{\displaystyle O}{\|}}{C}-} \rightleftharpoons \underset{\text{enol form}}{>C=C<^{OH}} \qquad (9.41)$$

This type of structural isomerism is called **tautomerism** (from the Greek *tauto,* the same, and *meros,* part). The two forms of the aldehyde or ketone are called **tautomers.**

EXAMPLE 9.6 Write formulas for the keto and enol forms of acetone.

Solution

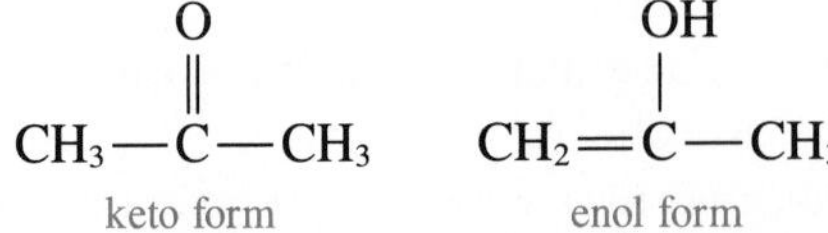

$$\underset{\text{keto form}}{CH_3-\overset{\overset{\displaystyle O}{\|}}{C}-CH_3} \qquad \underset{\text{enol form}}{CH_2=\overset{\overset{\displaystyle OH}{|}}{C}-CH_3}$$

PROBLEM 9.23 Draw the structural formula for the enol form of

a. cyclopentanone. b. acetaldehyde.

Tautomers are structural isomers, *not* contributors to a resonance hybrid. They readily equilibrate, and we indicate that fact by the equilibrium symbol $\rightleftharpoons$ between their structures.

To be capable of existing in an enol form, a carbonyl compound must have a hydrogen atom attached to the carbon atom adjacent to the carbonyl group. This hydrogen is called an **α-hydrogen** and is attached to the **α-carbon atom** (from the first letter of the Greek alphabet, α, or alpha).

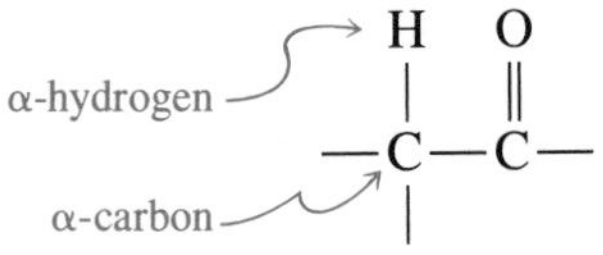

α-hydrogen
α-carbon

Most simple aldehydes and ketones exist mainly in the keto form. Acetone, for example, is 99.9997% in the keto form, with only 0.0003% of the enol present. The main reason for the greater stability of the keto form is that the C=O plus C—H bond energy present in the keto form is greater than the C=C plus O—H bond energy of the enol form. We have already encountered some molecules, however, that have mainly the enol structure—the *phenols*. In this case, the resonance stabilization of the aromatic ring is greater than the usual energy difference that favors the keto over the enol form. Aromaticity would be destroyed if the molecule existed in the keto form; therefore, the enol form is preferred.

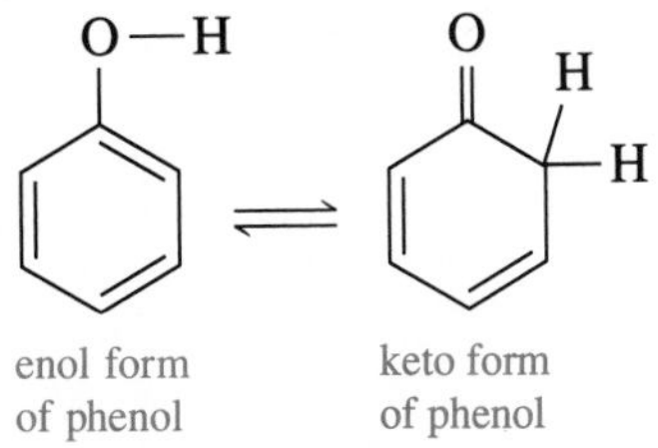

(9.42)

enol form of phenol — keto form of phenol

Carbonyl compounds that do not have an α-hydrogen cannot form enols and exist only in the keto form. Examples include

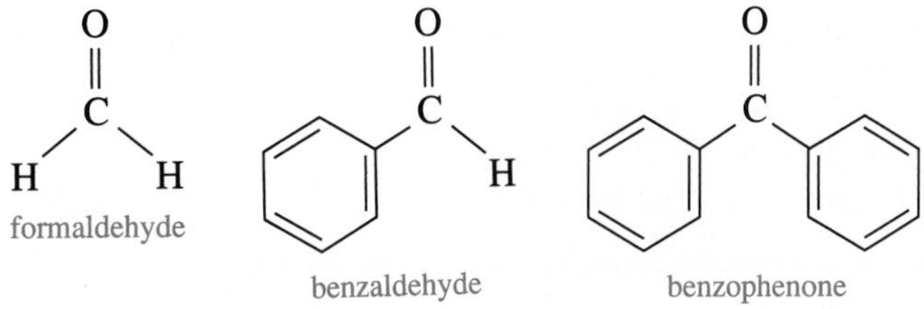

formaldehyde — benzaldehyde — benzophenone

A Word About . . .

18. Tautomerism and Photochromism

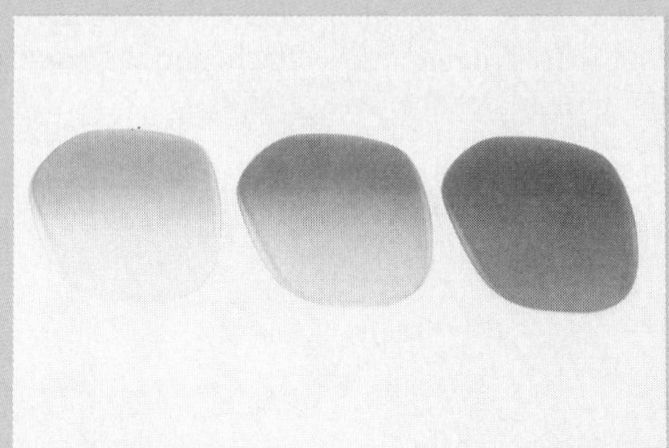

The concept of tautomerism can be expanded beyond keto and enol forms to include any pair or group of isomers that can be easily interconverted by the relocation of an atom and/or bonds. For example, imines and enamines (unsaturated amines) are tautomers whose relationship is similar to that of keto and enol forms.

an imine ⇌ an enamine

keto ⇌ enol

For some pairs of tautomers, one can be converted to the other photochemically—that is, by the absorption of light. Irradiation of the following pale yellow phenol-imine causes the hydrogen atom to shift from the oxygen to the nitrogen, with appropriate rebonding.

a phenol-imine
pale yellow
(both rings aromatic)

$\underset{\text{dark}}{\overset{h\nu}{\rightleftharpoons}}$

a keto-enamine
red
(only one ring aromatic)

If the keto-enamine product of this photochemical reaction is allowed to remain in the dark, it gradually changes back to the more stable phenol-imine. Of what use can such a cycle of reactions, with no net change, possibly be?

Note that, in this example, one tautomer is pale yellow, the other red. This phenomenon, in which two compounds undergo a reversible photochemical color change, is called **photochromism.** Photochromic substances have many practical uses. One thinks immediately of glasses that, when exposed to sunlight, become darker because their lenses are impregnated with a photochromic material. When the sunlight dims or when one goes indoors, the colored photochromic substance gradually changes back to its colorless form. Photochromic substances can be used for data storage and display (as in digital watches), for chemical switches in computers, for micro images (microfilm and microfiche), for protection against sudden light flashes, for camouflage, and in many other creative ways.

9.16 Acidity of α-Hydrogens; the Enolate Anion

The α-hydrogen in a carbonyl compound is more acidic than normal hydrogens bonded to a carbon atom. Table 9.2 shows the pK_a values for a typical aldehyde and ketone, as well as for reference compounds. The result of placing a carbonyl group adjacent to methyl protons is truly striking, an increase in their acidity of over 30 powers of 10! (Compare acetaldehyde or acetone with propane.) Indeed, these compounds are almost as acidic as the O—H protons in alcohols. Why is this?

There are two reasons. First, the carbonyl carbon carries a partial positive charge. Bonding electrons are displaced toward the carbonyl carbon and away from the α-hydrogen, making it easy for a base to remove the α-hydrogen as a proton (that is, without its bonding electrons).

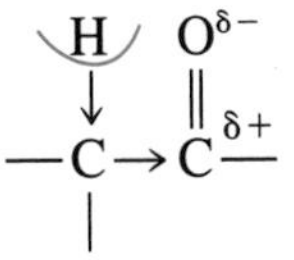

$$\begin{array}{c} \text{H} \quad \text{O}^{\delta-} \\ \downarrow \qquad \| \;\; \delta+ \\ -\text{C}\rightarrow\text{C}- \\ | \end{array}$$

Second and more important, the resulting anion is stabilized by resonance.

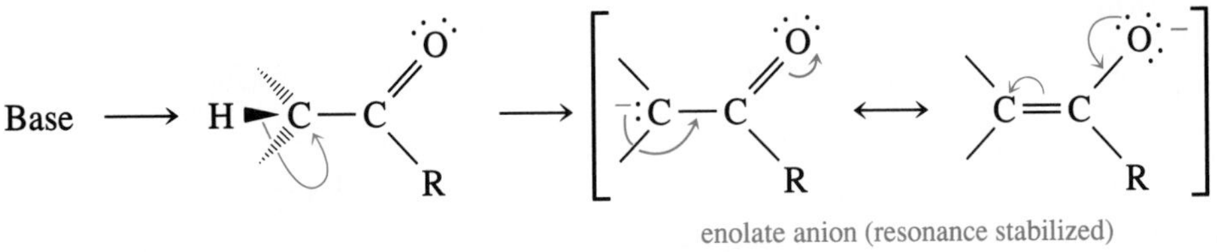

$$\text{Base} \longrightarrow \text{H}-\text{C}-\text{C}(=\ddot{\text{O}}:)\text{R} \longrightarrow \left[\bar{:}\text{C}-\text{C}(=\ddot{\text{O}})\text{R} \longleftrightarrow \text{C}=\text{C}(-\ddot{\text{O}}:^{-})\text{R} \right] \quad (9.43)$$

enolate anion (resonance stabilized)

The anion is called an **enolate anion.** The negative charge is distributed between the α-carbon and the carbonyl oxygen atom.

TABLE 9.2 Acidity of α-hydrogens

Compound	*Name*	pK_a
$CH_3CH_2CH_3$	propane	~50
$CH_3C(=O)CH_3$	acetone	19
$CH_3C(=O)H$	acetaldehyde	17
CH_3CH_2OH	ethanol	15.9

EXAMPLE 9.7 Draw the formula for the enolate anion of acetone.

Solution

$$\left[\ddot{C}H_2-\overset{\overset{:\ddot{O}}{\|}}{C}-CH_3 \longleftrightarrow CH_2{=}\overset{\overset{:\ddot{O}:^-}{|}}{C}-CH_3\right] \quad \text{or} \quad \left[CH_2\overset{\cdots}{=}\overset{\overset{O}{\|}}{C}-CH_3\right]^-$$

An enolate anion is a resonance hybrid of two contributing structures that differ *only* in the arrangement of the electrons.

PROBLEM 9.24 Draw the resonance contributors to the enolate ion of

a. cyclopentanone. b. acetaldehyde.

9.17 Deuterium Exchange in Carbonyl Compounds

Even though its concentration is very low, the presence of the enol form of ordinary aldehydes and ketones can be demonstrated experimentally. For example, the α-hydrogens can be exchanged for deuterium by placing the carbonyl compound in a solvent such as D_2O or CH_3OD that contains O—D bonds. The exchange is catalyzed by acid or base. *Only the α-hydrogens exchange*, as illustrated by the following examples:

$$\text{cyclohexanone} \xrightarrow[\text{CH}_3\text{OD (excess)}]{\text{Na}^{+\,-}\text{OCH}_3} \text{2,2,6,6-tetradeuteriocyclohexanone} \qquad (9.44)$$

$$\underset{\text{butanal}}{CH_3CH_2CH_2\overset{\overset{O}{\|}}{C}H} \xrightarrow[D^+]{D_2O} \underset{\text{2,2-dideuteriobutanal}}{CH_3CH_2CD_2\overset{\overset{O}{\|}}{C}H} \qquad (9.45)$$

The mechanism of base-catalyzed exchange of the α-hydrogens (eq. 9.44) involves two steps.

$$\text{cyclohexanone} + {}^-OCH_3 \rightleftharpoons [\text{enolate anion}] + CH_3OH \;\xrightleftharpoons{CH_3OD}\; \text{2-deuteriocyclohexanone} + CH_3O^- \qquad (9.46)$$

enolate anion

The base (methoxide ion) removes an α-proton to form the enolate anion. Reprotonation, but with CH_3OD, replaces the α-hydrogen with deuterium. With excess CH_3OD, all four α-hydrogens are eventually exchanged.

The mechanism of acid-catalyzed exchange of the α-hydrogens (eq. 9.45) also involves several steps. The keto form is first protonated and, by loss of an α-hydrogen, converted to its enol.

$$\underset{\text{keto form}}{CH_3CH_2CH_2\overset{\overset{\ddot{O}:}{\|}}{C}H} \xrightleftharpoons{D^+} CH_3CH_2\underset{H}{\overset{H}{C}}-\overset{\overset{+\ddot{O}-D}{\|}}{C}H \xrightleftharpoons{-H^+} \underset{\text{enol form}}{CH_3CH_2CH=\overset{\overset{:\ddot{O}-D}{|}}{C}H} \quad (9.47)$$

In the reversal of these equilibria, the enol then adds D^+ at the α-carbon.

$$CH_3CH_2CH=\overset{\overset{:\ddot{O}-D}{|}}{C}H \;(+\;\overset{+}{D}) \longrightarrow CH_3CH_2\overset{\overset{D}{|}}{C}H-\overset{\overset{:\ddot{O}}{\|}}{C}H + D^+ \quad (9.48)$$

Repetition of this sequence results in exchange of the other α-hydrogen.

PROBLEM 9.25 Identify the hydrogens that are readily exchanged for deuterium in

a. 2-methylcyclopentanone.
b. *t*-butyl methyl ketone.

9.18 The Aldol Condensation

Enolate anions may act as carbon nucleophiles. They can add to the carbonyl group of another aldehyde or ketone molecule in a reaction called the **aldol condensation,** an extremely useful carbon–carbon bond-forming reaction.

The simplest example of an aldol condensation is the combination of two acetaldehyde molecules, which occurs when a solution of acetaldehyde is treated with catalytic amounts of aqueous base.

$$\underset{\text{acetaldehyde}}{CH_3\overset{\overset{O}{\|}}{C}H + CH_3\overset{\overset{O}{\|}}{C}H} \xrightleftharpoons{OH^-} \underset{\substack{\text{3-hydroxybutanal}\\ \text{(an aldol)}}}{CH_3\overset{\overset{OH}{|}}{C}H-CH_2\overset{\overset{O}{\|}}{C}H} \quad (9.49)$$

The product is called an **aldol** (so named because the product is both an *ald*ehyde and an alcoh*ol*).

The aldol condensation of acetaldehyde occurs according to the following three-step mechanism:

$$\textbf{Step 1.}\quad \overset{\alpha}{CH_3}-\overset{\overset{\ddot{O}:}{\|}}{C}-H + OH^- \rightleftharpoons \underset{\text{enolate anion}}{\overset{-}{\ddot{C}}H_2-\overset{\overset{\ddot{O}:}{\|}}{C}-H} + HOH \quad (9.50)$$

$$\textbf{Step 2.}\quad CH_3-\overset{\overset{\ddot{O}:}{\|}}{C}H + \underset{\text{nucleophile}}{\overset{-}{\ddot{C}}H_2-\overset{\overset{\ddot{O}:}{\|}}{C}H} \rightleftharpoons \underset{\text{an alkoxide ion}}{CH_3\overset{\overset{:\ddot{O}:^-}{|}}{C}H-CH_2\overset{\overset{\ddot{O}:}{\|}}{C}H} \quad (9.51)$$

$$\text{Step 3.}\quad \underset{}{CH_3\overset{\overset{:\ddot{O}:^-}{|}}{C}H}-CH_2\overset{\overset{\ddot{O}:}{\|}}{C}H + HOH \rightleftharpoons \underset{\text{aldol}}{CH_3\overset{\overset{:\ddot{O}H}{|}}{C}H-\overset{\alpha}{C}H_2\overset{\overset{\ddot{O}:}{\|}}{C}H} + OH^- \tag{9.52}$$

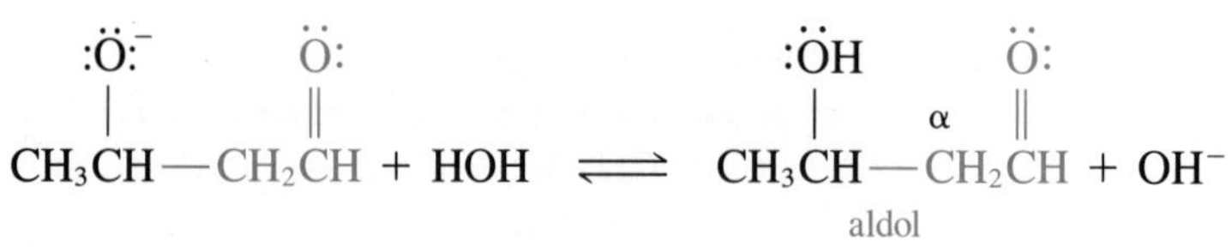

In Step 1, the base removes an α-hydrogen to form the enolate anion. In Step 2, this anion adds to the carbonyl carbon of another acetaldehyde molecule, forming a new carbon–carbon bond. Ordinary bases convert a small fraction of the carbonyl compound to the enolate anion, so that a substantial fraction of the aldehyde is still present in the un-ionized carbonyl form needed for this step. In Step 3, the alkoxide ion formed in Step 2 accepts a proton from the solvent, thus regenerating the hydroxide ion needed for the first step.

In the aldol condensation, the α-carbon of one aldehyde molecule becomes connected to the carbonyl carbon of another aldehyde molecule.

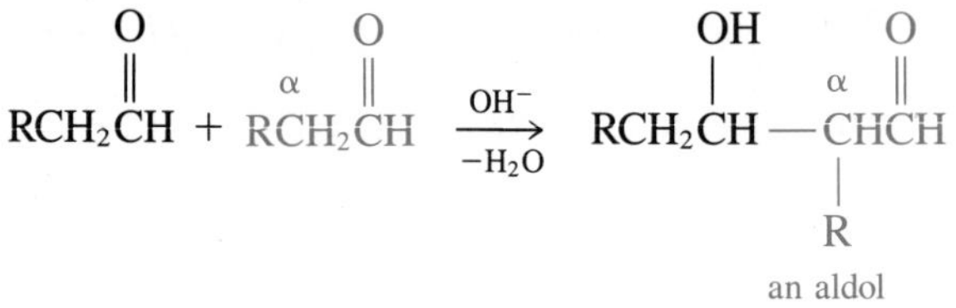

$$RCH_2\overset{\overset{O}{\|}}{C}H + R\overset{\alpha}{C}H_2\overset{\overset{O}{\|}}{C}H \xrightarrow[-H_2O]{OH^-} \underset{\text{an aldol}}{RCH_2\overset{\overset{OH}{|}}{C}H-\underset{\underset{R}{|}}{\overset{\alpha}{C}H}\overset{\overset{O}{\|}}{C}H} \tag{9.53}$$

Aldols are therefore 3-hydroxyaldehydes. ***Since it is always the α-carbon that acts as a nucleophile, the product always has just one carbon atom between the aldehyde and alcohol carbons,*** regardless of how long the carbon chain is in the starting aldehyde.

EXAMPLE 9.8 Give the structure of the aldol that is obtained by treating propanal ($CH_3CH_2CH{=}O$) with base.

Solution Rewriting eq. 9.53 with $R{=}CH_3$, the product is

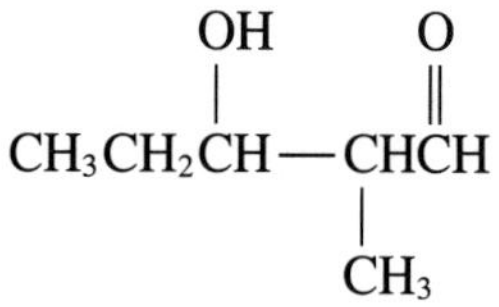

$$CH_3CH_2\overset{\overset{OH}{|}}{C}H-\underset{\underset{CH_3}{|}}{C}H\overset{\overset{O}{\|}}{C}H$$

PROBLEM 9.26 Write out the steps in the mechanism for formation of the product in Example 9.8.

9.19 *The Mixed Aldol Condensation*

The aldol condensation is very versatile in that the enolate anion of *one* carbonyl compound can be made to add to the carbonyl carbon of *another*, provided that the reaction partners are carefully selected. Consider, for example, the reaction between acetaldehyde and benzaldehyde, when treated with base. Only acetaldehyde can form an enolate anion (benzaldehyde has no α-hydrogen). If the enolate ion of acetaldehyde adds to the benzaldehyde carbonyl group, a mixed aldol condensation occurs.

$$C_6H_5\overset{O}{\overset{\|}{C}}H + CH_3\overset{\alpha}{}\overset{O}{\overset{\|}{C}}H \underset{}{\overset{OH^-}{\rightleftharpoons}} \underset{\text{a mixed aldol}}{C_6H_5\overset{OH}{\overset{|}{C}}H-CH_2\overset{O}{\overset{\|}{C}}H} \xrightarrow[-H_2O]{\text{heat}} \underset{\text{cinnamaldehyde}}{C_6H_5-CH{=}CH\overset{O}{\overset{\|}{C}}H} \quad (9.54)$$

In this particular example, the resulting mixed aldol eliminates water on heating to give **cinnamaldehyde** (the flavor constituent of cinnamon).

EXAMPLE 9.9 Write the structure of the mixed aldol obtained from acetone and formaldehyde.

Solution Of the two reactants, only acetone has α-hydrogens.

$$H-\overset{O}{\overset{\|}{C}}-H + CH_3\overset{O}{\overset{\|}{C}}CH_3 \xrightarrow{\text{base}} H-\underset{H}{\underset{|}{\overset{OH}{\overset{|}{C}}}}-CH_2\overset{O}{\overset{\|}{C}}CH_3$$

PROBLEM 9.27 Using eqs. 9.50 through 9.52 as a guide, write out the steps in the mechanism for eq. 9.54.

PROBLEM 9.28 Write the structure of the mixed aldol obtained from propanal and benzaldehyde. What structure is obtained from dehydration of this mixed aldol?

9.20 Commercial Syntheses via the Aldol Condensation

Aldols are useful in synthesis. For example, acetaldehyde is converted commercially to crotonaldehyde, 1-butanol, and butanal using the aldol condensation.

$$\underset{\text{acetaldehyde}}{2\ CH_3\overset{O}{\overset{\|}{C}}H} \xrightarrow{OH^-} \underset{\text{aldol}}{CH_3\overset{OH}{\overset{|}{C}}HCH_2\overset{O}{\overset{\|}{C}}H} \xrightarrow[-H_2O]{H^+} \underset{\text{crotonaldehyde}}{CH_3CH{=}CH\overset{O}{\overset{\|}{C}}H} \xrightarrow[\text{catalyst}]{H_2}$$

$$\underset{\text{butanal}}{CH_3CH_2CH_2\overset{O}{\overset{\|}{C}}H} \quad \text{or} \quad \underset{\text{1-butanol}}{CH_3CH_2CH_2CH_2OH} \quad (9.55)$$

The particular product obtained in the hydrogenation step depends on the catalyst and reaction conditions.

Butanal is the starting material for the synthesis of the mosquito repellent "6-12" (2-ethylhexane-1,3-diol). The first step is an aldol condensation, and the second step is reduction of the aldehyde group to a primary alcohol.

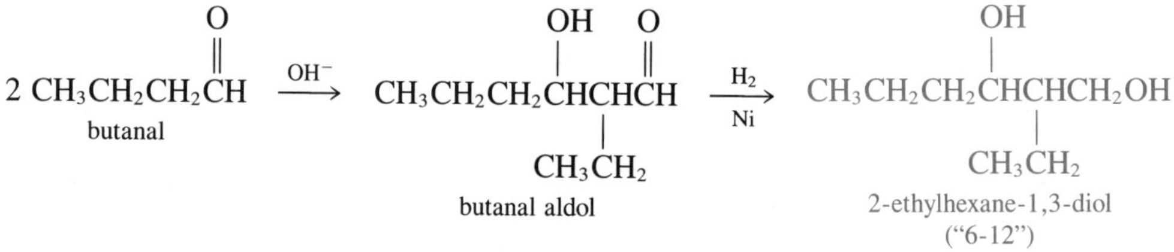

$$2\,CH_3CH_2CH_2\overset{\overset{\displaystyle O}{\|}}{C}H \xrightarrow{OH^-} CH_3CH_2CH_2\overset{\overset{\displaystyle OH}{|}}{C}H\underset{\underset{\displaystyle CH_3CH_2}{|}}{C}H\overset{\overset{\displaystyle O}{\|}}{C}H \xrightarrow[Ni]{H_2} CH_3CH_2CH_2\overset{\overset{\displaystyle OH}{|}}{C}H\underset{\underset{\displaystyle CH_3CH_2}{|}}{C}HCH_2OH \quad (9.56)$$

butanal — butanal aldol — 2-ethylhexane-1,3-diol ("6-12")

The aldol condensation is also used in nature to build up (and, in the case of *reverse* aldol condensations, to break down) carbon chains.

PROBLEM 9.29 2-Ethylhexanol, used commercially in the manufacture of plasticizers and synthetic lubricants, is synthesized from butanal via its aldol. Devise a route to it.

A Word About . . .

19. Quinones, Dyes, and Electron Transfer

Quinones constitute a unique class of carbonyl compounds. They are cyclic, conjugated diketones. The simplest example is 1,4-benzoquinone. All quinones are colored. Many are naturally occurring plant pigments, which often exhibit special biological activity.

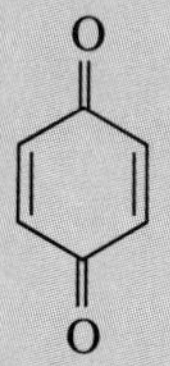

1,4-benzoquinone
mp 118°C, yellow

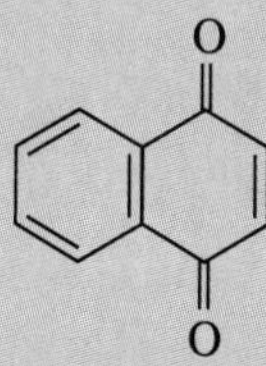

1,4-naphthoquinone
mp 129°C, yellow

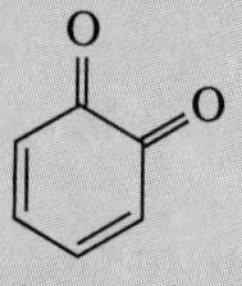

1,2-benzoquinone
mp 70°C, red

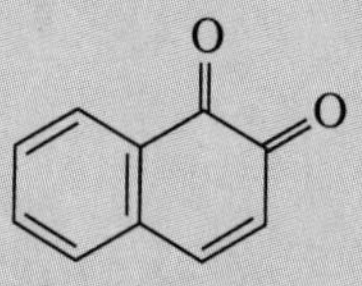

1,2-naphthoquinone
mp 146°C, yellow-red

Alizarin is a quinone dye that was known in ancient Egypt, Persia, and India. It was extracted from the madder root (an herb) and fixed to cloth by various metal ions (for example, aluminum). The red coats used by the British army during the American Revolution were dyed with alizarin. In the mid-nineteenth century, madder was a substantial agricultural crop in Europe, with as many as 70,000 tons in annual production. Later, organic chemists determined alizarin's structure and were able to work out a commercial synthesis that dramatically reduced the price. Many alizarin-type pigments are still used today.

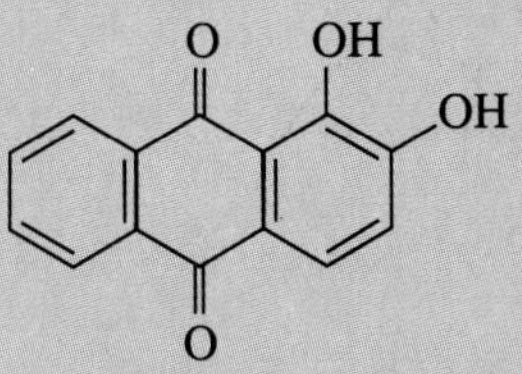

alizarin
mp 290°C, orange-red

Lawsone is a quinone pigment extracted from the tropical henna shrub (*Lawsonia inermis*). It dyes wool and silk orange and can tint hair red. The Islamic prophet Muhammad is said to have dyed his beard with henna.

Juglone was first isolated from walnut shells (*Juglans regia*), where it occurs as the colorless hydroquinone (1,4,5-trihydroxynaphthalene). In air, it is oxidized to the quinone. Juglone, which also occurs in pecans, stains skin brown.

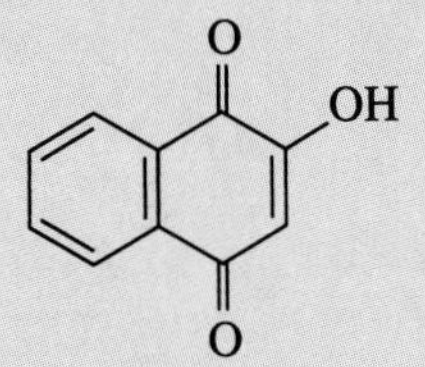

lawsone
mp 192°C, red-brown

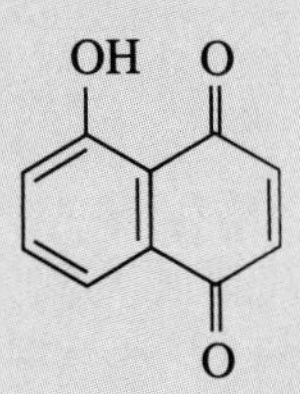

juglone
mp 155°C, yellow

Perhaps the most important property of quinones is their *reversible reduction* to hydroquinones.

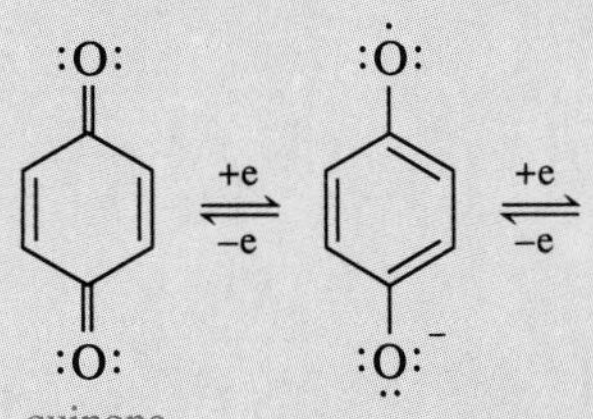

$$\text{quinone} \underset{-e}{\overset{+e}{\rightleftharpoons}} \text{radical anion} \underset{-e}{\overset{+e}{\rightleftharpoons}} \text{dianion}$$

quinone radical anion

$$\text{dianion} \underset{-2H^+}{\overset{+2H^+}{\rightleftharpoons}} \text{hydroquinone}$$

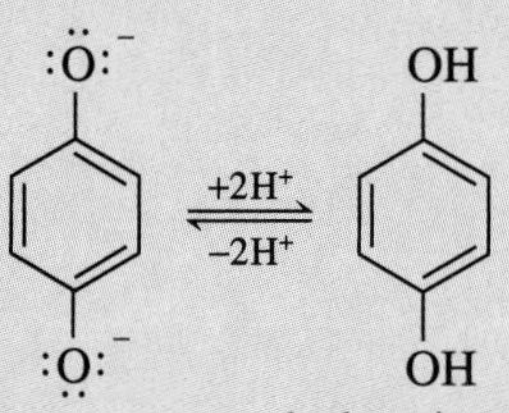

dianion hydroquinone

Virtually all quinones undergo this reaction. The reduction involves stepwise addition of two electrons to give first a radical anion and then a dianion. It is this property that permits quinones to play an important role in reversible biochemical oxidation-reduction (electron transport) reactions. A group of enzymes called **coenzymes Q** (also known as *ubiquinones* because of their ubiquitous, or widespread, occurrence in animal and plant cells) participate in electron transport in mitochondria, the granular bodies in the cell that are involved in the metabolism of lipids, carbohydrates, and proteins.

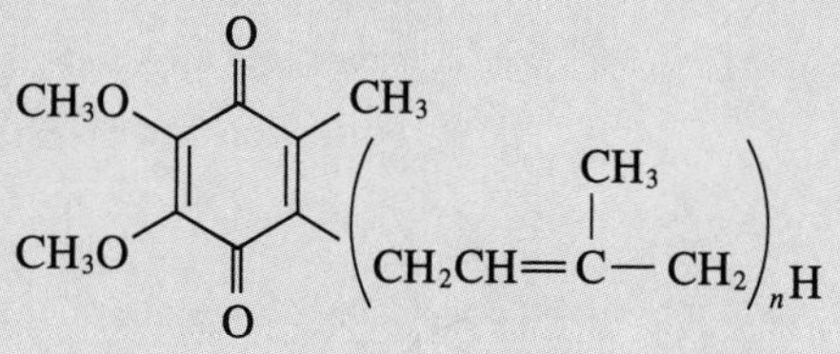

CH_3O, CH_3O, CH_3, $(CH_2CH{=}C(CH_3){-}CH_2)_nH$

coenzymes Q
($n = 10$ is common)

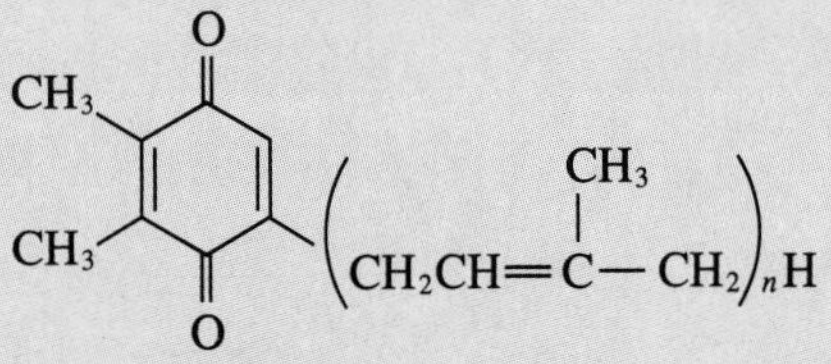

CH_3, CH_3, $(CH_2CH{=}C(CH_3){-}CH_2)_nH$

plastoquinones
($n = 9$ is common)

In plant tissues the **plastoquinones** perform a similar function in photosynthesis. The long isoprenoid carbon chain in these quinones is no doubt necessary to promote fat solubility of these coenzymes.

Vitamin K (Figure 9.1) is a quinone that is required for the normal clotting of blood.

REACTION SUMMARY

1. Preparation of Aldehydes and Ketones

a. Oxidation of Alcohols (Sec. 9.4)

$$\underset{1^\circ\text{ alcohol}}{R{-}CH_2{-}OH} \xrightarrow{\text{PCC}} \underset{\text{aldehyde}}{R{-}\overset{O}{\overset{\|}{C}}{-}H}$$

$$\underset{2^\circ\text{ alcohol}}{R{-}\overset{OH}{\overset{|}{C}H}{-}R} \xrightarrow{\text{PCC}} \underset{\text{ketone}}{R{-}\overset{O}{\overset{\|}{C}}{-}R}$$

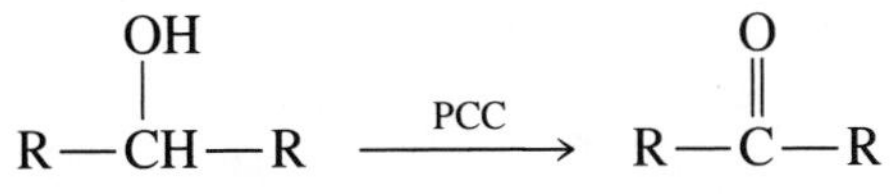

b. Friedel-Crafts Acylation (Sec. 9.4)

$$C_6H_6 + R-\overset{O}{\overset{\|}{C}}-Cl \xrightarrow{AlCl_3} C_6H_5-\overset{O}{\overset{\|}{C}}-R$$

c. Hydration of Alkynes (Sec. 9.4)

$$R-C\equiv C-H \xrightarrow[Hg^{+2}]{H_3O^+} R-\overset{O}{\overset{\|}{C}}-CH_3$$

2. Reactions of Aldehydes and Ketones

a. Formation and Hydrolysis of Acetals (Sec. 9.8)

$$\underset{\text{alcohol}}{ROH} + \underset{\text{carbonyl}}{R'-\overset{O}{\overset{\|}{C}}-R''} \underset{}{\overset{H^+}{\rightleftharpoons}} \underset{\text{hemiacetal}}{R'-\overset{OH}{\overset{|}{\underset{OR}{\underset{|}{C}}}}-R''} \underset{H^+}{\overset{ROH}{\rightleftharpoons}} \underset{\text{acetal}}{R'-\overset{OR}{\overset{|}{\underset{OR}{\underset{|}{C}}}}-R''} + H_2O$$

b. Addition of Grignard Reagents (Sec. 9.10)

$$R'-MgX + R-\overset{O}{\overset{\|}{C}}-R \longrightarrow R-\overset{OMgX}{\overset{|}{\underset{R'}{\underset{|}{C}}}}-R \xrightarrow{H_3O^+} R-\overset{OH}{\overset{|}{\underset{R'}{\underset{|}{C}}}}-R$$

Formaldehyde gives 1° alcohols, other aldehydes give 2° alcohols, ketones give 3° alcohols.

c. Formation of Cyanohydrins (Sec. 9.11)

$$HC\equiv N + \underset{R=H,\ \text{alkyl}}{R-\overset{O}{\overset{\|}{C}}-R} \overset{\text{NaOH (catalyst)}}{\rightleftharpoons} \underset{\text{cyanohydrin}}{R-\overset{OH}{\overset{|}{\underset{CN}{\underset{|}{C}}}}-R}$$

d. Addition of Nitrogen Nucleophiles (Sec. 9.12 and Table 9.1)

$$R'-\ddot{N}H_2 + \underset{R=H,\ \text{alkyl}}{R-\overset{O}{\overset{\|}{C}}-R} \longrightarrow R-\overset{N-R'}{\overset{\|}{C}}-R + H_2O$$

The product is an imine when R′ is an alkyl group.

e. Reduction to Alcohols (Sec. 9.13)

$$\underset{R=H,\ \text{alkyl}}{R-\overset{O}{\overset{\|}{C}}-R} \xrightarrow[\text{or } H_2\text{, catalyst, heat}]{LiAlH_4 \text{ or } NaBH_4} \underset{\text{alcohol}}{R-\overset{OH}{\overset{|}{\underset{H}{\underset{|}{C}}}}-R}$$

f. Oxidation to Carboxylic Acids (Sec. 9.14)

$$\underset{\text{aldehyde}}{R{-}\overset{\overset{\displaystyle O}{\|}}{C}{-}H} \xrightarrow[O_2 \text{ or } Ag^{2+},\ NaOH]{CrO_3,\ H_2SO_4,\ H_2O \text{ or}} \underset{\text{carboxylic acid}}{R{-}\overset{\overset{\displaystyle O}{\|}}{C}{-}OH}$$

g. Aldo Condensation (Sec. 9.18)

$$\underset{\text{aldehyde}}{2\ RCH_2CH{=}O} \xrightarrow{\text{base}} \underset{\text{aldol}}{RCH_2\overset{\overset{\displaystyle OH}{|}}{C}H\underset{\underset{\displaystyle R}{|}}{C}HCH{=}O}$$

ADDITIONAL PROBLEMS

9.30. Name each of the following compounds.

a. $CH_3CH_2\overset{\overset{O}{\|}}{C}CH_2CH_3$ **b.** $CH_3(CH_2)_4CH{=}O$ **c.** $(C_6H_5)_2C{=}O$

d. Br–C_6H_4–CH=O (para) **e.** cyclopentanone **f.** $(CH_3)_3CCH{=}O$

g. dicyclobutyl ketone (cyclobutyl–$\overset{\overset{O}{\|}}{C}$–cyclobutyl) **h.** $CH_3CH{=}CH\overset{\overset{O}{\|}}{C}CH_3$ **i.** $CH_2Br\overset{\overset{O}{\|}}{C}CH_3$

9.31. Write a structural formula for each of the following:

a. 2-octanone **b.** 4-methylpentanal
c. *m*-chlorobenzaldehyde **d.** 3-methylcyclohexanone
e. 2-butenal **f.** benzyl phenyl ketone
g. *p*-tolualdehyde **h.** *p*-benzoquinone
i. 2,2-dibromohexanal **j.** 1-phenyl-2-butanone

9.32. Give an example of each of the following:

a. acetal **b.** hemiacetal
c. cyanohydrin **d.** imine
e. oxime **f.** phenylhydrazone
g. enol **h.** aldehyde with no α-hydrogen
i. enolate **j.** hydrazone

9.33. Write an equation for the synthesis of 2-pentanone by
a. oxidation of an alcohol. **b.** hydration of an alkyne.

9.34. Write an equation, using the Friedel-Crafts reaction, for the preparation of methyl 1-naphthyl ketone.

9.35. Write an equation for the reaction, if any, of *p*-bromobenzaldehyde with each of the following reagents, and name the organic product.

a. Tollens' reagent **b.** hydroxylamine
c. CrO_3, H^+ **d.** ethylmagnesium bromide, then H_3O^+
e. phenylhydrazine **f.** methylamine (CH_3NH_2)

g. cyanide ion **h.** excess methanol, dry HCl
i. ethylene glycol, H^+ **j.** lithium aluminum hydride

9.36. What simple chemical test can distinguish between the members of the following pairs of compounds?
a. hexanal and 2-hexanone
b. benzyl alcohol and benzaldehyde
c. cyclopentanone and 2-cyclopentenone

9.37. Use the structures shown in Figure 9.1 to write equations for the following reactions of natural products:
a. cinnamaldehyde + Tollens' reagent
b. vanillin + hydroxylamine
c. carvone + sodium borohydride
d. camphor + (1) methylmagnesium bromide and (2) H_3O^+

9.38. The boiling points of the isomeric carbonyl compounds heptanal, 4-heptanone, and 2,4-dimethyl-3-pentanone are 155°C, 144°C, and 124°C, respectively. Suggest a possible explanation for the observed order of boiling points.

9.39. Complete each of the following equations.
a. butanal + excess ethanol, H^+ $\longrightarrow$
b. $CH_3CH(OCH_3)_2$ + H_2O, H^+ $\longrightarrow$

c. (tetrahydropyran ring, O, OCH_3) + H_2O, H^+ $\longrightarrow$

d. (tetrahydropyran ring, O, OH) + excess CH_3OH, H^+ $\longrightarrow$

9.40. Write an equation for the reaction of each of the following with methylmagnesium bromide, followed by hydrolysis with aqueous acid.
a. acetaldehyde **b.** acetophenone **c.** formaldehyde **d.** cyclohexanone

9.41. Using a Grignard reagent and the appropriate aldehyde or ketone, show how each of the following can be prepared:
a. 1-pentanol **b.** 3-pentanol
c. 2-methyl-2-butanol **d.** 1-cyclopentylcyclopentanol
e. 1-phenyl-1-propanol **f.** 3-butene-2-ol

9.42. Complete the equation for the reaction of
a. cyclohexanone + $Na^{+\,-}C\equiv CH$ $\longrightarrow$ $\xrightarrow[H^+]{H_2O}$
b. cyclopentanone + HCN $\xrightarrow{KOH}$
c. 2-butanone + NH_2OH $\xrightarrow{H^+}$
d. benzaldehyde + benzylamine $\longrightarrow$
e. propanal + phenylhydrazine $\longrightarrow$

9.43. Give the structure of each product.

a. $CH_3\overset{O}{\overset{\|}{C}}-C_6H_5$ $\xrightarrow[2.\ H_2O,\ H^+]{1.\ LiAlH_4}$

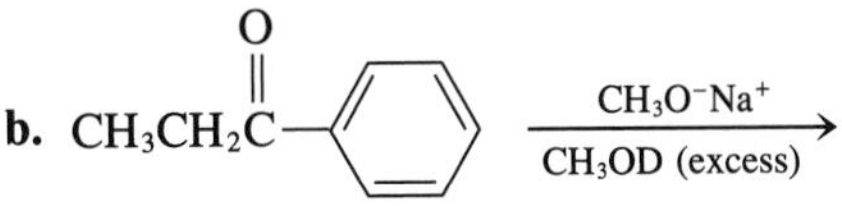

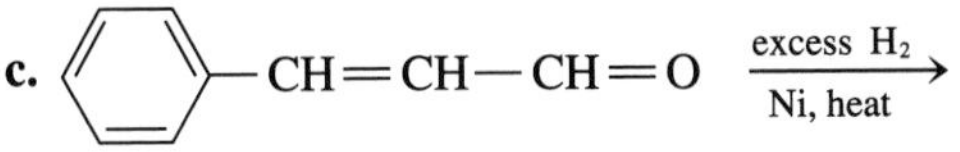

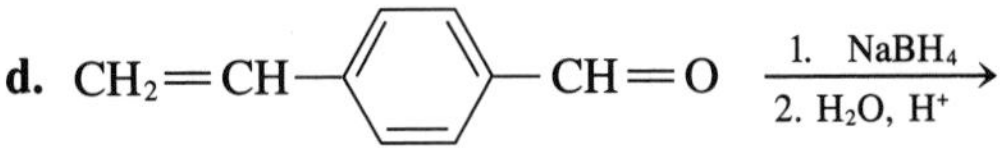

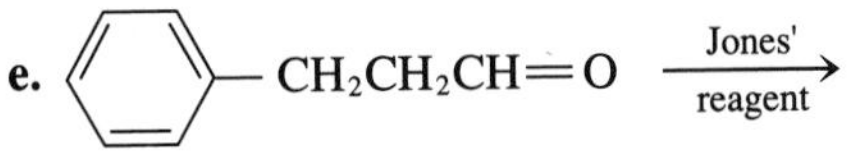

b. $CH_3CH_2\overset{O}{\overset{\|}{C}}-C_6H_5 \xrightarrow[CH_3OD\ (excess)]{CH_3O^-Na^+}$

c. $C_6H_5-CH{=}CH-CH{=}O \xrightarrow[Ni,\ heat]{excess\ H_2}$

d. $CH_2{=}CH-C_6H_4-CH{=}O \xrightarrow[2.\ H_2O,\ H^+]{1.\ NaBH_4}$

e. $C_6H_5-CH_2CH_2CH{=}O \xrightarrow[reagent]{Jones'}$

f. $CH_3CH{=}CHCHO \xrightarrow{Ag_2O}$

9.44. Write the structural formulas for all possible enols of
a. 2-butanone. **b.** phenylacetaldehyde. **c.** 2,4-pentanedione.

9.45. How many hydrogens are replaced by deuterium when each of the following compounds is treated with NaOD in D_2O?
a. 3-methylcyclopentanone **b.** 2-methylbutanal

9.46. Write out the steps in the mechanism for the aldol condensation of butanal (the first step in the synthesis of the mosquito repellent "6-12," eq. 9.56).

9.47. Treatment of the unsaturated diketone

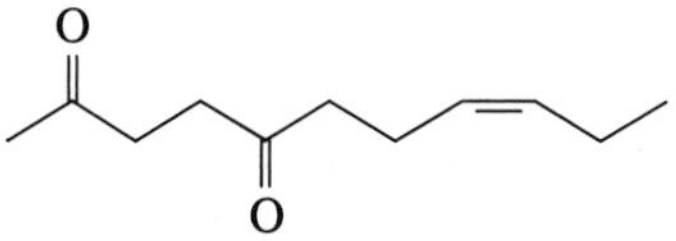

with sodium hydroxide in ethanol gives *jasmone,* the fragrant component of the jasmine flower, used in perfumes.

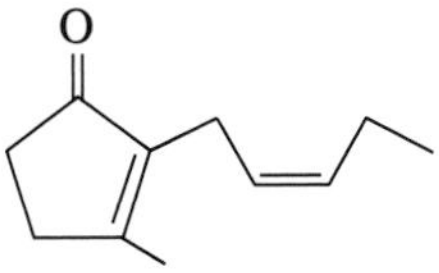

jasmone

Explain how this reaction takes place.

9.48. Excess benzaldehyde reacts with acetone and base to give a yellow crystalline product, $C_{17}H_{14}O$. Deduce its structure and explain how it is formed.

9.49. The final steps in the synthesis of two oral contraceptives, Enovid and Norlutin, are shown. For each step, supply the missing reagent, and tell what general type of reaction is involved.

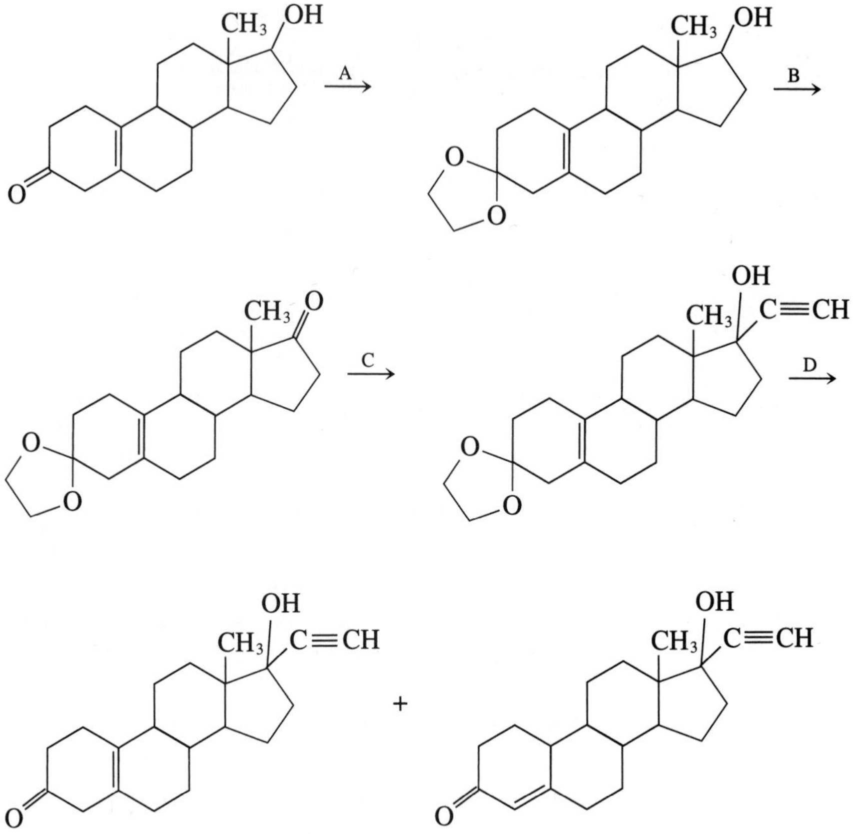

$\xrightarrow{A}$ $\xrightarrow{B}$ $\xrightarrow{C}$ $\xrightarrow{D}$

Enovid + Norlutin

Explain why the carbonyl group in the starting material was converted to an acetal before the rest of the synthesis proceeded, since in the final products, that carbonyl group is to be there. Explain how the carbon–carbon double bond ends up where it does in Norlutin.

9.50. Lily aldehyde, used in perfumes, can be made starting with a mixed aldol condensation. Show how its carbon skeleton could be assembled.

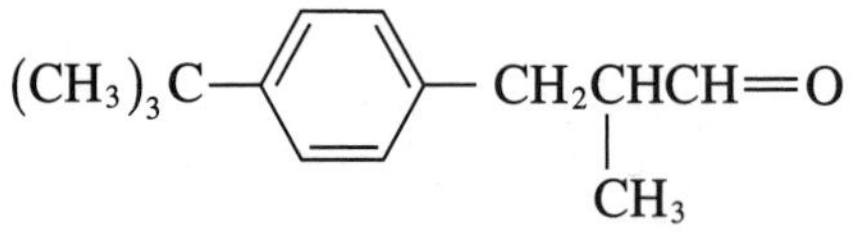

$(CH_3)_3C-C_6H_4-CH_2CH(CH_3)CH{=}O$

lily aldehyde

10 Carboxylic Acids and Their Derivatives

10.1 Introduction

Carboxylic acids are the most important organic acids; their functional group is the **carboxyl group.** This name is a contraction of the parts: the *carb*onyl and hydr*oxyl* groups. The general formula for a carboxylic acid can be written in expanded or abbreviated forms.

$$-\mathrm{C}\overset{\textstyle\mathrm{O}}{\underset{\textstyle\mathrm{OH}}{\lessgtr}} \qquad \mathrm{R}-\mathrm{C}\overset{\textstyle\mathrm{O}}{\underset{\textstyle\mathrm{OH}}{\lessgtr}} \quad \text{or} \quad \mathrm{RCOOH} \quad \text{or} \quad \mathrm{RCO_2H}$$

carboxyl group — three ways to write a carboxylic acid

In this chapter, we will describe the structure, acidity, preparation, and reactions of carboxylic acids. We will also discuss some common related classes of compounds called **acid derivatives,** in which the —OH group of an acid is replaced by other functions (—OR, halogen, or others).

10.2 Nomenclature of Acids

Because of their abundance in nature, carboxylic acids were among the earliest classes of compounds studied by organic chemists. It is not surprising, then, that many of them have common names. These names usually come from some Latin or Greek word that indicates the original source of the acid. Table 10.1 lists the first ten unbranched carboxylic acids, with their common and IUPAC names.

To obtain the IUPAC name of a carboxylic acid, we replace the final *e* in the name of the corresponding alkane with the suffix *-oic* and add the word *acid*. Substituted acids are named in two ways. In the IUPAC system, *the chain is numbered beginning with the carboxyl carbon atom,* and substituents are located in the usual way. If the common name of the acid is used, substituents are located with Greek letters, beginning with the α-carbon atom.

TABLE 10.1 Aliphatic carboxylic acids

Carbon atoms	*Formula*	*Source*	*Common name*	*IUPAC name*
1	HCOOH	ants (Latin, *formica*)	formic acid	methanoic acid
2	CH_3COOH	vinegar (Latin, *acetum*)	acetic acid	ethanoic acid
3	CH_3CH_2COOH	milk (Greek, *protos pion,* first fat)	propionic acid	propanoic acid
4	$CH_3(CH_2)_2COOH$	butter (Latin, *butyrum*)	butyric acid	butanoic acid
5	$CH_3(CH_2)_3COOH$	valerian root (Latin, *valere,* to be strong)	valeric acid	pentanoic acid
6	$CH_3(CH_2)_4COOH$	goats (Latin, *caper*)	caproic acid	hexanoic acid
7	$CH_3(CH_2)_5COOH$	vine blossom (Greek, *oenanthe*)	enanthic acid	heptanoic acid
8	$CH_3(CH_2)_6COOH$	goats (Latin, *caper*)	caprylic acid	octanoic acid
9	$CH_3(CH_2)_7COOH$	pelargonium (an herb with stork-shaped seed capsules; Greek, *pelargos,* stork)	pelargonic acid	nonanoic acid
10	$CH_3(CH_2)_8COOH$	goats (Latin, *caper*)	capric acid	decanoic acid

IUPAC and common naming systems should not be mixed.

$$\overset{\beta}{\overset{3}{CH_3}}-\overset{\alpha}{\overset{2}{CH}}(Br)-\overset{1}{CO_2H}$$

2-bromopropanoic acid
(α-bromopropionic acid)

$$\overset{3}{CH_2}=\overset{2}{C}H\overset{1}{C}O_2H$$

propenoic acid
(acrylic acid)

$$\overset{4}{CH_3}\overset{\gamma}{\overset{3}{C}}H(OH)\overset{\beta}{\overset{2}{C}}H_2\overset{\alpha}{\overset{1}{C}}O_2H$$

3-hydroxybutanoic acid
(β-hydroxybutyric acid)

The carboxyl group has priority over alcohol, aldehyde, or ketone functionality in naming. In the latter cases, the prefix *oxo-* is used to locate the carbonyl group of the aldehyde or ketone, as in these examples:

$$H\overset{3}{C}(=O)-\overset{2}{C}H_2\overset{1}{C}O_2H$$

3-oxopropanoic acid

$$\overset{5}{CH_3}\overset{4}{C}(=O)\overset{3}{C}H_2\overset{2}{C}H(Br)\overset{1}{C}O_2H$$

2-bromo-4-oxopentanoic acid

PROBLEM 10.1 Write the structure for

a. 3-chlorobutanoic acid.
b. 2-hydroxy-2-methylpropanoic acid.
c. 2-butynoic acid.
d. 5-methyl-6-oxohexanoic acid.

PROBLEM 10.2 Give an IUPAC name for

a. C_6H_5–$CH_2CH_2CO_2H$. b. Cl_2CHCO_2H.

c. $CH_3CH{=}CHCO_2H$. d. $(CH_3)_3CCO_2H$.

When the carboxyl group is attached to a ring, the ending *-carboxylic acid* is added to the name of the parent cycloalkane.

COOH

cyclopentanecarboxylic acid

Cl 3 2 1 H
H 4 COOH

trans-3-chlorocyclobutanecarboxylic acid

Aromatic acids are named by attaching the suffix *-oic acid* or *-ic acid* to an appropriate prefix derived from the aromatic hydrocarbon.

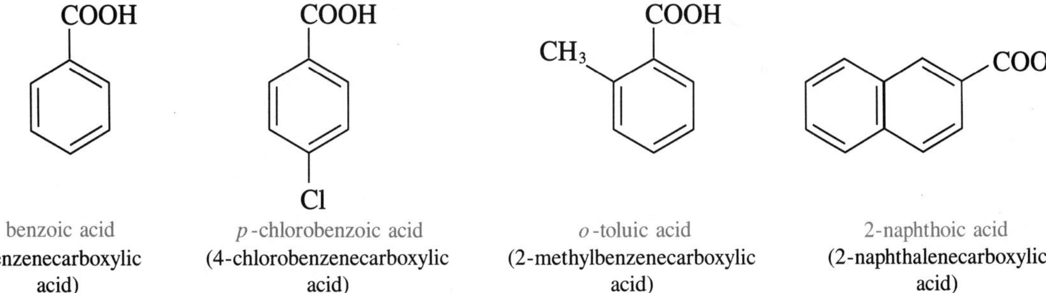

benzoic acid (benzenecarboxylic acid)

p-chlorobenzoic acid (4-chlorobenzenecarboxylic acid)

o-toluic acid (2-methylbenzenecarboxylic acid)

2-naphthoic acid (2-naphthalenecarboxylic acid)

PROBLEM 10.3 Write the structure for

a. *trans*-4-methylcyclohexanecarboxylic acid.
b. *m*-nitrobenzoic acid.

PROBLEM 10.4 Give the correct name for

a. (cyclopropyl)–COOH. b. CH_3–C_6H_4–COOH.

Aliphatic dicarboxylic acids are given the suffix *-dioic acid* in the IUPAC system. For example,

$$\overset{1}{HO_2C}-\overset{2}{CH_2}\overset{3}{CH_2}-\overset{4}{CO_2H} \qquad HO_2C-C{\equiv}C-CO_2H$$

butanedioic acid — butynedioic acid

Many dicarboxylic acids occur in nature and go by their common names, which are based on their source. Table 10.2 lists some common aliphatic

TABLE 10.2 Aliphatic dicarboxylic acids

Formula	*Common name*	*Source*	*IUPAC name*
HOOC—COOH	oxalic acid	plants of the *oxalic* family (for example, sorrel)	ethanedioic acid
HOOC—CH_2—COOH	malonic	apple (Gk. *malon*)	propanedioic
HOOC—$(CH_2)_2$—COOH	succinic	amber (L. *succinum*)	butanedioic
HOOC—$(CH_2)_3$—COOH	glutaric	gluten	pentanedioic
HOOC—$(CH_2)_4$—COOH	adipic	fat (L. *adeps*)	hexanedioic
HOOC—$(CH_2)_5$—COOH	pimelic	fat (Gk. *pimele*)	heptanedioic

diacids.* The most important commercial compound in this group is adipic acid, used to manufacture nylon.

The two butenedioic acids played an important role in the discovery of *cis–trans* isomerism and are usually known by their common names **maleic**** and **fumaric† acid.**

HOOC(H)C=C(H)COOH (cis) and HOOC(H)C=C(COOH)H (trans)

maleic acid
(*cis*-2-butenedioic acid)

fumaric acid
(*trans*-2-butenedioic acid)

The three benzenedicarboxylic acids are generally known by their common names.

phthalic acid — $C_6H_4(CO_2H)_2$ (1,2)

isophthalic acid — $C_6H_4(CO_2H)_2$ (1,3)

terephthalic acid — $C_6H_4(CO_2H)_2$ (1,4)

All three are important commercial chemicals, used to make polymers and other useful materials.

Finally, it is useful to have a name for an **acyl group.** Particular acyl groups are named from the corresponding acid by changing the *-ic* ending to *-yl.*

* The first letter of each word in the sentence "Oh my, such good apple pie" gives, in order, the first letters of the common names of these acids and can help you to remember them.

** From the Latin *malum* (apple). Malic acid (2-hydroxybutanedioic acid), found in apples, can be dehydrated on heating to give maleic acid.

† Found in fumitory, an herb of the genus *fumaria*.

$$R{-}\overset{\overset{\displaystyle O}{\|}}{C}{-} \quad H{-}\overset{\overset{\displaystyle O}{\|}}{C}{-} \quad CH_3{-}\overset{\overset{\displaystyle O}{\|}}{C}{-} \quad CH_3CH_2\overset{\overset{\displaystyle O}{\|}}{C}{-} \quad C_6H_5{-}\overset{\overset{\displaystyle O}{\|}}{C}{-}$$

an acyl group; formyl (methanoyl); acetyl (ethanoyl); propanoyl; benzoyl

PROBLEM 10.5 Write the formula for

a. 4-acetylbenzoic acid.
b. benzoyl bromide.
c. butanoyl chloride.
d. formylcyclopentane.

10.3 Physical Properties of Acids

The first members of the carboxylic acid series are colorless liquids with sharp or unpleasant odors. Acetic acid, which constitutes about 4 to 5% of vinegar, gives vinegar its characteristic odor and flavor. Butyric acid gives rancid butter its disagreeable odor, and the goat acids (caproic, caprylic, and capric in Table 10.1) smell like goats. Table 10.3 lists some physical properties of selected carboxylic acids.

Carboxylic acids are polar. Like alcohols, they form hydrogen bonds with themselves or with other molecules. They therefore have rather high boiling points for their molecular weights—higher even than those of comparable alcohols. For example, acetic acid and propyl alcohol, which have the same formula weights (60), boil at 118°C and 97°C, respectively. Carboxylic acids form dimers, with the units firmly held together by *two* hydrogen bonds.

$$R{-}C\begin{matrix} \nearrow\!\!\!\!\!\!\!= O \cdots H{-}O \searrow \\ \searrow O{-}H \cdots O \nearrow\!\!\!\!\!\!\!= \end{matrix}C{-}R$$

Hydrogen bonding also explains the water-solubility of the lower-molecular-weight carboxylic acids.

Table 10.3 Physical properties of some carboxylic acids

Name	bp, °C	mp, °C	Solubility, g/100 g H_2O at 25°C
formic	101	8	miscible (∞)
acetic	118	17	miscible (∞)
propanoic	141	−22	miscible (∞)
butanoic	164	−8	miscible (∞)
hexanoic	205	−1.5	1.0
octanoic	240	17	0.06
decanoic	270	31	0.01
benzoic	249	122	0.4 (but 6.8 at 95°C)

10.4 Acidity and Acidity Constants

Carboxylic acids dissociate in water, yielding a **carboxylate anion** and a hydronium ion.

$$R-C(=O)OH + H\ddot{O}H \rightleftharpoons R-C(=O)O^- + H-\overset{H}{\underset{..}{O^+}}-H \qquad (10.1)$$

carboxylate ion   hydronium ion

Their **acidity constant** K_a is given by the expression

$$K_a = \frac{[RCO_2^-][H_3O^+]}{[RCO_2H]} \qquad (10.2)$$

(Before proceeding further, it would be a good idea for you to review Secs. 7.6 and 7.7.)

Table 10.4 lists the acidity constants for some carboxylic and other acids. In comparing data in this table, remember that the larger the value of K_a or the smaller the value of pK_a, the stronger the acid.

EXAMPLE 10.1 Which is the stronger acid, formic or acetic, and by how much?

Solution Formic acid is stronger; it has the larger K_a. The ratio of acidities is

$$\frac{2.1 \times 10^{-4}}{1.8 \times 10^{-5}} = 1.16 \times 10^{1} = 11.6$$

This means that formic acid is 11.6 times stronger than acetic acid.

PROBLEM 10.6 Using the data given in Table 10.4, determine which is the stronger acid, acetic or chloroacetic, and by how much.

Before we can explain the acidity differences in Table 10.4, we must examine the structural features that make carboxylic acids acidic.

10.5 Resonance in the Carboxylate Ion

You might wonder why carboxylic acids are so much more acidic than alcohols or phenols, since all three classes ionize by losing H^+ from a hydroxyl group. One answer lies in a comparison of the possibilities for charge delocalization in the resulting anions. Let us use a specific example.

From Table 10.4 we see that acetic acid is approximately 10^{11}, or one hundred thousand million, times stronger an acid than ethanol.

$$CH_3CH_2OH \rightleftharpoons CH_3CH_2O^- + H^+ \qquad K_a = 10^{-16} \qquad (10.3)$$

ethoxide ion

$$CH_3\overset{O}{\overset{\|}{C}}-OH \rightleftharpoons CH_3\overset{O}{\overset{\|}{C}}-O^- + H^+ \qquad K_a = 10^{-5} \qquad (10.4)$$

acetate ion

In ethoxide ion, *the negative charge is localized on a single oxygen atom*. In

TABLE **10.4** The ionization constants of some acids

Name	*Formula*	K_a	pK_a
formic	$HCOOH$	2.1×10^{-4}	3.68
acetic	CH_3COOH	1.8×10^{-5}	4.74
propanoic	CH_3CH_2COOH	1.4×10^{-5}	4.85
butanoic	$CH_3CH_2CH_2COOH$	1.6×10^{-5}	4.80
chloroacetic	$ClCH_2COOH$	1.5×10^{-3}	2.82
dichloroacetic	$Cl_2CHCOOH$	5.0×10^{-2}	1.30
trichloroacetic	CCl_3COOH	2.0×10^{-1}	0.70
2-chlorobutanoic	$CH_3CH_2CHClCOOH$	1.4×10^{-3}	2.85
3-chlorobutanoic	$CH_3CHClCH_2COOH$	8.9×10^{-5}	4.05
benzoic acid	C_6H_5COOH	6.6×10^{-5}	4.18
o-chlorobenzoic	*o*-$Cl—C_6H_4COOH$	12.5×10^{-4}	2.90
m-chlorobenzoic	*m*-$Cl—C_6H_4COOH$	1.6×10^{-4}	3.80
p-chlorobenzoic	*p*-$Cl—C_6H_4COOH$	1.0×10^{-4}	4.00
p-nitrobenzoic	*p*-$NO_2—C_6H_4COOH$	4.0×10^{-4}	3.40
phenol	C_6H_5OH	1.0×10^{-10}	10.00
ethanol	CH_3CH_2OH	1.0×10^{-16}	16.00
water	HOH	1.8×10^{-16}	15.74

acetate ion, on the other hand, *the negative charge can be delocalized through resonance*.

$$\left[CH_3—C(=\ddot{O}:)(—\ddot{O}:^-) \longleftrightarrow CH_3—C(—\ddot{O}:^-)(=\ddot{O}:) \right] \quad \text{or} \quad \left[CH_3—C\genfrac{}{}{0pt}{}{\cdots O}{\cdots O} \right]^-$$

resonance in a carboxylate ion

The negative charge is spread *equally* over the two oxygens, so that each oxygen in the carboxylate ion carries only half the negative charge. The acetate ion is stabilized by resonance compared to the ethoxide ion. This stabilization drives the equilibrium more to the right in eq. 10.4 than in eq. 10.3. Consequently, more H^+ is formed from acetic acid than from ethanol. That is, acetic acid is a stronger acid than ethanol.

EXAMPLE 10.2 Phenoxide ions are also stabilized by resonance (Sec. 7.7). Why aren't phenols as strong acids as carboxylic acids?

Solution Charge delocalization is not as great in phenoxide ions as in carboxylate ions because the contributors to the resonance hybrid are not equivalent. Some of them put the negative charge on carbon instead of on oxygen and disrupt aromaticity. In the carboxylate ion, both contributors are identical and have the negative charge on oxygen, a more electronegative atom than carbon.

Physical data support the importance of resonance in carboxylate ions. In formic acid molecules, the two carbon-oxygen bonds have different lengths. But in sodium formate, both carbon-oxygen bonds of the formate ion are identical, and their length is between those of normal double and single carbon-oxygen bonds.

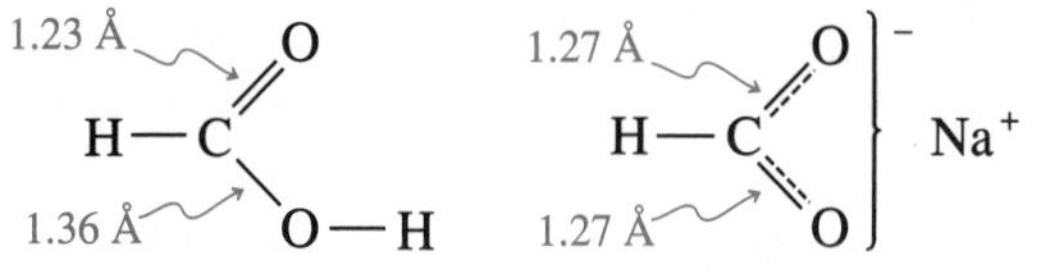

formic acid: H—C(=O)—O—H, with C=O 1.23 Å and C—O 1.36 Å

sodium formate: $[H—C(O)(O)]^- Na^+$, with both carbon-oxygen bonds 1.27 Å

10.6 Effect of Structure on Acidity; the Inductive Effect Revisited

The data in Table 10.4 show that even among carboxylic acids (where the ionizing functional group is kept constant), acidities can vary depending on what other groups are attached to the molecule. Compare, for example, the K_a of acetic acid with those of mono-, di-, and trichloroacetic acids, and note that the acidity varies by a factor of 10,000.

The most important factor operating here is the **inductive effect** of the groups close to the carboxyl group. This effect relays charge through bonds, by displacing bonding electrons toward electronegative atoms, or away from electropositive atoms. Recall that *electron-withdrawing groups enhance acidity, and electron-releasing groups reduce acidity* (see Sec. 7.7).

Let us examine the carboxylate ions formed when acetic acid and its chloro derivatives ionize:

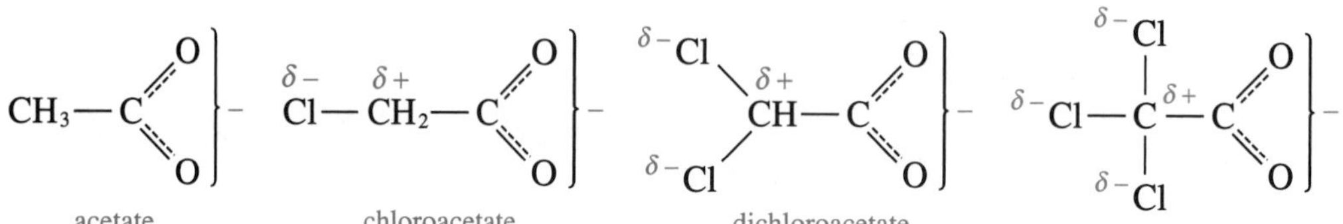

$$[CH_3—CO_2]^- \qquad [\overset{\delta-}{Cl}—\overset{\delta+}{CH_2}—CO_2]^- \qquad [(\overset{\delta-}{Cl})_2\overset{\delta+}{CH}—CO_2]^- \qquad [(\overset{\delta-}{Cl})_3\overset{\delta+}{C}—CO_2]^-$$

acetate, chloroacetate, dichloroacetate, trichloroacetate

Because chlorine is more electronegative than carbon, the C—Cl bond is polarized with the chlorine partially negative and the carbon partially positive. Thus, electrons are pulled away from the carboxylate end of the ion toward the chlorine. The effect tends to spread the negative charge over more atoms than in acetate ion itself and thus stabilizes the ion. The more chlorines, the greater the effect and the greater the strength of the acid.

EXAMPLE 10.3 Explain the acidity order in Table 10.4 for butanoic acid and its 2- and 3-chloro derivatives.

Solution The 2-chloro substituent increases the acidity of butanoic acid substantially, due to its inductive effect. In fact, the effect is about the same as for chloroacetic and acetic acids. The 3-chloro substituent exerts a similar *but much smaller* effect, because the C—Cl bond is now farther away from the carboxylate group. *Inductive effects fall off rapidly with distance.*

PROBLEM 10.7 Account for the relative acidities of benzoic acid and its *ortho, meta,* and *para* chloro derivatives (Table 10.4).

We saw in Example 10.1 that formic acid is a substantially stronger acid than acetic acid. This suggests that the methyl group is more electron-releasing (hence anion destabilizing and acidity-reducing) than hydrogen. This observation is consistent with what we have already learned about carbocation stabilities—that alkyl groups are more effective than hydrogen atoms at releasing electrons to, and therefore stabilizing, a positive carbon atom (see Sec. 3.10).

10.7 Conversion of Acids to Salts

Carboxylic acids, when treated with a strong base, form salts. For example,

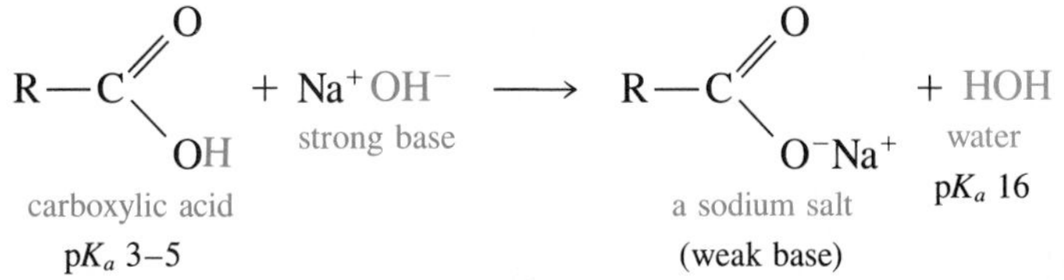

$$\underset{\substack{\text{carboxylic acid} \\ \text{p}K_a\ 3\text{–}5}}{R{-}C(=O)OH} + \underset{\text{strong base}}{Na^+OH^-} \longrightarrow \underset{\substack{\text{a sodium salt} \\ \text{(weak base)}}}{R{-}C(=O)O^-Na^+} + \underset{\substack{\text{water} \\ \text{p}K_a\ 16}}{HOH} \qquad (10.5)$$

The salt can be isolated by evaporating the water. As we will see in Chapter 15, carboxylate salts of certain acids are useful as soaps and detergents.

Salts are named as shown in the following examples:

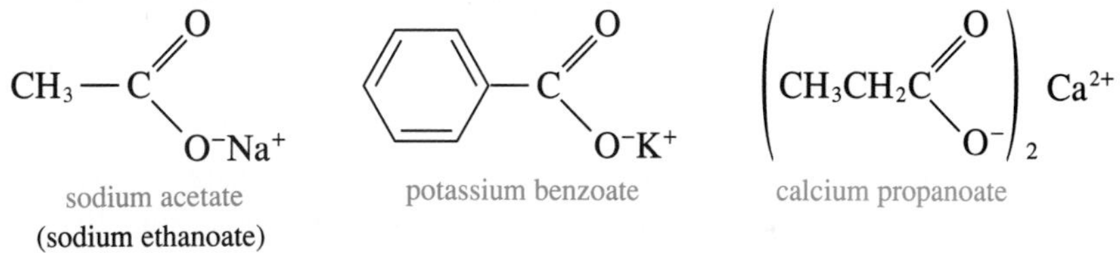

$$\underset{\substack{\text{sodium acetate} \\ \text{(sodium ethanoate)}}}{CH_3{-}C(=O)O^-Na^+} \qquad \underset{\text{potassium benzoate}}{C_6H_5{-}C(=O)O^-K^+} \qquad \underset{\text{calcium propanoate}}{\left(CH_3CH_2C(=O)O^-\right)_2 Ca^{2+}}$$

The cation is named first, followed by the name of the carboxylate ion, which is obtained by changing the *-ic* ending of the acid to *-ate*.

EXAMPLE 10.4 Name the following carboxylate salt:

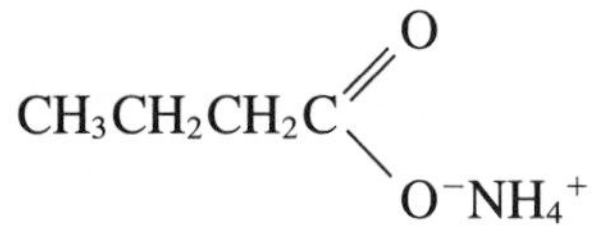

$$CH_3CH_2CH_2C(=O)O^-NH_4^+$$

Solution The salt is ammonium butanoate (IUPAC) or ammonium butyrate (common).

PROBLEM 10.8 Write an equation, analogous to eq. 10.5, for the preparation of potassium 3-bromopropanoate from the corresponding acid.

10.8 Preparation of Acids

Organic acids can be prepared in many ways, four of which will be described here: (1) oxidation of primary alcohols or aldehydes, (2) oxidation of alkyl side chains on aromatic rings, (3) reaction of Grignard reagents with carbon dioxide, and (4) hydrolysis of alkyl cyanides (nitriles).

10.8a Oxidation of Primary Alcohols or Aldehydes The oxidation of primary alcohols (eq. 7.34) and aldehydes (eq. 9.35) to carboxylic acids has already been mentioned. It is easy to see that these are oxidation reactions because going from an alcohol to an aldehyde to an acid requires replacement of C—H bonds by C—O bonds.

$$\underset{\substack{\text{alcohol}\\\text{(one C—O bond)}}}{R-CH_2-OH} \longrightarrow \underset{\substack{\text{aldehyde}\\\text{(two C—O bonds)}}}{R(H)C{=}O} \longrightarrow \underset{\substack{\text{acid}\\\text{(three C—O bonds)}}}{R-C(=O)OH} \tag{10.6}$$

The most commonly used oxidizing agents for these purposes are potassium permanganate ($KMnO_4$), chromic acid (CrO_3), nitric acid (HNO_3), and with aldehydes only, silver oxide (Ag_2O). For specific examples, see eqs. 7.37, 9.36, 9.37, and 9.40.

10.8b Oxidation of Aromatic Side Chains Aromatic acids can be prepared by oxidizing an alkyl side chain on an aromatic ring.

$$\underset{\text{toluene}}{C_6H_5-CH_3} \xrightarrow[\text{heat}]{KMnO_4} \underset{\text{benzoic acid}}{C_6H_5-C(=O)OH} \tag{10.7}$$

This reaction illustrates the striking stability of aromatic rings; it is the alkane-like methyl group, not the aromatic ring, that is oxidized. The reaction involves attack of the oxidant at a C—H bond adjacent to the benzene ring. Longer side chains are also oxidized to a carboxyl group.

$$C_6H_5-CH_2CH_2CH_3 \xrightarrow[\text{heat}]{KMnO_4} C_6H_5-CO_2H \tag{10.8}$$

If no C—H bond is in the benzylic position, however, the aromatic ring is oxidized, although only under severe reaction conditions.

$$(CH_3)_3C-C_6H_5 \xrightarrow[\text{heat}]{KMnO_4} (CH_3)_3C\,CO_2H \tag{10.9}$$

With oxidants other than potassium permanganate, this reaction is commercially important. For example, **terephthalic acid,** one of the two raw materials needed to manufacture Dacron, is produced in this way, using a cobalt catalyst for the air oxidation.

$$\underset{p\text{-xylene}}{CH_3-C_6H_4-CH_3} \xrightarrow[CH_3CO_2H]{O_2,\,Co(III)} \underset{\text{terephthalic acid}}{HOOC-C_6H_4-COOH} \tag{10.10}$$

Phthalic acid, used for making plasticizers, resins, and dyestuffs, is manufactured by similar oxidations, starting with *o*-xylene.

$$\text{o-xylene } (C_6H_4(CH_3)_2) \xrightarrow[CH_3CO_2H]{O_2, Co(III)} \text{phthalic acid } (C_6H_4(COOH)_2) \tag{10.11}$$

o-xylene — phthalic acid

10.8c Reaction of Grignard Reagents with Carbon Dioxide As we saw previously, Grignard reagents add to the carbonyl groups of aldehydes or ketones to give alcohols. In a similar way, they add irreversibly to the carbonyl group of carbon dioxide to give acids, after protonation of the intermediate carboxylate salt with a mineral acid like aqueous HCl.

$$R—MgX + O=C=O \longrightarrow R—\overset{\overset{O}{\|}}{C}—OMgX \xrightarrow{HX} R—\overset{\overset{O}{\|}}{C}—OH + Mg^{2+}X_2^{-} \tag{10.12}$$

This reaction gives good yields and is an excellent laboratory method for preparing both aliphatic and aromatic acids. Note that the acid obtained has one more carbon atom than the alkyl or aryl halide from which the Grignard reagent is prepared, so the reaction provides a way to increase the length of a carbon chain.

EXAMPLE 10.5 Show how $(CH_3)_3CBr$ can be converted to $(CH_3)_3CCO_2H$.

Solution

$$(CH_3)_3CBr \xrightarrow[\text{ether}]{Mg} (CH_3)_3CMgBr \xrightarrow[2.\ H_3O^+]{1.\ CO_2} (CH_3)_3CCO_2H$$

PROBLEM 10.9 Show how cyclohexyl chloride can be converted to cyclohexane carboxylic acid.

PROBLEM 10.10 Devise a synthesis of butanoic acid from 1-propanol.

PROBLEM 10.11 Explain why the first step in Eq. 10.12 is irreversible (see Sec. 9.10 for help).

10.8d Hydrolysis of Cyanides (Nitriles) The carbon-nitrogen triple bond of organic cyanides can be hydrolyzed to a carboxyl group. The reaction requires either acid or base. In acid, the nitrogen atom of the cyanide is converted to an ammonium ion.

$$\underset{\text{a cyanide, or nitrile}}{R—C\equiv N} + 2\,H_2O \xrightarrow{HCl} \underset{\text{an acid}}{R—\overset{\overset{O}{\|}}{C}—OH} + \underset{\text{ammonium ion}}{\overset{+}{N}H_4} + Cl^- \tag{10.13}$$

In base, the nitrogen is converted to ammonia and the organic product is the carboxylate salt, which must be neutralized in a separate step to give the acid.

$$R{-}C{\equiv}N + 2\,H_2O \xrightarrow{NaOH} \underset{\text{a carboxylate salt}}{R{-}\overset{\overset{O}{\|}}{C}{-}O^- Na^+} + \underset{\text{ammonia}}{NH_3} \quad (10.14)$$

$$\downarrow H^+$$

$$R{-}\overset{\overset{O}{\|}}{C}{-}OH$$

Alkyl cyanides are generally made from the corresponding alkyl halide (usually primary) and sodium cyanide by an S_N2 displacement. For example,

$$\underset{\substack{\text{propyl bromide}\\ \text{(1-bromopropane)}}}{CH_3CH_2CH_2Br} \xrightarrow{NaCN} \underset{\substack{\text{butyronitrile}\\ \text{(butanenitrile)}}}{CH_3CH_2CH_2CN} \xrightarrow[H^+]{H_2O} \underset{\substack{\text{butyric acid}\\ \text{(butanoic acid)}}}{CH_3CH_2CH_2CO_2H} + NH_4^+ \quad (10.15)$$

PROBLEM 10.12 Why is it *not* possible to convert bromobenzene to benzoic acid by the nitrile method? How could this conversion be accomplished?

Organic cyanides are commonly named after the corresponding acid, by changing the *-ic* or *-oic* suffix to *-onitrile* (hence butyronitrile in eq. 10.15). In the IUPAC system, the suffix *-nitrile* is added to the name of the hydrocarbon with the same number of carbon atoms (hence butanenitrile in eq. 10.15). Sometimes these cyanides are named as alkyl cyanides.

EXAMPLE 10.6 Name CH_3CN in three ways.

Solution Acetonitrile (it gives acetic acid on hydroylsis), ethanenitrile (IUPAC), or methyl cyanide.

Note that with the hydrolysis of nitriles, as with the Grignard method, the acid obtained has one more carbon atom than the alkyl halide from which the cyanide is prepared. Consequently, both methods provide ways of increasing the length of a carbon chain.

PROBLEM 10.13 Write equations for synthesizing phenylacetic acid from benzyl bromide by two routes.

10.9 Carboxylic Acid Derivatives

Carboxylic acid derivatives are compounds in which the hydroxyl part of the carboxyl group is replaced by various other groups. All acid derivatives can be hydrolyzed to the corresponding acid. In the remainder of this chapter, we will consider the preparation and reactions of the more important of these acid derivatives. Their general formulas are as follows:

$R{-}\overset{\overset{\large O}{\|}}{C}{-}OR'$ — ester

$R{-}\overset{\overset{\large O}{\|}}{C}{-}X$ (X is usually Cl or Br) — acyl halide

$R{-}\overset{\overset{\large O}{\|}}{C}{-}O{-}\overset{\overset{\large O}{\|}}{C}{-}R$ — acid anhydride

$R{-}\overset{\overset{\large O}{\|}}{C}{-}NH_2$ — primary amide

Esters and amides occur widely in nature. Anhydrides, however, are uncommon in nature, and acyl halides are strictly creatures of the laboratory.

10.10 Esters

Esters are derived from acids by replacing the —OH group by an —OR group. They are named in a manner analogous to that for carboxylic acid salts. The R part of the —OR group is named first, followed by the name of the acid, with the *-ic* ending changed to *-ate*.

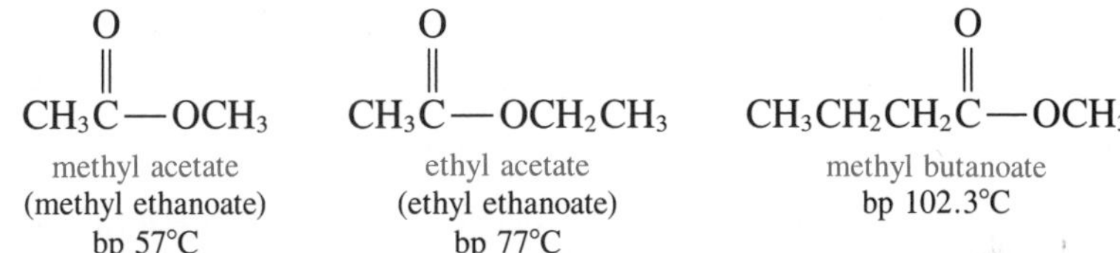

methyl acetate (methyl ethanoate) bp 57°C

ethyl acetate (ethyl ethanoate) bp 77°C

methyl butanoate bp 102.3°C

Notice the different names of the following pair of isomeric esters, where the R and R′ groups are interchanged.

$CH_3\overset{\overset{\large O}{\|}}{C}{-}O{-}C_6H_5$

phenyl acetate
bp 195.7°C

$C_6H_5{-}\overset{\overset{\large O}{\|}}{C}{-}OCH_3$

methyl benzoate
bp 196.6°C

EXAMPLE 10.7 Name $CH_3CH_2CO_2CH(CH_3)_2$.

Solution The related acid is $CH_3CH_2CO_2H$, so the last part of the name is *propanoate* (change the *-ic* of propanoic to *-ate*). The alkyl group that replaces the hydrogen is *isopropyl,* or *2-propyl,* so the correct name is *isopropyl propanoate*, or *2-propyl propanoate*.

PROBLEM 10.14 Write the IUPAC name for

a. $H{-}\overset{\overset{\large O}{\|}}{C}{-}OCH_3$ **b.** $CH_3CH_2\overset{\overset{\large O}{\|}}{C}{-}OCH_2CH_2CH_3$

PROBLEM 10.15 Write the structure of

a. 3-pentyl ethanoate.
b. ethyl 2-methylpropanoate.

Many esters are rather pleasant-smelling substances, which are responsible for the flavor and fragrance of many fruits and flowers. Among the more common are pentyl acetate (bananas), octyl acetate (oranges), ethyl butanoate (pineapples), and pentyl butanoate (apricots). Natural flavors can be exceedingly complex. For example, no fewer than 53 esters have been identified among the volatile constituents of Bartlett pears! Mixtures of esters are used as perfumes and artificial flavors.

10.11 Preparation of Esters; Fischer Esterification

When a carboxylic acid and an alcohol are heated in the presence of an acid catalyst (usually HCl or H_2SO_4), an equilibrium is established with the ester and water.

$$\underset{\text{acid}}{R-\overset{\overset{\displaystyle O}{\|}}{C}-OH} + \underset{\text{alcohol}}{HO-R'} \underset{}{\overset{H^+}{\rightleftharpoons}} \underset{\text{ester}}{R-\overset{\overset{\displaystyle O}{\|}}{C}-OR'} + H_2O \qquad (10.16)$$

The process is called **Fischer esterification,** after Emil Fischer (page 164), who developed the method. Although the reaction is an equilibrium, it can be used to make esters in high yield by shifting the equilibrium to the right. This can be accomplished in several ways. If either the alcohol or the acid is inexpensive, a large excess can be used. Alternatively, the ester and/or water may be removed as formed (by distillation, for example), thus driving the reaction forward.

PROBLEM 10.16 Following eq. 10.16, write an equation for the preparation of propyl butanoate from the correct acid and alcohol.

10.12 The Mechanism of Acid-catalyzed Esterification; Nucleophilic Acyl Substitution

We can ask the following simple mechanistic question about Fischer esterification: Is the water molecule formed from the hydroxyl group of the acid and the hydrogen of the alcohol (as shown in color in eq. 10.16), or from the hydrogen of the acid and the hydroxyl group of the alcohol? This question may seem rather trivial, but the answer provides a key to understanding much of the chemistry of acids, esters, and their derivatives.

This question was resolved using isotopic labeling. For example, Fischer esterification of benzoic acid with methanol that had been enriched with the ^{18}O isotope of oxygen gave labeled methyl benzoate.*

$$C_6H_5-\overset{\overset{\displaystyle O}{\|}}{C}-OH + H^{18}OCH_3 \overset{H^+}{\rightleftharpoons} \underset{\text{methyl benzoate}}{C_6H_5-\overset{\overset{\displaystyle O}{\|}}{C}-{}^{18}OCH_3} + HOH \qquad (10.17)$$

* ^{18}O is oxygen with two additional neutrons in its nucleus. It is two mass units heavier than ^{16}O. ^{18}O can be distinguished from ^{16}O by mass spectrometry (see Chapter 12).

None of the ^{18}O appeared in the water. Thus it is clear that *the water was formed using the hydroxyl group of the acid and the hydrogen of the alcohol.* In other words, in Fischer esterification, the —OR group of the alcohol replaces the —OH group of the acid.

How can we explain this experimental fact? A mechanism consistent with this result is as follows (the oxygen atom of the alcohol is shown in color so that its path can be traced):

$$R-C(=O)-OH \underset{①}{\overset{H^+}{\rightleftharpoons}} R-C(=\overset{+}{O}H)-OH \overset{②}{\rightleftharpoons} R-C(OH)(OH)-\overset{+}{O}(H)R' \underset{③}{\overset{-H^+}{\rightleftharpoons}} R-C(OH)(OH)-OR'$$

nucleophilic addition

$$\overset{④}{\rightleftharpoons}{}^{H^+} \quad R-C(OH)(OR')-\overset{+}{O}H_2 \underset{⑤}{\overset{-H_2O}{\rightleftharpoons}} R-C(=\overset{+}{O}H)-OR' \underset{⑥}{\overset{-H^+}{\rightleftharpoons}} R-C(=O)-OR' \qquad (10.18)$$

elimination step

Let us go through this mechanism, which looks more complicated than it really is, one step at a time.

Step 1. The carbonyl group of the acid is reversibly protonated. This step explains how the acid catalyst works. Protonation increases the positive charge on the carboxyl carbon and enhances its reactivity toward nucleophiles (recall the similar effect of acid catalysts with aldehydes and ketones, eq. 9.9).

Step 2. *This is the crucial step.* The alcohol, as a nucleophile, attacks the carbonyl carbon of the protonated acid. This is the step in which the new C—O bond (the ester bond) is formed.

Steps 3 and 4. These steps are equilibria in which oxygens lose or gain a proton. Such acid-base equilibria are reversible and rapid and go on constantly in any acidic solution of an oxygen-containing compound. In step 4, it doesn't matter which —OH group is protonated since these groups are equivalent.

Step 5. This is the step in which water, one product of the overall reaction, is eliminated. For this step to occur, an —OH group must be protonated to improve its leaving-group capacity. (This step is similar to the reverse of step 2.)

Step 6. This deprotonation step gives the ester and regenerates the acid catalyst. (This step is similar to the reverse of Step 1.)

Some other features of the mechanism in eq. 10.18 are worth examining. The reaction begins with an acid, in which the carboxyl carbon is trigonal and sp^2-hybridized. The end product is an ester; the ester carbon is also trigonal and sp^2-hybridized. However, *the reaction proceeds through a neutral tetrahedral intermediate* (shown in a box in eq. 10.18 and in color in eq. 10.19), in which the carbon atom has four groups attached to it and is thus sp^3-hybridized. If we omit all of the proton-transfer steps in eq. 10.18, we can focus on this feature of the reaction:

$$\underset{sp^2}{\mathrm{R{-}\overset{\overset{\displaystyle O}{\|}}{C}{-}OH}} + \mathrm{R'OH} \rightleftharpoons \underset{sp^3}{\mathrm{R{-}\overset{\overset{\displaystyle OH}{|}}{\underset{\underset{\displaystyle R'O}{|}}{C}}{-}OH}} \rightleftharpoons \underset{sp^2}{\mathrm{R{-}\overset{\overset{\displaystyle O}{\|}}{C}{-}OR'}} + \mathrm{HOH} \qquad (10.19)$$

The net result of this process is substitution of the —OR′ group of the alcohol for the —OH group of the acid. Hence the reaction is referred to as **nucleophilic acyl substitution.** But the reaction is not a direct substitution. Instead, it occurs in two steps: (1) nucleophilic addition, followed by (2) elimination. We will see in the next and subsequent sections of this chapter that this is a general mechanism for nucleophilic substitutions at the carbonyl carbon atoms of acid derivatives.

PROBLEM 10.17 Following eq. 10.18, write out the steps in the mechanism for the acid-catalyzed preparation of ethyl acetate from ethanol and acetic acid. In the United States, this method is used commercially to produce more than 100 million pounds of ethyl acetate annually, mainly for use as a solvent in the paint industry, but also as a solvent for nail polish and various glues.

10.13 *Lactones*

Hydroxy acids contain both functional groups required for ester formation. If these groups can come in contact through bending of the chain, they may react with one another to form **cyclic esters** called **lactones.** For example,

$$\underset{\displaystyle OH}{\overset{\gamma\;\;\beta\;\;\alpha}{\overset{4\;\;\;3\;\;\;2\;\;\;1}{\mathrm{CH_2CH_2CH_2CO_2H}}}} \xrightarrow[\text{or heat}]{H^+} \underset{\gamma\text{-butyrolactone}}{\text{(five-membered ring: C1(=O), C2, C3, C4, O)}} + H_2O \qquad (10.20)$$

Most common lactones have five- or six-membered rings, although lactones with smaller or larger rings are known. Two examples of six-membered lactones from nature are **coumarin,** which is responsible for the pleasant odor of newly mown hay, and **nepatalactone,** the compound in catnip that excites cats. **Erythromycin,** widely used as an antibiotic, is an example of a macrocyclic lactone.*

---

*The R and R′ groups in erythromycin are carbohydrate units (see Chapter 16).

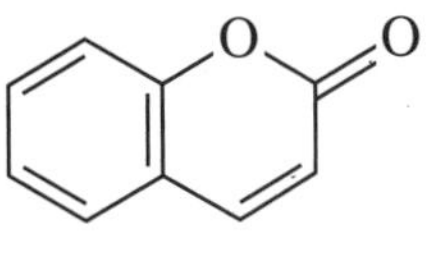

coumarin

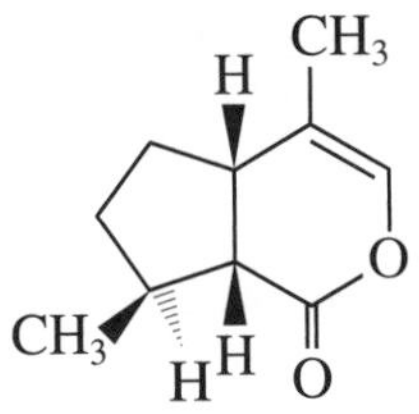

nepatalactone

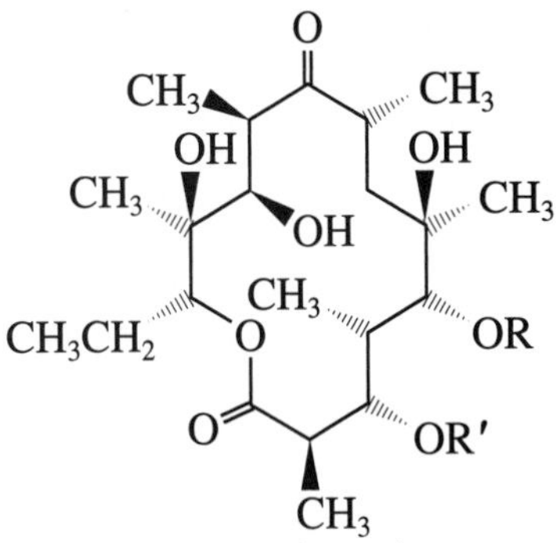

erythromycin

PROBLEM 10.18 Write the steps in the mechanism for the acid-catalyzed reaction in eq. 10.20.

10.14 Saponification of Esters

Esters are commonly hydrolyzed with base. The reaction is called **saponification** (from the Latin *sapon,* soap) because this type of reaction is used to make soaps from fats (Chapter 15). The general reaction is as follows:

$$\underset{\text{ester}}{R{-}C(=O){-}OR'} + \underset{\text{nucleophile}}{Na^+OH^-} \xrightarrow[H_2O]{\text{heat}} \underset{\text{salt of an acid}}{R{-}C(=O){-}O^-Na^+} + \underset{\text{alcohol}}{R'OH} \quad (10.21)$$

The mechanism is another example of a nucleophilic acyl substitution. It involves nucleophilic attack by hydroxide ion, a strong nucleophile, on the carbonyl carbon of the ester.

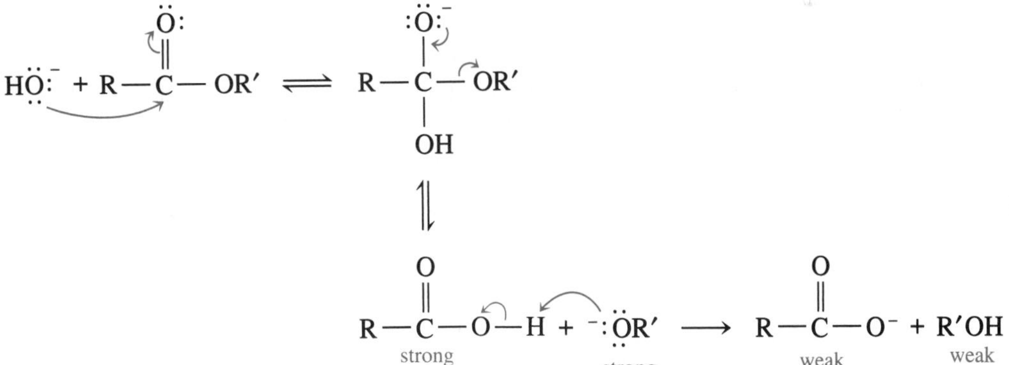

$$HO^- + R{-}C(=O){-}OR' \rightleftharpoons R{-}C(O^-)(OH){-}OR'$$

$$\rightleftharpoons$$

$$\underset{\substack{\text{strong}\\ \text{acid}\\ (pK_a\,5)}}{R{-}C(=O){-}O{-}H} + \underset{\substack{\text{strong}\\ \text{base}}}{^-{:}OR'} \longrightarrow \underset{\substack{\text{weak}\\ \text{base}}}{R{-}C(=O){-}O^-} + \underset{\substack{\text{weak}\\ \text{acid}\\ (pK_a\,16)}}{R'OH} \quad (10.22)$$

The key step is nucleophilic addition to the carbonyl group. The reaction proceeds via a tetrahedral intermediate, but the reactant and the product are trigonal. *Saponification is not reversible;* in the final step, the strongly basic alkoxide ion removes a proton from the acid to form a carboxylate ion and an alcohol molecule—a step that proceeds completely in the forward direction.

Saponification is especially useful for breaking down an unknown ester, perhaps isolated from a natural source, into its component acid and alcohol for structure determination.

PROBLEM 10.19 Following eq. 10.21, write an equation for the saponification of methyl benzoate.

10.15 Ammonolysis of Esters

Ammonia converts esters to amides.

$$\underset{\text{ester}}{R-C(=O)-OR'} + \ddot{N}H_3 \longrightarrow \underset{\text{amide}}{R-C(=O)-NH_2} + R'OH \qquad (10.23)$$

For example,

$$\underset{\text{methyl benzoate}}{C_6H_5-C(=O)-OCH_3} + \ddot{N}H_3 \xrightarrow{\text{ether}} \underset{\text{benzamide}}{C_6H_5-C(=O)-NH_2} + CH_3OH \qquad (10.24)$$

The reaction mechanism is very much like that of saponification. The unshared electron pair on the ammonia nitrogen initiates nucleophilic attack on the ester carbonyl group.

PROBLEM 10.20 Using eq. 10.22 as a guide, write out the steps in the mechanism for eq. 10.24. (Use NH_3 as the initial nucleophile instead of HO^-.)

10.16 Reaction of Esters with Grignard Reagents

Esters react with two equivalents of a Grignard reagent to give tertiary alcohols. The reaction proceeds by nucleophilic attack of the Grignard reagent on the ester carbonyl group. The initial product, a ketone, reacts further in the usual way to give the tertiary alcohol.

$$\underset{\text{ester}}{R-C(=O)-OR'} + 2\,R''MgBr \xrightarrow{\text{overall}} R-C(OMgBr)(R'')-R'' \xrightarrow[H^+]{H_2O} \underset{\text{tertiary alcohol}}{R-C(OH)(R'')-R''} \qquad (10.25)$$

$$\text{ester} \xrightarrow{R''MgBr} R-C(O-MgBr)(R'')-OR' \xrightarrow{-R'OMgBr} \underset{\text{ketone}}{R-C(=O)-R''} \xrightarrow{R''MgBr} R-C(OMgBr)(R'')-R''$$

This method is useful for making tertiary alcohols in which at least two of three alkyl groups attached to the hydroxyl-bearing carbon atom are identical.

PROBLEM 10.21 Using eq. 10.25 as a guide, write the structure of the tertiary alcohol that is obtained from

$$\text{(cyclopropyl)}-\overset{\overset{\displaystyle O}{\|}}{C}-OCH_3 + \text{excess } C_6H_5-MgBr$$

10.17 Reduction of Esters

Esters can be reduced to primary alcohols by lithium aluminum hydride.

$$\underset{\text{ester}}{R-\overset{\overset{\displaystyle O}{\|}}{C}-OR'} \xrightarrow[\text{ether}]{LiAlH_4} \underset{\text{primary alcohol}}{RCH_2OH} + R'OH \qquad (10.26)$$

The mechanism is similar to the hydride reduction of aldehydes and ketones (eq. 9.32).

$$\underset{\text{ester}}{R-\overset{\overset{\displaystyle O}{\|}}{C}-OR'} \xrightarrow{H-\bar{A}lH_3} R-\overset{\overset{\displaystyle O-\bar{A}lH_3}{|}}{\underset{\underset{\displaystyle H}{|}}{C}}-OR' \xrightarrow{-\bar{A}lH_3(OR')}$$

$$\underset{\text{aldehyde}}{R-\overset{\overset{\displaystyle O}{\|}}{C}-H} \xrightarrow{H-\bar{A}lH_2(OR')} R-\overset{\overset{\displaystyle O-\bar{A}lH_2(OR')}{|}}{\underset{\underset{\displaystyle H}{|}}{C}}-H \xrightarrow[H^+]{H_2O} \underset{1^\circ \text{ alcohol}}{RCH_2OH} + R'OH \qquad (10.27)$$

The intermediate aldehyde is not usually isolable and reacts rapidly with additional hydride to produce the alcohol.

It is possible to reduce an ester carbonyl group without reducing a C=C bond in the same molecule. For example,

$$\underset{\text{ethyl 2-butenoate}}{CH_3CH=CH\overset{\overset{\displaystyle O}{\|}}{C}-OCH_2CH_3} \xrightarrow[\text{2. } H_2O,\ H^+]{\text{1. } LiAlH_4} \underset{\text{2-buten-1-ol}}{CH_3CH=CHCH_2OH} + CH_3CH_2OH \qquad (10.28)$$

10.18 The Need for Activated Acyl Compounds

As we have seen, most reactions of carboxylic acids, esters, and related compounds involve, as the first step, nucleophilic attack on the carbonyl carbon atom. Examples are Fischer esterification, saponification and ammonolysis of esters, and the first stage of the reaction of esters with Grignard reagents or lithium aluminum hydride. All of these reactions can be summarized by a single mechanistic equation:

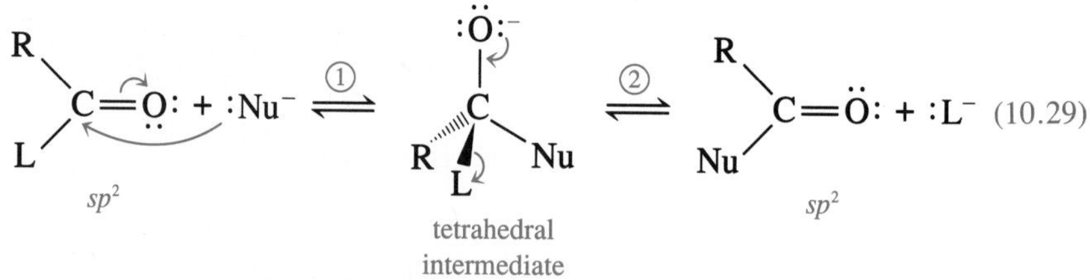

$$\underset{sp^2}{\text{R(L)C=}\ddot{\text{O}}\text{:}} + \text{:Nu}^- \overset{①}{\rightleftharpoons} \underset{\text{tetrahedral intermediate}}{\text{R(L)(Nu)C–}\ddot{\text{O}}\text{:}^-} \overset{②}{\rightleftharpoons} \underset{sp^2}{\text{R(Nu)C=}\ddot{\text{O}}\text{:}} + \text{:L}^- \quad (10.29)$$

The carbonyl carbon, initially trigonal, is attacked by a nucleophile, Nu:, to form a **tetrahedral intermediate** (step 1). Loss of a leaving group L (step 2) then regenerates the carbonyl group with its trigonal carbon atom. The net result is the replacement of L by Nu.

Biochemists look at eq. 10.29 in a slightly different way. They refer to the overall reaction as **acyl transfer.** The acyl group is transferred from L in the starting material to Nu in the product.

Regardless of how we consider the reaction, one important feature that can affect the rate of both steps is the nature of the leaving group. ***The rates of both steps in a nucleophilic acyl substitution reaction are enhanced by increasing the electron-withdrawing properties of the leaving group.*** Step 1 is favored because the more electronegative L is, the more positive the carbonyl carbon becomes, and therefore the more susceptible it is to nucleophilic attack. Step 2 is also facilitated because the more electronegative L is, the better leaving group it becomes.

In general, *esters are less reactive toward nucleophiles than are aldehydes or ketones* because the positive charge on the carbonyl carbon in esters can be delocalized to the oxygen atom.

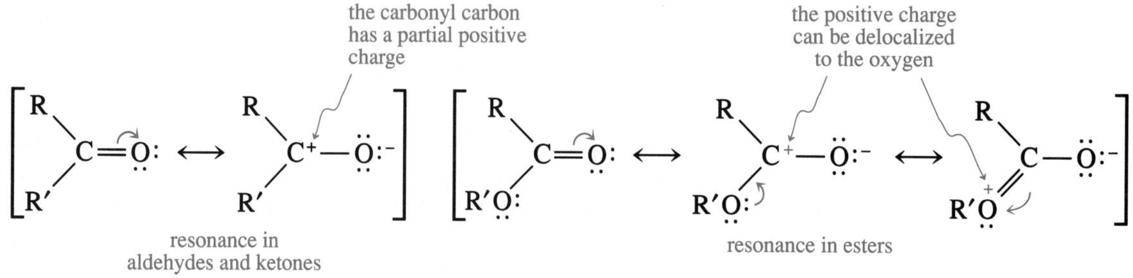

$$\left[\text{RR'C=}\ddot{\text{O}}\text{:} \longleftrightarrow \text{RR'C}^+\text{–}\ddot{\text{O}}\text{:}^-\right] \qquad \left[\text{R(R'}\ddot{\text{O}}\text{)C=}\ddot{\text{O}}\text{:} \longleftrightarrow \text{R(R'}\ddot{\text{O}}\text{)C}^+\text{–}\ddot{\text{O}}\text{:}^- \longleftrightarrow \text{R(R'}\overset{+}{\text{O}}\text{=)C–}\ddot{\text{O}}\text{:}^-\right]$$

the carbonyl carbon has a partial positive charge

the positive charge can be delocalized to the oxygen

resonance in aldehydes and ketones

resonance in esters

Consequently, the carbonyl carbon is less positive in esters than it is in aldehydes or ketones and therefore less susceptible to nucleophilic attack.

Now let us examine some of the ways in which the carboxyl group can be modified to *increase* its reactivity toward nucleophiles.

10.19 Acyl Halides

Acyl halides are among the most reactive of carboxylic acid derivatives. *Acyl chlorides* are more common and less expensive than bromides or iodides. They are usually prepared from acids by reaction with thionyl chloride or phosphorus pentachloride (compare with Sec. 7.11).

$$\text{R—C}(\text{=O})\text{—OH} + \underset{\text{thionyl chloride}}{\text{SOCl}_2} \longrightarrow \text{R—C}(\text{=O})\text{—Cl} + \text{HCl}\uparrow + \text{SO}_2\uparrow \quad (10.30)$$

$$\text{R—C}(\text{=O})\text{—OH} + \underset{\text{phosphorus pentachloride}}{\text{PCl}_5} \longrightarrow \text{R—C}(\text{=O})\text{—Cl} + \text{HCl} + \underset{\text{phosphorus oxychloride}}{\text{POCl}_3} \quad (10.31)$$

PROBLEM 10.22 Rewrite eq. 10.30 to show the preparation of benzoyl chloride.

Acyl halides react rapidly with most nucleophiles. For example, they are rapidly hydrolyzed by water.

$$\underset{\text{acetyl chloride}}{\text{CH}_3\text{—C}(\text{=O})\text{—Cl}} + \text{HOH} \xrightarrow{\text{rapid}} \underset{\text{acetic acid}}{\text{CH}_3\text{—C}(\text{=O})\text{—OH}} + \underset{\text{(fumes)}}{\text{HCl}} \quad (10.32)$$

For this reason, acyl halides have irritating odors. Benzoyl chloride, for example, is a lachrymator (tear gas).

PROBLEM 10.23 Explain why acyl halides may be irritating to the nose.

Acyl halides react rapidly with alcohols to form esters.

$$\underset{\text{benzoyl chloride}}{\text{C}_6\text{H}_5\text{—C}(\text{=O})\text{—Cl}} + \text{CH}_3\text{OH} \xrightarrow[\text{temp.}]{\text{room}} \underset{\text{methyl benzoate}}{\text{C}_6\text{H}_5\text{—C}(\text{=O})\text{—OCH}_3} + \text{HCl} \quad (10.33)$$

Indeed, the most common way to prepare an ester *in the laboratory* is to convert an acid to its acid chloride, then react the latter with an alcohol. Even though two steps are necessary (compared with one step for Fischer esterification), the method may be preferable, especially if either the acid or the alcohol is expensive. (Recall that Fischer esterification is an equilibrium reaction and must often be carried out with a large excess of one of the reactants.)

Acyl halides react rapidly with ammonia to form amides.

$$\underset{\text{acetyl chloride}}{\text{CH}_3\text{C}(\text{=O})\text{—Cl}} + 2\,\text{NH}_3 \longrightarrow \underset{\text{acetamide}}{\text{CH}_3\text{C}(\text{=O})\text{—NH}_2} + \text{NH}_4^+\text{Cl}^- \quad (10.34)$$

The reaction is much more rapid than the ammonolysis of esters. Two equivalents of ammonia are required, however—one to form the amide and one to neutralize the hydrogen chloride.

EXAMPLE 10.8 Explain why acyl chlorides are more reactive than esters toward nucleophiles.

Solution The electronegativity order is Cl > OR. Therefore, the carbonyl carbon is more positive in acyl halides than in esters and more reactive toward nucleophiles. Also, Cl^- is a better leaving group (weaker nucleophile) than RO^-.

Acyl halides are used to synthesize aromatic ketones, through Friedel-Crafts acylation of aromatic rings (review Sec. 4.10d).

PROBLEM 10.24 Devise a synthesis of 4-methylphenyl propyl ketone from toluene and butanoic acid as starting materials.

10.20 Acid Anhydrides

Acid anhydrides are derived from acids by removing water from two carboxyl groups and connecting the fragments.

$$\underset{\text{two acid molecules}}{R-\overset{\overset{\displaystyle O}{\|}}{C}-OH \quad HO-\overset{\overset{\displaystyle O}{\|}}{C}-R} \qquad \underset{\text{an acid anhydride}}{R-\overset{\overset{\displaystyle O}{\|}}{C}-O-\overset{\overset{\displaystyle O}{\|}}{C}-R}$$

The most important commercial aliphatic anhydride is **acetic anhydride ($R = CH_3$).** About 1 million tons are manufactured annually, mainly to react with alcohols to form acetates. The two most common uses are in making cellulose acetate (rayon) and aspirin.

The name of an anhydride is obtained by naming the acid from which it is derived and replacing the word *acid* with *anhydride.*

$$\underset{\text{ethanoic anhydride or acetic anhydride}}{CH_3-\overset{\overset{\displaystyle O}{\|}}{C}-O-\overset{\overset{\displaystyle O}{\|}}{C}-CH_3}$$

PROBLEM 10.25 Write the structural formula for

a. butanoic anhydride.
b. benzoic anhydride.

Anhydrides are prepared by dehydration of acids. Dicarboxylic acids with appropriately spaced carboxyl groups lose water on heating to form cyclic anhydrides with five- and six-membered rings. For example,

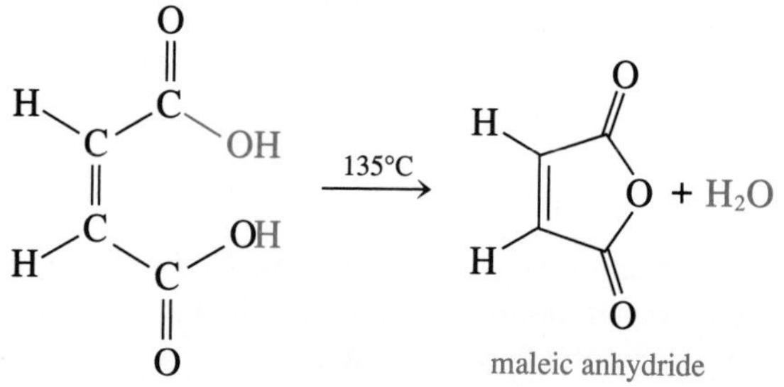

$$\underset{\text{maleic acid}}{HC(COOH){=}CH(COOH)} \xrightarrow{135°C} \underset{\text{maleic anhydride}}{C_4H_2O_3} + H_2O \qquad (10.35)$$

PROBLEM 10.26 Predict and name the product of the following reaction:

$$\text{C}_6\text{H}_4(\text{COOH})_2 \text{ (ortho)} \xrightarrow{\text{heat}}$$

PROBLEM 10.27 Do you expect fumaric acid (page 292) to form a cyclic anhydride on heating? Explain.

Anhydrides can also be prepared from acid chlorides and carboxylate salts in a reaction that occurs by a nucleophilic acyl substitution mechanism. This is a good method for preparing anhydrides derived from two different carboxylic acids, called **mixed anhydrides.**

$$CH_3CH_2CH_2{-}\overset{\overset{\displaystyle O}{\|}}{C}{-}Cl + Na^+\,{}^-O{-}\overset{\overset{\displaystyle O}{\|}}{C}{-}CH_3 \longrightarrow \underset{\text{butanoic ethanoic anhydride}}{CH_3CH_2CH_2{-}\overset{\overset{\displaystyle O}{\|}}{C}{-}O{-}\overset{\overset{\displaystyle O}{\|}}{C}{-}CH_3} + NaCl \qquad (10.36)$$

Anhydrides undergo nucleophilic acyl substitution reactions. They are more reactive than esters, but less reactive than acyl halides, toward nucleophiles. Some typical reactions of acetic anhydride are the following:

$$\underset{\substack{\text{acetic anhydride}\\ \text{bp 139.5°C}}}{CH_3{-}\overset{\overset{\displaystyle O}{\|}}{C}{-}O{-}\overset{\overset{\displaystyle O}{\|}}{C}{-}CH_3} \begin{cases} \xrightarrow{HO-H} \underset{\text{acid}}{CH_3\overset{\overset{\displaystyle O}{\|}}{C}{-}OH} + CH_3\overset{\overset{\displaystyle O}{\|}}{C}{-}OH \\ \xrightarrow{RO-H} \underset{\text{ester}}{CH_3\overset{\overset{\displaystyle O}{\|}}{C}{-}OR} + CH_3\overset{\overset{\displaystyle O}{\|}}{C}{-}OH \\ \xrightarrow{NH_2-H} \underset{\text{amide}}{CH_3\overset{\overset{\displaystyle O}{\|}}{C}{-}NH_2} + CH_3\overset{\overset{\displaystyle O}{\|}}{C}{-}OH \end{cases} \qquad (10.37)$$

Water hydrolyzes an anhydride to the corresponding acid. Alcohols give esters, and ammonia gives amides. In each case, one equivalent of acid is also produced.

PROBLEM 10.28 Write an equation for the reaction of acetic anhydride with 1-butanol.

PROBLEM 10.29 Write equations for the reactions of maleic anhydride with

a. water. b. 1-butanol. c. ammonia.

The reaction of acetic anhydride with **salicylic acid** (*o*-hydroxybenzoic acid) is used to synthesize **aspirin.** In this reaction, the phenolic hydroxyl group is **acetylated** (converted to its acetate ester).

Annual aspirin production in the United States is more than 50 million pounds, enough to produce over 50 billion standard 5-grain tablets. Aspirin is widely used, either by itself or mixed with other drugs, as an analgesic and antipyretic. It is not without dangers, however. Repeated use may cause gastrointestinal bleeding, and a large single dose (10 to 20g) can cause death.

$$C_6H_4(OH)(CO_2H) + CH_3\overset{O}{\overset{\|}{C}}-O-\overset{O}{\overset{\|}{C}}CH_3 \longrightarrow C_6H_4(O\overset{O}{\overset{\|}{C}}CH_3)(CO_2H) + CH_3CO_2H \quad (10.38)$$

salicylic acid; acetic anhydride; acetylsalicylic acid (aspirin)

PROBLEM 10.30 Methyl salicylate is the chief component of oil of wintergreen. It is used to flavor gum and candy and in rubbing liniments, where its mild irritating action on skin provides a counterirritant for sore muscles. Write an equation to show how methyl salicylate can be prepared from salicylic acid and methanol.

A Word About . . .

20. Thioesters, Nature's Acyl-Activating Groups

Acyl transfer plays an important role in many biochemical processes. However, acyl halides and anhydrides are far too corrosive to be cell constituents—they are hydrolyzed quite rapidly by water and are therefore incompatible with cellular fluid. Most ordinary esters, on the other hand, react too slowly with nucleophiles for acyl transfer to be carried out efficiently at body temperatures. Consequently, other groups have evolved to activate acyl groups in the cell. The most important of these is **coenzyme A** (the A stands for acetylation, one of the functions of this enzyme). Coenzyme A is a complex *thiol* (Figure 10.1). It is usually abbreviated by the symbol **CoA—SH**. Though its structure is made up of three parts—adenosine diphosphate (ADP), pantothenic acid (a vitamin), and 2-aminoethanethiol—it is the thiol group that gives coenzyme A its most important functions.

Coenzyme A can be converted to **thioesters,** which are the active acyl-transfer agents in the cell. Of the thioesters that coenzyme A forms, the acetyl ester, called **acetyl-coenzyme A** and abbreviated as

$$CH_3\overset{O}{\overset{\|}{C}}-S-CoA$$

is the most important. Acetyl-CoA reacts with many nucleophiles to transfer the acetyl group.

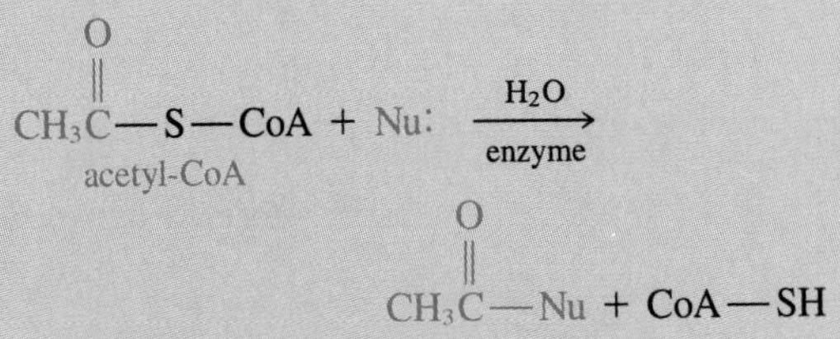

$$\mathrm{CH_3\overset{\overset{\large O}{\|}}{C}{-}S{-}CoA} + \mathrm{Nu{:}} \xrightarrow[\text{enzyme}]{H_2O} \mathrm{CH_3\overset{\overset{\large O}{\|}}{C}{-}Nu} + \mathrm{CoA{-}SH}$$

acetyl-CoA

The reactions are usually enzyme-mediated and occur rapidly at ordinary cell temperatures.

Why are thioesters superior to ordinary esters as acyl-transfer agents? Part of the answer lies in the acidity difference between alcohols and thiols (Sec. 7.17). Since thiols are approximately 1 million times stronger acids than are alcohols, their conjugate bases, $^-$SR, are a million times weaker bases than $^-$OR. Thus the —SR group of thioesters is a much better leaving group, in nucleophilic substitutions, than is the —OR group of ordinary esters. Thioesters are not so reactive that they hydrolyze in cellular fluid, but they are appreciably more reactive than simple esters. Nature makes use of this fact.

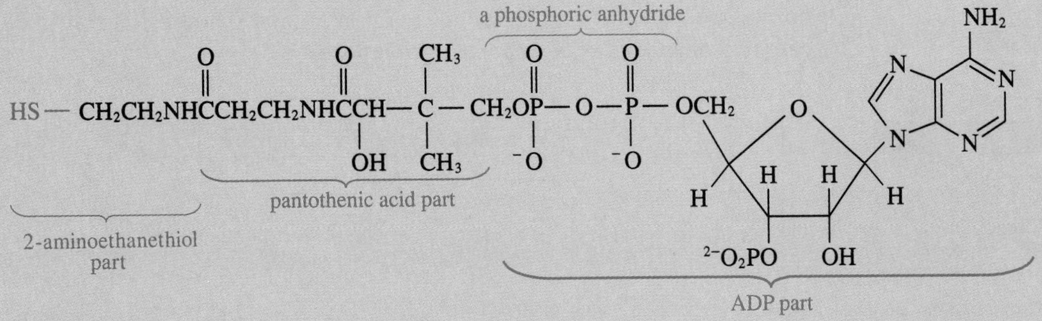

HS— $\mathrm{CH_2CH_2NH\overset{O}{\overset{\|}{C}}CH_2CH_2NH\overset{O}{\overset{\|}{C}}CH(OH){-}C(CH_3)_2{-}CH_2O\overset{O}{\overset{\|}{P}}(O^-){-}O{-}\overset{O}{\overset{\|}{P}}(O^-){-}OCH_2}$ (ribose ring with H, H, H, H; $^{2-}O_3PO$, OH; adenine with NH_2)

2-aminoethanethiol part; pantothenic acid part; a phosphoric anhydride; ADP part

FIGURE 10.1 Coenzyme A.

10.21 *Amides*

Amides are the least reactive of the common carboxylic acid derivatives. They occur widely in nature. The most important amides are the proteins, whose chemistry we will discuss in Chapter 17. Here we will concentrate on just a few properties of simple amides.

Primary amides have the general formula $RCONH_2$. They can be prepared by the reaction of ammonia with esters (eq. 10.23), with acyl halides (eq. 10.34), or with acid anhydrides (eq. 10.36). Amides can also be prepared by heating the ammonium salts of acids.

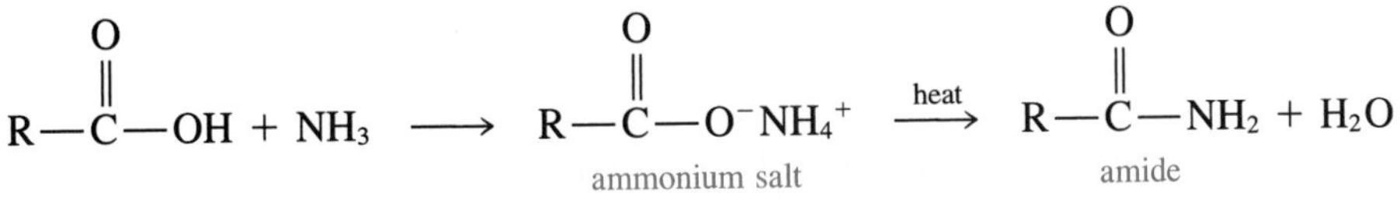

$$\mathrm{R{-}\overset{\overset{\large O}{\|}}{C}{-}OH} + \mathrm{NH_3} \longrightarrow \underset{\text{ammonium salt}}{\mathrm{R{-}\overset{\overset{\large O}{\|}}{C}{-}O^-NH_4^+}} \xrightarrow{\text{heat}} \underset{\text{amide}}{\mathrm{R{-}\overset{\overset{\large O}{\|}}{C}{-}NH_2}} + \mathrm{H_2O} \qquad (10.39)$$

Amides are named by replacing the *-ic* or *-oic* ending of the acid name, either the common or the IUPAC name, with the *-amide* ending.

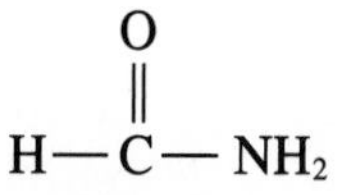

$$\underset{\text{formamide (methanamide)}}{H-\overset{O}{\overset{\|}{C}}-NH_2} \quad \underset{\text{acetamide (ethanamide)}}{CH_3-\overset{O}{\overset{\|}{C}}-NH_2} \quad \underset{\text{butanamide}}{CH_3CH_2CH_2\overset{O}{\overset{\|}{C}}-NH_2} \quad \underset{\text{benzamide (benzenecarboxamide)}}{C_6H_5-\overset{O}{\overset{\|}{C}}-NH_2}$$

formamide (methanamide) acetamide (ethanamide) butanamide 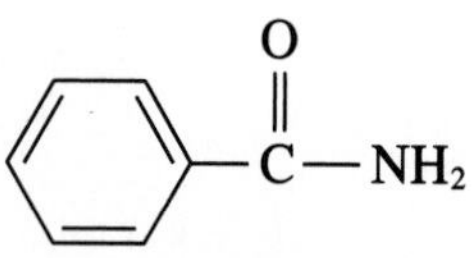 benzamide (benzenecarboxamide)

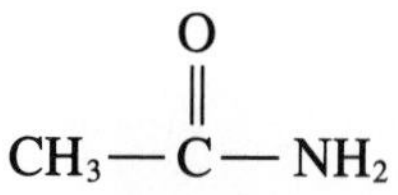

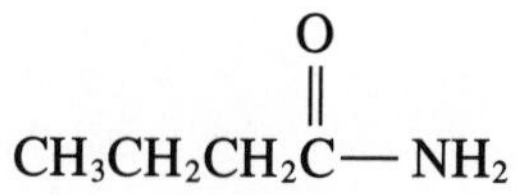

PROBLEM 10.31 a. Name $(CH_3)_2CHCONH_2$.
b. Write the structure of 1-methylcyclopropancarboxamide.

The above examples are all primary amides. Secondary and tertiary amides, in which one or both of the hydrogens on the nitrogen atom are replaced by organic groups, are described in the next chapter.

Amides have a planar geometry. Even though the carbon–nitrogen bond is normally written as a single bond, rotation around that bond is restricted, because resonance is very important in amides.

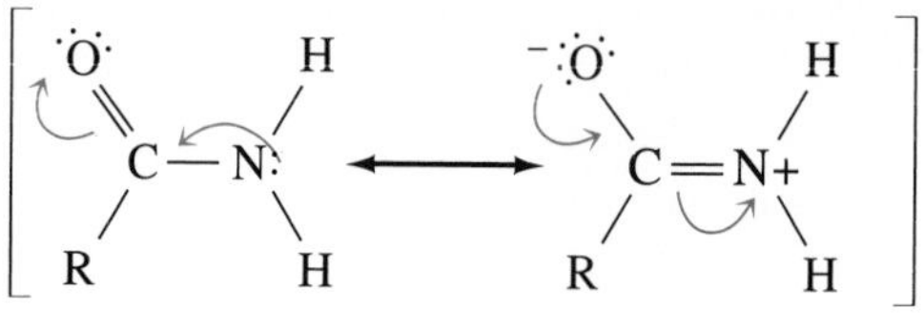

$$\left[R-\overset{\ddot{O}:}{\overset{\|}{C}}-\ddot{N}H_2 \longleftrightarrow R-\overset{^-:\ddot{O}:}{\overset{|}{C}}=N^+H_2 \right]$$

amide resonance

The dipolar contributor is so important that the carbon-nitrogen bond behaves much like a double bond. Consequently, the nitrogen and the carbonyl carbon, and the two atoms attached to each of them, lie in the same plane, and rotation at the C—N bond is restricted. Indeed, the C—N bond in amides is only 1.32 Å long—much shorter than the usual carbon-nitrogen single bond (which is about 1.47 Å).

As the dipolar resonance contributor suggests, amides are highly polar and form strong hydrogen bonds.

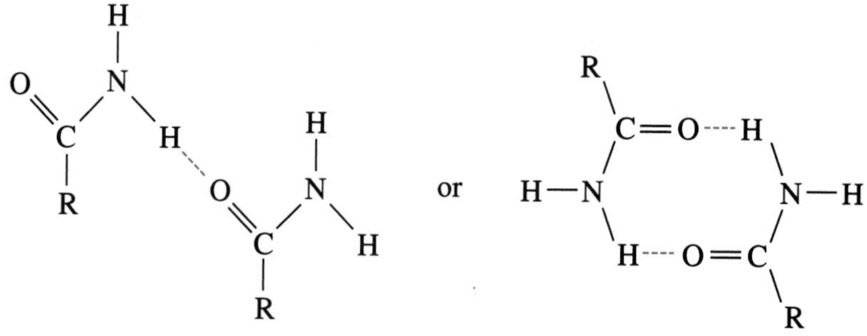

$$R-\overset{O}{\overset{\|}{C}}-N(H)-H\cdots O=\overset{|}{C}(R)-NH_2 \quad \text{or} \quad \text{cyclic dimer: } R-C(=O\cdots H-N(H)-)\ \ (-N(H)-H\cdots O=)C-R$$

They have exceptionally high boiling points for their molecular weights, although alkyl substitution on the nitrogen lowers the boiling and melting points by decreasing the hydrogen-bonding possibilities.

	$H-\overset{\overset{O}{\|\|}}{C}-NH_2$	$H-\overset{\overset{O}{\|\|}}{C}-N(CH_3)_2$	$CH_3\overset{\overset{O}{\|\|}}{C}-NH_2$	$CH_3\overset{\overset{O}{\|\|}}{C}-N(CH_3)_2$
	formamide	*N,N*-dimethylformamide	acetamide	*N,N*-dimethylacetamide
bp	210°C	153°C	222°C	165°C
mp	2.5°C	−60.5°C	81°C	−20°C

PROBLEM 10.32 Show that hydrogen bonding is possible for acetamide, but not for *N,N*-dimethylacetamide.

Like other acid derivatives, amides react with nucleophiles. For example, they can be hydrolyzed by water.

$$\underset{\text{amide}}{R-\overset{\overset{O}{\|}}{C}-NH_2} + H-OH \xrightarrow[OH^-]{H^+ \text{ or}} \underset{\text{acid}}{R-\overset{\overset{O}{\|}}{C}-OH} + NH_3 \qquad (10.40)$$

The reactions are slow, however. Prolonged heating and acid or base catalysis are usually necessary.

PROBLEM 10.33 Using eq. 10.40 as a model, write an equation for the hydrolysis of acetamide.

Amides can be reduced by lithium aluminum hydride to give amines.

$$\underset{\text{amide}}{R-\overset{\overset{O}{\|}}{C}-NH_2} \xrightarrow[\text{ether}]{LiAlH_4} \underset{\text{amine}}{RCH_2NH_2} \qquad (10.41)$$

This is an excellent way to make primary amines, whose chemistry is discussed in the next chapter.

PROBLEM 10.34 Using eq. 10.41 as a model, write an equation for the reduction of acetamide with $LiAlH_4$.

PROBLEM 10.35 Outline steps for the synthesis of benzylamine

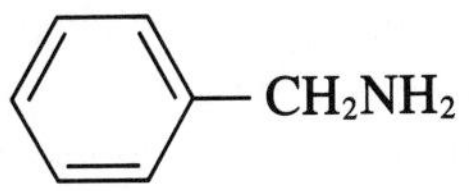

$$C_6H_5-CH_2NH_2$$

starting with benzoic acid.

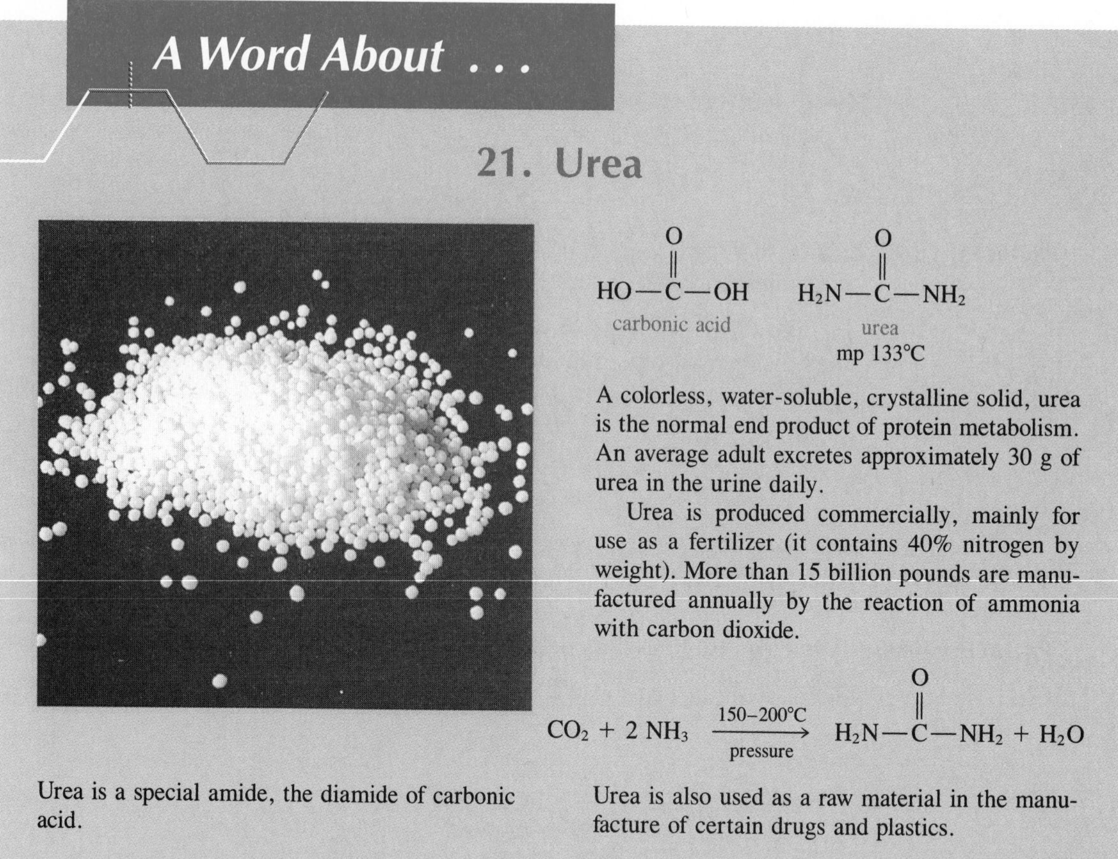

A Word About . . .

21. Urea

$$HO-\overset{\overset{O}{\|}}{C}-OH \qquad H_2N-\overset{\overset{O}{\|}}{C}-NH_2$$

carbonic acid — urea, mp 133°C

A colorless, water-soluble, crystalline solid, urea is the normal end product of protein metabolism. An average adult excretes approximately 30 g of urea in the urine daily.

Urea is produced commercially, mainly for use as a fertilizer (it contains 40% nitrogen by weight). More than 15 billion pounds are manufactured annually by the reaction of ammonia with carbon dioxide.

$$CO_2 + 2\,NH_3 \xrightarrow[\text{pressure}]{150\text{–}200^\circ C} H_2N-\overset{\overset{O}{\|}}{C}-NH_2 + H_2O$$

Urea is a special amide, the diamide of carbonic acid.

Urea is also used as a raw material in the manufacture of certain drugs and plastics.

10.22 *A Summary of Carboxylic Acid Derivatives*

We have studied a rather large number of reactions in this chapter. However, most of them can be summarized in a single chart, shown in Table 10.5.

The four types of acid derivatives are listed at the left of the chart in order of decreasing reactivity toward nucleophiles. Three common nucleophiles are listed across the top. Note that the main organic product in each column is the same, regardless of which type of acid derivative we start with. For example, **hydrolysis** gives the corresponding organic acid, whether we start with an acyl halide, acid anhydride, ester, or amide. Similarly, **alcoholysis** gives an ester, and **ammonolysis** gives an amide. Note also that the *other* reaction product is generally the same from a given acid derivative (horizontally across the table), regardless of the nucleophile. For example, starting with an ester, RCO_2R', we obtain as the second product the alcohol $R''OH$, regardless of whether the reaction type is hydrolysis, alcoholysis, or ammonolysis.

All of the reactions in Table 10.5 take place via attack of the nucleophile on the carbonyl carbon of the acid derivative, as described in eq. 10.29. Indeed, most of the reactions from Secs. 10.11 through 10.20 occur by that same mechanism. We can sometimes use this idea to predict new reactions.

TABLE 10.5 Reactions of acid derivatives with certain nucleophiles

Acid derivative (decreasing reactivity ↓)	*Nucleophile* HOH (hydrolysis)	R′OH (alcoholysis)	NH_3 (ammonolysis)
$R{-}\overset{O}{\overset{\Vert}{C}}{-}Cl$ acyl halide	$R{-}\overset{O}{\overset{\Vert}{C}}{-}OH + HCl$	$R{-}\overset{O}{\overset{\Vert}{C}}{-}OR' + HCl$	$R{-}\overset{O}{\overset{\Vert}{C}}{-}NH_2 + NH_4^+Cl^-$
$R{-}\overset{O}{\overset{\Vert}{C}}{-}O{-}\overset{O}{\overset{\Vert}{C}}{-}R$ acid anhydride	$2\ R{-}\overset{O}{\overset{\Vert}{C}}{-}OH$	$R{-}\overset{O}{\overset{\Vert}{C}}{-}OR' + RCO_2H$	$R{-}\overset{O}{\overset{\Vert}{C}}{-}NH_2 + RCO_2H$
$R{-}\overset{O}{\overset{\Vert}{C}}{-}O{-}R''$ ester	$R{-}\overset{O}{\overset{\Vert}{C}}{-}OH + R''OH$	$R{-}\overset{O}{\overset{\Vert}{C}}{-}OR' + R''OH$ (ester interchange)	$R{-}\overset{O}{\overset{\Vert}{C}}{-}NH_2 + R''OH$
$R{-}\overset{O}{\overset{\Vert}{C}}{-}NH_2$ amide	$R{-}\overset{O}{\overset{\Vert}{C}}{-}OH + NH_3$	—	—
Main organic product	acid	ester	amide

For example, the reaction of esters with Grignard reagents (eq. 10.25) involves nucleophilic attack of the Grignard reagent on the ester carbonyl group. Keeping in mind that all acid derivatives are susceptible to nucleophilic attack, it is understandable that acyl halides also react with Grignard reagents to give teriary alcohols. The first steps involve ketone formation as follows:

$$R{-}\overset{O}{\overset{\Vert}{C}}{-}Cl + R'MgX \longrightarrow R{-}\overset{O^-\overset{+}{MgX}}{\overset{|}{\underset{\underset{R'}{|}}{C}}}{-}Cl \longrightarrow R{-}\overset{O}{\overset{\Vert}{C}}{-}R' + MgXCl \qquad (10.42)$$

The ketone can sometimes be isolated, but usually it reacts with a second mole of Grignard reagent to give a tertiary alcohol.

$$R{-}\overset{O}{\overset{\Vert}{C}}{-}R' + R'MgX \longrightarrow R{-}\overset{O^-\overset{+}{MgX}}{\overset{|}{\underset{\underset{R'}{|}}{C}}}{-}R' \xrightarrow{H_3O^+} R{-}\overset{OH}{\overset{|}{\underset{\underset{R'}{|}}{C}}}{-}R' \qquad (10.43)$$

PROBLEM 10.36 Modified metal hydrides, such as lithium tri-*t*-butoxyaluminum hydride, $Li^{+}\,^{-}Al[OC(CH_3)_3]_3H$, react with acyl chlorides to give aldehydes. Using eq. 10.27 as a guide, write an equation for the reaction of benzoyl chloride with this reagent, and suggest a possible reaction mechanism (simply treat lithium tri-*t*-butoxy aluminum hydride as a source of hydride, H^-).

10.23 The α-Hydrogen of Esters; the Claisen Condensation

In this final section, we describe an important reaction of esters that resembles the aldol condensation of aldehydes and ketones (Sec. 9.18). It makes use of the **α-hydrogen** of an ester.

Being adjacent to a carbonyl group, **the α-hydrogens of an ester are weakly acidic ($pK_a \sim 23$) and can be removed by a strong base. The product is an ester enolate.**

$$\text{H}-\overset{|}{\underset{|}{\text{C}}}{}_{\alpha}-\text{C}(=\text{O})\text{OR} \xrightleftharpoons{\text{base}} \left[-\overset{\ominus}{\ddot{\text{C}}}-\text{C}(=\ddot{\text{O}}\text{:})\text{OR} \longleftrightarrow -\text{C}=\text{C}(-\ddot{\text{O}}\text{:}^{\ominus})\text{OR} \right] \quad (10.44)$$

resonance contributors to an ester enolate

Common bases used for this purpose are sodium alkoxides or sodium hydride. The ester enolate, once formed, can act as a carbon nucleophile and add to the carbonyl group of another ester molecule. This reaction is called the **Claisen condensation.** It is a way of making **β-keto esters.** We can use ethyl acetate as an example to see how the reaction works.

Treatment of ethyl acetate with sodium ethoxide in ethanol produces the β-keto ester **ethyl acetoacetate:**

$$\underset{\text{ethyl acetate}}{CH_3\overset{O}{\overset{\|}{C}}-OCH_2CH_3} + \underset{\text{ethyl acetate}}{H-\overset{\alpha}{CH_2}-\overset{O}{\overset{\|}{C}}-OCH_2CH_3} \xrightarrow[\text{in ethanol}\\ \text{2. } H_3O^+]{\text{1. } NaOCH_2CH_3}$$

$$\underset{\substack{\text{ethyl acetoacetate}\\ \text{(ethyl 3-oxobutanoate)}}}{CH_3\overset{O}{\overset{\|}{C}}-CH_2-\overset{O}{\overset{\|}{C}}-OCH_2CH_3} + CH_3CH_2OH \quad (10.45)$$

The first part of the Claisen condensation takes place in three steps.

Step 1. $$CH_3\overset{O}{\overset{\|}{C}}-OCH_2CH_3 + \underset{\text{sodium ethoxide}}{Na^{+}\,^{-}OCH_2CH_3} \rightleftharpoons$$

$$\underset{\text{ester enolate}}{Na^{+}\,^{-}CH_2\overset{O}{\overset{\|}{C}}OCH_2CH_3} + CH_3CH_2OH \quad (10.46)$$

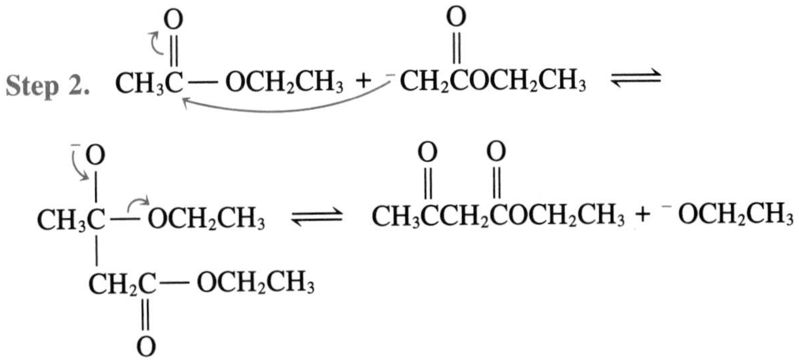

Step 2. $CH_3\overset{O}{\overset{\|}{C}}-OCH_2CH_3 + {}^{-}CH_2\overset{O}{\overset{\|}{C}}OCH_2CH_3 \rightleftharpoons$

$$CH_3\underset{CH_2\overset{\|}{\underset{O}{C}}-OCH_2CH_3}{\overset{{}^{-}O}{\overset{|}{C}}}-OCH_2CH_3 \rightleftharpoons CH_3\overset{O}{\overset{\|}{C}}CH_2\overset{O}{\overset{\|}{C}}OCH_2CH_3 + {}^{-}OCH_2CH_3 \quad (10.47)$$

Step 3. $CH_3\overset{O}{\overset{\|}{C}}CH_2\overset{O}{\overset{\|}{C}}OCH_2CH_3 + {}^{-}OCH_2CH_3 \longrightarrow$

$$CH_3\overset{O}{\overset{\|}{C}}\cdots\overset{\ominus}{C}H\cdots\overset{O}{\overset{\|}{C}}OCH_2CH_3 + CH_3CH_2OH \quad (10.48)$$

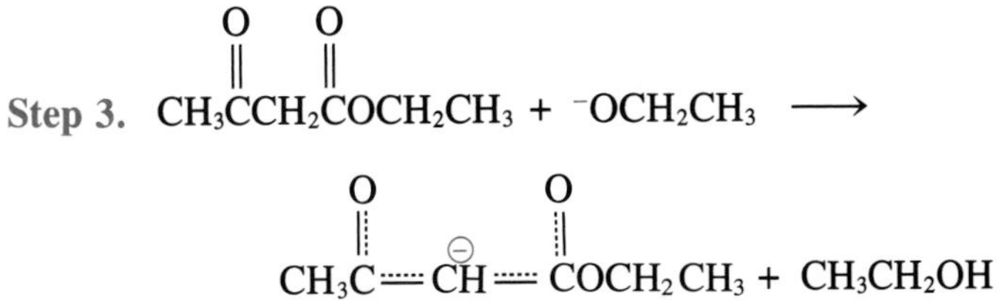

enolate ion of a β-keto ester

In Step 1, the base (sodium ethoxide) removes an α-hydrogen from the ester to form an ester enolate. In Step 2, this ester enolate, acting as a nucleophile, adds to the carbonyl group of a second ester molecule, displacing ethoxide ion. This step follows the mechanism in eq. 10.29 and proceeds through a tetrahedral intermediate. These first two steps of the reaction are completely reversible.

Step 3 drives the equilibria forward. In this step, the β-keto ester is converted to its enolate anion. The methylene (CH_2) hydrogens in ethyl acetoacetate are *α to two carbonyl groups* and hence are appreciably more acidic than ordinary α-hydrogens. They have a pK_a of 12 and are easily removed by the base (ethoxide ion) to form a resonance-stabilized β-keto enolate ion, *with the negative charge delocalized to both carbonyl oxygen atoms.*

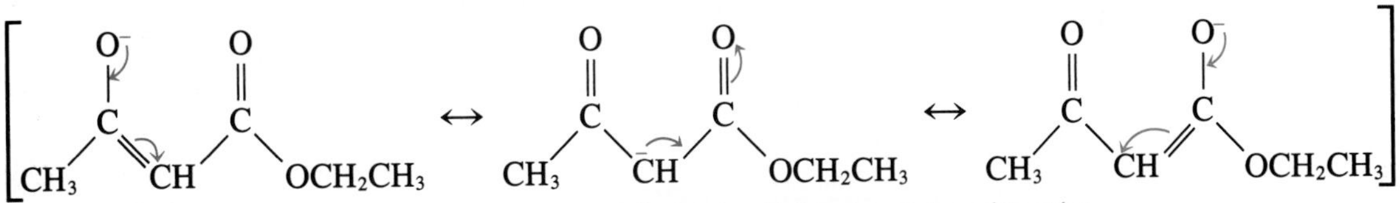

$$\left[CH_3-\overset{O^-}{\overset{|}{C}}=CH-\overset{O}{\overset{\|}{C}}-OCH_2CH_3 \leftrightarrow CH_3-\overset{O}{\overset{\|}{C}}-\overset{-}{C}H-\overset{O}{\overset{\|}{C}}-OCH_2CH_3 \leftrightarrow CH_3-\overset{O}{\overset{\|}{C}}-CH=\overset{O^-}{\overset{|}{C}}-OCH_2CH_3 \right]$$

resonance contributors to ethyl acetoacetate enolate anion

To complete the Claisen condensation, the solution is acidified, to regenerate the β-keto ester from its enolate anion.

EXAMPLE 10.9 Identify the product of the Claisen condensation of ethyl propanoate:

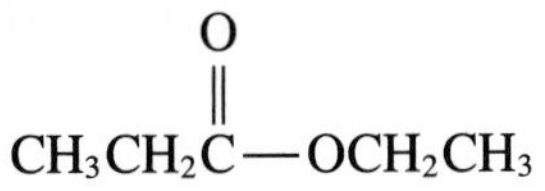

$$CH_3CH_2\overset{O}{\overset{\|}{C}}-OCH_2CH_3$$

Solution The product is

$$CH_3CH_2\overset{\overset{O}{\|}\beta}{C}-\overset{\alpha}{\underset{\underset{CH_3}{|}}{CH}}-\overset{\overset{O}{\|}}{C}OCH_2CH_3$$

The α-carbon of one ester molecule displaces the —OR group and becomes joined to the carbonyl carbon of the other ester. The product is always a β-keto ester.

PROBLEM 10.37 Using eqs. 10.46 through 10.48 as a model, write out the steps in the mechanism for the Claisen condensation of ethyl propanoate.

The Claisen condensation, like the aldol condensation, is useful for making new carbon-carbon bonds. The resulting β-keto esters can be converted to a variety of useful products. For example, ethyl acetate can be converted to ethyl butanoate by the following sequence.

$$\underset{\text{ethyl acetate}}{2\ CH_3\overset{\overset{O}{\|}}{C}-OCH_2CH_3} \xrightarrow[NaOCH_2CH_3]{\text{Claisen}} \underset{\text{ethyl acetoacetate}}{CH_3\overset{\overset{O}{\|}}{C}CH_2\overset{\overset{O}{\|}}{C}OCH_2CH_3} \xrightarrow{NaBH_4} \underset{\text{ethyl 3-hydroxybutanoate}}{CH_3\overset{\overset{OH}{|}}{C}HCH_2\overset{\overset{O}{\|}}{C}OCH_2CH_3} \xrightarrow[-H_2O]{H^+}$$

$$\underset{\text{ethyl 2-butenoate}}{CH_3CH=CH\overset{\overset{O}{\|}}{C}OCH_2CH_3} \xrightarrow[Pt]{H_2} \underset{\text{ethyl butanoate}}{CH_3CH_2CH_2\overset{\overset{O}{\|}}{C}OCH_2CH_3} \quad (10.49)$$

In this way, the acetate chain is lengthened by two carbon atoms. Nature makes use of a similar process, catalyzed by various enzymes, to construct the long-chain carboxylic acids that are components of fats and oils (Chapter 15).

REACTION SUMMARY

1. Preparation of Acids

a. From Alcohols or Aldehydes (Sec. 10.8)

$$RCH_2OH \xrightarrow{CrO_3,\ H_2SO_4,\ H_2O} RCO_2H \xleftarrow[\text{or } O_2 \text{ or } Ag^+]{CrO_3,\ H_2SO_4,\ H_2O} RCH=O$$

b. From Alkyl Benzenes (Sec. 10.8)

$$ArCH_3 \xrightarrow[\text{or } O_2,\ Co^{+3}]{KMnO_4} ArCO_2H$$

c. From Grignard Reagents (Sec. 10.8)

$$RMgX + CO_2 \longrightarrow RCO_2MgX \xrightarrow{H_3O^+} RCO_2H$$

d. From Nitriles (Sec. 10.8)

$$RC\equiv N + 2\ H_2O \xrightarrow{H^+ \text{ or } HO^-} RCO_2H + NH_3$$

2. Reactions of Acids

a. Acid–Base (Sec. 10.7)

$$RCO_2H \rightleftharpoons RCO_2^- + H^+ \quad \text{(ionization)}$$

$$RCO_2H + NaOH \longrightarrow RCO_2^-Na^+ + H_2O \quad \text{(salt formation)}$$

b. Preparation of Esters (Secs. 10.11 and 10.13)

$$RCO_2H + R'OH \xrightarrow{H^+} RCO_2R' + H_2O$$

c. Preparation of Acid Chlorides (Sec. 10.19)

$$RCO_2H + SOCl_2 \longrightarrow RCOCl + HCl + SO_2$$

$$RCO_2H + PCl_5 \longrightarrow RCOCl + HCl + POCl_3$$

d. Preparation of Anhydrides (Sec. 10.20)

$$R{-}\overset{\overset{\displaystyle O}{\|}}{C}{-}Cl + Na^{+}\,{}^{-}O{-}\overset{\overset{\displaystyle O}{\|}}{C}{-}R' \longrightarrow R{-}\overset{\overset{\displaystyle O}{\|}}{C}{-}O{-}\overset{\overset{\displaystyle O}{\|}}{C}{-}R' + NaCl$$

e. Preparation of Amides (Secs. 10.20–10.22)

$$RCO_2^-NH_4^+ \xrightarrow{\text{heat}} RCONH_2 + H_2O$$

Also see reactions of esters, acid chlorides, and anhydrides.

3. Reactions of Carboxylic Acid Derivatives

a. Saponification of Esters (Sec. 10.14)

$$RCO_2R' + NaOH \longrightarrow RCO_2^-Na^+ + R'OH$$

b. Ammonolysis of Esters (Sec. 10.15)

$$RCO_2R' + NH_3 \longrightarrow RCONH_2 + R'OH$$

c. Esters with Grignard Reagents (Sec. 10.16)

$$RCO_2R' \xrightarrow{R''MgX} R{-}\underset{\underset{\displaystyle R''}{|}}{\overset{\overset{\displaystyle R''}{|}}{C}}{-}OMgX \xrightarrow{H_3O^+} R{-}\underset{\underset{\displaystyle R''}{|}}{\overset{\overset{\displaystyle R''}{|}}{C}}{-}OH$$

$$+ R'OH$$

d. Reduction of Esters (Sec. 10.17)

$$RCO_2R' + LiAlH_4 \longrightarrow RCH_2OH + R'OH$$

e. Nucleophilic Acyl Substitution Reactions of Acid Chlorides and Anhydrides (Secs. 10.20 and 10.21)

$$R{-}\overset{\overset{\displaystyle O}{\|}}{C}{-}Cl \text{ or } R{-}\overset{\overset{\displaystyle O}{\|}}{C}{-}O{-}\overset{\overset{\displaystyle O}{\|}}{C}{-}R \begin{cases} \xrightarrow{H_2O} RCO_2H + HCl \text{ (or } RCO_2H) \\ \xrightarrow{R'OH} RCO_2R' + HCl \text{ (or } RCO_2H) \\ \xrightarrow{NH_3} RCONH_2 + NH_4Cl \text{ (or } RCO_2H) \end{cases}$$

f. Hydrolysis of Amides (Sec. 10.22)

$$RCONH_2 + H_2O \xrightarrow{H^+ \text{ or } HO^-} RCO_2H + NH_3$$

g. Reduction of Amides (Sec. 10.22)

$$RCONH_2 \xrightarrow{LiAlH_4} RCH_2NH_2$$

h. Claisen Condensation (Sec. 10.23)

$$2\ RCH_2CO_2R' \xrightarrow[2.\ H_3O^+]{1.\ R'O^-Na^+} RCH_2\overset{\overset{O}{\|}}{C}\underset{\underset{R}{|}}{C}HCO_2R' + R'OH$$

ADDITIONAL PROBLEMS

10.38. Write a structural formula for each of the following acids:

a. 3-methylpentanoic acid
b. 2,2-dichlorobutanoic acid
c. 4-hydroxyhexanoic acid
d. *p*-toluic acid
e. cyclobutanecarboxylic acid
f. 2-propanoylbenzoic acid
g. phenylacetic acid
h. 2-naphthoic acid
i. 2,3-dimethyl-3-butenoic acid
j. 3-oxobutanoic acid
k. 2,2-dimethylbutanedioic acid
l. α-methyl-γ-butyrolactone

10.39. Name each of the following acids:

a. $(CH_3)_2CHCH_2CH_2COOH$
b. $CH_3CHBrCH(CH_3)COOH$
c. (benzene ring with COOH and ortho NO_2)
d. (cyclohexane ring)—COOH
e. $CH_2{=}CHCOOH$
f. $CH_3CH(C_6H_5)COOH$
g. CH_3CF_2COOH
h. $HC{\equiv}CCH_2CO_2H$

10.40. *Ibuprofen,* an anti-inflammatory agent used to treat rheumatoid arthritis and other diseases, is chemically called 2-(*p*-isobutylphenyl)propanoic acid. Draw its structure.

10.41. Which will have the higher boiling point? Explain your reasoning.

a. CH_3CH_2COOH or $CH_3CH_2CH_2CH_2OH$
b. $CH_3CH_2CH_2CH_2COOH$ or $(CH_3)_3CCOOH$

10.42. In each of the following pairs of acids, which would be expected to be the stronger, and why?

a. CH_2ClCO_2H and CH_2BrCO_2H
b. *o*-$BrC_6H_4CO_2H$ and *m*-$BrC_6H_4CO_2H$
c. CCl_3CO_2H and CF_3CO_2H
d. $C_6H_5CO_2H$ and *p*-$CH_3OC_6H_4CO_2H$
e. $ClCH_2CH_2CO_2H$ and $CH_3CHClCO_2H$

10.43. Write a balanced equation for the reaction of

a. chloroacetic acid with potassium hydroxide.
b. decanoic acid with calcium hydroxide.

10.44. Give equations for the synthesis of

a. $CH_3CH_2CH_2CO_2H$ from $CH_3CH_2CH_2CH_2OH$.
b. $CH_3CH_2CH_2CO_2H$ from $CH_3CH_2CH_2OH$ (two ways).

c. 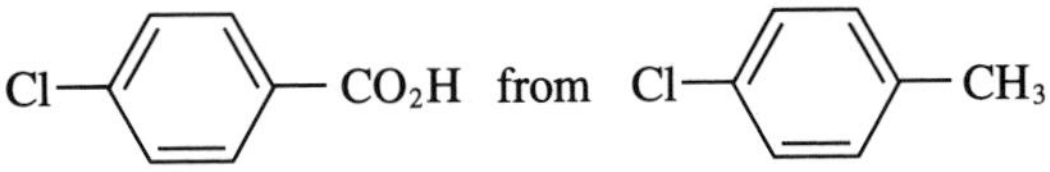 Cl—C_6H_4—CO_2H from Cl—C_6H_4—CH_3

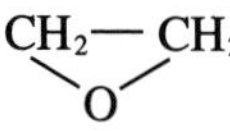

d. 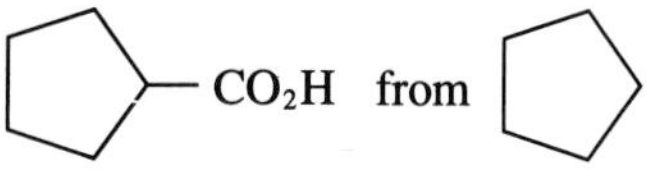 (cyclopentyl)—CO_2H from (cyclopentane)

e. $CH_3OCH_2CO_2H$ from CH_2—CH_2 (bridged by O) (two steps).

f. 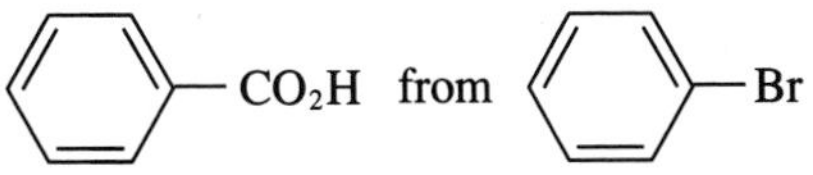 C_6H_5—CO_2H from C_6H_5—Br

10.45. The Grignard route for the synthesis of $(CH_3)_3CCO_2H$ from $(CH_3)_3CBr$ (Example 10.5) is far superior to the nitrile route. Explain why.

10.46. Write a structure for each of the following compounds:

a. sodium 2-chloropropanoate
b. calcium acetate
c. isopropyl acetate
d. ethyl formate
e. phenyl benzoate
f. benzonitrile
g. propanoic anhydride
h. *m*-toluamide
i. 4-chlorobutanoyl chloride
j. 3-formylcyclopentanecarboxylic acid

10.47. Name each of the following compounds:

a. 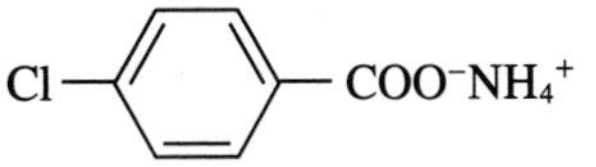 Cl—C_6H_4—$COO^-NH_4^+$

b. $[CH_3(CH_2)_2CO_2^-]_2Ca^{2+}$

c. $(CH_3)_2CHCOOC_6H_5$

d. $CF_3CO_2CH_3$

e. $HCONH_2$

f. $CH_3CH_2—\overset{\overset{\large O}{\|}}{C}—O—\overset{\overset{\large O}{\|}}{C}—CH_2CH_3$

10.48. Write out each step in the Fischer esterification of benzoic acid with methanol (you may wish to use eq. 10.18 as a model).

10.49. Write an equation for the reaction of ethyl benzoate with

a. hot aqueous sodium hydroxide.
b. ammonia (heat).
c. propylmagnesium bromide (two equivalents), then H_3O^+.
d. lithium aluminum hydride (two equivalents), then H_3O^+.

10.50. Write out all the steps in the mechanism for

a. saponification of $CH_3CH_2CO_2CH_3$.
b. ammonolysis of $CH_3CH_2CO_2CH_3$.

10.51. Explain each difference in reactivity toward nucleophiles.

a. Esters are less reactive than ketones.
b. Benzoyl chloride is less reactive than cyclohexanecarbonyl chloride.

10.52. Identify the Grignard reagent and the ester that would be used to prepare

a.
$$CH_3CH_2CH_2-\underset{\underset{C_6H_5}{|}}{\overset{\overset{OH}{|}}{C}}-CH_2CH_2CH_3$$

b. $CH_3CH_2CH_2C(C_6H_5)_2OH$

10.53. Write an equation for
a. hydrolysis of acetyl chloride.
b. reaction of benzoyl chloride with methanol.
c. esterification of 1-pentanol with acetic anhydride.
d. ammonolysis of butanoyl bromide.
e. Fischer esterification of pentanoic acid with ethanol.
f. 2-methylpropanoyl chloride + ethylbenzene + $AlCl_3$.
g. succinic acid + heat (235°C).
h. phthalic anhydride + methanol (1 equiv.) + H^+.
i. phthalic anhydride + methanol (excess) + H^+.
j. adipol chloride + ammonia (excess).

10.54. Complete the equation for each of the following reactions:
a. $CH_3CH_2CH_2CO_2H + PCl_5 \rightarrow$
b. $CH_3(CH_2)_8CO_2H + SOCl_2 \rightarrow$
c. 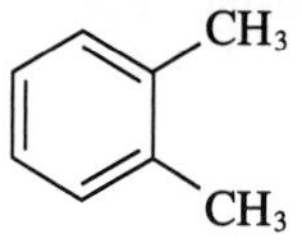 (benzene ring with two adjacent CH_3 groups) $+ KMnO_4 \longrightarrow$
d. 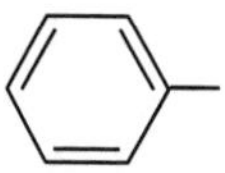 (benzene ring)$-CO_2^-NH_4^+$ + heat $\longrightarrow$
e. $CH_3(CH_2)_5CONH_2 + LiAlH_4 \rightarrow$
f. 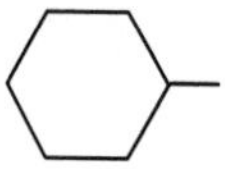 (cyclohexane ring)$-CO_2CH_2CH_3 + LiAlH_4 \longrightarrow$

10.55. Write the important resonance contributors to the structure of propanamide and tell which atoms will lie in a single plane.

10.56. Considering the relative reactivities of ketones and esters toward nucleophiles, which of the following products seems the more likely?

$$CH_3\overset{\overset{O}{\|}}{C}CH_2CH_2CO_2CH_3 \xrightarrow{NaBH_4} CH_3\overset{\overset{O}{\|}}{C}CH_2CH_2CH_2OH \text{ or } CH_3\overset{\overset{OH}{|}}{C}HCH_2CH_2CO_2CH_3$$

10.57. Mandelic acid, which has the formula $C_6H_5CH(OH)COOH$, can be isolated from bitter almonds (called *Mandel* in German). It is sometimes used in medicine to treat urinary infections. Devise a two-step synthesis of mandelic acid from benzaldehyde, using the latter's cyanohydrin (see Sec. 9.11) as an intermediate.

10.58. Consider the structure of the catnip ingredient nepatalactone (page 305).
a. Show with dotted lines that the structure is composed of two isoprene units.
b. Circle the stereogenic centers and determine their configurations (*R* or *S*).

10.59. When maleic acid is heated at reflux with a little concentrated hydrochloric acid, it is gradually converted to fumaric acid. Explain how this isomerization might occur.

10.60. Write the structure of the Claisen condensation product of ethyl phenylacetate, and show the steps in its formation.

10.61. Diethyl adipate, when heated with sodium ethoxide, gives the product shown, by an *intra*molecular Claisen condensation.

$$\underset{\text{diethyl adipate}}{CH_3CH_2O\overset{\overset{O}{\|}}{C}-(CH_2)_4-\overset{\overset{O}{\|}}{C}OCH_2CH_3} \xrightarrow{NaOCH_2CH_3} \text{ethyl 2-oxocyclopentanecarboxylate}$$

(product: cyclopentanone ring bearing $\overset{\overset{O}{\|}}{C}OCH_2CH_3$ at C-2)

ethyl 2-oxocyclopentanecarboxylate

Write out the steps in a plausible mechanism for the reaction.

10.62. Analogous to the mixed aldol condensation (Sec. 9.19), mixed Claisen condensations are possible. Predict the structure of the product obtained when a mixture of ethyl benzoate and ethyl acetate is heated with sodium ethoxide in ethanol.

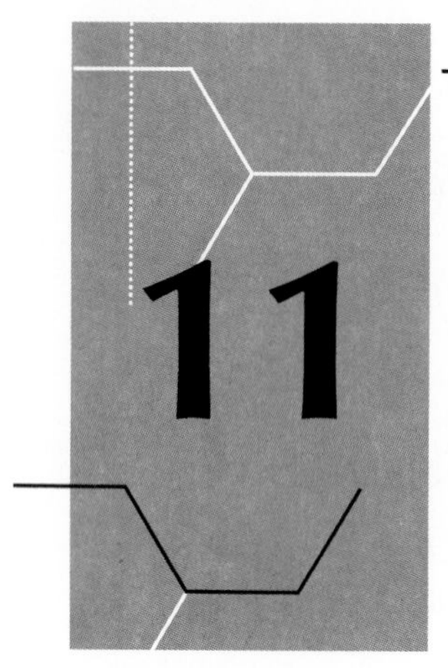

11 Amines and Related Nitrogen Compounds

11.1 Introduction

In this chapter, we discuss the last of the major, simple functionalities—the **amines.**

Amines are organic relatives of ammonia, derived by replacing one, two, or all three hydrogens of ammonia with organic groups. Like ammonia, amines are bases. In fact, *amines are the most important type of organic base that occurs in nature.* In this chapter, we will first describe the structure, preparation, chemical properties, and uses of some simple amines. Later in the chapter, we will discuss a few natural and synthetic amines with important biological activity.

11.2 Classification and Structure of Amines

The relation between ammonia and amines is illustrated by the following structures:

$H-\ddot{N}-H$ (with H below N)	$R-\ddot{N}-H$ (with H below N)	$R-\ddot{N}-R$ (with H below N)	$R-\ddot{N}-R$ (with R below N)
ammonia	primary amine	secondary amine	tertiary amine

For convenience, amines are classified as **primary, secondary,** or **tertiary,** depending on whether one, two, or three organic groups are attached to the nitrogen. The *R* groups in these structures may be alkyl or aryl, and when two or more *R* groups are present, they may be identical to or different from one another. In some secondary and tertiary amines, the nitrogen may be part of a ring.

PROBLEM 11.1 Classify each of the following amines as primary, secondary, or tertiary:

a. $(CH_3)_3CNH_2$ b. (five-membered ring containing N–H) c. $CH_3-C_6H_4-NH_2$ (para) d. $(CH_3)_2N-C_6H_5$

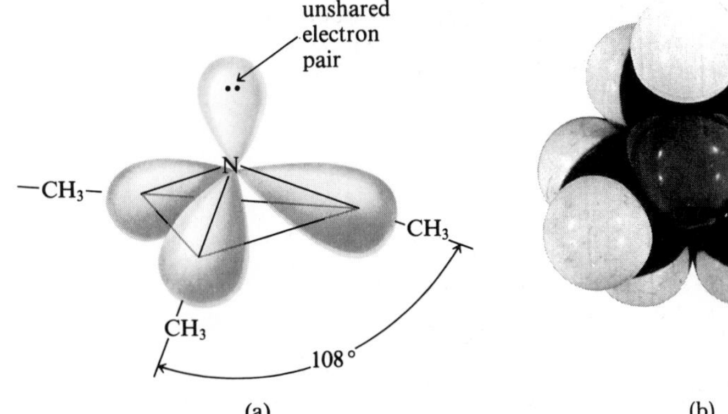

FIGURE 11.1 (a) An orbital view of the pyramidal bonding in trimethylamine. (b) Top view of a space-filling model of trimethylamine. The center ball represents the orbital with the unshared electron pair.

The nitrogen atom in amines is trivalent. In addition, the nitrogen carries an unshared electron pair. The nitrogen orbitals are therefore sp^3-hybridized, and the overall geometry is pyramidal (nearly tetrahedral), as shown for trimethylamine in Fig. 11.1. From this geometry, one might think that an amine with three different groups attached to the nitrogen would be chiral, with the unshared electron pair acting as the fourth group. This is true in principle, but in practice the two enantiomers usually interconvert rapidly through inversion, via an "umbrella-in-the-wind" type of process, and are not resolvable.

$$N(R_1)(R_2)(R_3) \rightleftharpoons \left[R_1{-}N(R_2)(R_3) \right]^{\ddagger} \rightleftharpoons N(R_1)(R_2)(R_3) \qquad (11.1)$$

planar transition state

11.3 Nomenclature of Amines

Amines can be named in several different ways. Commonly, simple amines are named by specifying the alkyl groups attached to the nitrogen and adding the suffix *-amine*.

$CH_3CH_2NH_2$	$(CH_3CH_2)_2NH$	$(CH_3CH_2)_3N$
ethylamine	diethylamine	triethylamine
(primary)	(secondary)	(tertiary)

In the IUPAC system, the **amino group, —NH_2,** is named as a substituent, as in the following examples:

$CH_3CH_2NH_2$
aminoethane

$\overset{1}{C}H_3\overset{2}{C}H(NH_2)\overset{3}{C}H_2\overset{4}{C}H_2\overset{5}{C}H_3$
2-aminopentane

cis-1,3-diaminocyclobutane

In this system, secondary or tertiary amines are named by using a prefix that includes all but the longest carbon chain, as in

$$\overset{}{CH_3NH}\overset{1}{C}H_2\overset{2}{C}H_2\overset{3}{C}H_3 \qquad CH_3\underset{}{\overset{CH_2CH_3}{N}}-\overset{1}{C}H_2\overset{2}{C}H_2\overset{3}{C}H_3$$

1-methylamino**propane** 1-(ethylmethylamino)**propane** dimethylamino**cyclohexane**

Recently, *Chemical Abstracts* (CA) introduced a system for naming amines that is rational and easy to use. In this system, amines are named as **alkanamines.** For example,

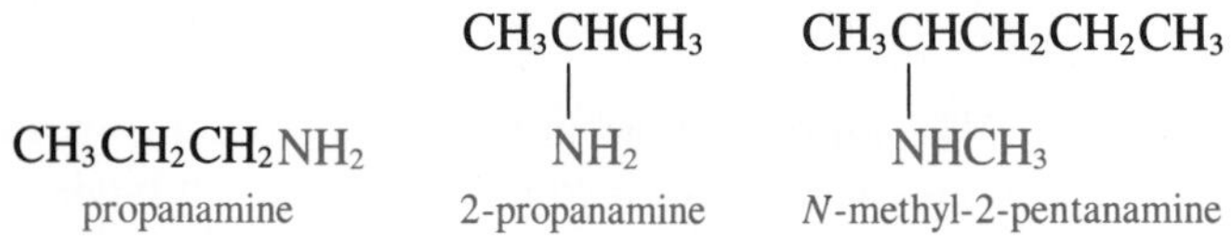

EXAMPLE 11.1 Name cyclohexyl—$N(CH_3)_2$ by the *Chemical Abstracts* system.

Solution The largest alkyl group attached to nitrogen is used as the root of the name. The compound is *N,N*-dimethylcyclohexanamine.

PROBLEM 11.2 Name $CH_3CH_2\underset{N(CH_3)_2}{CH}CH_2CH_3$ by the CA system.

When other functional groups are present, the amino group is named as a substituent:

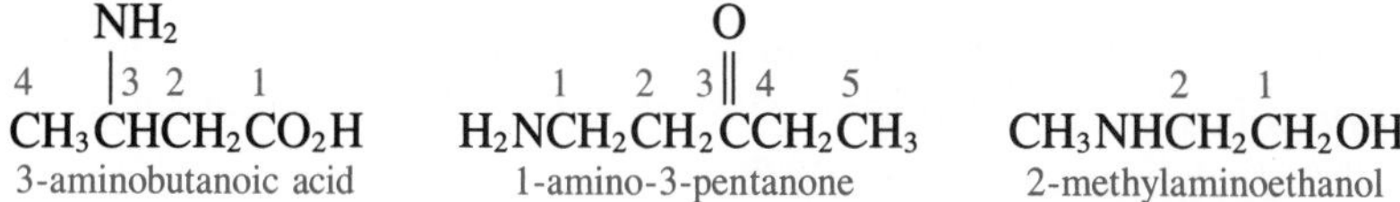

Aromatic amines are named as derivatives of aniline. In the CA system, aniline is called benzenamine; these CA names are shown in parentheses.

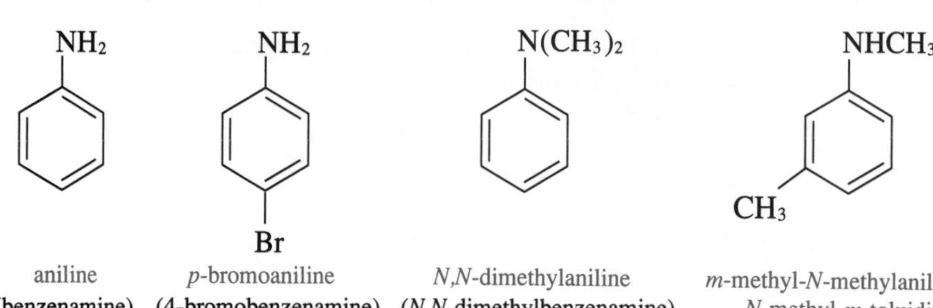

aniline (benzenamine) *p*-bromoaniline (4-bromobenzenamine) *N,N*-dimethylaniline (*N,N*-dimethylbenzenamine) *m*-methyl-*N*-methylaniline, or *N*-methyl-*m*-toluidine (*N*-methyl-3-methylbenzenamine)

EXAMPLE 11.2 Give an acceptable name for the following compounds:

a. $(CH_3)_2CHCH_2NH_2$ b. $CH_3NHCH_2CH_3$

c. 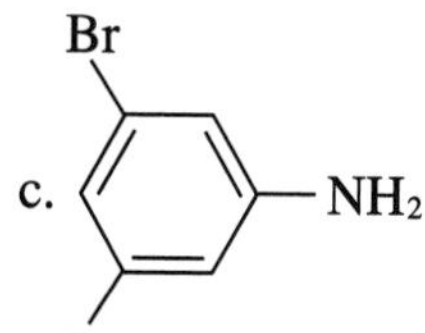

d. 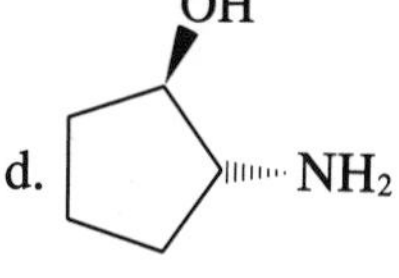

Solution

a. isobutylamine (common); 1-amino-2-methylpropane (IUPAC); 2-methylpropanamine (CA).
b. ethylmethylamine (common); methylaminoethane (IUPAC); *N*-methylethanamine (CA).
c. 3,5-dibromoaniline (common, IUPAC); 3,5-dibromobenzenamine (CA).
d. *trans*-2-aminocyclopentanol (only name).

PROBLEM 11.3 Give an acceptable name for the following compounds:

a. $(CH_3)_3CNH_2$ b. $H_2NCH_2CH_2OH$

c. O_2N— —NH_2

PROBLEM 11.4 Write the structure for

a. dipropylamine. b. 3-aminohexane.
c. pentamethylaniline. d. *N,N*-dimethyl-2-butanamine.

11.4 *Physical Properties of Amines*

Table 11.1 lists the boiling points of some common amines. Methylamine and ethylamine are gases, but primary amines with three or more carbons are liquids. Primary amines boil well above alkanes with comparable molecular weights, but below comparable alcohols, as shown in Table 11.2. Intermolecu-

TABLE 11.1 The boiling points of some simple amines

Name	*Formula*	*bp, °C*
ammonia	NH_3	−33.4
methylamine	CH_3NH_2	−6.3
dimethylamine	$(CH_3)_2NH$	7.4
trimethylamine	$(CH_3)_3N$	2.9
ethylamine	$CH_3CH_2NH_2$	16.6
propylamine	$CH_3CH_2CH_2NH_2$	48.7
butylamine	$CH_3CH_2CH_2CH_2NH_2$	77.8
aniline	$C_6H_5NH_2$	184.0

TABLE 11.2 A comparison of alkane, amine, and alcohol boiling points*

alkane	CH_3CH_3 (30) bp −88.6°C	$CH_3CH_2CH_3$ (44) bp −42.1°C
amine	CH_3NH_2 (31) bp −6.3°C	$CH_3CH_2NH_2$ (45) bp +16.6°C
alcohol	CH_3OH (32) bp +65.0°C	CH_3CH_2OH (46) bp +78.5°C

* Molecular weights are given in parentheses.

lar N—H···N hydrogen bonds are important and raise the boiling points of primary and secondary amines but are not as strong as the O—H···O bonds of alcohols. The reason for this is that nitrogen is not as electronegative as oxygen.

PROBLEM 11.5 Explain why the tertiary amine $(CH_3)_3N$ boils so much lower than its primary isomer $CH_3CH_2CH_2NH_2$.

All three classes of amines can form hydrogen bonds with the —OH group of water (that is, O—H···N). Thus, most simple amines with up to five or six carbon atoms are either completely or appreciably soluble in water.

Now we will describe some ways that amines can be prepared.

11.5 Preparation of Amines; Alkylation of Ammonia and Amines

Ammonia reacts with alkyl halides to give amines via a two-step process. The first step is a nucleophilic substitution reaction (S_N2).

$$\underset{\text{ammonia}}{H_3N\!:} + R{-}X \longrightarrow \underset{\text{alkylammonium halide}}{R{-}\overset{+}{N}H_3} + X^- \tag{11.2}$$

The free amine can then be obtained from its salt by treatment with a strong base.

$$R{-}\overset{+}{N}H_3\,X^- + NaOH \longrightarrow \underset{\text{primary amine}}{RNH_2} + H_2O + Na^+X^- \tag{11.3}$$

Primary, secondary, and tertiary amines can be similarly alkylated.

$$\underset{\text{primary amine}}{R\ddot{N}H_2} + R{-}X \longrightarrow R_2\overset{+}{N}H_2X^- \xrightarrow{NaOH} \underset{\text{secondary amine}}{R_2NH} \tag{11.4}$$

$$\underset{\text{secondary amine}}{R_2\ddot{N}H} + R{-}X \longrightarrow R_3\overset{+}{N}HX^- \xrightarrow{NaOH} \underset{\text{tertiary amine}}{R_3N} \tag{11.5}$$

$$R_3\ddot{N} + R{-}X \longrightarrow R_4N^+ X^- \qquad (11.6)$$

tertiary amine — quaternary ammoniun salt

Unfortunately, mixtures of products are often obtained in these reactions because the starting ammonia or amine and the alkylammonium ion formed in the S_N2 step can equilibrate, as in the following equation:

$$NH_3 + R\overset{+}{N}H_3\,X^- \rightleftharpoons NH_4^+\,X^- + RNH_2 \qquad (11.7)$$

So, in the reaction of ammonia with an alkyl halide (eq. 11.2), some primary amine is formed (eq. 11.7), and it may be further alkylated (eq. 11.4) to give a secondary amine, and so on. By adjusting the ratio of the reactants, however, a good yield of one desired amine may be obtained. For example, with a large excess of ammonia, the primary amine is the major product.

Aromatic amines can often be alkylated selectively.

$$C_6H_5NH_2 \xrightarrow{CH_3I} C_6H_5NHCH_3 \xrightarrow{CH_3I} C_6H_5N(CH_3)_2 \qquad (11.8)$$

aniline — *N*-methylaniline — *N,N*-dimethylaniline

The alkylation can be intramolecular, as in the following final step in a laboratory synthesis of nicotine:

$$\text{3-pyridyl}{-}CH(Br){-}CH_2{-}CH_2{-}CH_2{-}\ddot{N}H{-}CH_3 \xrightarrow{-HBr} \text{nicotine} \qquad (11.9)$$

nicotine

EXAMPLE 11.3 Write an equation for the synthesis of benzylamine, $C_6H_5{-}CH_2NH_2$

Solution

$$C_6H_5{-}CH_2X + 2\,\ddot{N}H_3 \longrightarrow C_6H_5{-}CH_2\ddot{N}H_2 + NH_4^+X^-$$

(X = Cl, Br, or I)

Use of excess ammonia helps prevent further substitution.

PROBLEM 11.6 Complete equations for the following reactions.

a. $CH_3CH_2CH_2CH_2Br + 2\ NH_3 \longrightarrow$
b. $CH_3CH_2I + 2(CH_3CH_2)_2NH \longrightarrow$
c. $(CH_3)_3N + CH_3I \longrightarrow$
d. $CH_3CH_2CH_2NH_2 + C_6H_5{-}CH_2Br \longrightarrow$

PROBLEM 11.7 Give a synthesis of $C_6H_5{-}NHCH_2CH_3$ from aniline.

11.6 Preparation of Amines; Reduction of Nitrogen Compounds

All bonds to the nitrogen atom in amines are either N—H or N—C bonds. Nitrogen in ammonia or amines is therefore in a reduced form. It is not surprising, then, that organic compounds in which a nitrogen atom is present in a more oxidized form can be reduced to amines by appropriate reducing agents. Several examples of this useful synthetic approach to amines are described here.

The best route to aromatic primary amines is by reduction of the corresponding nitro compounds, which are in turn prepared by electrophilic aromatic nitration. The nitro group is easily reduced, either catalytically with hydrogen or by chemical reducing agents.

$$CH_3{-}C_6H_4{-}NO_2 \xrightarrow[\text{1. } SnCl_2,HCl \quad \text{2. } NaOH,H_2O]{3H_2,\text{ Ni catalyst} \quad \text{or}} CH_3{-}C_6H_4{-}NH_2 + 2H_2O \tag{11.10}$$

p-nitrotoluene *p*-toluidine

EXAMPLE 11.4 Devise a synthesis of *p*-chloroaniline, $Cl{-}C_6H_4{-}NH_2$, from chlorobenzene.

Solution Chlorobenzene is first nitrated; —Cl is an *o,p*-directing group, so the major product is *p*-chloronitrobenzene. This product is then reduced.

$$C_6H_5Cl \xrightarrow[H_2SO_4]{HONO_2} Cl{-}C_6H_4{-}NO_2 \xrightarrow[Ni]{H_2} Cl{-}C_6H_4{-}NH_2$$

PROBLEM 11.8 Give a synthesis for H_2N–C_6H_3(–NH_2)–CH_3 (2,4-diaminotoluene structure) from toluene.

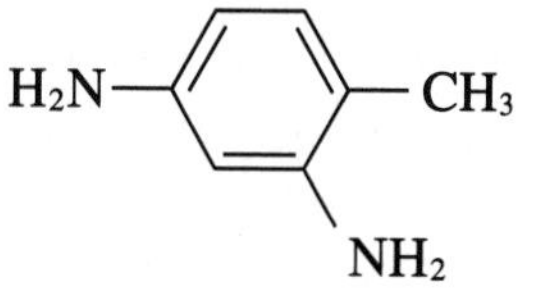

As described in the previous chapter (eq. 10.41), *amides can be reduced to amines with lithium aluminum hydride.*

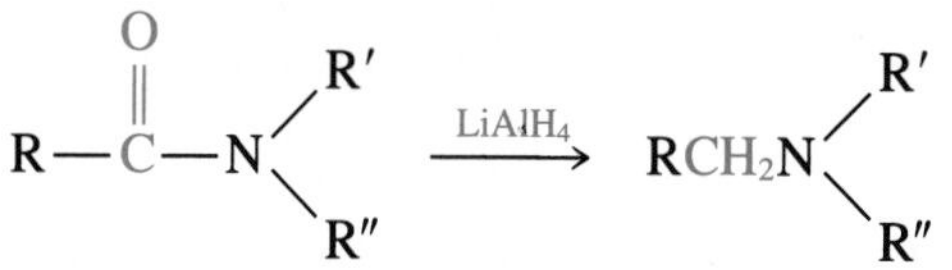

$$R-\overset{O}{\overset{\|}{C}}-N\begin{matrix}R'\\R''\end{matrix} \xrightarrow{LiAlH_4} RCH_2N\begin{matrix}R'\\R''\end{matrix} \quad \text{(R' and R'' may be H or organic groups)} \tag{11.11}$$

Depending on the structures of R′ and R″, we can obtain primary, secondary, or tertiary amines in this way.

EXAMPLE 11.5 Complete the equation $CH_3\overset{O}{\overset{\|}{C}}NHCH_2CH_3 \xrightarrow{LiAlH_4}$

Solution The C=O group is reduced to CH_2. The product is the secondary amine $CH_3CH_2NHCH_2CH_3$.

PROBLEM 11.9 Show how $CH_3CH_2N(CH_3)_2$ can be synthesized from an amide.

Reduction of nitriles (cyanides) gives primary amines.

$$R-C\equiv N \xrightarrow[\text{or } H_2,\ Ni]{LiAlH_4} RCH_2NH_2 \tag{11.12}$$

EXAMPLE 11.6 Complete the equation $NCCH_2CH_2CH_2CH_2CN \xrightarrow[\text{Ni catalyst}]{\text{excess } H_2}$

Solution Both CN groups are reduced. The product $H_2N-(CH_2)_6-NH_2$, or 1,6-diaminohexane, is one of two raw materials for the manufacture of nylon (p. 407).

PROBLEM 11.10 Devise a synthesis of $C_6H_5-CH_2CH_2NH_2$ from $C_6H_5-CH_2Br$

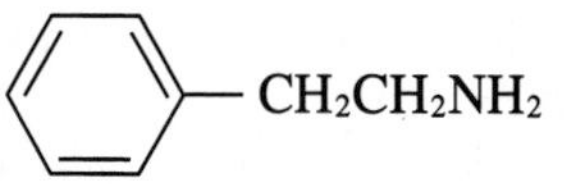

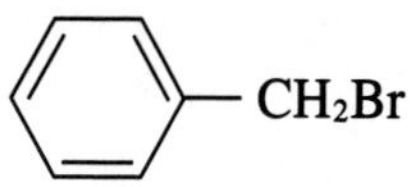

Aldehydes and ketones undergo ***reductive amination*** when treated with ammonia, primary, or secondary amines, to give primary, secondary, or tertiary amines, respectively. The most commonly used laboratory reducing agent for this purpose is the metal hydride sodium cyanoborohydride, $NaBH_3(CN)$.

$$\text{>C=\ddot{O}: + R\ddot{N}H_2} \underset{}{\overset{-H_2O}{\rightleftharpoons}} \left[\text{>C=NR}\right] \xrightarrow{NaBH_3CN} \text{>CHNHR} \qquad (11.13)$$

aldehyde or ketone; primary amine; imine; secondary amine

The reaction involves nucleophilic attack on the carbonyl groups, leading to an imine (in the case of ammonia or primary amines; compare with eq. 9.30) or an iminium ion with secondary amines. The reducing agent then reduces the C=N bond.

PROBLEM 11.11 Using eq. 11.13 as a guide, devise a synthesis of 3-aminopentane from 3-pentanone.

Now that we know several ways to make amines, let us examine some of their properties.

11.7 The Basicity of Amines

The unshared pair of electrons on the nitrogen atom dominates the chemistry of amines. Because of this electron pair, amines are both basic and nucleophilic.

Aqueous solutions of amines are basic because of the following equilibrium:

$$\text{>N: + H—\ddot{O}H} \rightleftharpoons \text{>N^+—H + {}^-:\ddot{O}H} \qquad (11.14)$$

amine; ammonium ion; hydroxide ion

EXAMPLE 11.7 Write an equation that shows why aqueous solutions of ethylamine are basic.

Solution

$$CH_3CH_2\ddot{N}H_2 + H_2O \rightleftharpoons CH_3CH_2\overset{+}{N}H_3 + OH^-$$

ethylamine; ethylammonium ion

Amines are more basic than water. They accept a proton from water, producing hydroxide ion, so their solutions are basic.

PROBLEM 11.12 Write an equation representing the equilibrium in an aqueous solution of trimethylamine.

An amine and its ammonium ion (eq. 11.14) *are related as a* base *and its* conjugate acid. For example, RNH_3^+ is the conjugate acid of the primary amine RNH_2. It is convenient, when comparing basicities of different amines, to compare instead the acidity constants (pKa's) of their conjugate acids. Eq. 11.15 expresses this acidity for a primary alkylammonium ion.

$$\underset{\text{conjugate acid}}{R\overset{+}{N}H_3} + H_2O \rightleftharpoons \underset{\text{base}}{RNH_2} + H_3O^+ \tag{11.15}$$

$$K_a = \frac{[RNH_2][H_3O^+]}{[RNH_3{}^+]}$$

The larger the K_a *(or the smaller the* pK_a*) the stronger* $R\overset{+}{N}H_3$ *is as an acid, or the weaker* RNH_2 *is as a base.*

EXAMPLE 11.8 The pK_a's of $NH_4{}^+$ and $CH_3\overset{+}{N}H_3$ are 9.30 and 10.64, respectively. Which is the stronger base, NH_3 or CH_3NH_2?

Solution $NH_4{}^+$ is the stronger acid (lower pK_a). Therefore, NH_3 is the *weaker* base, and CH_3NH_2 is the *stronger*.

Table 11.3 lists some amine basicities. Alkylamines are approximately 10 times as basic as ammonia. Recall that alkyl groups are electron-donating relative to hydrogen $R \rightarrow \ddot{N}\langle^{H}_{H}$. This electron-donating effect stabilizes the am-

TABLE 11.3 Basicities of some common amines, expressed as pK_a of the corresponding ammonium ions.

	Formula		
Name	*Amine*	*Ammonium Ion*	pK_a *of the ammonium ions*
ammonia	$\ddot{N}H_3$	$\overset{+}{N}H_4$	9.30
methylamine	$CH_3\ddot{N}H_2$	$CH_3\overset{+}{N}H_3$	10.64
dimethylamine	$(CH_3)_2\ddot{N}H$	$(CH_3)_2\overset{+}{N}H_2$	10.71
trimethylamine	$(CH_3)_3\ddot{N}$	$(CH_3)_3\overset{+}{N}H$	9.77
ethylamine	$CH_3CH_2\ddot{N}H_2$	$CH_3CH_2\overset{+}{N}H_3$	10.67
propylamine	$CH_3CH_2CH_2\ddot{N}H_2$	$CH_3CH_2CH_2\overset{+}{N}H_3$	10.58
aniline	$C_6H_5\ddot{N}H_2$	$C_6H_5\overset{+}{N}H_3$	4.62
N-methylaniline	$C_6H_5\ddot{N}HCH_3$	$C_6H_5\overset{+}{N}H_2(CH_3)$	4.85
N,N-dimethylaniline	$C_6H_5\ddot{N}(CH_3)_2$	$C_6H_5\overset{+}{N}H(CH_3)_2$	5.04
p-chloroaniline	p-$ClC_6H_4\ddot{N}H_2$	p-$ClC_6H_4\overset{+}{N}H_3$	3.98

monium ion (positive charge) relative to the free amine (eq. 11.14). Hence it decreases the acidity of the ammonium ion, or increases the basicity of the amine. In general, *electron-donating groups increase the basicity of amines, and electron-withdrawing groups decrease their basicity.*

PROBLEM 11.13 Do you expect $ClCH_2CH_2NH_2$ to be a stronger or weaker base than $CH_3CH_2NH_2$? Explain.

Aromatic amines are much weaker bases than aliphatic amines or ammonia. For example, aniline is less basic than cyclohexylamine by nearly a million times.

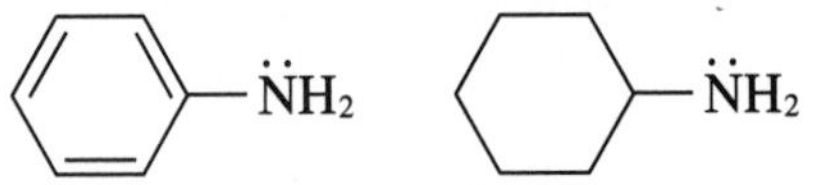

	aniline	cyclohexylamine
pK_a of ammonium ion	4.62	9.8

The reason for this huge difference is the resonance delocalization of the unshared electron pair that is possible in aniline but not in cyclohexylamine.

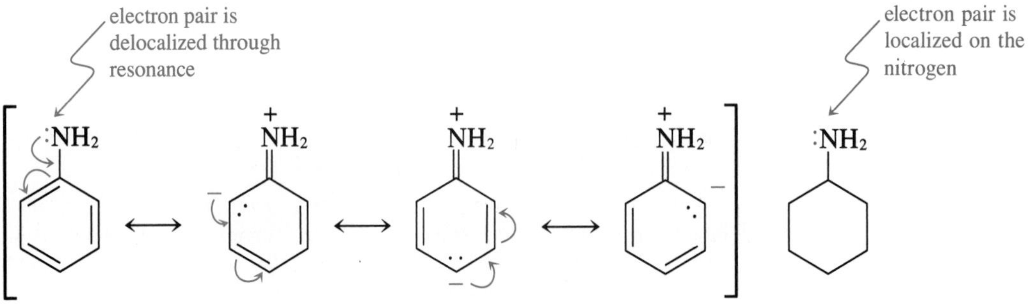

electron pair is delocalized through resonance

electron pair is localized on the nitrogen

resonance structures of aniline

cyclohexylamine

Resonance stabilizes the unprotonated form of aniline. This shifts the equilibrium in eq. 11.15 to the right, increasing the acidity of the anilinium ion or decreasing the basicity of aniline. Another way to describe the situation is to say that the unshared electron pair in aniline is delocalized and therefore less available for donation to a proton than is the electron pair in cyclohexylamine.

PROBLEM 11.14 Compare the basicities of the last four amines in Table 11.3, and explain the reasons for the observed basicity order.

PROBLEM 11.15 Place the following amines in order of increasing basicity: aniline, *p*-nitroaniline, *p*-toluidine.

11.8 Comparison of the Basicity and Acidity of Amines and Amides

Both amines and amides have nitrogens with an unshared electron pair. There is a huge difference, however, in their basicities. *Aqueous solutions of amines are basic; aqueous solutions of amides are essentially neutral.* Why this striking difference?

The answer lies in their structures, as illustrated in the following comparison of a primary amine with a primary amide:

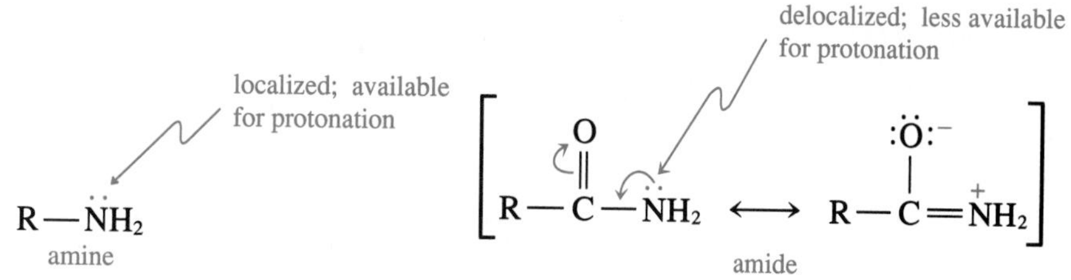

localized; available for protonation

delocalized; less available for protonation

$$\underset{\text{amine}}{R{-}\ddot{N}H_2} \qquad \underset{\text{amide}}{\left[R{-}\overset{\overset{\displaystyle O}{\|}}{C}{-}\ddot{N}H_2 \longleftrightarrow R{-}\overset{\overset{\displaystyle :\ddot{O}:^-}{|}}{C}{=}\overset{+}{N}H_2\right]}$$

In the amine, the electron pair is mainly localized on the nitrogen. In the amide, the electron pair is delocalized to the carbonyl oxygen. The effect of this delocalization is seen in the low pK_a values for the conjugate acids of amides, compared with those for the conjugate acids of amines, for example:

	$CH_3CH_2\overset{+}{N}H_3$	$CH_3\overset{\overset{\displaystyle \overset{+}{:O}{-}H}{\|}}{C}\ddot{N}H_2$
conjugate acid of	ethylamine	acetamide
pK_a	10.67	−0.6

Notice that amides are protonated on the carbonyl oxygen, not on nitrogen. This is because protonation on oxygen gives a resonance-stabilized cation, while protonation on nitrogen does not.

Primary and secondary amines and amides have N—H bonds, and one might expect that they would on occasion behave as acids (proton donors).

$$R{-}\ddot{N}H_2 \rightleftharpoons R{-}\ddot{\underset{\cdot\cdot}{N}}H^- + H^+ \qquad K_a \cong 10^{-40} \tag{11.16}$$

Primary amines are exceedingly weak acids, much weaker than alcohols. Their pK_a is about 40, compared with about 16 for alcohols. The main reason for the difference is that nitrogen is much less electronegative than oxygen and thus cannot stabilize a negative charge nearly as well.

Amides, on the other hand, are much *stronger acids than amines;* in fact, their pK_a (about 15) is comparable to that of alcohols:

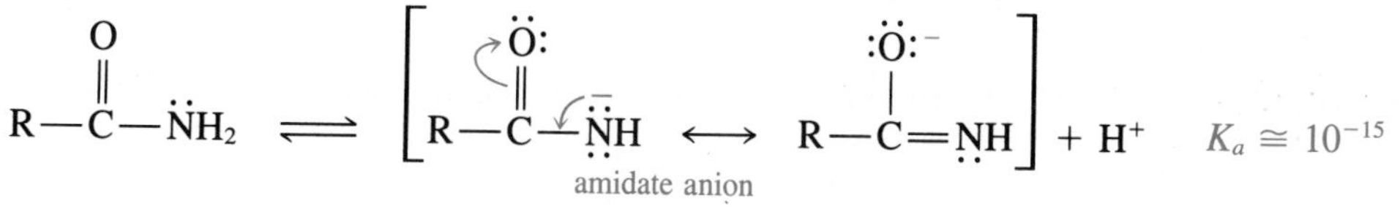

$$R{-}\overset{\overset{\displaystyle O}{\|}}{C}{-}\ddot{N}H_2 \rightleftharpoons \underset{\text{amidate anion}}{\left[R{-}\overset{\overset{\displaystyle \ddot{O}:}{\|}}{C}{-}\ddot{\underset{\cdot\cdot}{N}}H \longleftrightarrow R{-}\overset{\overset{\displaystyle :\ddot{O}:^-}{|}}{C}{=}\underset{\cdot\cdot}{N}H\right]} + H^+ \qquad K_a \cong 10^{-15} \tag{11.17}$$

One reason is that the negative charge of the **amidate anion** can be delocalized through resonance. Another reason is that the nitrogen in an amide carries a partial positive charge, making it easy to lose the attached proton, which is also positive.

It is important to understand these differences between amines and amides, not only because they involve important chemical principles, but also because they help us understand the chemistry of certain natural products, such as peptides and proteins.

PROBLEM 11.16 Place the following compounds in order of increasing basicity; in order of increasing acidity.

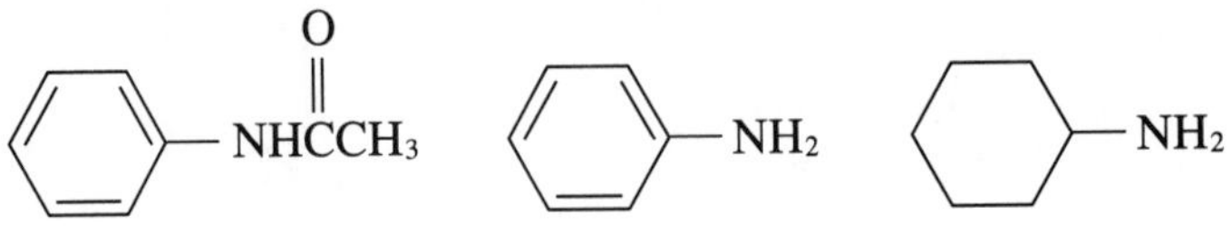

$$C_6H_5\text{—}NH\overset{O}{\overset{\|}{C}}CH_3 \qquad C_6H_5\text{—}NH_2 \qquad C_6H_{11}\text{—}NH_2$$

acetanilide — aniline — cyclohexylamine

11.9 Reaction of Amines with Strong Acids; Amine Salts

Amines react with strong acids to form **alkylammonium salts.** An example of this reaction for a primary amine and HCl is as follows:

$$R\text{—}\ddot{N}H_2 + HCl \longrightarrow R\overset{+}{N}H_3\ Cl^- \qquad (11.18)$$

primary amine — an alkylammonium chloride

EXAMPLE 11.9 Complete the following acid-base reactions, and name the products.

a. $CH_3CH_2NH_2 + HI \longrightarrow$ b. $(CH_3)_3N + HBr \longrightarrow$

Solution

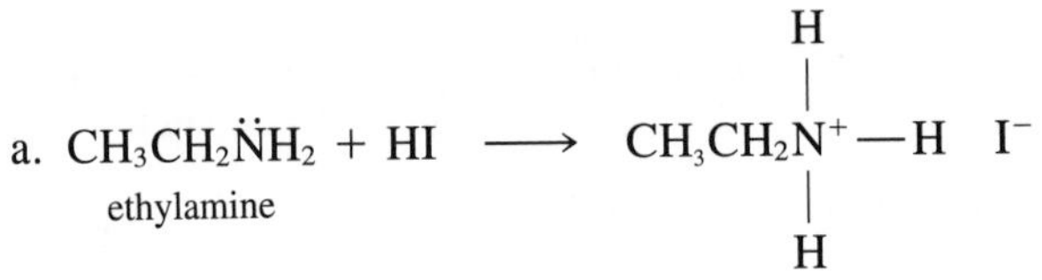

a. $CH_3CH_2\ddot{N}H_2 + HI \longrightarrow CH_3CH_2N^+H_3\ I^-$ (N bonded to three H)

ethylamine → ethylammonium iodide

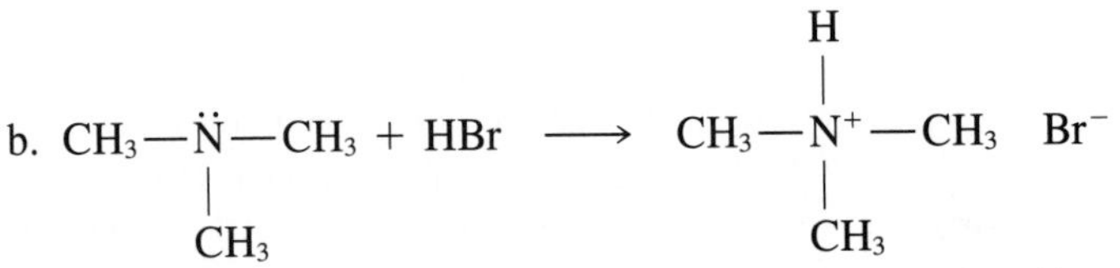

b. $CH_3\text{—}\ddot{N}(CH_3)\text{—}CH_3 + HBr \longrightarrow CH_3\text{—}N^+(H)(CH_3)\text{—}CH_3\ Br^-$

trimethylamine → trimethylammonium bromide

PROBLEM 11.17 Complete the following equation, and name the product.

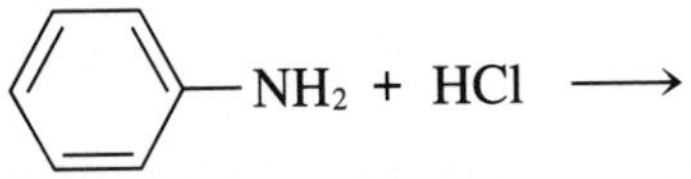

$$C_6H_5\text{—}NH_2 + HCl \longrightarrow$$

This type of reaction is used to separate or extract amines from neutral or acidic water-insoluble substances. Consider, for example, a mixture of *p*-

toluidine and p-nitrotoluene, which might arise from a preparation of the amine that for some reason does not go to completion (eq. 11.10). The amine can be separated from the unreduced nitro compound by the following scheme:

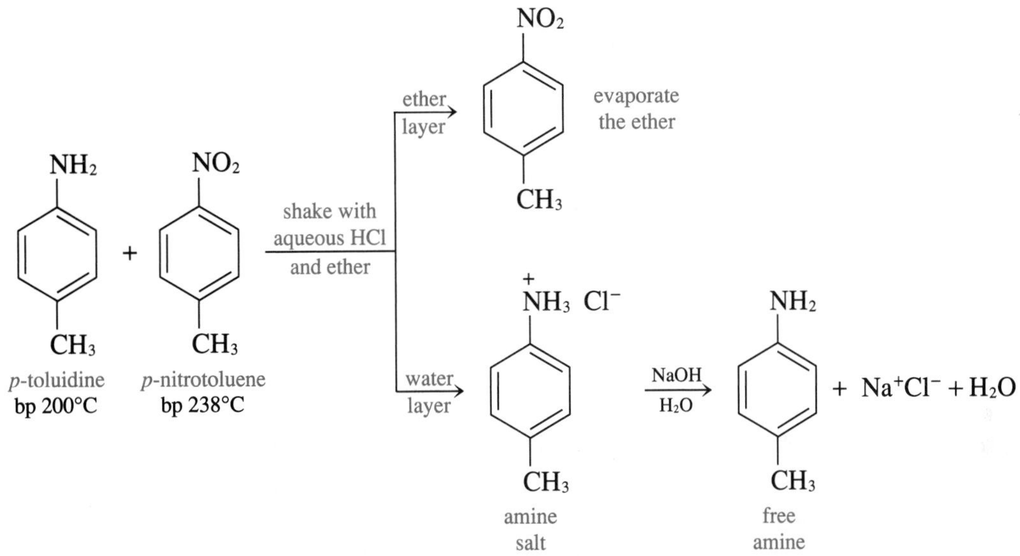

p-toluidine (bp 200°C) + p-nitrotoluene (bp 238°C) —shake with aqueous HCl and ether→ ether layer: p-nitrotoluene (evaporate the ether); water layer: $CH_3C_6H_4\overset{+}{N}H_3\ Cl^-$ (amine salt) $\xrightarrow[H_2O]{NaOH}$ $CH_3C_6H_4NH_2$ (free amine) + Na^+Cl^- + H_2O (11.19)

The mixture, neither component of which is water soluble, is dissolved in an inert, low-boiling solvent such as ether and is shaken with aqueous hydrochloric acid. The amine reacts to form a salt, which is ionic and dissolves in the water layer. The nitro compound does not react and remains in the ether layer. The two layers are then separated. The nitro compound can be recovered by evaporating the ether. The amine can be recovered from its salt by making the aqueous layer alkaline with a strong base such as NaOH.

There are many natural and synthetic amine salts of biological interest. Two examples are squalamine, an antimicrobial steroid recently isolated from the dogfish shark (see Chapter 15 for more about steroids), and (+)-methamphetamine hydrochloride, the addictive and toxic stimulant commonly known as "ice."*

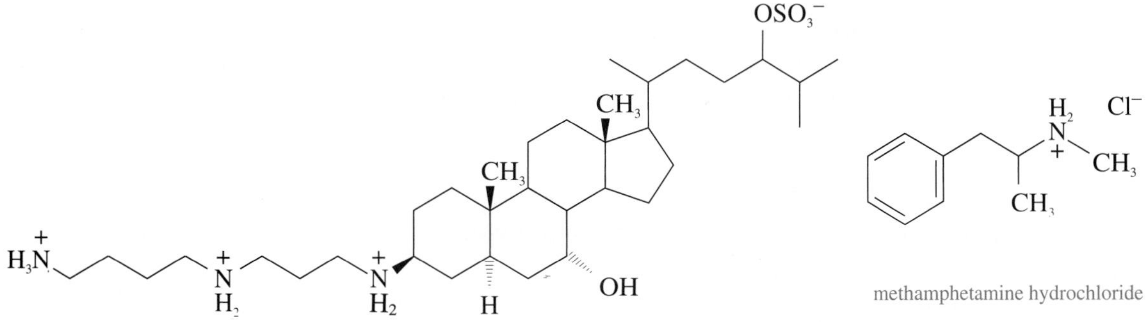

squalamine

methamphetamine hydrochloride

*For further reading about squalamine and "ice", see R. Stone, *Science* **1993,** *259,* 1125, and A. K. Cho, *Science* **1990,** *249,* 631–634.

11.10 Chiral Amines as Resolving Agents

Amines also form salts with organic acids. This reaction is used to resolve enantiomeric acids (Sec. 5.13). For example, (*R*)- and (*S*)-lactic acids can be resolved by reaction with a chiral amine such as (*S*)-1-phenylethylamine:

$$\left[\begin{array}{c} \text{(R)-lactic acid} \\ + \\ \text{(S)-lactic acid} \end{array}\right] + \text{(S)-1-phenylethylamine} \longrightarrow \left[\begin{array}{c} (R,S)\text{ salt} \\ + \\ (S,S)\text{ salt} \end{array}\right] \quad (11.20)$$

(*R*)-lactic acid: CO_2H, H, HO, CH_3 on C; (*S*)-lactic acid: CO_2H, HO, H, CH_3 on C; (*S*)-1-phenylethylamine: NH_2, CH_3, H, C_6H_5 on C; (*R,S*) salt: CO_2^- / $\overset{+}{N}H_3$; (*S,S*) salt: CO_2^- / $\overset{+}{N}H_3$

The salts are diastereomers, not enantiomers, and can be separated by ordinary methods, such as fractional crystallization. Once separated, each salt can be treated with a strong acid, such as HCl, to liberate one enantiomer of lactic acid. For example,

$$(R,S)\text{ salt} + HCl \longrightarrow \text{(R)-lactic acid} + \text{(S)-1-phenylethylammonium chloride } (\overset{+}{N}H_3\ Cl^-) \quad (11.21)$$

The chiral amine can be recovered for reuse by treating its salt with sodium hydroxide (as in the last step of eq. 11.19).

Numerous chiral amines are available from natural products and can be used to resolve acids. Conversely, some chiral acids are available to resolve amine enantiomers.

So far we have considered reactions in which amines act as bases. Now we will examine some reactions in which they act as nucleophiles.

11.11 Acylation of Amines with Acid Derivatives

Amines are nitrogen nucleophiles. They react with the carbonyl group of acid derivatives (acyl halides, anhydrides, and esters) by nucleophilic acyl substitution (Sec. 10.12).

Looked at from the viewpoint of the amine, we can say that the N—H bond in primary and secondary amines can be **acylated** by acid derivatives. For example, primary and secondary amines react with acyl halides to form amides (compare with eq. 10.34).

$$\underset{\text{acyl halide}}{R{-}\overset{\overset{\displaystyle O}{\|}}{C}{-}Cl} + \underset{\text{primary amine}}{H_2\ddot{N}{-}R'} \longrightarrow \underset{\text{secondary amide}}{R{-}\overset{\overset{\displaystyle O}{\|}}{C}{-}NHR'} + HCl \qquad (11.22)$$

$$\underset{\text{acyl halide}}{R{-}\overset{\overset{\displaystyle O}{\|}}{C}{-}Cl} + \underset{\text{secondary amine}}{H\ddot{N}R'R''} \longrightarrow \underset{\text{tertiary amide}}{R{-}\overset{\overset{\displaystyle O}{\|}}{C}{-}NR'R''} + HCl \qquad (11.23)$$

If the amine is inexpensive, two equivalents are used—one to form the amide and the second to neutralize the HCl. Alternatively, an inexpensive base may be added for the latter purpose. This can be sodium hydroxide (especially if R is *aromatic*), or a tertiary amine; having no N—H bonds, tertiary amines cannot be acylated, but they can neutralize the HCl.

EXAMPLE 11.10 Using eq. 10.29 as a guide, write out the steps in the mechanism for eq. 11.22.

Solution

$$R{-}\overset{\overset{\displaystyle \ddot{O}:}{\|}}{C}{-}Cl + H_2\ddot{N}R' \longrightarrow R{-}\overset{\overset{\displaystyle :\ddot{O}:^-}{|}}{\underset{\underset{\displaystyle H{-}\overset{+}{N}{-}R'\atop \displaystyle H}{|}}{C}}{-}Cl$$

$$\downarrow -Cl^-$$

$$\underset{\text{amide}}{R{-}\overset{\overset{\displaystyle O}{\|}}{C}{-}\underset{\underset{\displaystyle H}{|}}{\ddot{N}}{-}R'} \xleftarrow{-H^+} R{-}\overset{\overset{\displaystyle O}{\|}}{C}{-}\underset{\underset{\displaystyle H}{|}}{\overset{\overset{\displaystyle H}{|}}{N^+}}{-}R' \quad H_2\ddot{N}R'$$

The first step involves nucleophilic addition to the carbonyl group. Elimination of HCl completes the substitution reaction.

Acylation of amines is put to practical use. For example, the insect repellent "Off" is the amide formed in the reaction of *m*-toluyl chloride and diethylamine.

$$\underset{m\text{-toluyl chloride}}{m\text{-}CH_3C_6H_4{-}\overset{\overset{\displaystyle O}{\|}}{C}{-}Cl} + \underset{\text{diethylamine}}{(CH_3CH_2)_2NH} \xrightarrow{NaOH} \underset{\substack{N,N\text{-diethyl-}m\text{-toluamide}\\ \text{(the insect repellent Off)}}}{m\text{-}CH_3C_6H_4{-}\overset{\overset{\displaystyle O}{\|}}{C}{-}N(CH_2CH_3)_2} + Na^+Cl^- + H_2O \qquad (11.24)$$

PROBLEM 11.18 Write out the steps in the mechanism for the synthesis of Off (eq. 11.24).

The antipyretic (fever-reducing substance) acetanilide is an amide made from aniline and acetic anhydride.

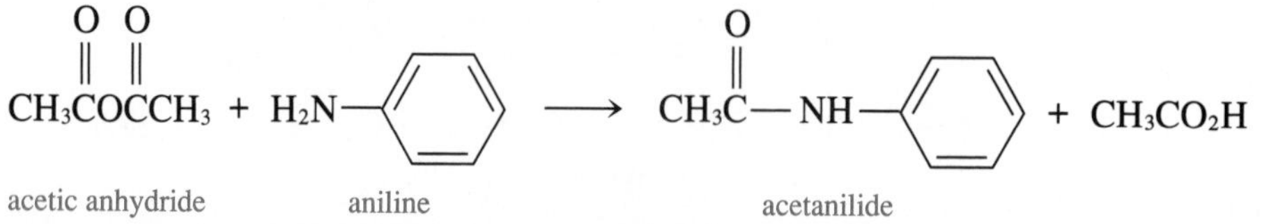

$$\underset{\text{acetic anhydride}}{CH_3\overset{O}{\overset{\|}{C}}O\overset{O}{\overset{\|}{C}}CH_3} + \underset{\text{aniline}}{H_2N-C_6H_5} \longrightarrow \underset{\text{acetanilide}}{CH_3\overset{O}{\overset{\|}{C}}-NH-C_6H_5} + CH_3CO_2H \qquad (11.25)$$

PROBLEM 11.19 Write out the steps in the mechanism of the preparation of acetanilide from aniline and acetic anhydride (eq. 11.25).

PROBLEM 11.20 Complete the following equation:

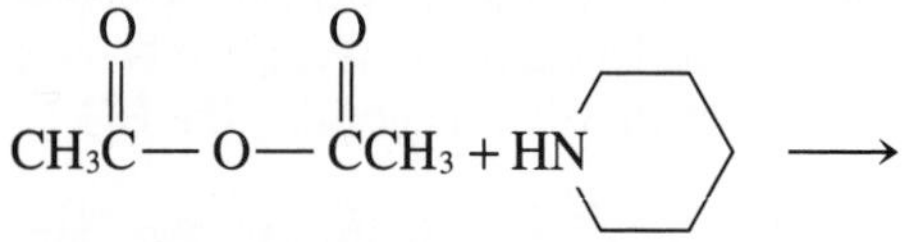

$$CH_3\overset{O}{\overset{\|}{C}}-O-\overset{O}{\overset{\|}{C}}CH_3 + HN(\text{piperidine ring}) \longrightarrow$$

A classic laboratory test for distinguishing among primary, secondary, and tertiary amines, the **Hinsberg test,** is based on their reaction with **benzenesulfonyl chloride,** the acid chloride of benzenesulfonic acid.

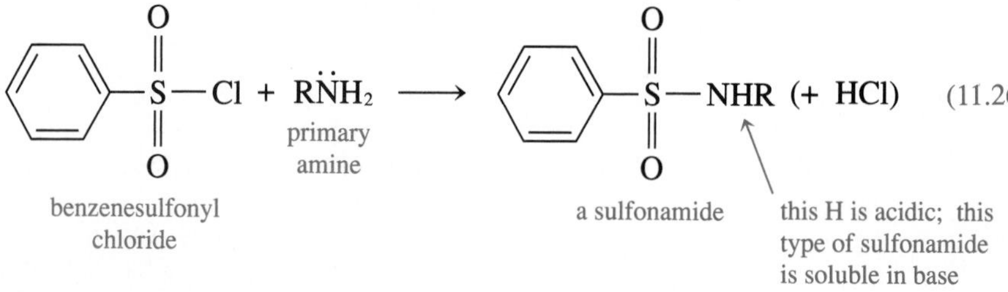

$$\underset{\text{benzenesulfonyl chloride}}{C_6H_5-\overset{O}{\overset{\|}{\underset{\underset{O}{\|}}{S}}}-Cl} + \underset{\text{primary amine}}{R\ddot{N}H_2} \longrightarrow \underset{\text{a sulfonamide}}{C_6H_5-\overset{O}{\overset{\|}{\underset{\underset{O}{\|}}{S}}}-NHR} \ (+\ HCl) \qquad (11.26)$$

this H is acidic; this type of sulfonamide is soluble in base

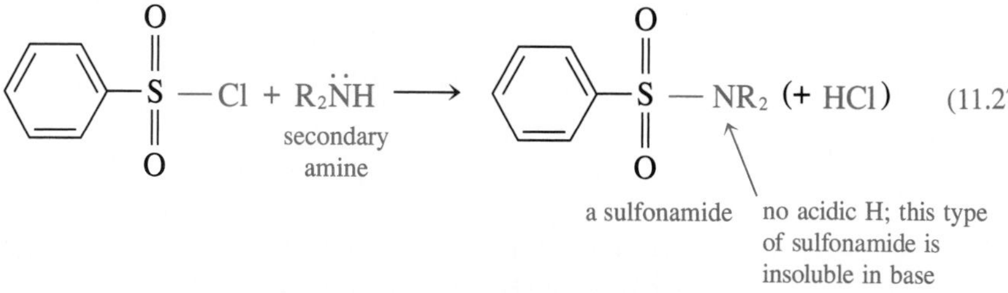

$$C_6H_5-\overset{O}{\overset{\|}{\underset{\underset{O}{\|}}{S}}}-Cl + \underset{\text{secondary amine}}{R_2\ddot{N}H} \longrightarrow \underset{\text{a sulfonamide}}{C_6H_5-\overset{O}{\overset{\|}{\underset{\underset{O}{\|}}{S}}}-NR_2} \ (+\ HCl) \qquad (11.27)$$

no acidic H; this type of sulfonamide is insoluble in base

Primary amines form sulfonamides that have an N—H bond. This proton is acidic, and such sulfonamides are soluble in aqueous sodium hydroxide. Sulfonamides derived from secondary amines have no N—H bond, are insoluble in aqueous base. Tertiary amines cannot eliminate the HCl with sulfonyl chlorides (they have no N—H bond that can be acylated) and show no net reaction.

PROBLEM 11.21 How could you distinguish between this pair of isomers?

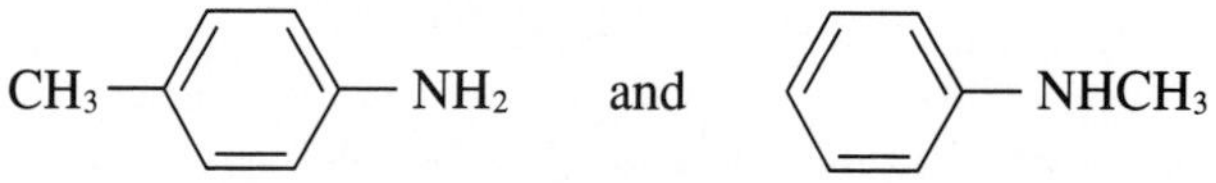

$$CH_3-C_6H_4-NH_2 \quad \text{and} \quad C_6H_5-NHCH_3$$

A Word About . . .

22. Sulfanilamide and Sulfa Drugs

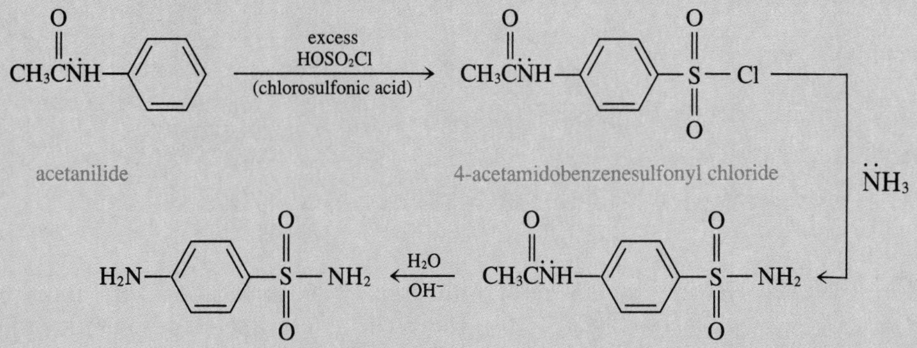

$$CH_3\overset{O}{\overset{\|}{C}}\ddot{N}H-C_6H_4 \xrightarrow[\text{(chlorosulfonic acid)}]{\text{excess } HOSO_2Cl} CH_3\overset{O}{\overset{\|}{C}}\ddot{N}H-C_6H_4-SO_2-Cl$$

acetanilide → 4-acetamidobenzenesulfonyl chloride

$$\xrightarrow{\ddot{N}H_3} CH_3\overset{O}{\overset{\|}{C}}\ddot{N}H-C_6H_4-SO_2-NH_2 \xrightarrow[OH^-]{H_2O} H_2N-C_6H_4-SO_2-NH_2$$

4-acetamidobenzenesulfonamide → sulfanilamide

Sulfa drugs were among the first antibiotics. Developed during World War II, they saved thousands of lives by preventing infection in the wounded. Even today they still see use, although in many instances they have been replaced by more effective and safer antibiotics.

Sulfanilamide is the parent compound of all the sulfa drugs. Its synthesis, which begins with **acetanilide** (eq. 11.25), illustrates several important reaction types that we have already studied. The first step is an electrophilic aromatic substitution, very much like sulfonation (Sec. 4.10c), but using chlorosulfonic acid (the acid chloride of sulfuric acid) as the sulfonating agent. The product is an aromatic sulfonyl chloride. Like other acid chlorides, it reacts with ammonia or amines to form amides—in this case, sulfonamides. The product contains two types of amide substituents, a carboxamide and a sulfonamide. In the third and final step of the synthesis, the carboxamide group is *selectively* hydrolyzed; the sulfonamide group is unaffected.

Sulfanilamide acts by binding to an enzyme site that is normally occupied by ***p*-aminobenzoic acid (PABA),** a compound with a similar structure and shape. PABA is used to synthesize folic acid, a compound essential for normal bacterial growth. By binding to the enzyme site, sulfanilamide blocks folic acid synthesis, and thus inhibits growth of the infecting bacteria.

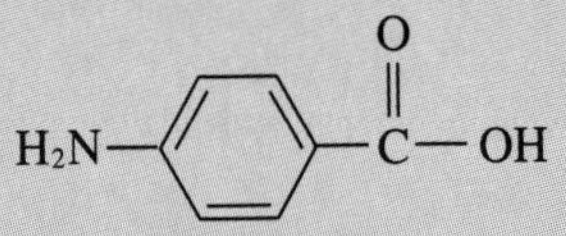

$$H_2N-C_6H_4-\overset{O}{\overset{\|}{C}}-OH$$

p-aminobenzoic acid (PABA)

One problem with sulfanilamide is its low solubility. It can crystallize out in the kidneys or urine, causing discomfort and tissue damage. Within a short time of its discovery, over 5000 analogs of sulfanilamide were synthesized, to correct the deficiencies of the parent drug. The structure was modified mainly by using various amines in place of ammonia, in the second step of the synthesis. **Sulfadiazine,** perhaps the most common of these analogs, is used to treat meningitis, dysentery, and urinary infections.

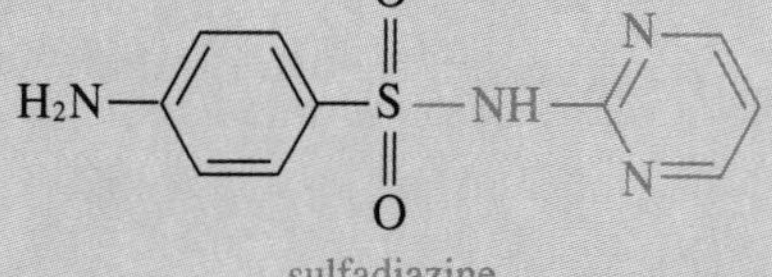

$$H_2N-C_6H_4-SO_2-NH-C_4H_3N_2$$

sulfadiazine

Sulfa drugs were discovered by accident. It was found that a dye called **prontosil,** used to stain bacteria, was an effective antibacterial agent. It was later discovered that the active agent was not prontosil itself, but sulfanilamide, formed from it in the cell. This discovery was then exploited rapidly and systematically, through the preparation of many variations on the basic structure in order to improve the pharmacological properties. Some of these variations led to new drugs with entirely new activity. This pattern of serendipitous discovery, followed by scientific development that engenders another new discovery, is not unusual in drug research.

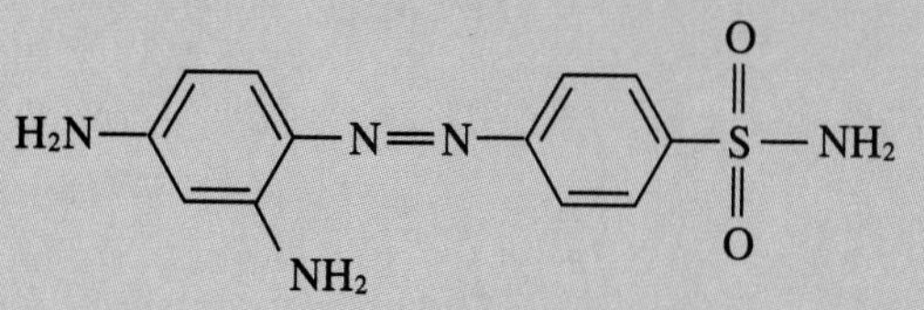

$$H_2N{-}C_6H_3(NH_2){-}N{=}N{-}C_6H_4{-}S(=O)_2{-}NH_2$$

prontosil

11.12 Quaternary Ammonium Compounds

Tertiary amines react with primary or secondary alkyl halides by an S_N2 mechanism (eq. 11.6). The products are **quaternary ammonium salts,** in which all four hydrogens of ammonium ion are replaced by organic groups. For example,

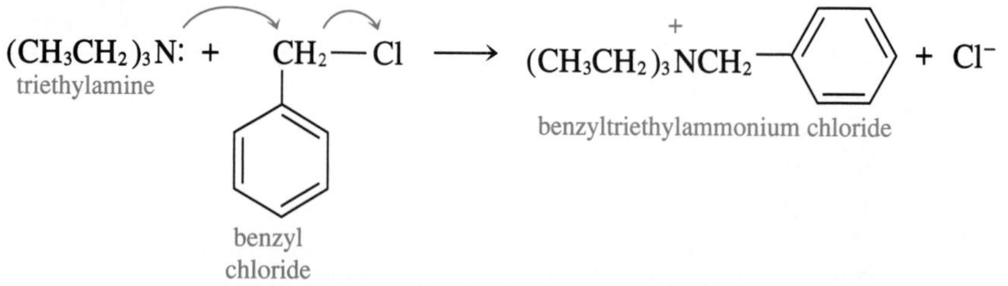

$$\underset{\text{triethylamine}}{(CH_3CH_2)_3N\!:} + \underset{\text{benzyl chloride}}{C_6H_5CH_2{-}Cl} \longrightarrow \underset{\text{benzyltriethylammonium chloride}}{(CH_3CH_2)_3\overset{+}{N}CH_2{-}C_6H_5} + Cl^- \qquad (11.28)$$

Quaternary ammonium compounds are important in biological processes. One of the most common natural quaternary ammonium ions is **choline,** which is present in phospholipids (Sec. 15.7).

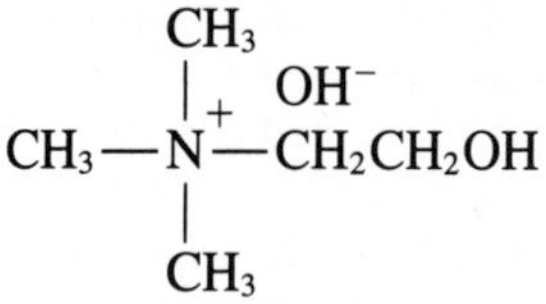

$$\underset{\text{choline}}{(CH_3)_3\overset{+}{N}{-}CH_2CH_2OH\ \ OH^-} \qquad \underset{\text{acetylcholine}}{(CH_3)_3\overset{+}{N}{-}CH_2CH_2{-}O{-}\overset{O}{\overset{\|}{C}}{-}CH_3\ \ OH^-}$$

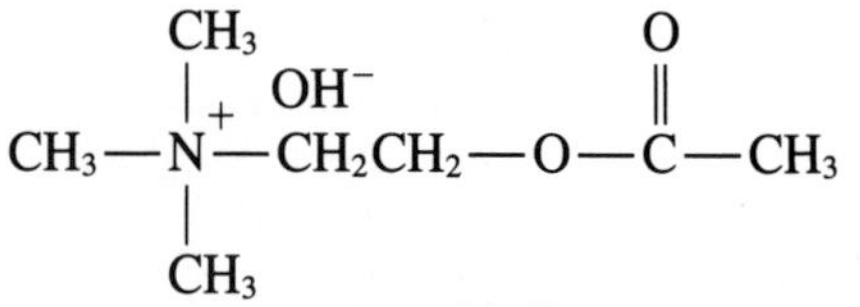

Choline not only is involved in various metabolic processes, but is also the precursor of **acetylcholine,** a compound that plays a key role in the transmission of nerve impulses.

11.13 Aromatic Diazonium Compounds

Primary aromatic amines react with nitrous acid at 0°C to yield ***aryldiazonium ions****. The process is called* ***diazotization****.*

$$\text{C}_6\text{H}_5\text{—NH}_2 + \text{HONO} + \text{H}^+\text{Cl}^- \xrightarrow[\text{aqueous solution}]{0\text{–}5^\circ\text{C}} \text{C}_6\text{H}_5\text{—N}_2^+\text{Cl}^- + 2\,\text{H}_2\text{O} \quad (11.29)$$

aniline; nitrous acid; benzenediazonium chloride

Diazonium compounds are extremely useful synthetic intermediates. Before we describe their chemistry, let us try to understand the steps in eq. 11.29. First we need to examine the structure of nitrous acid.

Nitrous acid decomposes rather rapidly at room temperature. It is therefore prepared as needed by treating an aqueous solution of sodium nitrite with a strong acid at ice temperature. At that temperature, nitrous acid solutions are reasonably stable.

$$\text{Na}^+\text{NO}_2^- + \text{H}^+\text{Cl}^- \xrightarrow{0\text{–}5^\circ\text{C}} \text{H—O—}\ddot{\text{N}}\text{=}\ddot{\text{O}}\text{:} + \text{Na}^+\text{Cl}^- \quad (11.30)$$

sodium nitrite; nitrous acid

The reactive species in reactions of nitrous acid is the **nitrosonium ion** NO^+. It is formed by protonation of the nitrous acid, followed by loss of water (compare with eq. 4.21):

$$\text{H}\ddot{\text{O}}\text{—}\ddot{\text{N}}\text{=}\ddot{\text{O}}\text{:} + \text{H}^+ \rightleftharpoons \text{H}\overset{\oplus}{\ddot{\text{O}}}(\text{H})\text{—}\ddot{\text{N}}\text{=}\ddot{\text{O}}\text{:} \rightleftharpoons \text{H}_2\text{O} + \text{:}\overset{\oplus}{\text{N}}\text{=}\ddot{\text{O}}\text{:} \quad (11.31)$$

nitrosonium ion

How do the two nitrogens, one from the amine and one from the nitrous acid, become bonded to one another, as they appear in diazonium ions? This happens in the first step of diazotization (eq. 11.29), which involves nucleophilic attack of the primary amine on the nitrosonium ion, followed by proton loss.

$$\text{Ar}\ddot{\text{N}}\text{H}_2 + \text{:}\overset{+}{\text{N}}\text{=}\ddot{\text{O}}\text{:} \longrightarrow \text{Ar}\overset{+}{\text{N}}\text{H}_2\text{—}\ddot{\text{N}}\text{=}\ddot{\text{O}}\text{:} \rightleftharpoons \text{Ar}\ddot{\text{N}}\text{H—}\ddot{\text{N}}\text{=}\ddot{\text{O}}\text{:} + \text{H}^+ \quad (11.32)$$

a primary nitrosamine

Protonation of the oxygen in the resulting nitrosamine, followed by elimination of water then gives the aromatic diazonium ion.

$$\text{Ar}\ddot{\text{N}}\text{H—}\ddot{\text{N}}\text{=}\ddot{\text{O}}\text{:} + \text{H}^+ \longrightarrow \text{Ar}\overset{+}{\text{N}}\text{H=}\ddot{\text{N}}\text{—}\ddot{\text{O}}\text{H} \xrightarrow{-\text{H}_2\text{O}} \text{Ar}\overset{+}{\text{N}}\equiv\text{N:} \quad (11.33)$$

aryldiazonium ion

Notice that in the final product there are no N—H bonds; both hydrogens of the amino group are lost, the first in eq. 11.32 and the second in eq. 11.33. Therefore, *only primary amines can be diazotized.* (Secondary and tertiary amines do react with nitrous acid, but their reactions are less important in synthesis.)

Solutions of aryldiazonium ions are moderately stable and can be kept at 0°C for several hours. They are useful in synthesis because the **diazonio group** (—N_2^+) can be replaced by nucleophiles; the other product is nitrogen gas.

$$Ar{-}\overset{+}{N}{\equiv}N\colon + Nu\colon^- \longrightarrow Ar{-}Nu + N_2 \qquad (11.34)$$

Specific useful examples are shown in eq. 11.35. The nucleophile always takes the position on the benzene ring that was occupied by the diazonio group.

CN Cl
H KCN $\overset{+}{N_2}$ HCl OH
$Cu_2(CN)_2$ Cu_2Cl_2
H_3PO_2 H_2O, H^+ (11.35)
I KI HBr Br
Cu_2Br_2
HBF_4,
heat
F

Conversion of diazonium compounds to aryl chlorides, bromides, or cyanides is usually accomplished using cuprous salts, and is known as the **Sandmeyer reaction.** Since a CN group is easily converted to a CO_2H group (eq. 10.13), this provides another route to aromatic carboxylic acids. The reaction with KI gives aryl iodides, usually not easily accessible by direct electrophilic iodination. Similarly, direct aromatic fluorination is difficult, but aromatic fluorides can be prepared from diazonium compounds and tetrafluoroboric acid, HBF_4.

Phenols can be prepared by adding diazonium compounds to hot aqueous acid. This reaction is important because there are not many ways to introduce an —OH group directly on an aromatic ring.

Finally, we sometimes use the orienting effect of a nitro or amino group and afterwards remove this substituent from the aromatic ring. This can be done by diazotization followed by reduction. A common reducing agent for this purpose is **hypophosphorous acid,** H_3PO_2.

Here are some examples of ways that diazonium compounds can be used in synthesis:

EXAMPLE 11.11 How can *m*-dibromobenzene be prepared?

Solution It *cannot* be prepared by direct electrophilic bromination of bromobenzene, because the Br group is *o,p*-directing (Sec. 4.12). But we can take advantage of the *m*-directing effect of a nitro group and then convert the nitro group to a bromine atom, as follows:

$$C_6H_5NO_2 \xrightarrow[FeBr_3]{Br_2} m\text{-}BrC_6H_4NO_2 \xrightarrow[HCl]{SnCl_2} m\text{-}BrC_6H_4NH_2 \xrightarrow[0°]{HONO} m\text{-}BrC_6H_4N_2^+ \xrightarrow[Cu_2Br_2]{HBr} m\text{-}C_6H_4Br_2$$

EXAMPLE 11.12 How can *o*-toluic (*o*-methylbenzoic) acid be prepared from *o*-toluidine (*o*-methylaniline)?

Solution

$$o\text{-}CH_3C_6H_4NH_2 \xrightarrow[2.\ Cu_2(CN)_2,\ KCN]{1.\ HONO} o\text{-}CH_3C_6H_4CN \xrightarrow{H_3O^+} o\text{-}CH_3C_6H_4CO_2H$$

o-toluidine → *o*-toluic acid

EXAMPLE 11.13 Design a route to 1,3,5-tribromobenzene from aniline.

Solution First brominate; the amino group is *o,p*-directing and ring-activating. Then remove the amino group by diazotization and reduction.

$$C_6H_5NH_2 \xrightarrow{3Br_2} 2,4,6\text{-}Br_3C_6H_2NH_2 \xrightarrow[2.\ H_3PO_2]{1.\ HONO} 1,3,5\text{-}C_6H_3Br_3$$

PROBLEM 11.22 Design a synthesis of each of the following compounds, using a diazonium ion intermediate.

a. *m*-bromochlorobenzene from benzene
b. *m*-nitrophenol from *m*-nitroaniline
c. 2,4-difluorotoluene from toluene
d. 3,5-dibromotoluene from *p*-toluidine

11.14 *Diazo Coupling; Azo Dyes*

Being positively charged, aryldiazonium ions are electrophiles. They are *weak* electrophiles, however, because the positive charge can be delocalized through resonance.

EXAMPLE 11.14 Write the resonance contributors for the benzenediazonium ion that show how the nitrogen furthest from the benzene ring can become electrophilic.

Solution

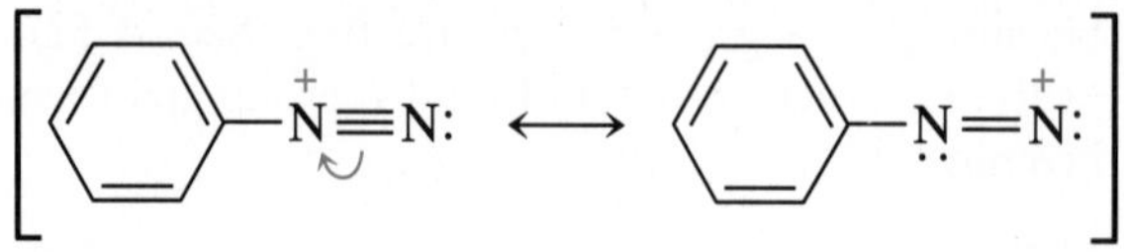

$$\left[C_6H_5-\overset{+}{N}{\equiv}N: \longleftrightarrow C_6H_5-\ddot{N}{=}\overset{+}{N}: \right]$$

In the second contributor, the nitrogen at the right has only six electrons; it can react as an electrophile.

PROBLEM 11.23 Draw resonance contributors which show that the positive charge in benzenediazonium ion can also be delocalized to the ortho and para carbons of the benzene ring. (CAREFUL! These contributors have two positive charges and one negative charge.)

Aryldiazonium ions react with strongly activated aromatic rings (phenols and aromatic amines) to give **azo compounds.** For example,

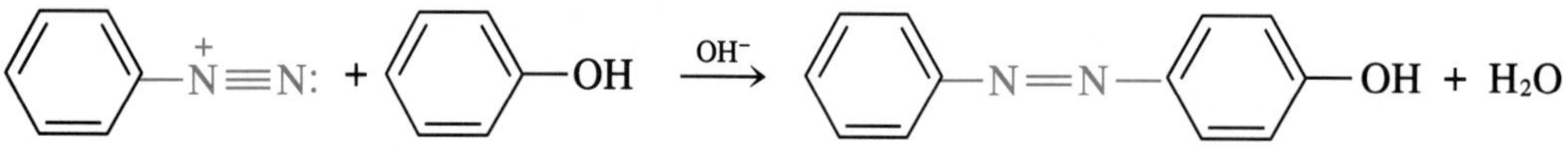

$$C_6H_5-\overset{+}{N}{\equiv}N: + C_6H_5-OH \xrightarrow{OH^-} C_6H_5-N{=}N-C_6H_4-OH + H_2O \quad (11.36)$$

benzenediazonium ion — phenol — *p*-hydroxyazobenzene, yellow leaflets, mp 155–157°C

The nitrogen atoms are retained in the product. This electrophilic aromatic substitution reaction is called **diazo coupling,** because in the product, two aromatic rings are coupled by the azo, or —N=N—, group. *Para* coupling is preferred, as in eq. 11.36, but if the *para* position is blocked by another substituent, *ortho* coupling can occur. *All azo compounds are colored, and many are used commercially as dyes for cloth and in color photography.**

EXAMPLE 11.15 Write out the steps in the mechanism for eq. 11.36.

Solution The reaction is an electrophilic aromatic substitution. The phenoxide ion, formed by dissociation of phenol, is readily attacked, even though the diazonium ion is a weak electrophile.

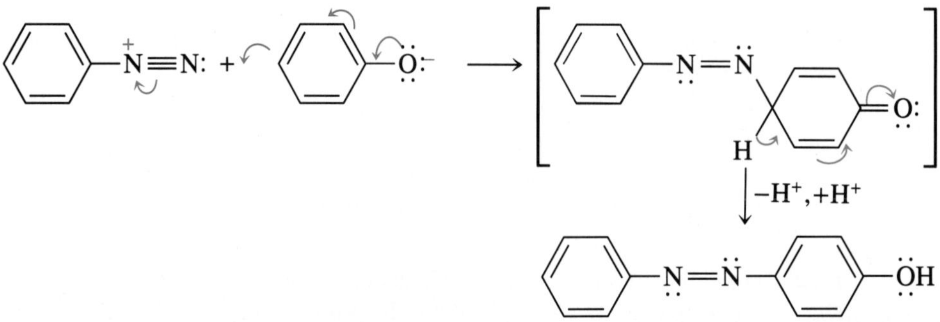

$$C_6H_5-\overset{+}{N}{\equiv}N: + C_6H_5-\ddot{O}:^- \longrightarrow \left[C_6H_5-\ddot{N}{=}\ddot{N}-C_6H_5(H){=}\ddot{O}: \right] \xrightarrow{-H^+,\,+H^+} C_6H_5-\ddot{N}{=}\ddot{N}-C_6H_4-\ddot{O}H$$

* For an interesting discussion of the diazo copying process, see the article by B. Osterby, *J. Chem. Ed.* **1989,** *66,* 1206–1208.

PROBLEM 11.24 Methyl orange is an azo dye used as an indicator in acid-base titrations. (It is yellow-orange above pH 4.5 and red below pH 3.) Show how it can be synthesized from *p*-aminobenzenesulfonic acid (sulfanilic acid) and *N,N*-dimethylaniline.

$$(CH_3)_2N-C_6H_4-N{=}N-C_6H_4-SO_3^-Na^+$$

methyl orange

At this point, we have completed a survey of the main functional groups in organic chemistry. By now, all of the structures in the table inside the front cover of this book should seem familiar to you. In the next chapter, we will describe some modern techniques that help us to assign a structure to a particular molecule. After that, we will conclude with a series of chapters on important commercial and biological applications of organic chemistry.

REACTION SUMMARY

1. Alkylation of Ammonia and Amines to Form Amines (Sec. 11.5)

$$R{-}X + 2\,NH_3 \longrightarrow R{-}NH_2 + NH_4^+X^-$$

$$C_6H_5{-}NH_2 \xrightarrow{R-X} C_6H_5{-}NHR \xrightarrow{R-X} C_6H_5{-}NR_2$$

2. Reduction Routes to Amines (Sec. 11.6)

a. Catalytic or Chemical Reduction of the Nitro Group

$$C_6H_5{-}NO_2 \xrightarrow[\substack{\text{or}\\ 1.\ SnCl_2,\ HCl\\ 2.\ NaOH,\ H_2O}]{H_2,\text{Ni catalyst}} C_6H_5{-}NH_2$$

b. Hydride Reduction of Amides and Nitriles

$$R{-}\overset{O}{\overset{\|}{C}}{-}NR'R'' \xrightarrow{LiAlH_4} R{-}CH_2{-}NR'R''$$

$$R{-}C{\equiv}N\!: \xrightarrow{LiAlH_4} R{-}CH_2{-}NH_2$$

c. Reductive Amination of Aldehydes and Ketones

$$R{-}\overset{O}{\overset{\|}{C}}{-}R' \xrightarrow[NaBH_3CN]{R''NH_2} R{-}\underset{H}{\underset{|}{\overset{NHR''}{\overset{|}{C}}}}{-}R'$$

3. Amines as Bases (Secs. 11.7 and 11.9)

$$R{-}NH_2 + H{-}OH \longrightarrow R{-}\overset{+}{N}H_3 + {}^{-}OH$$

$$R{-}NH_2 + H{-}Cl \longrightarrow R{-}\overset{+}{N}H_3 + Cl^-$$

4. Amines as Nucleophiles

a. Acylation of Amines (Sec. 11.11)

Secondary and Tertiary Amides from Primary and Secondary Amines

$$\underset{\text{(primary amine)}}{R{-}NH_2} \xrightarrow[\text{or } R'C(=O)Cl]{(R'C(=O))_2O} \underset{\text{(secondary amide)}}{R'{-}\overset{O}{\overset{\|}{C}}{-}NHR}$$

$$\underset{\text{(secondary amine)}}{R_2NH} \xrightarrow[\text{or } R'C(=O)Cl]{(R'C(=O))_2O} \underset{\text{(tertiary amide)}}{R'{-}\overset{O}{\overset{\|}{C}}{-}NR_2}$$

b. Hinsberg Test: Sulfonamides (Sec. 11.11)

$$\underset{\text{(primary amine)}}{R-NH_2} + C_6H_5{-}SO_2Cl \longrightarrow \underset{\text{(soluble in aqueous NaOH)}}{C_6H_5{-}SO_2NHR}$$

$$\underset{\text{(secondary amine)}}{R_2NH} + C_6H_5{-}SO_2Cl \longrightarrow \underset{\text{(insoluble in aqueous NaOH)}}{C_6H_5{-}SO_2NR_2}$$

c. Alkylation of Amines: Quaternary Ammonium Salts (Sec. 11.12)

$$R_3N + R'X \longrightarrow R_3\overset{+}{N}{-}R'X^-$$

5. Aryldiazonium Salts: Formation and Reactions (Secs. 11.13 and 11.14)

a. Formation from Aniline and Nitrous Acid (Sec. 11.13)

$$ArNH_2 + HONO \xrightarrow{HX} \underset{\text{(aryldiazonium salt)}}{ArN_2^+X^-}$$

b. Reactions to Form Substituted Benzenes (Sec. 11.13)

$$ArN_2^+ + H_2O \xrightarrow{\text{heat}} ArOH + N_2 + H^+ \quad \text{(phenols)}$$

$$ArN_2^+ + HX \xrightarrow{Cu_2X_2} ArX \quad (X{=}Cl, Br)$$

$$ArN_2^+ + KI \longrightarrow ArI$$

$$ArN_2^+ + KCN \xrightarrow{Cu_2(CN)_2} ArCN$$

$$ArN_2^+ + HBF_4 \longrightarrow ArF$$

$$ArN_2^+ + H_3PO_2 \longrightarrow ArH$$

c. Diazo Coupling (Sec. 11.14)

$$ArN_2^+X^- + C_6H_5OH \longrightarrow p\text{-}HOC_6H_4{-}N{=}N{-}Ar + HX$$

(azo compound)

ADDITIONAL PROBLEMS

11.25. Give an example of each of the following:

a. a primary amine
b. a cyclic tertiary amine
c. a secondary aromatic amine
d. a quaternary ammonium salt
e. an aryldiazonium salt
f. an azo compound
g. a primary amide
h. a sulfonamide

11.26. Write a structural formula for each of the following compounds:

a. *m*-bromoaniline
b. *sec*-butylamine
c. 2-aminopentane
d. dimethylpropylamine
e. *N*-methylbenzylamine
f. 1,2-diaminopropane
g. *N,N*-dimethylaminocyclohexane
h. tetraethylammonium bromide
i. triphenylamine
j. *o*-toluidine
k. 2-methyl-2-butanamine
l. *N,N*-dimethyl-3-hexanamine

11.27. Write a correct name for each of the following compounds:

a. $Cl{-}C_6H_4{-}NH_2$ (para)

b. $CH_3NHCH_2CH_2CH_3$

c. $(CH_3CH_2)_2NCH_3$

d. $(CH_3)_4N^+Cl^-$

e. $CH_3CH(OH)CH_2CH_2NH_2$

f. $H_2N{-}C_6H_{10}{=}O$ (4-aminocyclohexanone ring)

g. $Br{-}C_6H_4{-}N_2^+Cl^-$ (para)

h. $H_3C{-}C_6H_4{-}NHCH_3$ (meta)

i. $C_5H_9{-}NH_2$ (cyclopentyl)

j. $H_2N(CH_2)_6NH_2$

11.28. Draw the structures for, name, and classify as primary, secondary, or tertiary the eight isomeric amines with the molecular formula $C_4H_{11}N$.

11.29. Explain why the difference in the boiling points of isobutane (2-methylpropane; bp −10.2°C) and trimethylamine (bp 2.9°C) is much smaller than the difference in the boiling points of butane (bp −0.5°C) and propylamine (bp 48.7°C). All four compounds have nearly identical formula weights.

11.30. Place the following substances, which have nearly identical formula weights, in order of increasing boiling point: 1-aminobutane, 1-butanol, methyl propyl ether, pentane.

11.31. Give equations for the preparation of the following amines from the indicated precursor.

a. *N,N*-diethylaniline from aniline **b.** *m*-chloroaniline from benzene
c. *p*-chloroaniline from benzene **d.** 1-aminohexane from 1-bromopentane

11.32. Complete the following equations:

a. $C_5H_9{-}NH_2 + CH_2{=}CHCH_2Br \xrightarrow{\text{heat}}$ (cyclopentyl–NH_2)

b. $CH_3\overset{O}{\overset{\|}{C}}Cl + H_2NCH_2CH_2CH(CH_3)_2 \longrightarrow A \xrightarrow{LiAlH_4} B$

c. $CH_3O\overset{O}{\overset{\|}{C}}{-}C_6H_5 \xrightarrow[H^+]{HONO_2} C \xrightarrow[\text{excess}]{LiAlH_4} D$

d. $C_6H_5{-}CH_2Br \xrightarrow{NaCN} E \xrightarrow{LiAlH_4} F \xrightarrow{(CH_3\overset{O}{\overset{\|}{C}})_2O} G$

e. (cyclohexanone) $C_6H_{10}{=}O + CH_3CH_2CH_2NH_2 \xrightarrow{NaBH_3CN} H$

11.33. Adamantadine hydrochloride is used (under the name SYMMETREL®) to treat influenza, a viral respiratory tract illness. Suggest how it might by synthesized from the hydrocarbon adamantane.

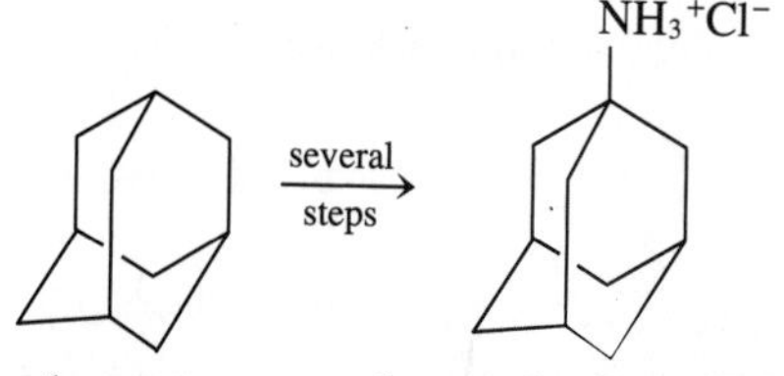

11.34. Tell which is the stronger base and why.

a. aniline or *p*-cyanoaniline **b.** aniline or diphenylamine

11.35. Write out a scheme similar to eq. 11.19 to show how you could separate a mixture of *p*-toluidine, *p*-methylphenol, and *p*-xylene.

$CH_3{-}C_6H_4{-}NH_2$ *p*-toluidine

$CH_3{-}C_6H_4{-}OH$ *p*-methylphenol

$CH_3{-}C_6H_4{-}CH_3$ *p*-xylene

11.36. Draw the important contributors to the resonance hydrid structure of *p*-nitroaniline.

11.37. Write an equation for the reaction of
a. *p*-toluidine with hydrochloric acid.
b. triethylamine with sulfuric acid.
c. dimethylammonium chloride with sodium hydroxide.
d. *N,N*-diethylaniline with methyl iodide.
e. cyclopentylamine with acetic anhydride.

11.38. Write out the steps in the mechanism for the following reaction:

$$CH_3CH_2NH_2 + CH_3\overset{\overset{O}{\|}}{C}O\overset{\overset{O}{\|}}{C}CH_3 \longrightarrow CH_3CH_2NH\overset{\overset{O}{\|}}{C}CH_3 + CH_3COOH.$$

Explain why only one of the hydrogens of the amine is replaced by an acetyl group, even if a large excess of acetic anhydride is used.

11.39. Explain why compound A can be separated into its *R*- and *S*-enantiomers, but compound B cannot.

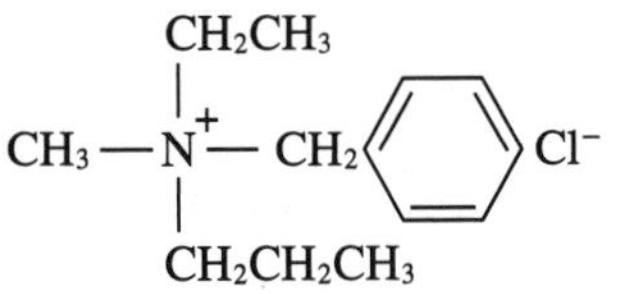

$$CH_3-\overset{+}{N}(CH_2CH_3)(CH_2CH_2CH_3)-CH_2C_6H_5\ Cl^- \qquad CH_3-\ddot{N}(CH_2CH_3)(CH_2CH_2CH_3)$$

compound A 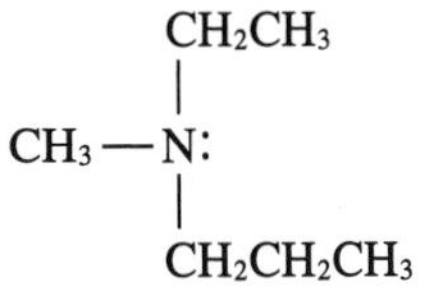 compound B

11.40. Give the priority order of groups in compound A (Problem 11.39), and draw a Fischer projection formula for its *R*-isomer.

11.41. Write equations that show the different behavior of aniline, *N*-methylaniline, and *N,N*-dimethylaniline in the Hinsberg test (Sec. 11.11).

11.42. Answer the following questions with regard to the sulfanilamide synthesis (page 343).
a. Why can we not use aniline in place of acetanilide in the first step (reaction with chlorosulfonic acid)?
b. What would the structure of the final product be if we were to use $H_2\ddot{N}$–(2-pyrimidinyl) in place of ammonia in the second step?
c. Write equations for the mechanism of the last step.

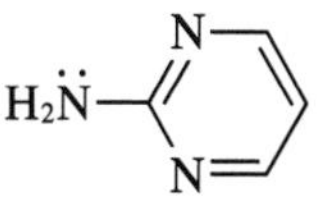

11.43. Choline (Sec. 11.12) can be prepared by the reaction of trimethylamine with ethylene oxide. Write an equation for the reaction, and show its mechanism.

11.44. Acetylcholine (Sec. 11.12) is synthesized in the cell body of neurons. The enzyme choline acetyltransferase catalyzes its synthesis from acetyl-CoA (page 312) and choline. Write an equation for the reaction, using the formula $CH_3\overset{\overset{O}{\|}}{C}-S-CoA$ for acetyl-CoA.

11.45. Primary aliphatic amines RNH_2 react with nitrous acid in the same way that primary arylamines $ArNH_2$ do, to form diazonium ions. But alkyldiazonium ions RN_2^+ are much less stable than aryldiazonium ions ArN_2^+ and readily lose nitrogen even at 0°C. Explain the difference.

11.46. Write an equation for the reaction of $CH_3-C_6H_4-N_2^+HSO_4^-$ with

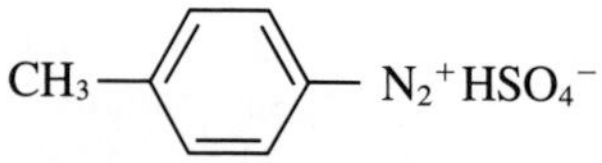

a. KCN and cuprous cyanide. **b.** aqueous acid, heat.
c. HCl and cuprous chloride. **d.** potassium iodide.
e. *p*-methylphenol and OH^-. **f.** *N,N*-dimethylaniline and base.
g. hypophosphorous acid. **h.** fluoroboric acid, then heat.

11.47. Show how diazonium ions could be used to synthesize
a. *p*-bromobenzoic acid from *p*-bromoaniline.
b. *m*-iodochlorobenzene from benzene.
c. *m*-iodoacetophenone from benzene.
d. 3-cyano-4-methylbenzenesulfonic acid from toluene.

11.48. Congo red is used as a direct dye for cotton. Write equations to show how it can be synthesized from benzidine and 1-aminonaphthalene-4-sulfonic acid.

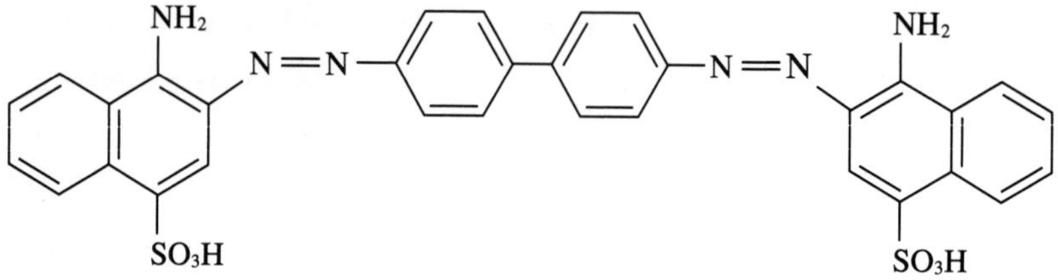

NH₂ N=N N=N NH₂ SO₃H SO₃H

Congo red

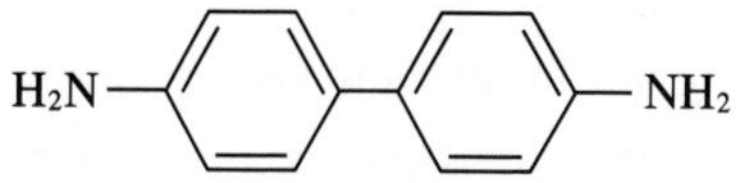

H_2N NH_2

benzidine

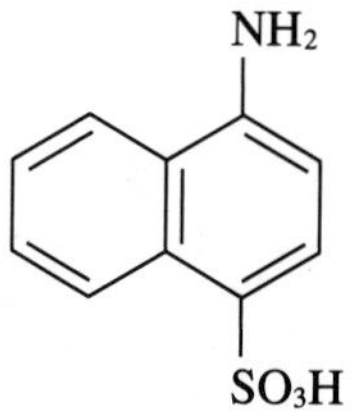

NH_2 SO_3H

1-aminonaphthalene-4-sulfonic acid

11.49. Show how prontosil (page 344) could be synthesized from *p*-aminobenzenesulfonic acid and *m*-dinitrobenzene.

11.50. Sunset yellow is a food dye that can be used to color Easter eggs. Write an equation for an azo coupling reaction that will give this dye.

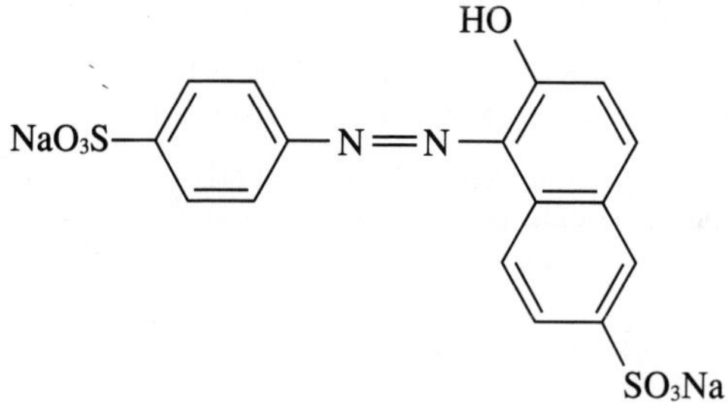

HO NaO_3S N=N SO_3Na

sunset yellow

12 Spectroscopy and Structure Determination

12.1 Introduction

In the early years of organic chemistry, determining the structure of a new compound was often a formidable task. The first step, of course, was an elemental analysis. Knowing the percentage of each element present allowed the empirical formula to be calculated; the molecular formula was then either the same as or a multiple of that formula. Elemental analysis is also an important criterion of the *purity* of a compound.

But how are the atoms arranged? What functional groups are present? And what about the carbon skeleton? Is it acyclic or cyclic, are there branches and where are they located, are benzene rings present? All these questions and more had to be answered by chemical means. Reactions such as ozonolysis or saponification could be used to convert complex molecules to simpler ones whose structures were easier to determine. To identify functional groups, various chemical tests could be applied (such as the bromine or permanganate tests for unsaturation, or the Tollens' silver mirror test for an aldehyde group, or the Hinsberg test for the class of amine).

Once the functionality was known, reactions whose chemistry was well understood could be used to convert the unknown compound in one or more steps to a compound whose structure was already known. For example, if the compound was an aldehyde suspected to have the same R group as a known acid, it could be oxidized. If the physical properties (bp, mp, specific rotation if chiral, and so on) and chemical reactions of the acid obtained from the aldehyde

$$RCH{=}O \xrightarrow{KMnO_4} RCO_2H \qquad (12.1)$$

agreed with those of the known acid, it could safely be concluded that the two R groups *were* the same, and the structure of the aldehyde also became known. If they did *not* agree, one had to do some rethinking about the suspected structure. Ultimate structure proof came through synthesis of the unknown from compounds whose structures were already known, by reactions whose outcome was unambiguous. Gradually, over the years, a vast network of compounds with known structures was built up and catalogued in reference books.

These methods—which often required weeks, months, even years—are still used in appropriate situations. But since the 1940s, various types of spectroscopy have simplified and speeded up the process of structure determination greatly. Automated instruments have been developed that permit us to determine and record spectroscopic properties often with little more effort than pushing a button. And these spectra, if properly interpreted, yield a great deal of structural information.

Spectroscopic methods have many advantages. Usually only a very small sample of material is required, and it can often be recovered if necessary. The methods are rapid, sometimes requiring only a few minutes. And frequently we can obtain more detailed structural information from spectra than from ordinary laboratory methods.

In this chapter, we will describe some of the more important spectroscopic techniques used today and how they can be applied to structural problems. But first, let us examine some principles that form the basis of most of these techniques.

12.2 *Principles of Spectroscopy*

Equation 12.2 describes the relationship between the energy of light, (or any other form of radiation) ***E***, and its **frequency,** ν (Greek nu, pronounced "new").

$$E = h\nu \tag{12.2}$$

The equation says that there is a direct relationship between the frequency of light and its energy: the higher the frequency, the higher the energy. The proportionality constant between the two is known as **Planck's constant, *h*.** Because the frequency of light and its wavelength are *inversely* proportional, the equation can also be written

$$E = hc/\lambda, \text{ because } \nu = c/\lambda \tag{12.3}$$

where λ is the **wavelength** of light and c is the speed of light. In this form, the equation tells us that the shorter the wavelength of light, the higher its energy.

Molecules can exist at various energy levels. For example, the bonds in a given molecule may stretch, bend, or rotate; electrons may move from one orbital to another; and so on. These processes are quantized; that is, bonds may stretch, bend, or rotate only with certain frequencies (or energies; the two are proportional), and electrons may only jump between orbitals with well-defined energy differences. It is these energy (or frequency) differences that we measure by various types of spectra.

The idea behind most forms of spectroscopy is very simple and is expressed schematically in Figure 12.1. A molecule at some energy level, E_1, is exposed to radiation. The radiation passes through the molecule to a detector. As long as the molecule does not absorb the radiation, the amount of radiation detected

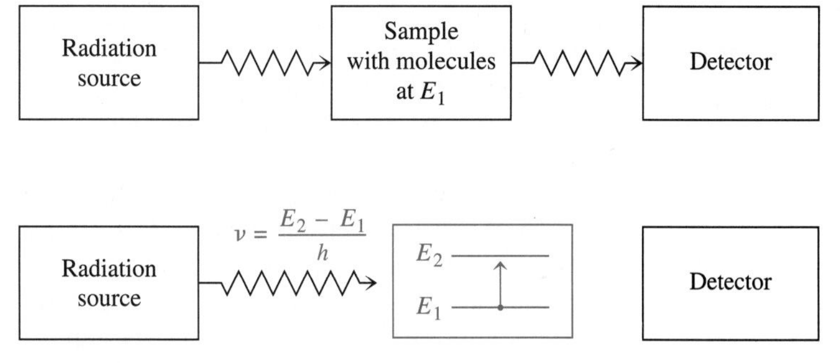

FIGURE 12.1 Radiation passes through the sample unchanged, except when its frequency corresponds to the energy difference between two energy states of the molecule.

will be equal to the amount of radiation emitted by the source (top part of Figure 12.1). At a frequency that corresponds to some molecular energy transition, from E_1 to E_2, the radiation will be absorbed by the molecule and will *not* appear at the detector (bottom part of Figure 12.1). The spectrum, then, consists of a record or plot of the amount of energy (radiation) received by the detector as the input energy is gradually varied.

Some transitions require more energy than others, so we must use radiation of the appropriate frequency to determine them. In this chapter, we will discuss three types of spectroscopy that depend on such transitions. They are nuclear magnetic resonance (NMR), infrared (IR), and ultraviolet-visible (UV-vis) spectroscopy. Table 12.1 summarizes the regions of the electromagnetic spectrum in which transitions for these three types of spectroscopy can be observed. We will begin with NMR spectroscopy and nuclear spin transitions, which require exceedingly small amounts of energy.

TABLE 12.1 Types of spectroscopy and the electromagnetic spectrum

		Region of the spectrum			
Type of spectroscopy	*Radiation source*	*Frequency (hertz)*	*Wavelength (meters)*	*Energy (kcal/mol)*	*Type of transition*
nuclear magnetic resonance	radio waves	$60-600 \times 10^6$ (depends on magnet strength of the instrument)	5–0.5	$6-60 \times 10^{-6}$	nuclear spin
infrared	infrared light	$0.2-1.2 \times 10^{14}$	$15.0-2.5 \times 10^{-6}$	2–12	molecule vibrations
visible-ultraviolet (electronic)	visible or ultraviolet light	$0.375-1.5 \times 10^{15}$	$8-2 \times 10^{-7}$	37–150	electronic states

12.3 Nuclear Magnetic Resonance Spectroscopy (NMR)

The kind of spectroscopy that has had by far the greatest impact on the determination of organic structures is nuclear magnetic resonance (NMR) spectroscopy. Commercial instruments became available in the late 1950s, and since then, NMR spectroscopy has become an indispensable tool for the organic chemist. Let us look briefly at the theory and then see what practical information we can obtain from an NMR spectrum.

Certain nuclei behave as though they are spinning. Because nuclei are charged and a spinning charge creates a magnetic field, these spinning nuclei behave like tiny magnets. The most important nuclei for organic structure determination are ^{1}H (ordinary hydrogen) and ^{13}C, a stable, nonradioactive isotope of ordinary carbon. Although ^{12}C and ^{16}O are present in most organic compounds, they do not possess a spin and do not give NMR spectra.

When nuclei with spin are placed between the poles of a powerful magnet, they align their magnetic fields *with* or *against* the field of the magnet. Nuclei aligned with the applied field have a slightly lower energy than those aligned against the field (Figure 12.2). By applying energy in the radiofrequency range, it is possible to excite nuclei in the lower energy state to the higher energy spin state (we sometimes say that the spins "flip").

The energy gap between the two spin states depends on the strength of the applied magnetic field; the stronger the field, the larger the energy gap. Instruments currently in use have magnetic fields that range from about 1.4 to 14 tesla (T) (by comparison, the earth's magnetic field is only about 0.0007 T). At these field strengths, the energy gap corresponds to a radiofrequency of 60 to 600 MHz (megahertz; 1 MHz = 10^6 Hz or 10^6 cycles per second). Translated to energy units to which chemists are more accustomed, the energy gap between the spin states is only $6–60 \times 10^{-6}$ kcal/mol. Even though this gap is exceedingly small, modern technology permits its detection with great accuracy.

12.3a. Measuring an NMR Spectrum A ^{1}H NMR* spectrum is usually obtained in the following way. A sample of the compound being studied (usually only a few milligrams) is dissolved in some inert solvent that does not contain

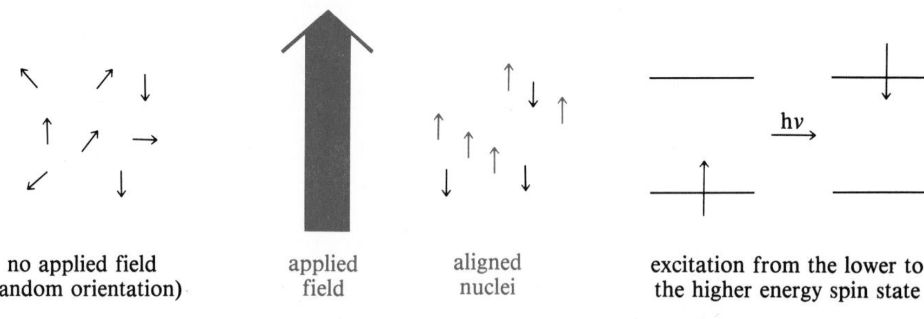

FIGURE 12.2 Orientation of nuclei in an applied field, and excitation of nuclei from the lower to the higher energy spin state.

*The term *proton* is often used interchangeably with *hydrogen* or ^{1}H in discussing NMR spectra, even though the hydrogens are covalently bound (and not H^+). This is an incorrect, but common usage.

^1H nuclei. Examples of such solvents are CCl_4, or solvents with the hydrogens replaced by deuterium, such as $CDCl_3$ (deuteriochloroform) and CD_3COCD_3 (hexadeuterioacetone). A small amount of a reference compound is also added (we will say more about this in the next section). The solution, in a thin glass tube, is placed in the center of a radiofrequency (rf) coil, between the pole faces of a powerful magnet. The nuclei align themselves with or against the field. Continuously increasing amounts of energy can then be applied to the nuclei by the rf coil. When this energy corresponds exactly to the energy gap between the lower and higher energy spin states, it is absorbed by the nuclei. At this point, the nuclei are said to be in resonance with the applied frequency—hence the term **nuclear magnetic resonance.** A plot of the energy absorbed by the sample against the applied frequency of the rf coil gives an NMR spectrum.

In practice, it is usually easier to apply a *constant* rf frequency and slightly vary the strength of the applied magnetic field. One then measures exactly the strength of the magnetic field that corresponds to the applied radiofrequency. The spectra given in this book were obtained in this way. The **applied magnetic field** increases as we go from left to right in the recorded spectra.

12.3b Chemical Shifts and Peak Areas Not all ^1H nuclei flip their spins at precisely the same radiofrequency because they may differ in chemical (and, more particularly, electronic) environment. We will return to this point, but first let us examine some spectra.

Figure 12.3 shows the ^1H NMR spectrum of *p*-xylene. The spectrum is very simple and consists of two peaks. The positions of the peaks are measured in δ

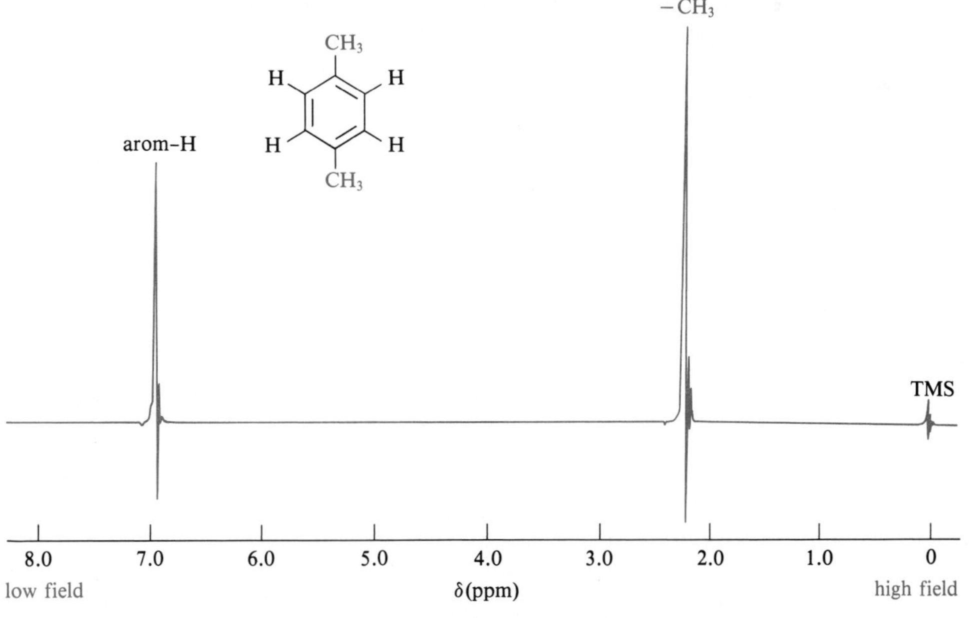

FIGURE **12.3** ^1H NMR spectrum of *p*-xylene.

(delta) units from the peak of a reference compound, which is **tetramethylsilane (TMS),** $(CH_3)_4Si$. The reasons for selecting TMS as a reference compound are (1) all 12 of its hydrogens are equivalent, so it shows only one sharp NMR signal, which serves as a reference point; (2) its ^1H signals appear at higher field than do most ^1H signals in organic compounds, thus making it easy to identify the TMS peak; and (3) TMS is inert, so it does not react with most organic compounds, and it is low-boiling and can be removed easily at the end of a measurement.

Most organic compounds have peaks *downfield* (at lower field) from TMS and are given positive δ values. A δ value of 1.00 means that a peak appears 1 part per million (ppm) downfield from the TMS peak. If the spectrum is measured at 60 MHz (60×10^6Hz), then 1 ppm is 60 Hz (one-millionth of 60 MHz) downfield from TMS. If the spectrum is run at 100 MHz, a δ value of 1 ppm is 100 Hz downfield from TMS, and so on. *The **chemical shift** of a particular kind of ^1H signal is its δ value with respect to TMS.* It is called a *chemical* shift because it depends on the chemical environment of the hydrogens. The chemical shift is independent of the instrument on which it is measured.

$$\text{Chemical shift} = \delta = \frac{\text{distance of peak from TMS, in Hz}}{\text{spectrometer frequency in MHz}}\ \text{ppm} \qquad (12.4)$$

In the spectrum of *p*-xylene, we see a peak at δ 2.20 and another at δ 6.95. It seems reasonable that these peaks are caused by the two different "kinds" of ^1H nuclei in the molecule: the methyl hydrogens and the aromatic ring hydrogens. How can we tell which is which?

One way is to integrate the area under each peak. *The **peak area** is directly proportional to the number of ^1H nuclei responsible for the particular peak.* All commercial NMR spectrometers are equipped with electronic integrators that print out these areas. Thus, we find that the areas of the peaks at δ 2.20 and δ 6.95 in the *p*-xylene spectrum give a ratio of 3 : 2 (or 6 : 4). These areas allow us to assign the peak at δ 2.20 to the six methyl hydrogens and the peak at δ 6.95 to the four aromatic ring hydrogens.

EXAMPLE 12.1 How many peaks do you expect to see in the NMR spectrum of each of the following compounds? If you expect several peaks, what will their relative areas be?

a. $(CH_3)_4C$: $CH_3—C(CH_3)_2—CH_3$

b. p-$C_6H_4(CO_2CH_3)_2$ (benzene ring with CO_2CH_3 at top and bottom)

c. $BrCH_2—C(CH_3)_2—CH_2Br$

Solution

a. All twelve ^1H nuclei are equivalent and appear as a single peak.

b. The four aromatic hydrogens are equivalent, and the six methyl hydrogens on the ester functions are equivalent. There will be two peaks in the spectrum, with an area ratio of 4 : 6 (or 2 : 3).

c. There are two kinds of hydrogens, CH_3—C and CH_2—Br. There will be two peaks, with the area ratio 6 : 4 (or 3 : 2).

PROBLEM 12.1 Which of the following compounds show only a single peak in their NMR spectrum?

a. CH_3OCH_3 b. $CH_3CH_2OCH_2CH_3$ c. (cyclohexane)

PROBLEM 12.2 Each of the following compounds shows more than one peak in its NMR spectrum. What will the area ratios be?

a. CH_3OH b. $CH_3\overset{O}{\overset{\|}{C}}OCH_3$ c. $CH_3CH_2\overset{O}{\overset{\|}{C}}CH_2CH_3$

PROBLEM 12.3 How could ^1H NMR spectroscopy be used to distinguish 1,1-dichloroethane from 1,2-dichloroethane?

A more general way to assign peaks is to compare chemical shifts with those of similar protons in a known reference compound. For example, benzene has six equivalent hydrogens and shows a single peak in its ^1H NMR spectrum, at δ 7.24. Other aromatic compounds also show a peak in this region. We can conclude that most aromatic ring hydrogens will have chemical shifts at about δ 7. Similarly, most CH_3—Ar hydrogens appear at δ 2.2–2.5 (see Figure 12.3).

The chemical shifts of ^1H nuclei in various chemical environments have been determined by measuring the ^1H NMR spectra of a large number of compounds with known, relatively simple structures. Table 12.2 gives the chemical shifts for several common types of ^1H nuclei.

EXAMPLE 12.2 Using the data in Table 12.2, describe the expected ^1H NMR spectrum of

a. $CH_3\overset{O}{\overset{\|}{C}}—OCH_3$ b. $Cl_2CH—\overset{CH_3}{\underset{CH_3}{\overset{|}{\underset{|}{C}}}}—CH_2Cl$

Solution a. The spectrum will consist of two peaks, equal in area, at about δ 2.3 (for the $CH_3\overset{O}{\overset{\|}{C}}$— hydrogens) and δ 3.6 (for the —OCH_3 hydrogens).

b. The spectrum will consist of three peaks, with relative areas 6 : 2 : 1 at δ 0.9 (the two methyls), δ 3.5 (the —CH_2—Cl hydrogens), and δ 5.8 (the —$CHCl_2$ hydrogen).

TABLE 12.2 Typical ^{1}H chemical shifts (relative to tetramethylsilane)

Type of ^1H	δ (ppm)	Type of ^1H	δ (ppm)
C—CH_3	0.85–0.95	—CH_2—F	4.3–4.4
C—CH_2—C	1.20–1.35	—CH_2—Br	3.4–3.6
C—CH(—C)—C	1.40–1.65	—CH_2—I	3.1–3.3
		CH_2=C	4.6–5.0
CH_3—C=C	1.6–1.9	—CH=C	5.2–5.7
CH_3—Ar	2.2–2.5	Ar—H	6.6–8.0
CH_3—C(—)=O	2.1–2.6	—C≡C—H	2.4–2.7
CH_3—N<	2.1–3.0	—C(=O)—H	9.5–9.7
CH_3—O—	3.5–3.8	—C(=O)—OH	10–13
—CH_2—Cl	3.6–3.8	R—OH	0.5–5.5
—$CHCl_2$	5.8–5.9	Ar—OH	4–8

PROBLEM 12.4 Describe the expected ^1H NMR spectrum of

a. $CH_3C(=O)OH$ b. $(CH_3)_2C{=}CH_2$

PROBLEM 12.5 An ester is suspected of being either $(CH_3)_3CC(=O)OCH_3$ or $CH_3C(=O)$—$OC(CH_3)_3$. Its ^1H NMR spectrum consists of two peaks at δ 0.9 and δ 3.6 (relative areas 3 : 1). Which compound is it? Describe the spectrum that would be expected if it had been the other ester.

Now let us return to the point mentioned at the beginning of this section, the factors that influence chemical shifts. One important factor is the **electronegativity of groups in the immediate environment of the ^1H nuclei. Electron-withdrawing groups generally cause a downfield chemical shift.** Compare, for example, the following chemical shifts from Table 12.2:

$—CH_3$ $—CH_2Cl$ $—CHCl_2$

~ 0.9 ~ 3.7 ~ 5.8

Electrons in motion near a ^1H nucleus create a small magnetic field in its microenvironment that tends to shield the nucleus from the externally applied magnetic field. Chlorine is an electron-withdrawing group. Withdrawal of electron density by the chlorines therefore "deshields" the nucleus, allowing it to flip its spin at a lower applied external field. The more chlorines, the larger the effect.

EXAMPLE 12.3 Predict the order of chemical shifts of the various ^1H signals for 1-bromopropane.

Solution $\overset{3}{C}H_3\overset{2}{C}H_2\overset{1}{C}H_2Br$

The hydrogens at C1 will be at lowest field because they are closest to the electron-withdrawing Br atom. The methyl hydrogens will be at highest field because they are farthest from the Br, and the peak for the C2 hydrogens will appear between the other two. The inductive effect falls off rapidly with distance, as seen by the actual chemical shift values.

δ 1.06 1.81 3.47

$CH_3—CH_2—CH_2—Br$

PROBLEM 12.6 Explain the following chemical shifts:

δ	0.23	3.05	2.68	2.16
	CH_4	CH_3Cl	CH_3Br	CH_3I

A second factor that influences chemical shifts is the presence of pi electrons. Hydrogens attached to a carbon that is part of a multiple bond or aromatic ring usually appear downfield from hydrogens attached to saturated carbons. Compare these values from Table 12.2:

$C—CH_2—C$	$CH_2{=}C$	$—CH{=}C$	H—(benzene ring)
δ 1.2–1.35	4.6–5.0	5.2–5.7	6.6–8.0

The reasons for this effect are complex, but it is useful in assigning structures.

PROBLEM 12.7 Describe the ^1H NMR spectrum of *trans*-2,2,5,5-tetramethyl-3-hexene.

12.3c Spin–Spin Splitting Many compounds give spectra that show more complex peaks than just single peaks **(singlets)** for each type of hydrogen. Let us examine some spectra of this type to see what additional structural information they convey.

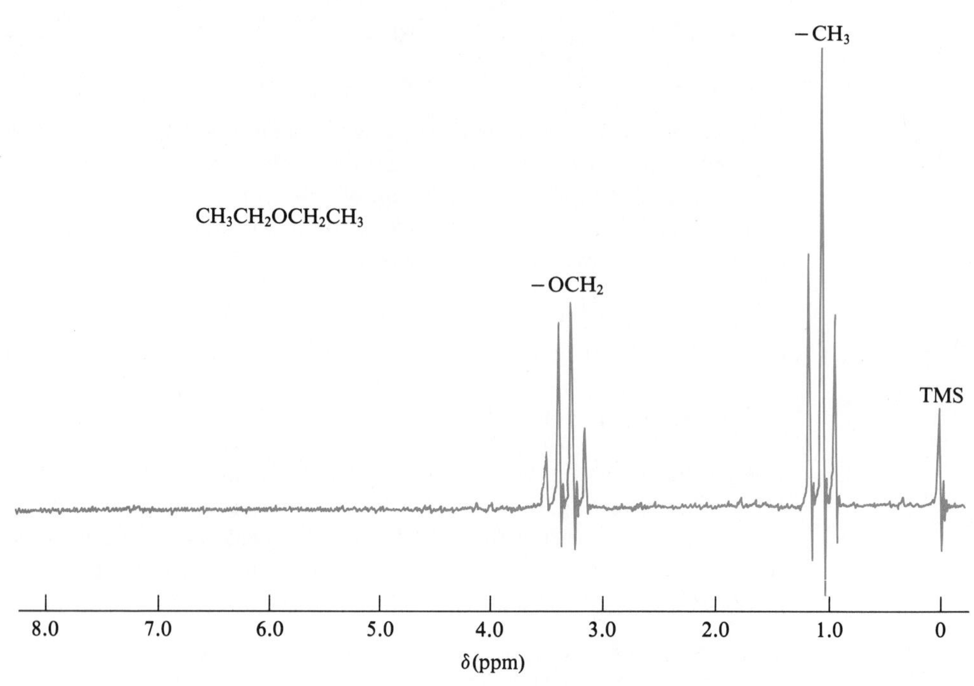

FIGURE 12.4 ^{1}H NMR spectrum of diethyl ether, showing spin–spin splitting.

Figure 12.4 above shows the ^{1}H NMR spectrum of diethyl ether, $CH_3CH_2OCH_2CH_3$. From the information given in Table 12.2, we might have expected the ^{1}H NMR spectrum of diethyl ether to consist of two lines: one in the region of δ 0.9 for the six equivalent CH_3 hydrogens and one at about δ 3.5 for the four equivalent CH_2 hydrogens adjacent to the oxygen atom, with relative areas 6 : 4. Indeed, in Figure 12.4 we see absorptions in each of these regions, with the expected total area ratio. But we do not see singlets! Instead, the methyl signal is split into three peaks, a **triplet,** with relative areas 1 : 2 : 1; and the methylene signal is split into four peaks, a **quartet,** with relative areas 1 : 3 : 3 : 1. These **spin–spin splittings,** as they are called, tell us quite a bit about molecular structure, and they arise in the following way.

We know that each ^{1}H nucleus in the molecule acts as a tiny magnet. When we run a ^{1}H NMR spectrum, each hydrogen "feels" not only the very large applied magnetic field but also a tiny field due to its neighboring hydrogens. During the time that we sweep through one signal, the ^{1}H nuclei on neighboring carbons can be in either the lower or the higher energy spin state, with nearly equal probabilities (nearly equal because, as we have said, the energy difference between the two states is exceedingly small). So the magnetic field of the nuclei whose peak we sweep through is perturbed slightly by the tiny fields of its neighboring ^{1}H nuclei.

We can predict the splitting pattern by **the $n + 1$ rule: if a ^{1}H nucleus or a set of equivalent ^{1}H nuclei has n ^{1}H neighbors with a substantially different chemical shift, its NMR signal will be split into $n + 1$ peaks.** In diethyl ether, each CH_3 hydrogen has *two* ^1H neighbors (on the CH_2 group). Therefore, the CH_3 signal is split into $2 + 1 = 3$ peaks. At the same time, each CH_2 hydrogen has *three* ^1H neighbors (on the CH_3 group). The CH_2 signal is therefore split into $3 + 1 = 4$ peaks. Let us see why this rule works and why the split peaks have the area ratios they do.

Consider first the system

```
 |    |
—C—C—
 |    |
 Ha   Hb
```

H_a has *one* nonequivalent neighbor, H_b. At the time we pass through the H_a signal, H_b can be in either the lower or the higher energy spin state. Because these two possibilities are nearly equal, the H_a signal will be split into two *equal* peaks: a **doublet.** The same is true for the peak due to H_b.

Now consider the system

```
      Hb
 |    |
—C—C—
 |    |
 Ha   Hb
```

H_a has *two* neighbors, H_b. At the time we pass through the H_a signal, there are three possibilities for the two H_b nuclei:

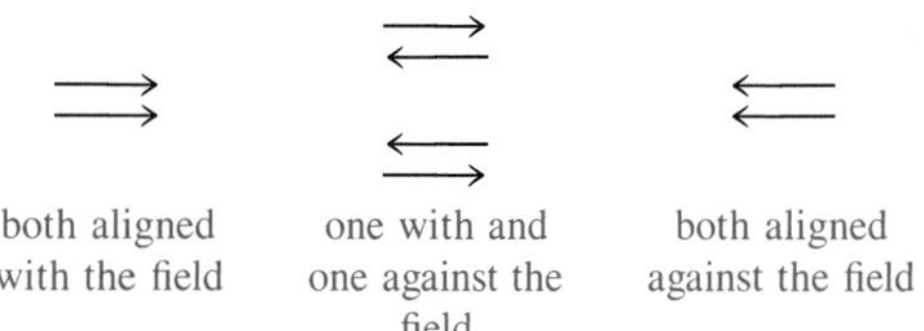

Both can be in the lower energy state, both can be in the higher energy state, or one can be in each state, with this latter arrangement being possible in two ways. Hence the H_a signal will be a *triplet* with relative areas 1 : 2 : 1. The H_b signal, on the other hand, will be a *doublet* because of the two possible spin states for H_a.

EXAMPLE 12.4 Explain why the signal of H_a with *three* neighboring nuclei H_b is a *quartet* with relative areas 1 : 3 : 3 : 1.

Solution The system is

```
      Hb
 |    |
—C—C—Hb      or     >CH—CH3
 |    |
 Ha   Hb
```

The possibilities for the spin states of the three H_b nuclei are

$$\begin{matrix}\rightarrow \\ \rightarrow \\ \leftarrow\end{matrix} \quad \begin{matrix}\rightarrow \\ \leftarrow \\ \leftarrow\end{matrix}$$

$$\begin{matrix}\rightarrow \\ \rightarrow \\ \rightarrow\end{matrix} \quad \begin{matrix}\rightarrow \\ \leftarrow \\ \rightarrow\end{matrix} \quad \begin{matrix}\leftarrow \\ \rightarrow \\ \leftarrow\end{matrix} \quad \begin{matrix}\leftarrow \\ \leftarrow \\ \leftarrow\end{matrix}$$

$$\begin{matrix}\leftarrow \\ \rightarrow \\ \rightarrow\end{matrix} \quad \begin{matrix}\leftarrow \\ \leftarrow \\ \rightarrow\end{matrix}$$

Hence the H_a signal will appear as four peaks (a quartet), with the area ratio 1 : 3 : 3 : 1. The H_b peak will appear as a doublet, because the possible alignments of H_a are $\longrightarrow$ or $\longleftarrow$.

PROBLEM 12.8 Use the data in Table 12.1 to predict the ^1H NMR spectrum of CH_3CHCl_2. Give the approximate chemical shifts *and* the splitting patterns of the various peaks.

^1H nuclei that split one another's signals are said to be **coupled.** The extent of the coupling, or the number of hertz by which the signals are split, is called the **coupling constant** (abbreviated J). A few typical coupling constants are shown in Table 12.3. ***Spin–spin splitting falls off rapidly with distance.*** Whereas hydrogens on adjacent carbons may show appreciable splitting (J = 6–8 Hz), hydrogens farther apart hardly "feel" each other's presence (J = 0–1 Hz). As seen in Table 12.3, coupling constants can even be used at

TABLE 12.3 Some typical coupling constants

Group	*J (Hz)*	*Group*	*J (Hz)*
—C(H)—C(H)—	6–8	H–C$_6$H$_4$–H (benzene ring)	*ortho:* 6–10 *meta:* 1–3 *para:* 0–1
—C(H)—C—C(H)—	0–1	(H)(H)C=C(R$_1$)(R$_2$)	0–3
(H)(R$_1$)C=C(R$_2$)(H)	12–18	(H)(R$_1$)C=C(H)(R$_2$)	6–12

times to distinguish between *cis-trans* isomers or between positions of substituents on a benzene ring.

Chemically equivalent ^{1}H nuclei do not split each other. For example, $BrCH_2CH_2Br$ shows only a sharp singlet in its ^{1}H NMR spectrum for all four hydrogens. Even though they are on adjacent carbons, the hydrogens do not split each other because they have identical chemical shifts.

PROBLEM 12.9 Describe the ^{1}H NMR spectrum of

a. ICH_2CH_2Cl b. $ClCH_2CH_2Cl$

Not all ^{1}H NMR spectra are simple; they may sometimes be quite complex. This complexity can arise when adjacent hydrogens have nearly the same, but not identical, chemical shifts. An example is phenol (see Figure 12.5). We can easily distinguish the aromatic hydrogens (δ 6.6–7.4) from the hydroxyl hydrogen (δ 5.85), but the splitting pattern of the complex **multiplet** seen for the aromatic hydrogens cannot be analyzed with the simple $n + 1$ rule. Such spectra can, however, be thoroughly analyzed by specially designed computer programs.

In summary, then, ^{1}H NMR spectroscopy can give us the following kinds of structural information:

1. **The number of signals and their chemical shifts can be used to identify the kinds of chemically different ^{1}H nuclei in the molecule.**
2. **The peak areas tell us how many ^{1}H nuclei of each kind are present.**
3. **The spin–spin splitting pattern gives us information about the number of nearest ^{1}H neighbors that a particular kind of ^{1}H nucleus may have.**

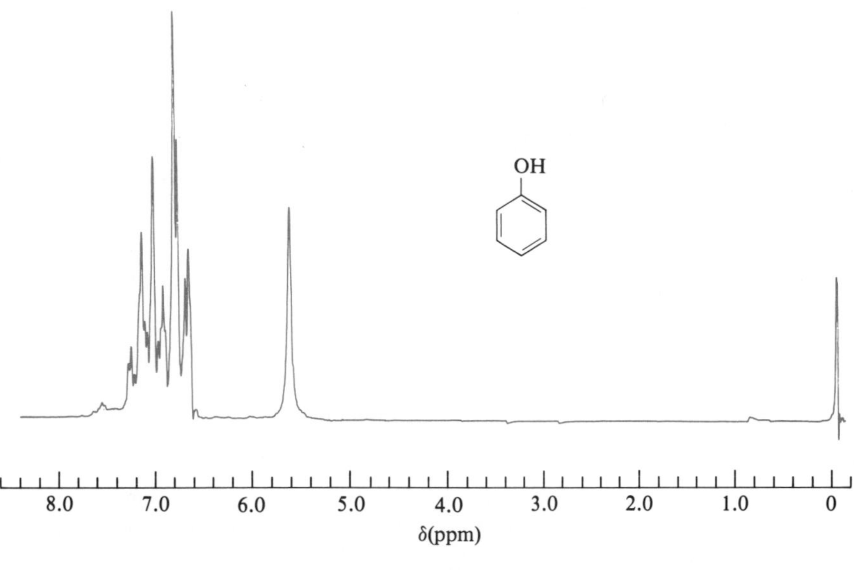

FIGURE 12.5
The ^{1}H NMR spectrum of phenol. Note the complexity in the aromatic ^{1}H region (δ 6.6–7.4). Reprinted with permission from University Science Books.

12.4 ^{13}C NMR Spectroscopy

Whereas ^1H NMR spectroscopy gives information about the arrangement of hydrogens in a molecule, ^{13}C NMR spectroscopy gives information about the carbon skeleton. The ordinary isotope of carbon, carbon-12, does not have a nuclear spin, but carbon-13 does. ^{13}C constitutes only 1.1% of naturally occurring carbon atoms. Also, the energy gap between the higher and lower spin states of ^{13}C is very small. For these two reasons, ^{13}C NMR spectrometers must be exceedingly sensitive. Nevertheless, the use of such instruments has become fairly routine in recent years.

^{13}C spectra differ from ^1H spectra in several ways. ^{13}C chemical shifts occur over a wider range than those of ^1H nuclei. They are measured against the same reference compound, TMS, whose methyl carbons are all equivalent and give a sharp signal. Chemical shifts for ^{13}C are reported in δ units, but the usual range is about 0 to 200 ppm downfield from TMS (instead of the smaller range of 0 to 10 ppm observed for ^1H). This wide range of chemical shifts tends to simplify ^{13}C spectra relative to ^1H spectra.

Because of the low natural abundance of ^{13}C, the chance of finding two adjacent ^{13}C atoms in the same molecule is small. Hence ^{13}C–^{13}C spin–spin splitting is ordinarily not seen. This feature simplifies ^{13}C spectra. However, ^{13}C–^1H spin–spin splitting can occur. A spectrum can be run in such a way as to show this splitting or not, as desired. Figure 12.6 shows the ^{13}C NMR spectrum of 2-butanol measured with and without ^{13}C–^1H splitting. The spectrum without ^1H splitting (called a **proton-decoupled spectrum**) shows four sharp singlets, one for each type of carbon atom. The hydroxyl-bearing carbon occurs at the lowest field (δ 69.3), and the two methyl carbons are well separated (δ 10.8 and 22.9). In the spectrum *with* ^{13}C–^1H splitting, the $n + 1$ rule applies. The signal for each type of ^{13}C nucleus is split by the ^1H nuclei bonded directly to it. Both CH_3 signals are quartets (three hydrogens, therefore $n + 1 = 4$), the CH_2 carbon is a triplet, and the CH carbon is a doublet.

EXAMPLE 12.5 Describe the ^{13}C spectrum of CH_3CH_2OH.

Solution The spectrum without ^{13}C–^1H splitting consists of two lines because there are two nonequivalent carbons (their signals come at δ 18.2 and 57.8). With ^{13}C–^1H splitting, the signal at δ 18.2 is a quartet, and the one at δ 57.8 is a triplet.

PROBLEM 12.10 Describe the main features of the ^{13}C spectrum of $CH_3CH_2CH_2OH$.

PROBLEM 12.11 How many peaks would you expect to see in the ^1H-decoupled ^{13}C NMR spectrum of

a. cyclopentane?
b. 3-pentanone?
c. 2-methyl-1-propanol?
d. *cis*-1,3-dimethylcyclopentane?

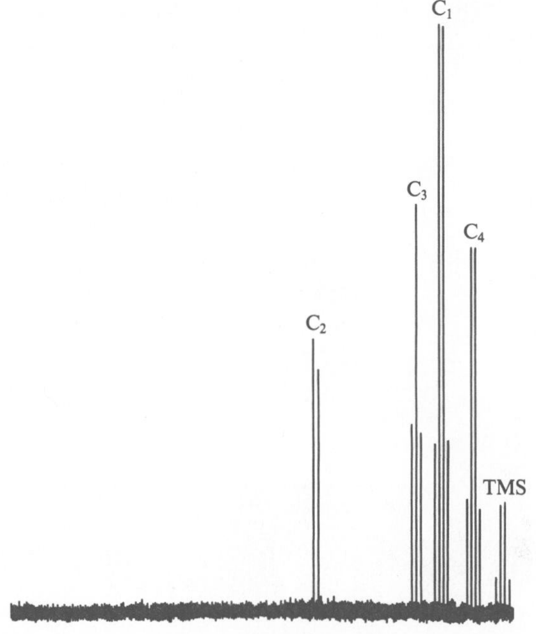

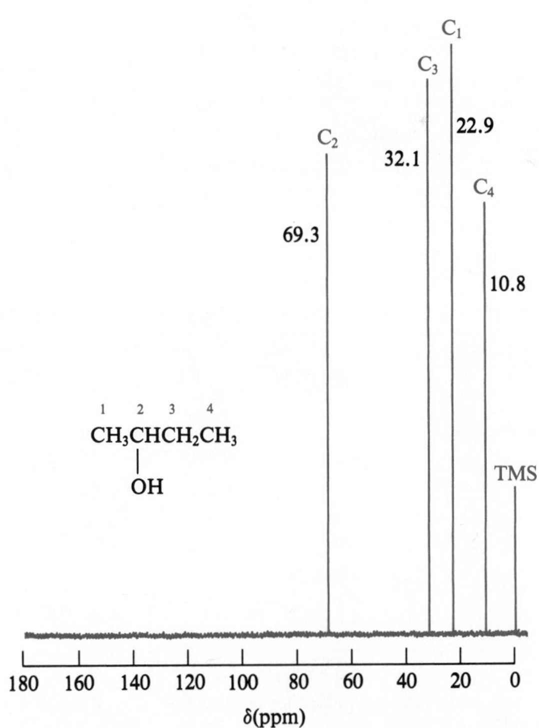

FIGURE 12.6 The ^{13}C NMR spectrum of 2-butanol without (bottom) and with (top) ^{13}C–^{1}H coupling. δ values are shown in the lower spectrum. Reprinted with permission from University Science Books.

A Word About . . .

23. NMR in Biology and Medicine

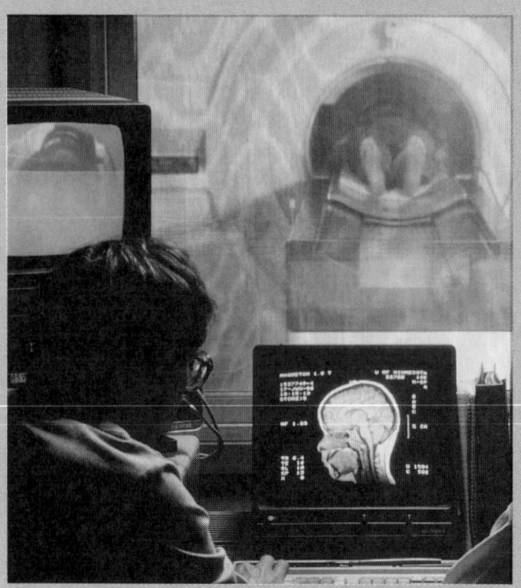

In this chapter, we have presented only the bare bones of NMR spectroscopy, with examples of organic structure determination. But with the application of computer technology, particularly minicomputers and microprocessors, to NMR instrumentation, the field has developed rapidly and added many sophisticated techniques.* It is possible, for example, to measure spectra at variable temperatures (−180 to +200°C) and to measure the rates of processes such as rotations about single bonds, "flipping" of cyclohexane chair conformations, nitrogen inversion in amines, and reactions of free radicals. With other techniques, it is possible to obtain not only ^{13}C spectra of *all* the carbons in a molecule, but separate spectra of *only* CH_3 groups (or only CH_2 or only CH groups), which helps enormously in determining the structures of complex biological molecules. Other techniques allow one to locate groups in the same molecule that are close in space even though they may seem to be far apart if we count by bonds, a real help in studying conformations of complex molecules.

The potential of NMR spectroscopy for solving biological and medical problems is now being realized. For example, instruments are now sufficiently sensitive that one can study intact bodily fluids such as urine, blood plasma, seminal fluid, cerebrospinal fluid, and eye fluid. **Creatinine,** for example, is a normal organic protein metabolite excreted in the urine, where it is present at millimolar levels. The singlets due to its CH_3 (δ 3.1) and CH_2 (δ 4.2) groups are easily observed at these low concentrations with a high resolution (500 MHz) spectrometer, even in the presence of all the other constituents of urine. Its concentration, which can be determined in just a few minutes, provides a useful indicator of kidney function. NMR spectroscopy has been used to detect inherited metabolic diseases by studying the urine of newborn infants.

creatinine

Other bodily fluids also provide useful medical information. Monitoring low-density and high-density lipoproteins in blood plasma can help in understanding heart disease. Studies of seminal fluid could be of use for problems of infertility. Neurologists are examining NMR spectra of cerebrospinal fluid in infants for new leads on brain disorders.**

Other biologically important nuclei with spin are ^{31}P, ^{23}Na, and ^{19}F. Instrumentation has been developed for studying NMR spectra of intact

* For excellent descriptions of these methods, see Lambert, J. B.; Shurvell, H. F.; Lightner, D. A.; Cooks, R. G. *Introduction to Organic Spectroscopy,* Macmillan: New York, 1987.

** For further information, see Bell, J. D.; Brown, J. C. C.; and Sadler, P. J., "NMR Spectroscopy of Body Fluids," *Chemistry in Britain* **1988,** 1021–1024.

human and animal body parts. For example, ^{31}P spectra of human forearm muscle (the forearm is placed directly in the magnetic field) taken before, during, and after exercise has allowed the monitoring of several phosphorus-containing components of muscle tissue—for example, adenosine triphosphate (ATP), phosphocreatine, and inorganic phosphate. By comparing concentration changes of these components in normal people with those in patients with muscle disorders, the nature of the disorders and methods of treatment can be devised.*

In **topical NMR,** the NMR magnet is brought to the object being studied, instead of vice versa. A magnetic probe of some sort placed on the surface induces resonances in molecules close to the surface, and useful ^{1}H, ^{13}C or ^{31}P spectra of molecules within living bodies can be obtained in this way. The method has been used, for example, to monitor effects of various drugs on metabolic processes.

Magnetic resonance imaging (MRI) is a new technique that has been used clinically in hospitals only since the mid-1980s. The method allows one to obtain internal images of whole-body parts and has several advantages over x-rays. For one, it is far less hazardous, since it does not cause radiation damage. For another, it gives good images of soft tissues, which are more difficult to obtain with x-rays. How does it work?

A large percentage of the human body is water. Most MRI uses the ^{1}H spectra of water in various tissues (although ^{31}P has also been used) to create images. The magnet must be very large, with about a 25-cm-diameter gap to accommodate a human head, and an even larger one for whole-body work. Fortunately, the magnetic field need not be as uniform as in high-resolution NMR for structure determination, so that construction of such large magnets becomes feasible. A gradient field, instead of a uniform field, is essential for imaging. The field strength can also be appreciably lower than for laboratory NMR work. This is fortunate as well because, as you can imagine, the large magnets required for imaging are quite expensive.

The time required for MRI varies from about thirty seconds to ten minutes, with about two minutes being fairly typical. Imaging can be done at preselected planar sections through the body or in three dimensions. The picture at the beginning of this "A Word About" shows a cross-section of a human head obtained by MRI.

The principal use of MRI is, of course, in *diagnostic medicine,* but it can also be used in *food science* to assess correct conditions for picking, storing, or marketing food; in *agriculture* to study seed germination; in the *building industry* to study water concentration and distribution in timbers; and in many other areas.**

Nuclear magnetic resonance was discovered by physicists and developed by chemists and is now being applied by biologists and others. The benefits to humanity could not have been foreseen in the beginning—another example of the wisdom of investing in fundamental research without immediate regard for practical applications.

*For more details, see Radda, G. K., "The Use of NMR Spectroscopy for the Understanding of Disease," *Science* **1986,** *233,* 640–645.

**For more about MRI see the book by Morris, P. G. *Nuclear Magnetic Resonance in Medicine and Biology,* Clarendon Press: Oxford, England, 1986.

12.5 Infrared Spectroscopy

Even though NMR spectroscopy is a powerful tool for deducing structures, it is usually supplemented by other spectroscopic methods that provide additional structural information. One of the more important of these is **infrared spectroscopy.**

Infrared frequency is usually expressed in units of **wavenumber,** defined as *the number of waves per centimeter*. Ordinary instruments scan the range of about 700 to 5000 cm^{-1}. This frequency range corresponds to energies of about

2 to 12 kcal/mol (Table 12.1). This amount of energy is sufficient to affect bond vibrations (motions such as *bond stretching* or *bond bending*) but is appreciably less than would be needed to break bonds. These motions are exemplified for a CH_2 group in Figure 12.7.

Particular types of bonds usually stretch within certain rather narrow frequency ranges. ***Infrared spectroscopy is particularly useful for determining the types of bonds that are present in a molecule.*** Table 12.4 gives the ranges of stretching frequencies for some bonds commonly found in organic molecules.

The infrared spectrum of a compound can easily be obtained in a few minutes. A small sample of the compound is placed in an instrument with an infrared radiation source. The spectrometer automatically scans the amount of radiation that passes through the sample over a given frequency range and records on a chart the percentage of radiation that is transmitted. Radiation absorbed by the molecule appears as a band in the spectrum.

Figure 12.8 on page 374 shows two typical infrared spectra. Both spectra show C—H stretching bands near 3000 cm^{-1} and a C=O stretching band near 1700 cm^{-1}. **Functional group bands** will appear in the same range regardless of the details of the molecular structure. Both cyclopentanone and cyclohexanone have C—H and C=O bonds and therefore have similar spectra in the functional group region of the spectrum (1500 to 4000 cm^{-1}).

FIGURE 12.7
Stretching and bending motions of a CH_2 group that require energies in the infrared region.

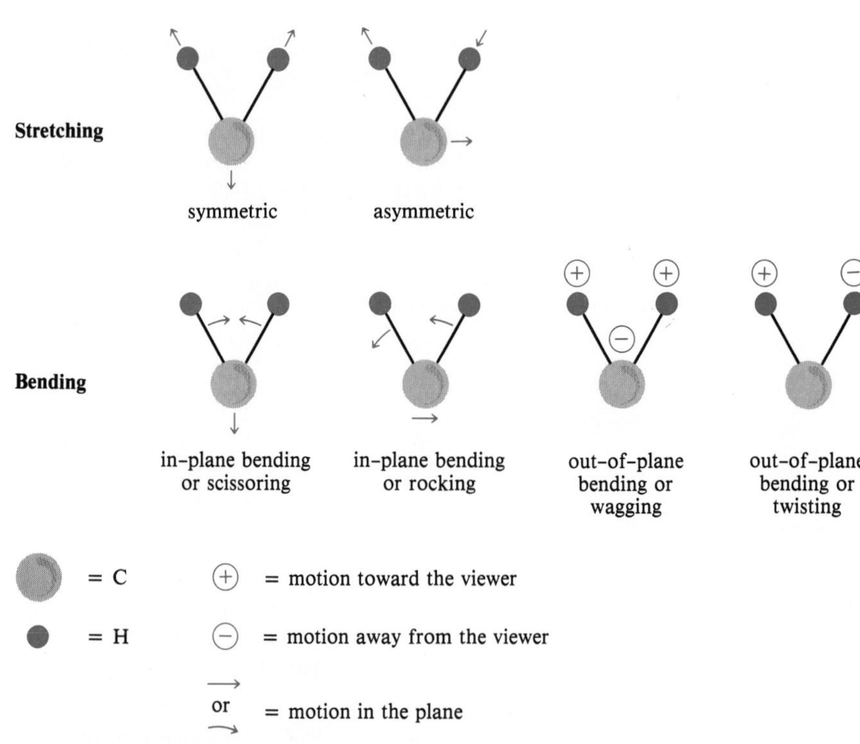

Table 12.4 Infrared stretching frequencies of some typical bonds

Bond type	*Group*	*Class of compound*	*Frequency range (cm^{-1})*
single bonds to hydrogen	C—H	alkanes	2850–3000
	=C—H	alkenes and aromatic compounds	3030–3140
	≡C—H	alkynes	3300
	O—H	alcohols and phenols	3500–3700 (free) 3200–3500 (hydrogen-bonded)
		carboxylic acids	2500–3000
	N—H	amines	3200–3600
	S—H	thiols	2550–2600
double bonds	C=C	alkenes	1600–1680
	C=N	imines, oximes	1500–1650
	C=O	aldehydes, ketones, esters, acids	1650–1780
triple bonds	C≡C	alkynes	2100–2260
	C≡N	nitriles	2200–2400

The spectra of cyclopentanone and cyclohexanone differ in the low frequency or **fingerprint region,** from 700 to 1500 cm^{-1}. Bands in this region result from combined bending and stretching motions of the atoms and *are unique for each particular compound.*

EXAMPLE 12.6 Using infrared spectroscopy, how could you quickly distinguish between the structural isomers benzyl alcohol and anisole?

C_6H_5—CH_2OH benzyl alcohol C_6H_5—OCH_3 anisole

Solution Benzyl alcohol's infrared spectrum will show a band in the O—H stretching frequency region (3200 to 3700 cm^{-1}); the spectrum of anisole will not have a band in that frequency region.

PROBLEM 12.12 How could you use infrared spectroscopy to distinguish between the isomers 1-hexyne and 1,3-hexadiene?

To summarize, infrared spectra can be used to tell what types of bonds may be present in a molecule (by using the functional group region) and to tell whether two substances are identical or different (by using the fingerprint region).

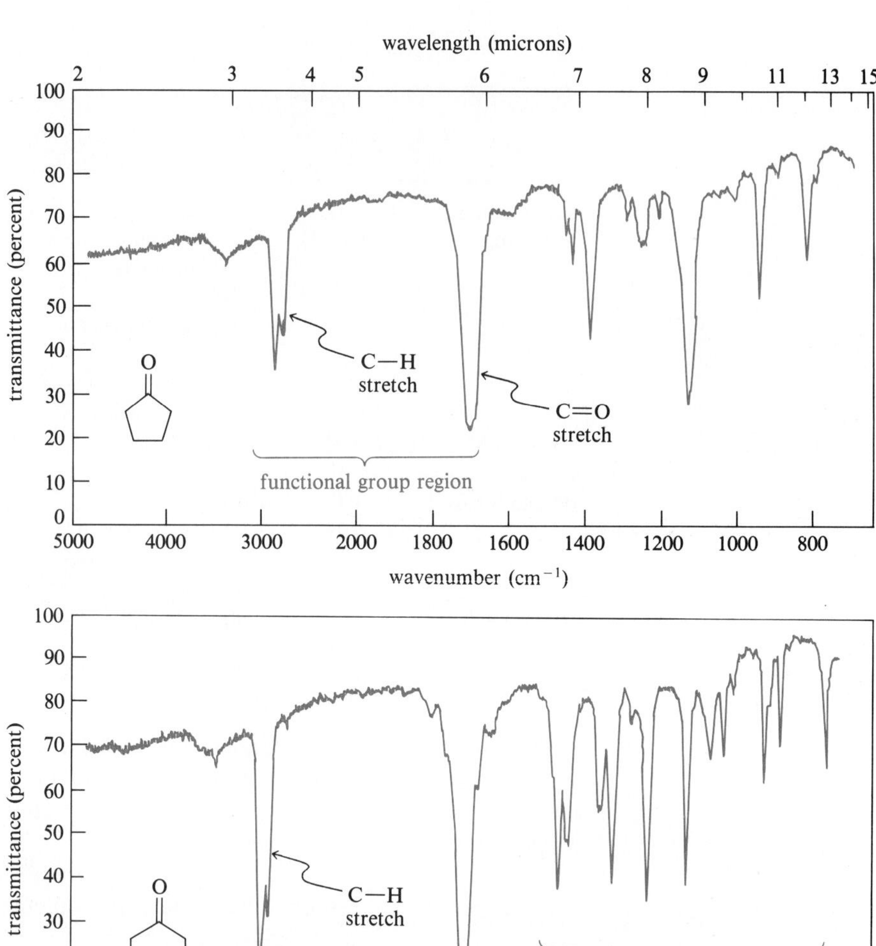

FIGURE 12.8
The infrared spectra of two similar ketones. The spectra have similar functional group bands but differ in the fingerprint region.

12.6 Visible and Ultraviolet Spectroscopy

The visible region of the spectrum (visible to the human eye, that is) corresponds to light with wavelengths of 400 to 800 **nanometers** (nm; 1 nm = 10^{-9} meters). Ultraviolet light has a shorter wavelength, about 200 to 400 nm, whereas infrared light has a longer wavelength, above 2500 nm. Older units for reporting these spectra are **millimicrons** (mμ = 1 nm) or **angstrom units** (Å; 10Å = 1 nm). Table 12.5 summarizes these units.

The amounts of energy associated with light are 37 to 75 kcal/mol for the visible region and 75 to 150 kcal/mol for the ultraviolet region (Table 12.1). These energies are much larger than those involved in infrared spectroscopy (2

TABLE 12.5 Units for visible-ultraviolet spectra

visible (vis)	400–800 nm (or mμ)	4000–8000 Å
ultraviolet (uv)	200–400 nm (or mμ)	2000–4000 Å

to 12 kcal/mol). They correspond to the amounts of energy needed to cause an electron to jump from a filled molecular orbital to a higher-energy, vacant molecular orbital. Such electron jumps are called **electronic transitions.**

Figure 12.9 shows a typical ultraviolet absorption spectrum. Unlike infrared spectra, visible-ultraviolet spectra are quite broad and generally show only a small number of peaks. The peaks are reported as the wavelengths where maxima occur. The conjugated, unsaturated ketone whose spectrum is shown in Figure 12.9 has an intense absorption at $\lambda_{max} = 232$ nm and a much weaker absorption at $\lambda_{max} = 330$ nm. The band at shorter wavelength corresponds to a pi electron transition, whereas the longer-wavelength, weaker intensity band corresponds to a transition of the nonbonding electrons on the carbonyl oxygen atom.

The intensity of an absorption band can be expressed quantitatively. Band intensity depends on the particular molecular structure and also on the number of absorbing molecules in the light path. **Absorbance,** which is the log of the ratio of light intensities entering and leaving the sample, is given by the equation

$$A = \epsilon c l \text{ (Beer's law)} \tag{12.5}$$

where ϵ is the **molar absorptivity** (sometimes called the **extinction coefficient),** c is the concentration of the solution in moles per liter, and l is the length in centimeters of the sample through which the light passes. The value of ϵ for any peak in the spectrum of a compound is a constant characteristic of that particular molecular structure. For example, the values of ϵ for the peaks in the spectrum of the unsaturated ketone shown in Figure 12.9 are $\lambda_{max} = 232$ nm ($\epsilon = 12{,}600$) and $\lambda_{max} = 330$ nm ($\epsilon = 78$).

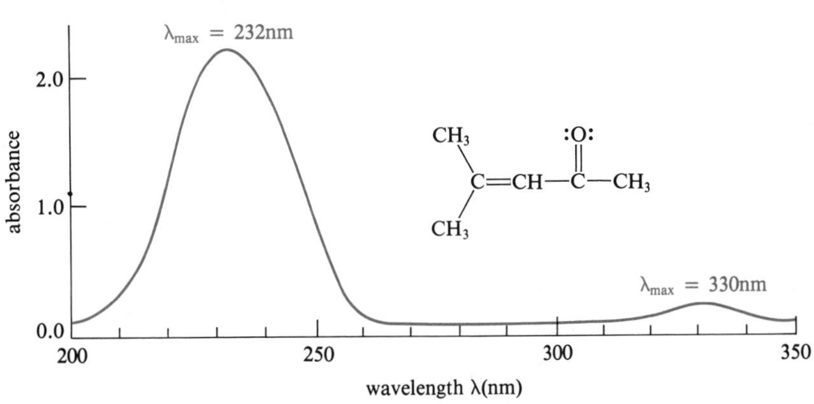

FIGURE 12.9 The absorption spectrum of 4-methyl-3-penten-2-one.

EXAMPLE 12.7 What is the effect of doubling the concentration of a particular absorbing sample on A? on ϵ?

Solution The observed absorbance A will be doubled since A is directly proportional to c. The value of ϵ, however, is a function of molecular structure and is a constant, independent of the concentration.

PROBLEM 12.13 A particular solution of $(CH_3)_2C{=}CH{-}\overset{\overset{\displaystyle O}{\|}}{C}{-}CH_3$, the ketone whose spectrum is shown in Figure 12.9, placed in a 1-cm absorption cell shows a peak at $\lambda_{max} = 232$ nm with an observed absorbance $A = 2.2$. Calculate the concentration of the solution, using the value of ϵ given in the text.

Visible-ultraviolet spectra are most commonly used to detect conjugation. In general, molecules with no double bonds or with only one double bond do not absorb in the visible-ultraviolet region (200 to 800 nm). Conjugated systems do absorb there, however, and the greater the conjugation, the longer the wavelength of maximum absorption, as seen in the following examples:

$CH_2{=}CH{-}CH{=}CH_2$
$\lambda_{max} = 220$ nm
($\epsilon = 20{,}900$)

$CH_2{=}CH{-}CH{=}CH{-}CH{=}CH_2$
$\lambda_{max} = 257$ nm
($\epsilon = 35{,}000$)

$CH_2{=}CH{-}CH{=}CH{-}CH{=}CH{-}CH{=}CH_2$
$\lambda_{max} = 287$ nm
($\epsilon = 52{,}000$)

$\lambda_{max} = 255$ nm
($\epsilon = 215$)

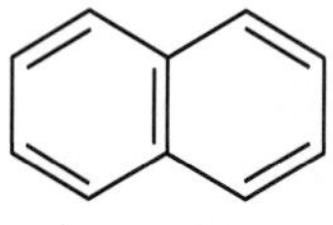

$\lambda_{max} = 314$ nm
($\epsilon = 289$)

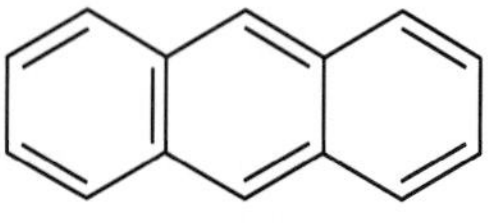

$\lambda_{max} = 380$ nm
($\epsilon = 9000$)

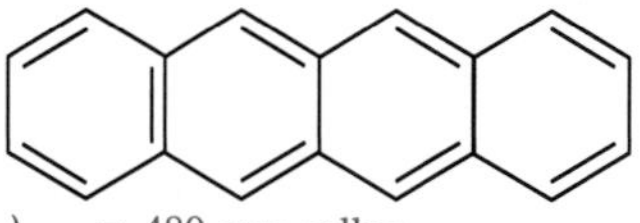

$\lambda_{max} = 480$ nm: yellow
($\epsilon = 12{,}500$)

PROBLEM 12.14 Which of the following aromatic compounds do you expect to absorb at the longer wavelength?

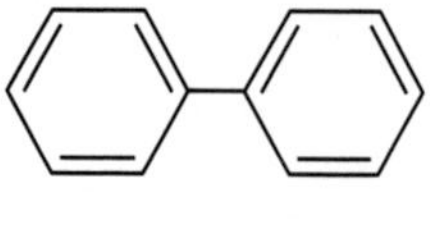

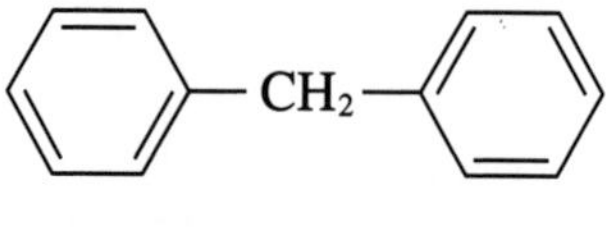

CH_2

PROBLEM 12.15 Naphthalene is colorless, but its isomer azulene is blue. Which compound has the lower-energy pi electronic transition?

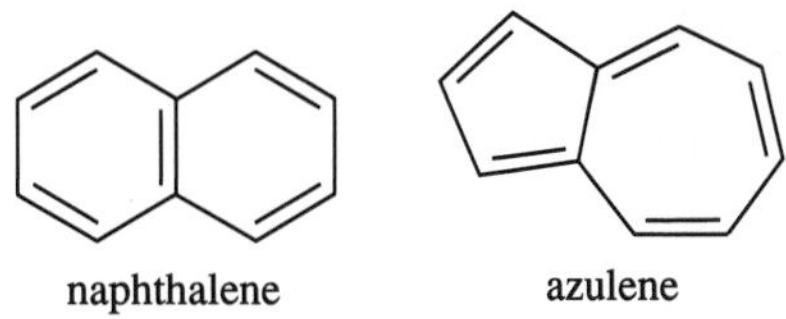

12.7 Mass Spectrometry

Mass spectrometry differs from the other types of spectroscopy discussed in this chapter, in that it does not depend on transitions between energy states. Instead, a mass spectrometer converts molecules to ions, sorts them according to their mass-to-charge ratio (m/z), and determines the relative amounts of each ion present. A small sample of the substance is introduced into a high-vacuum chamber, where it is vaporized and bombarded with high-energy electrons. These bombarding electrons eject an electron from the molecule M, to give a **cation radical** called the **molecular ion** $M^{+\cdot}$ (sometimes referred to as the **parent ion**).

$$M + e^- \longrightarrow \underset{\text{molecular ion}}{M^{+\cdot}} + 2\,e^- \qquad (12.6)$$

Methanol, for example, forms a molecular ion in the following way:

$$e^- + CH_3\ddot{\underset{\cdot\cdot}{O}}H \longrightarrow \underset{\text{methanol molecular ion } (m/z = 32)}{[CH_3\dot{\underset{\cdot\cdot}{O}}H]^+} + 2\,e^- \qquad (12.7)$$

The beam of these parent ions then passes between the poles of a powerful magnet, which deflects the beam. The extent of the deflection depends on the mass of the ion. Since $M^{+\cdot}$ has a mass that is essentially identical to the mass of the molecule M (the mass of the ejected electron is trivial compared to the mass of the rest of the molecule), **mass spectrometers can be used to determine molecular weights.**

Frequently, mass spectra show a peak one or two mass units *higher* than the molecular weights. How can this be? Recall that the isotope ^{13}C (one mass unit higher than ordinary ^{12}C) has a natural abundance of about 1.1%. This gives rise to an $(M + 1)^{+\cdot}$ peak in carbon compounds. The intensity of this peak relative to the $M^{+\cdot}$ peak is approximately 1.1% times the number of carbons in the compound (because the chance of finding a ^{13}C atom in a compound is proportional to the number of carbon atoms present).

PROBLEM 12.16 An alkane shows an $M^{+\cdot}$ peak at m/z 142. What is its molecular formula? What will be the relative intensities of the 143/142 peaks?

Other isotopic peaks can also be useful. For example, chlorine consists of a mixture of ^{35}Cl (75%) and ^{37}Cl (25%), and bromine consists of a 50 : 50 mixture of ^{79}Br and ^{81}Br. A monochloroalkane will therefore show two parent ion peaks, two mass units apart and in an intensity ratio of 3 : 1. Monobromoalkanes also show two parent ion peaks two mass units apart, but in a 1 : 1 intensity ratio. These isotopic peaks can be used to obtain structural information, as in the following example:

EXAMPLE 12.8 A bromoalkane shows two equal-intensity parent ion peaks at m/e 136 and 138. Deduce its molecular formula.

Solution Only one Br can be present (the molecular weight is not high enough for two bromines). Subtract 136−79 (or 138−81) to get 57 for the mass of the carbons and hydrogens. Dividing 57 by 12 (the mass of carbon), we get 4, with 9 mass units left over for the hydrogens. The formula is C_4H_9Br.

PROBLEM 12.17 A compound containing only C, H, and Cl shows parent ion peaks at m/e 74 and 76 in a ratio of 3 : 1. Suggest possible structures for the compound.

If bombarding electrons have enough energy, they produce not only parent ions but also fragments called **daughter ions.** That is, the original molecular ion breaks into smaller fragments, some of which are ionized and get sorted on an m/z basis by the spectrometer. A prominent peak in the mass spectrum of methanol, for example, is the $M^+ - 1$ peak at $m/z = 31$. This peak arises through loss of a hydrogen atom from the molecular ion.

$$\underset{m/z\,=\,32}{H_3C\text{—}\overset{\cdot +}{\ddot{O}}\text{—}H} \longrightarrow \underset{m/z\,=\,31}{H_2C\text{=}\overset{+}{\ddot{O}}\text{—}H} + H\cdot \qquad (12.8)$$

You will recognize this daughter ion as protonated formaldehyde, a resonance-stabilized carbocation.

A mass spectrum consists, then, of a series of signals of varying intensities at different m/z ratios. In practice, most of the ions are singly charged ($z = 1$), so that we can readily obtain their masses, m. Figure 12.10 shows a mass spectrum printed as the computer output of a mass spectrometer. It is the mass spectrum of a typical ketone, 4-octanone. The peak at $m/z = 128$ is the most intense high mass peak in the spectrum and corresponds to the molecular weight of the ketone. In addition, we see certain prominent daughter ion peaks. For example, the peaks at $m/z = 85$ and $m/z = 71$ correspond in mass to $C_4H_9CO^+$ and $C_3H_7CO^+$, respectively. This suggests that one easy fragmentation path for the parent ion is to break the carbon-carbon bond adjacent to the carbonyl group. Ion fragmentation paths depend on ion structure, and the interpretation of mass spectral fragmentation patterns can give significant information about molecular structure.

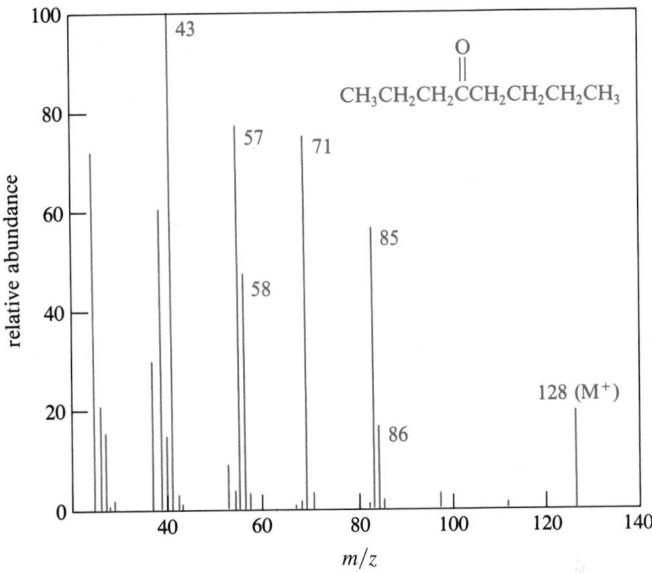

FIGURE 12.10
The mass spectrum of 4-octanone.

EXAMPLE 12.9 The most intense peak (called the *base peak*) in Figure 12.10 occurs at m/z = 43. Suggest how it might arise.

Solution This peak corresponds to the m/z for $C_3H_7^+$, suggesting that the daughter ion $C_3H_7CO^+$ loses carbon monoxide to give $C_3H_7^+$. This explanation becomes more plausible when we consider that the spectrum also contains an intense peak at m/z = 57, corresponding to the analogous process for $C_4H_9CO^+$. We can summarize these conclusions in the following "family tree" of ions:

$$C_8H_{16}O^{+\cdot}\ \text{M}^+\ (128) \xrightarrow{-C_3H_7\cdot} C_4H_9CO^+\ (85) \xrightarrow{-CO} C_4H_9^+\ (57)$$

$$C_8H_{16}O^{+\cdot}\ \text{M}^+\ (128) \xrightarrow{-C_4H_9\cdot} C_3H_7CO^+\ (71) \xrightarrow{-CO} C_3H_7^+\ (43)$$

PROBLEM 12.18 In what ways will the mass spectrum of 4-heptanone be similar to and in what ways will it differ from the mass spectrum in Figure 12.10?

The types of spectroscopy that we have described here are routinely used in research laboratories. With modern instrumentation, each type of spectrum can be obtained in a few minutes to an hour, including time to prepare the sample. Interpretation of the spectra may take longer, but investigators with experience can often deduce the structure of even complex molecules from their spectra alone in a relatively short time.

ADDITIONAL PROBLEMS

12.19. Draw the structure of a compound with each of the following molecular formulas that will show only one peak in its ¹H NMR spectrum.

a. C_5H_{10} **b.** $C_3H_6Br_2$ **c.** C_4H_6
d. $C_{12}H_{18}$ **e.** C_2H_6O **f.** C_5H_{12}

12.20. Tell how many chemically different types of ¹H nuclei are present in

a. $(CH_3)_3C—CH_2CH_3$ **b.** $(CH_3)_2CHNHCH_2CH_3$

c. 1,2-dimethylcyclohexene (CH_3, CH_3 on the double-bonded ring carbons)

d. $H_3C—CH(OH)—CH_3$

12.21. The ¹H NMR spectrum of a compound, C_4H_9Br, consists of a single sharp peak. What is its structure? The spectrum of an isomer of this compound consists of a doublet at δ 3.2, a complex pattern at δ 1.9, and a doublet at δ 0.9, with relative areas 2 : 1 : 6. What is its structure?

12.22. The chemical shifts of the ¹H nuclei in 2,2-dimethylpropane and TMS are δ 0.95 and δ 0.0, respectively. From these data, what can you deduce about the relative electronegativities of carbon and silicon?

12.23. How could you distinguish between the following pairs of isomers by ¹H NMR spectroscopy?

a. CH_3CCl_3 and $CH_2ClCHCl_2$ **b.** $CH_3CH_2CH_2OH$ and $(CH_3)_2CHOH$

c. $CH_3—C(=O)—OCH_3$ and $H—C(=O)—OCH_2CH_3$

d. $C_6H_5—CH_2—CH{=}O$ and $C_6H_5—C(=O)—CH_3$

12.24. The ¹H NMR spectrum of methyl *p*-toluate consists of a singlet at δ 2.35, a singlet at δ 3.82, and two doublets, at δ 7.15 and δ 7.87; relative areas are 3 : 3 : 2 : 2. Draw the structure and determine which hydrogens are responsible for each peak, as far as you can tell (use Table 12.2).

12.25. Using the information in Table 12.2, sketch the ¹H NMR spectrum of each of the following compounds. Be sure to show all splitting patterns.

a. $CH_3CH_2—C(=O)—H$ **b.** $(CH_3)_2CHOCH(CH_3)_2$

c. $Cl_2C{=}CH(CH_3)$ **d.** $CH_3O—C_6H_4—OCH_3$ (para)

12.26. The ¹H NMR spectrum of a compound, $C_3H_3Cl_5$, consists of a triplet at δ 4.5 and a doublet at δ 6.0 (J = 7 Hz), with relative areas 1 : 2. What is its structure?

12.27. A compound is known to be the methyl ester of a toluic acid, but the orientation of the two substituents ($—CH_3$ and $—CO_2CH_3$) on the aromatic ring is not known. The ¹³C NMR spectrum shows seven peaks. Which isomer is it? What will the ¹H NMR spectrum look like?

12.28. A compound, C_3H_6O, has no bands in the infrared region around 3500 or 1720 cm^{-1}. What structures can be eliminated by these data? Suggest a possible structure, and tell how you could determine whether it is correct.

12.29. From Table 12.4, deduce the order of ease with which C—H bonds in alkanes, alkenes, and alkynes are stretched. Suggest an explanation for the observed order.

12.30. A very dilute solution of ethanol in carbon tetrachloride shows a sharp infrared band at 3580 cm^{-1}. As the solution is made more concentrated, a new, rather broad band appears at 3250 to 3350 cm^{-1}. Eventually the sharp band disappears and is replaced entirely by the broad band. Explain.

12.31. For the following pairs of compounds, give at least one major peak in the infrared spectrum of one member of the pair that will enable you to distinguish it from the other member.

a. $CH_3-C(=O)-CH_2CH_3$ and $CH_3OCH_2CH_3$

b. 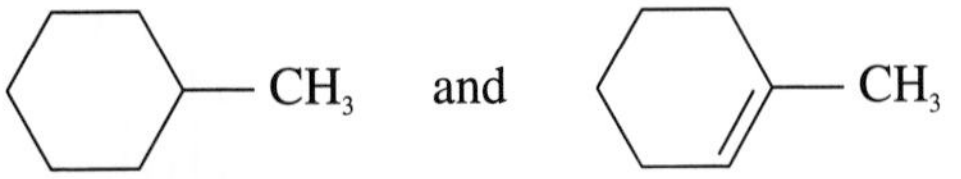 (cyclohexane)—CH_3 and (cyclohexene)—CH_3

c. $CH_3OCH_2CH_3$ and $CH_3CH_2CH_2OH$

d. $CH_3CH_2CH(=O)$ and $CH_3CH_2C(=O)-OH$

12.32. What features will be similar in the infrared spectra of the following compounds, and how will their infrared spectra differ?

$CH_3C(=O)CH_2CH_2CH_2OH$ and $CH_3CH_2C(=O)CH_2CH_2OH$?

12.33. How could you use infrared spectroscopy to distinguish between the following pairs of isomers?

a. $CH_3C(=O)CH_2CH_3$ and $CH_3CH(OH)CH{=}CH_2$

b. 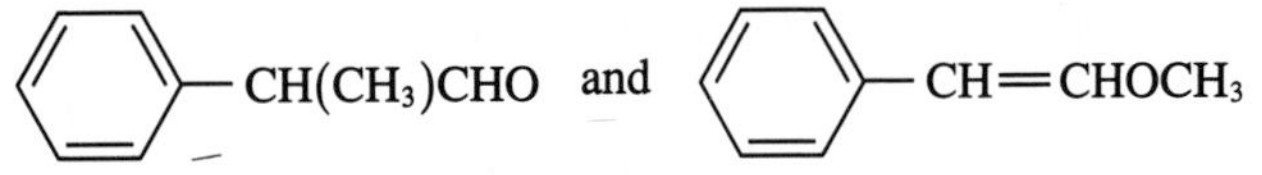 (benzene)—$CH(CH_3)CHO$ and (benzene)—$CH{=}CHOCH_3$

c. $(CH_3CH_2)_3N$ and $(CH_3CH_2CH_2)_2NH$

12.34. A compound, $C_5H_{10}O$, has an intense infrared band at 1725 cm^{-1}. Its ^1H NMR spectrum consists of a quartet at δ 2.7 and a triplet at δ 0.9, with relative areas 2:3. What is its structure?

12.35. A compound, $C_5H_{10}O_3$, has a strong infrared band at 1745 cm^{-1}. Its ^1H NMR spectrum consists of a quartet at δ 4.15 and a triplet at δ 1.20; relative areas are 2:3. What is the correct structure?

12.36 Compound A, $C_4H_{10}O$, is oxidized with PCC (see Sec. 7.13) to Compound B, C_4H_8O, which gives a positive Tollens' test and has a strong IR band at 1725 cm^{-1}.

The ^{1}H NMR spectrum of Compound A is given below. What are the structures of Compound A and Compound B?

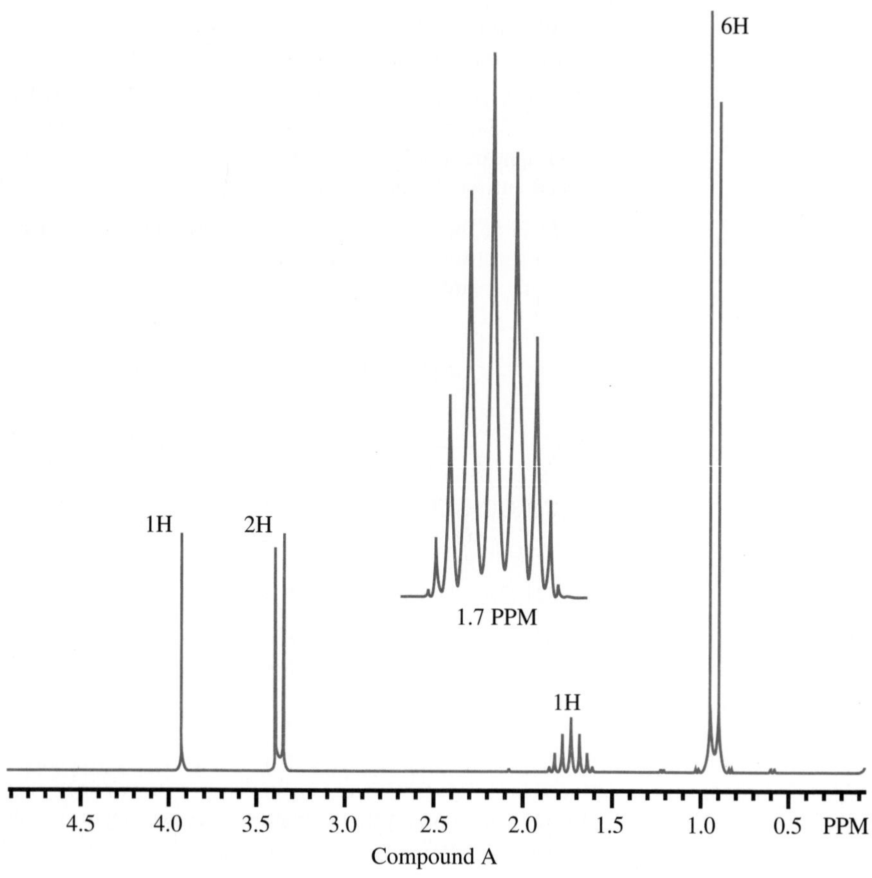

12.37. You are oxidizing cyclohexanol with CrO_3 to obtain cyclohexanone (eq. 7.33). How could you use infrared spectroscopy to tell that the reaction was complete and that the product was free of starting material?

12.38. Which of the following compounds are not likely to absorb ultraviolet radiation in the range 200 to 400 nm?

a.

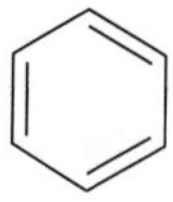

b. $CH_3CH_2CH_2OH$

c.

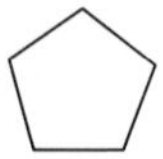

d.

e. $CH_3CH_2OCH_2CH_3$

f. $CH_2{=}CHCH_2CH_2CH{=}CH_2$

12.39. The unsaturated aldehydes $CH_3(CH{=}CH)_nCH{=}O$ have ultraviolet absorption spectra that depend on the value of n; the λ_{max} values are 220, 270, 312, and 343 nm as n changes from 1 to 4. Explain.

12.40. The λ_{max} for *cis*-1,2-diphenylethene is at shorter wavelength (280 nm) than for *trans*-1,2-diphenylethene (295 nm). Suggest an explanation.

12.41. A sample of cyclohexane is suspected of being contaminated with benzene, from which it had been prepared by hydrogenation. At λ = 255 nm, benzene has the molar absorptivity ϵ = 215, whereas cyclohexane does not absorb at that wavelength (ϵ = 0). An ultraviolet spectrum of the contaminated cyclohexane (obtained in a 1.0-cm cell) shows an absorbance A = 0.43 at 255 nm. Calculate the concentration of benzene in the cyclohexane.

12.42. Write a formula for the molecular ion of *n*-propanol.

12.43. The mass spectrum of 1-pentanol shows an intense daughter ion peak at m/z = 31. Explain how this peak might arise.

12.44. An alcohol, $C_5H_{12}O$, shows a daughter ion peak at m/z = 59 in its mass spectrum. An isomeric alcohol shows no daughter ion at m/z = 59 but does have a peak at m/z = 45. Suggest possible structures for each isomer. How could you confirm these structures by ^1H NMR spectroscopy? by ^{13}C NMR spectroscopy?

12.45. A hydrocarbon shows a parent ion peak in its mass spectrum at m/z = 102. Its ^1H NMR spectrum shows peaks at δ 2.7 and δ 7.4, with relative areas 1 : 5. What is the correct structure?

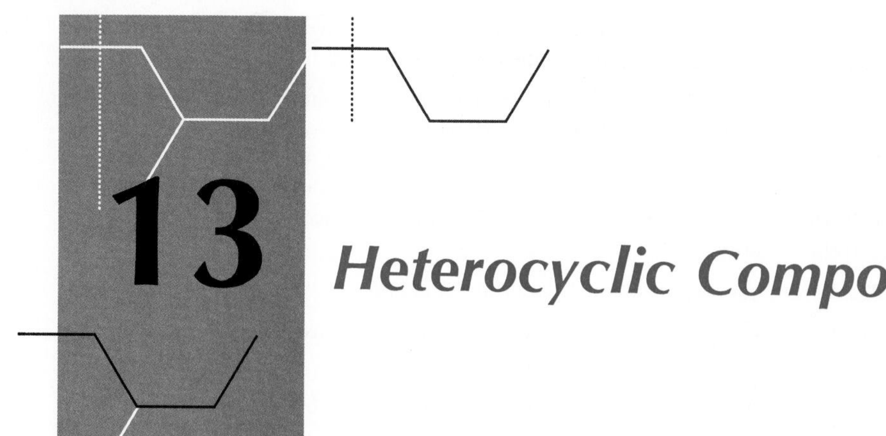

13 Heterocyclic Compounds

13.1 Introduction

From an organic chemist's viewpoint, heteroatoms are atoms other than carbon or hydrogen that may be present in organic compounds. The most common heteroatoms are oxygen, nitrogen, and sulfur. In heterocyclic compounds, one or more of these heteroatoms replaces carbon in a ring.

Heterocycles form the largest class of organic compounds. In fact, most natural products and drugs contain heterocyclic rings; indeed, well over half of all organic chemical publications deal in one way or another with heterocycles.

Heterocycles can be divided into two subgroups: **nonaromatic** and **aromatic.** We have already encountered a few nonaromatic heterocycles—ethylene oxide and other cyclic ethers (Chapter 8), cyclic acetals (Chapter 9), cyclic esters (lactones, Chapter 10), and cyclic amines (Chapter 11). In general, these nonaromatic heterocycles behave a great deal like their acyclic analogs with the same functional groups and do not require special discussion.

Much more important are the **aromatic heterocycles;** in this chapter, we will focus our attention on them. We begin with an important six-membered aromatic nitrogen heterocycle, **pyridine.**

13.2 Pyridine: Bonding and Basicity

Pyridine has a structure similar to that of benzene, except that **one CH unit is replaced by a nitrogen atom.**

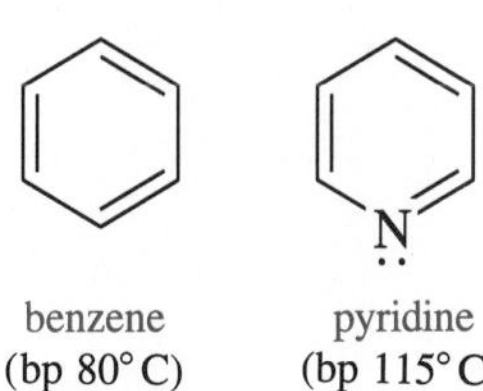

benzene (bp 80° C) pyridine (bp 115° C)

The orbital pictures for benzene and pyridine are similar (review Sec. 4.5 and Figure 4.2). The nitrogen atom, like the carbons, is sp^2-hybridized, with one electron in a p orbital perpendicular to the ring plane.

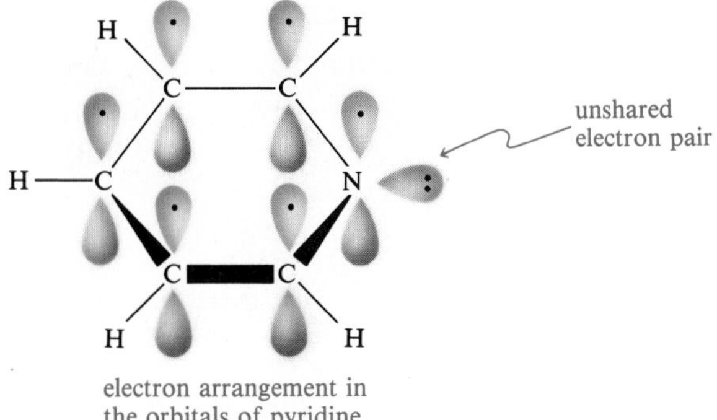

electron arrangement in the orbitals of pyridine

Thus, the nitrogen contributes *one* electron to the six electrons that form the aromatic pi cloud above and below the ring plane. On the other hand, the unshared electron pair on nitrogen lies in the ring plane (just like the C—H bonds) in an sp^2 orbital.

Pyridine is a resonance hybrid of Kekulé-type structures.

Because of the similarities in bonding, pyridine resembles benzene in shape. It is planar, with nearly perfect hexagonal geometry. It is aromatic, and tends to undergo substitution rather than addition reactions.

But the substitution of nitrogen for carbon changes many of the properties. Like benzene, pyridine is miscible with most organic solvents, but unlike benzene, pyridine is also completely miscible with water! One explanation lies in its hydrogen-bonding capability.

Another reason is that pyridine is much more polar than benzene. The nitrogen atom is electron-withdrawing compared with carbon; hence there is a shift of electrons away from the ring carbons and toward the nitrogen, making it partially negative and the ring carbons partially positive.

This polarity enhances the solubility of pyridine in polar solvents like water.

Pyridine is a weakly basic tertiary amine, with $pK_a = 5.29$. It is a much weaker base than aliphatic amines ($pK_a \cong 10$; see Table 11.3), mainly because of the different hybridization of the nitrogen (sp^2 in pyridine, sp^3 in aliphatic amines). The greater s-character ($\frac{1}{3}s$ in pyridine, $\frac{1}{4}s$ in aliphatic amines) means that the unshared electron pair is held closer to the nitrogen nucleus in pyridine, decreasing its basicity.

Pyridine reacts with strong acids to form **pyridinium salts.**

$$C_5H_5N: + H^+Cl^- \longrightarrow C_5H_5\overset{+}{N}\text{—}H \;\; Cl^- \qquad (13.1)$$

pyridinium chloride

For this reason, pyridine is often used as a scavenger in acid-producing reactions (for example, in the reaction of thionyl chloride with alcohols, eq. 7.31).

PROBLEM 13.1 Write an equation for the reaction of pyridine with

a. cold sulfuric acid.
b. methyl iodide (see Sec. 11.12).

13.3 Substitution in Pyridine

Though aromatic, *pyridine is very resistant to electrophilic aromatic substitution* and undergoes reaction only under drastic conditions. For example, nitration or bromination requires high temperatures and strong acid catalysis.

Pyridine (positions 1–4 labeled: N = 1)

$\xrightarrow[\text{300°C, Fe}]{KNO_3,\ HNO_3,\ \text{fuming } H_2SO_4}$ 3-nitropyridine (NO_2)

$\xrightarrow[\text{300°C}]{Br_2,\ H_2SO_4, SO_3}$ 3-bromopyridine (Br)

(13.2)

One reason for this sluggishness is that electron withdrawal by the nitrogen makes the ring partially positive and therefore not receptive to attack by electrophiles, which are also positive. A second reason is that, under the acidic conditions for these reactions, most of the pyridine is protonated and present as the positive pyridinium ion, which is even more unlikely to be attacked by electrophiles than neutral pyridine.

When substitution does occur, electrophiles attack pyridine mainly at C3. The cationic intermediate (review Sec. 4.10) is *least unfavorable* in this case, because it does not put a positive charge on the electron-deficient nitrogen (especially bad if the nitrogen is protonated).

EXAMPLE 13.1 Draw all contributors to the resonance hybrid for electrophilic attack at C3 of pyridine.

Solution

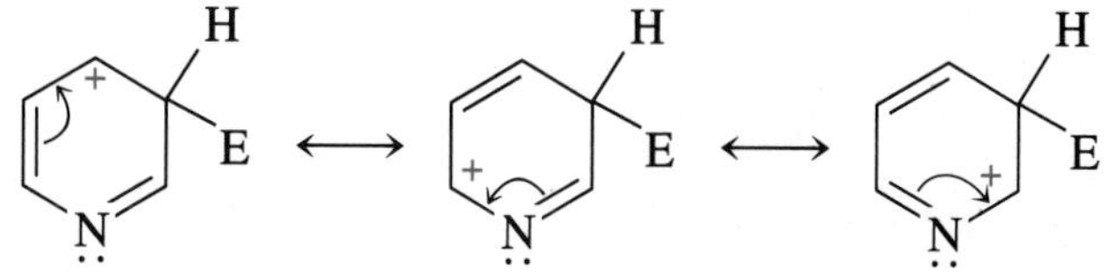

For substitution at C3, the "pyridinonium ion" positive charge is delocalized to C2, C4, and C6, but not to the nitrogen.

PROBLEM 13.2 Repeat Example 13.1, but for electrophilic substitution at C2 or C4 of pyridine. Explain why substitution at C3 (eq. 13.2) is preferred.

Although resistant to electrophilic substitution, *pyridine undergoes nucleophilic substitution.* The pyridine ring is partially positive (due to electron withdrawal by the nitrogen) and is therefore susceptible to attack by nucleophiles. Here are two examples:

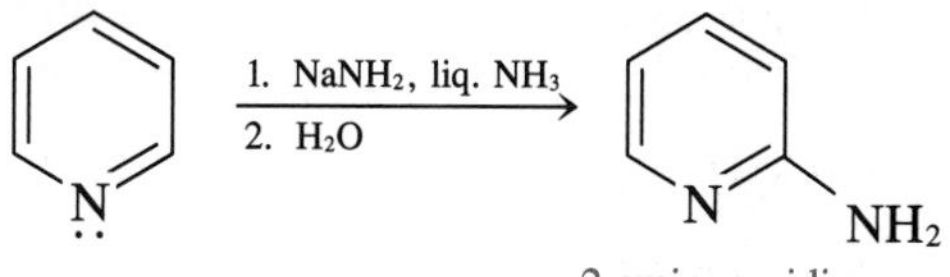

1. $NaNH_2$, liq. NH_3
2. H_2O

2-aminopyridine (13.3)

$NaOCH_3, CH_3OH$

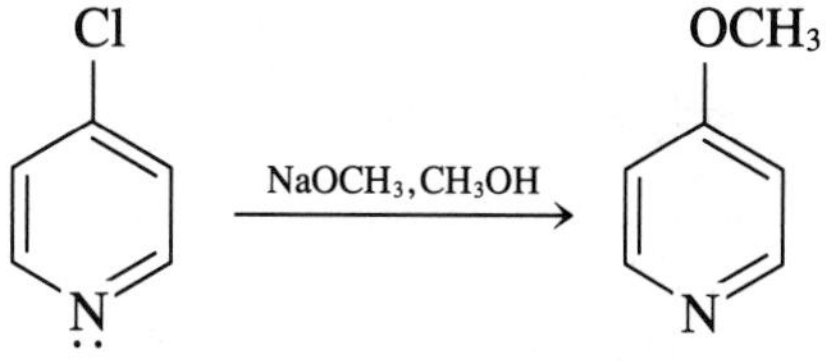

4-chloropyridine 4-methoxypyridine (13.4)

EXAMPLE 13.2 Write a mechanism for eq. 13.3.

Solution Attack of amide ion at C2 gives an anionic intermediate with the negative charge mainly on nitrogen.

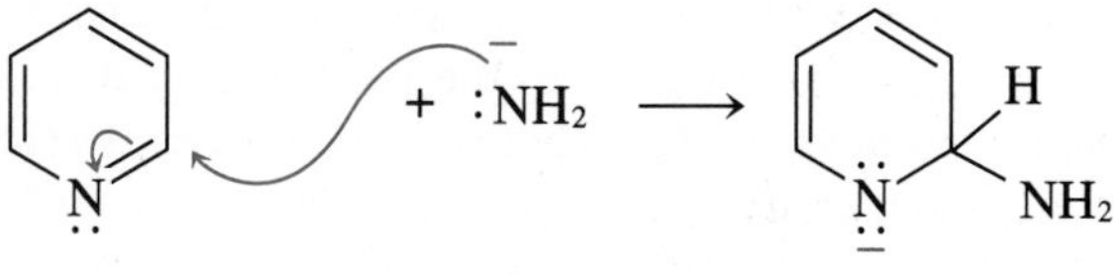

$+ \; :\!NH_2^-$

You can think of this as nucleophilic addition to a C=N bond, analogous to addition of nucleophiles to a C=O bond (Sec. 9.7).

To restore aromaticity, hydride ion is displaced. The hydride then attacks the amino group to give hydrogen gas and an amide-type anion (cf. eq. 11.17).

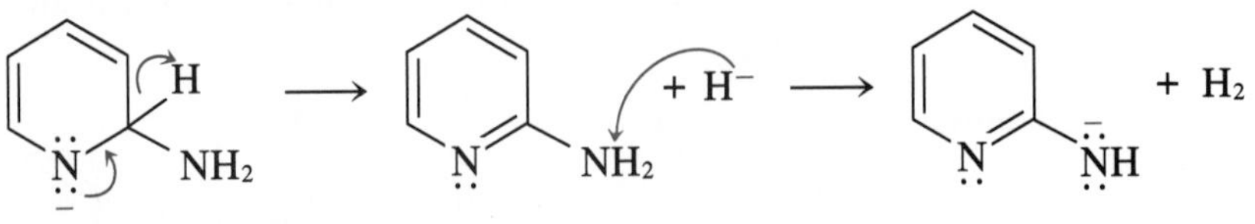

$$\text{(ring, H, } NH_2) \longrightarrow \text{(ring, } NH_2) + H^- \longrightarrow \text{(ring, } \ddot{N}H^-) + H_2$$

In the final step, this ion is protonated by water.

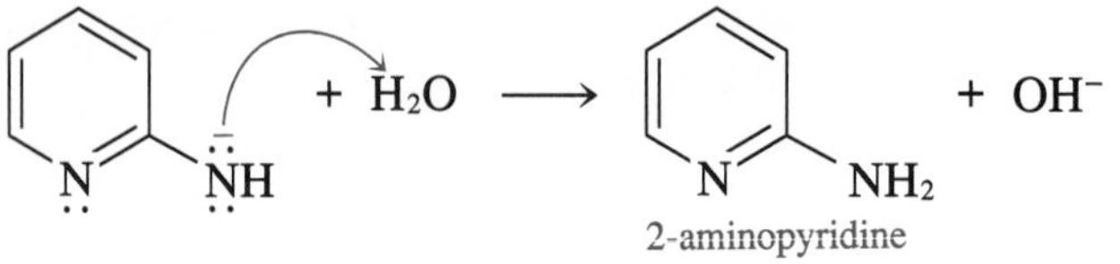

$$\text{(ring, } \ddot{N}H^-) + H_2O \longrightarrow \text{(ring, } NH_2) + OH^-$$

2-aminopyridine

PROBLEM 13.3 Write a mechanism for eq. 13.4.

Pyridine and alkylpyridines are found in coal tar. The monomethyl pyridines (called **picolines**) undergo side-chain oxidation to carboxylic acids (review Sec. 10.8b). For example, 3-picoline gives **nicotinic acid** (or **niacin**), a vitamin essential in the human diet, to prevent the disease *pellagra*.

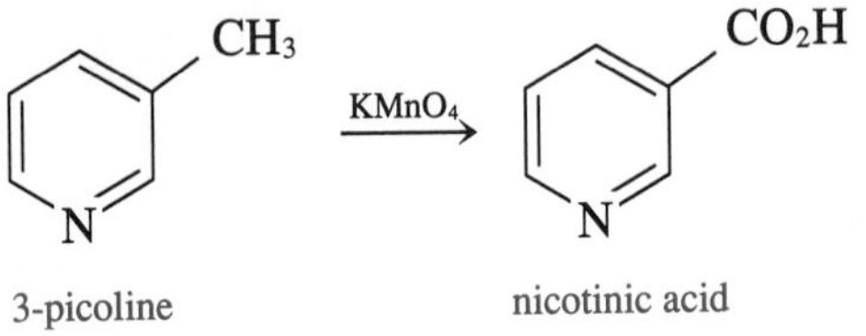

$$\text{3-picoline } (CH_3) \xrightarrow{KMnO_4} \text{nicotinic acid } (CO_2H) \tag{13.5}$$

Pyridine can be reduced to the fully saturated secondary amine **piperidine.**

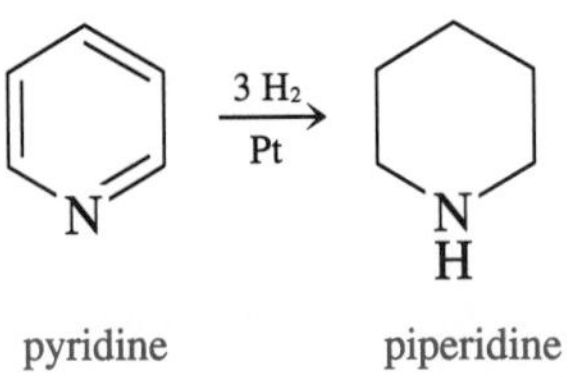

$$\text{pyridine} \xrightarrow[Pt]{3\ H_2} \text{piperidine (NH)} \tag{13.6}$$

The pyridine and piperidine rings are found in many natural products. Examples include **nicotine** (the major alkaloid in tobacco, used as an agricultural insecticide and highly toxic to humans), **pyridoxine** (vitamin B_6, a coenzyme), and **coniine** (the toxic principle of poison hemlock, taken by Socrates).

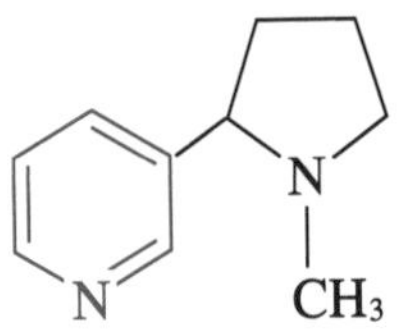

nicotine 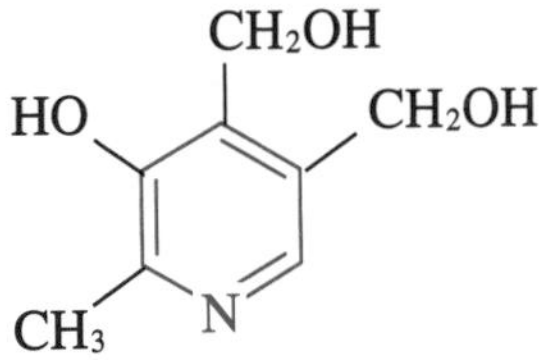 pyridoxine 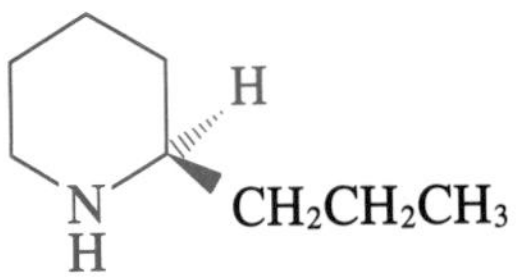 (+)-coniine

PROBLEM 13.4 Naturally occurring coniine is the (+)-isomer shown. What is its configuration, *R* or *S*?

PROBLEM 13.5 Naturally occurring nicotine is the *(S)-(−)* isomer. Locate the stereogenic center, and draw the three-dimensional structure.

PROBLEM 13.6 Nicotine contains two nitrogens, one in a pyridine ring and one in a pyrrolidine ring. It reacts with *one* equivalent of HCl to form a crystalline salt, $C_{10}H_{15}N_2Cl$. Draw its structure. Nicotine also reacts with *two* equivalents of HCl to form another crystalline salt, $C_{10}H_{16}N_2Cl_2$. Draw its structure.

13.4 Other Six-Membered Heterocycles

The pyridine ring can be fused with benzene rings to produce polycyclic aromatic heterocycles. The most important examples are **quinoline** and **isoquinoline,** analogs of naphthalene (Sec. 4.14) but with N in place of CH at C1 or C2.

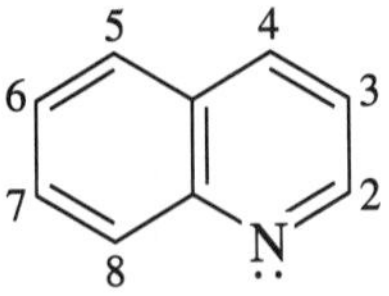

quinoline
bp 237°C

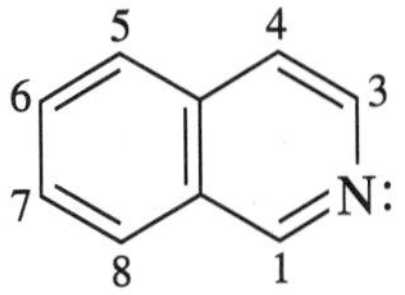

isoquinoline
bp 243°C, mp 26.5°C

Electrophilic substitution in these amines occurs in the carbocyclic ring, illustrating the inactivity toward electrophiles of the pyridinoid vis-à-vis the benzenoid rings.

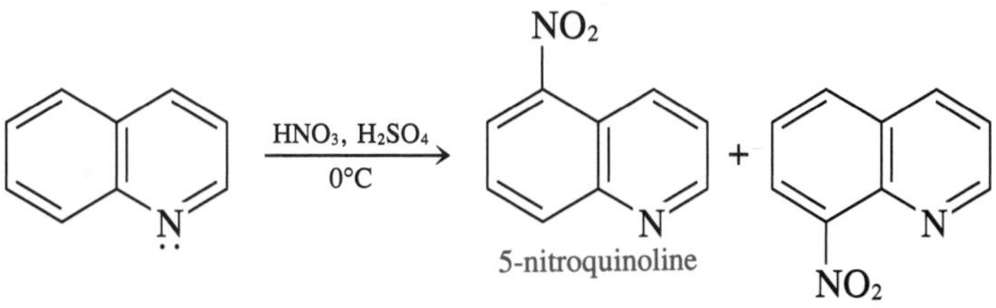

$$\text{quinoline} \xrightarrow[0°C]{HNO_3,\ H_2SO_4} \text{5-nitroquinoline} + \text{8-nitroquinoline} \quad (13.7)$$

The stability of the pyridine ring is also illustrated by its resistance to oxidation. Thus, when quinoline is treated with potassium permanganate, the benzenoid ring is oxidized.

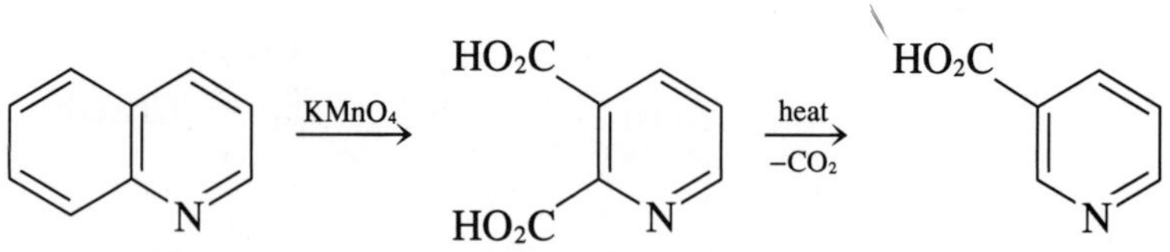

(13.8)

quinoline quinolinic acid nicotinic acid

The quinoline and isoquinoline rings occur in many natural products. Good examples are **quinine** (which occurs in cinchona bark and is used to treat malaria) and **papaverine** (present in opium and used as a muscle relaxant).

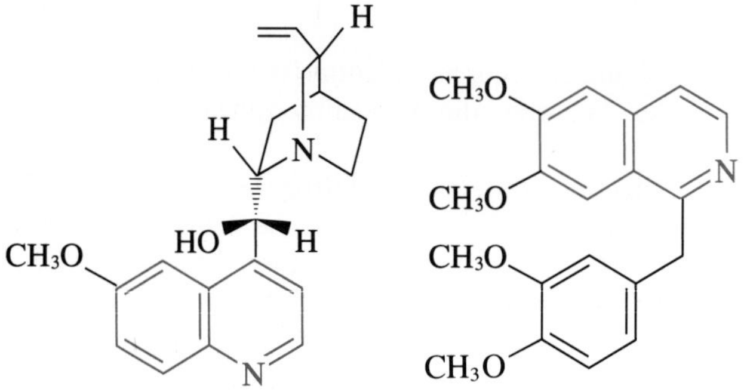

quinine papaverine

It is logical to ask: If we can replace one CH with N in a benzene ring to create pyridine, can we replace more than one of the benzene CH groups with nitrogens? The answer is yes. For example, there are *three* **diazines.**

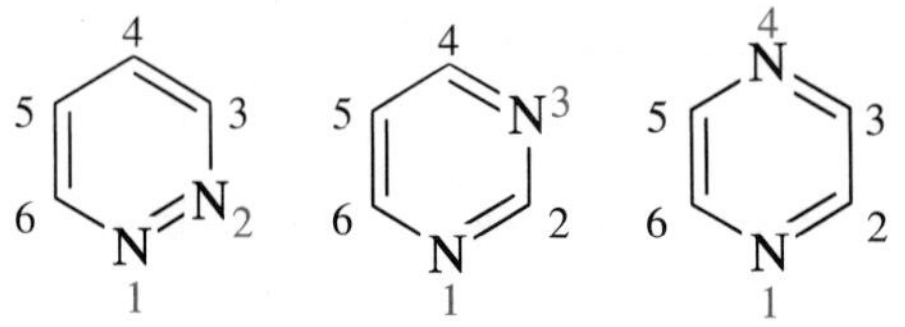

pyridazine (bp 208° C) pyrimidine (bp 134° C) pyrazine (bp 118° C)

Of these, the most important are the **pyrimidines,** derivatives of which (**cytosine, thymine,** and **uracil**) are important bases in nucleic acids (DNA and RNA; see Chapter 18).

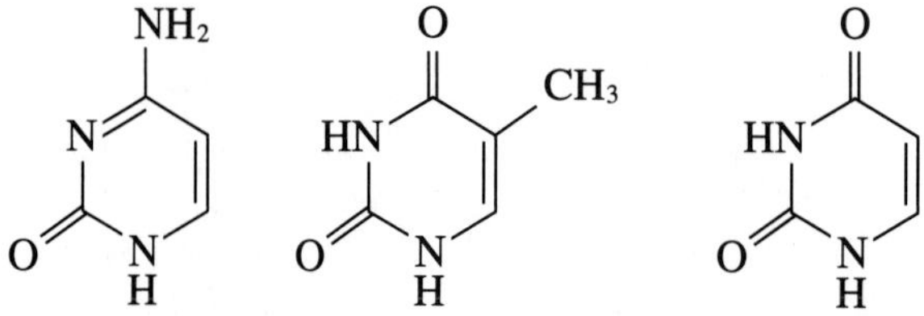

cytosine thymine uracil

Triazines and tetrazines are also known, but neither pentazine nor hexazine (which would really not be a heterocycle, since it would contain only one element and be an allotrope of nitrogen) is known.

Similar analogs of naphthalene with more than one nitrogen are also known. The **pteridine** ring, with nitrogens replacing C1, C3, C5, and C8 in naphthalene, is present in many natural products, such as the butterfly wing pigment **xanthopterin** and the blood-forming vitamin **folic acid.** The analog of folic acid, but with an NH_2 group in place of the OH group on the pteridine ring and a methyl group on the first nitrogen in the side chain, is useful in cancer chemotherapy (it is called **methotrexate**).

pteridine part — 4-aminobenzoic acid part — glutamic acid part

NH_2 group here in methotrexate

methyl group here in methotrexate

xanthopterin

folic acid

To summarize, six-membered heterocycles with nitrogen (one or more) in place of the CH groups in benzenoid aromatics are also aromatic. Each nitrogen atom contributes one electron to the 6π aromatic system. The nitrogens are also basic, because of unshared electron pairs located in the ring plane. The nitrogen, being electron-withdrawing, deactivates the ring toward electrophilic aromatic substitution but activates it toward nucleophilic aromatic substitution. These heterocyclic aromatic rings are present in many natural products.

If we replace a benzene CH group with oxygen (instead of nitrogen), we obtain an aromatic cation called a **pyrylium ion.**

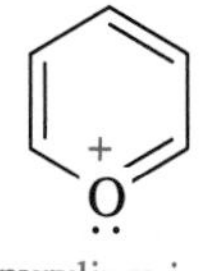

pyrylium ion

EXAMPLE 13.3 Describe the bonding in a pyrylium ion.

Solution The oxygen is sp^2-hybridized, with only five electrons (hence the $+1$ formal charge). Two of these, the unshared electron pair, are in an sp^2 orbital that lies in the ring plane. Two others are in sp^2 orbitals that form the sigma bonds with the adjacent carbon atoms, also sp^2-hybridized. The fifth electron is in a p orbital perpendicular to the ring plane; it overlaps with similar orbitals on each ring carbon atom (as in benzene) to form the aromatic six-electron pi cloud above and below the ring plane.

The red and blue colors of many flowers are due to **anthocyanines,** compounds in which a pyrylium ring is present. For example, the pigment responsible for the color of red roses is

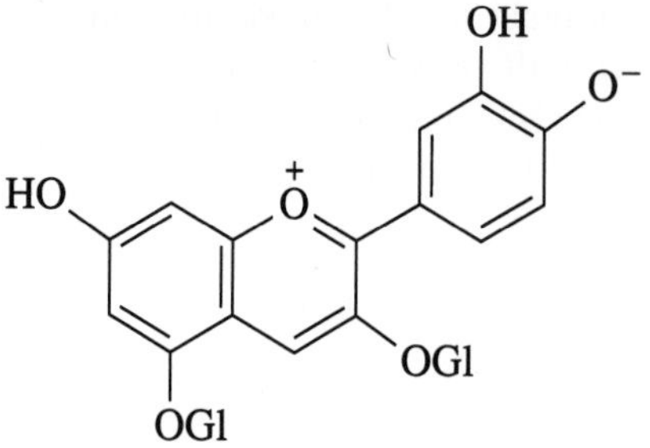

red rose pigment
(Gl = Glucose)

The glucose units solubilize the pigment in the aqueous cellular fluid. The color of blue cornflowers is due to the same pigment, but complexed with metallic ions, such as Fe^{3+} or Al^{3+}. Other flower pigments have the same basic structures, but with fewer, more, or differently located hydroxyl groups.

Now let us examine rather different types of heteroaromatic compounds: those with five-membered rings.

13.5 Five-Membered Heterocycles: Furan, Pyrrole, and Thiophene

Furan, pyrrole, and thiophene are important five-membered ring heterocycles with one heteroatom.

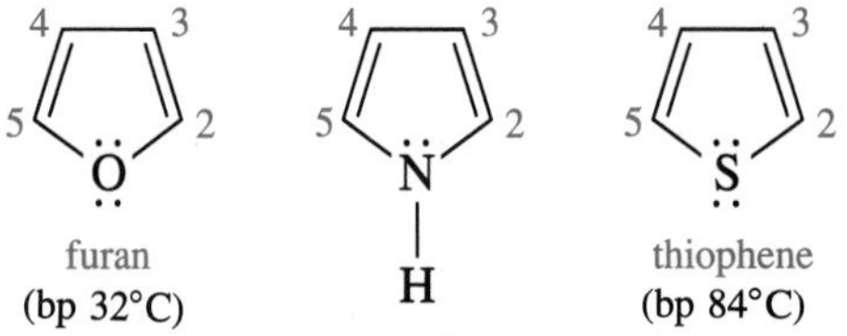

furan (bp 32°C) pyrrole (bp 131°C) thiophene (bp 84°C)

Numbering begins with the heteroatom and proceeds around the ring.

The most important commercial source of furans is **furfural** (2-furaldehyde), obtained by heating oat hulls, corn cobs, or straw with strong acid. These naturally occurring materials are polymers of a five-carbon sugar, which is dehydrated by the acid to furfural.

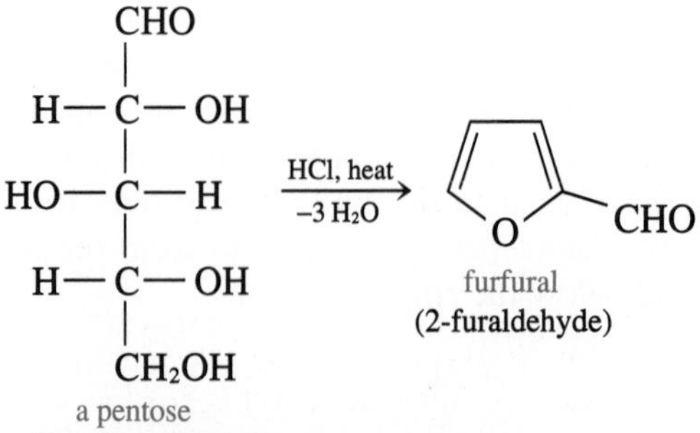

a pentose (5-carbon sugar) $\xrightarrow[-3\,H_2O]{HCl,\ heat}$ furfural (2-furaldehyde) (13.9)

Pyrrole is obtained commercially by distillation of coal tar or from furan, ammonia, and a catalyst. Thiophene is obtained by heating a mixture of butanes and butenes with sulfur.

As drawn, the structures of these heterocycles look as if they ought to be dienes, but in fact, these ring systems are aromatic; they behave like benzene in many ways, particularly in their tendency to undergo electrophilic aromatic substitution. The reasons for this behavior will become clearer if we examine the bonding in these molecules.

Furan has a planar, pentagonal structure in which each ring atom is sp^2-hybridized.

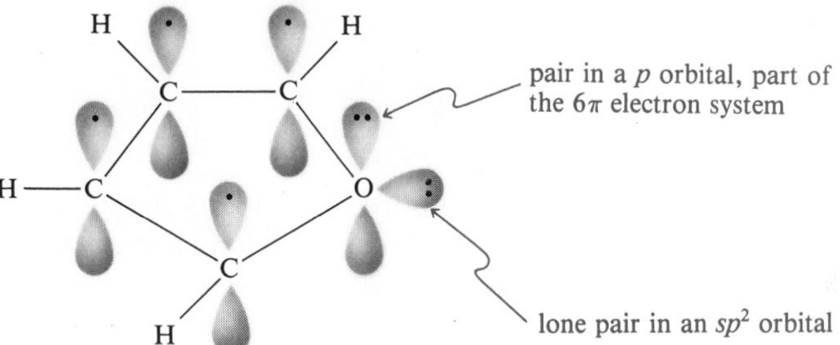

orbital structure of furan

Each ring atom uses two of these orbitals to form sigma bonds with its neighbors. Each carbon also uses one sp^2 orbital to form a sigma bond in the ring plane with a hydrogen atom and has one electron in a p orbital perpendicular to the ring plane. Now look at the oxygen. It has an unshared electron pair in an sp^2 orbital in the ring plane and **two electrons in a *p* orbital perpendicular to the ring plane.** These two electrons overlap with those in the p orbitals on the carbons to form a 6π electron cloud above and below the ring plane, just as in benzene. The bonding in pyrrole and thiophene is similar to that in furan.

The important difference between five- and six-membered aromatic heterocycles is that, **in the five-membered heterocycles, the heteroatom contributes *two* electrons to the aromatic 6π systems, whereas in six-membered heterocycles, the heteroatom contributes only one electron to that system.** This difference has important consequences for the chemical behavior of the two types of heterocycles.

Before we consider those consequences, let us examine the bonding in furan in another way. We can write these heterocycles as a resonance hybrid in which the electron pair from the heteroatom is delocalized to all ring atoms.

contributors to the resonance hybrid structure of furan

Notice that four of these structures are dipolar and place a *negative* charge on the ring carbons. As we will see, this enhances their susceptibility to attack by electrophiles.

EXAMPLE 13.4 Although pyrrole is an amine, it is an exceedingly weak base. It has a pK_a of -4.4, nearly 10^{10} times weaker than pyridine ($pK_a = 5.29$). Suggest an explanation.

Solution In pyrrole, the unshared electron pair on nitrogen is part of the aromatic 6π electron system.

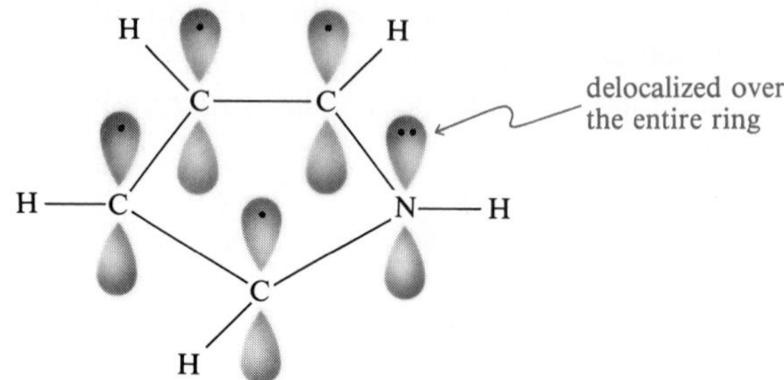

Protonation of the nitrogen would destroy the aromatic system, thus losing its resonance energy. Hence pyrrole is a very weak base; in very strong acids, *it is protonated on carbon rather than on the nitrogen.*

$$\text{pyrrole (N–H)} + H^+ \rightleftharpoons \text{2H-pyrrolium (}\overset{+}{N}\text{–H, } CH_2 \text{ at C-2)}$$

$pK_a = -4.4$

In pyridine, the unshared electron pair is *not* part of the aromatic pi system and is available for adding a proton (eq. 13.1).

PROBLEM 13.7 Pyrrole is insoluble in water, but its saturated analog, pyrrolidine, is completely miscible with water. Suggest an explanation.

CH_2-CH_2
CH_2 CH_2
N̈
|
H

pyrrolidine

13.6 *Electrophilic Substitution in Furan, Pyrrole, and Thiophene*

Furan, pyrrole, and thiophene are all much more reactive than benzene toward electrophilic substitution. Each reacts predominantly at the 2-position (and if that position is already substituted, at the 5-position). Here are typical examples:

$$\text{pyrrole} + HNO_3 \xrightarrow{0°C} \text{2-nitropyrrole} + H_2O \qquad (13.10)$$

2-nitropyrrole

$$\text{furan} + Br_2 \xrightarrow[0°C]{\text{ether}} \text{2-bromofuran} + HBr \qquad (13.11)$$

2-bromofuran

$$\text{2-methylthiophene} + CH_3\overset{O}{\overset{\|}{C}}Cl \xrightarrow{SnCl_4} \text{2-acetyl-5-methylthiophene} + HCl \qquad (13.12)$$

2-methylthiophene

2-acetyl-5-methylthiophene

The reason for predominant attack at C2 (instead of the other possibility, C3) becomes clear if we examine the carbocation intermediate in each case:

Attack of electrophile at C2 (X = NH, O, or S):

$$+ E^+ \longrightarrow [\ \cdots \longleftrightarrow \cdots \longleftrightarrow \cdots\ ] \longrightarrow \text{2-E product} + H^+ \qquad (13.13)$$

Attack of electrophile at C3:

$$+ E^+ \longrightarrow [\ \cdots \longleftrightarrow \cdots\ ] \longrightarrow \text{3-E product} + H^+ \qquad (13.14)$$

Attack at C2 is preferred because, in the carbocation intermediate, the positive charge can be delocalized over *three* atoms, whereas attack at C3 allows delocalization of the charge over only two positions.

PROBLEM 13.8 Write out the steps in the mechanism for bromination of furan (eq. 13.11).

A Word About . . .

24. Porphyrins: What Makes Blood Red and Grass Green?

Pyrrole rings form the building blocks of several biologically important pigments. The **porphyrins** are macrocyclic compounds that contain four pyrrole rings linked by one-carbon bridges. The molecules are flat and have a conjugated system of 18π electrons, shown in color in the parent molecule, **porphine.**

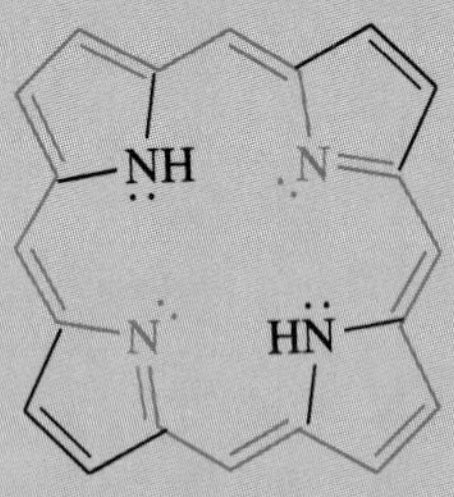

porphine
red crystals

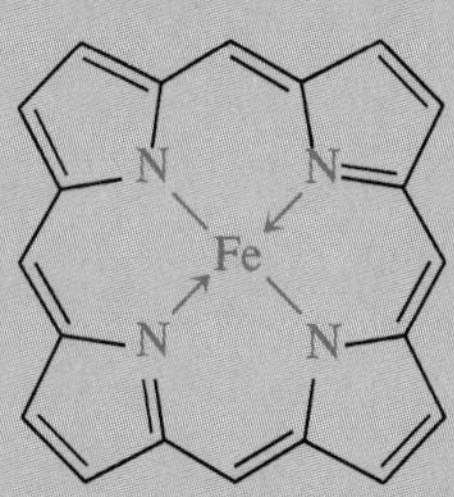

the Fe^{2-} porphine complex
brown cubic crystals

Porphyrins are exceptionally stable and highly colored. They form complexes with metallic ions. In these complexes, the hydrogens bonded to two of the nitrogens are absent, and each of the four nitrogens donates an electron pair to the metal, which sits in the middle of the structure.

Porphine itself does not occur in nature, but analogous compounds with various side chains on the pyrrole rings are some of the most important life-sustaining compounds of nature. One example is **heme,** the iron-porphyrin complex responsible for the red color of arterial blood.

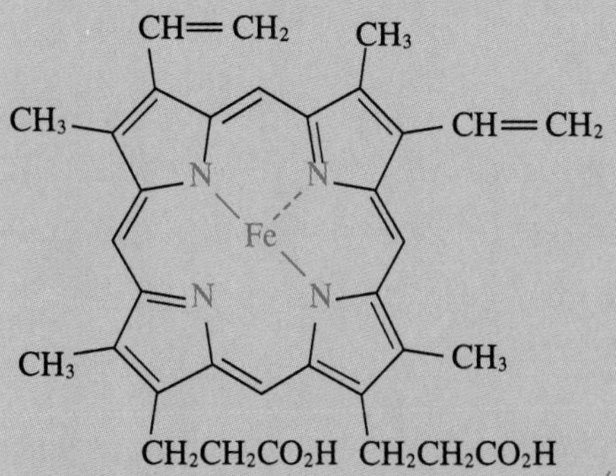

heme (brown needles with violet sheen)

Heme is present in red blood cells as a complex with a protein called **globin.** The complex is called **hemoglobin.** This complex is responsible for binding molecular oxygen and transporting it to sites where it is needed. The iron atom is complexed with the four porphyrin nitrogens but also has two additional coordination sites, above and below the plane of the porphyrin ring. One of these sites is occupied by an imidazole ring from a histidine unit in the protein, but the second site is available for **reversible binding to oxygen.** Carbon monoxide is toxic because it, too, can bind to this site, preventing oxygen transport to the lungs and elsewhere and ultimately leading to death by suffocation. Fortunately, CO binding can be reversed if oxygen is given quickly to a patient poisoned by the gas.

Heme is also associated with **myoglobin,** the oxygen-transport protein of muscle, where it performs a similar function.

The green color of plants is due to **chlorophyll,** a magnesium complex of a modified porphyrin. This pigment is present in chloroplasts, one or more of which can be present in cells, depending on the plant. Chlorophyll-*a* is an ester of the long-chain alcohol **phytol,** which helps solubilize the pigment in the chloroplasts.

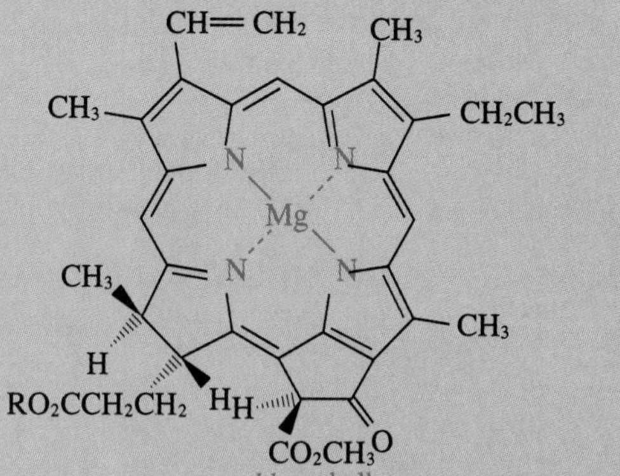

chlorophyll-*a*
blue-black crystals (mp 117–120°C)

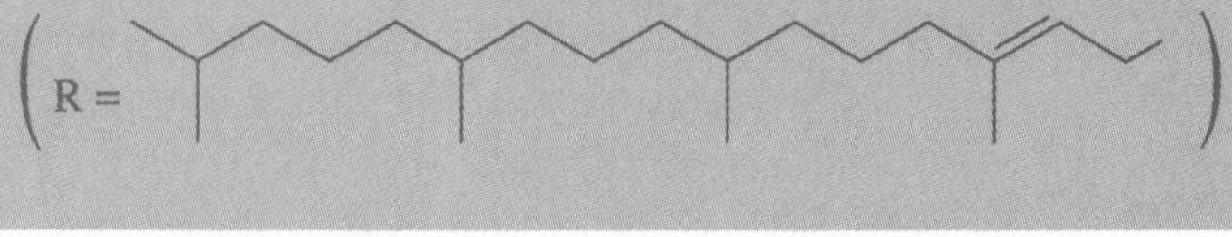

Chlorophyll-*b*, another plant pigment, differs from chlorophyll-*a* only in the replacement of one ring methyl group by an aldehyde group. The mechanism of photosynthesis (the conversion of CO_2 and H_2O by plants to carbohydrates) is quite complex and involves many steps. The first of these is the absorption of sunlight by chlorophyll and its conversion into chemical energy. Thus, chlorophyll is, in a sense, a solar energy converter.

13.7 Other Five-Membered Heterocycles: Azoles

It is possible to introduce a second heteroatom (and even a third and fourth) into five-membered heterocycles. The most important of these are the **azoles,** in which the second heteroatom, located at position 3, is nitrogen.

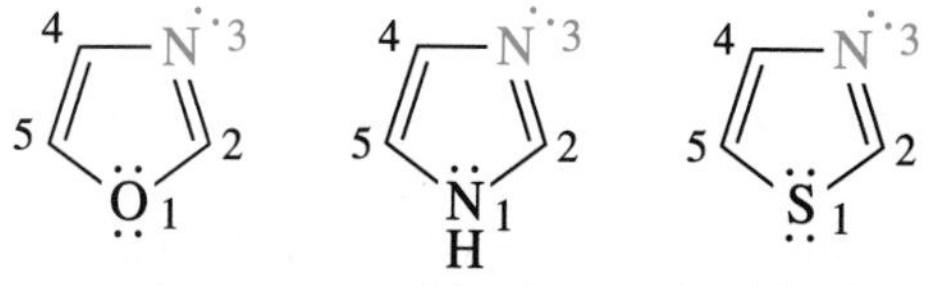

oxazole imidazole thiazole

Analogs in which the two heteroatoms are adjacent are also known.

As with pyridine, these heterocycles can be thought of as derived by replacing an aromatic CH group by N. *The unshared electron pair on this nitrogen (at position 3) is therefore* ***not*** *part of the 6π aromatic system,* as seen in the following orbital picture for imidazole:

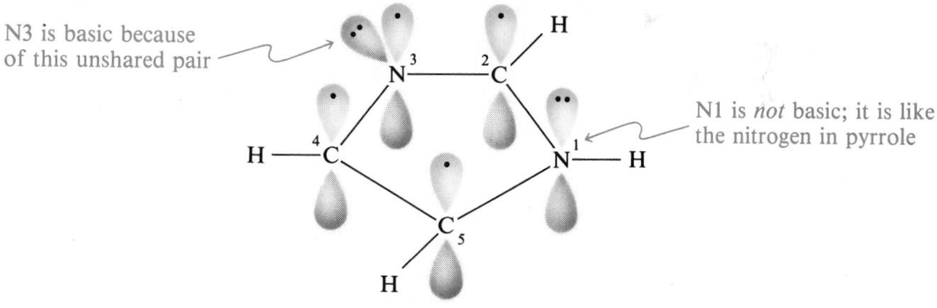

N3 is basic because of this unshared pair

N1 is *not* basic; it is like the nitrogen in pyrrole

bonding in imidazole

Consequently, the N3 nitrogen is basic and can be protonated. Imidazole is even more basic than pyridine (pK_a = 7.0; compare with pK_a 5.2 for pyridine), mainly because the positive charge in the imidazolium ion can be delocalized equally over both nitrogens.

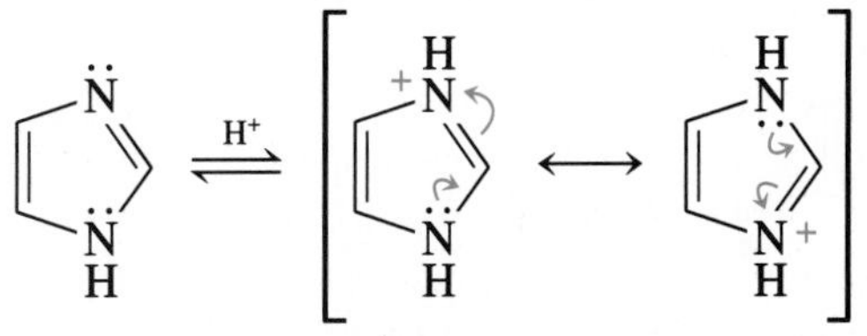

(13.15)

resonance in the imidazolium ion

These ring systems occur in nature. For example, the imidazole skeleton is present in the amino acid **histidine,** where it plays an important role in the reactions of many enzymes. Decarboxylation of histidine gives **histamine,** a toxic substance present in combination with proteins in body tissues. It is released as a consequence of allergic hypersensitivity or inflammation (for example, in hay fever sufferers). Many **antihistamines,** compounds that counteract the effects of histamine, have been developed. One of the better known of these is the drug **benadryl** (diphenylhydramine).

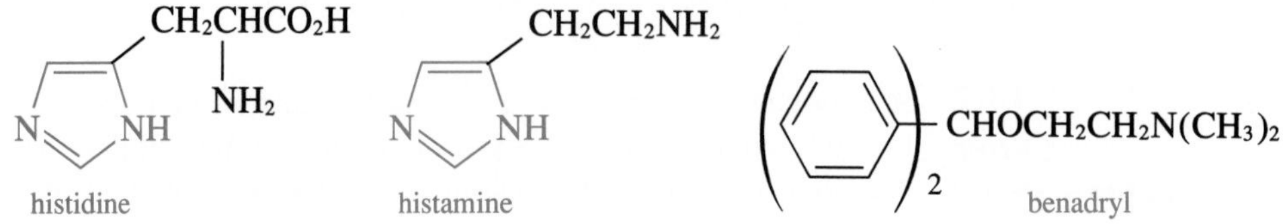

CH_2CHCO_2H / NH_2 / N, NH

histidine

$CH_2CH_2NH_2$ / N, NH

histamine

$(C_6H_5)_2$ $CHOCH_2CH_2N(CH_3)_2$

benadryl
(an antihistamine)

The thiazole ring occurs in **thiamin** (vitamin B_1), a coenzyme required for certain metabolic processes and hence essential to life (thiamin also contains a pyrimidine ring). In its reduced form, the tetrahydrothiazole ring appears in penicillins, important antibiotics.

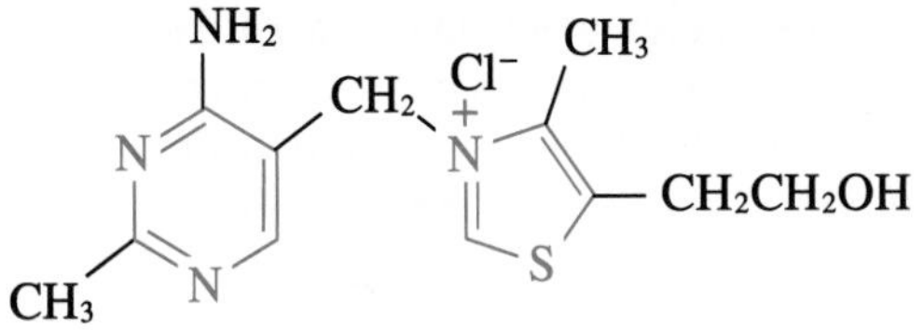

NH_2, CH_2, Cl^-, CH_3, $\overset{+}{N}$, N, N, S, CH_2CH_2OH, CH_3

thiamin
(vitamin B_1)

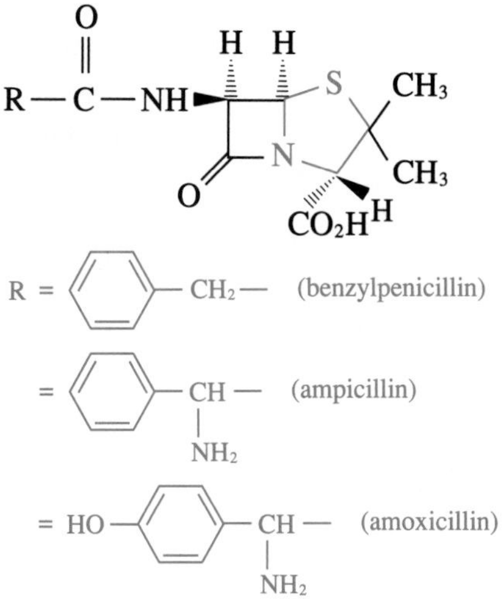

O, R—C—NH, H H, S, CH_3, N, CH_3, O, CO_2H, H

R = C_6H_5—CH_2— (benzylpenicillin)

= C_6H_5—CH(NH_2)— (ampicillin)

= HO—C_6H_4—CH(NH_2)— (amoxicillin)

13.8 Fused-Ring Five-Membered Heterocycles: Indoles and Purines

Another aromatic or heteroaromatic ring can be fused to the double bonds of five-membered heterocycles. For example, **indole** has a benzene ring fused to the C2–C3 bond of pyrrole.

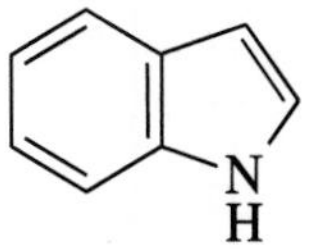

N
H

indole

The indole ring system, which is prevalent in a number of important natural products, is usually biosynthesized from the amino acid **tryptophan,** one of the building blocks of proteins. Indole itself and its 3-methyl derivative, **skatole,** are formed during protein decay.

Decarboxylation of tryptophan gives **tryptamine.** Several compounds with this skeleton have a profound effect on the brain and nervous system. One example is **serotonin** (5-hydroxytryptamine), a neurotransmitter and vasoconstrictor active in the central nervous system.

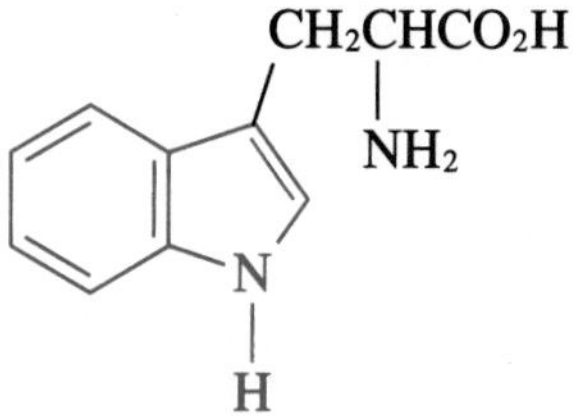

CH_2CHCO_2H / NH_2 / N / H

tryptophan

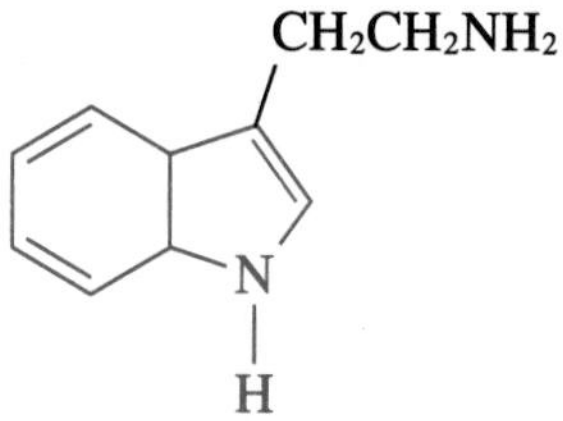

$CH_2CH_2NH_2$ / N / H

tryptamine

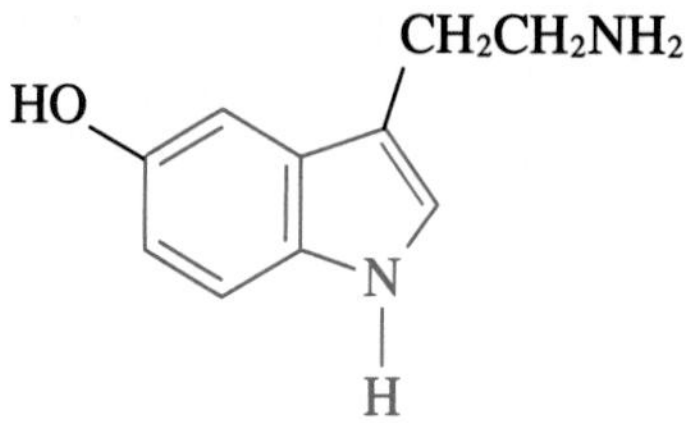

HO / $CH_2CH_2NH_2$ / N / H

serotonin

The tryptamine skeleton is disguised but present (shown in color) in more complex molecules. **Reserpine,** present in Indian snake root (*Rauwolfia serpintina*), which grows wild on the foothills of the Himalayas, has been used medically for centuries. It lowers blood pressure and is used to calm schizophrenics and improve their accessibility to psychiatric treatment. **Lysergic acid** is present in the fungus *ergot,* which grows on rye and other grains. Conversion of the carboxyl group to its diethylamide gives the extremely potent hallucinogen LSD.

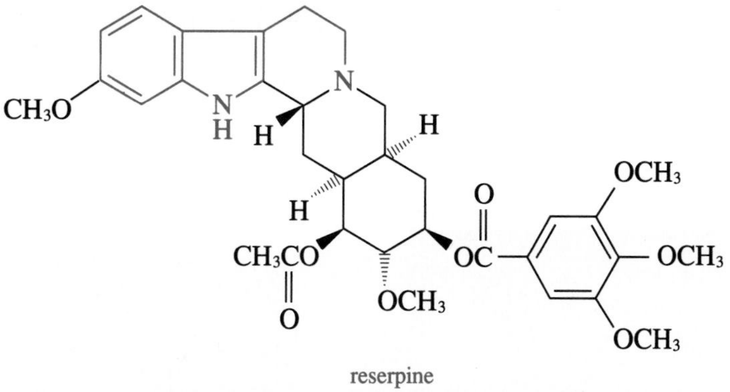

CH_3O / N / H / H / N / H / H / CH_3CO / O / OCH_3 / O / OC / OCH_3 / OCH_3 / OCH_3

reserpine

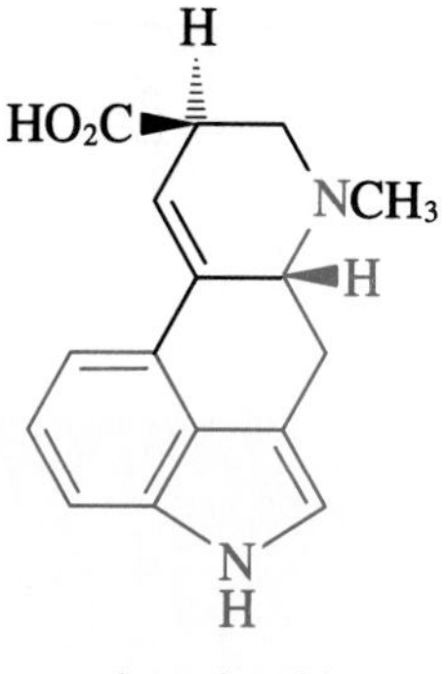

H / HO_2C / NCH_3 / H / N / H

lysergic acid

The **purines** are another biologically important class of fused-ring heterocycles. They contain a pyrimidine ring fused to an imidazole ring.

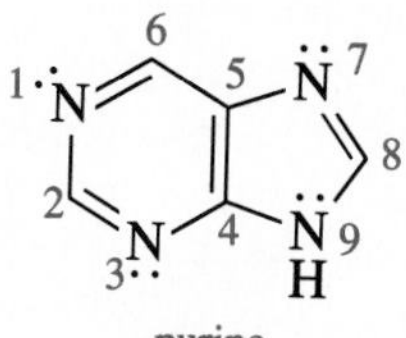

6 / 1 N / 5 / N 7 / 8 / 2 / N 3 / 4 / N 9 / H

purine
(mp 217°C)

Uric acid is present in the urine of all carnivores and is the main product of nitrogen metabolism in the excrement of birds and reptiles. The disease gout results from deposition of sodium urate (the salt of uric acid) in joints and tendons. **Caffeine,** present in coffee, tea, and cola beverages, and **theobromine** (in cocoa) are also purines.

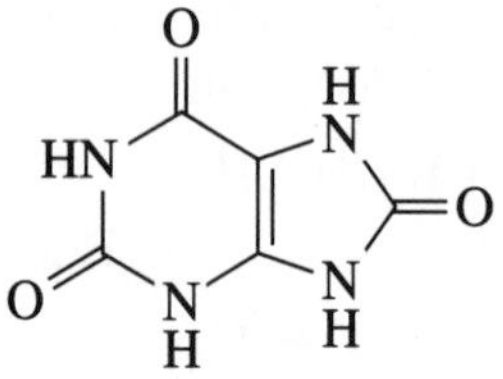

uric acid

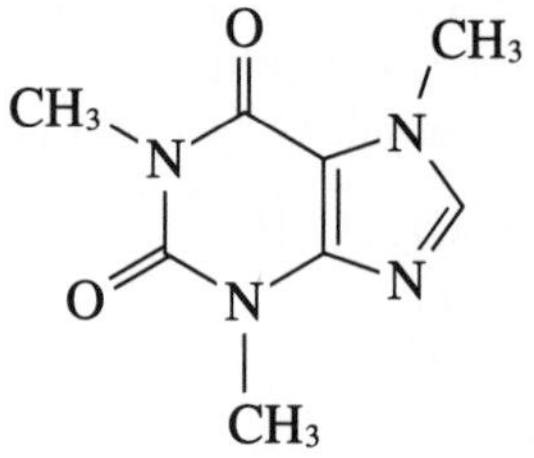

caffeine

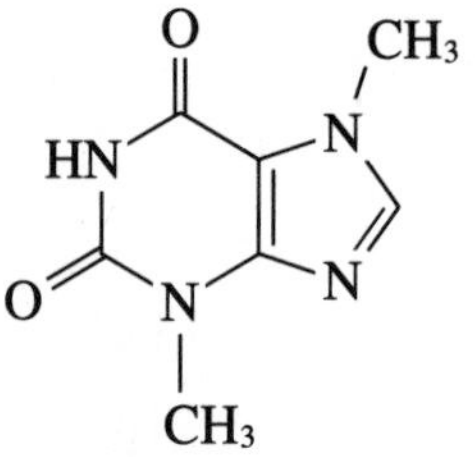

theobromine

Perhaps the most important purines in nature are **adenine** and **guanine,** two of the nitrogen bases that are present in nucleic acids (DNA and RNA; for further details, see Chapter 18).

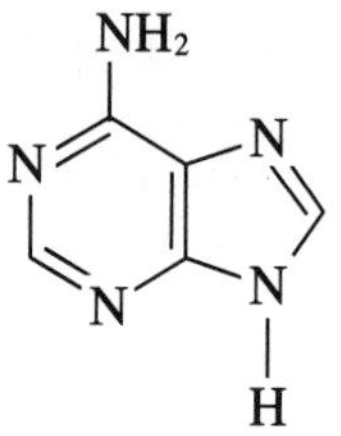

adenine

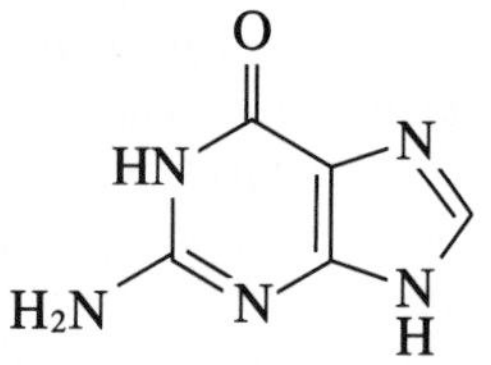

guanine

Many nitrogen heterocycles play a role in medicine. One leading actor in this field is morphine.

A Word About . . .

25. Morphine and Other Nitrogen-Containing Drugs

Morphine (named after Morpheus, the Greek god of dreams) is the major alkaloid present in opium. An **alkaloid** is any basic, nitrogen-containing plant product, often with a complex structure and significant pharmacological properties. Quinine, papaverine, and caffeine are examples of alkaloids already mentioned in this chapter.

Opium is the dried sap of the unripe seed capsule of the poppy *Papaver somniferum;* its medical properties have been known since ancient times. Morphine was not isolated in pure form until 1805; its correct structure was not established until 1925, and it was not synthesized in the laboratory until 1952.

Pain is a major problem in medicine, and relief of pain has long been a medical goal. Morphine is an **analgesic,** a substance that relieves pain without causing unconsciousness. It was used for the large-scale relief of pain from battle wounds during the American Civil War (largely as a consequence of the invention, at about that time, of the hypodermic syringe). But morphine has serious side effects. It is addictive and also can cause nausea, a decrease in blood pressure, and a depressed breathing rate that can be fatal to the very young or the severely debilitated.

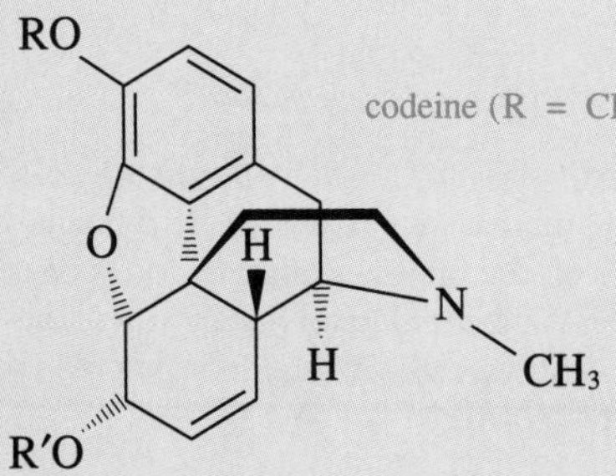

morphine (R = R′ = H)
heroin (R = R′ = $-\overset{\overset{\displaystyle O}{\|}}{C}CH_3$)
codeine (R = CH_3, R′ = H)

The first attempts to find a substance with morphine's benefits but without its side effects involved minor modification of its structure. Acetylation with acetic anhydride gave its diacetyl derivative **heroin,** which is a good analgesic with less of a respiratory depressant effect than morphine. But heroin is severely addictive, and its abuse has become a serious problem. Partial methylation of morphine gave **codeine,** which is useful as an antitussive (anticough) agent. Unfortunately, it is less than one-tenth as effective as morphine as an analgesic.

Many compounds similar to various parts of the morphine structure have been synthesized and tested for their analgesic properties. Two of these are shown here. Their structural similarity to morphine is evident in the colored parts of the structures.

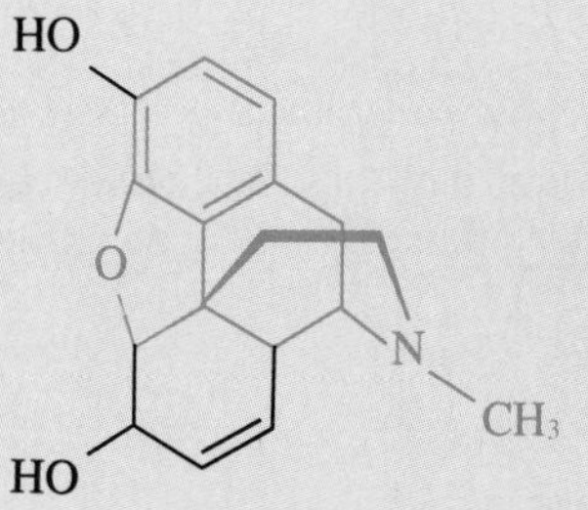

morphine

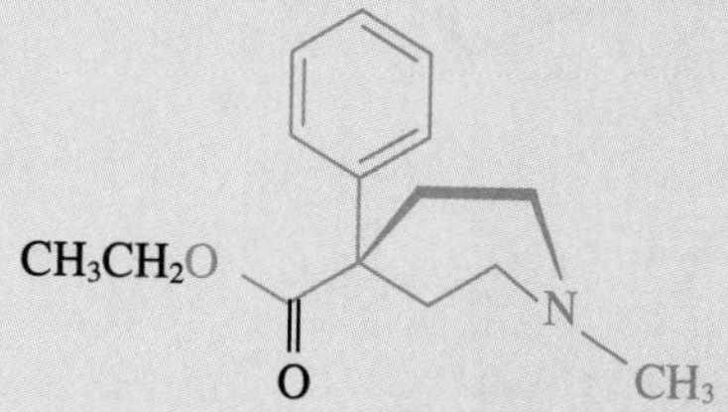

demerol
(meperidine)

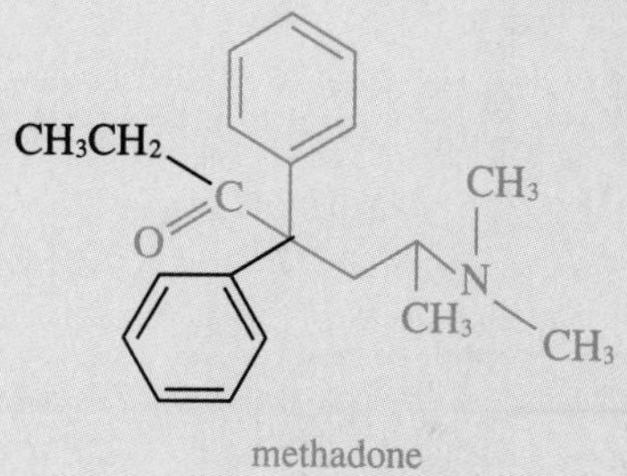

methadone

Demerol is an effective analgesic with a relatively simple structure compared to that of morphine. Notice that it still retains the piperidine ring present in morphine. **Methadone,** which retains the nitrogen of morphine but is no longer heterocyclic, was synthesized and used as an analgesic by the Germans during World War II, when natural sources of morphine were in short supply. Later it was used in substitution therapy for heroin addiction, but it, too, is addictive. The search for a perfect analgesic still goes on.

The pain associated with surgery or injury can sometimes be treated with local or regional anesthetics, many of which are nitrogen-containing drugs. **Cocaine,** an alkaloid found in the plant *Erythroxylum coca,* was one of the first anesthetics of this type. It constricts blood vessels, thus producing a bloodless surgical area.

But cocaine is addictive and has other undesirable properties. Its medical use (it is also used illegally for nonmedical purposes) has been supplanted largely by **procaine hydrochloride** (Novocain).

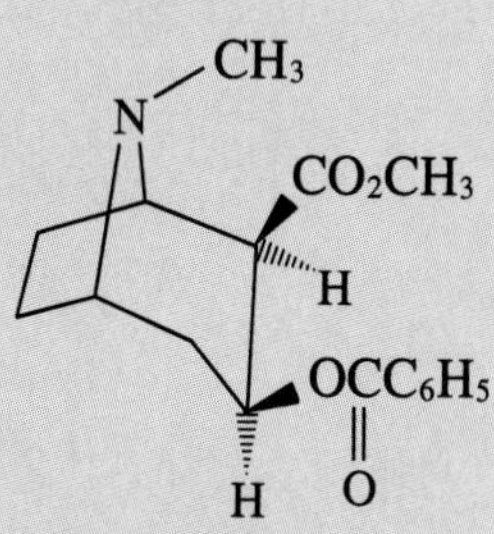

cocaine
mp 98°C

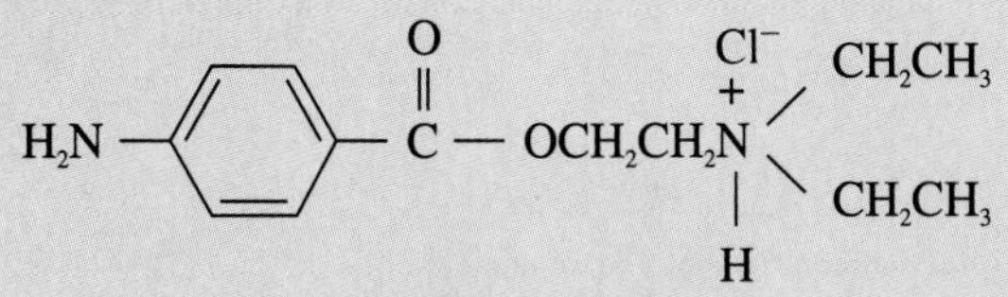

procaine hydrochloride
(mp 153–156°C)

Procaine is less toxic, is easier to synthesize and sterilize, and has a desirably shorter period of action than cocaine. It is usually injected into a nerve to anesthetize a small region of the body. It acts by inhibiting nerve impulse transmission by acetylcholine. Procaine hydrochloride is a widely used drug, sold to dentists and to doctors of human and veterinary medicine under at least 27 different trade names.

Benzocaine (ethyl *p*-aminobenzoate), which has a very simple structure, is used as a mild topical anesthetic in ointments for burns, insect bites, and open wounds. Note that procaine and benzocaine are different esters of the same acid, *p*-aminobenzoic acid.

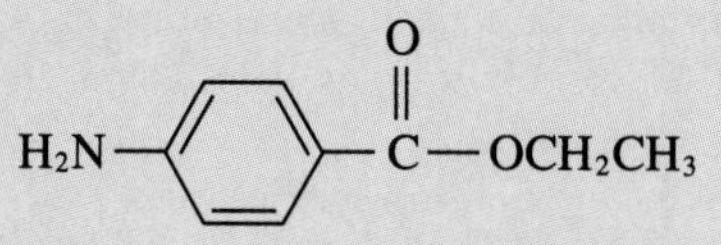

benzocaine
(mp 88–90°C)

Fentanyl, a very short-acting drug about 100 times as potent as morphine, is extremely important in medicine today. It is used as an anesthetic in perhaps 70% of all surgical procedures in the United States.

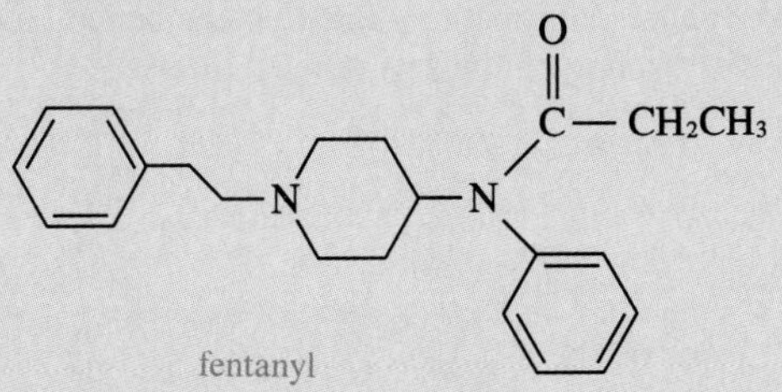

fentanyl

There are, of course, many types of pain. Sometimes a mild tranquilizer can be medically useful. Two of the most commonly prescribed contain seven-membered heterocyclic rings; they are the well-known twins of modern psychiatry: **librium** and **valium.**

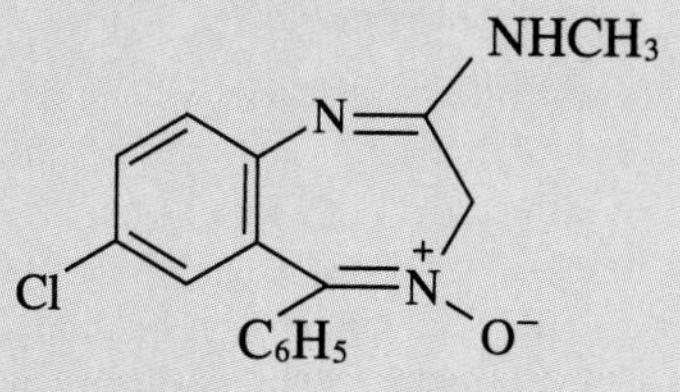

librium

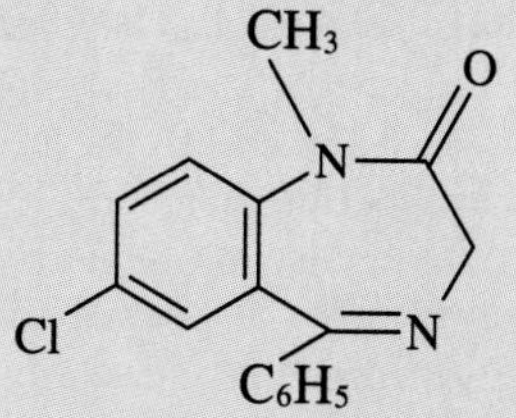

valium

REACTION SUMMARY **Reactions of Pyridine and Related Six-Membered Ring Aromatic Heterocycles**

1. Protonation (Sec. 13.2)

pyridine + HX ⟶ pyridinium (N–H, +) X^-

X = Cl, Br, I, HSO_4

2. Electrophilic Aromatic Substitution (Sec. 13.3)

pyridine + E^+ ⟶ 3-E-pyridine + H^+

3. Nucleophilic Aromatic Substitution (Sec. 13.3)

pyridine + Nu–metal ⟶ 2-Nu-pyridine + metal–H

Examples: Nu—metal = H_2N—Na, Ph—Li

4-Cl-pyridine + Nu–metal ⟶ 4-Nu-pyridine + metal–Cl

Example: Nu—metal = CH_3O—Na

4. Alkyl Side Chain Oxidation (Sec. 13.3)

3-CH_3-pyridine $\xrightarrow{KMnO_4}$ 3-CO_2H-pyridine

5. Ring Reduction (Sec. 13.3)

pyridine $\xrightarrow{H_2}$ piperidine (N–H)

Electrophilic Aromatic Substitution Reactions of Five-Membered Ring Aromatic Heterocycles (Sec. 13.6)

$$\text{(five-membered ring with X)} + E^+ \longrightarrow \text{(2-substituted ring with X and E)} + H^+$$

X=O, S, N—H

ADDITIONAL PROBLEMS

13.9. In addition to the Kekulé-type contributors to the pyridine resonance hybrid shown on page 385, there are three minor dipolar contributors. Draw their structures. Do they suggest a reason why pyridine is deactivated (relative to benzene) toward reaction with electrophiles and a reason why substitution, when it does occur, takes place at the 3-position?

13.10. Although nitration of pyridine requires a temperature of 300°C (eq. 13.2), 2,6-dimethylpyridine is readily nitrated at 100°C. Write an equation for the reaction, and explain why milder conditions suffice.

13.11. Pyridine reacts with phenyllithium to give a good yield of 2-phenylpyridine. Write an equation and a mechanism for the reaction.

13.12. Oxidation of nicotine with $KMnO_4$ gives nicotinic acid. Write an equation for the reaction.

13.13. Write equations for the reaction of coniine (2-propylpiperidine) with
a. hydrochloric acid. **b.** methyl iodide (1 equivalent).
c. methyl iodide (2 equivalents). **d.** acetic anhydride.
e. benzenesulfonyl chloride.

13.14. Explain why nitration of quinoline (eq. 13.7) occurs mainly at C5 and C8.

13.15. Write an equation for each of the following reactions:
a. quinoline + HCl **b.** nitration of isoquinoline
c. quinoline + $NaNH_2$ **d.** isoquinoline + phenyllithium

13.16. What is the configuration (*R* or *S*) of the hydroxyl-bearing carbon atom in codeine?

13.17. In contrast with phenol, which exists almost entirely in the enol form, 2-hydroxypyridine exists mainly in the keto form. Draw the structures, and suggest a reason for the difference.

13.18. Write an equation for each of the following reactions:
a. piperidine + acetic anhydride
b. pyrrolidine + HBr
c. pyrimidine + HCl
d. isoquinoline + CH_3I
e. thiophene + HNO_3
f. furan + acetyl chloride + $AlCl_3$

13.19. Write equations to show how furfural (2-furaldehyde) could be converted to
a. 2-furylmethanol. **b.** 2-furoic acid.

13.20. Although electrophilic substitution occurs at C2 in pyrrole, it occurs predominantly at C3 in indole. Suggest an explanation.

13.21. Draw a molecular orbital picture of the bonding in oxazole (Sec. 13.7), using the bonding in imidazole as an example. Do you expect oxazole to be more or less basic than pyrrole? Explain.

13.22. Using structures for the parent compounds given in the text, write the formula for

a. 3-chlorothiophene.
b. 2,5-diethylpyrrole.
c. 5-hydroxyindole.
d. 4-hydroxyisoquinoline.
e. 5-hydroxyquinoline.
f. 2-methylimidazole.
g. 4-pyridinecarboxylic acid (isonicotinic acid).
h. 3-bromopyridinium bromide.
i. 4-aminopyrimidine.
j. 2,5-dimethylfuran.

13.23. Benadryl can be synthesized in two steps from diphenylmethane, according to the following sequence:

$$C_6H_5{-}CH_2{-}C_6H_5 \xrightarrow[\text{light}]{Br_2} \text{A} \xrightarrow{(CH_3)_2NCH_2CH_2OH} \text{benadryl}$$

a. What is the structure of A?
b. How can the aminoalcohol needed for the second step be synthesized from ethylene oxide?
c. Write a mechanism for each step.

13.24. The water solubility of morphine is only 0.2 g/L, but morphine hydrochloride has a water solubility of 57 g/L. Write equations to show how morphine could be extracted and isolated from opium.

13.25. Identify the stereogenic centers in amoxicillin, and assign an absolute configuration (*R* or *S*) for each center where the stereochemistry is shown.

13.26. Uric acid has four hydrogens. Which do you expect to be the most acidic, and why?

14 Synthetic Polymers

14.1 Introduction

Polymers, or **macromolecules,** as they are sometimes called, are large molecules that are built up by repetitive linking of many smaller units called **monomers.** Polymers can be **natural** or **synthetic.** The most important natural polymers are carbohydrates (starch, cellulose), proteins, and nucleic acids (DNA, RNA). We will study these **biopolymers** in the last three chapters of this book. In this chapter, we will focus on some of the most important *synthetic* polymers.

Synthetic polymers whose names may already be familiar to you include polyethylene (Sec. 3.16), Teflon, Styrofoam, nylon, Dacron, saran, and polyurethanes. There are many others. In the United States alone, annual synthetic polymer production exceeds 70 billion pounds. Synthetic polymers touch our daily lives perhaps to a greater extent than any other synthetic organic chemicals. They make up a substantial percentage of our clothing, appliances, vehicles, homes, packaging, toys, paints, plywood and fiberboard, and tires. It is virtually impossible to live in the modern world without, dozens of times each day, making use of these materials—*all of which were totally unknown less than a century ago*. Our standard of living could not possibly approach what it is today without products of the synthetic organic chemical industry.

14.2 Classification of Polymers

Synthetic polymers can be classified into two main types, depending on how they are made. **Chain-growth polymers** (also called *addition polymers)* are made by the addition of one monomer unit to another in a repetitive manner. Alkenes serve as monomers for the preparation of many important chain-growth polymers and a catalyst is required to initiate their polymerization.* It

---

*Other compounds, such as formaldehyde (Sec. 9.3 and Problem 14.36) and epoxides (see Word About #17), serve as monomers for the preparation of important chain-growth polymers.

adds to a carbon–carbon double bond to form a reactive intermediate and this intermediate then adds to the double bond of a second monomer unit to yield a new intermediate. The process continues until the polymer chain is built. Eventually the process is terminated in some way. *Chain-growth polymers retain all the atoms of the monomer units in the polymer.* A good example of a chain-growth polymer is polyethylene (Sec. 3.16).

$$n\text{CH}_2{=}\text{CH}_2 \xrightarrow[\text{initiator}]{\text{polymerization}} \text{+CH}_2\text{—CH}_2\text{+}_n \tag{14.1}$$

ethylene → polyethylene

Step-growth polymers (also called *condensation polymers*) are usually formed by a reaction between two different functional groups, with the loss of some small molecule, such as water. Thus, *a step-growth polymer does not contain all the atoms initially present in the monomers;* some atoms are lost in the small molecule that is eliminated. The monomer units are usually di- or polyfunctional, and the monomers usually appear in alternating order in the polymer chain. Perhaps the best known example of a step-growth polymer is the polyamide **nylon,** prepared from 1,6-diaminohexane (hexamethylenediamine) and hexanedioic acid (adipic acid).

$$\text{H}_2\text{N(CH}_2)_6\text{NH}_2 + \text{HO}\overset{\text{O}}{\overset{\|}{\text{C}}}(\text{CH}_2)_4\overset{\text{O}}{\overset{\|}{\text{C}}}\text{OH} \xrightarrow{200\text{–}300^\circ\text{C}}$$

1,6-diaminohexane (hexamethylenediamine); hexanedioic acid (adipic acid)

$$\left[\text{NH(CH}_2)_6\text{NH}\overset{\text{O}}{\overset{\|}{\text{C}}}(\text{CH}_2)_4\overset{\text{O}}{\overset{\|}{\text{C}}}\right]_n + 2n\,\text{H}_2\text{O} \tag{14.2}$$

nylon, a polyamide

Let us consider each of these ways of making polymers in greater detail.

14.3 Free-Radical Chain-Growth Polymerization

The free-radical chain mechanism described earlier for polyethylene (Sec. 3.15) is typical of chain-growth polymers. The overall reaction is

$$\underset{\text{L}}{\text{CH}_2{=}\underset{|}{\text{CH}}} \xrightarrow[\text{initiator}]{\text{radical}} \text{—CH}_2\underset{\text{L}}{\underset{|}{\text{CH}}}\left(\text{CH}_2\underset{\text{L}}{\underset{|}{\text{CH}}}\right)_n\text{CH}_2\underset{\text{L}}{\underset{|}{\text{CH}}}\text{—} \tag{14.3}$$

vinyl monomer → vinyl (or chain-growth) polymer

where L is some substituent. Table 14.1 lists some common commercial chain-growth polymers and their uses.

Free-radical chain-growth polymerization requires a radical initiator, of which **benzoyl peroxide** is an example. It decomposes at about 80°C to give benzoyloxy radicals. These radicals may initiate chains or may lose carbon dioxide to give phenyl radicals that can also initiate chains.

TABLE 14.1 Some commercial chain-growth (vinyl) polymers prepared by free-radical polymerization

Monomer name	*Formula*	*Polymer*	*Uses*
ethylene (ethene)	$CH_2{=}CH_2$	polyethylene	sheets and films, blow-molded bottles, injection-molded toys and housewares, wire and cable coverings, shipping containers
propylene (propene)	$CH_2{=}CHCH_3$	polypropylene	fiber products such as indoor-outdoor carpeting, car and truck parts, packaging, toys, housewares
styrene	$CH_2{=}CH{-}C_6H_5$	polystyrene	packaging and containers (Styrofoam), toys, recreational equipment, appliance parts, disposable food containers and utensils, insulation
acrylonitrile (propenenitrile)	$CH_2{=}CHCN$	polyacrylonitrile (Orlon, Acrilan)	sweaters and other clothing
vinyl acetate (ethenyl ethanoate)	$CH_2{=}CH{-}O\overset{O}{\overset{\|}{C}}CH_3$	polyvinyl acetate	adhesives, latex paints
methyl methacrylate (methyl 2-methylpropenoate)	$CH_2{=}C(CH_3){-}\overset{O}{\overset{\|}{C}}OCH_3$	polymethyl methacrylate (Plexiglas, Lucite)	objects that must be clear, transparent, and tough
vinyl chloride (chloroethene)	$CH_2{=}CHCl$	polyvinyl chloride (PVC)	plastic pipe and pipe fittings, films and sheets, floor tile, records, coatings
tetrafluoroethylene (tetrafluoroethene)	$CF_2{=}CF_2$	polytetrafluoroethylene (Teflon)	coatings for utensils, electric insulators

$$\text{C}_6\text{H}_5\text{-}\overset{\overset{\text{O}}{\|}}{\text{C}}\text{-O-O-}\overset{\overset{\text{O}}{\|}}{\text{C}}\text{-C}_6\text{H}_5 \xrightarrow{80°\text{C}} 2\,\text{C}_6\text{H}_5\text{-}\overset{\overset{\text{O}}{\|}}{\text{C}}\text{-O}\cdot \xrightarrow{-\text{CO}_2} 2\,\text{C}_6\text{H}_5\cdot \qquad (14.4)$$

weak bond (O—O)

benzoyl peroxide — benzoyloxy radicals — phenyl radicals

For simplicity, we can represent the initiator radicals by the symbol In·

Initiator radicals add to the carbon–carbon double bond of the vinyl monomer to produce a carbon radical.

Initiation step

$$\text{In}\cdot + \text{CH}_2\text{=}\underset{\underset{\text{L}}{|}}{\text{CH}} \longrightarrow \text{In—CH}_2\text{—}\underset{\underset{\text{L}}{|}}{\dot{\text{C}}\text{H}} \qquad (14.5)$$

monomer — carbon radical

Experience shows that the initiator usually adds to the *least* substituted carbon of the monomer, that is, to the CH_2 group. This gives a carbon radical adjacent to the substituent. There are two reasons for this preference: first, the terminal vinylic carbon is less hindered and therefore more easily attacked, and second, the substituent L usually can stabilize an adjacent radical by resonance.

The carbon radical formed in the initiation step then adds to another monomer molecule, and the adduct adds to another, and so on.

Propagation steps

$$\text{In CH}_2\underset{\underset{\text{L}}{|}}{\dot{\text{C}}\text{H}} \xrightarrow{\text{CH}_2\text{=CHL}} \text{In CH}_2\underset{\underset{\text{L}}{|}}{\text{CH}}\text{CH}_2\underset{\underset{\text{L}}{|}}{\dot{\text{C}}\text{H}} \xrightarrow{\text{CH}_2\text{=CHL}}$$

$$\text{In CH}_2\underset{\underset{\text{L}}{|}}{\text{CH}}\text{CH}_2\underset{\underset{\text{L}}{|}}{\text{CH}}\text{CH}_2\underset{\underset{\text{L}}{|}}{\dot{\text{C}}\text{H}} \longrightarrow \text{and so on} \qquad (14.6)$$

growing polymer chain

Chain propagation (eq. 14.6) occurs in the same sense as initiation (eq. 14.5), so that the monomer units are linked in a head-to-tail manner, with the substituent on alternate carbon atoms.

PROBLEM 14.1 In polystyrene, the chain grows in a strictly head-to-tail arrangement because the intermediate radical is stabilized through resonance. Draw the intermediate radical, and show how it is stabilized through resonance. (The structure of styrene is shown in Table 14.1.)

Chain propagation may continue until anywhere from a few hundred to several thousand monomer units are linked. The extent of reaction depends on several factors, some of which are the reaction conditions (temperature, pressure, solvent, concentration of monomer and catalyst, for example); the nature of the monomer, especially the substituent L; and the rates of competing reactions, which may terminate the chain. Two common chain-terminating reactions are **radical coupling** and **radical disproportionation.**

Termination steps

$$2 \sim CH_2\dot{C}H(L) \xrightarrow[\text{coupling}]{\text{radical}} \sim \overbrace{CH_2CH(L)-CH(L)CH_2}^{\text{head-to-head}} \sim \qquad (14.7)$$

$$2 \sim CH_2\dot{C}H(L) \xrightarrow[\text{disproportionation}]{\text{radical}} \underset{\text{alkane}}{\sim CH_2CH_2(L)} + \underset{\text{alkene}}{\sim CH{=}CHL} \qquad (14.8)$$

Radical coupling (eq. 14.7) gives rise to a head-to-head arrangement of two monomers. In radical disproportionation (eq. 14.8), one radical abstracts a hydrogen atom from the carbon adjacent to another radical site, producing a saturated and an unsaturated polymer.

EXAMPLE 14.1 What feature distinguishes propagation steps from termination steps?

Solution In propagation steps, one radical is destroyed, but another radical is created. In other words, the number of radicals on the left and on the right side of the equation for a propagation step is always the same. In termination steps, however, radicals are destroyed, and no new radicals are generated. Therefore, chain growth terminates.

If eqs. 14.5 through 14.8 told the whole story, all polymers formed by free-radical chain growth would be linear. But we know from physical measurements that many such polymers have branched chains, so something is missing from the scheme.

What is missing is a **chain-transfer reaction.** A growing polymer radical may abstract a hydrogen atom from another polymer chain.

Chain-transfer step

$$\sim CH_2\dot{C}H(L) + \sim CH_2CH(L) \sim \longrightarrow \sim CH_2CH_2(L) + \sim CH_2\dot{C}(L) \sim \qquad (14.9)$$

This step terminates one chain but initiates another chain somewhere along the length of the polymer, not at its end, so that, when the polymerization continues, a branch is produced.

$$\sim\sim CH_2\dot{C}(L)\sim\sim \xrightarrow{CH_2{=}CHL} \sim\sim CH_2C(L)(CH_2CH(L)\cdot)\sim\sim \qquad (14.10)$$

midchain radical from eq. 14.9

branch point

Chain transfer and radical disproportionation are similar in that both involve a hydrogen abstraction reaction.

The extent of chain branching in a particular polymer depends on the relative rates of the chain-propagating and chain-transfer steps. If radical addition is very fast compared with hydrogen abstraction, the polymer will be mainly linear. On the other hand, if the chain-transfer rate were, for example, 1/10 that of the addition rate, we might expect an average of 1 branch for every 10 linearly linked monomers.

Chain-transfer reactions can be used to control the molecular weight of a polymer. Certain reagents, such as thiols, have a hydrogen that is easily abstracted. The resulting RS· radical is not reactive enough to add to double bonds. Instead, it dimerizes to a disulfide.

$$\sim\sim CH_2\dot{C}H(L) + \underset{\text{thiol}}{RSH} \xrightarrow[\text{abstraction}]{\text{hydrogen}} \sim\sim CH_2CH_2(L) + RS\cdot \longrightarrow RSSR \qquad (14.11)$$

does not add to double bonds; dimerizes instead

Thiols are therefore chain terminators, and when added to a polymerization reaction mixture in small amounts, they limit the polymer chain length.

Free-radical chain-growth polymerization is a very fast reaction. A chain may grow to 1000 monomer units or more in less than a second! The polymer chains contain one or two groups derived from the initiator radical, but those groups make up only a small fraction of the polymer molecule, so that the polymer properties are determined largely by the particular monomer used.

Now let us consider two typical free-radical growth polymers: polystyrene and polyvinyl chloride.

Styrene is easily polymerized by benzoyl peroxide, and the product **polystyrene** has a molecular weight in the range of 1 to 3 million.

$$\underset{\text{styrene}}{CH_2{=}CH{-}C_6H_5} \xrightarrow{\text{benzoyl peroxide}} \underset{\text{polystyrene}}{\left(\!-CH_2{-}CH(C_6H_5)-\!\right)_n} \qquad (14.12)$$

Polystyrene is an amorphous, thermoplastic polymer. By **amorphous,** we mean that the polymer chains are irregularly arranged in a random manner; they are not regularly aligned, as in a crystal. By **thermoplastic,** we mean that the polymer melts or softens on heating and hardens again on cooling. Polystyrene can be molded or extruded to produce parts for housewares, toys, radio and television chassis, and bottles, jars, and containers of all kinds. **Styrofoam** is produced by including a low-boiling hydrocarbon such as pentane in the processing. When the polymer is heated, the pentane volatilizes, producing bubbles that expand the polymer into a foam. These foams are used for insulation, packaging, cups for hot drinks, egg cartons, and many other purposes. Annual U.S. production of polystyrene is more than 5 billion pounds.

Polystyrene can be modified in various ways. For example, it can be rigidified through **cross-linking,** by including small amounts of *p*-divinylbenzene with the monomer.

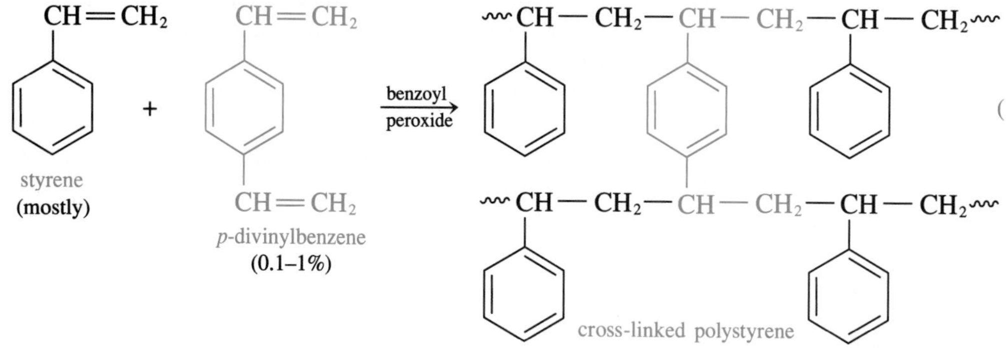

$CH{=}CH_2$ (on benzene ring) styrene (mostly) + $CH{=}CH_2$–C_6H_4–$CH{=}CH_2$ *p*-divinylbenzene (0.1–1%) $\xrightarrow[\text{peroxide}]{\text{benzoyl}}$ (14.13)

$\sim CH-CH_2-CH-CH_2-CH-CH_2\sim$

$\sim CH-CH_2-CH-CH_2-CH-CH_2\sim$

cross-linked polystyrene

The resulting polymer is more rigid and less soluble in organic solvents than ordinary polystyrene. By sulfonation, this cross-linked polymer can be converted to an ion-exchange resin used in water softeners. As hard water percolates through the resin, Ca^{2+} and Mg^{2+} ions are exchanged for Na^+ ions.

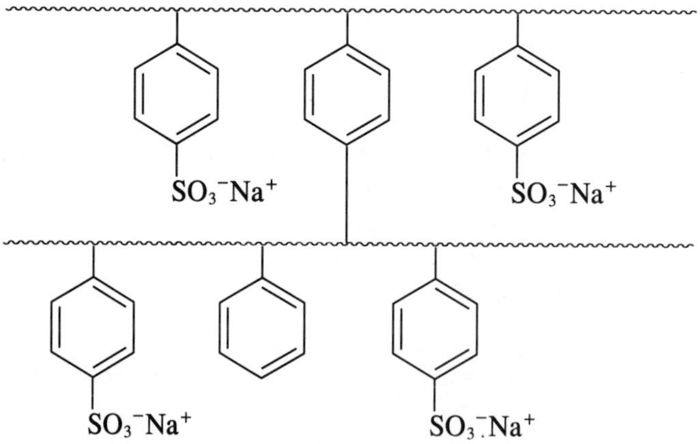

$SO_3^-Na^+$ $SO_3^-Na^+$

$SO_3^-Na^+$ $SO_3^-Na^+$

schematic structure of an ion-exchange resin

PROBLEM 14.2 Explain why sulfonation of polystyrene occurs in the positions shown in the ion-exchange resin formula.

Polyvinyl chloride (PVC) can be represented by the general formula

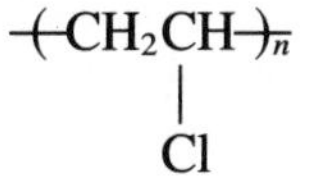

$$\text{+CH}_2\underset{\displaystyle\text{Cl}}{\underset{|}{\text{CH}}}\text{+}_n$$

polyvinyl chloride (PVC)

It has a head-to-tail structure. PVC is a hard polymer but can be softened by adding **plasticizers,** usually low-molecular-weight esters that act as lubricants between the polymer chains. A good example is bis-2-ethylhexyl phthalate.

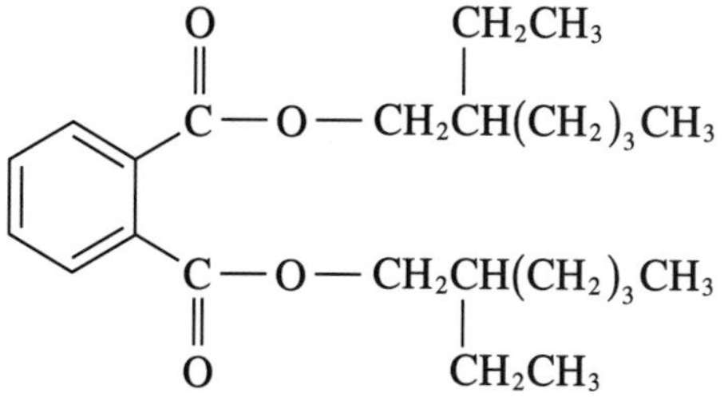

$$C_6H_4\left(\overset{\displaystyle O}{\overset{\|}{C}}-O-CH_2\overset{\displaystyle CH_2CH_3}{\overset{|}{C}H}(CH_2)_3CH_3\right)_2$$

bis-2-ethylhexyl phthalate
(a plasticizer)

PROBLEM 14.3 Bis-2-ethylhexyl phthalate can be prepared from 2-ethylhexanol (review Problem 9.29) and phthalic anhydride (review Problem 10.26). Write an equation for the reaction.

PVC is used to make floor tiles, vinyl upholstery (imitation leather), plastic pipes, plastic squeeze bottles, and so on. Annual U.S. production exceeds 10 billion pounds.

PROBLEM 14.4 Write the structural formula for a three-monomer segment of

a. polypropylene.
b. polyvinyl acetate.
c. poly(methyl methacrylate).
d. polyacrylonitrile.

PROBLEM 14.5 Using a three-monomer segment, write an equation for

a. the reaction of polystyrene with $Cl_2 + FeCl_3$.
b. the reaction of polyvinyl acetate with hot aqueous sodium hydroxide.

14.4 Cationic Chain-Growth Polymerization

Certain vinyl compounds are best polymerized via cationic rather than free-radical intermediates. The most common commercial example is isobutylene (2-methylpropene), which can be polymerized with Friedel-Crafts catalysts via tertiary carbocation intermediates.

Initiation

$$\mathrm{CH_2{=}C(CH_3)_2} \xrightarrow[\mathrm{BF_3;\ H^+}]{\mathrm{AlCl_3\ or}} \mathrm{CH_3{-}\overset{+}{C}(CH_3)_2} \tag{14.14}$$

isobutylene → tertiary carbocation

Propagation

$$\mathrm{CH_3{-}C^+(CH_3)_2 + CH_2{=}C(CH_3)_2 \longrightarrow CH_3C(CH_3)_2{-}CH_2{-}C^+(CH_3)_2 \longrightarrow}$$

$$\mathrm{CH_3{-}C(CH_3)_2{+}\!\left[CH_2{-}C(CH_3)_2\right]_n\!{-}CH_2{-}C^+(CH_3)_2} \tag{14.15}$$

Termination

$$\mathrm{(CH_3)_3C{+}\!\left[CH_2{-}C(CH_3)_2\right]_n\!{-}CH_2{-}C^+(CH_3)_2} \xrightarrow{-\mathrm{H^+}}$$

$$\mathrm{(CH_3)_3C{+}\!\left[CH_2{-}C(CH_3)_2\right]_n\!{-}CH_2{-}C({=}CH_2)CH_3} \tag{14.16}$$

polyisobutylene

Initiation gives the *tert*-butyl cation (eq. 14.14), which, in the propagation step, adds to the CH_2 carbon of the double bond in a Markovnikov manner to produce another tertiary carbocation, and so on (eq. 14.15). The chain is terminated by loss of a proton from a carbon atom adjacent to the positive carbon (eq. 14.16).

Polyisobutylenes prepared this way (n = about 50) are used as additives in lubricating oil and as adhesives in pressure-sensitive tape and removable paper labels. Higher-molecular-weight polymers are used in the manufacture of inner tubes for truck and bicycle tires.

14.5 Anionic Chain-Growth Polymerization

Alkenes with electron-withdrawing substituents can be polymerized via carbanionic intermediates. The catalyst may be an organometallic compound, for example, an alkyllithium.

Initiation $$\mathrm{CH_2{=}CHL + RLi \longrightarrow RCH_2\overset{..}{\underset{|\atop L}{C}}\overset{-}{H}\ \ \overset{+}{Li}} \tag{14.17}$$

Propagation

$$\underset{\displaystyle L}{\overset{}{\mathrm{RCH_2\underset{|}{C}H^-}}}\ \ \mathrm{Li^+} \xrightarrow{\mathrm{CH_2{=}CHL}} \mathrm{RCH_2\underset{\underset{L}{|}}{C}HCH_2\underset{\underset{L}{|}}{\bar{C}}H}\ \ \mathrm{Li^+},\ \text{and so on} \qquad (14.18)$$

Addition of the catalyst to the double bond gives a carbanion intermediate (eq. 14.17) in which the substituent L usually delocalizes the negative charge through resonance. Common L groups of this type are cyano, carbomethoxy, phenyl, and vinyl.

EXAMPLE 14.2 Draw the carbanion intermediate for anionic polymerization of acrylonitrile ($CH_2{=}CHCN$), and show how it is stabilized by resonance.

Solution

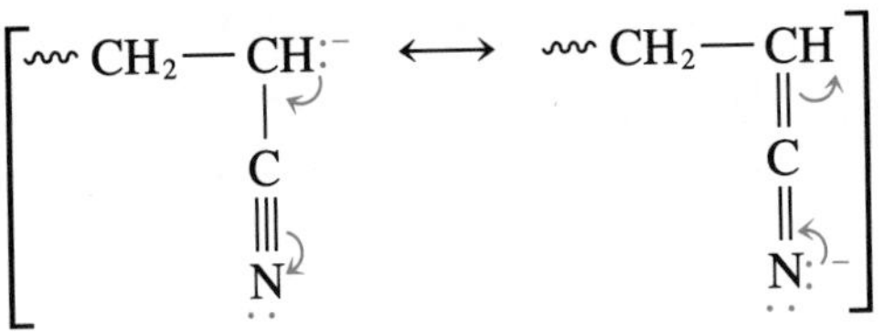

$$\left[\sim\mathrm{CH_2{-}\ddot{C}H^-}(\mathrm{-C{\equiv}\ddot{N}}) \longleftrightarrow \sim\mathrm{CH_2{-}CH{=}C{=}\ddot{N}:^-} \right]$$

PROBLEM 14.6 Methyl methacrylate (Table 14.1) can be polymerized by *n*-butyllithium at −78°C. Using eqs. 14.17 and 14.18 as a model, write a mechanism for the reaction. Show how the intermediate carbanion is resonance-stablized.

Anionic polymerizations are terminated by quenching the reaction mixture with a proton source (water or alcohol).

PROBLEM 14.7 Ethylene oxide can be polymerized by base to give carbowax, a water-soluble wax.

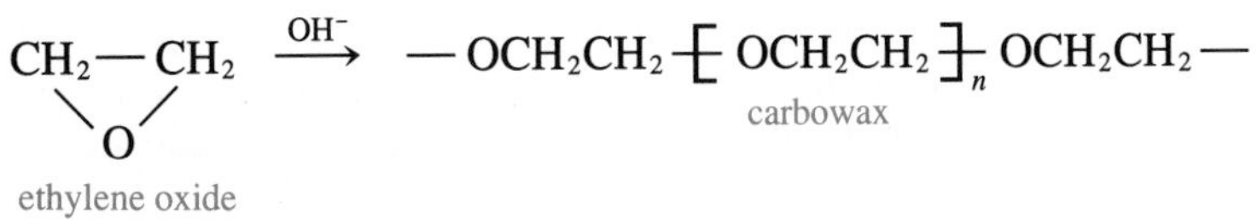

$$\underset{\text{ethylene oxide}}{\mathrm{CH_2{-}CH_2}\ (\text{with bridging O})} \xrightarrow{\mathrm{OH^-}} \underset{\text{carbowax}}{\mathrm{-OCH_2CH_2{+}OCH_2CH_2{+}_n\,OCH_2CH_2-}}$$

Suggest a mechanism for the reaction.

14.6 Stereoregular Polymers; Ziegler–Natta Polymerization

When a monosubstituted vinyl compound is polymerized, every other carbon atom in the chain becomes a stereogenic center:

$$\mathrm{CH_2{=}\underset{\underset{L}{|}}{C}H} \longrightarrow \mathrm{-CH_2{-}\underset{\underset{L}{|}}{\overset{*}{C}}H{-}CH_2{-}\underset{\underset{L}{|}}{\overset{*}{C}}H{-}CH_2{-}\underset{\underset{L}{|}}{\overset{*}{C}}H-} \qquad (14.19)$$

The carbons marked with an asterisk have four different groups attached and are therefore stereogenic centers. Three classes of such polymers are recognized:

atactic: stereocenters have random configurations
isotactic: all stereocenters have the same configuration
syndiotactic: stereocenters alternate in configuration

An atactic polymer is **stereorandom,** but an isotactic or syndiotactic polymer is **stereoregular.** These three classes of polymers, *even if derived from the same monomer,* will have different physical properties.

EXAMPLE 14.3 Draw a chain segment of isotactic polypropylene.

Solution For polypropylene, the group L in eq. 14.19 is $—CH_3$. With the chain extended in zigzag fashion, all methyl substituents occupy identical positions.

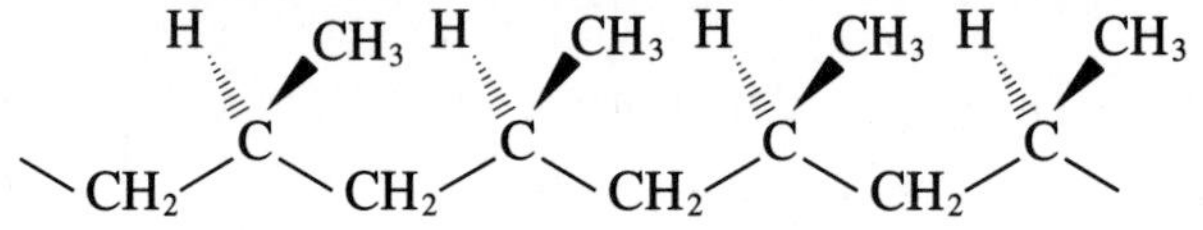

H CH₃ H CH₃ H CH₃ H CH₃ on C; —CH₂—C—CH₂—C—CH₂—C—CH₂—C—

PROBLEM 14.8 Using the definitions just given, draw a chain segment of

a. syndiotactic polypropylene.
b. atactic polypropylene.

Stereoregularity imparts certain favorable properties to polymers. Since free-radical polymerization usually results in an atactic polymer, the discovery in the 1950s by Ziegler and Natta* of mixed organometallic catalysts that produce stereoregular polymers was a landmark in polymer chemistry. One such catalyst system is a mixture of triethylaluminum (or other trialkylaluminums) and titanium tetrachloride. With this catalyst, for example, propylene gives a polymer that is more than 98% isotactic.

The mechanism of Ziegler–Natta catalysis is quite complex. A key step in the chain growth involves an alkyl–titanium bond and coordination of the monomer to the metal. The coordinated monomer then inserts into the carbon–titanium bond, and the process continues.

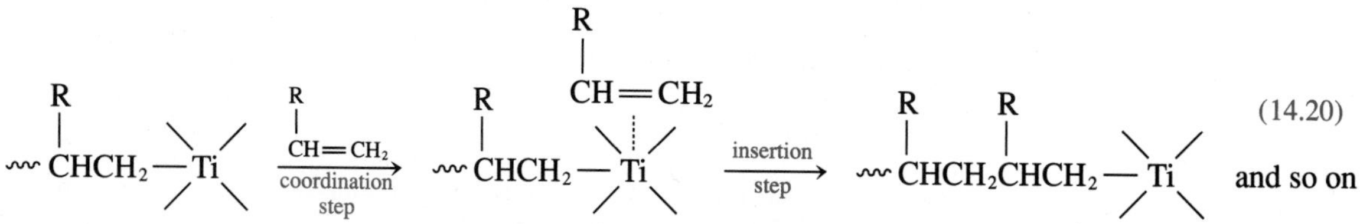

$$\sim\!\overset{R}{\overset{|}{C}}HCH_2\!-\!Ti \xrightarrow[\text{coordination step}]{RCH=CH_2} \sim\!\overset{R}{\overset{|}{C}}HCH_2\!-\!\overset{RCH=CH_2}{\overset{\vdots}{Ti}} \xrightarrow[\text{step}]{\text{insertion}} \sim\!\overset{R}{\overset{|}{C}}HCH_2\overset{R}{\overset{|}{C}}HCH_2\!-\!Ti \quad \text{and so on} \tag{14.20}$$

Because of the various ligands attached to the titanium atom, coordination and insertion occur in a stereoregular manner and can be controlled to give either an isotactic or a syndiotactic polymer.

---

* Karl Ziegler (Germany) and Giulio Natta (Italy) shared the 1963 Nobel Prize in chemistry for this discovery.

Commercial production of polypropylene is performed exclusively with Ziegler–Natta catalysts. A stereoregular, isotactic polymer that is highly crystalline is obtained. It is used for interior trim and battery cases in automobiles, for packaging (for example, containers for nested potato chips), and for furniture (such as plastic stacking chairs). It is also spun into fibers for ropes that float (an advantage for sailors and dockers), synthetic grass, carpet backings, and related materials.

Polyethylene obtained through Ziegler–Natta catalysis is linear, in contrast with the highly branched polyethylene obtained through free-radical processes. Linear polyethylene has a more crystalline structure, a higher density, and greater tensile strength and hardness than the branched polymer. It is used in thin-wall containers like those used to hold laundry bleach and detergents; in molded housewares such as mixing bowls, refrigerator containers, and toys; and in extruded plastic pipes and conduits.

14.7 Diene Polymers: Natural and Synthetic Rubber

Natural rubber is an unsaturated hydrocarbon polymer. It is obtained commercially from the milky sap (latex) of the rubber tree. Its chemical structure was deduced in part from the observation that, when latex is heated *in the absence of air,* it breaks down to give mainly a single unsaturated hydrocarbon product, **isoprene.**

$$\text{natural rubber} \xrightarrow{\text{heat}} \underset{\substack{\text{isoprene} \\ \text{(2-methyl-1,3-butadiene)}}}{CH_2{=}\underset{\underset{CH_3}{|}}{C}{-}CH{=}CH_2} \tag{14.21}$$

It is possible to synthesize a material that is nearly identical to natural rubber by treating isoprene with a Ziegler–Natta catalyst, such as a mixture of triethylaluminum, $(CH_3CH_2)_3Al$, and titanium tetrachloride, $TiCl_4$. The isoprene molecules add to one another by a **head-to-tail 1,4-addition.**

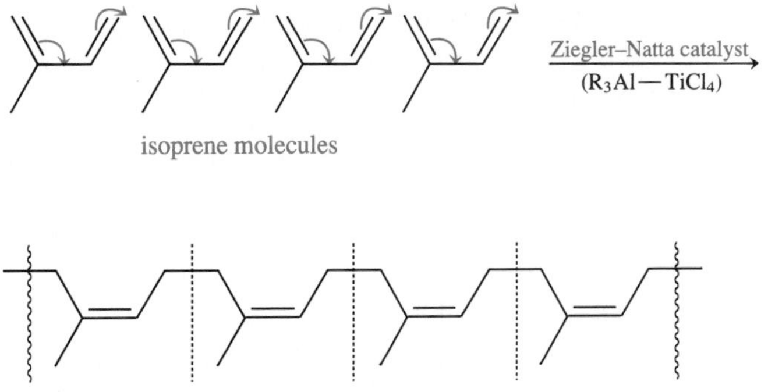

isoprene molecules $\xrightarrow[(R_3Al-TiCl_4)]{\text{Ziegler–Natta catalyst}}$ natural rubber segment (all Z) (14.22)

The double bonds in natural rubber are *isolated;* that is, they are separated from one another by three single bonds. They have a Z geometry.

PROBLEM 14.9 Gutta-percha is a less common form of natural rubber. It is also a 1,4-polymer of isoprene, but with E double bonds. Draw the structural formula for a three-monomer segment of gutta-percha.

Most rubber has a molecular weight in excess of 1,000,000, though the value varies with the source and method of processing. This corresponds to about 15,000 isoprene monomers per rubber molecule. Crude plantation rubber contains, in addition to polyisoprene, about 2.5 to 3.5% protein, 2.5 to 3.2% fats, 0.1 to 1.2% water, and traces of inorganic matter.

Although natural rubber has many useful properties, it also has some undesirable ones. Early manufactured rubber goods were often sticky and smelly, and they softened in warm weather and hardened in cold. Some of these weaknesses were overcome when Charles Goodyear invented **vulcanization,** a process of cross-linking polymer chains by heating rubber with sulfur. The cross-links add strength to the rubber and act as a kind of "memory" that helps the polymer recover its original shape after stretching.

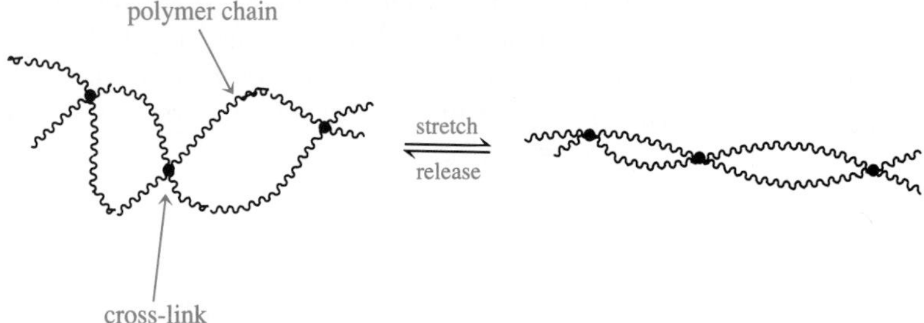

In spite of such improvements, there were still problems. For example, it was not uncommon years ago to have to check the air pressure in tires almost every time one purchased gasoline, because the rubber inner tubes were somewhat porous. Therefore, there was a need to develop **synthetic rubber,** a name given to polymers with properties similar to those of natural rubber but superior to and somewhat different chemically from it.

Many monomers or mixtures of monomers form **elastomers** (rubberlike substances) when they are polymerized. The largest scale commercial synthetic rubber is a **copolymer** of 25% styrene and 75% 1,3-butadiene, called SBR (styrene-butadiene rubber).

$$\underset{\text{styrene}}{n\,CH_2{=}CHC_6H_5} + \underset{\text{butadiene}}{3n\,CH_2{=}CH{-}CH{=}CH_2} \xrightarrow[\text{initiator}]{\text{free radical}} \qquad (14.23)$$

SBR

The structure is approximately as shown, although about 20% of the butadiene adds 1,2- instead of 1,4-. Unlike those in natural rubber, the double bonds in this polymer have E geometry. The dashed lines in the structure show the units from which the polymer is constructed. About two-thirds of SBR goes into tires. Its annual production exceeds that of natural rubber by a factor of two.

PROBLEM 14.10 Draw the structural formula for a three-monomer segment of poly(1,3-butadiene) in which

a. addition is 1,4 and double bonds are Z.
b. addition is 1,4 and double bonds are E.
c. addition is 1,2 for the middle unit and 1,4 for the outer units, with double bonds Z.

14.8 Copolymers

Most of the polymers we have described so far are **homopolymers,** polymers made from a single monomer. But the variety and utility of chain-growth polymerization can sometimes be enhanced (as we have just seen with SBR synthetic rubber) by using mixtures of monomers, to give **copolymers.** Figure 14.1 summarizes some of the ways that monomers can be arranged in homo- and copolymers. The copolymers depicted are limited to two different monomers (A and B); in principle, of course, the possibilities are unlimited.

The exact arrangement of monomers along a copolymer chain will depend on a number of factors. One of these is the relative reactivity of the two monomers. Let us assume that we polymerize a 1 : 1 mixture of A and B by a free-radical chain-growth process. Here are some of the possibilities:

1. A· reacts rapidly with B but slowly with A, and B· reacts rapidly with A but slowly with B. The polymer will then be alternating: —ABABAB—. Many copolymers tend toward this arrangement, though not always perfectly.
2. A and B are equally reactive toward radicals, and each reacts readily with A· or B·. The polymer will then be random: —AABABBA—.

FIGURE 14.1 Arrangements of monomers in polymers.

Homopolymers

```
                      AA—
                      |
—AAAAA—          —AAAAA—          —AAAAA—
                   |                  |
 linear            AA—            —AAAAA—
                branched         cross-linked
```

Copolymers

```
—ABABAB—     —AABABBA—     —AAAAABBBB—     —AAAAAAA—
alternating    random         block          |    |
                                          —BBB  BBB—
                                             graft
```

3. A is much more reactive than B toward all radicals. In this case, A will be consumed first, followed more slowly by B. We will obtain a mixture of two homopolymers, $—(A)_n—$ and $—(B)_m—$.

PROBLEM 14.11 1,1-Dichloroethene and vinyl chloride form a copolymer called *saran*, used in food packaging. The monomer units tend to alternate in the chain. Draw the structural formula for a 4-monomer segment of the chain.

Block and graft copolymers are made by special methods. If we first initiate polymerization of monomer A, then add some B, then add A again, and so on, we can obtain a block polymer with alternating segments of blocks of A units, then B units, and so on. This is particularly easy with anionic polymerizations, where there are no significant termination steps.

Graft polymers are made by taking advantage of functionality present in a homopolymer. For example, if a polymer contains double bonds (as in poly-1,3-butadiene), addition of a free-radical initiator R· and second monomer (such as styrene) will "graft" polystyrene chains onto the polybutadiene backbone.

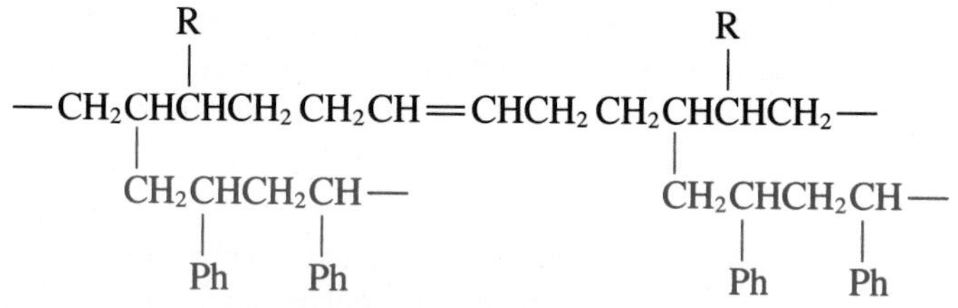

```
        R                                    R
        |                                    |
—CH2CHCHCH2 CH2CH=CHCH2 CH2CHCHCH2—
      |                                    |
      CH2CHCH2CH—                          CH2CHCH2CH—
         |     |                              |     |
         Ph    Ph                             Ph    Ph
```

poly-1,3-butadiene with polystyrene grafts

This particular graft polymer is used to make rubber soles for shoes.

14.9 Step-Growth Polymerization: Dacron and Nylon

Step-growth polymers are usually produced by a reaction between two monomers, each of which is at least difunctional. Many of them can be represented by overall equation 14.24,

$$\text{A}\sim\text{A} + \text{B}\sim\text{B} \longrightarrow —\text{A}\sim\text{A}—\text{B}\sim\text{B}—\text{A}\sim\text{A}—\text{B}\sim\text{B}— \quad (14.24)$$

where A∿A and B∿B are difunctional molecules with groups A and B that can react with one another. For example A might be an OH group, and B might be a CO_2H group, in which case A∿A would be a diol, B∿B would be a dicarboxylic acid and ∿A—B∿ would be an ester. The polymer would be a polyester.

Unlike chain-growth polymers, which grow by one monomer unit at a time, step-growth polymers are formed in steps (or leaps), often by reaction of one polymer molecule with another. The way this works is best illustrated with a specific example.

Consider the formation of a polyester from a diol and a diacid. In the first step, the product will be an ester, with an alcohol group at one end and an acid at the other (eq. 14.25).

$$\underset{\text{diol}}{HO{\sim}OH} + \underset{\text{diacid}}{HO_2C{\sim}CO_2H} \xrightarrow{-H_2O} \underset{\text{alcohol}}{HO{\sim}}\underset{\text{ester}}{O-\overset{\overset{\displaystyle O}{\|}}{C}}\underset{\text{acid}}{{\sim}CO_2H} \tag{14.25}$$

At the next stage, the alcohol-ester-acid can react with another diol, with another diacid, *or with another trifunctional molecule like itself*.

$$\underset{\text{alcohol-ester-acid}}{HO{\sim}O-\overset{\overset{\displaystyle O}{\|}}{C}{\sim}CO_2H} \begin{cases} \xrightarrow[-H_2O]{HO{\sim}OH} \underset{\text{diester-diol}}{HO{\sim}O-\overset{\overset{\displaystyle O}{\|}}{C}{\sim}\overset{\overset{\displaystyle O}{\|}}{C}-O{\sim}OH} \\ \xrightarrow[-H_2O]{HO_2C{\sim}CO_2H} \underset{\text{diester-diacid}}{HO_2C{\sim}\overset{\overset{\displaystyle O}{\|}}{C}-O{\sim}O-\overset{\overset{\displaystyle O}{\|}}{C}{\sim}CO_2H} \\ \xrightarrow[-H_2O]{HO{\sim}O-\overset{\overset{O}{\|}}{C}{\sim}CO_2H} \underset{\text{alcohol-triester-acid}}{HO{\sim}O-\overset{\overset{\displaystyle O}{\|}}{C}{\sim}\overset{\overset{\displaystyle O}{\|}}{C}-O{\sim}O-\overset{\overset{\displaystyle O}{\|}}{C}{\sim}CO_2H} \end{cases} \tag{14.26}$$

The consequences of these alternatives are different; each of the first two products contains three monomer units, but in the third alternative, we go directly from two-monomer fragments to a product with four monomer units. Since the reactivity of the —OH group or of the —CO_2H group in all these reactants is quite similar, there is no particular preference among them, and the rates of the various reactions will depend mainly on the concentrations of the particular reactants.

PROBLEM 14.12 How many monomer units will be present in the next product if the diester-diol and diester-diacid in eq. 14.26 react? Draw the structure of the product.

If we start with exactly one equivalent each of a diol and a diacid, we should, in principle, be able to form one giant polyester molecule. In practice, this does not happen. In fact, to form a polymer with an average of 100 monomer units or more, the reaction must go to at least 99% completion. Consequently, the starting materials for this type of polymerization must be exceedingly pure, their mole ratio must be controlled precisely, and the reaction must be forced to completion, usually by distilling or otherwise removing, as it is formed, the small molecule that is eliminated.

PROBLEM 14.13 What product will mainly be formed if a diacid is treated with a *large excess* of a diol? if a diol is treated with a *large excess* of a diacid? These reactions represent two extremes of what can happen as the ratio of reactant concentrations deviates from 1 : 1 in a step-growth polymerization.

Although many polyesters are known, the most common example is **Dacron, the polyester of terephthalic acid and ethylene glycol.**

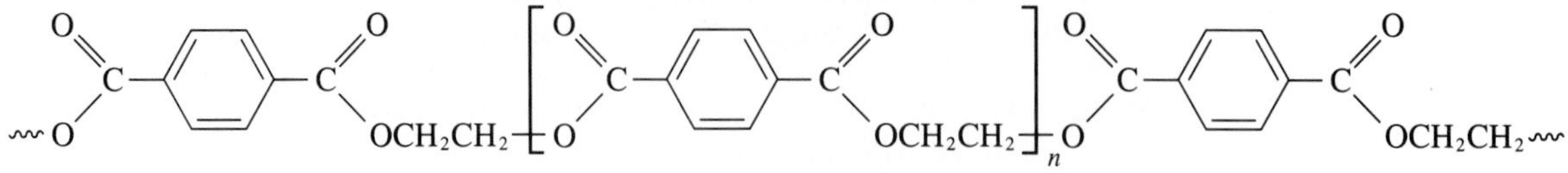

$$\sim O-C(=O)-C_6H_4-C(=O)-OCH_2CH_2\left[O-C(=O)-C_6H_4-C(=O)-OCH_2CH_2\right]_n O-C(=O)-C_6H_4-C(=O)-OCH_2CH_2\sim$$

the polyester Dacron,
poly(ethylene terephthalate)

The value of n is about 100 ± 20. The crude polyester can be spun into fibers for use in textiles. The fibers are highly resistant to wrinkling.

The same polyester can also be fabricated into a particularly strong film called **Mylar.** Mylar polyester film is used for the long-term protection of artwork and historical documents because of its transparency, strength, and inertness. Because of their extraordinary resistance to tear, polyesters are used to make magnetic tapes for the recording industry. In the United States, production of polyester fibers is almost 4 billion pounds per year.

A Word About . . .

26. Degradable Polymers

Widespread use of polymeric plastic materials that contribute so much to our standard of living has brought with it a pollution problem. What is to be done with these materials once they are no longer useful? Incineration represents one solution to the problem, but it can be accompanied by release of toxic substances into the environment. Another solution is recycling. For example poly(ethylene terephthalate) (PETE), which is widely used in the production of plastic soft drink bottles (look for the symbol PETE on the bottom of your next bottle of soda), can be recycled as polyester fiber for use in carpets or furniture stuffing. Also, milk jugs made from high-density poly(ethylene) (HDPE) can be recycled in a number of ways, for example, as plastic lumber.

A third solution to the pollution problem is to develop polymers that are degraded in the environment once they are no longer useful. This approach is quite difficult because it is often the durability of plastics that makes them so useful. It would be useless to produce a product that began to disintegrate as it came off the production line. Nonetheless, several commercially useful **degradable polymers** have been developed. For example, when ethylene is polymerized in the presence of carbon monoxide, a polyethylene-like polymer that contains carbonyl groups is produced. This material, which has been used in the production of trash bags and the plastic rings of beverage six packs, undergoes degradation upon exposure to ultraviolet light because of the photochemical properties of carbonyl groups.

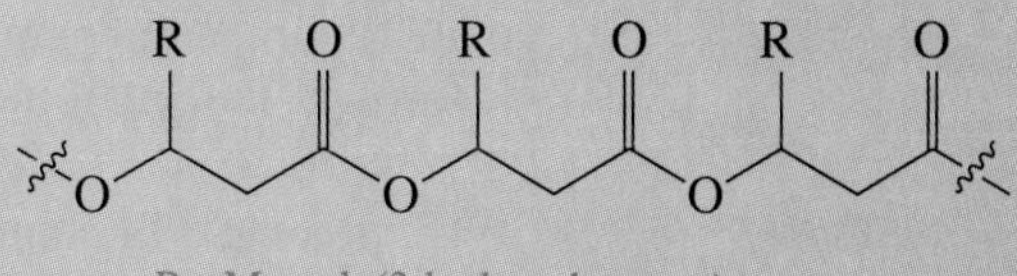

$$\sim O-CH(R)-CH_2-C(=O)-O-CH(R)-CH_2-C(=O)-O-CH(R)-CH_2-C(=O)\sim$$

R=Me poly(3-hydroxybutyrate)
R=Et poly(3-hydroxyvalerate)

ε-caprolactone lactic acid

Recently, natural biopolymers have been developed for use in the manufacturing of plastic containers. **Poly(hydroxyalkanoates)** such as poly(3-hydroxybutyrate) (PHB) and poly(3-hydroxyvalerate) (PHV) are naturally occuring polyesters produced by certain bacteria. PHB has many physical properties in common with poly(propylene) and a PHB-PHV copolymer (BIOPOL) has recently been used to manufacture plastic shampoo bottles. PHB-PHV is of special interest because it is biodegradable. Because it is a naturally occurring polymer, it is easily degraded by enzymes produced by soil microorganisms and therefore does not persist in the environment after disposal. Other biodegradable polymers, such as polyesters derived from ε-caprolactone and lactic acid, are also known and have been commercialized. Although it remains to be seen how widespread the use of biodegradable plastics will become, research and development of these materials is sure to continue as we try to deal with contemporary environmental issues.*

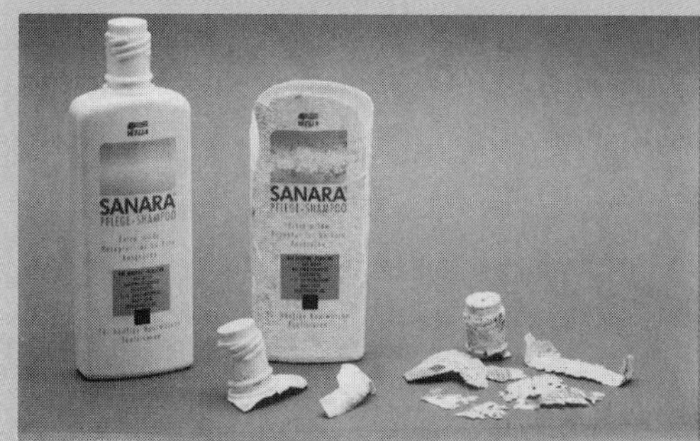

Shampoo bottles made of BIOPOL, after 0, 3, and 9 months in compost. The slower degradation of the printed areas is visible.

* For additional reading, see articles by J. D. Evans and S. K. Sikdar in *Chemtech* **1990,** 38, and H. Muller and D.Seebach in *Angew. Chem., Int. Engl. Ed.* **1993,** *32,* 477.

PROBLEM 14.14 *Kodel* is a polyester with the following structure:

$$\left[-\overset{\displaystyle O}{\overset{\|}{C}}-C_6H_4-\overset{\displaystyle O}{\overset{\|}{C}}-O-CH_2-C_6H_{10}-CH_2-O- \right]_n$$

From what two monomers is it made?

Nylons are polyamide step-growth polymers. The formula for **nylon-6,6,** so called because each monomer (diamine and diacid) has six carbon atoms, is shown in eq. 14.2. This polymer was first made by W. H. Carothers at the Du Pont Company in 1933 and was commercialized five years later.* When mixed, the two monomers form a polysalt which, on heating, loses water to form a polyamide. The molten polymer can be molded or spun into fibers.

* For an account of its discovery, the strange origin of its name, and a description of its properties and many uses, see the article by G. B. Kauffman in *J. Chem. Education* **1988,** *65,* 803–808.

A Word About . . .

27. Aramids, the Latest in Polyamides

Aromatic polyamides (called aramids) are being produced at a rapidly growing rate because of their special properties of heat resistance, low flammability, and exceptional strength. The best known is **Kevlar** (see Figure 14.2). Because of the aromatic rings, this type of polyamide has a much stiffer structure than do the nylons. It is being used in place of steel to make tire cord for radial tires, since a Kevlar fiber is five times stronger than an equal-weight steel fiber. Kevlar is used in lightweight personal body armor (such as bullet-resistant vests), which offers protection against handgun fire, shotgun pellets, and knife slashes.

Nomex has a structure similar to that of Kevlar but uses *meta*- instead of *para*-oriented monomers. It is used in flame-resistant clothing because its fibers char rather than melt when exposed to flame. It has wide applications, ranging from firefighters' coats to racing drivers' uniforms. It has also been used in flame-resistant building materials. Honeycomb core constructions of Nomex are used in many internal and external parts of military and commercial jets, helicopters, and space vehicles. For example, close to 25,000 square feet of Boeing 747's exterior structures include Nomex construction. The combination of strength and light weight has also made both Nomex and Kevlar popular in boat construction. For an account of the Kevlar story, see Tanner, D.; Fitzgerald, J. A.; and Phillips, B. R. *Angewandte Chemie International Edition in English, Advanced Materials* **1989,** *28,* 649–654.

$$\underset{p\text{-phenylenediamine}}{H_2N-C_6H_4-NH_2} + \underset{\text{terephthaloyl chloride}}{Cl-\overset{O}{\overset{\|}{C}}-C_6H_4-\overset{O}{\overset{\|}{C}}-Cl} \xrightarrow{\text{base}} \underset{\text{Kevlar}}{\left(-HN-C_6H_4-NH-\overset{O}{\overset{\|}{C}}-C_6H_4-\overset{O}{\overset{\|}{C}}- \right)_n}$$

FIGURE **14.2** Reaction used to synthesize Kevlar.

The second most important polyamide is **nylon-6,** made from **caprolactam.**

$$\text{caprolactam (seven-membered ring: C(=O)–NH–CH}_2\text{–CH}_2\text{–CH}_2\text{–CH}_2\text{–CH}_2\text{–)} \xrightarrow{250\text{–}270^\circ\text{C}} \left[\text{NHCH}_2\text{CH}_2\text{CH}_2\text{CH}_2\text{CH}_2\overset{\text{O}}{\overset{\|}{\text{C}}}\right]_n \quad \text{nylon-6} \tag{14.27}$$

Lactams are cyclic amides (compare with lactones, Sec. 10.13). On heating, the seven-membered ring of caprolactam opens as the amino group of one molecule reacts with the carbonyl group of the next, and so on, to produce the polyamide.

Nylons are extremely versatile polymers that can be processed to give materials as delicate as sheer fabrics, as long-wearing as carpets, as tough as molded automobile parts, or as useful as Velcro fasteners. In the United States, annual production of nylon fibers exceeds 2.5 billion pounds.

14.10 Polyurethanes and Other Step-Growth Polymers

A **urethane** (also called a **carbamate**) is a functional group that is simultaneously an ester and an amide, $\text{RNH}\overset{\text{O}}{\overset{\|}{\text{C}}}\text{OR}'$. Urethanes are commonly prepared from isocyanates and alcohols.

$$\underset{\text{isocyanate}}{\text{R—N=C=O}} + \underset{\text{alcohol}}{\text{R}'\text{OH}} \longrightarrow \underset{\text{urethane}}{\text{RNH}\overset{\text{O}}{\overset{\|}{\text{C}}}\text{OR}'} \tag{14.28}$$

The reaction is an example of a nucleophilic addition to the carbonyl group (compare with Secs. 9.7 and 10.12).

$$\text{R—}\ddot{\text{N}}\text{=C=}\ddot{\text{O}}\text{:} + \text{R}'\ddot{\text{O}}\text{H} \longrightarrow \text{R—}\ddot{\text{N}}\text{=}\overset{:\ddot{\text{O}}:^-}{\overset{|}{\text{C}}}\text{—}\underset{\text{H}}{\underset{|}{\overset{+}{\ddot{\text{O}}}}}\text{—R}' \quad (\text{H}^+) \longrightarrow \text{R—}\underset{\text{H}}{\underset{|}{\ddot{\text{N}}}}\text{—}\overset{:\ddot{\text{O}}}{\overset{\|}{\text{C}}}\text{—}\ddot{\text{O}}\text{R}' \tag{14.29}$$

PROBLEM 14.15 The highly effective, biodegradable insecticide Sevin is a urethane called 1-naphthyl-*N*-methylcarbamate. It is made from methyl isocyanate and 1-naphthol. Using eq. 14.28 as a guide, write an equation for its preparation.

Polyurethanes are made from *di*isocyanates and *di*ols. The most important commercial example is prepared from 2,4-tolylenediisocyanate (TDI) and ethylene glycol.

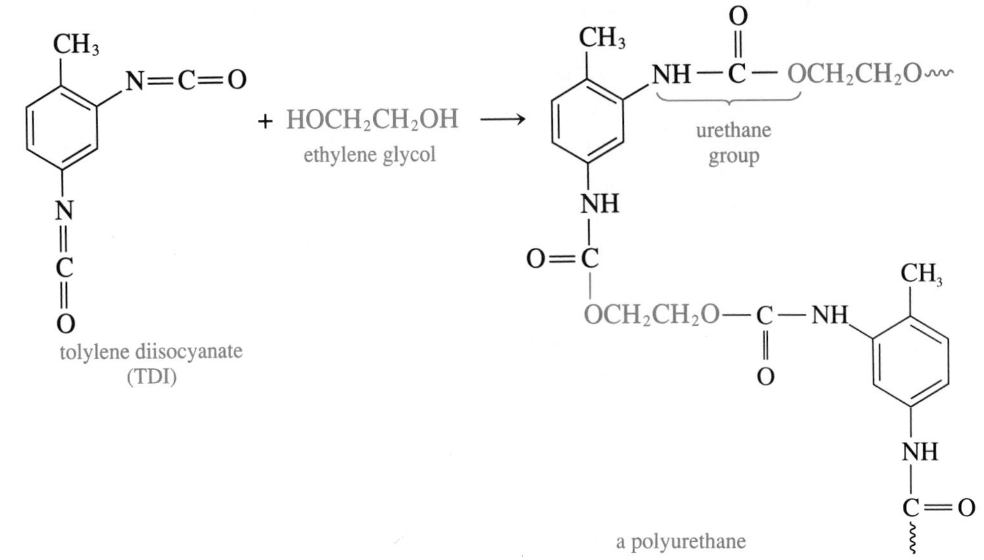
CH3
N=C=O
+ HOCH2CH2OH
ethylene glycol
N=C=O
tolylene diisocyanate
(TDI)
CH3
NH—C—OCH2CH2O
O
urethane
group
NH
O=C
OCH2CH2O—C—NH
O
CH3
NH
C=O
a polyurethane

(14.30)

This reaction is a little different from most step-growth polymerizations in that no small molecule is eliminated. Like other step-growth polymerizations, however, it does involve a reaction of two different difunctional monomers.

PROBLEM 14.16 Rewrite the partial polyurethane structure shown in eq. 14.30, but with one more monomer unit attached to each wavy bond.

The reaction in eq. 14.30 produces a polyurethane but does not produce a foam. To obtain a foam as the product, the polymerization is carried out with a little water present. The water reacts with isocyanate groups in the starting material or in the growing polymer to produce a **carbamic acid.** This acid spontaneously loses carbon dioxide, which creates bubbles as the polymer forms.

$$\underset{\text{isocyanate}}{\sim N{=}C{=}O} + \underset{\text{water}}{HOH} \longrightarrow \underset{\text{carbamic acid}}{\sim NH\overset{\overset{\large O}{\|}}{C}OH} \longrightarrow \underset{\text{amine}}{\sim NH_2} + CO_2\uparrow \qquad (14.31)$$

The amount of carbon dioxide formed, which determines the density of the foam, can be controlled by the amount of water used. The resulting amine can also react with isocyanate groups to form a urea, which may serve as a cross-link between polymer chains.

$$\underset{\text{isocyanate}}{\sim N{=}C{=}O} + \underset{\text{amine}}{H_2N\sim} \longrightarrow \underset{\text{urea link}}{\sim NH\overset{\overset{\large O}{\|}}{C}NH\sim} \qquad (14.32)$$

Polyurethanes have a wide range of applications. Polymers with relatively little cross-linking produce stretchable fibers (Spandex, Lycra) used for bathing suits. Polyurethane foams are used in furniture, mattresses, and car seats and as insulation in portable ice chests. Cross-linked polyurethanes form very tough surface coatings in paints and varnishes. The major component of the Jarvik-7 artificial heart is also a polyurethane.

Several commercially important step-growth polymers are based on reactions of formaldehyde. **Bakelite,** the oldest totally synthetic polymer, was invented by Leo Baekeland in 1907. It is prepared from phenol and formaldehyde. The polymer is highly cross-linked, with methylene groups *ortho* and/or *para* to the phenolic hydroxyl group.

phenol + $CH_2{=}O$ (formaldehyde) $\xrightarrow[-H_2O]{H^+,\ heat}$ segment of Bakelite (14.33)

EXAMPLE 14.4 Write a mechanism for the formation of Bakelite.

Solution The first step in this acid-catalyzed polymerization involves electrophilic aromatic substitution.

phenol + $CH_2{=}O$ + H^+ ⟶ (arenium ion with H and CH_2OH) $\xrightarrow{-H^+}$ o-(hydroxymethyl)phenol (CH_2OH)

The next step involves protonation of the alcohol, formation of a benzyl cation, and another electrophilic aromatic substitution. In this step, a molecule of water is eliminated.

o-(CH_2OH)phenol $\overset{H^+}{\rightleftharpoons}$ $CH_2{-}\overset{+}{O}H_2$ $\xrightarrow{-H_2O}$ CH_2^+ + phenol ⟶ (arenium ion, CH_2 bridge, H) $\xrightarrow{-H^+}$ CH_2-bridged bis(phenol)

Repetition at *ortho* and *para* positions gives the polymer.

Bakelite is a **thermosetting polymer.** Heating leads to further cross-linking, producing a hard, infusible material. This process cannot be reversed, and once setting has occurred, the polymer cannot be melted. Bakelite is used for molded plastic parts, such as appliance handles, and for applications requiring light materials that can withstand high temperatures, such as missile nose cones.

Urea and formaldehyde also form an important commercial polymer.

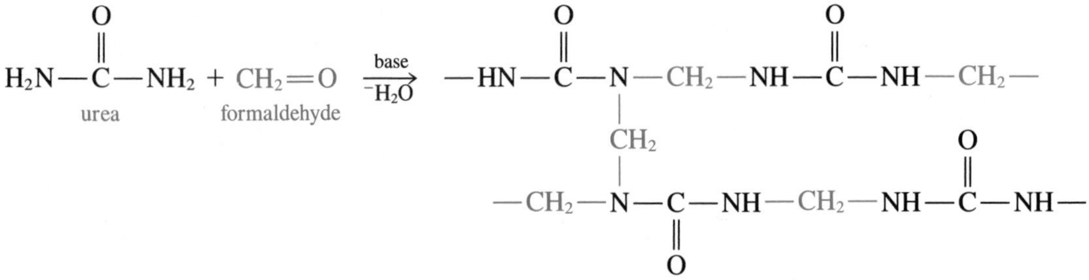

$$\underset{\text{urea}}{H_2N{-}\overset{\overset{\displaystyle O}{\|}}{C}{-}NH_2} + \underset{\text{formaldehyde}}{CH_2{=}O} \xrightarrow[-H_2O]{\text{base}} \text{—HN—}\overset{\overset{\displaystyle O}{\|}}{C}\text{—}\underset{\underset{\displaystyle —CH_2—N—\underset{\underset{\displaystyle O}{\|}}{C}—NH—CH_2—NH—\overset{\overset{\displaystyle O}{\|}}{C}—NH—}{|}}{\underset{CH_2}{\underset{|}{N}}}\text{—}CH_2\text{—NH—}\overset{\overset{\displaystyle O}{\|}}{C}\text{—NH—}CH_2\text{—} \tag{14.34}$$

urea-formaldehyde polymer

This kind of polymer is used in molded materials (electrical fittings and kitchenware), in laminates such as Formica, in plywood and particle board as adhesive, and in foams.

We have barely touched on polymer chemistry. The field is vast and constantly developing. No doubt you will see many new types of polymers reach the marketplace during your lifetime.*

REACTION SUMMARY

1. Chain-Growth Polymerization (Secs. 14.3–14.8)

a. Free Radical Chain-Growth Polymerization (Sec. 14.3)

$$CH_2{=}CH{-}L \xrightarrow[\text{initiator}]{\text{free radical}} {+}CH_2{-}\overset{\overset{\displaystyle L}{|}}{CH}{+}_n$$

L=H, alkyl, aryl, electron-donating group (OAc), electron-withdrawing group (CN)

b. Cationic Chain-Growth Polymerization (Sec 14.4)

$$CH_2{=}CH{-}L \xrightarrow[\text{initiator}]{\text{cationic}} {+}CH_2{-}\overset{\overset{\displaystyle L}{|}}{CH}{+}_n$$

L=carbocation stabilizing group (alkyl, Ph)

---

* For further reading, try "Practical Macromolecular Organic Chemistry," 3rd ed., by D. Braun, H. Cherdon, and W. Kern (Harwood Academic Publishers, New York, 1984, translated by K. J. Ivin) or "Industrial Organic Chemicals in Perspective" by H. A. Wittcoff and B. G. Reuben (John Wiley, New York, 1980, Vols. 1 and 2).

c. Anionic Chain-Growth Polymerization (Sec. 14.5)

$$CH_2{=}CH{-}L \xrightarrow[\text{initiator}]{\text{anionic}} \left(CH_2{-}\underset{L}{CH}\right)_n$$

L = carbanion stabilizing group (CN, Ph)

d. Ziegler–Natta Polymerization (Sec. 14.6)

$$CH_2{=}CH{-}CH_3 \xrightarrow[TiCl_4]{Et_3Al} \left(CH_2{-}\underset{CH_3}{CH}\right)_n \text{ isotactic}$$

$$CH_2{=}C(CH_3){-}CH{=}CH_2 \xrightarrow[TiCl_4]{Et_3Al} \left(CH_2{-}C(CH_3){=}CH{-}CH_2\right)_n \text{ all Z}$$

2. Step-Growth Polymerization (Secs. 14.9 and 14.10)

a. Preparation of Dacron (Sec. 14.9)

$$HO_2C{-}C_6H_4{-}CO_2H + HOCH_2CH_2OH \xrightarrow{\Delta} \left(\overset{O}{\overset{\|}{C}}{-}C_6H_4{-}\overset{O}{\overset{\|}{C}}{-}OCH_2CH_2O\right)_n + H_2O$$

Dacron

b. Preparation of Nylon (Secs. 14.2 and 14.9)

$$HO\overset{O}{\overset{\|}{C}}(CH_2)_4\overset{O}{\overset{\|}{C}}OH + H_2N(CH_2)_6NH_2 \xrightarrow{\Delta} \left(O\overset{O}{\overset{\|}{C}}(CH_2)_4\overset{O}{\overset{\|}{C}}NH(CH_2)_6NH\right)_n + H_2O$$

nylon-6,6

ADDITIONAL PROBLEMS

14.17. Define and give an example of the following terms:

a. homopolymer **b.** copolymer **c.** chain-growth polymerization
d. cross-linked polymer **e.** thermoplastic **f.** thermosetting
g. isotactic **h.** atactic **i.** chain transfer

14.18. Write out all the steps in the free-radical chain-growth polymerization of vinyl chloride.

14.19. Draw the structure of a chain segment of polyvinyl alcohol. This polymer cannot be made by polymerizing its monomer. Why not? It is usually made from polyvinyl acetate (see Problem 14.5b).

14.20. Although propylene can be polymerized with a free-radical initiator, the molecular weight of the polymer is never very high by this method because chain transfer from the methyl group of propylene keeps the chains short. Explain why this chain transfer occurs so easily.

14.21. Consider a polyethylene molecule containing 1000 monomer units initiated by one benzoyloxy radical (from benzoyl peroxide). What percentage of the molecular weight is due to the catalyst radical? Repeat the calculation for a similar polystyrene molecule.

14.22. Draw the expected structure of "propylene tetramer," made by the acid-catalyzed polymerization of propene.

14.23. Propylene oxide can be converted to a polyether by anionic chain-growth polymerization. What is the structure of the polymer, and how is it formed?

14.24. Superglue is made by the on-the-spot polymerization of methyl α-cyanoacrylate (methyl 2-cyanopropenoate) with a little water or base. Draw the structure of the repeating unit. Why is this monomer so susceptible to anionic polymerization?

14.25. Draw six-unit chain segments of isotactic, syndiotactic, and atactic polystyrene.

14.26. Can isobutylene polymerize in isotactic, syndiotactic, or atactic forms? Explain.

14.27. Explain how polyethylenes obtained by free-radical and by Ziegler-Natta polymerization differ in structure.

14.28. The ozonolysis of natural rubber gives levulinic aldehyde, $CH_3C(=O)CH_2CH_2CH{=}O$

Explain how this result is consistent with natural rubber's formula (eq. 14.22).

14.29. Write out the steps in a mechanism that explains the free-radical copolymerization of 1,3-butadiene and styrene to give the synthetic rubber SBR (eq. 14.23).

14.30. Neoprene is a synthetic rubber invented more than 50 years ago by the Du Pont Company. It is used to manufacture industrial hoses, drive belts, window gaskets, shoe soles, and packaging materials. Neoprene is a polymer of 2-chloro-1,3-butadiene. Assuming mainly 1,4-addition, draw the structure of the repeating unit in neoprene.

14.31. Draw the structural formula for a segment of alternating copolymer made from styrene and methyl methacrylate.

14.32. Draw the repeating unit in the polymer that results from each of the following step-growth polymerizations:

a. $Cl{-}C(=O)(CH_2)_6C(=O){-}Cl + H_2N(CH_2)_6NH_2$

b. 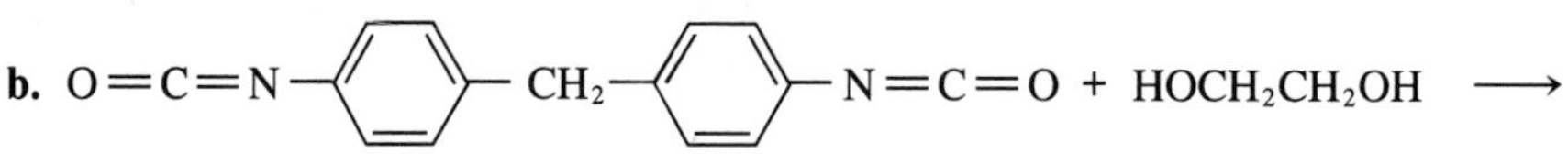 $O{=}C{=}N{-}C_6H_4{-}CH_2{-}C_6H_4{-}N{=}C{=}O + HOCH_2CH_2OH \longrightarrow$

c. $CH_3OC(=O)(CH_2)_4C(=O)OCH_3 + HOCH_2CH_2OH \xrightarrow{H^+}$

14.33. Lexan is a tough polycarbonate used to make molded articles. It is made from diphenyl carbonate and bisphenol A. Draw the structure of its repeating unit.

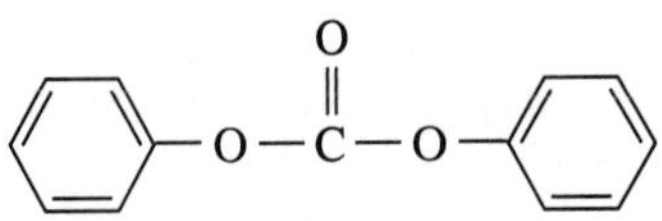

$C_6H_5{-}O{-}C(=O){-}O{-}C_6H_5$

diphenyl carbonate

$HO{-}C_6H_4{-}C(CH_3)_2{-}C_6H_4{-}OH$

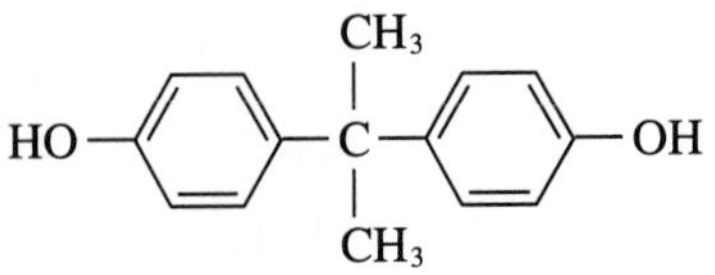

bisphenol A

14.34. Formaldehyde polymerizes in aqueous solution to give paraformaldehyde, $HO—(CH_2O)_n—H$. Although high molecular weights are achieved, the polymer "unzips" readily. However, if the polymer is treated with acetic anhydride, the resulting material (an important commercial polymer called **Delrin**) no longer "unzips." Explain the chemistry described.

14.35. Reaction of phthalic anhydride with glycerol gives a cross-linked polyester called a *glyptal resin*. Draw a partial structure, showing the cross-linking.

14.36. Using eq. 14.33 as a model, draw a segment of the polymer that will be formed from formaldehyde and *p*-methylphenol. Will the polymer be cross-linked? Explain.

14.37. To make polymers commercially, it is necessary to make the monomers inexpensively and on a large scale. One commercial method for making hexamethylenediamine (for nylon-6,6) starts with the 1,4-addition of chlorine to 1,3-butadiene. Suggest a possibility for the remaining steps.

14.38. Methyl methacrylate (methyl 2-methylpropenoate), the monomer for Plexiglas and Lucite, is made by reaction of acetone cyanohydrin (Sec. 9.11) with methanol and sulfuric acid. The reaction involves a methanolysis and a dehydration. Write an equation for the reaction and mechanistic equations for the two processes.

15 Lipids and Detergents

15.1 Introduction

In this chapter, we take up the first of four major classes of biologically important substances, the **lipids.** And because they are related to one type of lipid, we will also discuss here the chemistry of **detergents.**

Lipids (from the Greek *lipos,* fat) are constituents of plants or animals that are characterized by their solubility properties. In particular, *lipids are insoluble in water but soluble in nonpolar organic solvents,* such as ether. Lipids can be extracted from cells and tissues by organic solvents. This solubility property distinguishes lipids from three other major classes of natural products, carbohydrates, proteins, and nucleic acids, which in general are *not* soluble in organic solvents.

Lipids may vary considerably in chemical structure, even though they have similar solubility properties. Some are esters, some are hydrocarbons; some are acyclic and others are cyclic, even polycyclic. We will take up each structural type separately.

15.2 Fats and Oils; Triesters of Glycerol

Fats and oils are familiar parts of daily life. Common fats include butter, lard, and the fatty portions of meat. Oils come mainly from plants and include corn, cottonseed, olive, peanut, and soybean oils. Although fats are solids and oils are liquids, they have the same basic organic structure. ***Fats and oils are triesters of glycerol and are called triglycerides.*** When we boil a fat or oil with alkali and acidify the resulting solution, we obtain glycerol and a mixture of **fatty acids.** The reaction is called **saponification** (Sec. 10.14).

$$\begin{array}{l} CH_2-O-\overset{\overset{\large O}{\|}}{C}-R \\ | \\ CH-O-\overset{\overset{\large O}{\|}}{C}-R' \\ | \\ CH_2-O-\overset{\overset{\large O}{\|}}{C}-R'' \end{array} \xrightarrow[\text{2. } H^+]{\text{1. NaOH, } H_2O \text{ heat}} \begin{array}{l} CH_2OH \\ | \\ CHOH \\ | \\ CH_2OH \end{array} + \begin{array}{l} HO\overset{\overset{\large O}{\|}}{C}R \\ HO\overset{\overset{\large O}{\|}}{C}R' \\ HO\overset{\overset{\large O}{\|}}{C}R'' \end{array} \quad (15.1)$$

a triglyceride (fat or oil) — glycerol — three equivalents of fatty acids

The most common saturated and unsaturated fatty acids obtained in this way are listed in Table 15.1. Although exceptions are known, *most fatty acids are unbranched and contain an even number of carbon atoms.* If double bonds are present, they usually have the *cis* (or *Z*) configuration and are not conjugated.

EXAMPLE 15.1 Draw the structure of linoleic acid, showing the geometry at each double bond.

Solution Both double bonds have the *Z* configuration.

[Structure of linoleic acid drawn in fully extended zigzag conformation, with wedge/dash H atoms on each sp^3 carbon, two *cis* C=C double bonds, terminating in C(=O)OH]

The preferred conformation is fully extended, with a staggered arrangement at each C—C single bond.

TABLE 15.1 Common acids obtained from fats

	Common name	*Number of carbons*	*Structural formula*	*mp, °C*
Saturated	lauric	12	$CH_3(CH_2)_{10}COOH$	44
	myristic	14	$CH_3(CH_2)_{12}COOH$	58
	palmitic	16	$CH_3(CH_2)_{14}COOH$	63
	stearic	18	$CH_3(CH_2)_{16}COOH$	70
	arachidic	20	$CH_3(CH_2)_{18}COOH$	77
Unsaturated	oleic	18	$CH_3(CH_2)_7CH{=}CH(CH_2)_7COOH$ (*cis*)	13
	linoleic	18	$CH_3(CH_2)_4CH{=}CHCH_2CH{=}CH(CH_2)_7COOH$ (*cis, cis*)	−5
	linolenic	18	$CH_3CH_2CH{=}CHCH_2CH{=}CHCH_2CH{=}CH(CH_2)_7COOH$ (all *cis*)	−11

PROBLEM 15.1 Draw the structure of linolenic acid.

There are two types of triglycerides: **simple triglycerides,** in which all three fatty acids are identical, and **mixed triglycerides.**

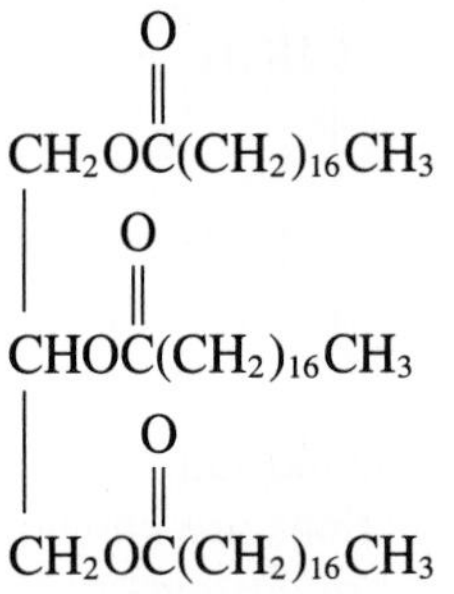

$$\begin{array}{l} CH_2OC(=O)(CH_2)_{16}CH_3 \\ | \\ CHOC(=O)(CH_2)_{16}CH_3 \\ | \\ CH_2OC(=O)(CH_2)_{16}CH_3 \end{array} \qquad \begin{array}{ll} CH_2{-}OC(=O)(CH_2)_{14}CH_3 & \text{ester of palmitic acid} \\ | & \\ CH{-}OC(=O)(CH_2)_{16}CH_3 & \text{ester of stearic acid} \\ | & \\ CH_2{-}OC(=O)(CH_2)_7CH{=}CH(CH_2)_7CH_3 & \text{ester of oleic acid} \end{array}$$

a simple triglyceride
(glyceryl tristearate or tristearin)

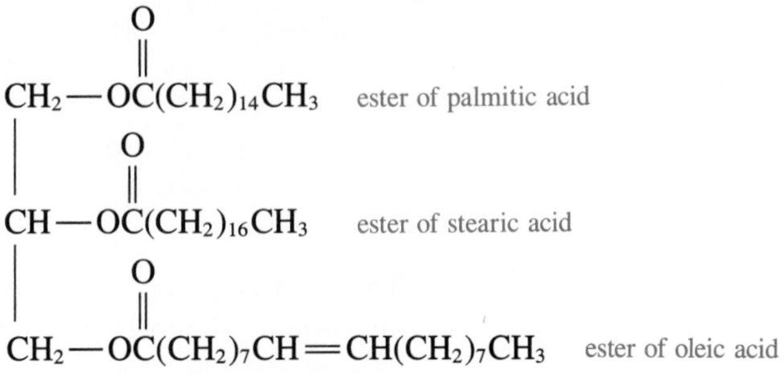

a mixed triglyceride
(glyceryl palmitostearoöleate)

EXAMPLE 15.2 Draw the structure of glyceryl stearopalmitoöleate, an isomer of the mixed triglyceride shown above.

Solution

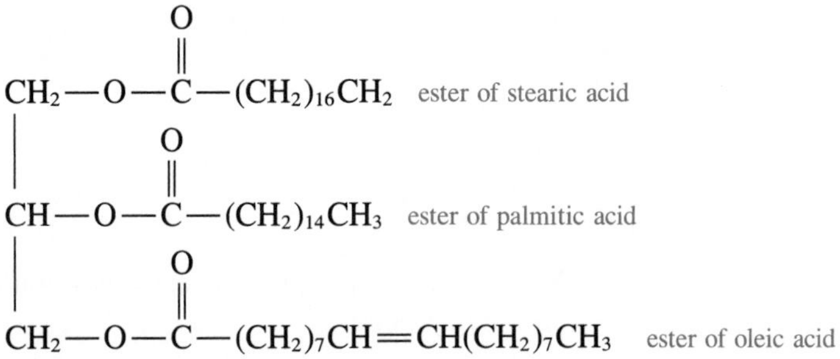

$$\begin{array}{ll} CH_2{-}O{-}C(=O){-}(CH_2)_{16}CH_2 & \text{ester of stearic acid} \\ | & \\ CH{-}O{-}C(=O){-}(CH_2)_{14}CH_3 & \text{ester of palmitic acid} \\ | & \\ CH_2{-}O{-}C(=O){-}(CH_2)_7CH{=}CH(CH_2)_7CH_3 & \text{ester of oleic acid} \end{array}$$

Notice that glyceryl palmitostearoöleate and glyceryl stearopalmitoöleate give the same saponification products.

PROBLEM 15.2 Draw the structure for

a. glyceryl trimyristate.
b. glyceryl palmitoöleostearate.

PROBLEM 15.3 What saponification products would be obtained from each of the triglycerides in Problem 15.2?

In general, a particular fat or oil consists, not of a single triglyceride, but of a complex mixture of triglycerides. For this reason, the composition of a fat or oil is usually expressed in terms of the percentages of the various acids obtained from it by saponification (Table 15.2). Some fats and oils give mainly one or two acids, with only minor amounts of other acids. Olive oil, for example, gives 83% oleic acid. Palm oil gives 43% palmitic acid and 43% oleic

TABLE 15.2 Fatty acid composition of some fats and oils (approximate)

	Saturated acids (%)					Unsaturated acids (%)	
Source	*C_{10} and less*	*C_{12} lauric*	*C_{14} myristic*	*C_{16} palmitic*	*C_{18} stearic*	*C_{18} oleic*	*C_{18} linoleic*
Animal Fats							
Butter	12	3	12	28	10	26	2
Lard	—	—	1	28	14	46	5
Beef tallow	—	0.2	3	28	24	40	2
Human	—	1	3	25	8	46	10
Vegetable Oils							
Olive	—	—	1	5	2	83	7
Palm	—	—	2	43	2	43	8
Corn	—	—	1	10	2	40	40
Peanut	—	—	—	8	4	60	25

acid, with lesser amounts of stearic and linoleic acids. Butterfat, on the other hand, gives at least 14 different acids on hydrolysis and is somewhat exceptional in that about 9% of these acids have fewer than 10 carbon atoms.

PROBLEM 15.4 From the data in Table 15.2, what can you say in general about the ratio of saturated to unsaturated acids in fats and oils?

What is it that makes some triglycerides solids (fats) and others liquids (oils)? The distinction is clear from their composition. *Oils contain a much higher percentage of unsaturated fatty acids than do fats.* For example, most vegetable oils (such as corn oil or soybean oil) give about 80% unsaturated acids on hydrolysis. For fats (such as beef tallow) the figure is much lower, just a little over 50%.

Table 15.1 shows that the melting points of unsaturated fatty acids are appreciably lower than those of saturated acids. Compare, for example, the melting points of stearic and oleic acids, which differ structurally by only one double bond. The same difference applies to triglycerides: *the more double bonds in the fatty acid portion of the triester, the lower its melting point.*

The reason for the effect of saturation or unsaturation on the melting point becomes apparent when we examine space-filling models. Figure 15.1 shows a model of a fully saturated triglyceride. The long, saturated chains have fully extended, staggered conformations. They can therefore pack together fairly regularly, as in a crystal. Consequently, saturated triglycerides are usually solids at room temperature.

The result of introducing just one *cis* double bond in one of the chains is shown in Figure 15.2. Clearly, the chains in this kind of molecule (and the molecules themselves, when many are close to one another) cannot align nicely in a crystalline array. The substance therefore remains a liquid. The more double bonds, the more disorderly the structure and the lower the melting point.

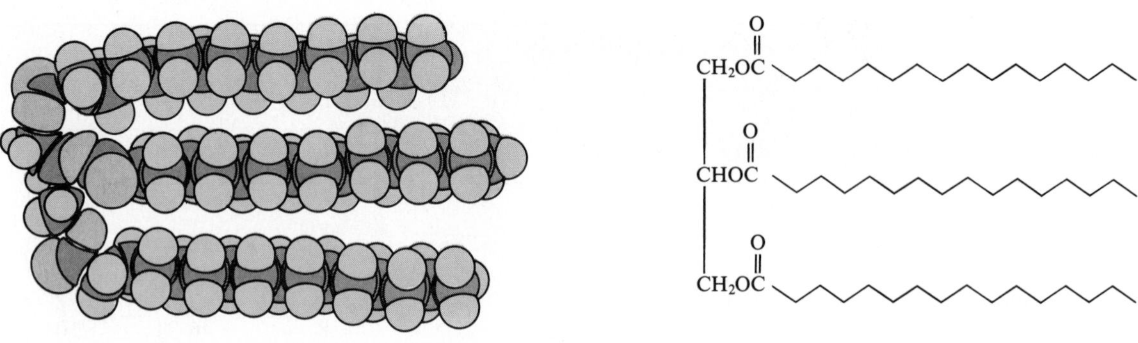

FIGURE 15.1 Space-filling and schematic models of glyceryl tripalmitate.

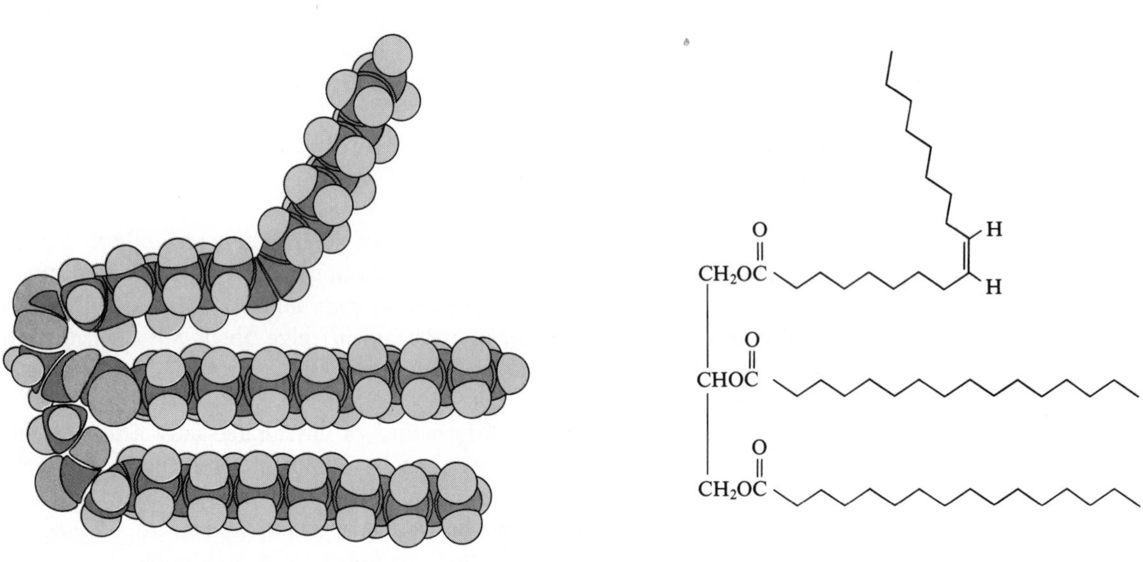

FIGURE 15.2 Space-filling and schematic models of glyceryl dipalmitoöleate.

15.3 Hydrogenation of Vegetable Oils

Vegetable oils, which are highly unsaturated, are converted into solid vegetable fats, such as Crisco, by catalytically hydrogenating some or all of the double bonds. This process, called **hardening,** is illustrated by the hydrogenation of glyceryl trioleate to glyceryl tristearate.

$$
\begin{array}{l}
\overset{\displaystyle O}{\overset{\|}{CH_2OC}}(CH_2)_7CH{=}CH(CH_2)_7CH_3 \\
\overset{\displaystyle O}{\overset{\|}{CHOC}}(CH_2)_7CH{=}CH(CH_2)_7CH_3 \\
\overset{\displaystyle O}{\overset{\|}{CH_2OC}}(CH_2)_7CH{=}CH(CH_2)_7CH_3
\end{array}
\xrightarrow[\text{Ni catalyst, heat}]{3\ H_2}
\begin{array}{l}
\overset{\displaystyle O}{\overset{\|}{CH_2OC}}(CH_2)_{16}CH_3 \\
\overset{\displaystyle O}{\overset{\|}{CHOC}}(CH_2)_{16}CH_3 \\
\overset{\displaystyle O}{\overset{\|}{CH_2OC}}(CH_2)_{16}CH_3
\end{array}
\quad (15.2)
$$

glyceryl trioleate (mp −17°C) → glyceryl tristearate (mp 55°C)

Margarine is made by hydrogenating cottonseed, soybean, peanut, or corn oil until the desired butterlike consistency is obtained. The product may be churned with milk and artificially colored to mimic butter's flavor and appearance.

15.4 Saponification of Fats and Oils; Soap

When a fat or oil is heated with alkali, the ester is converted to glycerol and the salts of fatty acids. The reaction is illustrated here with the saponification of glyceryl tripalmitate.

$$
\begin{array}{l}
\overset{\displaystyle O}{\overset{\|}{CH_2OC}}(CH_2)_{14}CH_3 \\
\overset{\displaystyle O}{\overset{\|}{CHOC}}(CH_2)_{14}CH_3 \\
\overset{\displaystyle O}{\overset{\|}{CH_2OC}}(CH_2)_{14}CH_3
\end{array}
+ 3\ Na^+OH^-
\xrightarrow{\text{heat}}
\begin{array}{l}
CH_2OH \\
| \\
CHOH \\
| \\
CH_2OH
\end{array}
+ 3\ CH_3(CH_2)_{14}CO_2^-Na^+
\quad (15.3)
$$

glyceryl tripalmitate (from palm oil); glycerol; sodium palmitate (a soap)

The salts (usually sodium) of long-chain fatty acids are **soaps.**

The conversion of animal fats (for example, goat tallow) into soap by heating with wood ashes (which are alkaline) is one of the oldest of chemical processes. Soap has been produced for at least 2300 years, having been known to the ancient Celts and Romans. Yet, as recently as the sixteenth and seventeenth centuries, soap was still a rather rare substance, used mainly in medicine. But by the nineteenth century, soap had come into such widespread use that the German organic chemist Justus von Liebig was led to remark that the quantity of soap consumed by a nation was an accurate measure of its wealth and civilization. At present, annual world production of ordinary soaps (not including synthetic detergents) is well over 6 million tons.

Soaps are made by either a batch process or a continuous process. In the batch process, the fat or oil is heated with a slight excess of alkali (NaOH) in an open kettle. When saponification is complete, salt is added to precipitate the soap as thick curds. The water layer, which contains salt, glycerol, and excess alkali, is drawn off, and the glycerol is recovered by distillation. The crude soap curds, which contain some salt, alkali, and glycerol as impurities, are purified by boiling with water and reprecipitating with salt several times. Finally, the curds are boiled with enough water to form a smooth mixture that, on standing, gives a homogeneous upper layer of soap. This soap may be sold without further processing, as a cheap industrial soap. Various fillers, such as sand or pumice, may be added, to make scouring soaps. Other treatments transform the crude soap to toilet soaps, powdered or flaked soaps, medicated or perfumed soaps, laundry soaps, liquid soaps, or (by blowing air in) floating soaps.

In the continuous process, which is more common today, the fat or oil is hydrolyzed by water at high temperatures and pressures in the presence of a catalyst, usually a zinc soap. The fat or oil and the water are introduced continuously into opposite ends of a large reactor, and the fatty acids and glycerol are removed as formed, by distillation. The acids are then carefully neutralized with an appropriate amount of alkali to make soap.

15.5 How Do Soaps Work?

Most dirt on clothing or skin adheres to a thin film of oil. If the oil film can be removed, the dirt particles can be washed away. A soap molecule consists of a long, hydrocarbon-like chain of carbon atoms with a highly polar or ionic group at one end (Figure 15.3). The carbon chain is **lipophilic** (attracted to or soluble in fats and oils), and the polar end is **hydrophilic** (attracted to or soluble in water). In a sense, soap molecules are "schizophrenic," having two different "personalities." Let us see what happens when we add soap to water.

When soap is shaken with water, it forms a colloidal dispersion—not a true solution. These soap solutions contain aggregates of soap molecules called **micelles.** The nonpolar, or lipophilic, carbon chains are directed toward the center of the micelle. The polar, or hydrophilic, ends of the molecule form the "surface" of the micelle that is presented to the water (Figure 15.4). In ordinary soaps, the outer part of each micelle is negatively charged, and the positive sodium ions congregate near the periphery of each micelle.

In acting to remove dirt, soap molecules surround and emulsify the droplets of oil or grease. The lipophilic "tails" of the soap molecules dissolve in the oil. The hydrophilic ends extend out of the oil droplet toward the water. In this

FIGURE 15.3 Sodium stearate, an ordinary soap.

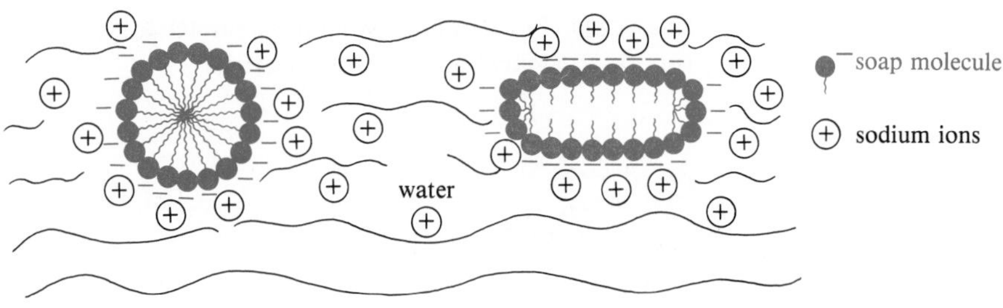

FIGURE 15.4 Soap molecules form micelles when "dissolved" in water.

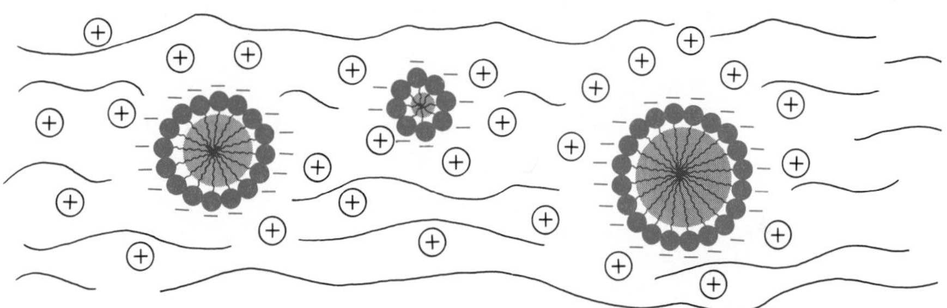

FIGURE 15.5 Oil droplets (shown in color) become emulsified by soap molecules.

way, the oil droplets are stabilized in the water solution because the negative surface charge of the droplets prevents their coalescence (as shown in Figure 15.5).

Another striking property of soap solutions is their unusually low surface tension, which gives a soap solution more "wetting" power than plain water has. As a consequence, soaps belong to a class of substances called **surfactants.** A combination of the emulsifying power and the surface action of soap solutions enables them to detach dirt, grease, and oil particles from the surface being cleaned and to emulsify them so that they can be washed away. These same principles of cleansing action apply to synthetic detergents.

15.6 Synthetic Detergents (Syndets)

Annual world production of synthetic detergents (sometimes called **syndets**) now exceeds that of ordinary soaps. Syndets evolved in response to two problems with the use of ordinary, unimproved soaps. First, being salts of weak acids, *soaps give rather alkaline solutions in water*. This is due to partial hydrolysis of the sodium salts.

$$\underset{\text{soap}}{R{-}\overset{\overset{\displaystyle O}{\|}}{C}{-}O^-Na^+} + H{-}OH \rightleftharpoons R{-}\overset{\overset{\displaystyle O}{\|}}{C}{-}OH + \underset{\text{alkali}}{Na^+OH^-} \quad (15.4)$$

Alkali can be harmful to certain fabrics. Yet ordinary soaps cannot function well in acid because the long-chain fatty acid will precipitate from the solution as a scum. For example, sodium stearate, a typical soap, is destroyed by conversion to stearic acid on acidification.

$$\underset{\text{sodium stearate (soluble)}}{C_{17}H_{35}C(=O)O^-Na^+} + H^+Cl^- \longrightarrow \underset{\text{stearic acid (insoluble)}}{C_{17}H_{35}C(=O)OH} \downarrow + Na^+Cl^- \qquad (15.5)$$

The second problem with ordinary soaps is that *they form insoluble salts with the calcium, magnesium, or ferric ions that may be present in hard water*.

$$\underset{\text{sodium stearate (soluble)}}{2\ C_{17}H_{35}C(=O)O^-Na^+} + Ca^{2+} \longrightarrow \underset{\text{calcium stearate (insoluble)}}{(C_{17}H_{35}COO^-)_2Ca^{2+}} \downarrow + 2\ Na^+ \qquad (15.6)$$

The insoluble salts are responsible for the "rings" around bathtubs or collars and for films that dull the look of clothing and hair.

These problems with ordinary soaps have been solved or diminished in several ways. For example, water can be "softened," either municipally or in individual households, to remove the offending calcium or magnesium ions. In softened water, these ions are replaced by sodium ions. If this water is also used for drinking, however, it may cause health problems, especially for people who have to limit their sodium ion intake.

Phosphates can also be added to soaps. Phosphates form soluble complexes with metal ions, thus keeping these ions from forming insoluble salts with the soap. But widespread use of phosphates in the past has created environmental problems. Because of their use in detergents, tremendous quantities of phosphates eventually found their way into lakes, rivers, and streams. Since they are fertilizers, these phosphates stimulated plant growth to such an extent that the plants exhausted the dissolved oxygen in the water, in turn causing fish to die. Phosphates are still used in detergents, but their use is now limited by law to levels that are unlikely to be harmful.

Another way to eliminate the problems associated with ordinary soaps is to design more effective detergents. These syndets must have several features. Like ordinary soaps, they must have a long lipophilic chain and a polar, or ionic, hydrophilic end. However, the polar end should not form insoluble salts with metal ions present in water and should not affect the acidity of water.

The first syndets were sodium salts of alkyl hydrogen sulfates. A long-chain alcohol was prepared by the **hydrogenolysis** of a fat or oil. For example, glyc-

$$\begin{array}{l} \quad\quad\quad\quad\;\; O \\ \quad\quad\quad\quad\;\; \| \\ CH_3(CH_2)_{10}C{-}OCH_2 \\ \quad\quad\quad\quad\;\; O \quad\;\; | \\ \quad\quad\quad\quad\;\; \| \quad\;\;\; | \\ CH_3(CH_2)_{10}C{-}OCH + 6\,H_2 \xrightarrow[\text{heat, pressure}]{\text{copper chromite}} 3\,CH_3(CH_2)_{10}CH_2OH + \begin{array}{c} HOCH_2 \\ | \\ HOCH \\ | \\ HOCH_2 \end{array} \\ \quad\quad\quad\quad\;\; O \quad\;\; | \\ \quad\quad\quad\quad\;\; \| \quad\;\;\; | \\ CH_3(CH_2)_{10}C{-}OCH_2 \end{array} \quad (15.7)$$

glyceryl trilaurate → 1-dodecanol (lauryl alcohol) + glycerol

eryl trilaurate can be reduced to a mixture of 1-dodecanol (from the acid part of the fat) and glycerol. Since glycerol is water soluble, whereas the long-chain alcohol is not, the two hydrogenolysis products can be easily separated. The long-chain alcohol was next treated with sulfuric acid to make the alkyl hydrogen sulfate, which was then neutralized with base.

$$\underset{\text{lauryl alcohol}}{CH_3(CH_2)_{10}CH_2OH} + \underset{\text{sulfuric acid}}{HOSO_2OH} \longrightarrow \underset{\text{lauryl hydrogen sulfate}}{CH_3(CH_2)_{10}CH_2OSO_2OH} + H_2O$$

$$\downarrow NaOH \quad (15.8)$$

$$\overbrace{CH_3CH_2CH_2CH_2CH_2CH_2CH_2CH_2CH_2CH_2CH_2CH_2}^{\text{lipophilic chain}}{-}\underbrace{O{-}\overset{\overset{\displaystyle O}{\|}}{\underset{\underset{\displaystyle O}{\|}}{S}}{-}O^-Na^+}_{\text{polar, hydrophilic end}} + H_2O$$

sodium lauryl sulfate

Sodium lauryl sulfate is an excellent detergent. Because it is a salt of a *strong* acid, its solutions are nearly neutral. Its calcium and magnesium salts do not precipitate from solution, so it is effective in hard as well as soft water. Unfortunately, its supply is too limited to meet the demand, so the need for other syndets persisted.

At present, *the most widely used syndets are straight-chain alkylbenzenesulfonates.* They are made in three steps, shown in eq. 15.9. A straight-chain alkene with 10 to 14 carbons is treated with benzene and a Friedel-Crafts catalyst ($AlCl_3$ or HF) to form an alkylbenzene. Sulfonation and neutralization of the sulfonic acid with base complete the process.

$$RCH{=}CHR' + C_6H_6 \xrightarrow[\text{catalyst}]{\text{Friedel-Crafts}} RCH(C_6H_5)CH_2R' \xrightarrow{H_2SO_4 \text{ or } SO_3} RCH(C_6H_4SO_3H)CH_2R' \xrightarrow{Na^+OH^-} RCH(C_6H_4SO_3^-Na^+)CH_2R' \quad (15.9)$$

(R and R′ are straight-chain alkyl groups; total 10–14 carbons)

lipophilic part: $RCHCH_2R'$ with the benzene ring; hydrophilic part: $SO_3^-Na^+$

a sodium alkylbenzenesulfonate

It is important that the alkyl chain in these detergents have no branches. The first alkylbenzenesulfonates had branched side chains and were found to be nonbiodegradable. They created severe pollution problems in the 1950s, causing foaming in sewage treatment plants, lakes, and rivers. But after about 1965, alkylbenzenesulfonates with unbranched side chains became available through further research. They are fully biodegradable and do not accumulate in the environment.

The soaps and syndets we have mentioned so far are *anionic* detergents; they have a lipophilic chain with a *negatively charged* polar end. But there are also *cationic, neutral,* and even *amphoteric* detergents, in which the polar portion of the molecule is *positive, neutral,* or *dipolar,* respectively. Here are some examples:

$$\left[R\overset{+}{N}(CH_3)_3\right]Cl^-$$

cationic detergent ($R = C_{16-18}$)

$$R{-}C_6H_4{-}O(CH_2CH_2O)_nH$$

neutral detergent ($R = C_{8-12}; n = 5-10$)

$$R{-}\overset{+}{N}(CH_3)_2{-}CH_2CO_2^-$$

amphoteric detergent ($R = C_{12-18}$)

The essential features of all these detergents are a lipophilic portion with a hydrocarbon chain of appropriate length to dissolve in oil or grease droplets and a polar portion to create a micelle surface that is attractive to water.

EXAMPLE 15.3 Design a synthesis of the cationic detergent shown above ($R = C_{16}$).

Solution

$$CH_3(CH_2)_{14}CH_2Cl + (CH_3)_3N{:} \longrightarrow \left[CH_3(CH_2)_{14}CH_2{-}\overset{+}{N}(CH_3)_2{-}CH_3 \quad Cl^-\right]$$

The reaction is an S_N2 displacement (see Table 6.1, entry 9, and eqs. 11.6 and 11.28).

PROBLEM 15.5 Design a synthesis of the neutral detergent shown above ($R = C_8$, $n = 5$), starting with *p*-octylphenol and ethylene oxide (review Sec. 8.9).

PROBLEM 15.6 Design a synthesis of the amphoteric detergent shown above, using an S_N2 displacement with $CH_3(CH_2)_{10}CH_2\ddot{N}(CH_3)_2$ as the nucleophile and an appropriate halide.

A Word About . . .

28. Commercial Detergents

The design and manufacture of a successful commercial detergent is a highly sophisticated process because, more and more, detergents are developed for a specific purpose and market. No one formulation fits every use. An inventory of the cleaning materials in a typical household might disclose half a dozen or more products designed to be most suitable for a specific job—to clean clothes, dishes, floors, automobiles, the human body (special hand and bath soaps and hair shampoo), and so on.

In Secs. 15.4 through 15.6 we described the structures of common surface active agents (*surfactants*). As we saw, these are organic molecules with a lipophilic portion and a polar portion, whose function is to emulsify and disperse oil and grease and to lower the surface tension of water to aid in wetting clothes, dishes, or other surfaces. Surfactants form an important part of all commercial detergents, but depending on the end use, these detergents will also contain a variety of other components, such as builders, bleaches, fabric softeners, enzymes, antiredeposition agents, and optical brighteners.

Builders are perhaps the second most important component of commercial detergents. They are added to remove calcium and magnesium ions ("hardness") from wash water. They may do this by chelation (forming a complex) or by exchanging these ions for sodium ions. Builders also raise pH to aid oil emulsification and buffer against pH changes. The most common builder is **sodium tripolyphosphate** ($5Na^+ \ P_3O_{10}^{5-}$), but because waste phosphates can be environmental pollutants, the amount that may be used is restricted by law; recently, sodium citrate, sodium carbonate, and sodium silicate have begun to replace it as builders. Zeolites (sodium aluminosilicates) are used as ion exchangers, especially for calcium ions.

Bleaches that contain chlorine (hypochlorites) have been used for years, sometimes separately and sometimes incorporated into the detergent. The chlorine acts as an oxidizing agent. Problems with residual chlorine and its odor have led more recently to the use of peroxides, especially sodium perborate ($NaBO_3$). Hydrolysis of the perborate produces hydrogen peroxide, which does the bleaching.

Fabric softeners are cationic surfactants that give clothes a soft feel. They may be incorporated in a laundry detergent or added to the wash separately.

Enzymes are added to detergents to deal with specific types of clothing stains. *Proteases* hydrolyze protein-based stains, and *amylases* convert starch-based stains to soluble materials that are easily washed away.

Antiredeposition agents are compounds added to laundry detergents to prevent the redeposition of soil on clothes. The most common examples are cellulose ethers or esters.

Optical brighteners are usually organic dyes that absorb ultraviolet light and fluoresce blue. In this way, a yellowed appearance of some "white" clothing can be avoided. In the mid-nineteenth century, "bluing" (a blue dye, such as ultramarine) used to be used for this purpose and is sometimes still used separately from the detergent. Brighteners that are incorporated in the detergent formulation are usually aromatic or heteroaromatic amines.

In addition to all these components, a modern commercial detergent may contain *antistatic agents* (cationic surfactants added to reduce static cling); *hydrotropes* (compounds added to liquid detergents to hold less-soluble surfactants and other compounds in solution); and, of course, *fragrances* and *perfumes* and inert fillers and formulation aids that keep powdered detergents free flowing.

Next time you do the laundry, bathe, or wash your hair, think about the complexity and diversity of the product you are using. It isn't just soap.

15.7 *Phospholipids*

Phospholipids constitute about 40% of cell membranes, the remaining 60% being proteins. Phospholipids are related structurally to fats and oils, except that one of the three ester groups is replaced by a phosphatidylamine.

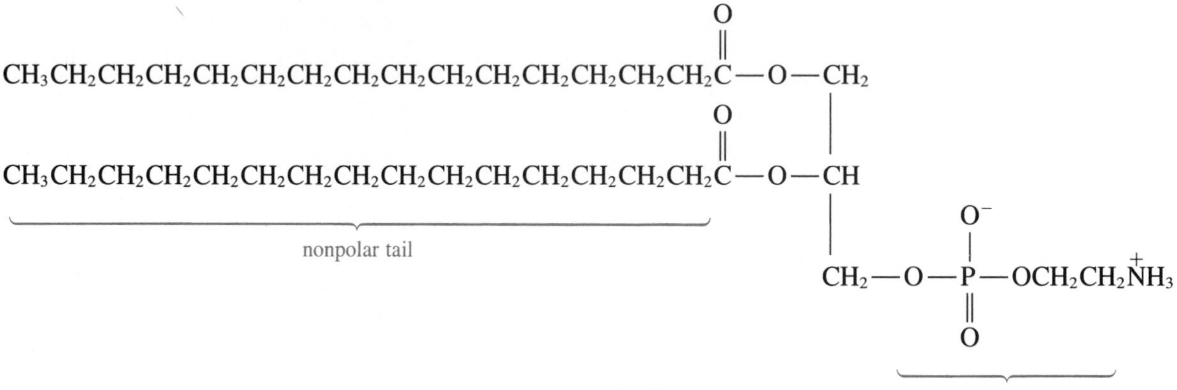

$$\begin{array}{l} CH_3CH_2CH_2CH_2CH_2CH_2CH_2CH_2CH_2CH_2CH_2CH_2CH_2CH_2CH_2\overset{\overset{O}{\|}}{C}-O-CH_2 \\ CH_3CH_2CH_2CH_2CH_2CH_2CH_2CH_2CH_2CH_2CH_2CH_2CH_2CH_2CH_2\overset{\overset{O}{\|}}{C}-O-CH \\ \qquad\qquad\qquad\qquad\qquad\qquad\qquad\qquad\qquad\qquad\qquad\qquad\qquad\qquad CH_2-O-\underset{\underset{O}{\|}}{\overset{\overset{O^-}{|}}{P}}-OCH_2CH_2\overset{+}{N}H_3 \end{array}$$

nonpolar tail (the two fatty acid chains); polar head (the phosphate–ethanolamine group)

a phospholipid

The fatty acid portions are usually palmityl, stearyl, or oleyl. The structure shown is a **cephalin;** the three protons on the nitrogen are replaced by methyl groups in the **lecithins.** Both types of phospholipids are widely distributed in the body, especially in the brain and nerve tissues.

Phospholipids arrange themselves in **bilayers** in membranes, with the two hydrocarbon "tails" pointing in and the polar phosphatidylamine ends constituting the membrane surface, as shown in Figure 15.6. Membranes play a key role in biology, controlling diffusion of substances into and out of cells.

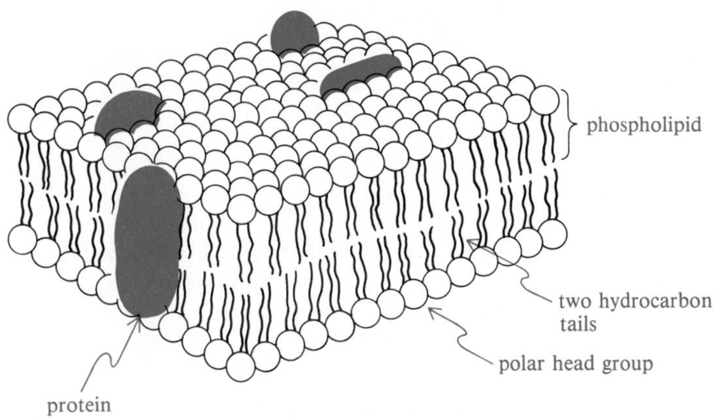

FIGURE 15.6 Schematic diagram of a cell membrane.

15.8 *Prostaglandins, Leukotrienes, and Lipoxins*

Prostaglandins are a group of compounds related to the unsaturated fatty acids. They were discovered in the 1930s, when it was found that human semen contained substances that could stimulate smooth muscle tissue, such as uterine muscle, to contract. On the assumption that these substances came from the prostate gland, they were named prostaglandins. We now know that prostaglandins are widely distributed in almost all human tissues, that they are biologically active in minute concentration, and that they have various effects on fat metabolism, heart rate, and blood pressure.

Prostaglandins have 20 carbon atoms. They are synthesized in the body by oxidation and cyclization of the 20-carbon unsaturated fatty acid **arachidonic acid.** Carbon-8 through carbon-12 of the chain are looped to form a cyclopentane ring, and an oxygen function (carbonyl or hydroxyl group) is always present at carbon-9. Various numbers of double bonds or hydroxyl groups may also be present elsewhere in the structure.

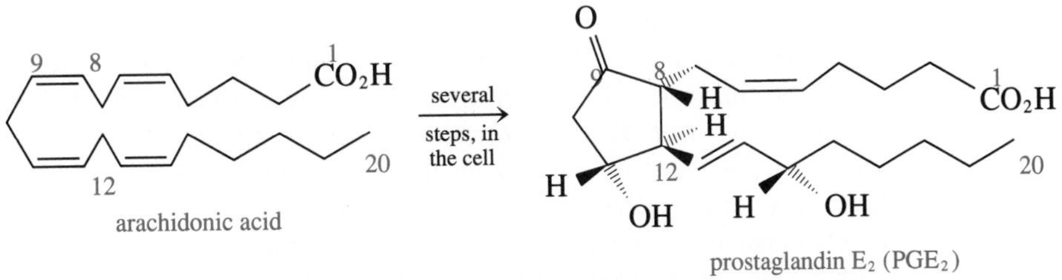

arachidonic acid $\xrightarrow[\text{steps, in the cell}]{\text{several}}$ prostaglandin E_2 (PGE_2) (15.10)

Prostaglandins have excited much interest in the medical community, where they may find use in the treatment of inflammatory diseases, such as asthma and rheumatoid arthritis; treatment of peptic ulcers; control of hypertension; regulation of blood pressure and metabolism; and inducing of labor and therapeutic abortions.

Enzymatic oxidation of arachidonic acid also leads to two important classes of *acyclic* products, **leukotrienes** and **lipoxins,** that arise from oxidation at C5 and/or C15.

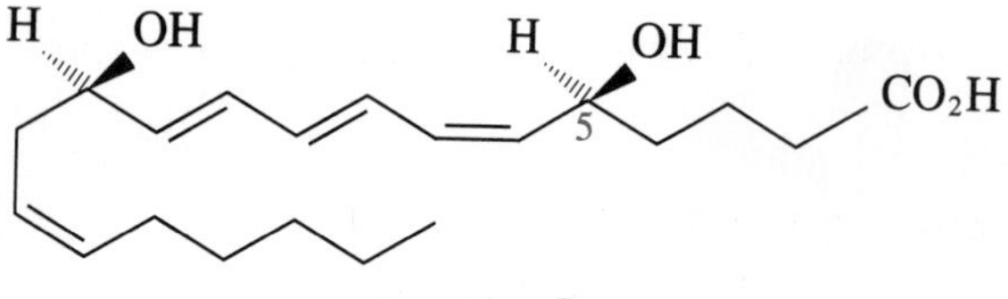

leucotriene B_4

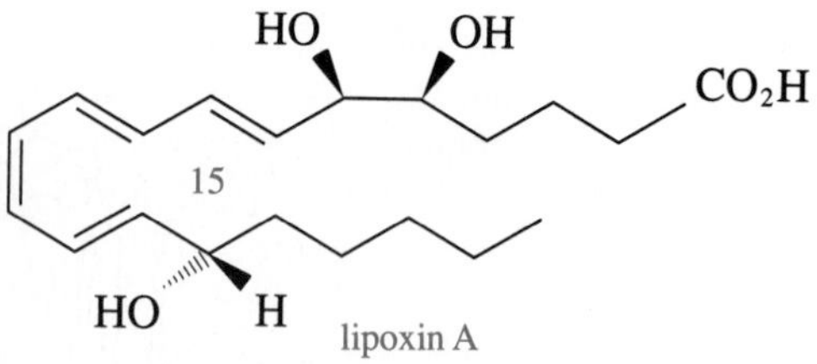

lipoxin A

These classes of compounds can regulate specific cellular responses important in inflammation and immune reactions and are the subject of intense current research.

15.9 *Waxes*

Waxes differ from fats and oils in that they are simple monoesters. The acid and alcohol portions of a wax molecule both have long saturated carbon chains.

$$CH_3(CH_2)_{13}CH_2\overset{O}{\overset{\|}{C}}—O(CH_2)_{15}CH_3$$

cetyl palmitate
(component of spermaceti, a wax in sperm whale oil)

$$C_{25\text{-}27}H_{51\text{-}55}\overset{O}{\overset{\|}{C}}—OC_{30\text{-}32}H_{61\text{-}65}$$

components of beeswax

Some plant waxes are simply long-chain saturated hydrocarbons (Sec. 2.8).

Waxes are more brittle, harder, and less greasy than fats. They are used to make polishes, cosmetics, ointments, and other pharmaceutical preparations, as well as candles and phonograph records. In nature, waxes coat the leaves and stems of plants that grow in arid regions, thus reducing evaporation. Similarly, insects with a high surface-area-to-volume ratio are often coated with a natural protective wax.

15.10 *Terpenes and Steroids*

Essential oils of many plants and flowers are obtained by distilling the plant with water. The water-insoluble oil that separates usually has an odor characteristic of the particular plant (rose oil, geranium oil, and others). Compounds isolated from these oils contain multiples of five carbon atoms (that is, 5, 10, 15, and so on) and are called **terpenes** (some compounds of this type were described in "A Word About Biologically Important Alcohols and Phenols", page 228). They are synthesized in the plant from acetate by way of an important biochemical intermediate, **isopentenyl pyrophosphate.** The five-carbon unit with a four-carbon chain and a one-carbon branch at C2 is called an **isoprene unit.**

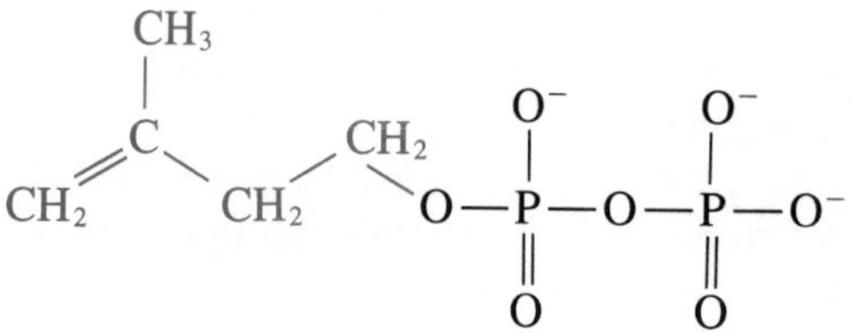

$$CH_2{=}C(CH_3)—CH_2—CH_2—O—\underset{\underset{O}{\|}}{\overset{O^-}{\overset{|}{P}}}—O—\underset{\underset{O}{\|}}{\overset{O^-}{\overset{|}{P}}}—O^-$$

isopentenyl pyrophosphate

$$C—\overset{C}{\overset{|}{C}}—C—C$$

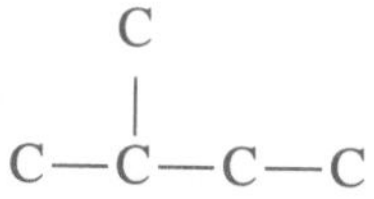

isoprene unit

Most terpene structures can be broken down into multiples of isoprene units. Terpenes contain various functional groups (C=C, OH, C=O) as part of their structures and may be acyclic or cyclic.

Compounds with a single isoprene unit (C_5) are relatively rare in nature, but compounds with two such units (C_{10}), called **monoterpenes,** are common. Examples include geraniol (page 228), **citronellal,** and **myrcene** (all acyclic) and **menthol** and **β-pinene** (both cyclic).

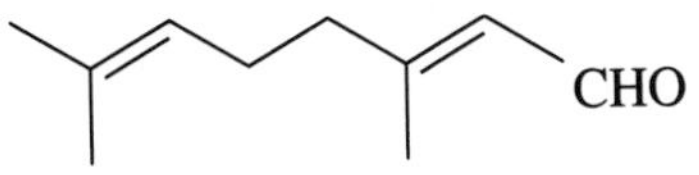

CHO

citronellal
(lemon oil)

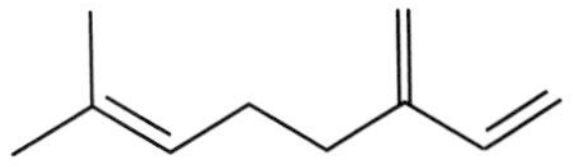

myrcene
(bay leaves)

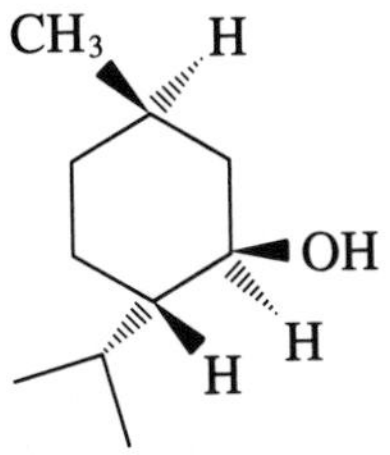

CH_3 H OH H H

menthol
(peppermint oil)

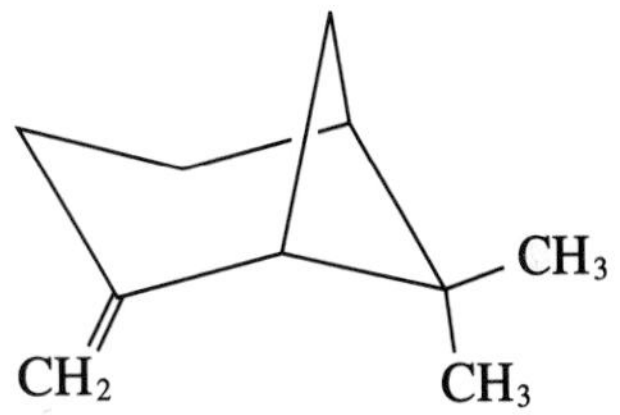

CH_3 CH_2 CH_3

β-pinene
(turpentine)

EXAMPLE 15.4 Mark off the isoprene units in citronellal and menthol.

Solution

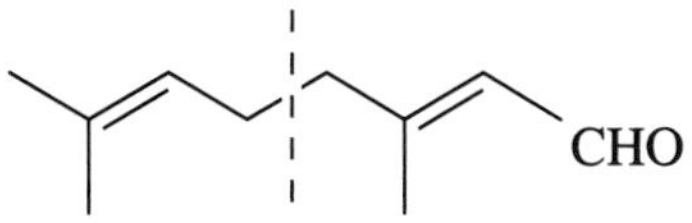

CHO

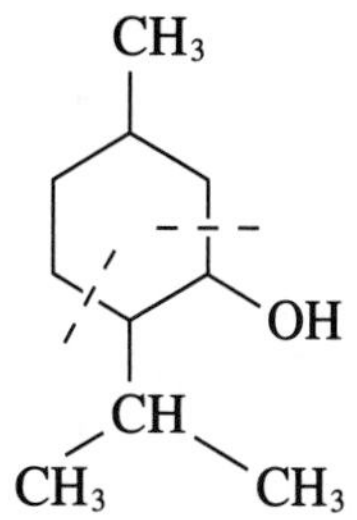

CH_3 OH CH CH_3 CH_3

The dashed lines divide the molecules into two isoprene units.

PROBLEM 15.7 There is another way to divide menthol into two isoprene units. Can you find it? Notice that both ways of dividing the structure join the isoprene units in a head-to-tail manner.

We have already seen examples of a **sesquiterpene** (C_{15}, farnesol, page 228), a **diterpene** (C_{20}, retinal, page 80), a **triterpene** (C_{30}, squalene, page 229), and even a **tetraterpene** (C_{40}, β-carotene, page 80).

PROBLEM 15.8 Draw the structures of farnesol, retinal, squalene, and β-carotene, and mark off the component isoprene units.

Steroids constitute a major class of lipids. They are related to terpenes in the sense that they are biosynthesized by a similar route. Through a truly remarkable reaction sequence, the acyclic triterpene squalene is converted *stereospecifically* to the tetracyclic steroid **lanosterol,** from which other steroids are subsequently synthesized.

squalene (C_{30})

1. O_2, enzyme
2. H^+, enzyme

(15.11)

lanosterol (C_{30})

PROBLEM 15.9 How many stereogenic centers are present in squalene? In lanosterol?

The common structural feature of steroids is a system of four fused rings. The A, B, and C rings are six-membered, and the D ring is five-membered, usually all fused in a *trans* manner.

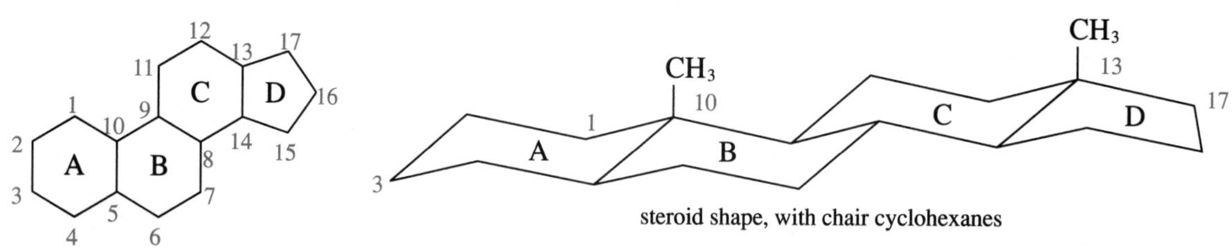

the steroid ring system, showing the numbering

steroid shape, with chair cyclohexanes

In most steroids, the six-membered rings are not aromatic, although there are exceptions. Usually there are methyl substituents attached to C-10 and C-13 (called "angular" methyl groups) and some sort of side chain attached to C-17.

Perhaps the best known steroid is **cholesterol.** With 27 carbons, it is biosynthesized from lanosterol by a sequence of reactions that includes removing three carbon atoms.

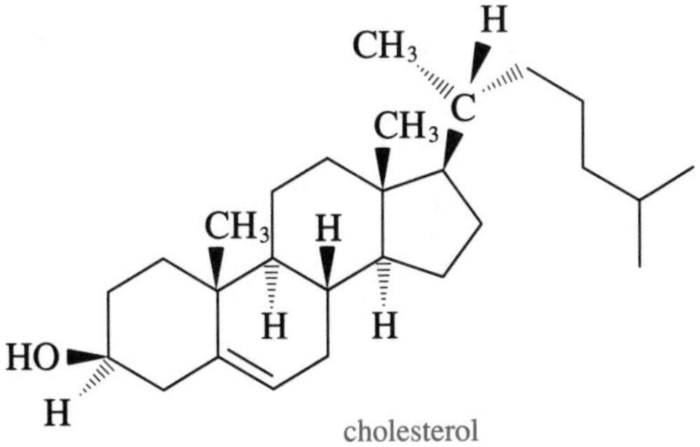

cholesterol

Cholesterol is present in all animal cells but is mainly concentrated in the brain and spinal cord. It is also the chief constituent of gallstones. The total amount of cholesterol present in an average human is about half a pound! There appears to be some connection between the concentration of blood serum cholesterol and coronary heart disease. Levels below 200 mg/dL seem desirable, whereas levels above 280 mg/dL may constitute a high risk.

Other steroids are common in animal tissues and play important biological roles. **Cholic acid,** for example, occurs in the bile duct, where it is present mainly in the form of various amide salts. These salts have a polar region (hydrophilic) and a largely hydrocarbon region (lipophilic) and function as emulsifying agents to facilitate the absorption of fats in the intestinal tract. They are, in a sense, biological soaps.

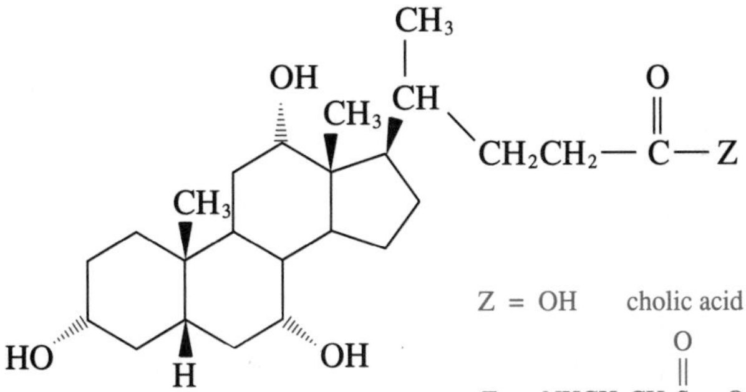

Z = OH cholic acid

$Z = NHCH_2CH_2SO_2-O^-Na^+$ a bile salt

The **sex hormones** are compounds, produced in the ovaries and testes, that control reproductive physiology and secondary sex characteristics. Those sex hormones that predominate in females are of two types. The **estrogens,** of which the most plentiful is **estradiol,** are essential for initiating changes during the menstrual cycle and for the development of female secondary sex characteristics. **Progesterone,** which prepares the uterus for implantation of the fertilized egg, also maintains a pregnancy and prevents further ovulation during that time. Progesterone is administered clinically to prevent abortion in difficult pregnancies. It differs structurally from estrogens, such as estradiol, in that the A ring is not aromatic.

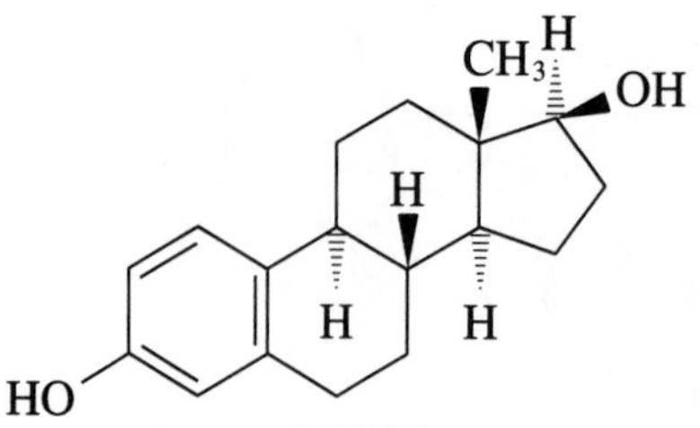

estradiol

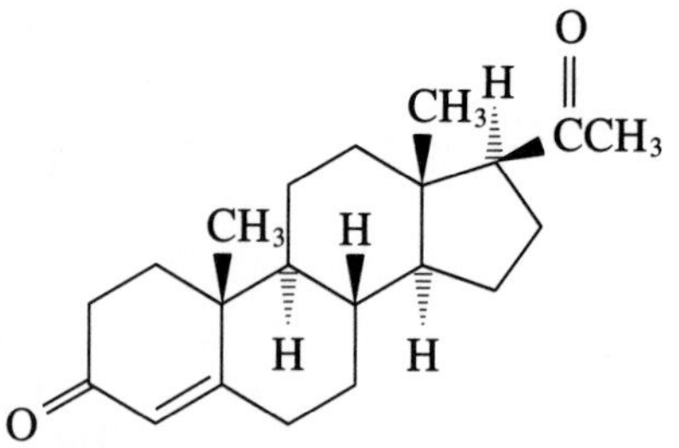

progesterone

Oral contraceptives, sometimes called "the pill," have structures similar to that of progesterone. An example is the acetylenic alcohol **norethindrone,** which prevents conception. On the other hand, **RU 486,** which also resembles progesterone, is a contragestive. It interferes with gestation of a fertilized ovum and, if taken in conjuction with prostaglandins, terminates a pregnancy within the first nine weeks of gestation more effectively and safely than surgical methods. It was discovered in France and is mainly available there.

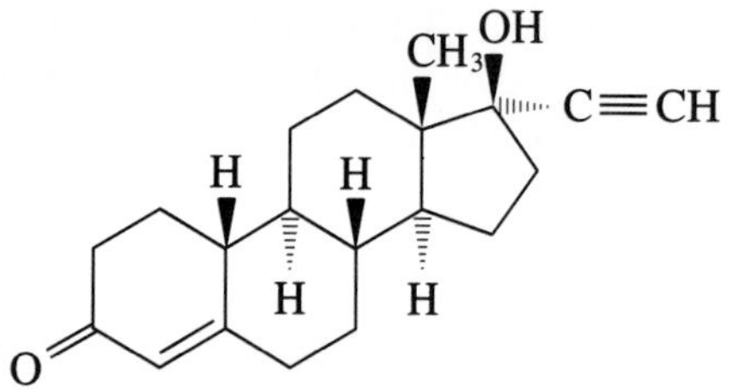

norethindrone (Norlutin)

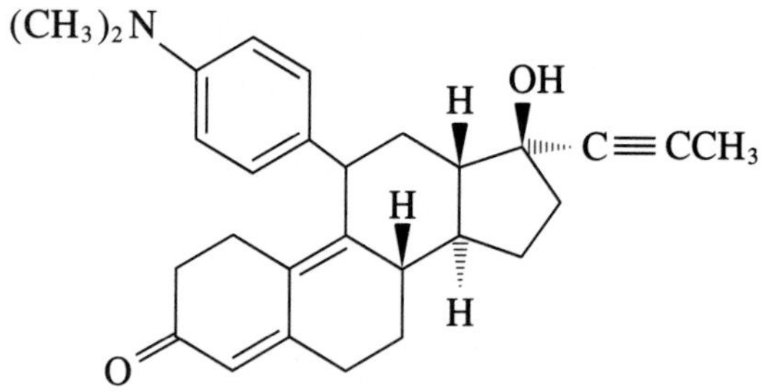

RU 486 (mifepristone)

Other steroid combinations show promise as "morning after" contraceptives, a current area of intense research.

Sex hormones that predominate in males are called **androgens.** They regulate the development of male reproductive organs and secondary sex characteristics, such as facial and body hair, deep voice, and male musculature. Two important androgens are **testosterone** and **androsterone.**

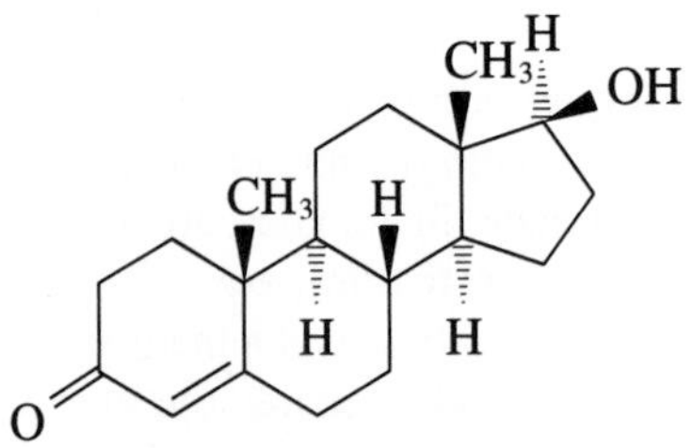

testosterone

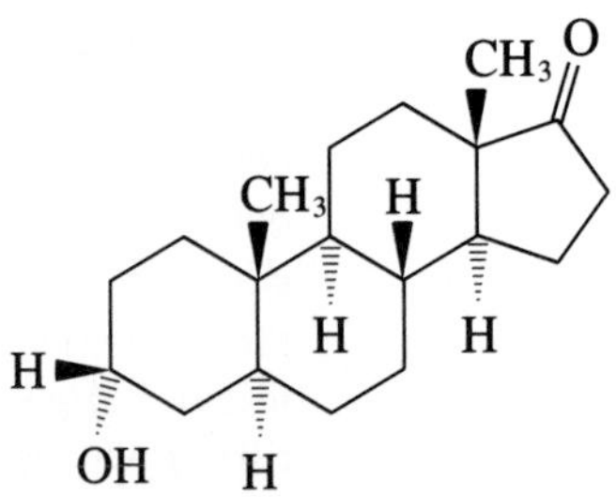

androsterone

Testosterone is an **anabolic** (muscle-building) steroid. Drugs based on its structure are sometimes administered to prevent withering of muscle in people recovering from surgery, starvation, or similar trauma. These same drugs, however, are sometimes illegally administered to healthy athletes and race horses to increase muscle mass and endurance. If taken in high doses, they can have serious side effects, including sexual malfunctions and liver tumors.

The only structural difference between testosterone and progesterone is the replacement of a hydroxyl group by an acetyl group at C-17 in the D ring. This great change in bioactivity, due to a seemingly minor change in structure, exemplifies the extreme specificity of biochemical reactions.

Another steroid bearing a resemblance to testosterone and progesterone is **cortisone,** a drug used in the treatment of arthritis.*

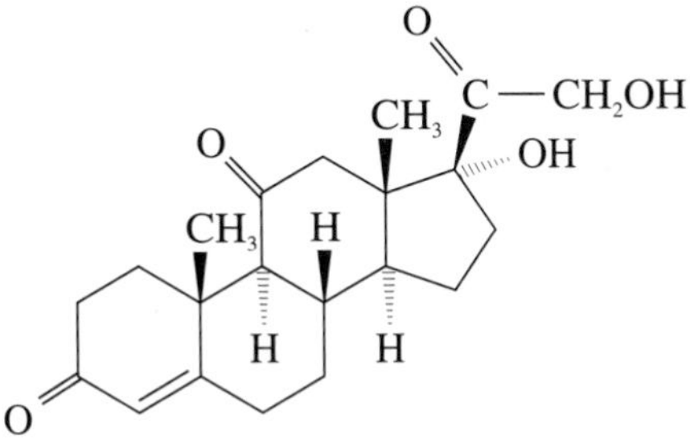

O, C—CH_2OH, CH_3, OH, O, CH_3, H, H, H, O

cortisone

REACTION SUMMARY

1. Saponification of a Triglyceride (Secs. 15.2 and 15.4)

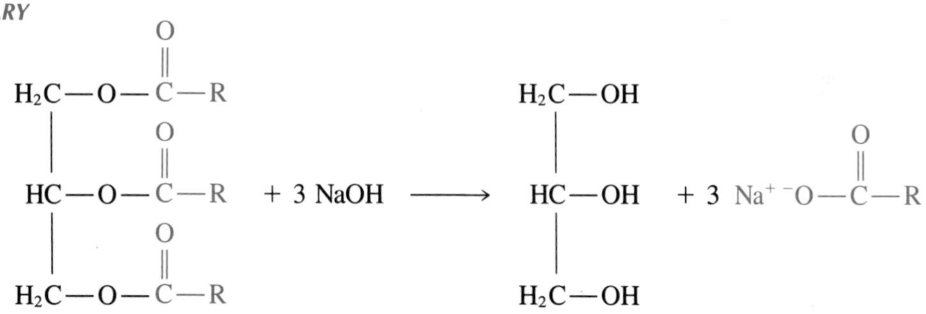

$$\begin{array}{l} H_2C-O-\overset{O}{\overset{\|}{C}}-R \\ \vert \\ HC-O-\overset{O}{\overset{\|}{C}}-R \\ \vert \\ H_2C-O-\overset{O}{\overset{\|}{C}}-R \end{array} + 3\ NaOH \longrightarrow \begin{array}{l} H_2C-OH \\ \vert \\ HC-OH \\ \vert \\ H_2C-OH \end{array} + 3\ Na^+\ {}^-O-\overset{O}{\overset{\|}{C}}-R$$

*A low-cost method for synthesizing cortisone was developed by the American chemist Percy Julian (1899–1975). Julian also discovered an economical way to produce sex hormones from soybean oil, and completed the first total synthesis of the glaucoma drug physostigmine (a nonaromatic nitrogen heterocycle). For a brief biography of Julian, see *Chemical and Engineering News,* Feb. 1, **1993,** p. 9.

2. Hydrogenation of a Triglyceride (Hardening) (Sec. 15.3)

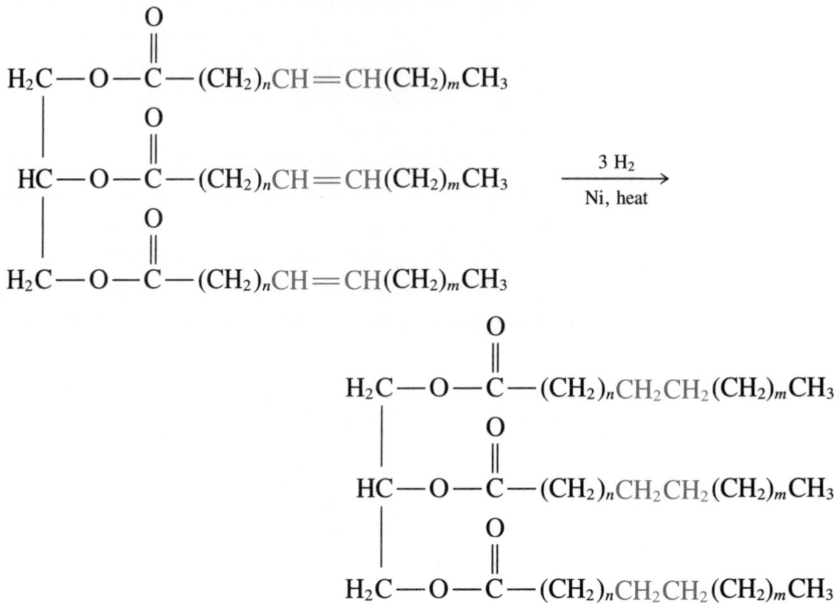

$$\begin{array}{l} H_2C-O-\overset{\overset{\displaystyle O}{\|}}{C}-(CH_2)_nCH=CH(CH_2)_mCH_3 \\ \;| \\ HC-O-\overset{\overset{\displaystyle O}{\|}}{C}-(CH_2)_nCH=CH(CH_2)_mCH_3 \\ \;| \\ H_2C-O-\overset{\overset{\displaystyle O}{\|}}{C}-(CH_2)_nCH=CH(CH_2)_mCH_3 \end{array} \xrightarrow[\text{Ni, heat}]{3\,H_2}$$

$$\begin{array}{l} H_2C-O-\overset{\overset{\displaystyle O}{\|}}{C}-(CH_2)_nCH_2CH_2(CH_2)_mCH_3 \\ \;| \\ HC-O-\overset{\overset{\displaystyle O}{\|}}{C}-(CH_2)_nCH_2CH_2(CH_2)_mCH_3 \\ \;| \\ H_2C-O-\overset{\overset{\displaystyle O}{\|}}{C}-(CH_2)_nCH_2CH_2(CH_2)_mCH_3 \end{array}$$

3. Hydrogenolysis of a Triglyceride (Sec. 15.6)

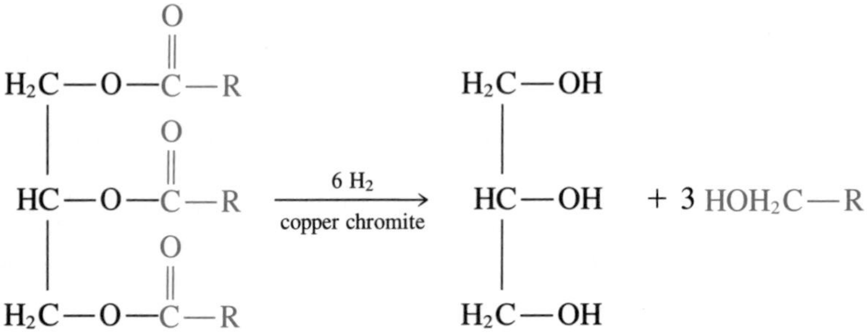

$$\begin{array}{l} H_2C-O-\overset{\overset{\displaystyle O}{\|}}{C}-R \\ \;| \\ HC-O-\overset{\overset{\displaystyle O}{\|}}{C}-R \\ \;| \\ H_2C-O-\overset{\overset{\displaystyle O}{\|}}{C}-R \end{array} \xrightarrow[\text{copper chromite}]{6\,H_2} \begin{array}{l} H_2C-OH \\ \;| \\ HC-OH \\ \;| \\ H_2C-OH \end{array} + 3\,HOH_2C-R$$

ADDITIONAL PROBLEMS

15.10. Using Table 15.1 as a guide, write the structural formula for

a. potassium stearate. **b.** calcium oleate.
c. glyceryl trilaurate. **d.** glyceryl palmitolauroöleate.
e. linoleyl myristate. **f.** methyl arachidate.

15.11. Write equations for the (a) saponification, (b) hydrogenation, and (c) hydrogenolysis of glyceryl trilinoleate.

15.12. Saponification of castor oil gives glycerol and mainly (80 to 90%) **ricinoleic acid,** also called 12-hydroxyoléic acid. Draw the structure of the main component of castor oil.

15.13. Complete the equation for each of the following reactions:

a. $C_{15}H_{31}\overset{\overset{\displaystyle O}{\|}}{C}O^-Na^+ + HCl \longrightarrow$ **b.** $C_{15}H_{31}\overset{\overset{\displaystyle O}{\|}}{C}O^-Na^+ + Mg^{2+} \longrightarrow$

15.14. Using eq. 15.9 as a model, write equations for the preparation of an alkylbenzenesulfonate synthetic detergent, starting with 1-dodecene and benzene.

15.15. A synthetic detergent widely used in dishwashing liquids has the structure $CH_3(CH_2)_{11}(OCH_2CH_2)_3OSO_3^-Na^+$. Write a series of equations showing how this detergent can be synthesized from $CH_3(CH_2)_{10}CH_2OH$ and ethylene oxide.

15.16. List important characteristics of a good syndet.

15.17. What difficulty might there be with a commercial detergent that includes nearly equal amounts of cationic and anionic surfactants?

15.18. Write the general structure for
a. a fat. **b.** a vegetable oil. **c.** a wax.
d. an ordinary soap. **e.** a synthetic detergent. **f.** a steroid.
g. a phospholipid. **h.** a terpene. **i.** an isoprene unit.

15.19. The central carbon of the glycerol unit in a phospholipid is stereogenic and, in nature, has the *R* configuration. Draw the structure of a cephalin that has this arrangement.

15.20. What is the configuration (*Z* or *E*) of each double bond in arachidonic acid? For the structure, see eq. 15.10.

15.21. Consider the structure of prostaglandin E_2 (page 445.)
a. How many stereogenic centers are present?
b. What is the configuration (*R* or *S*) of each?
c. What is the configuration of the double bonds (*Z* or *E*) in the two side chains?
d. Are the two side chains *cis* or *trans* to one another?

15.22. Write an equation for the saponification of cetyl palmitate, the main component of spermaceti.

15.23. When boiled with concentrated aqueous alkali, fats and oils dissolve, but waxes do not. Explain the difference.

15.24. Divide the structures of (a) myrcene and (b) β-pinene into their component isoprene units.

15.25. Notice (from your answer to Problem 15.8) that all the isoprene units in farnesol and retinal are joined in a head-to-tail manner, but that they are not all arranged that way in squalene and β-carotene. Where does the head-to-tail arrangement break down in these two terpenes? What does this suggest about the way these two terpenes might be biosynthesized?

15.26. Predict the products of the following reactions. (Consult the text for the structures of the starting materials).
a. cholesterol + acetic anhydride **b.** testosterone + $LiAlH_4$
c. cholesterol + peroxyacetic acid **d.** androsterone + chromic acid

15.27. Cortisone is a drug used in the treatment of arthritis. (See page 451 for the structure.)
a. Number all the carbon atoms according to the steroidal system.
b. Which carbons are stereogenic centers, and what is the configuration (*R* or *S*) of each?

16 Carbohydrates

16.1 Introduction

Carbohydrates occur in all plants and animals and are essential to life. Through photosynthesis, plants convert carbon dioxide to carbohydrates, mainly **cellulose, starch,** and **sugars.** Cellulose is the building block of rigid cell walls and woody tissue in plants, whereas starch is the chief storage form of carbohydrates for later use as a food or energy source. Some plants (cane and sugar beets) produce sucrose, ordinary table sugar. Another sugar, glucose, is an essential component of blood. Two other sugars, ribose and 2-deoxyribose, are components of the genetic materials RNA and DNA. Other carbohydrates are important components of coenzymes, antibiotics, cartilage, the shells of crustaceans, bacterial cell walls, and mammalian cell membranes.

In this chapter, we will describe the structures and a few reactions of the more important carbohydrates.

16.2 Definitions and Classification

The word ***carbohydrate*** arose because molecular formulas of these compounds can be expressed as *hydrates* of *carbon*. Glucose, for example, has the molecular formula $C_6H_{12}O_6$, which might be written as $C_6(H_2O)_6$. Although this type of formula is useless in studying the chemistry of carbohydrates, the old name persists.

We can now define carbohydrates more precisely in terms of their organic structures. *Carbohydrates are polyhydroxyaldehydes, polyhydroxyketones, or substances that give such compounds on hydrolysis.* The chemistry of carbohydrates is mainly the combined chemistry of two functional groups: the hydroxyl group and the carbonyl group.

Carbohydrates are usually classified according to their structure as **monosaccharides, oligosaccharides,** or **polysaccharides.** The term *saccharide* comes from Latin (*saccharum,* sugar) and refers to the sweet taste of some simple carbohydrates. The three classes of carbohydrates are related to each other through hydrolysis.

$$\text{Polysaccharide} \xrightarrow[H^+]{H_2O} \text{oligosaccharides} \xrightarrow[H^+]{H_2O} \text{monosaccharides} \quad (16.1)$$

For example, hydrolysis of starch, a polysaccharide, gives first maltose and then glucose.

$$\underset{\substack{\text{starch}\\ \text{(a polysaccharide)}}}{[C_{12}H_{20}O_{10}]_n} \xrightarrow[H^+]{n\,H_2O} \underset{\substack{\text{maltose}\\ \text{(a disaccharide)}}}{n\,C_{12}H_{22}O_{11}} \xrightarrow[H^+]{n\,H_2O} \underset{\substack{\text{glucose}\\ \text{(a monosaccharide)}}}{2n\,C_6H_{12}O_6} \quad (16.2)$$

Monosaccharides (or *simple sugars,* as they are sometimes called) *are carbohydrates that cannot be hydrolyzed to simpler compounds.* **Polysaccharides** contain many monosaccharide units—sometimes hundreds or even thousands. Usually, but not always, the units are identical. Two of the most important polysaccharides, starch and cellulose, contain linked units of the same monosaccharide, glucose. **Oligosaccharides** (from the Greek *oligos,* few) contain at least two and generally no more than a few linked monosaccharide units. They may be called **disaccharides, trisaccharides,** and so on, depending on the number of units, which may be the same or different. Maltose, for example, is a disaccharide made of two glucose units, but sucrose, another disaccharide, is made of two different monosaccharide units: glucose and fructose.

In the next section, we will describe the structures of monosaccharides. Later, we will see how these units are linked together to form oligosaccharides and polysaccharides.

16.3 Monosaccharides

Monosaccharides are classified according to the number of carbon atoms present (**triose, tetrose, pentose, hexose,** and so on) and according to whether the carbonyl group is present as an aldehyde (**aldose**) or as a ketone (**ketose**).

There are only two trioses: **glyceraldehyde** and **dihydroxyacetone.** Each has two hydroxyl groups, on separate carbon atoms, and one carbonyl group.

$$\begin{array}{ccc}
{}^{1}\text{CH}{=}\text{O} & {}^{1}\text{CH}_2\text{OH} & \text{CH}_2\text{OH} \\
| & | & | \\
{}^{2}\text{CHOH} & {}^{2}\text{C}{=}\text{O} & \text{CHOH} \\
| & | & | \\
{}^{3}\text{CH}_2\text{OH} & {}^{3}\text{CH}_2\text{OH} & \text{CH}_2\text{OH} \\
\text{glyceraldehyde} & \text{dihydroxyacetone} & \text{glycerol} \\
\text{(an aldose)} & \text{(a ketose)} &
\end{array}$$

Glyceraldehyde is the simplest aldose, and dihydroxyacetone is the simplest ketose. Each is related to glycerol in having a carbonyl group in place of one of the hydroxyl groups.

Other aldoses or ketoses can be derived from glyceraldehyde or dihydroxyacetone by adding carbon atoms, each with a hydroxyl group. In aldoses, the chain is numbered from the aldehyde carbon. In most ketoses, the carbonyl group is located at C-2.

aldoses: tetrose	aldoses: pentose	aldoses: hexose	ketoses: tetrose	ketoses: pentose	ketoses: hexose
¹CH=O	¹CH=O	¹CH=O	¹CH_2OH	¹CH_2OH	¹CH_2OH
²CHOH	²CHOH	²CHOH	²C=O	²C=O	²C=O
³CHOH	³CHOH	³CHOH	³CHOH	³CHOH	³CHOH
⁴CH_2OH	⁴CHOH	⁴CHOH	⁴CH_2OH	⁴CHOH	⁴CHOH
	⁵CH_2OH	⁵CHOH		⁵CH_2OH	⁵CHOH
		⁶CH_2OH			⁶CH_2OH

16.4 Chirality in Monosaccharides; Fischer Projection Formulas and D,L-Sugars

You will notice that glyceraldehyde, the simplest aldose, has one stereogenic carbon atom (C-2) and hence can exist in two enantiomeric forms.

CH=O / H►C◄OH / CH_2OH — *R*-(+)-glyceraldehyde, $[\alpha]^{25}_{D}$ +8.7(c = 2, H_2O)

CH=O / HO►C◄H / CH_2OH — *S*-(−)-glyceraldehyde, $[\alpha]^{25}_{D}$ −8.7(c = 2, H_2O)

The dextrorotatory form has the *R* configuration.

It was in connection with his studies on carbohydrate stereochemistry that Emil Fischer invented his system of projection formulas. Since we will be using these formulas here, it might be wise for you to review Secs. 5.8 through 5.10. Recall that, in a Fischer projection formula, *horizontal* lines show groups that project *above* the plane of the paper *toward* the viewer; *vertical* lines show groups that project *below* the plane of the paper *away* from the viewer. Thus, *R*-(+)-glyceraldehyde can be represented as

CH=O / H►C◄OH / CH_2OH ≡ CH=O / H—┼—OH / CH_2OH

R-(+)-glyceraldehyde

Fischer projection formula for *R*-(+)-glyceraldehyde

with the stereogenic center represented by the intersection of two crossed lines.

Fischer also introduced a stereochemical nomenclature that preceded the *R,S* system and is still in common use for sugars and amino acids. He used a small-capital D to represent the configuration of (+)-glyceraldehyde, with the hydroxyl group on the *right;* its enantiomer, with the hydroxyl group on the *left,* was designated L-(−)-glyceraldehyde. The most oxidized carbon (CHO) was placed at the top.

$$\begin{array}{ccc} & CHO & \\ H & + & OH \\ & CH_2OH & \end{array} \qquad \begin{array}{ccc} & CHO & \\ HO & + & H \\ & CH_2OH & \end{array}$$

D-(+)-glyceraldehyde L-(−)-glyceraldehyde

Fischer extended his system to other monosaccharides in the following way. If the stereogenic carbon *farthest* from the aldehyde or ketone group had the same configuration as D-glyceraldehyde (hydroxyl on the right), the compound was called a D-sugar. If the configuration at the remote carbon had the same configuration as L-glyceraldehyde (hydroxyl on the left), the compound was an L-sugar.

$$\begin{array}{ccc} & CH{=}O & \\ & | & \\ & (CHOH)_n & \\ H & + & OH \\ & CH_2OH & \end{array} \quad \begin{array}{ccc} & CH{=}O & \\ & | & \\ & (CHOH)_n & \\ HO & + & H \\ & CH_2OH & \end{array} \quad \begin{array}{ccc} & CH_2OH & \\ & | & \\ & C{=}O & \\ & | & \\ & (CHOH)_n & \\ H & + & OH \\ & CH_2OH & \end{array} \quad \begin{array}{ccc} & CH_2OH & \\ & | & \\ & C{=}O & \\ & | & \\ & (CHOH)_n & \\ HO & + & H \\ & CH_2OH & \end{array}$$

a D-aldose an L-aldose a D-ketose an L-ketose

Figure 16.1 shows the Fischer projection formulas for all the D-aldoses through the hexoses. Starting with D-glyceraldehyde, one CHOH unit at a time is inserted in the chain. This carbon, which adds a new stereogenic center to the structure, is shown in black. In each case, the new stereogenic center can have the hydroxyl group at the right or at the left in the Fischer projection formula (*R* or *S* absolute configuration).

EXAMPLE 16.1 Using Figure 16.1, write the Fischer projection formula for L-erythrose.

Solution L-Erythrose is the enantiomer of D-erythrose. Since both —OH groups are on the right in D-erythrose, they will both be on the left in its mirror image.

$$\begin{array}{ccc} & CH{=}O & \\ HO & + & H \\ HO & + & H \\ & CH_2OH & \end{array}$$

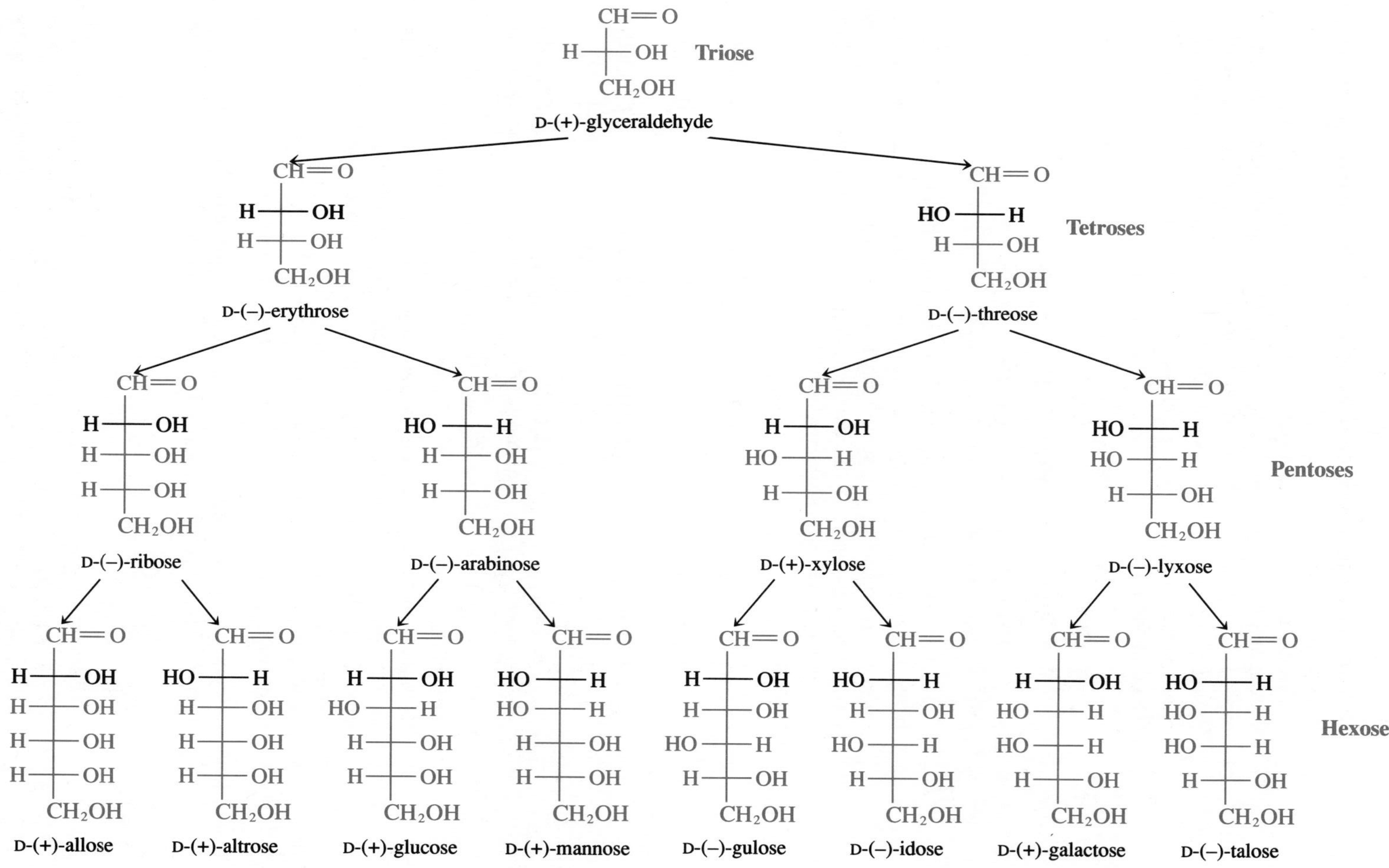

FIGURE 16.1 Fischer projection formulas and genealogy chart for the D-aldoses with up to six carbon atoms.

EXAMPLE 16.2 Convert the Fischer projection formula for D-erythrose to a three-dimensional structural formula.

Solution

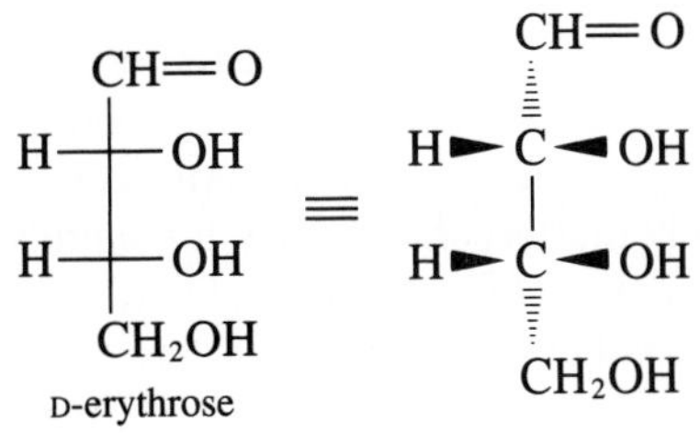

CH=O / H—OH / H—OH / CH_2OH ≡ CH=O / H►C◄OH / H►C◄OH / CH_2OH

D-erythrose

We can also write the structure as

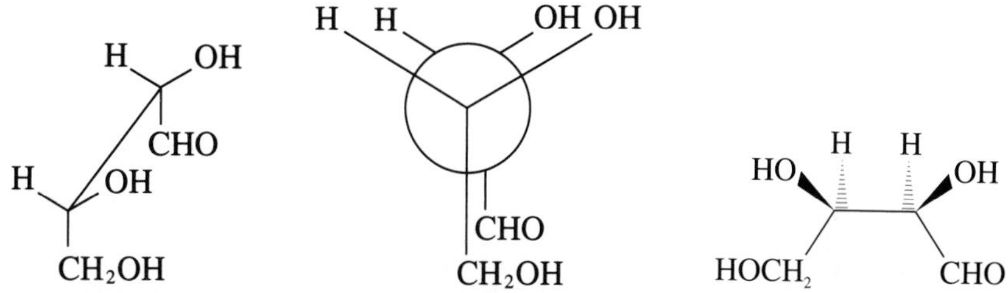

sawhorse | Newman | dash-wedge

and we can then rotate around the central C—C bond to more favorable staggered (instead of eclipsed) conformations, such as

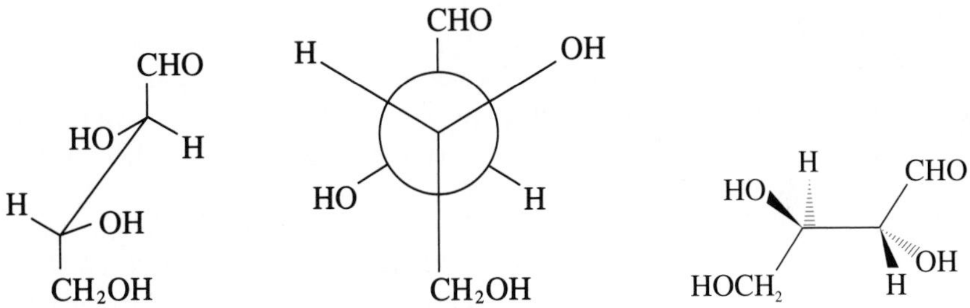

Molecular models may help you to follow these interconversions.

PROBLEM 16.1 Using Figure 16.1, write the Fischer projection formula for

a. L-threose. b. L-glucose.

PROBLEM 16.2 Convert the Fischer projection formula for D-threose to three-dimensional representations.

PROBLEM 16.3 How many D-aldoheptoses are possible?

How are the sugars with identical numbers of carbon atoms, shown horizontally across Figure 16.1, related to one another? Compare, for example, D-(−)-erythrose and D-(−)-threose. They have the same configuration at

C-3 (D, with the OH on the right), but opposite configurations at C-2. These sugars are stereoisomers, but *not* mirror images (*not* enantiomers). In other words, they are **diastereomers** (review Sec. 5.9). Similarly, there are four diastereomeric D-pentoses and eight diastereomeric D-hexoses.

A special name is given to diastereomers that differ in configuration *at only one stereogenic center;* they are called **epimers.** D-(−)-erythrose and D-(−)-threose are not only diastereomers, they are epimers. Similarly, D-glucose and D-mannose are epimers (at C-2), and D-glucose and D-galactose are epimers (at C-4). Each pair has the same configurations at all stereogenic centers except one.

PROBLEM 16.4 What pairs of D-pentoses are epimeric at C-3?

Notice that there is no direct relationship between configuration and the sign of optical rotation. Although all the sugars in Figure 16.1 are D-sugars, some are dextrorotatory (+) and others are levorotatory (−).

16.5 *The Cyclic Hemiacetal Structures of Monosaccharides*

The monosaccharide structures described so far are consistent with much of the known chemistry of these compounds, but they are oversimplified. Let us now examine the true structures of these compounds.

We learned earlier that alcohols undergo rapid and reversible addition to the carbonyl group of aldehydes and ketones, to form hemiacetals (review Sec. 9.8). This can happen *intramolecularly* when the hydroxyl and carbonyl groups are properly located in the same molecule (eqs. 9.14 and 9.15), which is the situation in many monosaccharides. *Monosaccharides exist mainly in cyclic, hemiacetal forms* and not in the acyclic aldo- or keto-forms we have depicted so far.

As an example, consider D-glucose. First, let us rewrite its Fischer projection formula in a way that brings the OH group at C-5 within bonding distance of the carbonyl group (as in eq. 9.14). This is shown in Figure 16.2. The Fischer projection is first converted to its three-dimensional (dash-wedge) structure, which is then turned on its side and bent around so that C-1 and C-6 are close to one another. Finally, rotation about the C-4–C-5 bond brings the hydroxyl oxygen at C-5 close enough for nucleophilic addition to the carbonyl carbon (C-1). Reaction then leads to the cyclic, hemiacetal structure shown at the bottom left of the figure.

The British carbohydrate chemist W. N. Haworth (Nobel Prize, 1937) introduced a useful way of representing the cyclic forms of sugars. In a **Haworth projection,** the ring is represented as if it were planar and viewed edge on, with the oxygen at the upper right. **The carbons are arranged clockwise numerically, with C-1 at the right.** Substituents attached to the ring lie above or below the plane. For example, the Haworth formula for D-glucose (Figure 16.2) is written

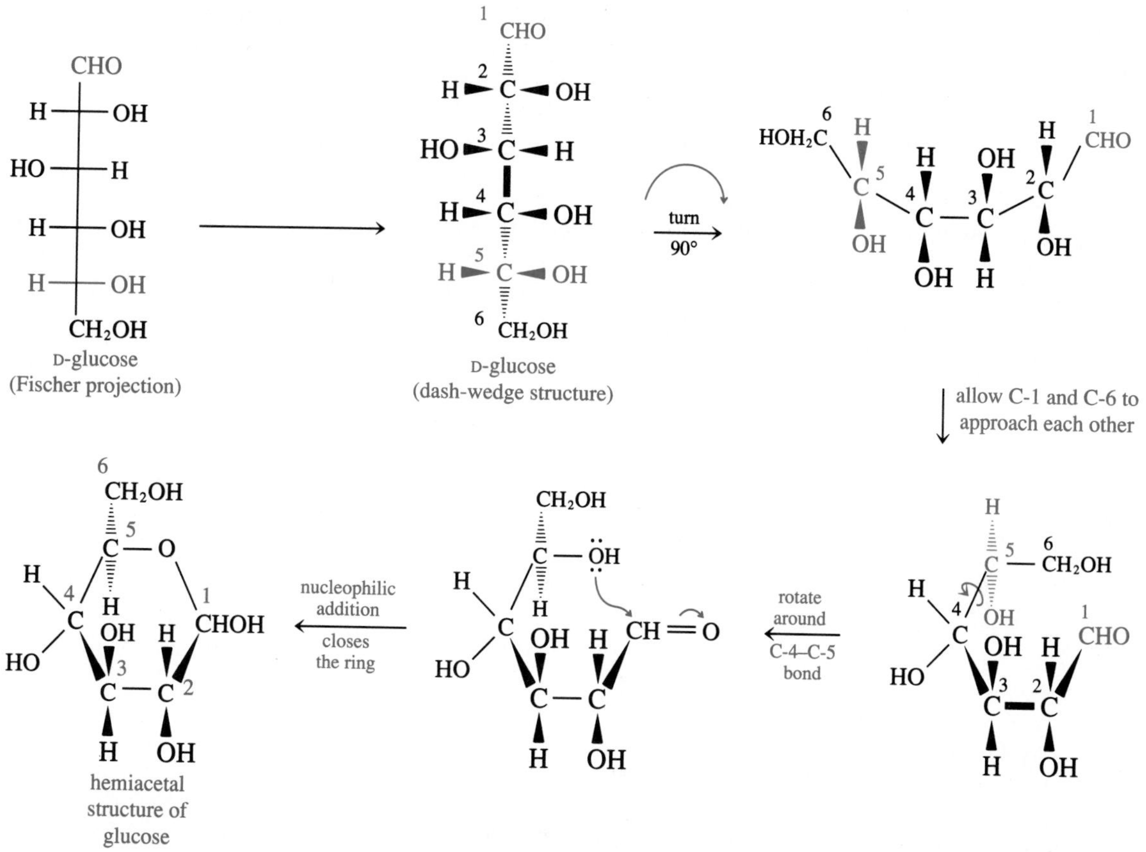

D-glucose
(Fischer projection)

D-glucose
(dash-wedge structure)

turn 90°

allow C-1 and C-6 to approach each other

rotate around C-4–C-5 bond

nucleophilic addition closes the ring

hemiacetal structure of glucose

FIGURE 16.2 Manipulation of the Fischer projection formula of D-glucose to bring the C-5 hydroxyl group in position for cyclization to the hemiacetal form.

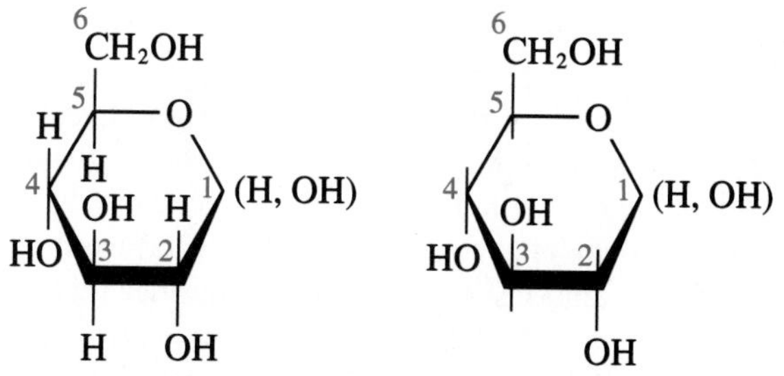

Haworth projection formulas for D-glucose

Sometimes, as in the structure at the right, the ring hydrogens are omitted for clarity.

In converting from one type of projection formula to another, notice that hydroxyl groups on the *right* in the Fischer projection are *down* in the Haworth projection (and conversely, hydroxyl groups on the *left* in the Fischer projection are *up* in the Haworth projection). For D-sugars, the terminal $—CH_2OH$ group is *up* in the Haworth projection; for L-sugars, it is down.

EXAMPLE 16.3 Draw the Haworth projection for the six-membered cyclic structure of D-mannose.

Solution Notice from Figure 16.1 that D-mannose differs from D-glucose *only* in the configuration at C-2. In the Fischer projection formula, the C-2 hydroxyl is on the *left;* therefore, it will be *up* in the Haworth projection. Otherwise, the structure is identical to that of D-glucose.

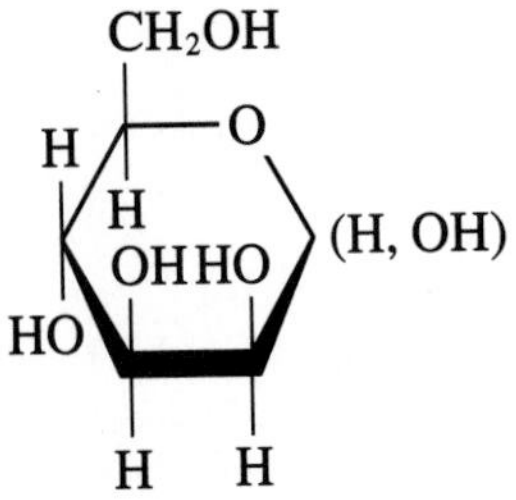

CH_2OH
H O
H
OH HO (H, OH)
HO
H H

D-mannose

PROBLEM 16.5 Draw the Haworth projection formula for the six-membered cyclic structure of D-galactose.

Now notice three important features of the hemiacetal structure of D-glucose. First, *the ring is heterocyclic,* with five carbons and an oxygen. Carbons 1 through 5 are part of the ring structure, but carbon 6 (the $—CH_2OH$ group) is a substituent on the ring. Next, C-1 is special. *C-1 is the hemiacetal carbon,* simultaneously an alcohol and an ether carbon (it carries a hydroxyl group, and it is also connected to C-5 by an ether link). In contrast, all the other carbons are monofunctional. C-2, C-3, and C-4 are secondary alcohol carbons; C-6 is a primary alcohol carbon; and C-5 is an ether carbon. These differences show up in the different chemical reactions of D-glucose. Finally, *C-1 in the cyclic, hemiacetal structure is a stereogenic center.* It has four different groups attached to it (H, OH, OC5, and C2) and can therefore exist in two configurations, *R* or *S*. Let us consider this last feature in greater detail.

16.6 Anomeric Carbons; Mutarotation

In the acyclic, aldehyde form of glucose, C-1 is achiral, but in the cyclic structures, this carbon becomes chiral. Consequently, *two* hemiacetal structures are possible, depending on the configuration at the new chiral center. The hemiacetal carbon, the carbon that forms the new stereogenic center, is called the **anomeric carbon.** Two monosaccharides that differ only in configuration at the anomeric center are **anomers** (a special kind of epimers). Anomers are called α or β, depending on the position of the hydroxyl group. For monosaccharides in the D-series, the hydroxyl group is "down" in the α anomer and "up" in the β anomer, when the structure is written in the usual way (eq. 16.3).

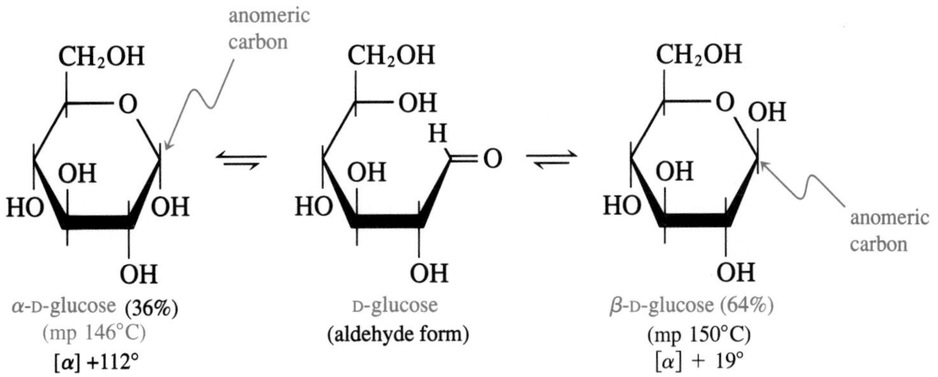

anomeric carbon

α-D-glucose (36%)
(mp 146°C)
[α] +112°

D-glucose
(aldehyde form)

β-D-glucose (64%)
(mp 150°C)
[α] + 19°

anomeric carbon

(16.3)

The α and β forms of D-glucose have identical configurations at every stereogenic center *except at C-1, the anomeric carbon.*

How do we know that monosaccharides exist mainly as cyclic hemiacetals? There is direct physical evidence. For example, if D-glucose is crystallized from methanol, the pure α form is obtained. On the other hand, crystallization of glucose from acetic acid gives the β form. The α and β forms of D-glucose are *diastereomers*. Being diastereomers, they have different physical properties, as shown under their structures in eq. 16.3, where we see that they have different melting points and different specific optical rotations.

The α and β forms of D-glucose interconvert in aqueous solution. For example, if crystalline α-D-glucose is dissolved in water, the specific rotation drops gradually from an initial value of +112° to an equilibrium value of +52°. Starting with the pure crystalline β form results in a gradual rise in specific rotation from an initial +19° to the *same equilibrium value* of +52°. These changes in optical rotation are called **mutarotation.** They can be explained by the equilibria shown in eq. 16.3. Recall that hemiacetal formation is a reversible equilibrium process (Sec. 9.8). Starting with either pure hemiacetal form, the ring can open to the acyclic aldehyde, which can then recyclize to give either the α or the β form. Eventually, an equilibrium mixture is obtained.

At equilibrium, an aqueous solution of D-glucose contains 35.5% of the α form and 64.5% of the β form. There is only about 0.003% of the open-chain aldehyde form present.

EXAMPLE 16.4 Show that the percentages of α- and β-D-glucose in aqueous solution at equilibrium can be calculated from the specific rotations of the pure α and β forms and the specific rotation of the solution at equilibrium.

Solution The equilibrium rotation is +52°, and the rotations of pure α and β forms are +112° and +19°, respectively. Assuming that no other forms are present, we can express these values graphically as follows:

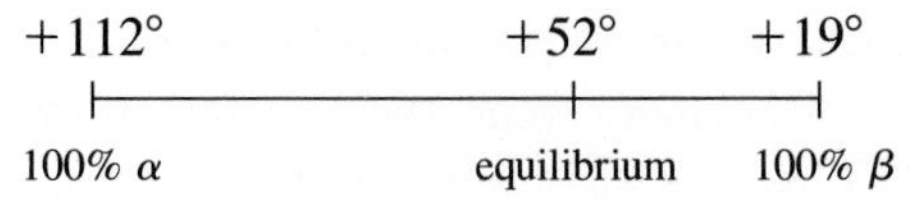

$+112°$ $\qquad\qquad$ $+52°$ $\qquad$ $+19°$

100% α $\qquad$ equilibrium $\qquad$ 100% β

The percentage of the β form at equilibrium is then

$$\frac{112 - 52}{112 - 19} \times 100 = \frac{60}{93} \times 100 = 64.5\%$$

The percentage of the α form at equilibrium is $100 - 64.5 = 35.5\%$.

16.7 Pyranose and Furanose Structures

The six-membered cyclic form of most monosaccharides is the preferred structure. These structures are called **pyranose** forms after the six-membered oxygen heterocycle **pyran.** The formula at the extreme left of eq. 16.3 is more completely named **α-D-glucopyranose,** the last part of the name showing the ring size.

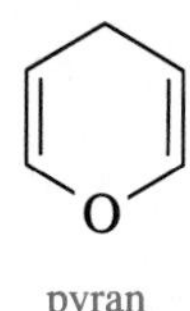

pyran

Pyranoses are formed by reaction of the hydroxyl group at C-5, with the carbonyl group. With some sugars, however, the hydroxyl group at C-4 reacts instead. In these cases, the cyclic hemiacetal that is formed has a *five*-membered ring. This type of cyclic monosaccharide is called a **furanose,** after the parent five-membered oxygen heterocycle **furan.**

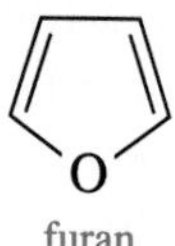

furan

For example, D-glucose could, in principle, exist in two furanose forms (α and β at C-1) through attack of the C-4 hydroxyl on the aldehyde carbon.

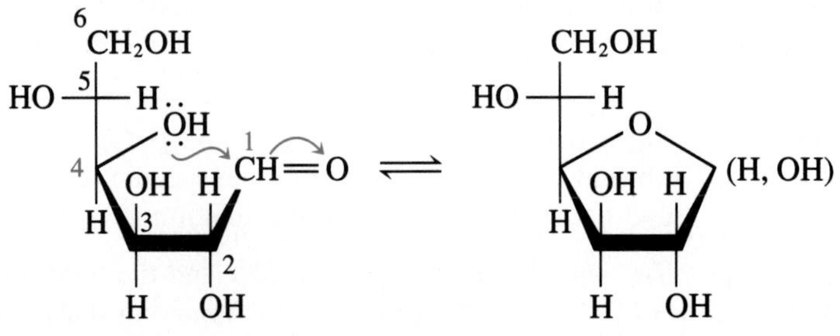

D-glucose $\rightleftharpoons$ α- and β-D-glucofuranose (16.4)

In practice, these forms are present to less than 1% in glucose solutions, but they are important with other monosaccharides. The ketose **D-fructose,** for example, exists in solution mainly in two furanose forms. The carbonyl carbon at C-2 and hydroxyl group at C-5 cyclize to give the furanose ring.

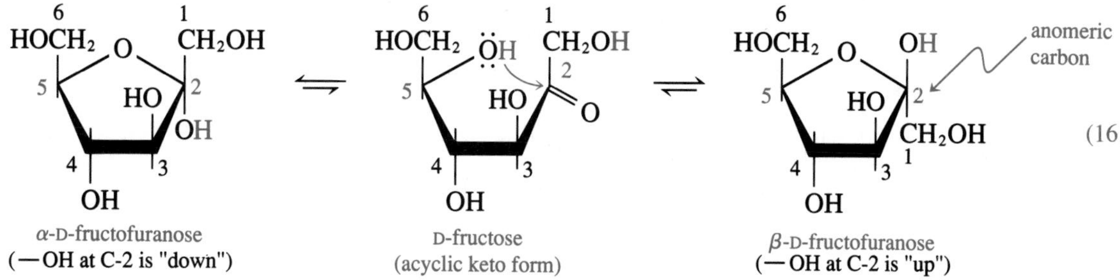

α-D-fructofuranose (—OH at C-2 is "down") ⇌ D-fructose (acyclic keto form) ⇌ β-D-fructofuranose (—OH at C-2 is "up") (16.5)

anomeric carbon

PROBLEM 16.6 Draw Haworth projections of the α and β forms of D-glucofuranose (eq. 16.4).

PROBLEM 16.7 D-Erythrose cannot exist in pyranose forms, but furanose cyclic forms are possible. Explain. Draw the structure for α-D-erythrofuranose.

16.8 Conformations of Pyranoses

Haworth projections depict pyranose rings as planar. However, as with cyclohexane, the rings generally prefer a chair conformation. Consequently, we can rewrite eq. 16.3 more accurately as eq. 16.6.

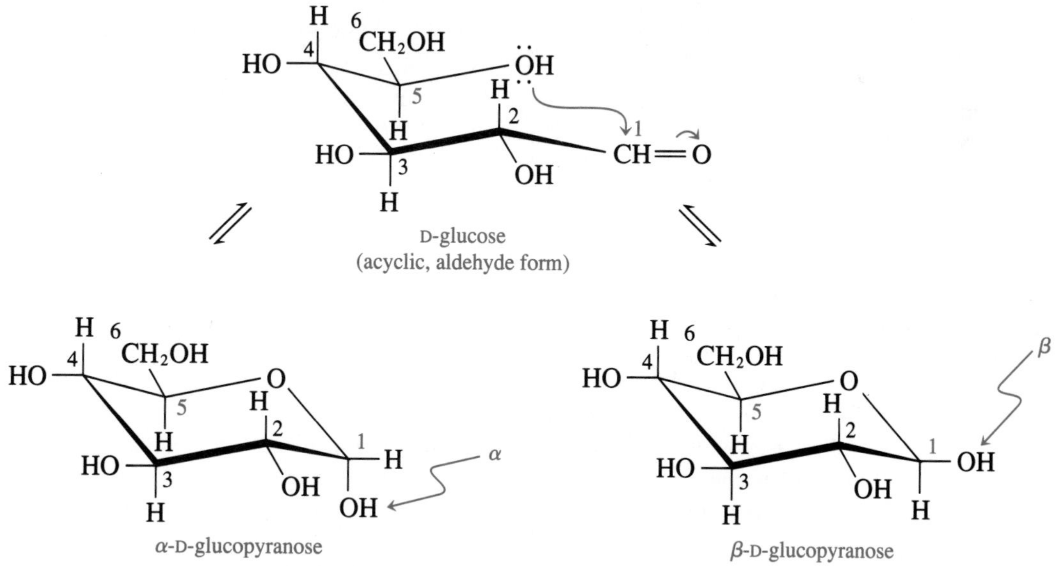

D-glucose (acyclic, aldehyde form) (16.6)

α-D-glucopyranose ⇌ D-glucose ⇌ β-D-glucopyranose

It is probably no accident that glucose is the most abundant natural monosaccharide because in D-glucose, the larger substituent at each ring carbon is equatorial. The only exception occurs at the anomeric carbon (C-1), where the hydroxyl group may be axial (in the α anomer) or equatorial (in the β anomer).

This difference provides one reason why the β form is preferred at equilibrium (eq. 16.3).

EXAMPLE 16.5 Draw the most stable chair conformation of α-D-mannopyranose.

Solution Recall from Example 16.3 that D-mannose differs from D-glucose only at C-2. Using the cyclic structure at the left of eq. 16.6 as a guide, we can write

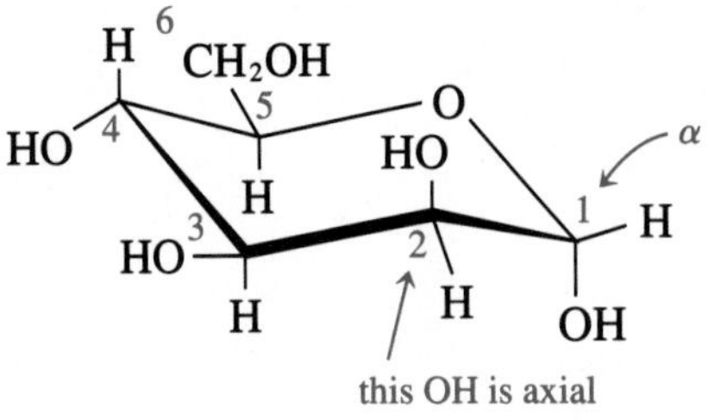

this OH is axial

α-D-mannopyranose

PROBLEM 16.8 D-Galactose differs from D-glucose only in the configuration at C-4. Draw the most stable chair conformation of β-D-galactopyranose.

Now that we have described the structures of monosaccharides, let us examine some of their common reactions.

16.9 *Esters and Ethers from Monosaccharides*

Monosaccharides contain hydroxyl groups. It is not surprising, then, that they undergo reactions typical of alcohols. For example, they can be converted to esters by reaction with acid halides or anhydrides. The conversion of β-D-glucose to its pentaacetate by reaction with excess acetic anhydride is typical; all five hydroxyl groups, including the hydroxyl at anomeric C-1, are esterified. (To clarify the structure, the ring H's are omitted.)

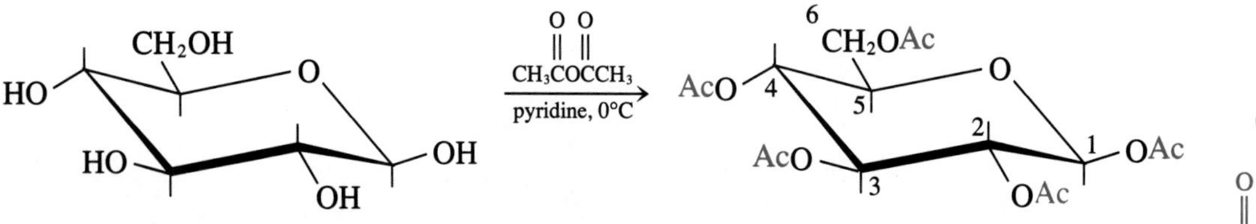

β-D-glucopyranose $\xrightarrow[\text{pyridine, 0°C}]{CH_3COCCH_3}$ β-D-glucopyranose pentaacetate (16.7)

Ac = $CH_3C(=O)$—

The hydroxyl groups can also be converted to ethers by treatment with an alkyl halide and a base (the Williamson synthesis, Sec. 8.6). Because sugars are sensitive to strong bases, the mild base silver oxide is preferred.

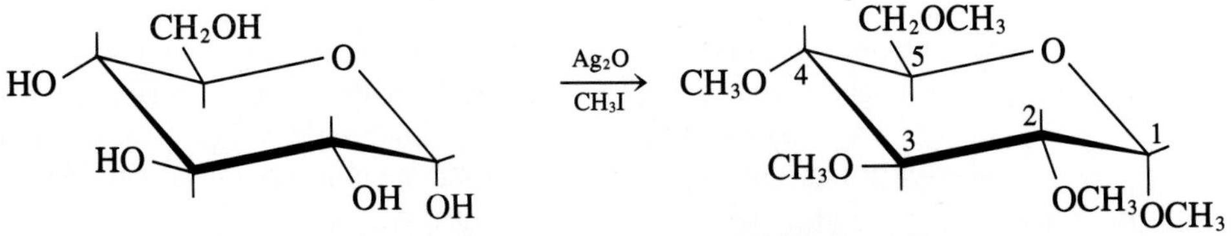

α-D-glucopyranose $\xrightarrow[CH_3I]{Ag_2O}$ α-D-glucopyranose pentamethyl ether (16.8)

Whereas sugars tend to be soluble in water and insoluble in organic solvents, the reverse is true for their esters and ethers. This often facilitates their purification and manipulation with organic reagents.

16.10 *Reduction of Monosaccharides*

The carbonyl group of aldoses and ketoses can be reduced by various reagents. The products are **polyols,** called **alditols.** For example, catalytic hydrogenation or reduction with sodium borohydride ($NaBH_4$) converts D-glucose to D-glucitol (also called sorbitol; review Sec. 9.13).

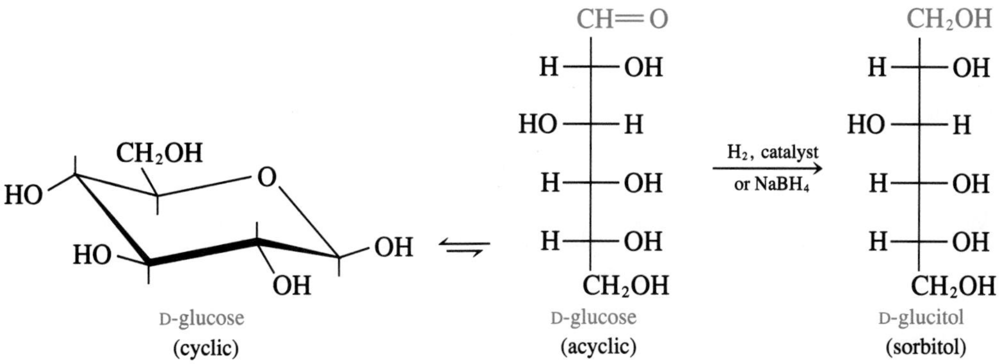

CH_2OH, HO, HO, O, OH, OH
D-glucose
(cyclic)

⇌

CH=O
H—OH
HO—H
H—OH
H—OH
CH_2OH
D-glucose
(acyclic)

$\xrightarrow[\text{or } NaBH_4]{H_2,\ \text{catalyst}}$

CH_2OH
H—OH
HO—H
H—OH
H—OH
CH_2OH
D-glucitol
(sorbitol)

(16.9)

Reaction occurs by reduction of the small amount of aldehyde in equilibrium with the cyclic hemiacetal. As that aldehyde is reduced, the equilibrium shifts to the right, so that eventually all of the sugar is converted. Sorbitol is used commercially as a sweetener and sugar substitute.

PROBLEM 16.9 D-Mannitol, which occurs naturally in olives, onions, and mushrooms, can be made by $NaBH_4$ reduction of D-mannose. Draw its structure.

16.11 *Oxidation of Monosaccharides*

Although aldoses exist primarily in cyclic hemiacetal forms, these structures are in equilibrium with a small but finite amount of the open-chain aldehyde. These aldehyde groups can be easily oxidized to acids (review Sec. 9.14). The products are called **aldonic acids.** For example, D-glucose is easily oxidized to D-gluconic acid.

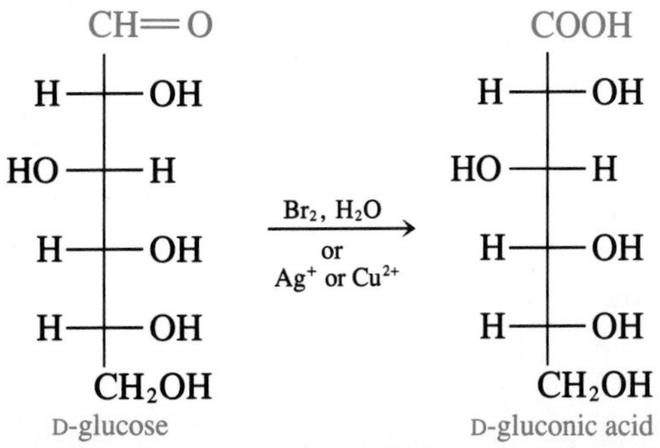

CH=O
H—OH
HO—H
H—OH
H—OH
CH_2OH
D-glucose

$\xrightarrow[\text{or } Ag^+ \text{ or } Cu^{2+}]{Br_2,\ H_2O}$

COOH
H—OH
HO—H
H—OH
H—OH
CH_2OH
D-gluconic acid

(16.10)

The oxidation of aldoses is so easy that they react with such mild oxidizing agents as Tollens' reagent (Ag^+ in aqueous ammonia), Fehling's reagent (Cu^{2+} complexed with tartrate ion), or Benedict's reagent (Cu^{2+} complexed with citrate ion). With Tollens' reagent, they give a silver mirror test (Sec. 9.14), and with the copper reagents, the blue solution gives a red precipitate of cuprous oxide, Cu_2O. A carbohydrate that reacts with Ag^+ or Cu^{2+} is called a **reducing sugar** because reduction of the metal accompanies oxidation of the aldehyde group.

$$\underset{\text{blue solution}}{RCH{=}O + 2\,Cu^{2+}} + 5\,OH^- \longrightarrow RC(=O)O^- + \underset{\text{red precipitate}}{Cu_2O} + 3\,H_2O \qquad (16.11)$$

PROBLEM 16.10 Write an equation for the reaction of D-mannose with Fehling's reagent (Cu^{2+}) to give D-mannonic acid.

Stronger oxidizing agents, such as aqueous nitric acid, attack the aldehyde group *and* the primary alcohol group, producing dicarboxylic acids called **aldaric acids.** For example, D-glucose gives D-glucaric acid.

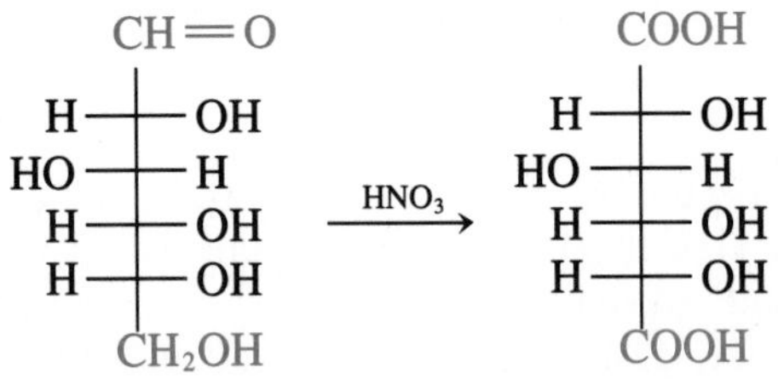

$$\begin{array}{c} CH{=}O \\ H-\!\!\!+\!\!\!-OH \\ HO-\!\!\!+\!\!\!-H \\ H-\!\!\!+\!\!\!-OH \\ H-\!\!\!+\!\!\!-OH \\ CH_2OH \\ \text{D-glucose} \end{array} \xrightarrow{HNO_3} \begin{array}{c} COOH \\ H-\!\!\!+\!\!\!-OH \\ HO-\!\!\!+\!\!\!-H \\ H-\!\!\!+\!\!\!-OH \\ H-\!\!\!+\!\!\!-OH \\ COOH \\ \text{D-glucaric acid} \end{array} \qquad (16.12)$$

PROBLEM 16.11 Write the structure of D-mannaric acid.

16.12 Formation of Glycosides from Monosaccharides

Because monosaccharides exist as cyclic hemiacetals, they can react with one equivalent of an alcohol to form acetals. An example is the reaction of β-D-glucose with methanol.

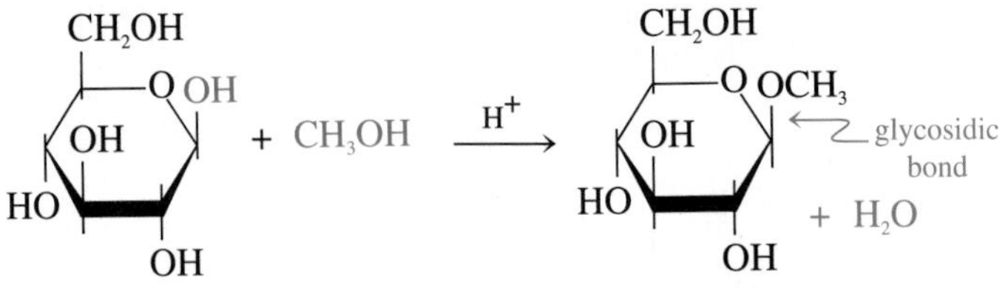

$$\underset{\beta\text{-D-glucopyranose}}{\text{(ring: }CH_2OH,\ O,\ OH,\ OH,\ HO,\ OH)} + CH_3OH \xrightarrow{H^+} \underset{\text{methyl }\beta\text{-D-glucopyranoside (mp 115–116°C)}}{\text{(ring: }CH_2OH,\ O,\ OCH_3,\ OH,\ HO,\ OH)} + H_2O \qquad (16.13)$$

glycosidic bond

Note that *only the —OH on the anomeric carbon is replaced by an —OR group*. Such acetals are called **glycosides,** and the bond from the anomeric carbon to the OR group is called the **glycosidic bond.** Glycosides are named from the corresponding monosaccharide by changing the *-e* ending to *-ide*. Thus, glucose gives glucosides, mannose gives mannosides, and so on.

EXAMPLE 16.6 Write a Haworth formula for ethyl α-D-mannoside.

Solution

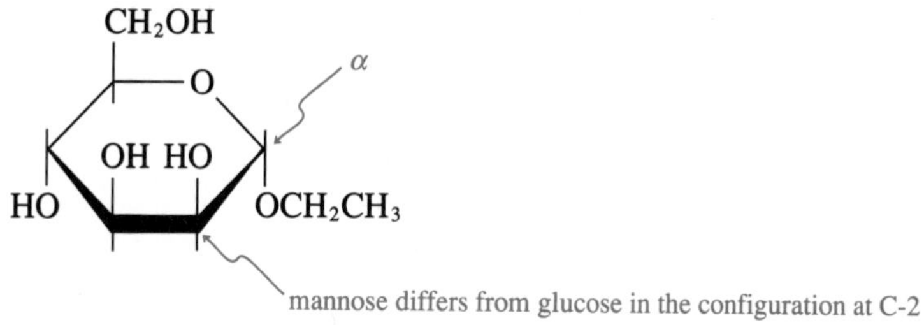

CH_2OH, O, α, OH HO, HO, OCH_2CH_3

mannose differs from glucose in the configuration at C-2

PROBLEM 16.12 Write an equation for the acid-catalyzed reaction of β-D-galactose with methanol.

The mechanism of glycoside formation is the same as that described in eq. 9.13 of Sec. 9.8. The acid catalyst can protonate any of the six oxygen atoms, since each has unshared electron pairs and is basic. However, *only protonation of the hydroxyl oxygen at C-1 leads, after water loss, to a resonance-stabilized carbocation.* In the final step, methanol can attack from either "face" of the six-membered ring, to give either the β-glycoside as shown, or the α-glycoside.

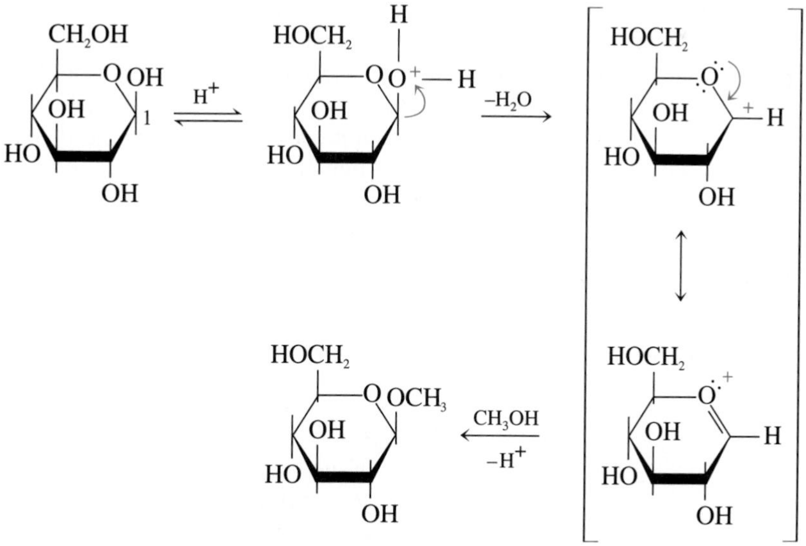

CH_2OH, O, OH, 1, OH, HO, OH $\underset{}{\overset{H^+}{\rightleftharpoons}}$ $HOCH_2$, O, O^+—H, H, OH, HO, OH $\xrightarrow{-H_2O}$ [$HOCH_2$, O, OH, + H, HO, OH ↔ $HOCH_2$, O^+, OH, H, HO, OH] $\xleftarrow[-H^+]{CH_3OH}$ $HOCH_2$, O, OCH_3, OH, HO, OH (16.14)

Naturally occurring alcohols or phenols often occur in cells combined as a glycoside with some sugar—most commonly, glucose. In this way, the many hydroxyl groups of the sugar portion of the glycoside solubilize compounds that would otherwise be incompatible with cellular protoplasm. An example is the bitter-tasting glucoside **salicin,** which occurs in willow bark and whose fever-reducing power was known to the ancients.

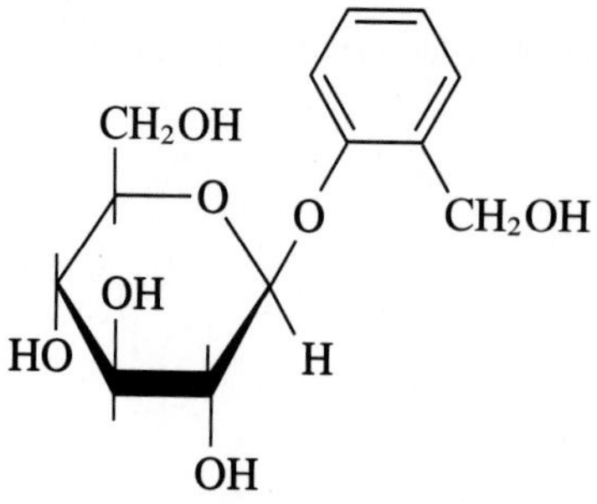

salicin
(the β-D-glucoside of salicyl alcohol)

The glycosidic bond is the key to understanding the structure of oligosaccharides and polysaccharides, as we will see in the following sections.

16.13 Disaccharides

The most common oligosaccharides are **disaccharides**. *In a disaccharide, two monosaccharides are linked by a glycosidic bond between the anomeric carbon of one monosaccharide unit and a hydroxyl group on the other unit.* In this section, we will describe the structure and properties of four important disaccharides.

16.13a Maltose

Maltose is the disaccharide obtained by the partial hydrolysis of starch. Further hydrolysis of maltose gives only D-glucose (eq. 16.2). Maltose must, therefore, consist of two linked glucose units. It turns out that the anomeric carbon of the left unit is linked to the C-4 hydroxyl group of the unit at the right as an acetal (glycoside). The configuration at the anomeric carbon of the left unit is α. In the crystalline form, the anomeric carbon of the right unit has the β configuration. Both units are pyranoses, and the right-hand unit fills the same role as the methanol in eq. 16.13.

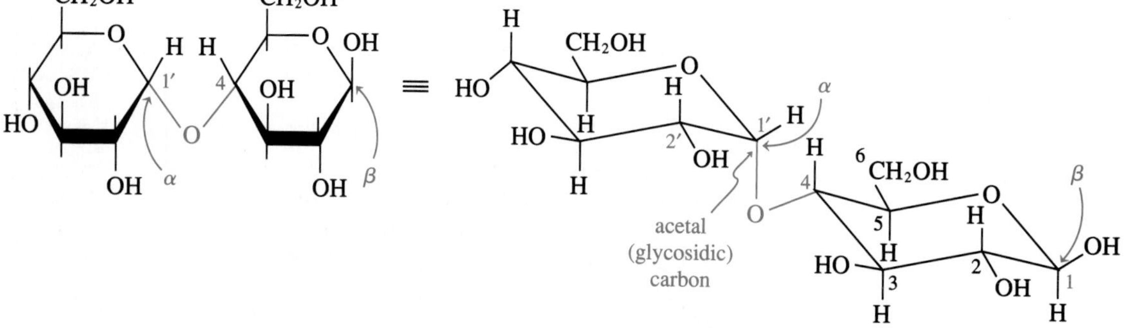

maltose
4-O-(α-D-glucopyranosyl)-β-D-glucopyranose

The systematic name for maltose, shown beneath the common name, describes the structure fully, including the name of each unit (D-glucose), the ring sizes (pyranose) the configuration at each anomeric carbon (α or β), and the location of the hydroxyl group involved in the glycosidic link (4-O).

The anomeric carbon of the right glucose unit in maltose is a hemiacetal. Naturally, when maltose is in solution, this hemiacetal function will be in equilibrium with the open-chain aldehyde form. Maltose therefore gives a positive Tollens' test and other reactions similar to those of the anomeric carbon in glucose.

PROBLEM 16.13 When crystalline maltose is dissolved in water, the initial specific rotation changes and gradually reaches an equilibrium value. Explain.

16.13b Cellobiose

Cellobiose is the disaccharide obtained by the partial hydrolysis of cellulose. Further hydrolysis of cellobiose gives only D-glucose. Cellobiose must therefore be an isomer of maltose. In fact, *cellobiose differs from maltose* only *in having the β configuration at C-1 of the left glucose unit*. Otherwise, all other structural features are identical, including a link from C-1 of the left unit to the hydroxyl group at C-4 in the right unit.

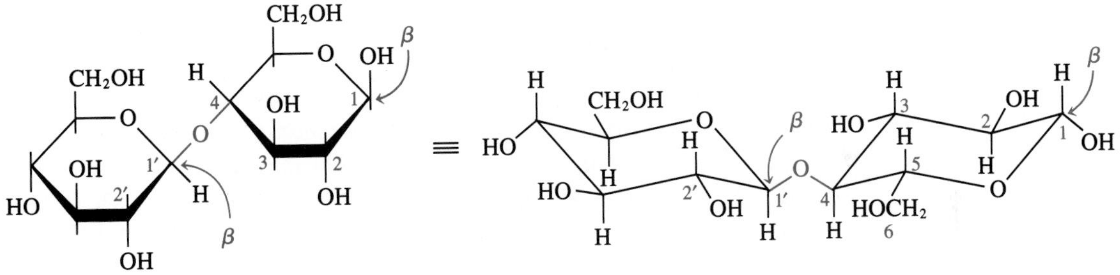

cellobiose
4-O-(β-D-glucopyranosyl)-β-D-glucopyranose

Note that, in the conformational formula for cellobiose, one ring-oxygen is drawn to the "rear" and one to the "front" of the molecule. This is the way the rings exist in the cellulose chain.

16.13c Lactose

Lactose is the major sugar in human and cow's milk (4 to 8% lactose). Hydrolysis of lactose gives equimolar amounts of D-galactose and D-glucose. The anomeric carbon of the galactose unit has the β configuration at C-1 and is linked to the hydroxyl group at C-4 of the glucose unit. The crystalline anomer that has the α configuration at the glucose unit is made commercially from cheese whey.

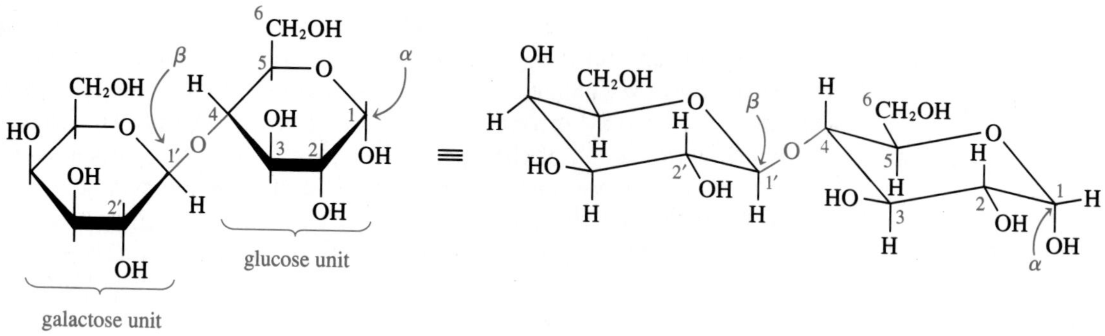

lactose
4-O-(β-D-galactopyranosyl)-α-D-glucopyranose

PROBLEM 16.14 Will lactose give a positive Fehling's test? Will it mutarotate?

Some human infants are born with a disease called *galactosemia*. They lack the enzyme that isomerizes galactose to glucose and cannot digest milk. If milk is excluded from such infants' diets, the disease symptoms caused by accumulation of galactose can be avoided.

16.13d Sucrose The most important commercial disaccharide is **sucrose,** ordinary table sugar. More than 100 million tons are produced annually worldwide. Sucrose occurs in all photosynthetic plants, where it functions as an energy source. It is obtained commercially from sugar cane and sugar beets, in which it constitutes 14 to 20% of the plant juices.

Hydrolysis of sucrose gives equimolar amounts of D-glucose and the ketose D-fructose. *Sucrose differs from the other disaccharides we have discussed in that the anomeric carbons of* ***both*** *units are involved in the glycosidic link.* That is, C-1 of the glucose unit is linked, via oxygen, to C-2 of the fructose unit. A further difference is that the fructose unit is in the furanose form.

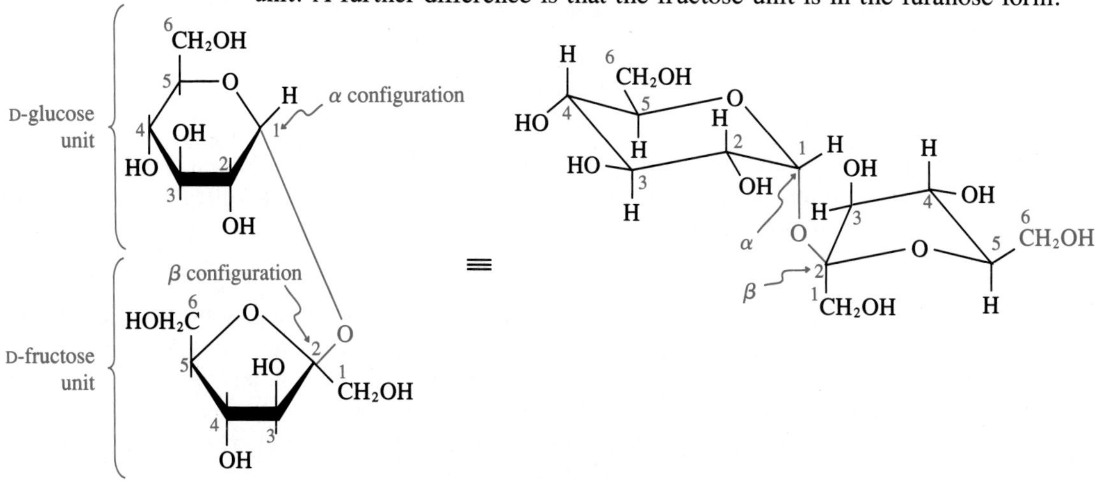

sucrose
α-D-glucopyranosyl-β-D-fructofuranoside
(or β-D-fructofuranosyl- α-D-glucopyranoside)

Since both anomeric carbons are linked in the glycosidic bond, neither monosaccharide unit has a hemiacetal group. Therefore, neither unit is in equilibrium with an acyclic form. Sucrose cannot mutarotate. And, because there is no free or potentially free aldehyde group, sucrose cannot reduce Tollens', Fehling's, or Benedict's reagent. Sucrose is therefore referred to as a *nonreducing sugar*, in contrast with the other disaccharides and monosaccharides we have discussed, which are reducing sugars.

PROBLEM 16.15 Although β-D-glucose is a reducing sugar, methyl β-D-glucopyranoside (eq. 16.13) is not. Explain.

Sucrose has an optical rotation of $[\alpha] = +66°$. When sucrose is hydrolyzed to an equimolar mixture of D-glucose and D-fructose, the optical rotation changes value and sign and becomes $[\alpha] = -20°$. This is because the equilibrium mixture of D-glucose anomers (α and β) has a rotation of $+52°$, but the mixture of fructose anomers has a strong negative rotation, $[\alpha] = -92°$. In the early days of carbohydrate chemistry, glucose was called **dextrose** (because it was dextrorotatory), and fructose was called **levulose** (because it was levorotatory). Because hydrolysis of sucrose inverts the sign of optical rotation (from + to −), enzymes that bring about sucrose hydrolysis are called **invertases,** and the resulting equimolar mixture of glucose and fructose is called **invert sugar.** A number of insects, including the honeybee, possess invertases. *Honey* is largely a mixture of D-glucose, D-fructose, and some unhydrolyzed sucrose. It also contains flavors from the particular flowers whose nectars are collected.

A Word About . . .

29. Sweetness and Sweeteners

Sweetness is literally a matter of taste. Although individuals vary greatly in their sensory perceptions, it is possible to make some quantitative comparisons of sweetness. For example, we can take some standard sugar solution (say 10% sucrose in water) and compare its sweetness with that of solutions containing other sugars or sweetening agents. If a 1% solution of some compound tastes as sweet as the 10% sucrose solution, we can say that the compound is 10 times sweeter than sucrose.

D-Fructose is the sweetest of the simple sugars—almost twice as sweet as sucrose. D-Glucose is almost as sweet as sucrose. On the other hand, sugars like lactose and galactose have less than 1% of the sweetness of sucrose.

Many synthetic sweeteners are known, perhaps the most familiar being **saccharin.** It was discovered in 1879 in the laboratory of Professor Ira Remsen at the Johns Hopkins University. Its structure has no relation whatever to that of the saccharides, but saccharin is about 300 times sweeter than sucrose. For most tastes, 0.5 grain (0.03 g) of saccharin is equivalent in sweetness to a heaping teaspoon (10 g) of sucrose. Saccha-

FIGURE 16.3
Synthesis of saccharin.

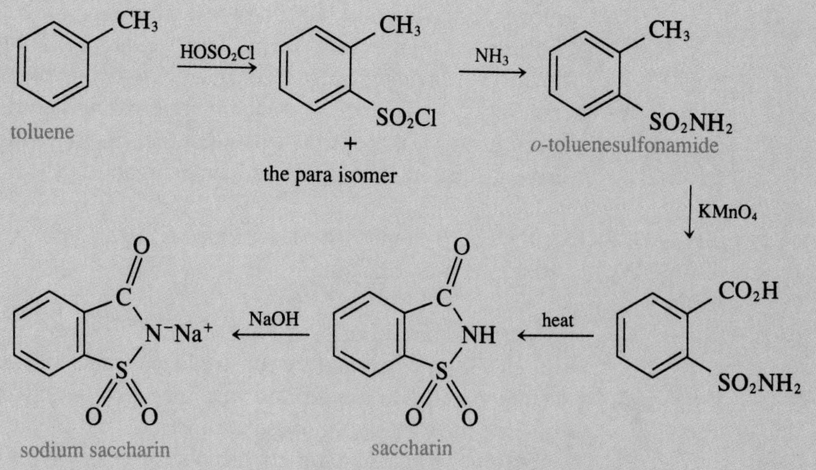

rin is made commercially from toluene, as shown in Figure 16.3.

Saccharin is very sweet yet has virtually no caloric content. It is useful as a sugar substitute for those who must restrict their sugar intake and also for those who wish to control their weight but still have a desire for sweets.

In 1981, **aspartame** became the first new sweetener to be approved by the U.S. Food and Drug Administration (FDA) in nearly 25 years. It is about 160 times sweeter than sucrose. Structurally, aspartame is the methyl ester of a dipeptide of two amino acids that occur naturally in proteins—aspartic acid and phenylalanine—and is sold under the trade name NutraSweet®. (Amino acids and peptides will be discussed in Chapter 17.)

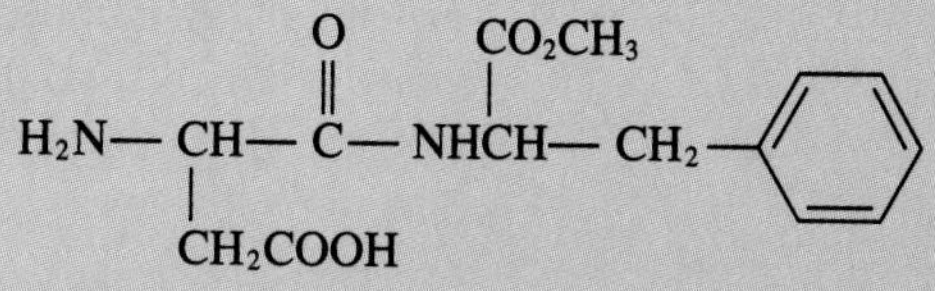

$$H_2N-\underset{\underset{CH_2COOH}{|}}{CH}-\overset{\overset{O}{||}}{C}-NH\overset{\overset{CO_2CH_3}{|}}{CH}-CH_2-C_6H_5$$

the methyl ester of
N-L-α-aspartyl-L-phenylalanine
(aspartame)

Aspartame has about the same caloric content as sucrose, but because of its intense sweetness, much less is used, so its energy value becomes insignificant.

16.14 Polysaccharides

Polysaccharides contain many linked monosaccharides and vary in chain length and molecular weight. Most polysaccharides give a single monosaccharide on complete hydrolysis. The monosaccharide units may be linked linearly, or the chains may be branched. In this section, we will describe a few of the more important polysaccharides.

16.14a Starch and Glycogen

Starch is the energy-storing carbohydrate of plants. It is a major component of cereals, potatoes, corn, and rice. It is the form in which glucose is stored by plants for later use.

Starch is made up of glucose units joined mainly by 1,4-α-glycosidic bonds, although the chains may have a number of branches attached through 1,6-α-glycosidic bonds. Partial hydrolysis of starch gives maltose, and complete hydrolysis gives only D-glucose.

Starch can be separated by various solution and precipitation techniques into two fractions: amylose and amylopectin. In **amylose,** which constitutes about 20% of starch, the glucose units (50 to 300) are in a continuous chain, with 1,4 linkages (Figure 16.4).

Amylopectin (Figure 16.5) is highly branched. Although each molecule may contain 300 to 5000 glucose units, chains with consecutive 1,4 links average only 25 to 30 units in length. These chains are connected at branch points by 1,6 linkages. Because of this highly branched structure, starch granules swell and eventually form colloidal systems in water.

Glycogen is the energy-storing carbohydrate of animals. Like starch, it is made of 1,4- and 1,6-linked glucose units. Glycogen has a higher molecular weight than starch (perhaps 100,000 glucose units). Its structure is even more branched than that of amylopectin, with a branch every 8 to 12 glucose units. Glycogen is produced from glucose that is absorbed from the intestines into the blood; transported to the liver, muscles, and elsewhere; and then polymerized enzymatically. Glycogen helps maintain the glucose balance in the body, by removing and storing excess glucose from ingested food and later supplying it to the blood when various cells need it for energy.

16.14b Cellulose **Cellulose** is an *unbranched* polymer of glucose joined by 1,4-β-glycosidic bonds. X-ray examination of cellulose shows that it consists of linear chains of cellobiose units, in which the ring oxygens alternate in "forward" and "backward" positions (Figure 16.6). These linear molecules,

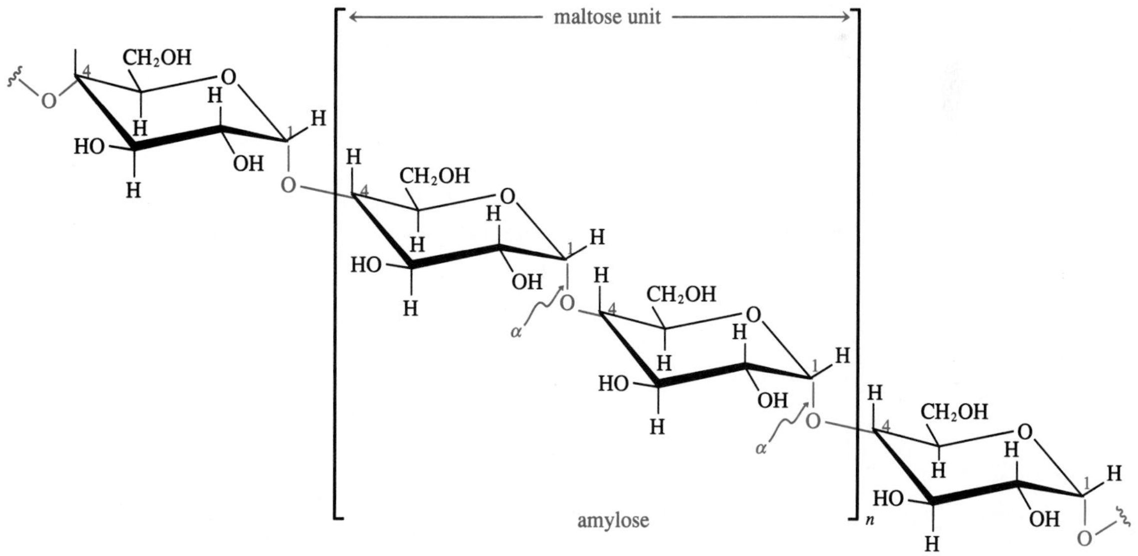

FIGURE 16.4 Structure of the amylose fraction of starch.

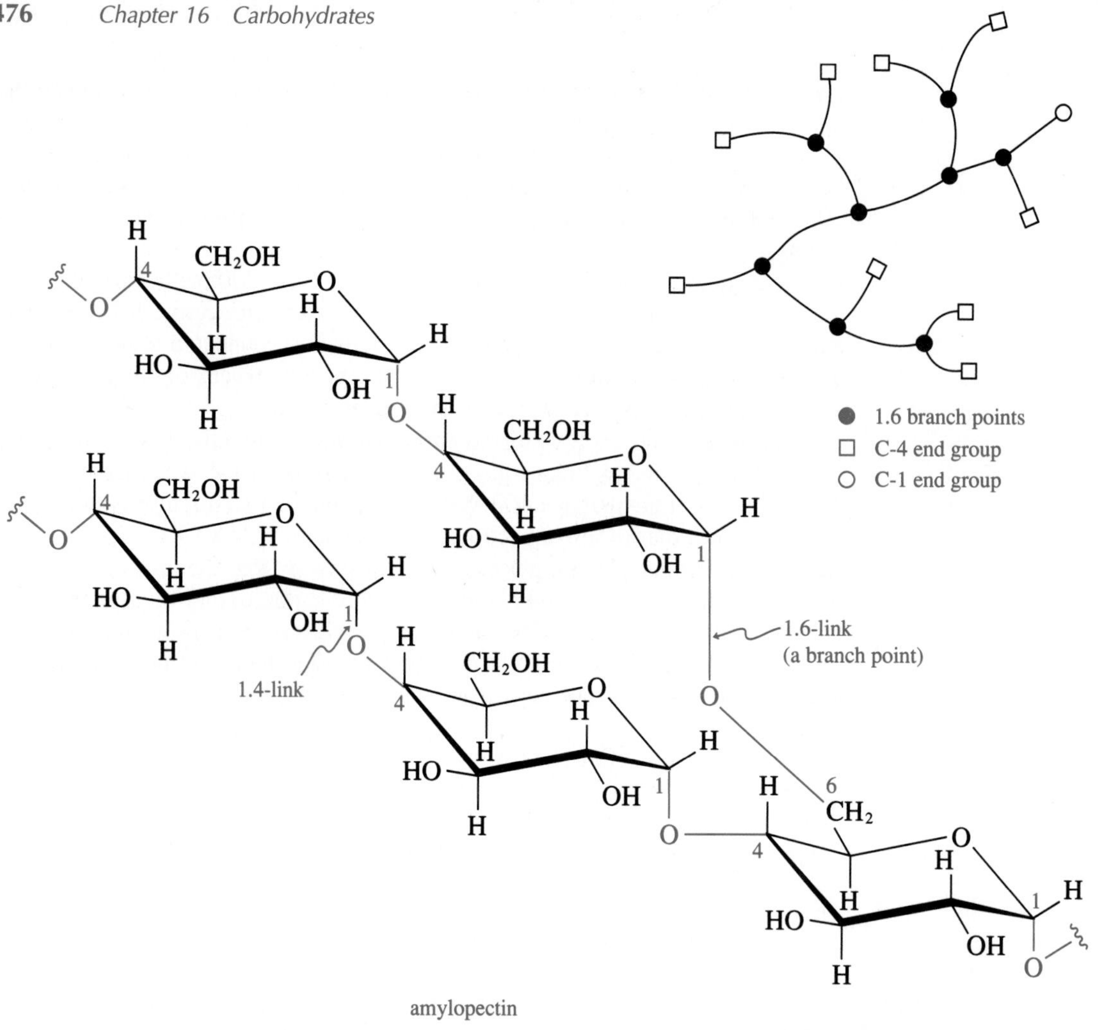

FIGURE 16.5 Structure of the amylopectin fraction of starch. (Adapted from Ferrier, R. J., and Collins, P. M., *Monosaccharide Chemistry*; Penguin Books, Ltd.; England. Used by permission.)

cellobiose unit

β

β

cellulose chain

FIGURE 16.6 Partial structure of a cellulose molecule showing the β linkages of each glucose unit.

containing an average of 5000 glucose units, aggregate to give fibrils bound together by hydrogen bonds between hydroxyls on adjacent chains. Cellulose fibers having considerable physical strength are built up from these fibrils, wound spirally in opposite directions around a central axis. Wood, cotton, hemp, linen, straw, and corncobs are mainly cellulose.

Although humans and other animals can digest starch and glycogen, they cannot digest cellulose. This is a truly striking example of the specificity of biochemical reactions. *The only chemical difference between starch and cellulose is the stereochemistry of the glucosidic link*—more precisely, the stereochemistry at C-1 of each glucose unit. The human digestive system contains enzymes that can catalyze the hydrolysis of α-glucosidic bonds, but it lacks the enzymes necessary to hydrolyze β-glucosidic bonds. Many bacteria, however, do contain β-glucosidases and can hydrolyze cellulose. Termites, for example, have such bacteria in their intestines and thrive on wood (cellulose) as their main food. Ruminants (cud-chewing animals such as cows) can digest grasses and other forms of cellulose because they harbor the necessary microorganisms in their rumen.

Cellulose is the raw material for several commercially important derivatives. Each glucose unit in cellulose contains three hydroxyl groups. These hydroxyl groups can be modified by the usual reagents that react with alcohols. For example, cellulose reacts with acetic anhydride to give **cellulose acetate.**

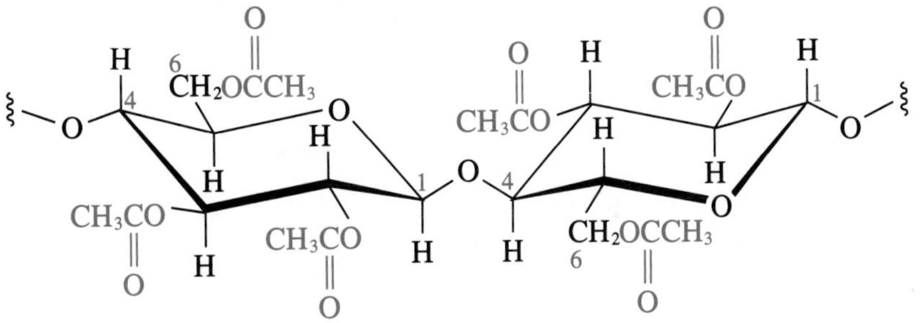

segment of a cellulose acetate molecule

Cellulose with about 97% of the hydroxyl groups acetylated is used to make acetate rayon.

Cellulose nitrate is another useful cellulose derivative. Like glycerol, cellulose can be converted with nitric acid to a nitrate ester (compare eq. 7.41). The number of hydroxyl groups nitrated per glucose unit determines the properties of the product. **Guncotton,** a highly nitrated cellulose, is an efficient explosive used in smokeless powders.

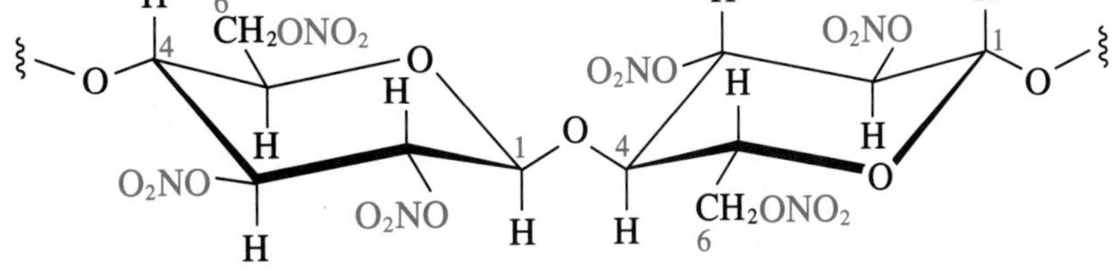

segment of a cellulose nitrate molecule

16.14c Other Polysaccharides **Chitin** is a nitrogen-containing polysaccharide that forms the shells of crustaceans and the exoskeletons of insects. It is similar to cellulose, except that the hydroxyl group at C-2 of each glucose unit is replaced by an acetylamino group, CH_3CONH-.

Pectins, which are obtained from fruits and berries, are polysaccharides used in making jellies. They are linear polymers of D-galacturonic acid, linked with 1,4-α-glycosidic bonds. D-Galacturonic acid has the same structure as D-galactose, except that the C-6 primary alcohol group is replaced by a carboxyl group.

Numerous other polysaccharides are known, such as gum arabic and other gums and mucilages, chondroitin sulfate (found in cartilage), the blood anticoagulant heparin (found in the liver and heart), and the dextrans (used as blood plasma substitutes).

Some saccharides have structures that differ somewhat from the usual polyhydroxyaldehyde or polyhydroxyketone pattern. In the final sections of this chapter, we will describe a few such modified saccharides that are important in nature.

16.15 *Sugar Phosphates*

Phosphate esters of monosaccharides are found in all living cells, where they are intermediates in carbohydrate metabolism. Some common **sugar phosphates** are the following:

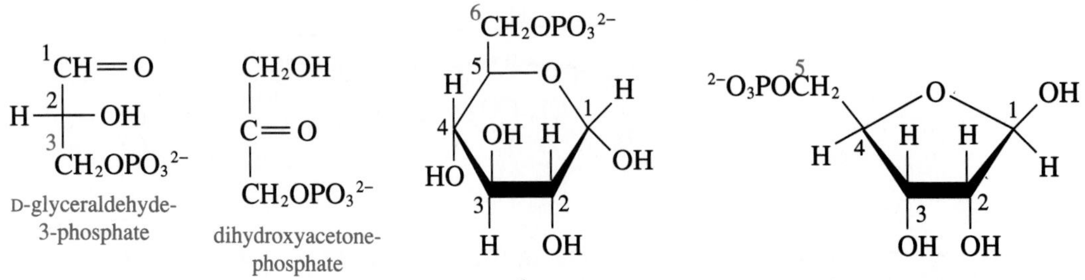

$^{1}CH{=}O$; H–^{2}C–OH; $^{3}CH_2OPO_3^{2-}$

D-glyceraldehyde-3-phosphate

CH_2OH–C=O–$CH_2OPO_3^{2-}$

dihydroxyacetone-phosphate

$^{6}CH_2OPO_3^{2-}$ (pyranose ring with H, O, H, OH, HO, OH, H, H, OH; carbons 1–5)

α-D-glucose-6-phosphate

$^{2-}O_3PO\overset{5}{C}H_2$ (furanose ring with O, OH, H, H, H, H, OH, OH; carbons 1–4)

β-D-ribose-5-phosphate

Phosphates of the five-carbon sugar ribose and its 2-deoxy analog are important in nucleic acid structures (DNA, RNA) and in other key biological compounds (Sec. 18.13).

16.16 *Deoxy Sugars*

In **deoxy sugars,** one or more of the hydroxyl groups is replaced by a hydrogen atom. The most important example is **2-deoxyribose,** the sugar component of DNA. It lacks the hydroxyl group at C-2 and occurs in DNA in the furanose form.

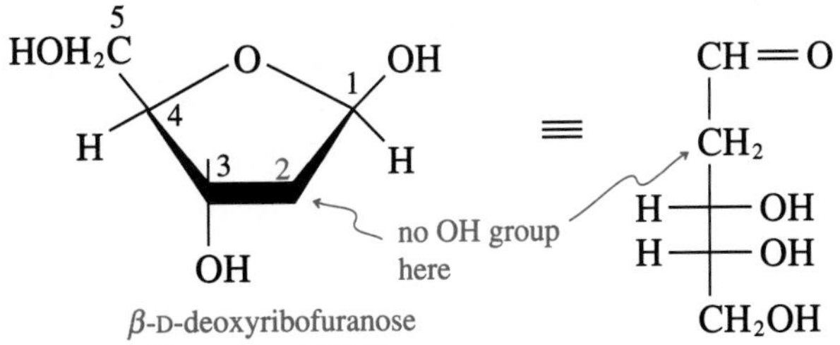

β-D-deoxyribofuranose
(the sugar of DNA)

16.17 Amino Sugars

In **amino sugars,** one of the sugar hydroxyl groups is replaced by an amino group. Usually the $—NH_2$ group is also acetylated. **D-Glucosamine** is one of the more abundant amino sugars.

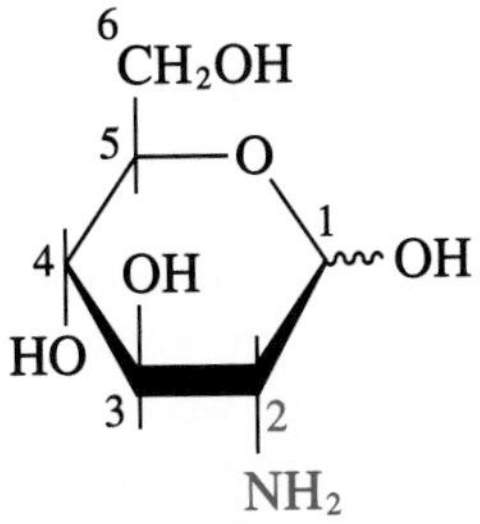

D-glucosamine
α mp 88°C
β [mp 110°C (decomposes)]

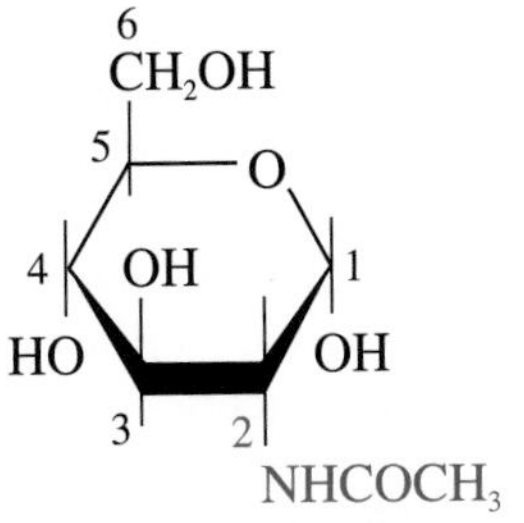

N-acetyl-α-D-glucosamine
[mp 211°C (decomposes)]

In its *N*-acetyl form, β-D-glucosamine is the monosaccharide unit of chitin, which forms the shells of lobsters, crabs, shrimp, and other shellfish.

16.18 Ascorbic Acid (Vitamin C)

L-Ascorbic acid (vitamin C) resembles a monosaccharide, but its structure has several unusual features. The compound has a five-membered unsaturated lactone ring with two hydroxyl groups attached to the doubly-bonded carbons. This **enediol** structure is relatively uncommon.

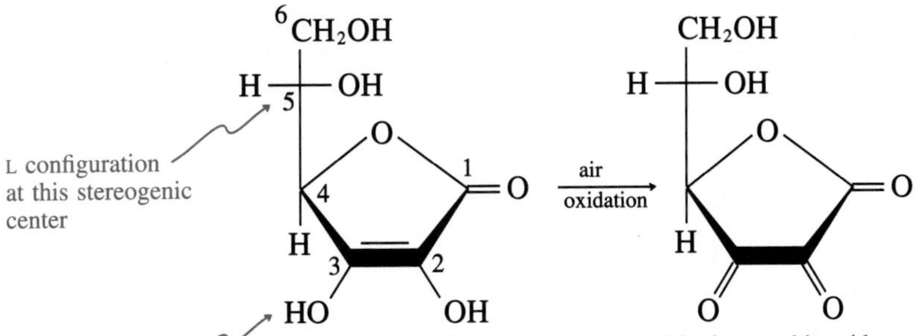

L-ascorbic acid
(vitamin C)
[mp 192°C (decomposes)]
pleasant, sharp-acid taste

dehydroascorbic acid

(16.15)

As a consequence of this structural feature, ascorbic acid is easily oxidized to dehydroascorbic acid. Both forms are biologically effective as a vitamin.

There is no carboxyl group in ascorbic acid, but it is nevertheless an acid with a pK_a of 4.17. The proton of the hydroxyl group at C-3 is acidic, because the anion resulting from its loss is resonance stabilized and similar to a carboxylate anion.

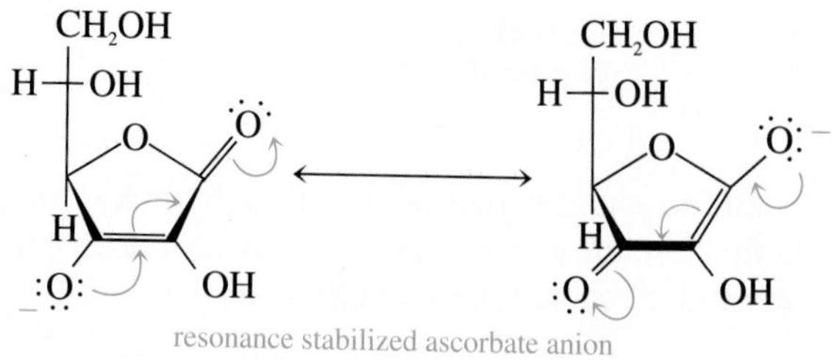

CH_2OH H–OH O O: H :O: OH ⟷ CH_2OH H–OH O :O:⁻ H :O OH

resonance stabilized ascorbate anion

Humans, monkeys, guinea pigs, and a few other vertebrates lack an enzyme that is essential for the biosynthesis of ascorbic acid from D-glucose. Hence ascorbic acid must be included in the diet of humans and these other species. Ascorbic acid is abundant in citrus fruits and tomatoes. Its lack in the diet causes scurvy, a disease that results in weak blood vessels, hemorrhaging, loosening of teeth, lack of ability to heal wounds, and eventual death. Ascorbic acid is needed for collagen synthesis (collagen is the structural protein of skin, connective tissue, tendon, cartilage, and bone). In the eighteenth century, British sailors were required to eat fresh limes (a vitamin C source) to prevent outbreaks of the dreaded scurvy; hence the nickname "limeys."

REACTION SUMMARY

1. Reactions of Monosaccharides

a. Mutarotation (Sec. 16.6)

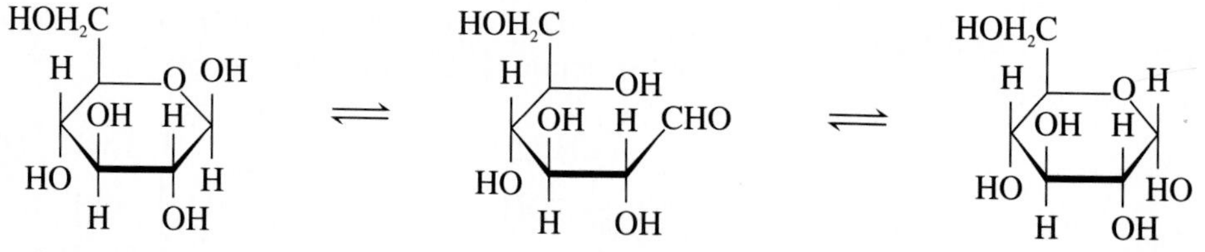

HOH_2C H O OH OH H HO H H OH ⇌ HOH_2C H OH OH H CHO HO H OH ⇌ HOH_2C H O H OH H HO HO H OH

β-anomer of glucose — acyclic form — α-anomer of glucose

b. Esterification (Sec 16.9)

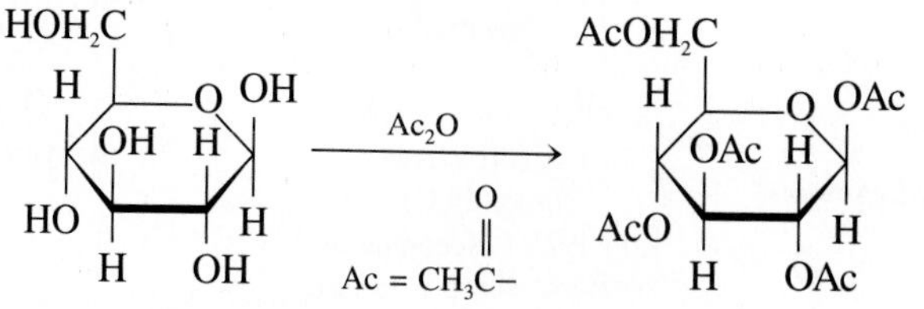

HOH_2C H O OH OH H HO H H OH $\xrightarrow{Ac_2O}$ $AcOH_2C$ H O OAc OAc H AcO H H OAc

$Ac = CH_3\overset{\overset{O}{\|}}{C}-$

c. Etherification (Sec. 16.9)

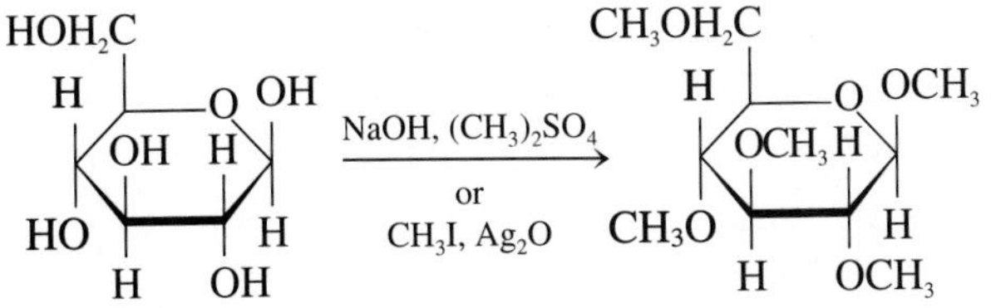

HOH$_2$C, H, O, OH, OH, H, HO, H, H, OH $\xrightarrow[\text{or } CH_3I,\ Ag_2O]{NaOH,\ (CH_3)_2SO_4}$ CH$_3$OH$_2$C, H, O, OCH$_3$, OCH$_3$, H, CH$_3$O, H, H, OCH$_3$

d. Reduction (Sec. 16.10)

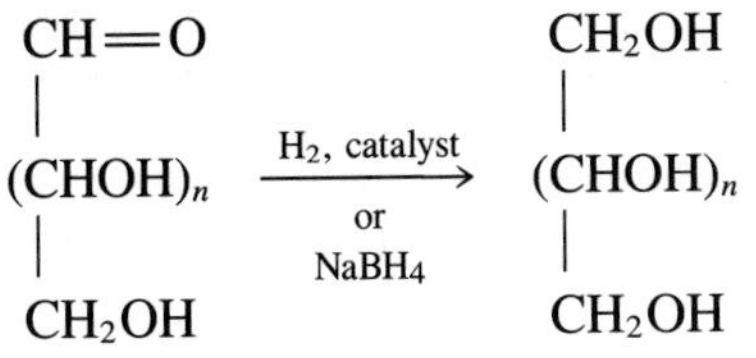

$$\begin{array}{c} CH{=}O \\ | \\ (CHOH)_n \\ | \\ CH_2OH \\ \text{aldose} \end{array} \xrightarrow[\text{or } NaBH_4]{H_2,\ \text{catalyst}} \begin{array}{c} CH_2OH \\ | \\ (CHOH)_n \\ | \\ CH_2OH \\ \text{alditol} \end{array}$$

e. Oxidation (Sec. 16.11)

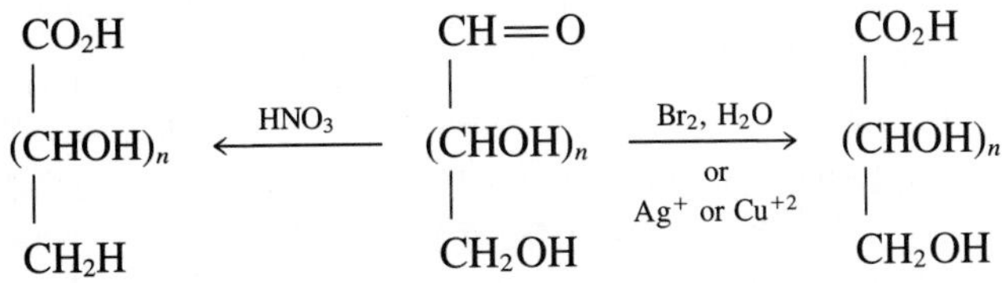

$$\begin{array}{c} CO_2H \\ | \\ (CHOH)_n \\ | \\ CH_2H \\ \text{aldaric acid} \end{array} \xleftarrow{HNO_3} \begin{array}{c} CH{=}O \\ | \\ (CHOH)_n \\ | \\ CH_2OH \\ \text{aldose} \end{array} \xrightarrow[\text{or } Ag^+ \text{ or } Cu^{+2}]{Br_2,\ H_2O} \begin{array}{c} CO_2H \\ | \\ (CHOH)_n \\ | \\ CH_2OH \\ \text{aldonic acid} \end{array}$$

f. Preparation and Hydrolysis of Glycosides (Sec. 16.12)

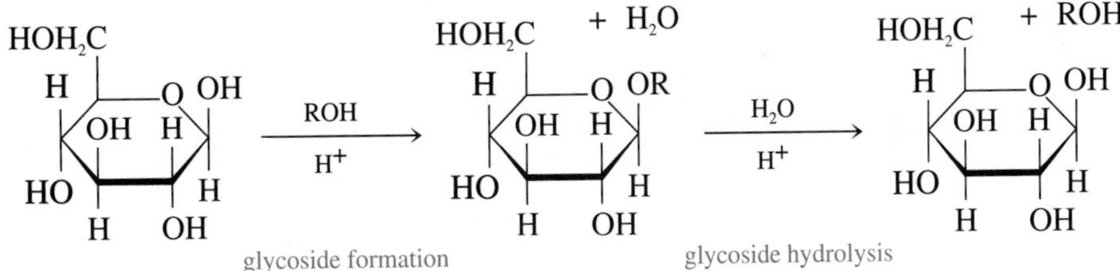

glycoside formation glycoside hydrolysis

2. Reactions of Polysaccharides

a. Hydrolysis (Sec. 16.2)

polysaccharide $\xrightarrow{H_3O^+}$ oligosaccharide $\xrightarrow{H_3O^+}$ monosaccharide

ADDITIONAL PROBLEMS

16.16. Define each of the following, and give the structural formula of one example.

a. aldohexose **b.** ketopentose **c.** monosaccharide
d. disaccharide **e.** polysaccharide **f.** furanose
g. pyranose **h.** glycoside **i.** anomeric carbon

16.17. Explain, using formulas, the difference between a D-sugar and an L-sugar.

16.18. Three of the four hydroxyl groups at the stereogenic centers in the Fischer projection of D-talose (Figure 16.1) are on the left, yet it is called a D-sugar. Explain.

16.19. What is the absolute configuration (*R* or *S*) of the stereogenic centers at C-2 and C-3 of D-erythrose? (Use Figure 16.1 for this and other problems if you have not yet memorized the structures of the monosaccharides.)

16.20. What is the absolute configuration (*R* or *S*) at each stereogenic center in the acyclic form of D-glucose? at the new stereogenic center in β-D-glucose?

16.21. What term would you use to describe the stereochemical relationship between D-gulose and D-idose?

16.22. Construct an array analogous to Figure 16.1 for the D-ketoses as high as the hexoses. Dihydroxyacetone should be at the head, in place of glyceraldehyde.

16.23. Using Figure 16.1 if necessary, write a Fischer projection formula and a Haworth projection formula for
a. methyl β-D-glucopyranoside. **b.** β-D-gulopyranose.
c. β-D-arabinofuranose. **d.** methyl β-L-glucopyranoside.

16.24. Draw the Fischer projection formula for
a. L-(−)-mannose. **b.** L-(+)-fructose.

16.25. At equilibrium in aqueous solution, D-ribose exists as a mixture containing 20% α-pyranose, 56% β-pyranose, 6% α-furanose, and 18% β-furanose forms. Draw Haworth formulas for each of these forms.

16.26. Write Fischer, Haworth, and conformational structures for β-D-allose.

16.27. The solubilities of α- and β-D-glucose in water at 25°C are 82 and 178 g/100 mL, respectively. Why are their solubilities not identical?

16.28. The specific rotations of pure α- and β-D-fructofuranose are $+21°$ and $-133°$, respectively. Solutions of each isomer mutarotate to an equilibrium specific rotation of $-92°$. Assuming that no other forms are present, calculate the equilibrium concentrations of the two forms.

16.29. D-Threose can exist in a furanose form but *not* in a pyranose form. Explain. Draw the β-furanose structure.

16.30. Draw the Fischer and Newman projection formulas for L-erythrose.

16.31. Starting with β-D-glucose and using acid (H^+) as a catalyst, write out all the steps in the mechanism for the mutarotation process. Use Haworth projections for the cyclic structures.

16.32. Oxidation of either D-erythrose or D-threose with nitric acid gives tartaric acid. In one case, the tartaric acid is optically active; in the other, it is optically inactive. How can these facts be used to assign stereochemical structures to erythrose and threose?

16.33. Write a structure for
a. D-galactonic acid **b.** D-galactaric acid.

16.34. Using complete structures, write out the reaction of D-galactose with
a. bromine water. **b.** nitric acid.
c. sodium borohydride. **d.** acetic anhydride.

16.35. Reduction of D-fructose with $NaBH_4$ gives a mixture of D-glucitol and D-mannitol. What does this result prove about the configurations of D-fructose, D-mannose, and D-glucose?

16.36. Although D-galactose contains five stereogenic centers, its oxidation with nitric acid gives an optically inactive dicarboxylic acid (called galactaric or mucic acid). What is the structure of this acid, and why is it optically inactive?

16.37. Write equations that clearly show the mechanism for the acid-catalyzed hydrolysis of

a. maltose to glucose.
b. lactose to galactose and glucose.
c. sucrose to fructose and glucose.

16.38. Write equations for the reaction of maltose with

a. methanol and H^+. **b.** Tollens' reagent.
c. bromine water. **d.** acetic anhydride.

16.39. Trehalose is a disaccharide that is the main carbohydrate component in the blood of insects. Its structure is

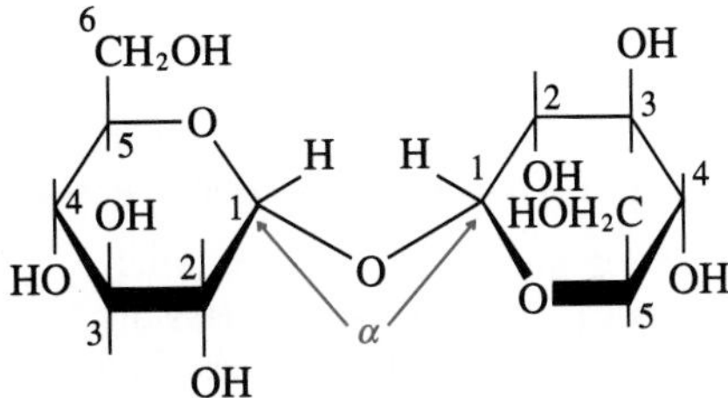

trehalose

a. What are its hydrolysis products?
b. Will trehalose give a positive or a negative test with Fehling's reagent? Explain.

16.40. Write an equation for the acid-catalyzed hydrolysis of salicin (Sec. 16.12). Notice that one of the products is structurally similar to aspirin (eq. 10.38), perhaps accounting for salicin's fever-reducing property.

16.41. Lactose exists in α and β forms, with specific rotations of +92.6° and +34°, respectively.

a. Draw their structures.
b. Solutions of each isomer mutarotate to an equilibrium value of +52°. What is the percentage of each isomer at equilibrium?

16.42. Write a balanced equation for the reaction of D-(+)-glucose (use either an acyclic or a cyclic structure, whichever seems more appropriate) with each of the following:

a. acetic anhydride (excess) **b.** bromine water
c. hydrogen, catalyst **d.** hydroxylamine (to form an oxime)
e. methanol, H^+ **f.** hydrogen cyanide (to form a cyanohydrin)
g. Fehling's reagent

16.43. Explain why sucrose is a nonreducing sugar but maltose is a reducing sugar.

16.44. Explain what type of reaction takes place in each of the five steps in the synthesis of sodium saccharin (Figure 16.3).

16.45. The last step in Figure 16.3 shows that saccharin is acidic and readily forms a sodium salt. How do you account for its acidity?

16.46. Using the descriptions in Sec. 16.14c, write a formula for
a. chitin. **b.** pectin.

16.47. L-Fucose is a component of bacterial cell walls. It is also called 6-deoxy-L-galactose. Using the description in Sec. 16.16, write its Fischer projection formula.

16.48. Daunosamine is an amino sugar that forms part of the structure of doxorubicin (adriamycin), a tetracycline anticancer agent. From its Haworth projection shown below, draw its conformational (chair) structure.

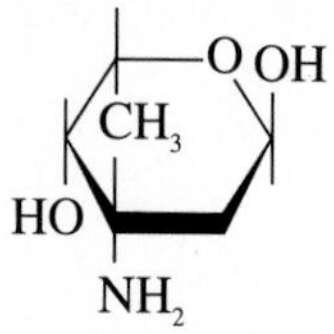

O OH
CH_3
HO
NH_2
daunosamine

Is daunosamine a D- or an L-sugar?

16.49. Write the main contributors to the resonance hybrid anion formed when ascorbic acid acts as an acid (loses the proton from the —OH group on C-3).

16.50. Hemicelluloses are noncellulose materials produced by plants and found in straw, wood, and other fibrous tissues. Xylans are the most abundant hemicelluloses. They consist of 1,4-β-linked D-xylopyranoses. Draw the structure for the repeating unit in xylans.

16.51. Inositols are hexahydroxycyclohexanes, with one hydroxyl group on each carbon atom of the ring. Although not strictly carbohydrates, they are obviously similar to pyranose sugars and do occur in nature. There are nine possible stereoisomers. Draw Haworth formulas for all possibilities (all are known), and tell which are chiral.

16.52. Sucralose is a chlorinated derivative of sucrose that is about 600 times sweeter than its parent. It is hydrolyzed only slowly at the mildly acidic pH of soft drinks. Draw the structure of the hydrolysis products.

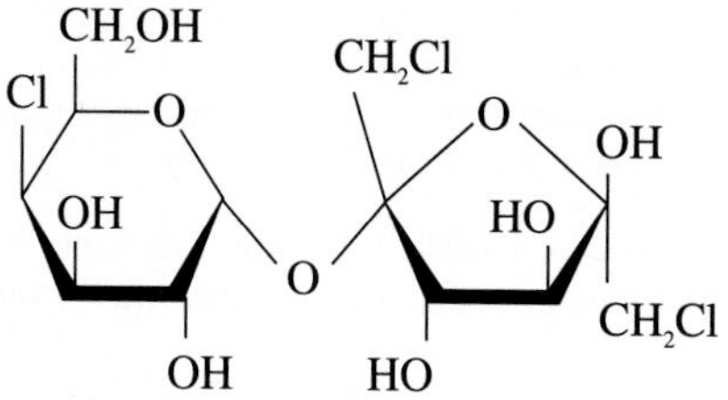

CH_2OH
CH_2Cl
Cl
O
O
OH
OH
HO
O
CH_2Cl
OH
HO
sucralose

16.53. Aspartame has the *S* configuration at each of its two stereogenic centers (see Word About #29). Draw a structure of aspartame that shows its stereochemistry.

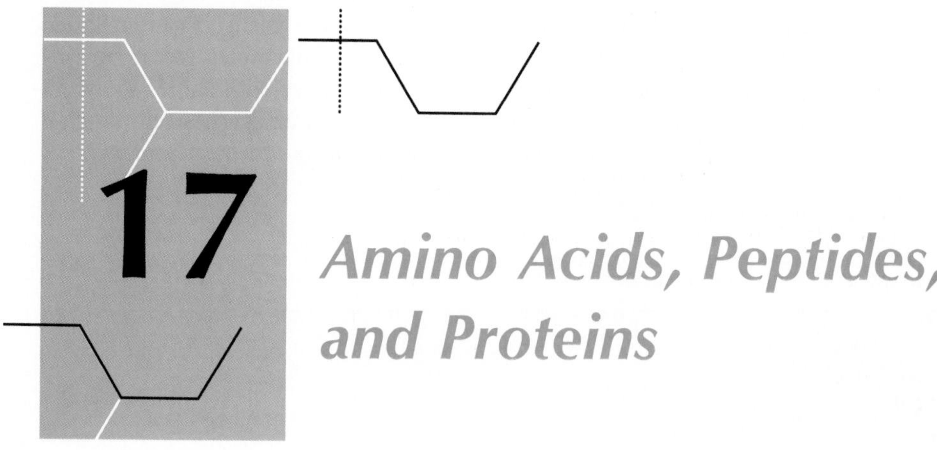

17 Amino Acids, Peptides, and Proteins

17.1 Introduction

Proteins are natural polymers composed of **amino acid** units joined one to another by amide (or peptide) bonds. Proteins are essential to the structure, function, and reproduction of living matter. In this chapter, we will first discuss the structure and properties of amino acids. We will next describe the properties of **peptides,** which consist of just a few amino acids linked together, and, finally, the structures of proteins.

17.2 Naturally Occurring Amino Acids

The amino acids obtained from protein hydrolysis are α-amino acids. That is, the amino group is on the α-carbon atom, the one adjacent to the carboxyl group.

$$\underset{\substack{|\\ NH_2}}{\overset{\alpha}{R-CH}}-C\begin{smallmatrix}\nearrow O\\ \searrow OH\end{smallmatrix}$$

an α-amino acid

With the exception of glycine, where R = H, α-amino acids have a stereogenic center at the α-carbon. All except glycine are therefore optically active. They have the L configuration relative to glyceraldehyde (Figure 17.1). Note that the Fischer convention, used with carbohydrates, is also applied to amino acids.

Table 17.1 lists the 20 α-amino acids commonly found in proteins. The amino acids are known by common names. Each also has a three-letter abbreviation based on this name, which is used when writing the formulas of peptides and proteins. The amino acids in Table 17.1 are grouped to emphasize structural similarities. Of the 20 amino acids listed in the table, 12 can be synthesized in the body from other foods. The other 8, those with names shown in color and referred to as **essential amino acids,** cannot be synthesized by adult humans and therefore must be included in the diet in the form of proteins.

Table 17.1 Names and formulas of the common amino acids

Name	*Abbreviation (Isoelectric point)*	*Formula*	*R*
A. One amino group and one carboxyl group			
1. glycine	Gly (6.0)	$H-CH-CO_2H$ | NH_2	
2. alanine	Ala (6.0)	$CH_3-CH-CO_2H$ | NH_2	
3. valine	Val (6.0)	$CH_3CH-CH-CO_2H$ | | CH_3 NH_2	R is hydrogen or an alkyl group.
4. leucine	Leu (6.0)	$CH_3CHCH_2-CH-CO_2H$ | | CH_3 NH_2	
5. isoleucine	Ile (6.0)	$CH_3CH_2CH-CH-CO_2H$ | | CH_3 NH_2	
6. serine	Ser (5.7)	$CH_2-CH-CO_2H$ | | OH NH_2	
7. threonine	Thr (5.6)	$CH_3CH-CH-CO_2H$ | | OH NH_2	R contains an alcohol function.
8. cysteine	Cys (5.0)	$CH_2-CH-CO_2H$ | | SH NH_2	
9. methionine	Met (5.7)	$CH_3S-CH_2CH_2-CH-CO_2H$ | NH_2	R contains sulfur.
10. proline	Pro (6.3)	$CH_2-CH-CO_2H$ | | CH_2 NH \\ / CH_2	The amino group is secondary and part of a ring.

TABLE 17.1 *(continued)*

Name	Abbreviation (Isoelectric point)	Formula	R
11. phenylalanine	Phe (5.5)	$C_6H_5-CH_2-CH(NH_2)-CO_2H$	
12. tyrosine	Tyr (5.7)	$HO-C_6H_4-CH_2-CH(NH_2)-CO_2H$	One hydrogen in alanine is replaced by an aromatic or heteroaromatic (indole) ring.
13. tryptophan	Trp (5.9)	$(\text{indol-3-yl})-CH_2-CH(NH_2)-CO_2H$	
B. One amino group and two carboxyl groups			
14. aspartic acid	Asp (3.0)	$HOOC-CH_2-CH(NH_2)-CO_2H$	
15. glutamic acid	Glu (3.2)	$HOOC-CH_2CH_2-CH(NH_2)-CO_2H$	
16. asparagine	Asn (5.4)	$H_2N-C(=O)-CH_2-CH(NH_2)-CO_2H$	
17. glutamine	Gln (5.7)	$H_2N-C(=O)-CH_2CH_2-CH(NH_2)-COOH$	
C. One carboxyl group and two basic groups			
18. lysine	Lys (9.7)	$CH_2(NH_2)CH_2CH_2CH_2-CH(NH_2)-CO_2H$	
19. arginine	Arg (10.8)	$(NH_2)(NH{=})C-NH-CH_2CH_2CH_2-CH(NH_2)-CO_2H$	The second basic group is a primary amine, a guanidine, or an imidazole.
20. histidine	His (7.6)	$CH{=}C(-CH_2-CH(NH_2)-CO_2H)-NH-CH{=}N-$ (imidazole ring)	

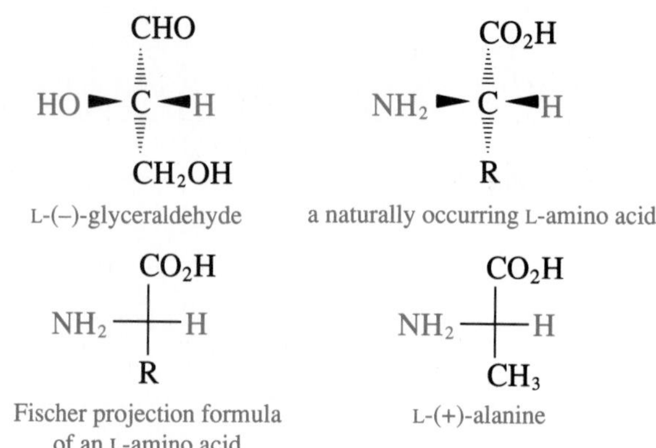

FIGURE 17.1 Naturally occurring α-amino acids have the L configuration.

A Word About . . .

30. Amino Acid Dating

The question of age is one of the first that archaeologists seek to answer when they find artifacts or skeletons at a "dig." Are the bones or the pot shards ancient or modern? If they are ancient, *how* ancient? Knowing the age helps to answer other questions, such as how the people lived, with what other groups they traded or had contact, whom they followed and who followed them, and so on.

Chemists have helped archaeologists with this problem. One of the best-known methods is carbon-14, or radioactive, dating, first proposed in 1947 by Willard F. Libby (Nobel Prize winner in 1960). The isotope ^{14}C decays with a half-life of 5730 years, which is sufficiently long for a steady-state equilibrium concentration to be established in the biosphere. A tiny but constant fraction (about $1.2 \times 10^{-10}\%$) of the carbon in live plants and animals is ^{14}C. After death, when ^{14}C is no longer taken in from the environment (as food, carbon dioxide, and so on), its concentration decreases. Knowing the decay rate of ^{14}C and comparing the ^{14}C content of the ancient material with that of modern allows the age of the ancient material to be calculated. The practical limit of the method is about 10 half-lives of ^{14}C, or about 50,000 years.

The extent of racemization of **amino acids** found in fossil bones, shells, and teeth offers another method of dating ancient material. In living systems, amino acids have the L configuration and are optically pure. Once death occurs, however, the biochemical reactions that prevent equilibration of the L and D forms are terminated, and gradual thermal equilibration of the two forms begins. This reaction can be used for dating because the amount of racemization is a function of the material's age.

$$\underset{\text{L form}}{NH_2-\overset{CO_2H}{\underset{R}{C}}-H} \rightleftharpoons \underset{\text{D form}}{H-\overset{CO_2H}{\underset{R}{C}}-NH_2}$$

Racemization rates differ for different amino acids. For example, the half-life at 25°C and pH 7 for aspartic acid is about 3000 years; for alanine, it is about 12,000 years. The racemization rate also depends on temperature. For example, the half-life at 0°C for aspartic acid increases to about 430,000 years. So it is necessary, for accurate dating, to know the temperature at which the material was stored. Fortunately, the temperature below ground at a given depth and climate often remains constant over long periods and can be estimated fairly accurately.

Accuracy can be improved by using a combination of dating methods or by calibrating one method with another. One advantage of amino acid dating is that the sample size can be very much smaller than is needed for carbon-14 dating. Also, a range of time spans can be covered by using different amino acids. Amino acid dating can be extended back in time well beyond the limits of ^{14}C dating, even to ice ages 100 to 400 thousand years ago!

17.3 The Acid–Base Properties of Amino Acids

The carboxylic acid and amine functional groups are *simultaneously* present in amino acids, and we might ask whether they are mutually compatible, since one group is acidic and the other is basic. Although we have represented the amino acids in Table 17.1 as having amino and carboxyl groups, these structures are oversimplified.

Amino acids with one amino group and one carboxyl group are better represented by a **dipolar ion structure.***

$$R-\underset{^+NH_3}{\underset{|}{CH}}-\overset{O}{\overset{\|}{C}}-O^-$$

dipolar structure of an α-amino acid

The amino group is protonated and present as an ammonium ion, whereas the carboxyl group has lost its proton and is present as a carboxylate anion. This dipolar structure is consistent with the saltlike properties of amino acids, which have rather high melting points (even the simplest, glycine, melts at 233°C) and relatively low solubilities in organic solvents.

Amino acids are amphoteric. They can behave as acids and donate a proton to a strong base, or they can behave as bases and accept a proton from a

---

* Such structures are sometimes called *zwitterions* (from a German word for hybrid ions).

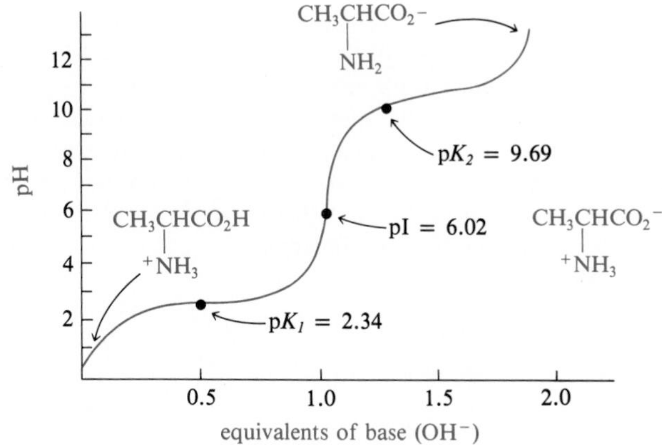

FIGURE 17.2 Titration curve for alanine, showing how its structure varies with pH.

strong acid. These behaviors are expressed in the following equilibria for an amino acid with one amino and one carboxyl group:

$$\underset{\substack{\text{amino acid}\\\text{at low pH}\\\text{(acid)}}}{\underset{|\atop {}^+NH_3}{RCHCO_2H}} \underset{H^+}{\overset{OH^-}{\rightleftharpoons}} \underset{\substack{\text{dipolar ion}\\\text{form}\\\text{(neutral)}}}{\underset{|\atop {}^+NH_3}{RCHCO_2^-}} \underset{H^+}{\overset{OH^-}{\rightleftharpoons}} \underset{\substack{\text{amino acid}\\\text{at high pH}\\\text{(base)}}}{\underset{|\atop NH_2}{RCHCO_2^-}} \quad (17.1)$$

Figure 17.2 shows a titration curve for alanine, a typical amino acid of this kind. At low pH (acidic solution), the amino acid is in the form of a substituted ammonium ion. At high pH (basic solution), it is present as a substituted carboxylate ion. At some intermediate pH (for alanine, pH 6.02), the amino acid is present as the dipolar ion. A simple rule to remember for any acidic site is that *if the pH of the solution is* less *than the* pK_a*, the proton is on; if the pH of the solution is* greater *than the* pK_a*, the proton is off.*

EXAMPLE 17.1 Starting with alanine hydrochloride (its structure at low pH in hydrochloric acid is shown in the lower left corner of the curve in Figure 17.2), write equations for its reaction with one equivalent of sodium hydroxide and then with a second equivalent of sodium hydroxide.

Solution

$$\underset{\text{ammonium salt}}{\underset{|\atop {}^+NH_3\ Cl^-}{CH_3CHCO_2H}} + Na^+OH^- \longrightarrow \underset{\text{dipolar ion}}{\underset{|\atop {}^+NH_3}{CH_3CHCO_2^-}} + Na^+Cl^- + H_2O \quad (17.2)$$

$$\underset{\text{dipolar ion}}{\underset{|\atop {}^+NH_3}{CH_3CHCO_2^-}} + Na^+OH^- \longrightarrow \underset{\text{carboxylate salt}}{\underset{|\atop NH_2}{CH_3CHCO_2^-Na^+}} + H_2O \quad (17.3)$$

The first equivalent of base removes a proton from the carboxyl group to give the dipolar ion, and the second equivalent of base removes a proton from the ammonium ion to give the sodium carboxylate.

PROBLEM 17.1 Starting with the sodium carboxylate salt of alanine, write equations for its reaction with one equivalent of hydrochloric acid and then with a second equivalent, and explain what each equivalent of acid does.

PROBLEM 17.2 Which group in the ammonium salt form of alanine is more acidic, the $-\overset{+}{N}H_3$ group or the $-CO_2H$ group?

PROBLEM 17.3 Which group in the carboxylate salt form of alanine is more basic, the $-NH_2$ group or the $-CO_2^-$ group?

Note from Figure 17.2 and from eq. 17.1 that the charge on an amino acid changes as the pH changes. At low pH, for example, the sign on alanine is positive, at high pH it is negative, and near neutrality the ion is dipolar. If placed in an electric field, the amino acid will therefore migrate toward the cathode (negative electrode) at low pH and toward the anode (positive electrode) at high pH (Figure 17.3). At some intermediate pH, called the **isoelectric point (pI),** the amino acid will be dipolar and have a net charge of zero. It will be unable to move toward either electrode. The isoelectric points of the various amino acids are listed in Table 17.1.

EXAMPLE 17.2 Write the structure of leucine

a. at the pI. b. at high pH. c. at low pH.

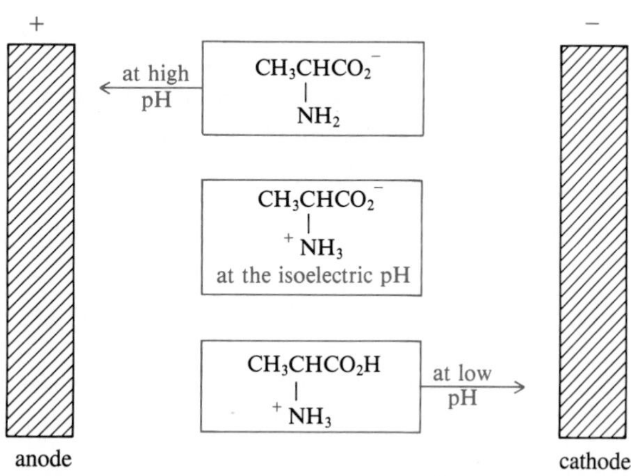

FIGURE 17.3 The migration of an amino acid (such as alanine) in an electric field depends on pH.

Solution

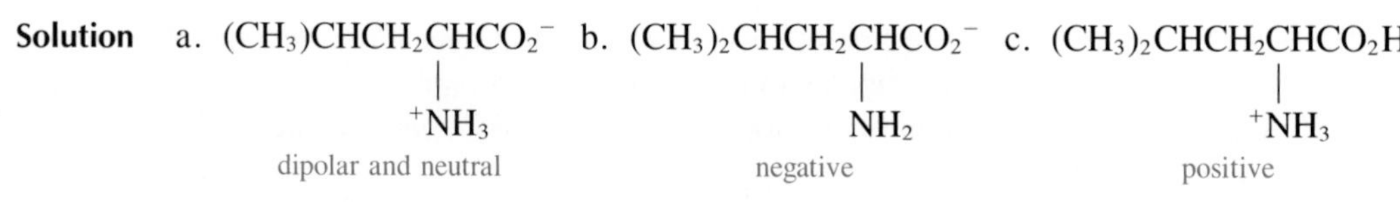

PROBLEM 17.4 Write the structure for the predominant form of each of the following amino acids at the indicated pH. If placed in an electric field, toward which electrode (+ or −) will each amino acid migrate?

a. methionine at its pI b. serine at low pH
c. phenylalanine at high pH

In general, amino acids with one amino group and one carboxyl group, and no other acidic or basic groups in their structure, have two pK_a values: one around 2 to 3 for proton loss from the carboxyl group and the other around 9 to 10 for proton loss from the ammonium ion. The isoelectric point is about halfway between the two pK_a values, near pH 6.

R is neutral

$$\underset{\underset{^+NH_3}{|}}{RCHCO_2H} \underset{}{\overset{pK_a = 2\text{–}3}{\rightleftharpoons}} \underset{\underset{^+NH_3}{|}}{RCHCO_2^-} \overset{pK_a = 9\text{–}10}{\rightleftharpoons} \underset{\underset{NH_2}{|}}{RCHCO_2^-} \tag{17.4}$$

net charge: low pH +1 → 0 → high pH −1

The situation is more complex with amino acids containing two acidic or two basic groups.

17.4 *The Acid-Base Properties of Amino Acids with More Than One Acidic or Basic Group*

Aspartic and glutamic acids (numbers 14 and 15 in Table 17.1) have two carboxyl groups and one amino group. In strong acid (low pH) all three of these groups are in their acidic form (protonated). As the pH is raised and the solution becomes more basic, each group in succession gives up a proton. The equilibria are shown for **aspartic acid,** with the three pK_a values over the equilibrium arrows:

$$\underset{\underset{^+NH_3}{|}}{HO_2CCH_2CHCO_2H} \overset{pK_a = 2.09}{\rightleftharpoons} \underset{\underset{^+NH_3}{|}}{HO_2CCH_2CHCO_2^-} \overset{pK_a = 3.86}{\rightleftharpoons} \underset{\underset{^+NH_3}{|}}{^-O_2CCH_2CHCO_2^-} \overset{pK_a = 9.82}{\rightleftharpoons} \underset{\underset{NH_2}{|}}{^-O_2CCH_2CHCO_2^-} \tag{17.5}$$

low pH → high pH

net charge: +1, 0, −1, −2

The isoelectric point for aspartic acid, the pH at which it is mainly in the neutral dipolar form, is 2.87 (in general, the pI is close to the average of the two pK_a's on either side of the neutral, dipolar species).

EXAMPLE 17.3 Which carboxyl group is the stronger acid in the most acidic form of aspartic acid?

Solution As shown at the extreme left of eq. 17.5, the first proton to be removed from the most acidic form of aspartic acid is the proton on the carboxyl group nearest the $^+NH_3$ substituent. The $^+NH_3$ group is electron-withdrawing due to its positive charge and enhances the acidity of the carboxyl group closest to it.

PROBLEM 17.5 Use eq. 17.5 to tell which is the least acidic group in aspartic acid and why.

The situation differs for amino acids with two basic groups and only one carboxyl group (numbers 18, 19, and 20 in Table 17.1). With **lysine,** for example, the equilibria are

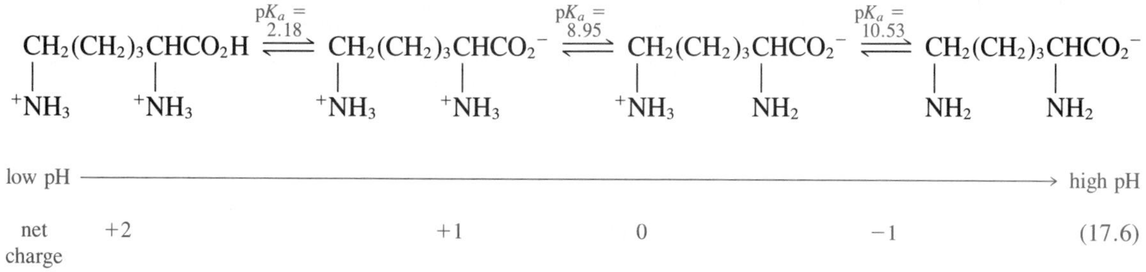

$$\underset{^+NH_3\qquad\qquad ^+NH_3}{CH_2(CH_2)_3CHCO_2H} \underset{}{\overset{pK_a = 2.18}{\rightleftharpoons}} \underset{^+NH_3\qquad\qquad ^+NH_3}{CH_2(CH_2)_3CHCO_2^-} \overset{pK_a = 8.95}{\rightleftharpoons} \underset{^+NH_3\qquad\qquad NH_2}{CH_2(CH_2)_3CHCO_2^-} \overset{pK_a = 10.53}{\rightleftharpoons} \underset{NH_2\qquad\qquad NH_2}{CH_2(CH_2)_3CHCO_2^-}$$

low pH ⟶ high pH

net charge +2 +1 0 −1 (17.6)

The pI for lysine comes in the basic region, at 9.74.

The second basic groups in arginine and histidine are not simple amino groups. They are a **guanidine** group and an **imidazole** ring, respectively, shown in color. The most protonated forms of these two amino acids are

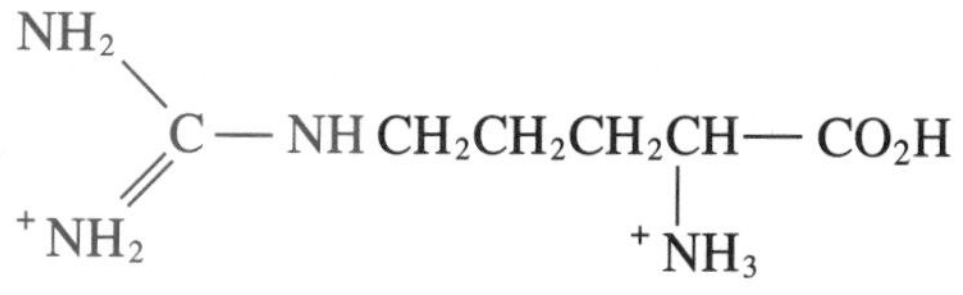

$$\underset{^+NH_2}{\overset{NH_2}{}}\!\!>C-NH\,CH_2CH_2CH_2\underset{^+NH_3}{CH}-CO_2H$$

arginine at pH 1

$$\text{(imidazolium ring: } \overset{+}{H}N-CH=C,\ CH=\ ,\ NH\text{)}\; C-CH_2\underset{^+NH_3}{CH}CO_2H$$

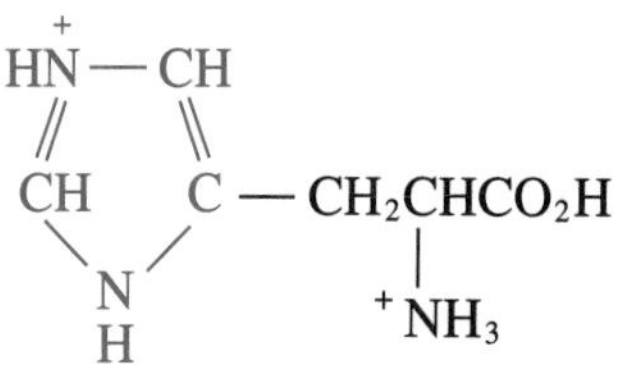

histidine at pH 1

PROBLEM 17.6 Arginine shows three pK_a's: at 2.17 (the —COOH group), at 9.04 (the $-\overset{+}{N}H_3$ group), and at 12.48 (the guanidinium ion). Write equilibria (similar to eq. 17.6) for its dissociation. At approximately what pH will the isoelectric point come, and what is the structure of the dipolar ion?

Table 17.2 summarizes the approximate pK_a values and isoelectric points for the three types of amino acids.

TABLE 17.2 Approximate acidity constants and isoelectric points (pI) for the three types of amino acids

Type		pK_a 1	pK_a 2	pK_a 3	pI
1 acidic and 1 basic group		2.3	9.4	—	6.0
2 acidic and 1 basic group		2.2	4.1	9.8	3.0
1 acidic and 2 basic groups	(Lys, Arg)	2.2	9.0	11.5	10.0
	(His)	1.8	6.0	9.2	7.6

17.5 Electrophoresis

As seen in eqs. 17.4 through 17.6, the charge on an amino acid depends on the pH of the solution. **Electrophoresis,** an important method for separating amino acids and proteins, takes advantage of these charge differences. It is based on the differential rates and directions of migration of amino acids or proteins in an electric field at a controlled pH.

EXAMPLE 17.4 Predict the direction of migration (toward the positive or negative electrode) of alanine in an electrophoresis apparatus at pH 5. Do the same for aspartic acid.

Solution A pH of 5 is *less* than the pI of alanine (~6). Therefore, the dipolar ions will be protonated (positive) and migrate toward the negative electrode. But pH 5 is *greater* than the pI of aspartic acid (~3). Therefore aspartic acid will exist mainly as the −1 ion (eq. 17.5) and migrate toward the positive electrode. A mixture of the two amino acids could therefore easily be separated in this way.

PROBLEM 17.7 Predict the direction of migration in an electrophoresis apparatus (toward the positive or negative electrode) of each component of the following amino acid mixtures:

a. glycine and lysine at pH 7
b. phenylalanine, leucine, and proline at pH 6

17.6 Reactions of Amino Acids

In addition to their acidic and basic behavior, amino acids undergo other reactions typical of carboxylic acids or amines. For example, the carboxyl group can be esterified (see Sec. 10.11).

$$\underset{\substack{|\\ {}^{+}NH_3}}{R-CH}-CO_2^- + R'OH + H^+ \xrightarrow{\text{heat}} \underset{\substack{|\\ {}^{+}NH_3}}{R-CH}-CO_2R' + H_2O \quad (17.7)$$

The amino group can be acylated to an amide. (see Sec. 11.11).

$$\underset{\underset{{}^{+}NH_3}{|}}{R-CH}-CO_2^- + R'-\overset{\overset{O}{\|}}{C}-Cl \xrightarrow{OH^-} \underset{\underset{\underset{\underset{O}{\|}}{R'C-NH}}{|}}{R-CH}-CO_2^- + H_2O + Cl^- \qquad (17.8)$$

These types of reactions are useful in temporarily modifying or protecting either of the two functional groups, especially during the controlled linking of amino acids to form peptides or proteins.

PROBLEM 17.8 Using eqs. 17.7 and 17.8 as models, write equations for the following reactions:

a. glutamic acid + CH_3OH + HCl $\longrightarrow$

b. proline + benzoyl chloride + NaOH $\longrightarrow$

c. phenylalanine + acetic anhydride $\xrightarrow{\text{heat}}$

17.7 The Ninhydrin Reaction

Ninhydrin is a useful reagent for detecting amino acids and determining the concentrations of their solutions. It is the hydrate of a cyclic triketone, and when it reacts with an amino acid, a violet dye is produced. The overall reaction, whose mechanism is complex and need not concern us in detail here, is as follows:

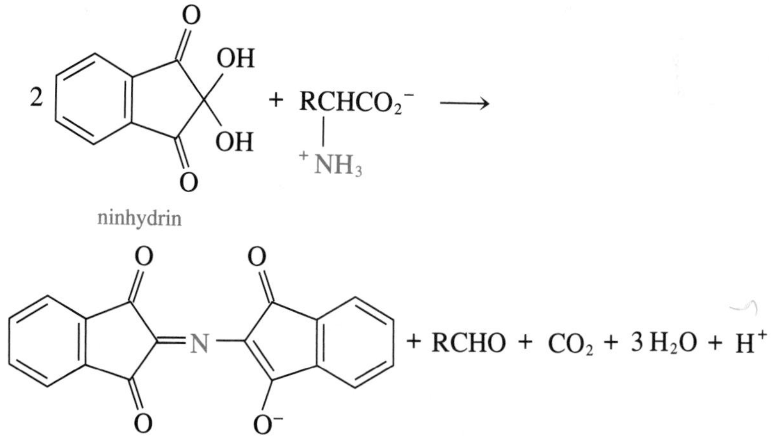

$$2\ \text{ninhydrin} + \underset{\underset{{}^{+}NH_3}{|}}{RCHCO_2^-} \longrightarrow \text{violet anion} + RCHO + CO_2 + 3\,H_2O + H^+ \qquad (17.9)$$

ninhydrin

violet anion

Only the nitrogen atom of the violet dye comes from the amino acid. The rest of the amino acid is converted to an aldehyde and carbon dioxide. Therefore, *the same violet dye is produced from all α-amino acids with a primary amino group,* and the intensity of its color is directly proportional to the concentration of the amino acid present. Only proline, which has a secondary amino group, reacts differently to give a yellow dye, but this, too, can be used for analysis.

PROBLEM 17.9 Write an equation for the reaction of alanine with ninhydrin.

17.8 Peptides

Amino acids are linked together in peptides and proteins by an amide bond between the carboxyl group of one amino acid and the α-amino group of another amino acid. Emil Fischer, who first proposed this structure, called this amide bond a **peptide bond.** A molecule containing only *two* amino acids (the shorthand aa is used for amino acid) joined in this way is called a **dipeptide:**

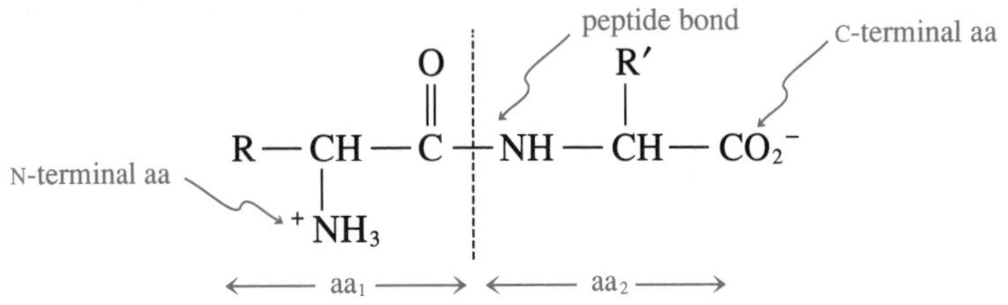

$$R{-}\underset{\substack{|\\ {}^{+}NH_3}}{CH}{-}\overset{\substack{O\\ \|}}{C}{-}NH{-}\overset{\substack{R'\\ |}}{CH}{-}CO_2^-$$

peptide bond

C-terminal aa

N-terminal aa

← aa₁ → ← aa₂ →

By convention, the peptide bond is written with the amino acid having a free $^{+}NH_3$ group at the left and the amino acid with a free CO_2^- group at the right. These amino acids are called, respectively, the **N-terminal amino acid** and the **C-terminal amino acid.**

EXAMPLE 17.5 Write the dipeptide structures that can be made by linking alanine and glycine with a peptide bond.

Solution There are two possibilities:

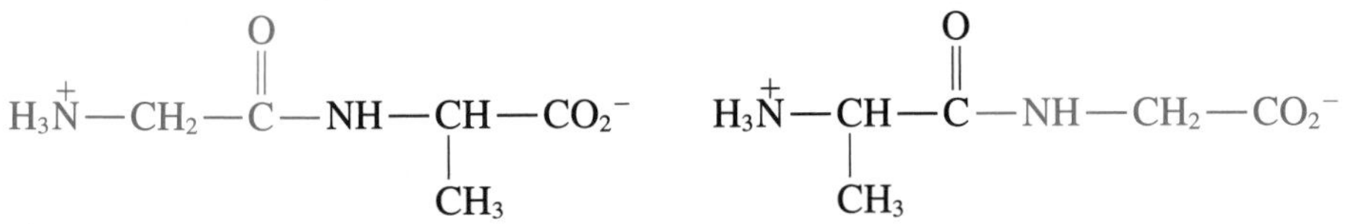

$$H_3\overset{+}{N}{-}CH_2{-}\overset{\substack{O\\ \|}}{C}{-}NH{-}\underset{\substack{|\\ CH_3}}{CH}{-}CO_2^- \qquad H_3\overset{+}{N}{-}\underset{\substack{|\\ CH_3}}{CH}{-}\overset{\substack{O\\ \|}}{C}{-}NH{-}CH_2{-}CO_2^-$$

glycyl**alanine** **alanyl**glycine

In glycylalanine, glycine is the N-terminal amino acid, and alanine is the C-terminal amino acid. In alanylglycine, these roles are switched. The two dipeptides are structural isomers.

We often write the formulas for peptides in a kind of shorthand by simply linking the three-letter abbreviations for each amino acid, *starting with the N-terminal one at the left*. For example, glycylalanine is Gly—Ala, and alanylglycine is Ala—Gly.

PROBLEM 17.10 In Example 17.5 the formulas for Gly—Ala and Ala—Gly are written in their dipolar forms. At what pH do you expect these structures to predominate? Draw the expected structure of Gly—Ala in solution at pH 3; at pH 9.

PROBLEM 17.11 Write the dipolar structural formula for

a. valylalanine. b. alanylvaline.

EXAMPLE 17.6 Consider the abbreviated formula Gly—Ala—Ser for a tripeptide. Which is the N-terminal amino acid, and which is the C-terminal amino acid?

Solution Such formulas always read from the N-terminal amino acid at the left to the C-terminal amino acid at the right. Glycine is the N-terminal amino acid, and serine is the C-terminal amino acid. Both the amino group *and* the carboxyl group of the middle amino acid, alanine, are tied up in peptide bonds.

PROBLEM 17.12 Write out the complete structural formula for Gly—Ala—Ser.

PROBLEM 17.13 Write out the *abbreviated* formulas for all possible tripeptide isomers of Gly—Ala—Ser.

The complexity that is possible in peptide and protein structures is truly astounding. For example, Problem 17.13 shows that there are 6 possible arrangements of 3 different amino acids in a tripeptide. For a tetrapeptide this number jumps to 24, and for an octapeptide (constructed from 8 different amino acids) there are 40,320 possible arrangements!

Now we must introduce one small additional complication before we consider the structures of particular peptides and proteins.

17.9 The Disulfide Bond

Aside from the peptide bond, the only other type of covalent bond between amino acids in peptides and proteins is the **disulfide bond.** It links two **cysteine** units. Recall that thiols are easily oxidized to disulfides (eq. 7.48). Two cysteine units can be linked by a disulfide bond.

$$
\begin{array}{c}
-NH-CH(CH_2SH)-C(=O)- \\
-NH-CH(CH_2SH)-C(=O)-
\end{array}
\underset{\text{reduction}}{\overset{\text{oxidation}}{\rightleftharpoons}}
\begin{array}{c}
-NH-CH(CH_2-S)-C(=O)- \\
| \\
-NH-CH(CH_2-S)-C(=O)-
\end{array}
\qquad (17.10)
$$

two cysteine units

—Cys—S—S—Cys—

disulfide bond

If the two cysteine units are in different parts of the *same* chain of a peptide or protein, a disulfide bond between them will form a "loop," or large ring. If the two units are on different chains, the disulfide bond will cross-link the two chains. We will see examples of both arrangements. Disulfide bonds can easily be broken by mild reducing agents (see "A Word About Hair" on page 231).

A Word About . . .

31. Some Naturally Occurring Peptides

Peptides with just a few linked amino acids per molecule have been isolated from living matter, where they often perform important roles in biology. Here are a few examples.

Bradykinin is a nonapeptide present in blood plasma and involved in regulating blood pressure. Several peptides found in the brain act as chemical transmitters of nerve impulses. One of these is the undecapeptide **substance P,** thought to be a transmitter of pain impulses. Notice that the C-terminal amino acid, methionine, is present as the primary amide. This is quite common in peptide chains and is symbolized by placing an NH_2 at the right-hand end of the formula.

Arg—Pro—Pro—Gly—Phe—Ser—
Pro—Phe—Arg

bradykinin

Arg—Pro—Lys—Pro—Gln—Gln—Phe—
Phe—Gly—Leu—Met—NH_2

substance P

Oxytocin and **vasopressin** are two cyclic nonapeptide hormones produced by the posterior pituitary gland. Oxytocin regulates uterine contraction and lactation and may be administered when it is necessary to induce labor at childbirth. Note that its structure includes two cysteine units joined by a disulfide bond and, once again, the C-terminal amino acid is present as the amide.

Vasopressin differs from oxytocin only in the substitution of Phe for Ile and Arg for Leu. Vasopressin regulates the excretion of water by the kidneys and also affects blood pressure. The disease *diabetes insipidus,* in which too much urine is excreted, is a consequence of vasopressin deficiency and can be treated by administering this hormone.

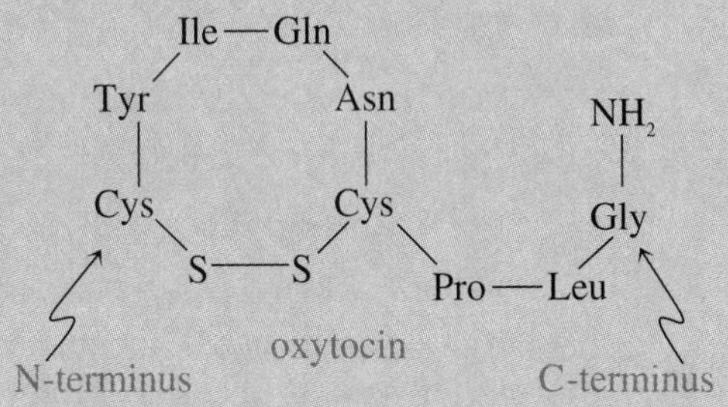

Ile—Gln
Tyr Asn NH_2
Cys Cys Gly
S—S Pro—Leu
N-terminus C-terminus

oxytocin

Cyclosporin A is another cyclic peptide that was first isolated from the fungus *Trichoderma polysporum.* Cyclosporin A has immunosuppressive activity and is used to help prevent organ rejection after transplant operations. Notice that cyclosporin A contains several unusual amino acids and that a number of the amide nitrogens are methylated.

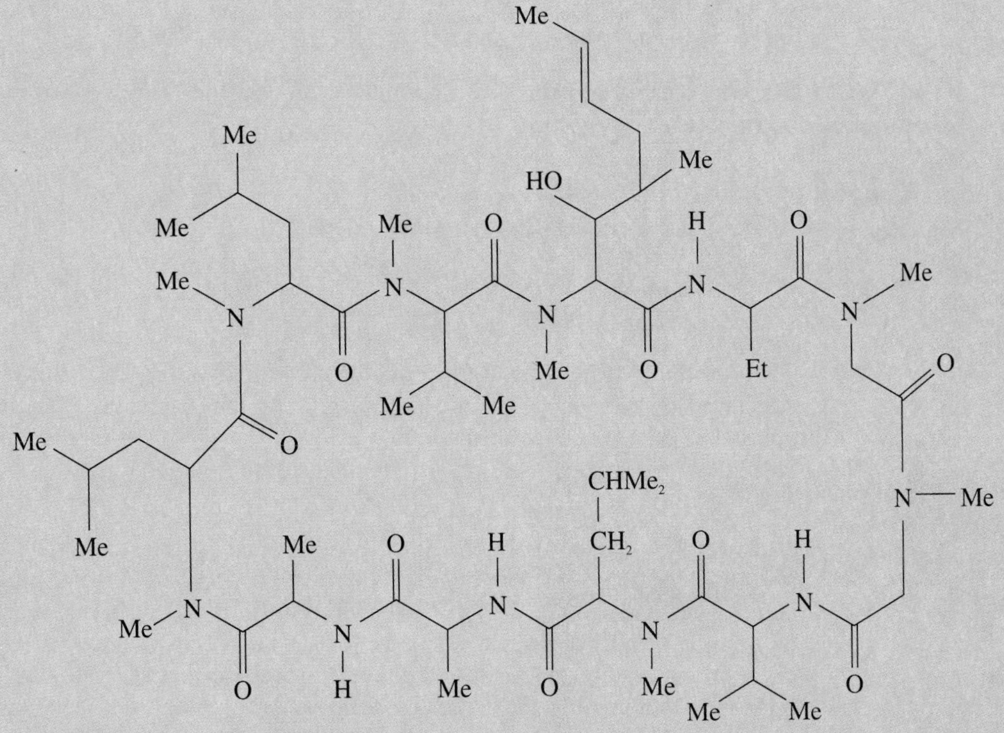

cyclosporin A

One area of current research is modifying the structures of natural, biologically important peptides (by replacing one amino acid by another or by altering side-chain structures or by replacing portions of the peptide chain with nonpeptidic structures) with the hope of developing new and useful drugs.

17.10 *Proteins*

Proteins are biopolymers composed of many amino acids connected to one another through amide (peptide) bonds. They play numerous roles in biological systems. For example, some proteins are major components of structural tissue (muscle, skin, nails, hair). Others transport molecules from one part of a living system to another. Yet others serve as catalysts for the many biological reactions needed to sustain life.

In the remainder of this chapter, we will describe the main features of peptide and protein structure. We will first examine what is called the primary structure of peptides and proteins, that is, how many amino acids are present and their sequence in the peptide or protein chain. We will then examine three-dimensional aspects of peptide and protein structure, usually referred to as secondary, tertiary, and quaternary structure.

17.11 *The Primary Structure of Proteins*

The backbone of proteins is a repeating sequence of one nitrogen and two carbon atoms.

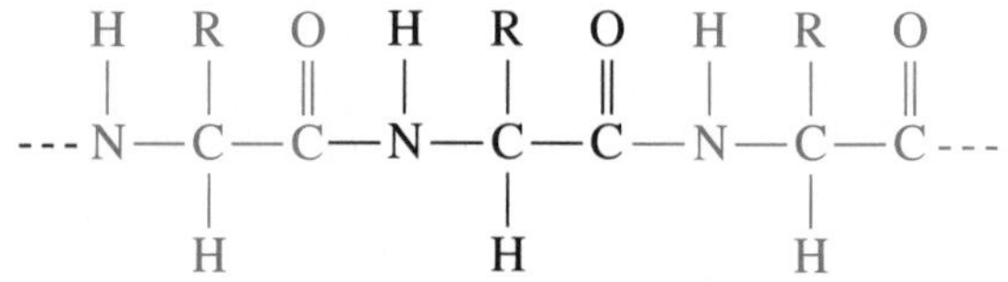

```
    H   R   O   H   R   O   H   R   O
    |   |   ‖   |   |   ‖   |   |   ‖
---N—C—C—N—C—C—N—C—C---
        |           |           |
        H           H           H
```

protein chain, showing amino acids linked by amide groups

Things we must know about a peptide or protein, if we are to write down its structure, are (1) which amino acids are present and how many of each there are and (2) the sequence of the amino acids in the chain. In this section, we will briefly describe ways to obtain this kind of information.

17.11a Amino Acid Analysis

Since peptides and proteins consist of amino acids held together by amide bonds, they can be hydrolyzed to their amino acid components (see Sec. 10.21). This hydrolysis is typically accomplished by heating the peptide or protein with 6M HCl at 110°C for 24 hours. Analysis of the resulting amino acid mixture requires a procedure that separates the amino acids from one another, identifies each amino acid present, and determines its amount.

An instrument called an **amino acid analyzer** performs these tasks automatically in the following way. The amino acid mixture from the complete hydrolysis of a few milligrams of the peptide or protein is placed at the top of a column packed with material that selectively absorbs amino acids. The packing is an insoluble resin that contains strongly acidic groups. These groups protonate the amino acids. Next, a buffer solution of known pH is pumped through the column. The amino acids pass through the column at different rates, depending on their structure and basicity, and are thus separated.

The column effluent is met by a stream of ninhydrin reagent. Therefore, the effluent is alternately violet or colorless, depending on whether or not an amino acid is being eluted from the column. The intensity of the color is automatically recorded as a function of the volume of effluent. Calibration with known amino acid mixtures allows each amino acid to be identified by the appearance time of its peak. Furthermore, the intensity of each peak gives a quantitative measure of the amount of each amino acid that is present. Figure 17.4 shows a typical plot obtained from an automatic amino acid analyzer.

PROBLEM 17.14 Show the products expected from complete hydrolysis of Gly—Ala—Ser (Problem 17.12).

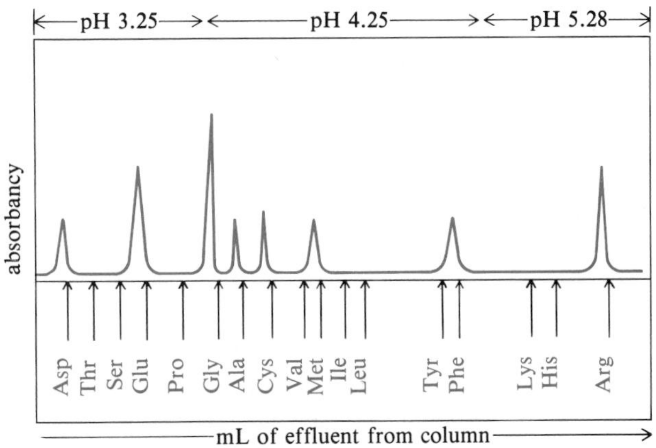

FIGURE 17.4 Different amino acids in a peptide hydrolyzate are separated on an ion-exchange resin. Buffers with different pH's elute the amino acids from the column. Each amino acid is identified by comparing the peaks with the standard elution profile shown near the bottom on the figure. The amount of each amino acid is proportional to the area under its peak.

17.11b Sequence Determination Frederick Sanger* devised a method for sequencing peptides based on the observation that the N-terminal amino acid differs from all others in the chain by having a free amino group. If that amino group were to react with some reagent prior to hydrolysis, then after hydrolysis, that amino acid would be labeled and could be identified. **Sanger's reagent** is 2,4-dinitrofluorobenzene, which reacts with the NH_2 group of amino acids and peptides to give yellow 2,4-dinitrophenyl (DNP) derivatives.

$$O_2N\text{-}C_6H_3(NO_2)\text{-}F + H_2N\text{—}\underset{\substack{|\\R_1}}{CH}\text{—}\overset{O}{\overset{\|}{C}}\text{----} \xrightarrow[\text{base}]{\text{mild}} \tag{17.11}$$

2,4-dinitrofluorobenzene; N-terminal amino acid

$$O_2N\text{-}C_6H_3(NO_2)\text{—}NH\text{—}\underset{\substack{|\\R_1}}{CH}\text{—}\overset{O}{\overset{\|}{C}}\text{----} + F^-$$

DNP-peptide, labeled at the N terminus

Hydrolysis of a peptide treated this way (eq. 17.11) would give the DNP derivative of the N-terminal amino acid; other amino acids in the chain would be unlabeled. In this way, the N-terminal amino acid could be identified.

---

* Frederick Sanger (Cambridge University, England) received *two* Nobel Prizes, the first in 1958 for his landmark work in amino acid sequencing and the second in 1980 for methodology in the base sequencing of RNA and DNA.

EXAMPLE 17.7 How might alanylglycine be distinguished from glycylalanine?

Solution Both dipeptides will give one equivalent each of alanine and glycine on hydrolysis. Therefore, we cannot distinguish between them without applying a sequencing method.

Treat the dipeptide with 2,4-dinitrofluorobenzene and *then* hydrolyze. If the dipeptide is alanylglycine, we will obtain DNP-alanine and glycine; if the dipeptide is glycylalanine, we will get DNP-glycine and alanine.

PROBLEM 17.15 Write out equations for the reactions described in Example 17.7.

Sanger used his method with great ingenuity to deduce the complete sequence of insulin, a protein hormone with 51 amino acid units. But the method suffers in that it identifies only the N-terminal amino acid.

An ideal method for sequencing a peptide or protein would have a reagent that clips off just one amino acid at a time from the end of the chain, and identifies it. Just such a method was devised by Pehr Edman (professor at the University of Lund in Sweden), and it is now widely used.

Edman's reagent is phenyl isothiocyanate, $C_6H_5N{=}C{=}S$. The steps in selectively labeling and releasing the N-terminal amino acid are shown in Figure 17.5. In the first step, the N-terminal amino acid acts as a nucleophile toward the C=S bond of the reagent to form a thiourea derivative. In the second step, the N-terminal amino acid is removed in the form of a heterocyclic compound, a phenylthiohydantoin. The specific phenylthiohydantoin that is formed can be identified by comparison with reference compounds prepared from the known amino acids. Then the two steps are repeated, to identify the next amino acid, and so on. The method has been automated, so currently amino acid "sequenators" can easily determine, in a day, the sequence of the first 20 or so amino acids in a peptide, starting at the N-terminal end. But the Edman method cannot be used indefinitely, due to the gradual buildup of impurities. It is most effective with peptides containing up to 20 to 25 amino acid units.

PROBLEM 17.16 Write the structure of the phenylhydantoin derived from the first cycle of Edman degradation of Phe—Ala—Ser.

17.11c Cleavage of Selected Peptide Bonds If a protein contains several hundred amino acid units, it is best to first partially hydrolyze the chain to smaller fragments that can be separated and subsequently sequenced by the Edman method. Certain chemicals or enzymes are used to cleave proteins at *particular* peptide bonds. For example, the enzyme *trypsin* (an intestinal digestive enzyme) specifically hydrolyzes polypeptides only at the carboxy end of arginine and lysine. A few of the many reagents of this type are listed in Table 17.3.

FIGURE 17.5 The Edman degradation of peptides.

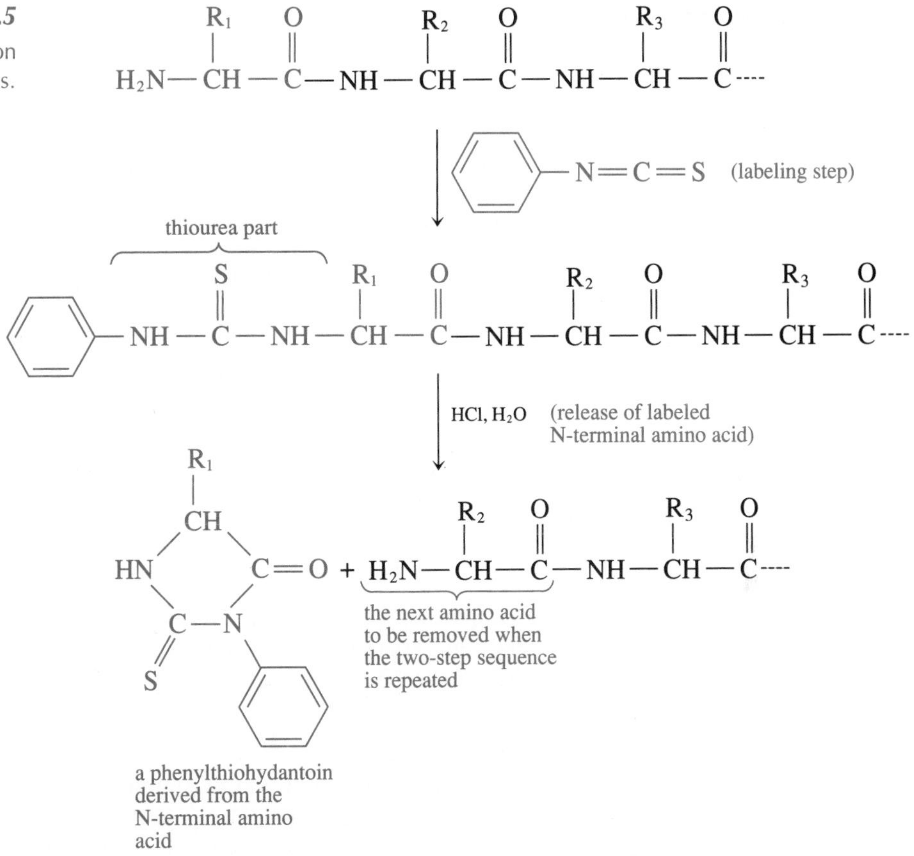

EXAMPLE 17.8 Consider the following peptide:

Ala—Gly—Tyr—Trp—Ser—Lys—Gly—Leu—Met—Gly

By referring to Table 17.3, determine what fragments will be obtained when this peptide is hydrolyzed with

a. trypsin. b. chymotrypsin. c. cyanogen bromide.

TABLE 17.3 Reagents for specific cleavage of polypeptides

Reagent	*Cleavage site*
trypsin	carboxyl side of Lys, Arg
chymotrypsin	carboxyl side of Phe, Tyr, Trp
cyanogen bromide (CNBr)	carboxyl side of Met
carboxypeptidase	a C-terminal amino acid

Solution

a. The enzyme trypsin will split the peptide on the carboxyl side of lysine, to give

Ala—Gly—Tyr—Trp—Ser—Lys and Gly—Leu—Met—Gly

b. The enzyme chymotrypsin will split the peptide on the carboxyl sides of tyrosine and tryptophan, to give

Ala—Gly—Tyr and Trp and Ser—Lys—Gly—Leu—Met—Gly

c. Cyanogen bromide will split the peptide on the carboxyl side of methionine, thus splitting off the C-terminal glycine and leaving the rest of the peptide untouched. (Carboxypeptidase would do the same thing, confirming that the C-terminal amino acid is glycine.)

PROBLEM 17.17 Determine what fragments will be obtained if bradykinin (its abbreviated formula is given on page 498) is hydrolyzed enzymatically with

a. trypsin. b. chymotrypsin.

During the past 15 years, the methods discussed here have been improved and expanded; the separation and sequencing of peptides and proteins can now be accomplished even if only minute amounts are available.

17.12 *The Logic of Sequence Determination*

A specific example will illustrate the reasoning that is used to fully determine the amino acid sequence in a particular peptide with 30 amino acid units.

First we hydrolyze the peptide completely, subject it to amino acid analysis, and find that it has the formula

$$Ala_2ArgAsnCys_2GlnGlu_2Gly_3His_2Leu_4LysPhe_3ProSerThrTyr_2Val_3$$

Using the Sanger method, we find that the N-terminal amino acid is Phe.

Since the chain is too long to degrade completely by the Edman method, we decide to simplify the problem by digesting the peptide with chymotrypsin. (We select chymotrypsin because the peptide contains three Phe's and two Tyr's and will undoubtedly be cleaved by that reagent.) When we carry out this cleavage, we get three fragment peptides. In addition, we get two equivalents of Phe and one of Tyr. After separation, we subject the three fragment peptides to Edman degradation and obtain their structures.

A. Leu—Val—Cys—Gly—Glu—Arg—Gly—Phe

B. Val—Asn—Gln—His—Leu—Cys—Gly—Ser—His—Leu—Val—Glu—Ala—Leu—Tyr

C. Thr—Pro—Lys—Ala (27 28 29 30)

We still cannot write a unique structure for the intact peptide, but we can say that the C-terminal amino acid must be Ala and that the last four amino acids must be in the sequence shown for fragment C. We deduce this because we know that Ala is *not* cleaved at its carboxyl end by chymotrypsin, yet it appears at the C-terminal end of one of the fragments. (Note that the C-terminal amino acids in fragments A and B are Phe and Tyr, both cleaved at the carboxyl ends by chymotrypsin.) That the C-terminal amino acid is Ala can be confirmed using carboxypeptidase. We can number the amino acids in fragment C as 27 through 30 in the chain.

What to do next? Cyanogen bromide is no help, because the peptide does not contain Met. But the peptide does contain Lys and Arg, so we go back to the beginning and digest the intact peptide with trypsin, which cleaves peptides on the carboxy side of these amino acid units. We obtain (not surprisingly) some Ala (the C-terminal amino acid) because it comes right after a Lys. We also obtain two peptides. One of them is relatively short, so we determine its sequence by the Edman method and find it to be

$$\text{D. } \overset{23}{\text{Gly}}\text{—}\overset{24}{\text{Phe}}\text{—}\overset{25}{\text{Phe}}\text{—}\overset{26}{\text{Tyr}}\text{—}\overset{27}{\text{Thr}}\text{—}\overset{28}{\text{Pro}}\text{—}\overset{29}{\text{Lys}}$$

Because the last three amino acids in fragment D *overlap* with 27, 28, and 29 of fragment C, we can number the rest of the chain, back to 23. We now note that amino acids 23 and 24 appear at the end of fragment A, so originally A must have been connected to C. The only place left for fragment B is in front of A. This leaves only one of the Phe's unaccounted for, and it must occupy the N-terminal position (recall the Sanger result). We can now write out the complete sequence!

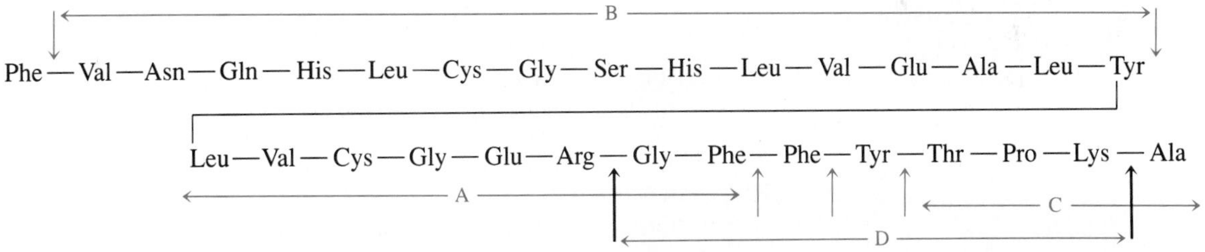

B

Phe—Val—Asn—Gln—His—Leu—Cys—Gly—Ser—His—Leu—Val—Glu—Ala—Leu—Tyr

Leu—Val—Cys—Gly—Glu—Arg—Gly—Phe—Phe—Tyr—Thr—Pro—Lys—Ala

A C

D

The colored vertical arrows show the cleavage points with chymotrypsin, and the black ones show the cleavage points with trypsin.

The peptide just used for illustration is the B chain of the protein hormone **insulin,** whose structure was first determined by Sanger and is shown schematically in Figure 17.6. Insulin consists of an A chain with 21 amino acid units and a B chain with 30 amino acid units. The two chains are joined by two disulfide bonds, and the A chain also contains a small disulfide loop.

PROBLEM 17.18 How could the A and B chains of insulin be separated chemically? (*Hint:* see eq. 7.48.)

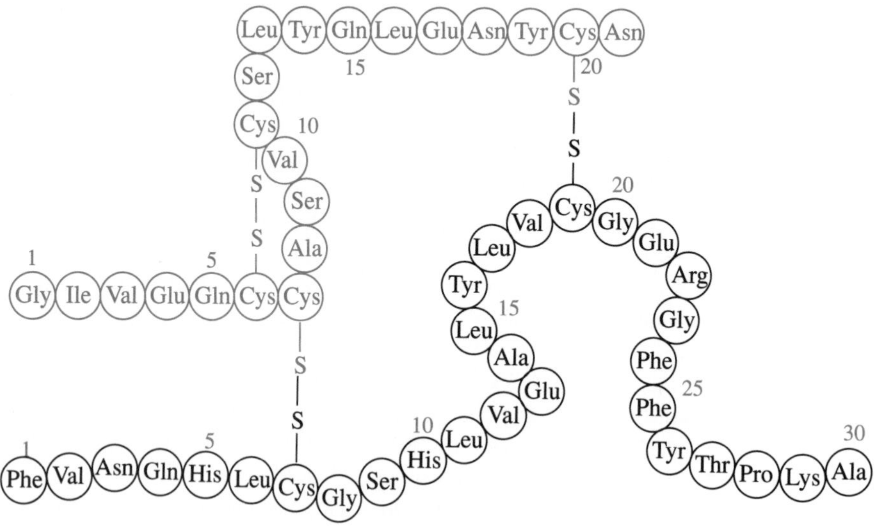

Figure 17.6 Primary structure of beef insulin. The A chain is in color, and the B chain, whose structure determination is described in the text, is shown in black.

A Word About . . .

32. Protein Sequencing and Evolution

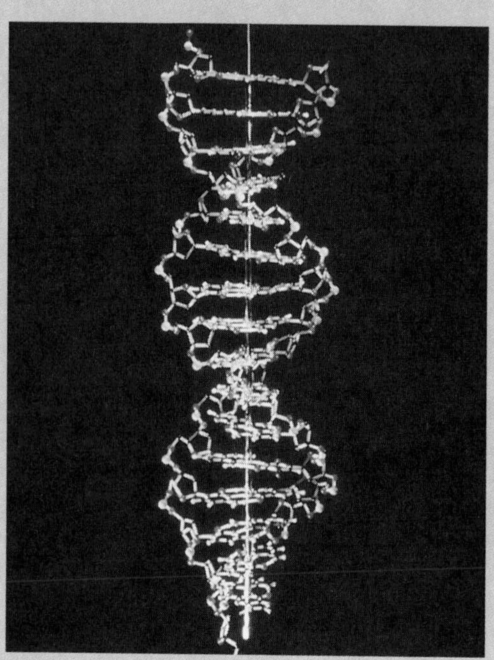

There are many reasons why it is important to determine the sequences of amino acids in proteins. First, we must know the detailed structures of proteins if we are to understand, at a molecular level, the way they function. The amino acid sequence is the link between the genetic message coded in DNA and the three-dimensional protein structure that forms the basis for biological function.

There are medical benefits to knowing amino acid sequences. Certain genetic diseases, such as sickle-cell anemia, can result from the change of a single amino acid unit in a protein. In this case, it is replacement of glutamic acid at position 6 in the β chain of hemoglobin by a valine unit. Thus, sequence determination is an important part of medical pathology. One future possible application of genetic engineering is devising ways to correct these amino acid sequence errors.

Another important reason for determining amino acid sequences is that they provide a chemical tool for studying our evolutionary history. Proteins resemble one another in amino acid sequence if they have a common evolutionary ancestry. Let's look at a specific example.

Cytochrome *c*, an enzyme important in the respiration of most plants and animals, is an electron-transport protein with 104 amino acid residues. It is involved in oxidation-reduction processes. In these reactions, cytochrome *c* must react with and transfer an electron from one enzyme complex (cytochrome reductase) to another (cytochrome oxidase).

Cytochrome *c* probably evolved more than 1.5 billion years ago, even before the evolutionary divergence of plants and animals. The function of this protein has been preserved all that time! We know this because the cytochrome *c* isolated from any eucaryotic microorganism (one that contains a cell nucleus) reacts in vitro with the cytochrome oxidase of every other species tested so far. For example, wheat germ cytochrome *c*, a plant-derived enzyme, reacts with human cytochrome *c* oxidase. Also, the three-dimensional structures of cytochrome *c* isolated from such diverse species as tuna and photosynthetic bacteria are very similar.

Although *the shape and functions are similar* for cytochrome *c* samples isolated from different sources, *the amino acid sequence varies somewhat* from one species to another. The amino acid sequence of cytochrome *c* isolated from humans differs from that of monkeys in *just 1* (out of 104) amino acid residues! On the other hand, cytochrome *c* from dogs, a more distant evolutionary relative of humans, differs from the human protein by 11 amino acid residues.

17.13 Peptide Synthesis

Once we know the amino acid sequence in a peptide or protein, we are in a position to synthesize it from its amino acid components. Why would we want to do this? There are several reasons. For example, we might wish to verify a particular peptide structure by comparing the properties of the synthetic and natural substances. Or we might wish to study the effect of substituting one amino acid for another on the biological properties of a peptide or protein. Such modified proteins could be valuable in treating disease or in understanding how the protein functions.

Many methods have been developed to link amino acids in a controlled manner. They require careful strategy. Amino acids are bifunctional. To link the carboxyl group of one amino acid to the amino group of a second amino acid, we must first prepare each compound by protecting the amino group of the first and the carboxyl group of the second.

$$\underset{aa_1}{H_2N-\underset{|}{\overset{R_1}{CH}}-CO_2H} \xrightarrow[\text{amino group}]{\text{protect the}} \boxed{P_1}-NH-\overset{R_1}{\underset{|}{CH}}-CO_2H \tag{17.12}$$

$$\underset{aa_2}{H_2N-\overset{R_2}{CH}-CO_2H} \xrightarrow[\text{carboxyl group}]{\text{protect the}} H_2N-\overset{R_2}{CH}-\overset{O}{\overset{\|}{C}}-\boxed{P_2} \tag{17.13}$$

In this way, we can control the linking of the two amino acids so that the carboxyl group of aa_1 combines with the amino group of aa_2.

$$\boxed{P_1}\!-\!NHCH(R_1)CO_2H + H_2N\!-\!CH(R_2)\!-\!C(=O)\!-\!\boxed{P_2} \xrightarrow{-H_2O} \boxed{P_1}\!-\!NHCH(R_1)\!-\!\boxed{C(=O)\!-\!NH}\!-\!CH(R_2)\!-\!C(=O)\!-\!\boxed{P_2} \quad (17.14)$$

peptide bond

doubly protected dipeptide

Later we will give specific examples of protecting groups and of a reagent that can be used to form the peptide bond.

EXAMPLE 17.9 What would happen if we tried to combine aa_1 with aa_2 without using protecting groups?

Solution Since each amino acid could react either as an amine or as a carboxylic acid, we could get not only aa_1—aa_2 but also aa_2—aa_1, aa_1—aa_1, and aa_2—aa_2. Furthermore, since the resulting dipeptides would still have a free amino and a free carboxyl group, we could also get trimers, tetramers, and so on. In other words, a mess.

After the peptide bond is formed, we must be able to *remove the protecting groups under conditions that do not hydrolyze the peptide bond.* Or, if more amino acids are to be added to the chain, we must be able to *selectively* remove one of the two protecting groups from the doubly protected dipeptide before joining the next amino acid to it. All of this can be quite a tricky and tedious process. Yet these methods were used by Vincent du Vigneaud* and his colleagues to synthesize oxytocin and vasopressin (page 498), the first naturally occurring polypeptides to be synthesized in the laboratory.

In 1965, R. Bruce Merrifield** developed a technique that revolutionized peptide synthesis. This **solid-phase technique** avoids many of the tedious aspects of previous methods and is now universally used. The principle is to *assemble the peptide chain while one end of it is chemically anchored to an insoluble inert solid.* In this way, excess reagents and by-products can be removed simply by washing and filtering the solid. The growing peptide chain does not need to be purified at any intermediate stage. When the peptide is fully constructed, it is cleaved chemically from the solid support.

---

* Vincent du Vigneaud (Cornell University) was awarded the 1955 Nobel Prize in chemistry for this achievement.

R. Bruce Merrifield (Rockefeller University) received the 1984 Nobel Prize in chemistry for his contribution, which not only revolutionized peptide synthesis but also affected many other areas of chemistry through the use of polymer-bound reagents. For an account of the history of this discovery, see the article by Merrifield in *Science* **1986, *232,* 341, and for a personal history, see *Chemistry in Britain* **1987,** 816.

Typically, the solid phase is a cross-linked polystyrene in which some (usually 1 to 10%) of the aromatic rings contain chloromethyl ($ClCH_2$—) groups.

$$\sim CH_2—CH—CH_2—CH—CH_2—CH—CH_2—CH—CH_2—CH—CH_2—CH\sim$$

(each CH bears a benzene ring; the first and last rings carry CH_2Cl groups)

CH_2Cl CH_2Cl

The polymer behaves chemically like benzyl chloride, an alkyl halide that is quite reactive in nucleophilic substitution reactions (S_N2).

The steps in a Merrifield synthesis are summarized for a dipeptide in Figure 17.7. In step 1, the polymer is first treated with an N-protected amino acid. The carboxylate ion acts as an oxygen nucleophile and displaces the chloride ion from the polymer, thus forming an ester link. *The first amino acid attached to the polymer will eventually become the C-terminal amino acid of the synthetic peptide.*

Many protecting groups are known, but in solid-phase peptide synthesis, the most frequently used N-protecting group is the ***t*-butoxycarbonyl (Boc)** group. The amino acid is protected by reaction with di-*t*-butyl dicarbonate.

$$[(CH_3)_3CO—C(=O)]_2O + H_3\overset{+}{N}—CH(R)—CO_2^- \xrightarrow{\text{base}} \underbrace{(CH_3)_3CO—\overset{O}{\overset{\|}{C}}}_{\boxed{P}}—NH—CH(R)—CO_2H \quad (17.15)$$

di-*t*-butyl dicarbonate

After the amino acid is attached to the polymer, the protecting group is removed. This is accomplished (step 2) by reaction with acid under mild conditions.

$$CH_3—C(CH_3)(CH_2—H)—O—\overset{O}{\overset{\|}{C}}—NHCH(R)—\underset{O}{\underset{\|}{C}}—O—CH_2—C_6H_4\text{—(polymer)} \xrightarrow[CH_2Cl_2]{CF_3CO_2H}$$

H^+

$$CH_3—\overset{CH_2}{\overset{\|}{C}}—CH_3 + CO_2 + H_2N—CH(R_1)—\underset{O}{\underset{\|}{C}}—O—CH_2—C_6H_4\text{—(polymer)} \quad (17.16)$$

2-methylpropene

ester group still intact

deprotected amino group

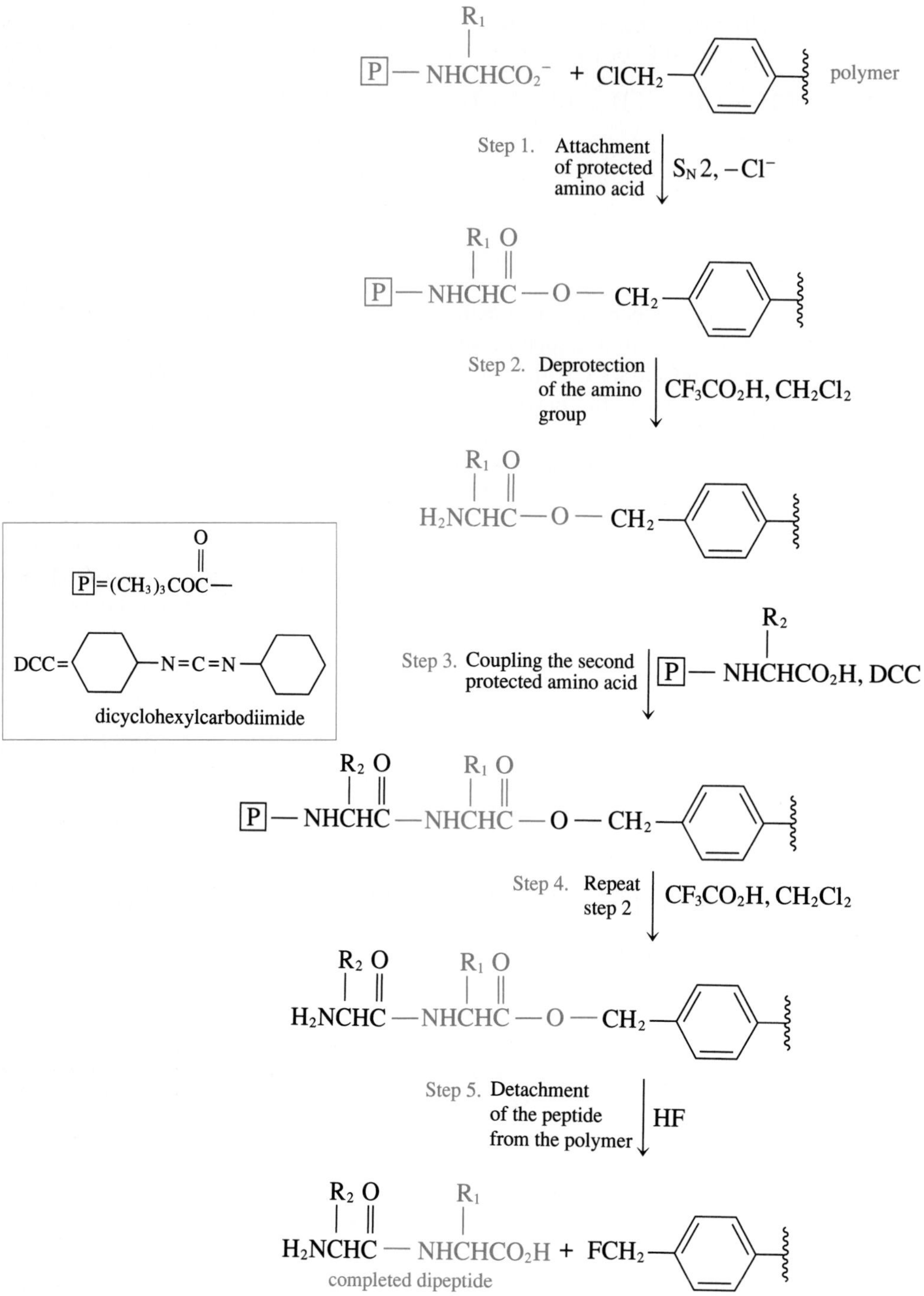

[P]—NHCH(R_1)CO_2^- + $ClCH_2$—C_6H_4—polymer

Step 1. Attachment of protected amino acid — S_N2, $-Cl^-$

[P]—NHCH(R_1)C(=O)—O—CH_2—C_6H_4—

Step 2. Deprotection of the amino group — CF_3CO_2H, CH_2Cl_2

$H_2NCH(R_1)C(=O)$—O—CH_2—C_6H_4—

[P] = $(CH_3)_3COC(=O)$—

DCC = C_6H_{11}—N=C=N—C_6H_{11}

dicyclohexylcarbodiimide

Step 3. Coupling the second protected amino acid — [P]—NHCH(R_2)CO_2H, DCC

[P]—NHCH(R_2)C(=O)—NHCH(R_1)C(=O)—O—CH_2—C_6H_4—

Step 4. Repeat step 2 — CF_3CO_2H, CH_2Cl_2

$H_2NCH(R_2)C(=O)$—NHCH(R_1)C(=O)—O—CH_2—C_6H_4—

Step 5. Detachment of the peptide from the polymer — HF

$H_2NCH(R_2)C(=O)$—NHCH(R_1)CO_2H + FCH_2—C_6H_4—

completed dipeptide

FIGURE 17.7 Solid-phase synthesis of a dipeptide.

The by-products of deprotection are gaseous (2-methylpropene and carbon dioxide) and are thus easily removed from the reaction mixture. The ester group that links the first amino acid to the polymer is *not* hydrolyzed under these conditions.

In step 3, the next N-protected amino acid is linked to the first one. This is accomplished with the aid of **dicyclohexylcarbodiimide (DCC).** DCC is able to link carboxyl and amino groups in a peptide bond; in the process, the DCC is converted to dicyclohexylurea.

$$\sim C(=O)-OH + H_2N\sim + C_6H_{11}-N=C=N-C_6H_{11} \longrightarrow \sim C(=O)-NH\sim + C_6H_{11}-NH-C(=O)-NH-C_6H_{11} \quad (17.17)$$

dicyclohexylcarbodiimide (DCC)

peptide bond

dicyclohexylurea

Steps 2 and 3 may be repeated to add a third amino acid, a fourth, and so on. Finally, when the desired amino acids have been connected in the proper sequence and the N-terminal amino group has been deprotected (step 4 in Figure 17.7), the complete polypeptide chain is detached from the polymer. This can be accomplished by treatment with anhydrous hydrogen fluoride, which cleaves the benzyl ester without hydrolyzing the amide bonds in the polypeptide (step 5 in Figure 17.7).

All operations in solid-phase peptide synthesis have been automated. The reactions occur in a single reaction vessel, with reagents and wash solvents automatically added from reservoirs by mechanical pumps. Working around the clock, the programmer can incorporate eight or more amino acids into a polypeptide in a day. Merrifield synthesized the nonapeptide bradykinin (page 498) in just 27 hours using this technique. And, in 1969, he used the automated synthesizer to prepare the enzyme ribonuclease (124 amino acid residues), the first enzyme to be prepared synthetically from its amino acid components. The synthesis, which required 369 chemical reactions and 11,391 steps, was completed in only six weeks. Automated peptide synthesis, though still not without occasional problems, is now a fairly routine matter.

PROBLEM 17.19 Write all the equations for the synthesis of Gly—Ala—Phe using Merrifield solid-phase technology.

We have seen how the primary structure of peptides and proteins can be determined and how peptides can be synthesized in the laboratory. Now let us examine some further details of protein structure.

17.14 Secondary Structure of Proteins

Because proteins consist of long chains of amino acids strung together, one might think that their shapes are rather amorphous, or "floppy" and ill-defined. This is incorrect. Many proteins have been isolated in pure crystalline form and are polymers with very well defined shapes. Indeed, even in solution, the shapes seem to be quite regular. Let us examine some of the structural features of peptide chains that are responsible for their definite shapes.

17.14a Geometry of the Peptide Bond We pointed out earlier that simple amides have a planar geometry, that the amide C—N bond is shorter than usual, and that rotation around that bond is restricted (Sec. 10.21). Bond planarity and restricted rotation, which are consequences of resonance, are also important in peptide bonds.

X-ray studies of crystalline peptides by Linus Pauling and his colleagues determined the precise geometry of peptide bonds. The characteristic dimensions, which are common to all peptides and proteins, are shown in Figure 17.8.

Things to notice about peptide geometry are as follows: (1) The amide group is flat; the carbonyl carbon, the nitrogen, and the four atoms connected to them all lie in a single plane. (2) The short amide C—N distance (1.32 Å, compared with 1.47 Å for the other C—N bond) and the 120° bond angles around that nitrogen show that it is essentially sp^2-hybridized and that the bond between it and the carbonyl carbon is like a double bond. (3) Although each amide group is planar, two adjacent amide groups need not be coplanar because of rotation about the other single bonds in the chain; that is, rotation can occur around the two single bonds to the —CHR— group.

The rather rigid geometry and restricted rotation of the peptide bond help to impart a definite shape to proteins.

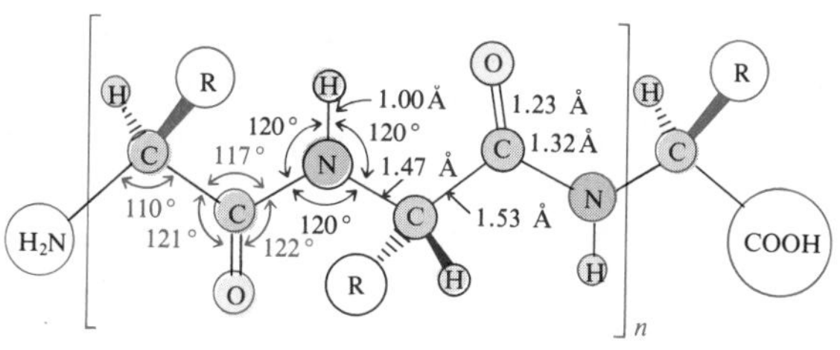

FIGURE 17.8 The characteristic bond angles and bond lengths in peptide bonds.

17.14b Hydrogen Bonding We pointed out earlier that amides readily form intermolecular hydrogen bonds between the carbonyl group and the N—H group, bonds of the type C=O···H—N. Such bonds are present and important in peptide chains. The chain may coil in such a way that the N—H of one peptide bond can hydrogen-bond with a carbonyl group of another peptide bond farther down the *same* chain, thus rigidifying the coiled structure. Alternatively, carbonyl groups and N—H groups on *different* peptide chains may hydrogen-bond, linking the two chains. Although a single hydrogen bond is relatively weak (perhaps only 5 kcal/mol of energy), the possibility of forming multiple intrachain or interchain hydrogen bonds makes this a very important factor in protein structure, as we will now see.

17.14c The α Helix and the Pleated Sheet X-ray studies of α-keratin, a structural protein present in hair, wool, horns, and nails, showed that some feature of the structure repeats itself every 5.4 Å. Using molecular models with the correct geometry of the peptide bond, Linus Pauling was able to suggest a structure that explains this and other features of the x-ray studies. Pauling proposed that the polypeptide chain coils about itself in a spiral manner to form a helix, held rigid by intrachain hydrogen bonds. The **α helix,** as it is called, is right-handed and has a pitch of 5.4 Å, or 3.6 amino acid units (Figure 17.9).

Note several features of the α helix. Proceeding from the N terminus (at the top of the structure as drawn in the figure), each carbonyl group points ahead or down toward the C terminus and is hydrogen-bonded to an N—H bond farther down the chain. The N—H bonds all point back toward the N terminus. All the hydrogen bonds are roughly aligned with the long axis of the helix. The very large number of hydrogen bonds (one for each amino acid unit) strengthen the helical structure. The R groups of the individual amino acid units are directed *outward* and do not disrupt the central core of the helix. It turns out that the α helix is a natural pattern into which many proteins fold. Figure 17.10 shows Professor Pauling seated by a scale model of the α helix.

The structural protein β-keratin, obtained from silk fibroin, shows a different repeating pattern (7 Å) in its x-ray structure. To explain the data, Pauling suggested a **pleated-sheet** arrangement of the peptide chain (Figure 17.11). In the pleated sheet, peptide chains lie side by side and are held together by *interchain* hydrogen bonds. Adjacent chains run in opposite directions. The repeating unit in each chain, which is stretched out compared with the α helix, is about 7 Å. In the pleated-sheet structure, the R groups of amino acid units in any one chain alternate above and below the mean plane of the sheet. If the R groups are large, there will be appreciable steric repulsion between them on adjacent chains. For this reason, the pleated-sheet structure is important *only* in proteins that have a high percentage of amino acid units with *small* R groups. In the β-keratin of silk fibroin, for example, 36% of the amino acid units are glycine (R = H) and another 22% are alanine (R = CH_3). Because this type of repulsion between R groups is not encountered in the α helix, the α helix is by far the more common structure of the two.

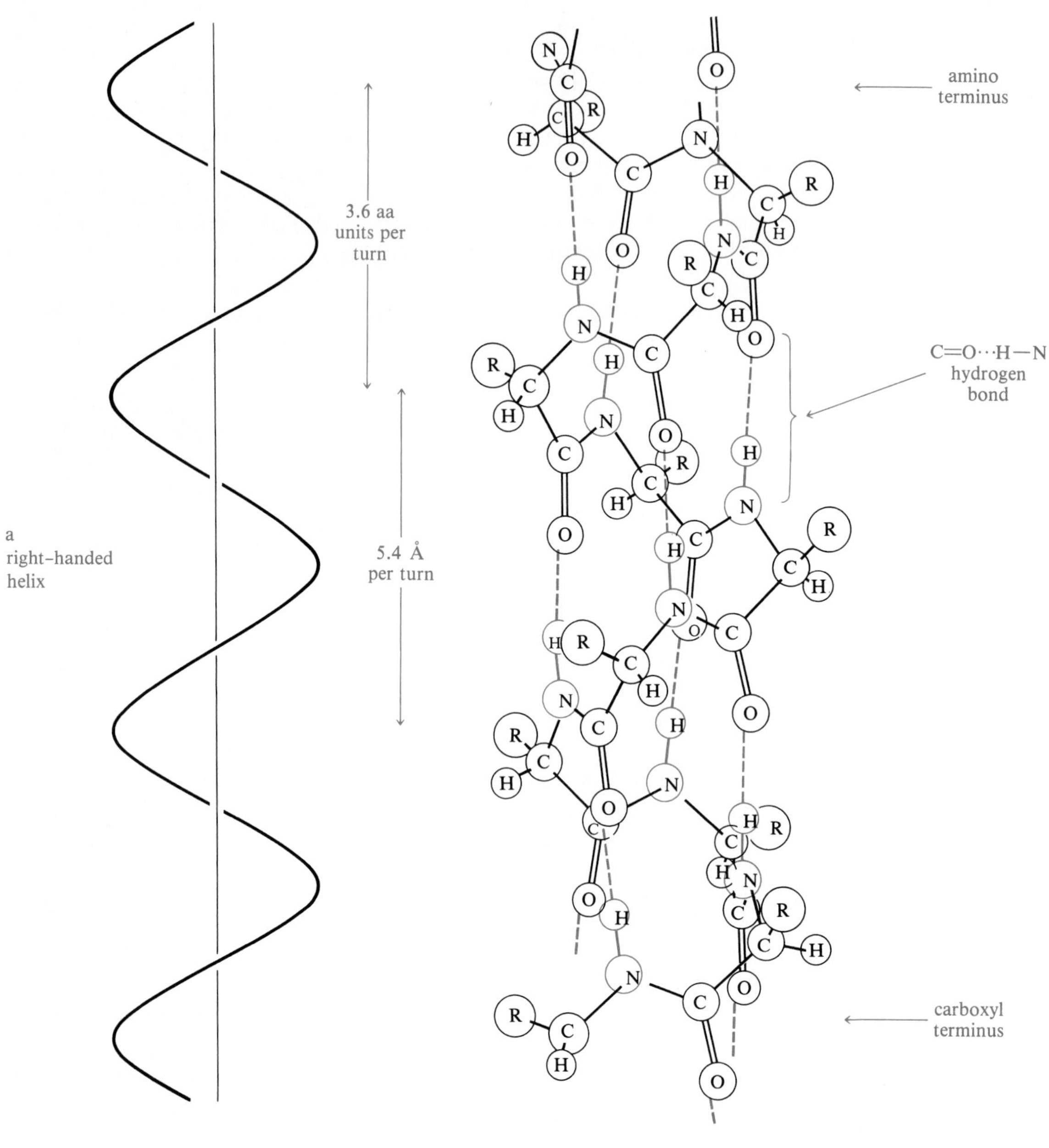

FIGURE 17.9 Segment of an α helix, showing three turns of the helix, with 3.6 amino acid units per turn. Hydrogen bonds are shown as dashed colored lines.

FIGURE 17.10

Linus Pauling (California Institute of Technology and Stanford University), who has made many contributions to our knowledge of organic structures. He did fundamental work on the theory of resonance, on the measurement of bond lengths and energies, and on the structure of proteins and the mechanism of antibody action. He received the Nobel Prize in chemistry in 1954 and the Nobel Peace Prize in 1962.

C terminus

possible source of strain if R groups are large

N terminus

~7 Å

N terminus

C terminus

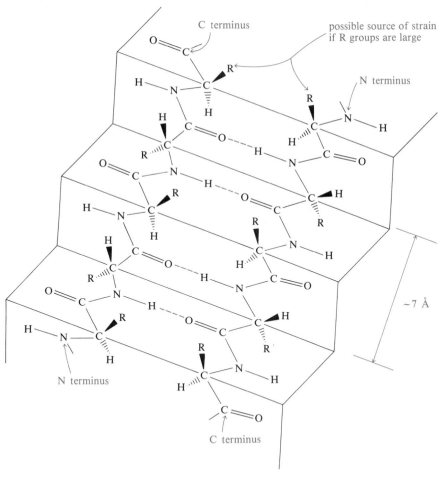

FIGURE 17.11 A segment of the pleated sheet structure of β-keratin. Adjacent chains run in opposite directions and are held together by hydrogen bonds (shown in color). R groups project above or below the mean plane of the sheet.

17.15 Tertiary Structure: Fibrous and Globular Proteins

We may well ask how materials as rigid as horses' hoofs, as springy as hair, as soft as silk, as slippery and shapeless as egg white, as inert as cartilage, and as reactive as enzymes can all be made of the same building blocks: amino acids and proteins. The key lies mainly in the amino acid makeup itself. So far we have focused on the protein backbone and its shape. But what about the diverse R groups of the various amino acids? How do they affect protein structure?

Some amino acids have nonpolar R groups, simple alkyl or aromatic groups. Others have highly polar R groups, with carboxylate or ammonium ions and hydroxyl or other polar groups. Still others have flat, rigid aromatic rings that may interact in specific ways. *Different R groups affect the gross properties of a protein.*

PROBLEM 17.20 Which amino acids in Table 17.1 have nonpolar R groups? highly polar groups? relatively flat R groups?

Proteins generally fall into one of two main classes: **fibrous** or **globular.** *Fibrous proteins* are animal structural materials and hence are water-insoluble. They fall into three general categories: the **keratins,** which make up protective tissue, such as skin, hair, feathers, claws, and nails; the **collagens,** which form connective tissue, such as cartilage, tendons, and blood vessels; and the **silks,** such as the fibroin of spider webs and cocoons.

Keratins and collagens have helical structures, whereas silks have pleated-sheet structures. A large fraction of the R groups attached to these frameworks are nonpolar, accounting for the insolubility of these proteins in water. In hair, three α helices are braided to form a rope, the helices being held together by disulfide cross-links. The ropes are further packed side by side in bundles that ultimately form the hair fiber. The α-keratin of more rigid structures, such as nails and claws, is similar to that of hair, except that there is a higher percentage of cysteine amino acid units in the polypeptide chain. Therefore, there are more disulfide cross-links, giving a firmer, less flexible overall structure.

To summarize, nonpolar R groups and disulfide cross-links, together with helical or sheetlike backbones, tend to give fibrous proteins their rather rigid, insoluble structures.

Globular proteins are very different from fibrous proteins. They tend to be water soluble and have roughly spherical shapes, as their name suggests. Instead of being structural, globular proteins perform various other biological functions. They may be **enzymes** (biological catalysts), **hormones** (chemical messengers that regulate biological processes), **transport proteins** (carriers of small molecules from one part of the body to another, such as hemoglobin, which transports oxygen in the blood), or **storage proteins** (which act as food stores; ovalbumin of egg white is an example).

Globular proteins have more amino acids with polar or ionic side chains than the water-insoluble fibrous proteins. An enzyme or other globular protein that carries out its function mainly in the aqueous medium of the cell will adopt a structure in which the nonpolar, hydrophobic R groups point in toward the center and the polar or ionic R groups point out toward the water.

Globular proteins are mainly helical, but they have folds that permit the overall shape to be globular. One of the 20 amino acids, proline, has a secondary amino group. Wherever a proline unit occurs in the primary peptide structure, there will be no N—H group available for intrachain hydrogen bonding.

$$\cdots\text{NH}-\overset{\overset{\displaystyle\text{R}}{|}}{\text{CH}}-\overset{\overset{\displaystyle\text{O}}{\|}}{\text{C}}-\text{N}(\text{proline ring})-\underset{\underset{\displaystyle\text{O}}{\|}}{\text{C}}-\text{NH}-\overset{\overset{\displaystyle\text{R}}{|}}{\text{CH}}-\overset{\overset{\displaystyle\text{O}}{\|}}{\text{C}}\cdots$$

no H for hydrogen bonding

proline unit

Proline units tend therefore to disrupt an α helix, and we frequently find proline units at "turns" in a protein structure.

Myoglobin, the oxygen-transport protein of muscle, is a good example of a globular protein (Figure 17.12). It contains 153 amino acid units, yet is extremely compact, with very little empty space in its interior. Approximately 75% of the amino acid units in myoglobin are part of eight right-handed α-helical sections. There are four proline units, and each occurs at or near "turns" in the structure. There are also three other "turns" caused by structural features of other R groups. The interior of myoglobin consists almost entirely of nonpolar R groups, such as those of leucine, valine, phenylalanine, and methionine. The only interior polar groups are two histidines. These perform a

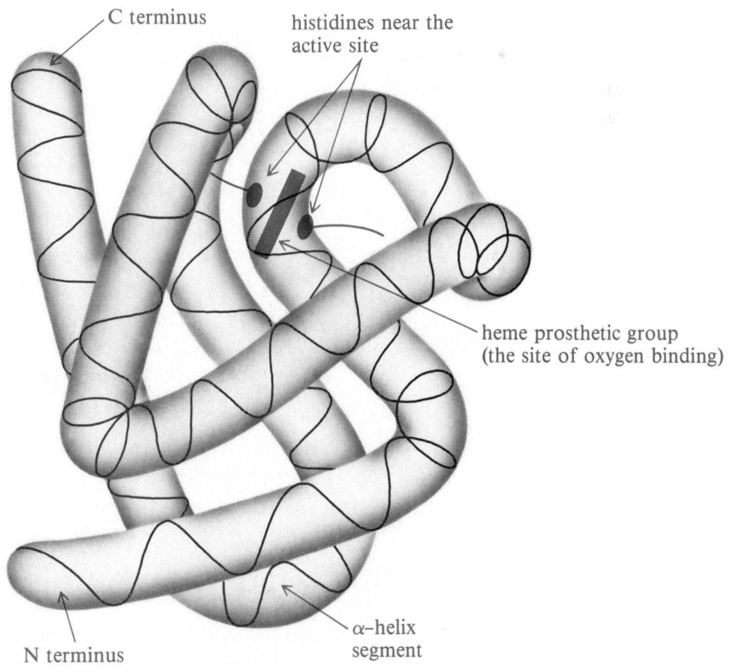

FIGURE 17.12 Schematic drawing of myoglobin. Each of the tubular sections is a segment of α helix, but the overall shape is globular.

necessary function at the *active site* of the protein, where the nonprotein portion, a molecule of the porphyrin *heme* (page 396) binds the oxygen. The outer surface of the protein includes many highly polar amino acids residues (lysine, arginine, glutamic acid, and so on).

To summarize this section, we see that the particular amino acid content of a peptide or protein influences its shape. These interactions are mainly a consequence of disulfide bonds and of the polarity or nonpolarity of the R groups, their shape, and their ability to form hydrogen bonds. When we refer to the **tertiary structure** of a protein, we refer to all the contributions of these factors to its three-dimensional structure.

A Word About . . .

33. Bacterial Cell Walls, Enzyme Inhibitors, and Antibiotics

Bacteria are protected by a fence-like cell wall. This wall, called a **peptidoglycan,** consists of polysaccharide strands cross-linked by peptide chains. The polysaccharide strands of the peptidoglycan cell wall are constructed from two monosaccharides, *N*-acetylglucosamine (G) and *N*-acetylmuramic acid (M). These are connected to one another by 1,4-β-glycosidic linkages (see Sec. 16.13). *N*-Acetylmuramic acid units of adjacent polysaccharide strands are crosslinked to one another by peptide chains as shown graphically in Figure 17.13.

Let us look in more detail at how bacteria assemble their cell walls as they grow. The first stage involves a complex series of enzyme-catalyzed reactions leading to construction, at the cell surface, of the polysaccharide strands with a pentapeptide (Ala—Gln—Lys—Ala—Ala) linked to the carboxyl group of the *N*-acetylmuramic acid units through a peptide bond.

Five glycine units are then attached to the side-chain amino group of the pentapeptide lysine residue. In the last step, the *N*-terminal amino group of the glycine pentapeptide of one peptidoglycan strand displaces the terminal alanine of the Ala—Gln—Lys—Ala—Ala pentapeptide of an adjacent strand as shown in Figure 17.14. This reaction is catalyzed by an enzyme called a **transpeptidase** and results in the bacteria being surrounded by an extremely

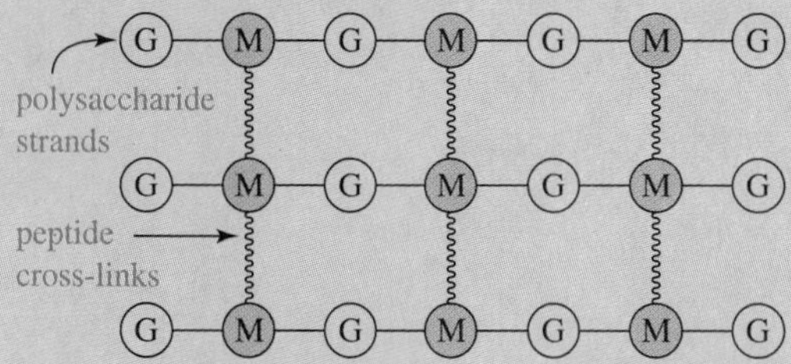

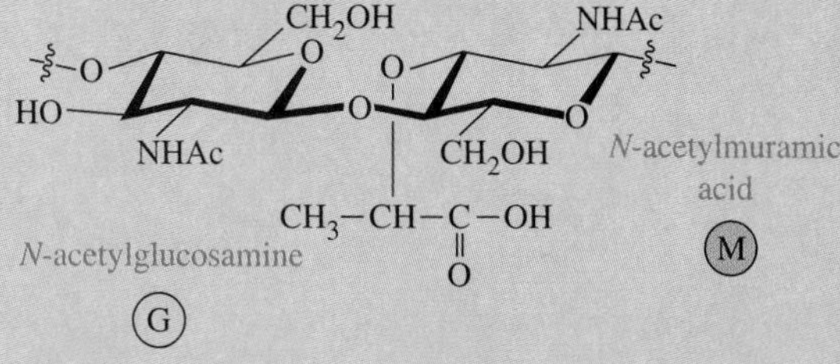

FIGURE 17.13 A schematic drawing of the bacterial cell wall and the structure of the polysaccharide repeating unit.

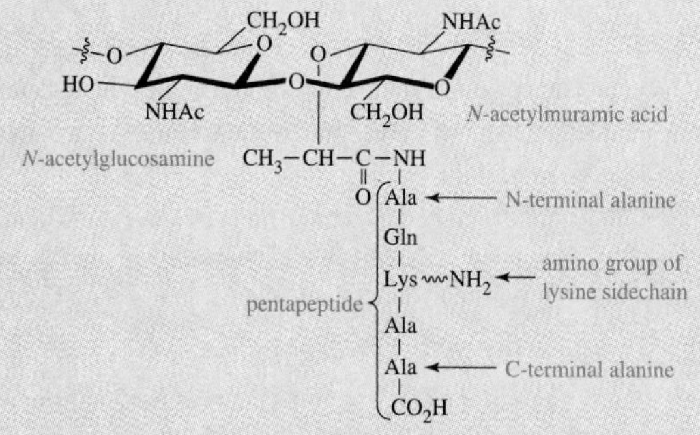

strong cell wall. The cross-linking process can be likened to vulcanization of rubber (Sec. 14.7), which also converts a linear polymer into a much more rigid and robust structure.

The cell wall is very important to bacteria because without strong cell walls, they are unable to withstand osmotic pressure and burst easily. Of course this leads to cell death. Since mammalian cells do not have such a cell wall, many compounds that interfere with cell wall biosynthesis will kill harmful bacteria without being toxic to the infected organism. Compounds that kill bacteria are call **antibiotics.** Although not all antibiotics inhibit cell wall biosynthesis, many do. Perhaps the most famous of these are the penicillins (see Sec. 13.7). These antibiotics bind to the transpeptidase enzyme and keep it from catalyzing the final cross-linking step in bacterial cell wall synthesis.

Compounds that keep enzymes from playing their roles as catalysts are known as **enzyme inhibitors** and the synthesis of inhibitors of specific enzymes now plays an important role in the discovery of new drugs.

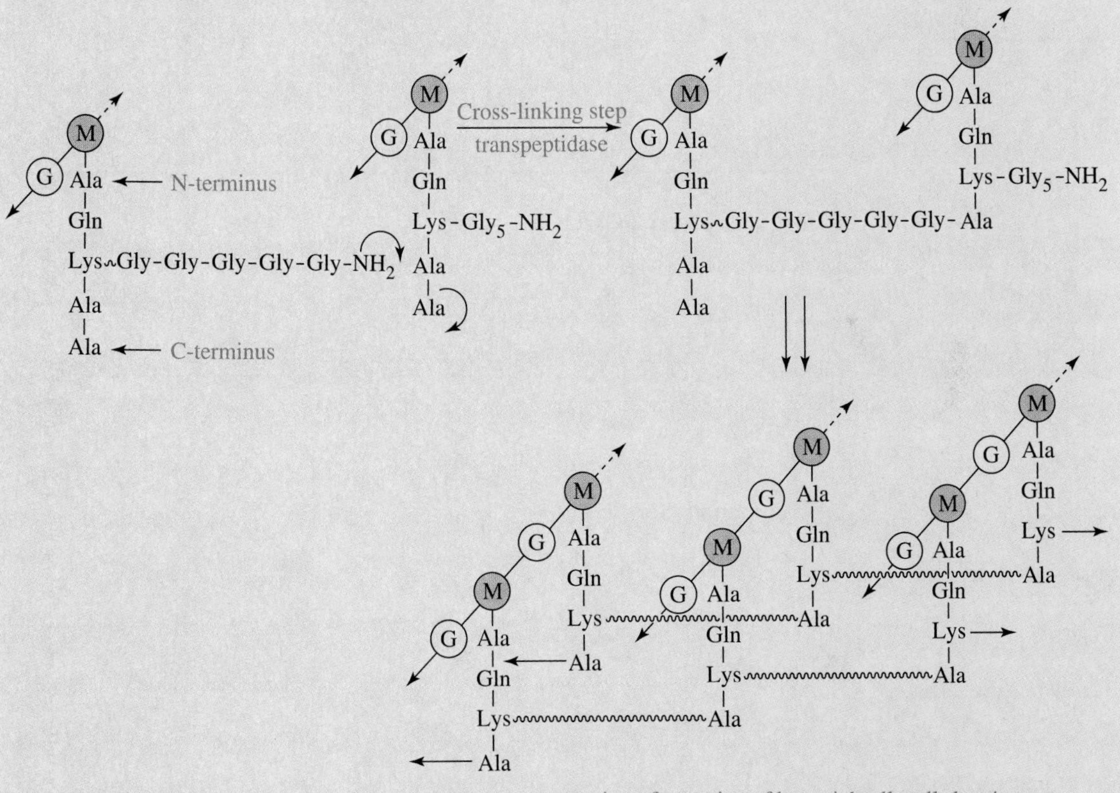

representation of a portion of bacterial cell wall showing four peptide cross-links along the polysaccharide backbone

FIGURE 17.14 A key step in bacterial biosynthesis involves the cross-linking of two peptidoglycan chains by an enzyme called a transpeptidase. Penicillin binds to the active site of transpeptidase and thus inhibits this step.

17.16 Quaternary Protein Structure

Some high-molecular-weight proteins exist as aggregates of several subunits. These aggregates are referred to as the **quaternary structure** of the protein. Aggregation helps to keep nonpolar portions of the protein surface from being exposed to the aqueous cellular environment.

Hemoglobin, the oxygen-transport protein of red cells, provides an example of such aggregation. It consists of four almost spherical units, two α units with 141 amino acids and two β units with 146 amino acids. The four units come together in a tetrahedral array, shown in Figure 17.15.

Many other proteins form similar aggregates. Some are active only in their aggregate state, whereas others are active only when the aggregate dissociates into subunits. Aggregation in quaternary structures, then, provides an additional control mechanism over biological activity.

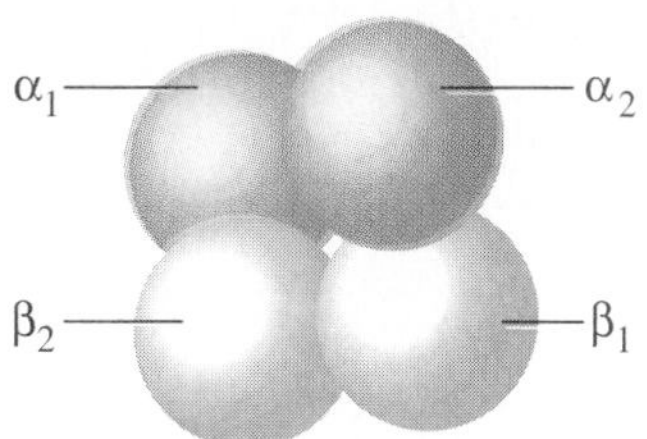

FIGURE 17.15 Schematic drawing of the four hemoglobin subunits.

REACTION SUMMARY

1. Reactions of Amino Acids

a. Acid–Base Reactions (Secs. 17.3 and 17.4)

$$\mathrm{R{-}CH(^{+}NH_3){-}CO_2H} \underset{H^+}{\overset{HO^-}{\rightleftharpoons}} \underset{\text{dipolar ion}}{\mathrm{R{-}CH(^{+}NH_3){-}CO_2^-}} \underset{H^+}{\overset{HO^-}{\rightleftharpoons}} \mathrm{R{-}CH(NH_2){-}CO_2^-}$$

b. Esterification (Sec 17.6)

$$\mathrm{R{-}CH(^{+}NH_3){-}CO_2^- + R'OH + H^+ \longrightarrow R{-}CH(^{+}NH_3){-}CO_2R' + H_2O}$$

c. Amide Formation (Sec. 17.6)

$$\mathrm{R{-}CH(^{+}NH_3){-}CO_2^- + R'{-}C({=}O){-}Cl \xrightarrow{2\ HO^-} R{-}CH(HN{-}C({=}O){-}R'){-}CO_2^- + 2\ H_2O + Cl^-}$$

d. Ninhydrin Reaction (Sec. 17.7)

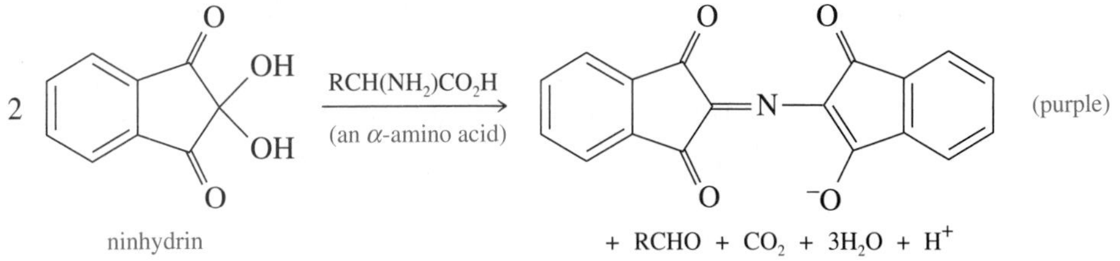

2 ninhydrin $\xrightarrow[\text{(an } \alpha\text{-amino acid)}]{RCH(NH_2)CO_2H}$ (purple) + RCHO + CO_2 + $3H_2O$ + H^+

2. Reactions of Proteins and Peptides

a. Hydrolysis (Sec. 17.11)

$$\text{proteins} \xrightarrow[H_2O,\ \Delta]{HCl} \text{peptides} \xrightarrow[H_2O,\ \Delta]{HCl} \alpha\text{-amino acids}$$

b. Sanger's Reagent (Sec. 17.11; used to identify the N-terminal amino acid of a peptide or protein)

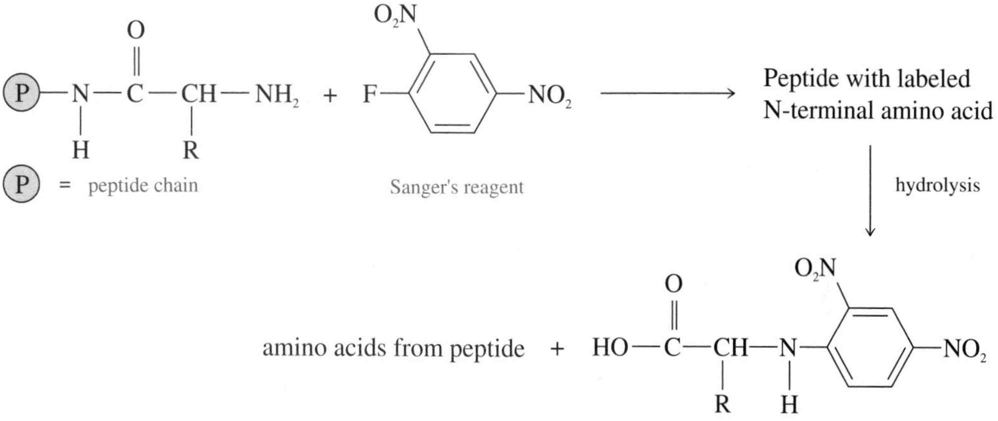

(P)—NH—C(=O)—CH(R)—NH_2 + F—C_6H_3(O_2N)—NO_2 ⟶ Peptide with labeled N-terminal amino acid

(P) = peptide chain

Sanger's reagent

hydrolysis

amino acids from peptide + HO—C(=O)—CH(R)—N(H)—C_6H_3(O_2N)—NO_2

labeled amino acid from N terminus of peptide

c. Edman degradation (Sec. 17.11; used to determine the amino acid sequence of a peptide)

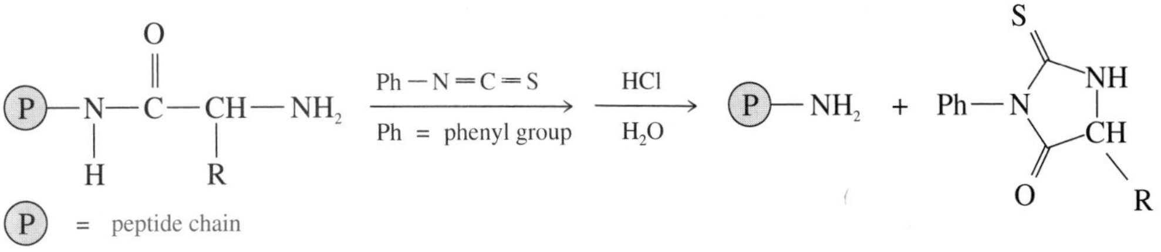

(P)—NH—C(=O)—CH(R)—NH_2 $\xrightarrow[\text{Ph = phenyl group}]{Ph-N=C=S}$ $\xrightarrow[H_2O]{HCl}$ (P)—NH_2 + Ph—N (cyclic: C(=S)—NH—CH(R)—C(=O))

(P) = peptide chain

d. Solid-Phase Peptide Synthesis (Sec. 17.13)

Details are given in Fig. 17.7.

ADDITIONAL PROBLEMS

17.21. Give a definition or illustration of each of the following terms:
a. peptide bond **b.** dipolar ion
c. dipeptide **d.** L configuration of amino acids
e. essential amino acid **f.** amino acid with a nonpolar R group
g. amino acid with a polar R group **h.** amphoteric compound
i. isoelectric point **j.** ninhydrin

17.22. Draw a Fischer projection for L-leucine. What is the priority order of the groups attached to the stereogenic center? What is the absolute configuration, *R* or *S*?

17.23. Write Fischer projection formulas for
a. L-tyrosine **b.** L-serine

17.24. Illustrate the amphoteric nature of amino acids by writing an equation for the reaction of alanine in its dipolar ion form with one equivalent of
a. hydrochloric acid. **b.** sodium hydroxide.

17.25. Write the formula for each of the following in its dipolar ion form:
a. valine **b.** serine **c.** proline **d.** tryptophan

17.26. Locate the most acidic proton in each of the following species, and draw the structure of the product formed by reaction with one equivalent of base (OH^-).

a. $HOOC{-}CH_2CH_2\underset{\substack{|\\ ^{+}NH_3}}{C}HCO_2H$ **b.** $HOCH_2{-}\underset{\substack{|\\ ^{+}NH_3}}{C}HCO_2^-$

c. $(CH_3)_2CH\underset{\substack{|\\ ^{+}NH_3}}{C}HCO_2H$ **d.** $NH_2(H_2\overset{+}{N}{=})C{-}NHCH_2CH_2CH_2\underset{\substack{|\\ NH_2}}{C}HCO_2^-$

17.27. What species is obtained by adding a proton to each of the following?

a. $CH_3\underset{\substack{|\\ OH}}{C}H{-}\underset{\substack{|\\ ^{+}NH_3}}{C}HCO_2^-$ **b.** $^{-}O_2CCH_2\underset{\substack{|\\ ^{+}NH_3}}{C}H{-}CO_2^-$

17.28. Protonated alanine, $CH_3CH(\overset{+}{N}H_3)CO_2H$, has a p$K_a$ of 2.34, whereas propanoic acid, $CH_3CH_2CO_2H$, has a pK_a of 4.85. Explain the increase in acidity due to replacing an α hydrogen with an $-\overset{+}{N}H_3$ substituent.

17.29. The pK_a's of glutamic acid are 2.19 (the α carboxyl group), 4.25 (the other carboxyl group), and 9.67 (the α ammonium ion). Write equations for the sequence of reactions that occurs when base is added to a strongly acidic (pH = 1) solution of glutamic acid.

17.30. The pK_a's of arginine are 2.17 for the carboxyl group, 9.04 for the ammonium ion, and 12.48 for the guanidinium ion. Write equations for the sequence of reactions that occurs when acid is gradually added to a strongly alkaline solution of arginine.

17.31. Draw the structure of histidine at pH 1, and show how the positive charge in the second basic group (the imidazole ring) can be delocalized.

17.32. Predict the direction of migration in an electrophoresis apparatus of each component in a mixture of asparagine, histidine, and aspartic acid at pH 6.

17.33. Write equations for the reaction of alanine with
a. $CH_3CH_2OH + HCl$. **b.** C_6H_5COCl + base. **c.** acetic anhydride.

17.34. Write equations for the following reactions:
a. serine + excess acetic anhydride →
b. threonine + excess benzoyl chloride →
c. glutamic acid + excess methanol + HCl →

17.35. Write the equations that describe what occurs when phenylalanine is treated with ninhydrin.

17.36. Write structural formulas for the following peptides:
a. alanylalanine **b.** valyltryptophan
c. tryptophanylvaline **d.** glycylalanylglycine

17.37. Write an equation for the hydrolysis of
a. leucylserine. **b.** serylleucine. **c.** valyltyrosylmethionine.

17.38. Write an equation for the acid-catalyzed hydrolysis of the artificial sweetener aspartame (page 474).

17.39. Write formulas that show how the structure of alanylglycine changes as the pH of the solution changes from 1 to 10. Estimate the pI (isoelectric point) of this dipeptide.

17.40. Use the three-letter abbreviations to write out all possible tetrapeptides containing one unit each of glycine, alanine, valine, and leucine. How many structures are possible?

17.41. Write the structure of the product expected from the reaction of glycylcysteine with a mild oxidizing agent, such as hydrogen peroxide (see Sec. 17.9).

17.42. Write equations for the following reactions of Sanger's reagent:
a. 2,4-dinitrofluorobenzene + glycine →
b. excess 2,4-dinitrofluorobenzene + lysine →

17.43. A pentapeptide was converted to its 2,4-dinitrophenyl (DNP) derivative, then completely hydrolyzed and analyzed quantitatively. It gave DNP-methionine, 2 moles of methionine, and 1 mole each of serine and glycine. The peptide was then partially hydrolyzed, the fragments were converted to their DNP derivatives, and each of them was hydrolyzed and analyzed quantitatively. Two tripeptides and two dipeptides isolated in this way gave the following products:
Tripeptide A: DNP-methionine and 1 mole each of methionine and glycine
Tripeptide B: DNP-methionine and 1 mole each of methionine and serine

Dipeptide C: DNP-methionine and 1 mole of methionine
Dipeptide D: DNP-serine and 1 mole of methionine
Deduce the structure of the original pentapeptide, and explain your reasoning.

17.44. Write the equations for the removal of one amino acid from the peptide alanylglycylvaline by the Edman method. What is the name of the remaining dipeptide?

17.45. Insulin (Figure 17.6), when subjected to the Edman degradation, gives *two* phenylthiohydantoins. From which amino acids are they derived? Draw their structures.

17.46. The following compounds are isolated as hydrolysis products of a peptide: Ala—Gly, Tyr—Cys—Phe, Phe—Leu—Try, Cys—Phe—Leu, Val—Tyr—Cys, Gly—Val, and Gly—Val—Tyr. Complete hydrolysis of the peptide shows that it contains one unit of each amino acid. What is the structure of the peptide, and what are its N- and C-terminal amino acids?

17.47. Simple pentapeptides called *enkephalins* are abundant in certain nerve terminals. They have opiate-like activity and are probably involved in organizing sensory information pertaining to pain. An example is *methionine enkephalin,* Tyr—Gly—Gly—Phe—Met. Write out its complete structure, including all the side chains.

17.48. Angiotensin II is an octapeptide with vasoconstrictor activity. Complete hydrolysis gives one equivalent each of Arg, Asp, His, Ile, Phe, Pro, Tyr, and Val. Reaction with Sanger's reagent gives, after hydrolysis,

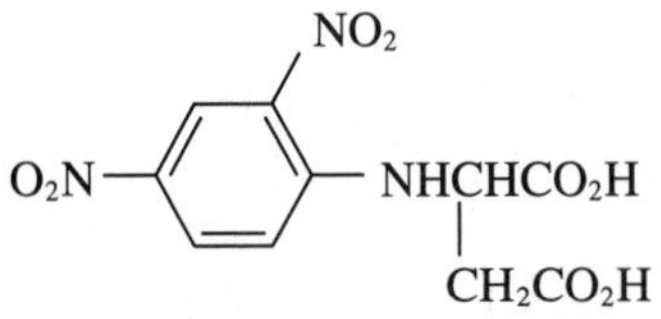

NO_2

O_2N—⟨benzene ring⟩—$NHCHCO_2H$

CH_2CO_2H

and seven amino acids. Treatment with carboxypeptidase gives Phe as the first released amino acid. Treatment with trypsin gives a dipeptide and a hexapeptide, whereas with chymotrypsin, two tetrapeptides are formed. One of these tetrapeptides, by Edman degradation, had the sequence Ile—His—Pro—Phe. From these data, deduce the complete sequence of angiotensin II.

17.49. *Endorphins* were isolated from the pituitary gland in 1976. They are potent pain relievers. β-Endorphin is a polypeptide containing 32 amino acid residues. Digestion of β-endorphin with trypsin gave the following fragments:

Lys
Gly—Gln
Asn—Ala—His—Lys
Asn—Ala—Ile—Val—Lys
Tyr—Gly—Gly—Phe—Leu—Met—Thr—Ser—Glu—Lys
Ser—Gln—Thr—Pro—Leu—Val—Thr—Leu—Phe—Lys

From these data only, what is the C-terminal amino acid of β-endorphin?
Treatment with cyanogen bromide gave the hexapeptide
Tyr—Gly—Gly—Phe—Leu—Met
and a 26-amino acid fragment. From these data only, what is the N-terminal amino acid of β-endorphin? Digestion of β-endorphin with chymotrypsin gave, among other fragments, a 15-unit fragment identified as
Leu—Met—Thr—Ser—Glu—Lys—Ser—Gln—Thr—Pro—Leu—Val—Thr—Leu—Phe
You should now be able to locate 22 of the 32 amino acid units. Write out as much as you can of the sequence. What further information do you need to complete the sequence?

17.50. The attachment of the N-protected C-terminal amino acid to the polymer in solid-phase peptide synthesis (Figure 17.7) is an S_N2 displacement reaction. What is the nucleophile? What is the leaving group? Write an equation that clearly shows the reaction mechanism.

17.51. The detachment of the peptide chain from the polymer in solid-phase peptide synthesis (Figure 17.7) occurs by an acid-catalyzed S_N2 mechanism. Write an equation to show this mechanism.

17.52. Write out all the steps in a Merrifield solid-phase synthesis of Leu—Pro.

17.53. Write the structure for glycylglycine, and show the resonance contributors to the peptide bond. At which bond is rotation restricted?

17.54. *Glucagon* is a polypeptide hormone secreted by the pancreas when the blood sugar level is low. It increases the blood sugar level by stimulating the breakdown of glycogen in the liver. The primary structure of glucagon is

His—Ser—Glu—Gly—Thr—Phe—Thr—Ser—Asp—Tyr—Ser—Lys—Tyr—Leu—
Asp—Ser—Arg—Arg—Ala—Gln—Asp—Phe—Val—Gln—Trp—Leu—
Met—Asn—Thr

What fragments would you expect to obtain from digestion of glucagon with:

a. trypsin **b.** chymotrypsin

17.55. In a globular protein, which of the following amino acid's side chains are likely to point toward the center of the structure? Which will point toward the surface when the protein is dissolved in water?

a. arginine **b.** phenylalanine **c.** isoleucine
d. glutamic acid **e.** asparagine **f.** tyrosine

18 Nucleotides and Nucleic Acids

18.1 Introduction

DNA, the double helix, and the genetic code—through the media's popularization of science, these have become household words. And they represent one of the greatest triumphs ever for chemistry and biology.

In this chapter, we will describe the structure of the nucleic acids, DNA and RNA. We will first look at their building blocks, the nucleosides and nucleotides, and then describe how these building blocks are linked to form giant nucleic acid molecules. Later we will consider the three-dimensional structures of these vital biopolymers and how the information they contain (the genetic code) was unraveled.

18.2 The General Structure of Nucleic Acids

Nucleic acids are linear, chainlike macromolecules that were first isolated from cell nuclei. **Hydrolysis of nucleic acids gives nucleotides,** which are the building blocks of nucleic acids, just as amino acids are the building blocks of proteins. A complete description of the primary structure of a nucleic acid requires knowledge of its nucleotide sequence, comparable to knowing the amino acid sequence in a protein.

Hydrolysis of a nucleotide gives 1 mole each of phosphoric acid and a nucleoside. The nucleoside can be hydrolyzed further, to 1 equivalent each of a sugar and a heterocyclic base.

$$\text{nucleic acid} \xrightarrow[\text{enzyme}]{H_2O} \underset{\text{(phosphate-sugar-heterocyclic base)}}{\text{nucleotide}}$$

$$\downarrow H_2O,\ OH^- \qquad (18.1)$$

$$\underset{\text{base}}{\text{heterocyclic}} + \text{sugar} \xleftarrow[H^+]{H_2O} \underset{\text{(sugar-base)}}{\text{nucleoside}} + H_3PO_4$$

The overall structure of the nucleic acid itself, then, is a macromolecule with a backbone of sugar molecules connected by phosphate links and with a base attached to each sugar unit.

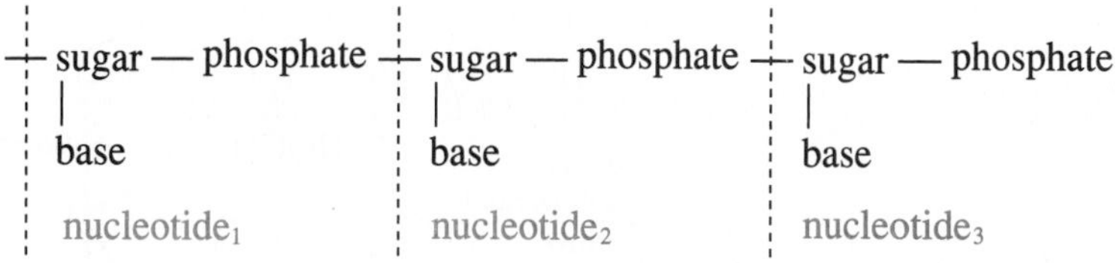
sugar — phosphate
base
nucleotide$_1$
sugar — phosphate
base
nucleotide$_2$
sugar — phosphate
base
nucleotide$_3$

schematic structure of a nucleic acid

18.3 *Components of Deoxyribonucleic Acid (DNA)*

Complete hydrolysis of DNA gives phosphoric acid, a single sugar, and a mixture of four heterocyclic bases. The sugar is **2-deoxy-D-ribose.**

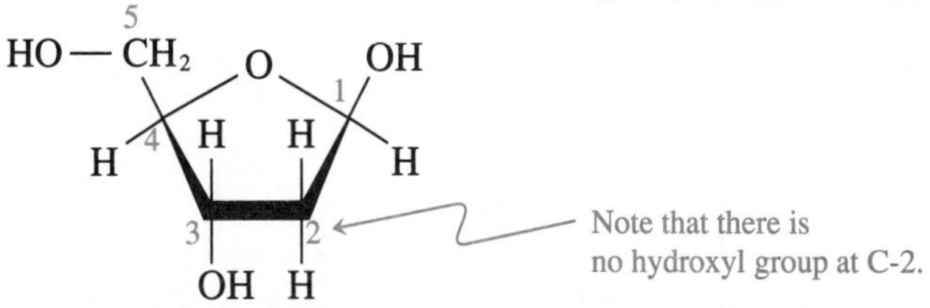
Note that there is
no hydroxyl group at C-2.

2-deoxy-D-ribose

The heterocyclic bases (Figure 18.1) fall into two categories, the pyrimidines (**cytosine** and **thymine;** review Sec. 13.4) and the purines (**adenine** and **guanine;** review Sec. 13.8). When we refer to these bases later, especially in connection with the genetic code, we will use the first letters of their names (capitalized) as abbreviations for their structures.

Now let us see how the sugar and bases are linked.

the pyrimidines | the purines

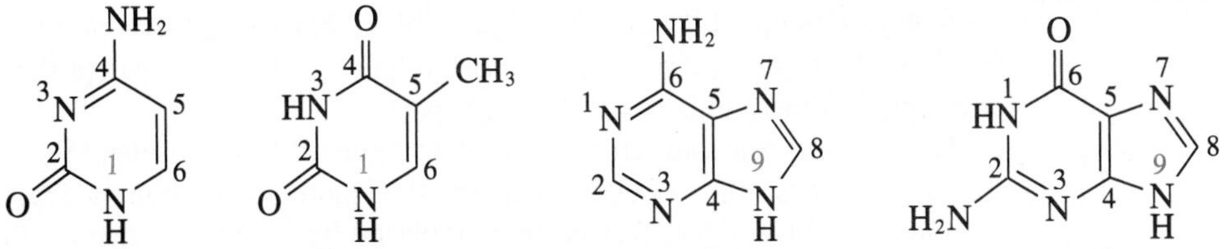

cytosine (C) | thymine (T) | adenine (A) | guanine (G)

FIGURE 18.1 The DNA bases.

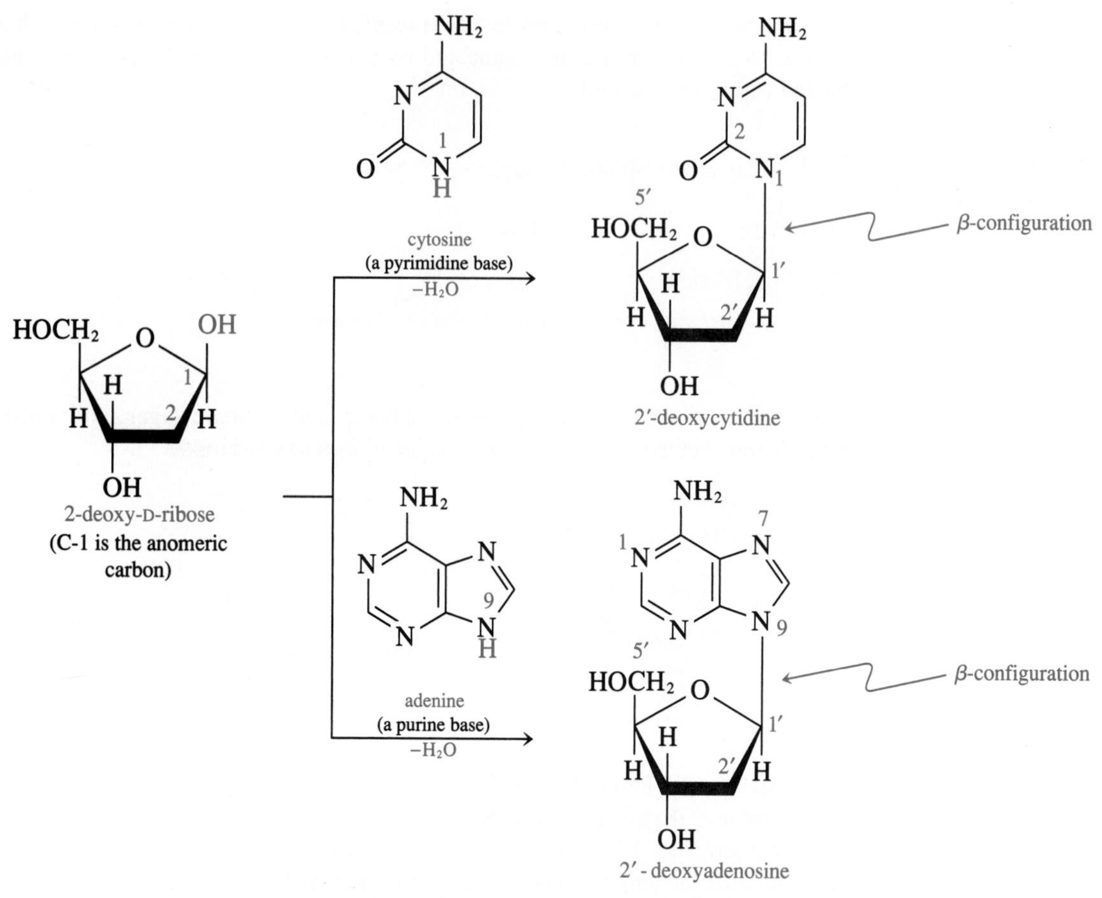

FIGURE 18.2 Schematic formation of nucleosides.

18.4 *Nucleosides*

A nucleoside is an *N-glycoside*. The pyrimidine or purine base is connected to the anomeric carbon (C-1) of the sugar. The pyrimidines are connected at N-1 and the purines at N-9 (Figure 18.2). Nucleoside structures are numbered in the same way as their component bases and sugars, except that primes are added to the numbers for the sugar part.

N-glycosides have structures similar to those of O-glycosides (Sec. 16.12). In O-glycosides, the —OH group on the anomeric carbon is replaced by —OR; in N-glycosides, that group is replaced by $—NR_2$.

EXAMPLE 18.1 Draw the structure of

a. the β-O-glycoside of 2-deoxy-D-ribose and methanol.
b. the β-N-glycoside of 2-deoxy-D-ribose and dimethylamine.

Solution a. 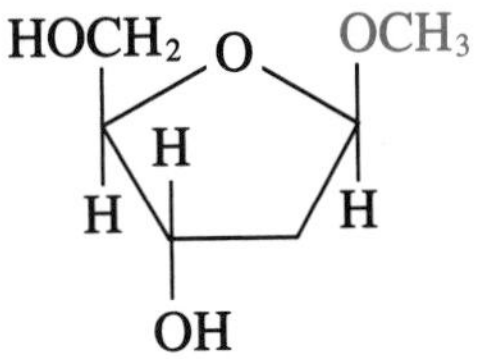 [structure: 5′-HOCH₂ furanose ring with OCH₃ at C-1, OH at C-3] b. 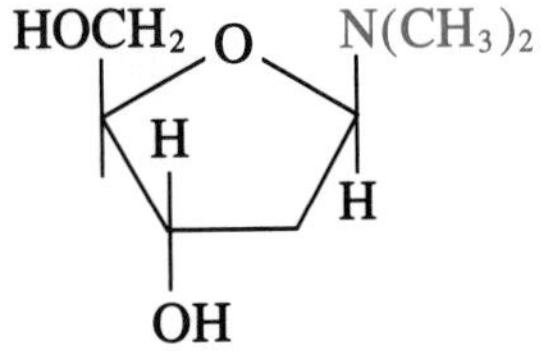 [structure: furanose ring with $N(CH_3)_2$ at C-1, OH at C-3]

Note the similarity between N- and O-glycosides.

PROBLEM 18.1 Figure 18.2 shows the structures of two DNA nucleosides. Draw the structures for the remaining two nucleosides of DNA: 2′-deoxythymidine and 2′-deoxyguanosine.

Because of their many polar groups, nucleosides are water soluble. Like other glycosides, they can be hydrolyzed readily by aqueous acid (or by enzymes) to the sugar and the heterocyclic base. For example,

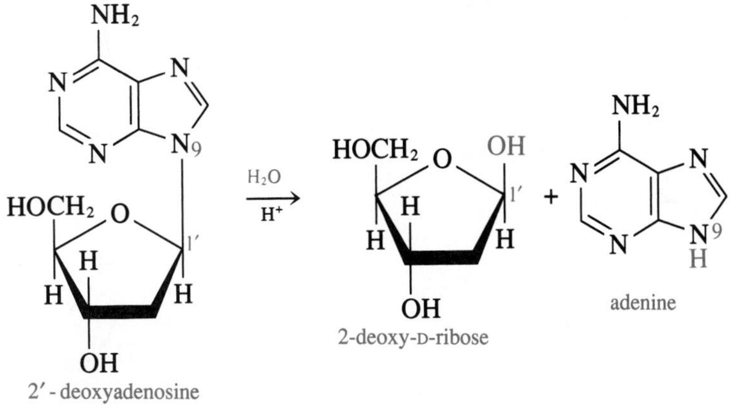

$$\text{2'-deoxyadenosine} \xrightarrow[H^+]{H_2O} \text{2-deoxy-D-ribose} + \text{adenine} \qquad (18.2)$$

PROBLEM 18.2 Using eq. 18.2 as a guide, write an equation for hydrolysis of

a. 2′-deoxythymidine. b. 2′-deoxyguanosine.

18.5 *Nucleotides*

Nucleotides are phosphate esters of nucleosides. A hydroxyl group in the sugar part of a nucleoside is esterified with phosphoric acid. In DNA nucleotides, either the 5′ or the 3′ hydroxyl group of 2-deoxy-D-ribose is esterified.

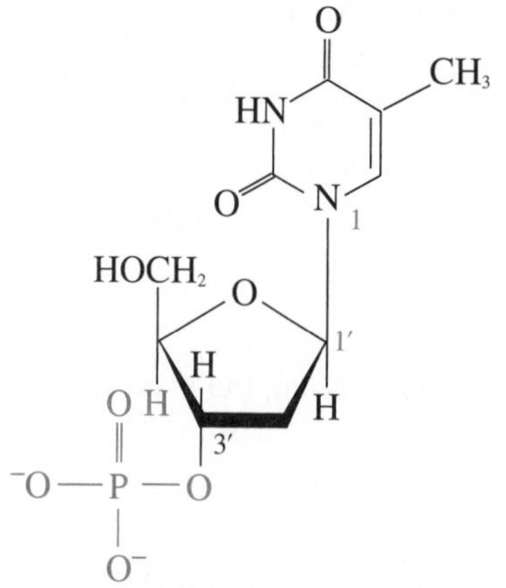

2′-deoxythymidine 3′-monophosphate

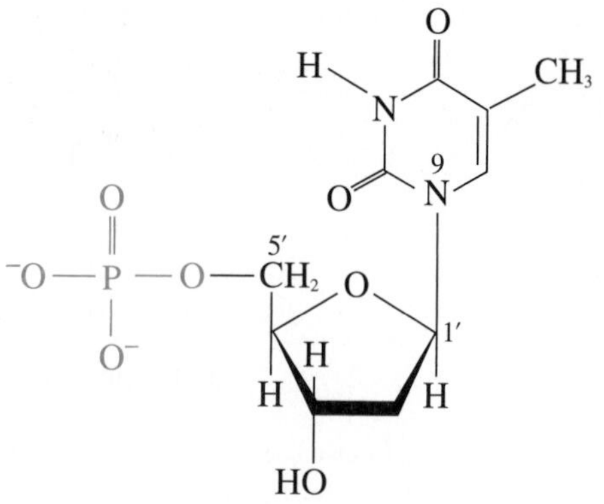

2′-deoxythymidine 5′-monophosphate

Nucleotides are named as the 3′- or 5′-monophosphate esters of a nucleoside, as shown above. These names are frequently abbreviated as shown in Table 18.1. In these abbreviations, the d stands for 2-deoxy-D-ribose, the next letter refers to the base, and MP stands for monophosphate. (Later we will see that some nucleotides are diphosphates, abbreviated DP, or triphosphates, TP.) Unless otherwise stated, the abbreviations usually refer to the 5′-phosphates.

TABLE 18.1 The common 2-deoxyribonucleotides:

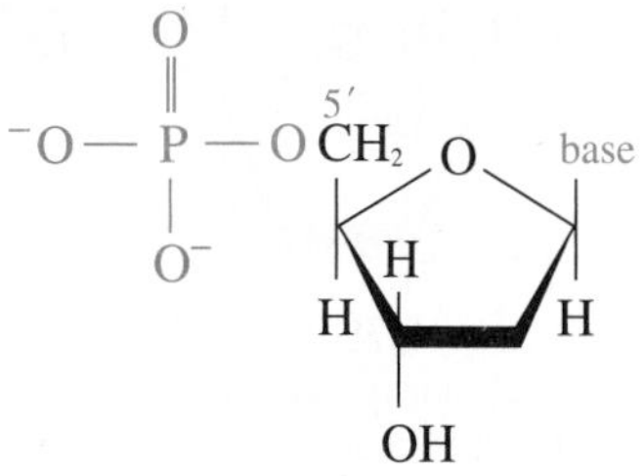

Base	*Monophosphate name*	*Abbreviation*
cytosine (C)	2′-deoxycytidine monophosphate	dCMP
thymine (T)	2′-deoxythymidine monophosphate	dTMP
adenine (A)	2′-deoxyadenosine monophosphate	dAMP
guanine (G)	2′-deoxyguanosine monophosphate	dGMP

EXAMPLE 18.2 Write the structure of dAMP.

Solution The letter d tells us that the sugar is 2-deoxy-D-ribose. The A stands for the base adenine, and the MP indicates a monophosphate. The structure is the same as that of the nucleotide shown in eq. 18.3.

PROBLEM 18.3 Write the structure for

a. dCMP. b. dGMP.

The phosphoric acid groups of nucleotides are acidic, and at pH 7, these groups exist mainly as dianions, as shown in the structures.

Nucleotides can be hydrolyzed by aqueous base (or by enzymes) to nucleosides and phosphoric acid. Phosphoric acid is sometimes abbreviated P_i, meaning inorganic phosphate.

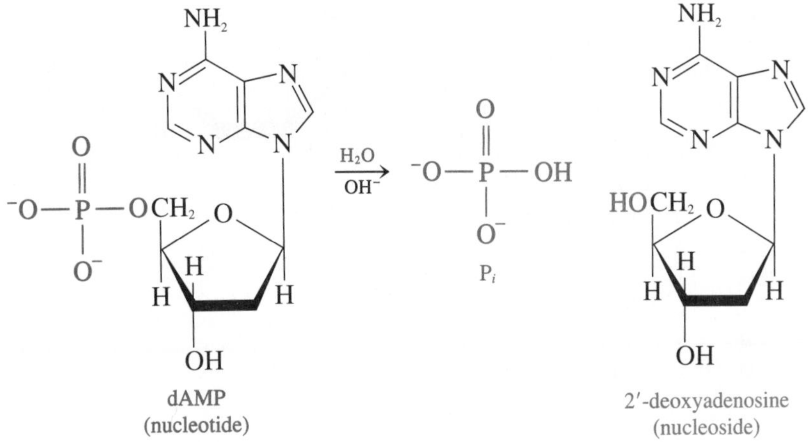

$$\text{dAMP (nucleotide)} \xrightarrow[OH^-]{H_2O} P_i + \text{2'-deoxyadenosine (nucleoside)} \qquad (18.3)$$

PROBLEM 18.4 Write a sequence of two equations showing the stepwise hydrolysis of dTMP first to its nucleoside, then to the sugar and free base.

Now let us see how the nucleotides are linked to one another in DNA.

18.6 *The Primary Structure of DNA*

In *deoxyribonucleic acid (DNA)*, 2-deoxy-D-ribose and phosphate units alternate in the backbone. The 3′ hydroxyl of one ribose unit is linked to the 5′ hydroxyl of the next ribose unit by a phosphodiester bond. The heterocyclic base is connected to the anomeric carbon of each deoxyribose unit by a β-N-glycosidic bond. Figure 18.3 shows a schematic drawing of a DNA segment.

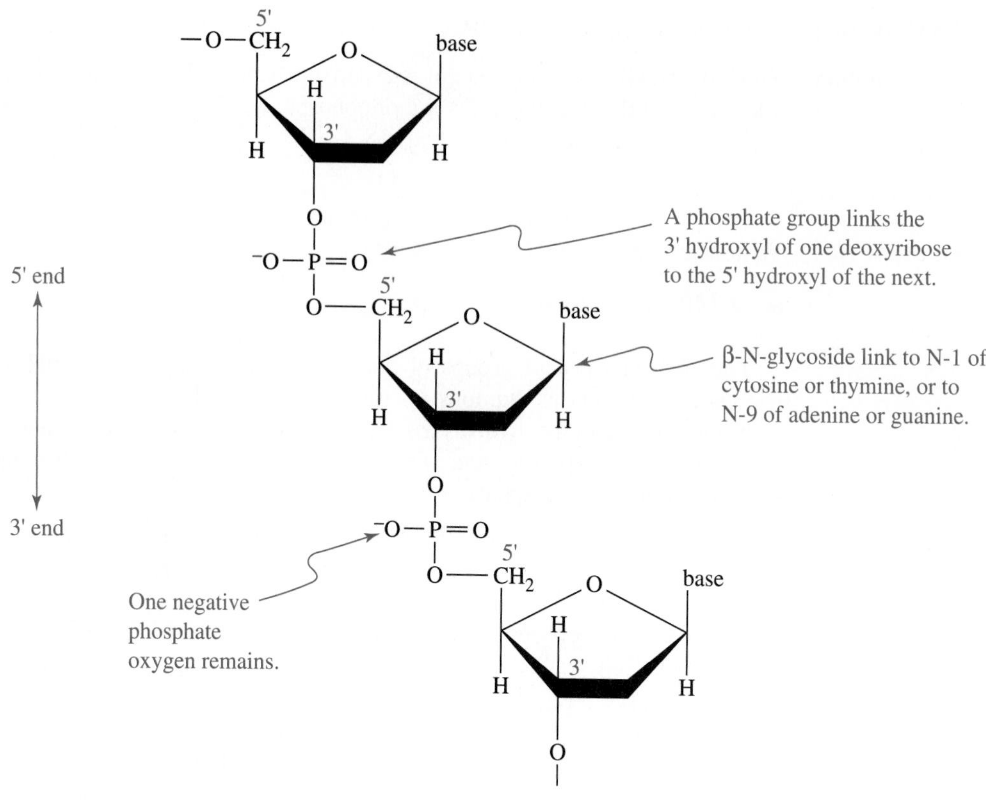
5' end
3' end
5'
base
3'
A phosphate group links the 3' hydroxyl of one deoxyribose to the 5' hydroxyl of the next.
β-N-glycoside link to N-1 of cytosine or thymine, or to N-9 of adenine or guanine.
One negative phosphate oxygen remains.

FIGURE 18.3 A segment of a DNA chain.

In DNA, there are no remaining hydroxyl groups on any deoxyribose unit. Each phosphate, however, still has one acidic proton that is usually ionized at pH 7, leaving a negatively charged oxygen, as shown in Figure 18.3. If this proton were present, the substance would be an acid; hence the name *nucleic acid*. A complete description of any particular DNA molecule, which may contain thousands or even millions of nucleotide units, would have to include the exact sequence of heterocyclic bases (A, C, G, and T) along the chain.

18.7 Sequencing Nucleic Acids

The problem of sequencing nucleic acids is in principle similar to that of sequencing proteins. At first, the job might appear to be easier because there are only 4 bases compared to 20 common amino acids. In fact, it is much more difficult. Even the smallest DNA molecule contains at least 5000 nucleotide units, and some DNA molecules may contain 1 million or more nucleotide units. To determine the exact base sequence in such a molecule is a task of considerable magnitude.

Without trying to discuss nucleic acid sequencing in detail, we can describe the strategy. The strategy basically relies on breaking the DNA into small identifiable fragments using a combination of enzymatic and chemical reactions. First, enzymes called **restriction endonucleases,** which split the DNA chain at known four-base sequences, are used to break the huge DNA molecule into smaller fragments with perhaps 100 to 150 nucleotide units. The purified fragments are then further degraded using four different and carefully controlled reaction conditions, which selectively split the chains at a particular base, A, G, C, or T. Each set of conditions gives a different group of yet smaller nucleotides that are subjected to gel electrophoresis (a technique similar to that used in separating peptides), which separates them based on the number of nucleotide units they contain. These experiments provide enough information such that the DNA sequence can be deciphered.

Progress in nucleic acid sequencing has been spectacular. In 1978, the longest known nucleic acid sequences (in RNA chains, which are shorter than DNA chains) were of about 200 nucleotide units. Later, the base sequence of DNA in a virus chromosome with 5375 nucleotide units was worked out by F. Sanger, who earned his second Nobel Prize in chemistry in 1980 for this achievement. In 1977, the Maxam-Gilbert* method for sequencing was introduced; and by 1985, base sequences of more than 170,000 nucleotide units became known. With present instrumentation, several thousand nucleotide bases can be sequenced in a day. Indeed, DNA sequencing, together with a knowledge of the genetic code (Sec. 18.12), is now sometimes used to sequence very large proteins.

A program of enormous magnitude—sequencing the entire human genome—now seems possible and is being undertaken on an international scale, with the hope of achieving the goal by the year 2000. This will require coding perhaps 3 billion base pairs, an impossible dream just a decade ago, but potential reality now.

A Word About . . .

34. DNA and Crime

DNA profiling, sometimes called **DNA fingerprinting,** is one of the most powerful new techniques of forensic science. Here's how it works.

A small quantity of DNA is obtained from some source associated with the crime—semen, blood, or hair roots, perhaps associated with a rape, murder, or other violent crime. The DNA is purified, cut with established restriction enzymes, and sequenced as described in Sec. 18.7. Functional genes—those that code for enzymes, hor-

* Walter Gilbert (Harvard University) shared a Nobel Prize in 1980 with F. Sanger and P. Berg.

mones, and other peptides common to all humans—do not vary from person to person, but these genes account for only about 5% of human DNA. The remaining DNA varies enormously from person to person, and in fact (except for identical twins) is characteristic of the individual. *The DNA in these genes can identify a single person.* And one tremendous advantage of the method is that the DNA sample can be very small, as little as a few micrograms.

There are two main uses of DNA profiling in dealing with a crime. One is to compare a suspect's profile with a sample from the scene of the crime. A few years ago, in one of the first applications of the DNA technique, this kind of pairwise comparison cleared one individual suspected of two murders and led to the conviction of another.

But DNA profiling also has the potential for use in crime investigation, through the accumulation of databases similar to those used for fingerprinting. Once that is done, a search of the database should answer such questions as: Does the assailant in this crime match up with one from a previous crime? Is there a match with an individual already on file?

Despite the tremendous potential of DNA profiling, it isn't without problems; a New York murder case that came to trial in 1989 identified some of them. As with all analytical methods, DNA typing must be done carefully and with proper controls; otherwise, errors are possible, especially with band shifting, when one lane in the gel electrophoresis step runs faster or slower than another. When comparing patterns from two DNA samples, it is essential to show that the bands match in order to conclude that the samples come from the same individual. When the job is done properly and a match is obtained, the odds for identity are very long indeed—usually better than one in a hundred million. But in their zeal to apply the method, some firms have been less careful than necessary about controls, and as a result of the New York trial, the method may be "on hold" for a few years. But it is hoped that guidelines soon will be set up by independent agencies (for example in the United States, the National Academy of Sciences) for appropriate practice of DNA profiling. Once that is done, the method should be a great boon in solving crimes.

18.8 Laboratory Synthesis of Nucleic Acids

It is important to be able to synthesize specific DNA segments in the laboratory, just as it is important to synthesize specific peptide segments. Short nucleotide chains of known sequence were needed, as we will see, to solve the genetic code. Longer chains of known sequence could, through genetic engineering techniques, be used to induce microorganisms to synthesize useful proteins—insulin, for example. Synthetic **oligonucleotides** of known sequence can be inserted into the DNA of microorganisms, where they can serve as templates for biological DNA synthesis.

The problems of laboratory DNA synthesis are similar to, but even greater than, those of peptide synthesis, mainly because nucleotides have more complicated structures than do amino acids. Each nucleotide has several functional groups that must be protected, and later deprotected, during synthesis. Despite these difficult chemical problems, methods for oligonucleotide synthesis have been developed in various laboratories. Using these methods, Khorana* and coworkers were able to synthesize a gene for the first time, by a combination of chemical and enzymatic means.

More recently, automated gene synthesizers have been developed that operate on principles similar to the Merrifield solid-phase technique for peptides. A protected nucleotide is covalently bonded to a polymer. Other protected nucleotides are then added sequentially to the chain, using a coupling reagent. Eventually, the protecting groups are removed, and the synthetic oligonucleotide is then detached from the solid support. Fairly rapid and reliable synthesis of oligonucleotides is now possible, and further progress in this area is likely to be rapid.

Let us now proceed from the primary DNA structure, the specific base sequences, to the secondary DNA structure: the double helix and the genetic code.

18.9 Secondary DNA Structure; the Double Helix

It has been known since 1938 that DNA molecules have a discrete shape, because x-ray studies of DNA threads showed a regular stacking pattern with some periodicity. A key observation by E. Chargaff (Columbia University) in 1950 provided an important clue to the structure. Chargaff analyzed the base content of DNA from many different organisms and found that the amounts of A and T are always equal and the amounts of G and C are also equal. For example, human DNA contains about 30% each of A and T and 20% each of G and C. Other DNA sources give different percentages, but the ratios of A to T and of G to C are always unity.

The meaning of these equivalences was not evident until 1953, when Watson and Crick, ** working together in Cambridge, England, proposed the double helix model for DNA. They received simultaneous supporting x-ray data for their proposal from Rosalind Franklin and Maurice Wilkins in London. The important features of their model are:

1. DNA consists of two helical polynucleotide chains coiled around a common axis.
2. The helices are right-handed, and the two strands run in opposite directions with regard to their 3′ and 5′ ends.
3. The purine and pyrimidine bases lie *inside* the helix, in planes perpendicular to the helical axis; the deoxyribose and phosphate groups form the outside of the helix.
4. The two chains are held together by purine-pyrimidine base pairs connected by hydrogen bonds. ***Adenine is always paired with thymine, and guanine is always paired with cytosine.***
5. The diameter of the helix is 20 Å. Adjacent base pairs are separated by 3.4 Å and oriented through a helical rotation of 36°. There are therefore 10

* Har Gobind Khorana (Massachusetts Institute of Technology) won the Nobel Prize in medicine, 1968.

** James D. Watson (Harvard University) and Francis H. Crick (Cambridge University) won the Nobel Prize in medicine in 1960.

base pairs for every turn of the helix (360°), and the structure repeats every 34 Å.

6. There is no restriction on the sequence of bases along a polynucleotide chain. The exact sequence carries the genetic information.

Figure 18.4 shows schematic models of the double helix. The key feature of the structure is the complementarity of the base pairing: **A—T** and **G—C.** Only purine–pyrimidine base pairs fit into the helical structure. There is not enough room for two purines and too much room for two pyrimidines, which would be too far apart to form hydrogen bonds. Of the purine–pyrimidine pairs, the hydrogen-bonding possibilities are best for A—T and G—C pairing.

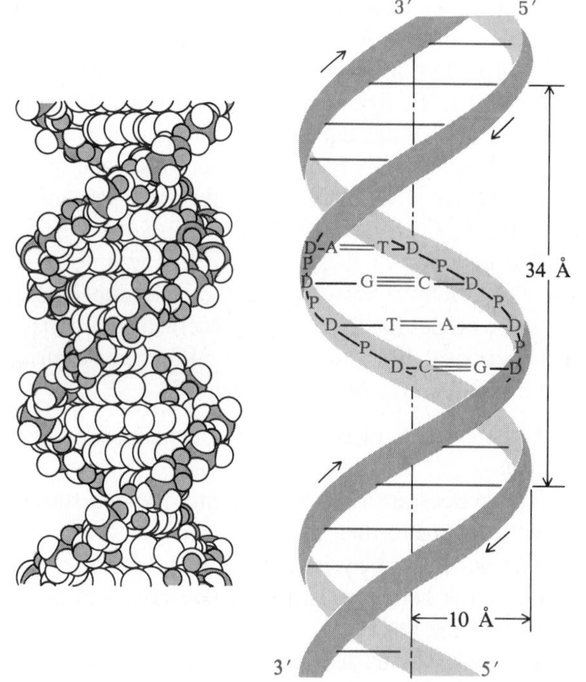

FIGURE 18.4
Model and schematic representations of the DNA double helix. The space-filling model at the left clearly shows the base pairs in the helix interior, in planes perpendicular to the main helical axis. The center drawing shows the structure more schematically, including the dimensions of the double helix. At the right is a schematic method for showing base pairing in the two strands.

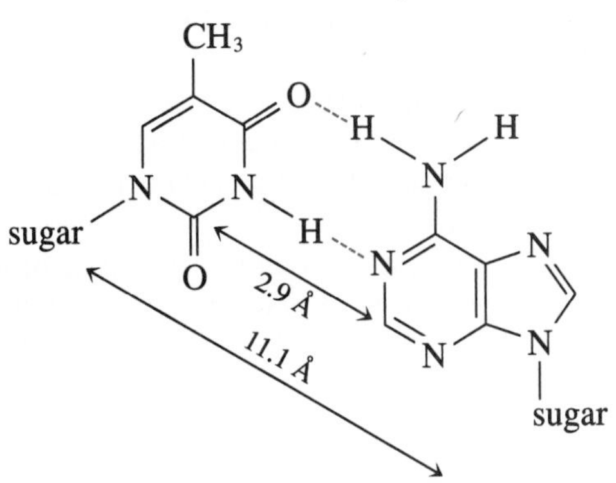

T — A base pair (2 H bonds)

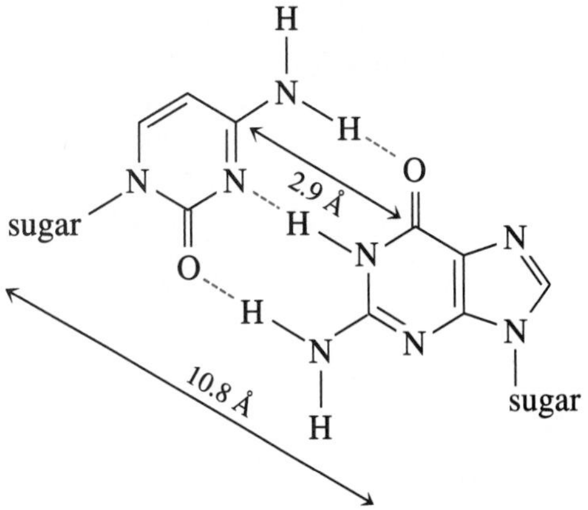

C — G base pair (3 H bonds)

The A — T pair is joined by two hydrogen bonds and the G — C pair by three. The geometries of the two pairs are nearly identical.

PROBLEM 18.5 Consider the following sequence of bases from one strand of DNA: — AGCCATGT — (written from 5′ to 3′). What will the sequence of bases on the other strand be?

We now know that, although the Watson–Crick model for the double helix is essentially correct, it is oversimplified. Helical conformations of DNA can now be classified into three general families, called the A-, B- and Z-forms. B-DNA, the predominant form, is the regular right-handed helix of Watson and Crick, with the base pairs essentially perpendicular to the helix axis. In the A-form, base pairs may be tilted by as much as 20° to the helix axis, and the sugar rings are puckered differently from the way they are in the B-form. And in the Z-form we see a 180° rotation of some of the bases about the C — N glycosidic bond, resulting in a *left*-handed helix.

The particular overall conformation adopted by a DNA molecule depends in part on the actual base sequence. For example, synthetic DNAs made of alternating purine-pyrimidine units have different conformations from DNAs made of blocks of purine bases followed by blocks of pyrimidine bases. Also, A — T and G — C base-pairing with different H-bonds from that originally proposed by Watson and Crick has been observed.

These variations in details of DNA structures lead, instead of to the rigid helical column shown in Figure 18.4, to DNA molecules with bends, hairpin loops, supercoils, single-stranded loops, and even cruciforms in which single intrastrand H-bonded loops are extruded from the double helix. These structural changes add flexibility to the way DNA molecules are able to recognize and interact with other cellular components to perform their functions.*

---

* For an excellent, readable article on this subject, see J. K. Barton, *Chem. and Eng. News* **1988,** Sept. 26, pp. 30–42.

18.10 DNA Replication

The beauty of the DNA double helix model was that it immediately suggested a molecular basis for transmitting information from one generation to the next: **DNA replication.** In 1954, Watson and Crick proposed that, as the two strands of a double helix separate, a new complementary strand is synthesized from nucleotides in the cell, using one strand as a template for the other. Figure 18.5 schematically depicts the process.

Though simple in principle, replication is quite a complex process in practice. The nucleotides must be present as triphosphates (not monophosphates),

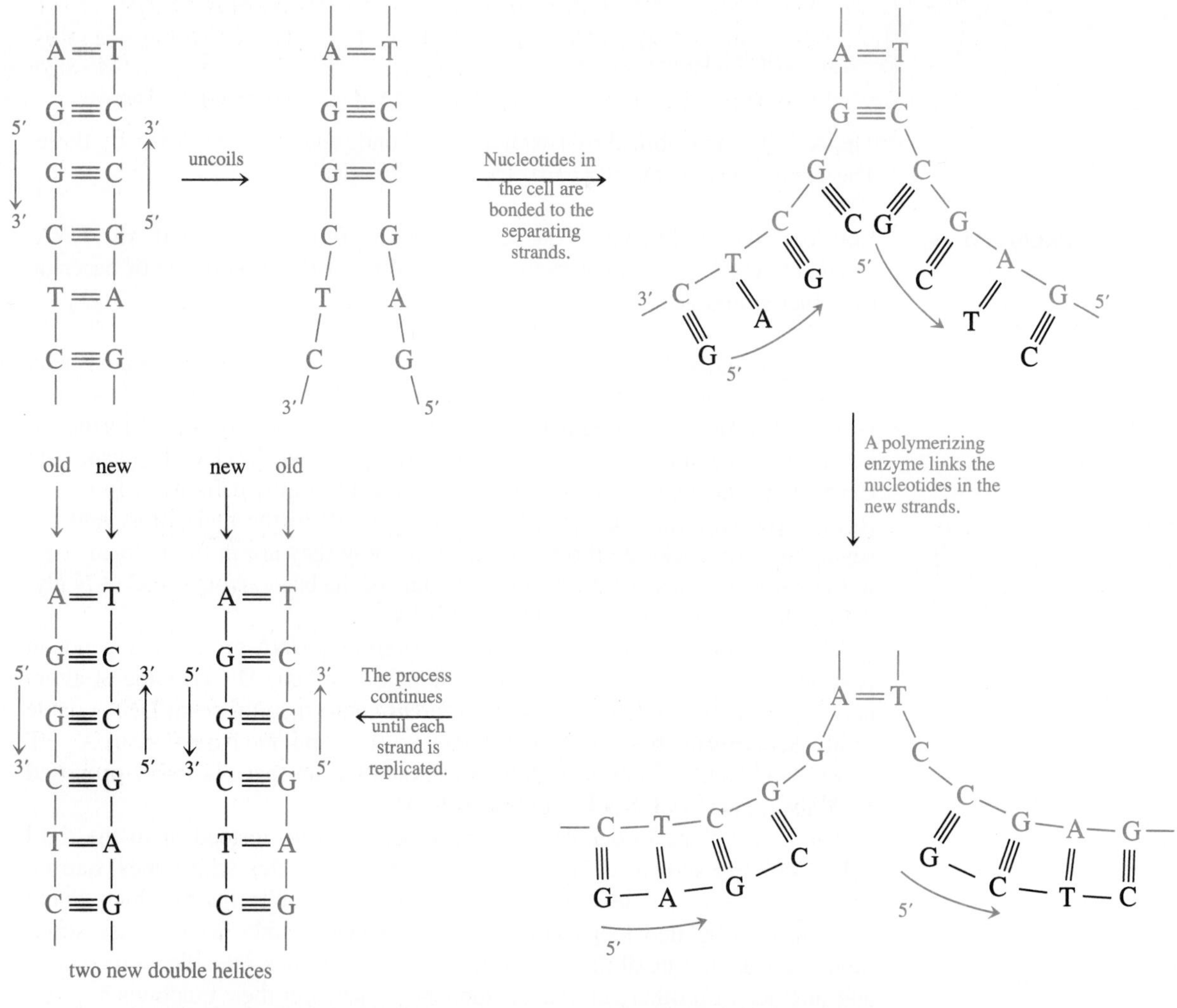

uncoils

Nucleotides in the cell are bonded to the separating strands.

A polymerizing enzyme links the nucleotides in the new strands.

The process continues until each strand is replicated.

old new new old

two new double helices

FIGURE 18.5 Schematic representation of DNA replication. As the double helix uncoils, nucleotides in the cell bond to the separate strands, following the base-pairing rules. A polymerizing enzyme links the nucleotides in the new strands to one another. Both new strands are assembled from the 5′ to the 3′ end.

an enzyme (DNA-polymerase) adds the nucleotides to a primer chain, other enzymes link DNA chains (DNA-ligase), there are specific places at which replication starts and stops, and so on. Our knowledge of the details of this process has increased considerably since the DNA double helix was proposed more than three decades ago, and as in all science, it seems that the more we know, the more we need to learn.

Before we turn to the role of DNA in protein synthesis, we must first consider another type of nucleic acid, RNA, which plays a key role in this process.

18.11 *Ribonucleic Acids; RNA*

Ribonucleic acids (RNA) differ from DNA in three important ways: (1) the sugar is D-ribose; (2) uracil replaces thymine as one of the four heterocyclic bases; and (3) most RNA molecules are single-stranded, although helical regions may be present by looping of the chain back on itself.

The RNA sugar D-ribose differs from the DNA sugar in that it has a hydroxyl group at C-2. Otherwise, the nucleosides and nucleotides of RNA have structures similar to those of DNA.

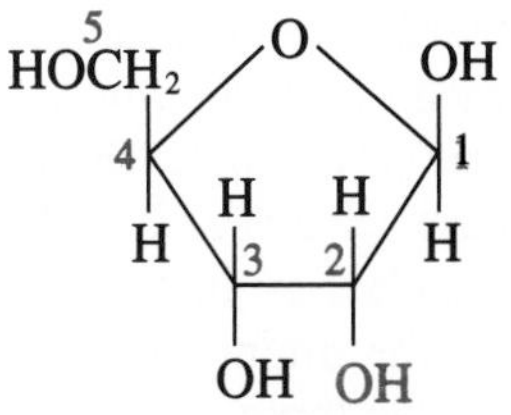

D-ribose

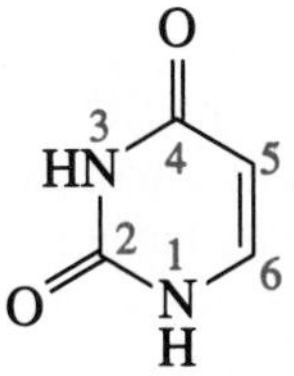

uracil

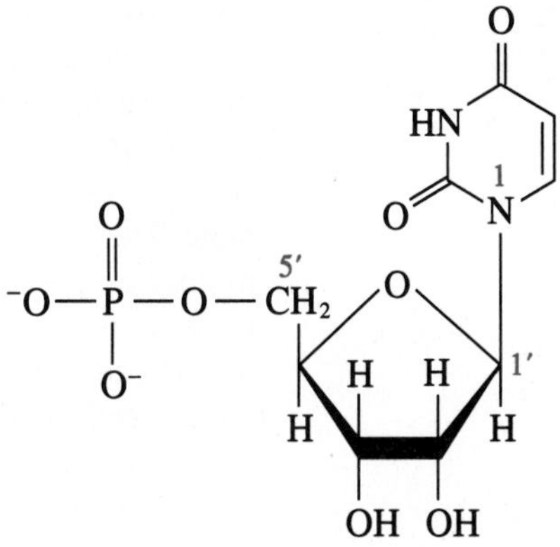

uridine-5′-monophosphate (UMP)

Uracil differs from thymine only in lacking the C-5 methyl group. Like thymine, it forms nucleotides at N-1, and their names are similar to those in Table 18.1. In the abbreviations of these names, the d (which stands for deoxyribose) is omitted because the sugar is ribose.

PROBLEM 18.6 Draw the full structure of

a. AMP.
b. the RNA trinucleotide UCG (written from 5′ to 3′).

Cells contain three major types of RNA. ***Messenger RNA*** *(mRNA) is involved in* ***transcription*** *of the genetic code and is the template for protein synthesis.* There is a specific *m*RNA for every protein synthesized by the cell. The base sequence of *m*RNA is complementary to the base sequence in a single strand of DNA, with U the complement of A.

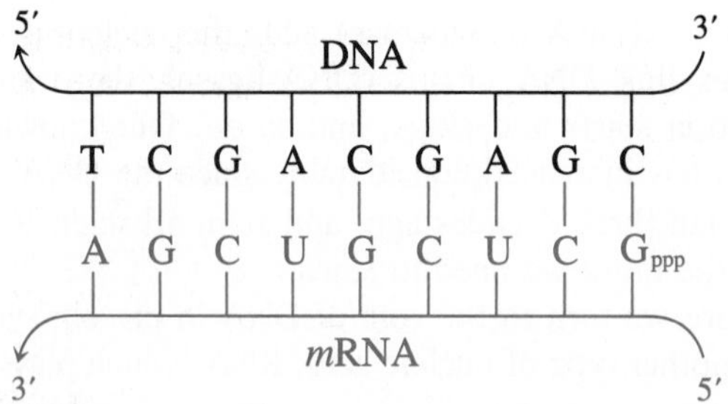

Transcription proceeds in the 3′-to-5′ direction along the DNA template. That is, the *m*RNA chain grows from its own 5′ end. The 5′ terminal nucleotide in *m*RNA is usually present as a triphosphate, not a monophosphate, and is commonly pppG or pppA. An enzyme called RNA-polymerase is essential for transcription. Usually only one strand of DNA is transcribed. It contains base sequences, called *promoter sites*, which initiate transcription. It also contains certain termination sequences, which signal the completion of transcription.

At the 3′ end of *m*RNA, there is usually a special sequence of about 200 successive nucleotide units of the same base, adenine. This sequence plays a role in transporting the *m*RNA from the cell nucleus to the *ribosomes,* the cellular structures where proteins are synthesized.

Transfer RNA (*t*RNA) *carries amino acids in an activated form to the ribosome for peptide bond formation,* in a sequence determined by the *m*RNA template. There is at least one *t*RNA for each of the 20 amino acids. Transfer RNA molecules are relatively small as nucleic acids go, with about 70 to 90 nucleotide units. Each *t*RNA has a three-base sequence, C—C—A, at the 3′ hydroxyl end, where the amino acid is attached as an ester. Each *t*RNA also has an **anticodon loop** quite remote from the amino acid attachment site. This loop contains seven nucleotides, the middle three of which are complementary to the three-base code word on the *m*RNA for that particular amino acid.

The third type of RNA is **ribosomal RNA** (*r*RNA). It comprises about 80% of the total cellular RNA (*t*RNA = 15%, *m*RNA = 5%) and is the main component of the ribosomes. Its molecular weight is large, and each molecule may contain several thousand nucleotide units.

Until recently, it was thought that all enzymes are proteins. But this dogma of biochemistry was recently overthrown by the discovery that some types of RNA can function as biocatalysts. They can cut, splice, and assemble themselves without outside help of conventional enzymes. This discovery* of **ribozymes,** as they are called, had a major impact on theories of the origin of life. The question was: Which came first in the primordial soup from which life began, the proteins or the nucleic acids? Proteins could be enzymes and catalyze the reactions needed for life, but they could not store genetic information. The reverse was thought to be true for nucleic acids. But, with the discovery of catalytic activity in certain types of RNA, it now seems almost certain that the

---

*Sidney Altman (Yale) and Thomas R. Cech (University of Colorado) received the 1989 Nobel Prize in chemistry for this discovery.

earth of 4 billion years ago was an RNA world, in which RNA molecules carried out all the processes of life without the help of proteins or DNA—even though the latter now contains the genetic code.

18.12 The Genetic Code and Protein Biosynthesis

It is beyond the scope of this book to give a detailed account of the genetic code, how it was unraveled, and how more than a hundred types of macromolecules must interact to translate that code into the synthesis of a protein. But we can present a few of the main concepts.

The **genetic code** is the relationship between the base sequence in DNA, or its RNA transcript, and the amino acid sequence in a protein. A three-base sequence, called a **codon,** corresponds to *one* amino acid. Because there are 4 bases in RNA (A, G, C, and U), there are $4 \times 4 \times 4 = 64$ possible codons. However, there are only 20 common amino acids in proteins. Each codon corresponds to only one amino acid, but the code is *degenerate;* that is, *several different codons may correspond to the same amino acid.* Of the 64 codons, 3 are codes for "stop" (UAA, UAG, and UGA). Each signals that the particular protein synthesis is complete. One codon, AUG, serves double duty. It is the initiator codon, but if it appears again after a chain has been initiated, it codes for the amino acid methionine. The entire code is summarized in Table 18.2.

TABLE 18.2 The genetic code; translation of the codons into amino acids

First base (5′ end)	*Second base*	*Third base (3′ end)* U	C	A	G
U	U	Phe	Phe	Leu	Leu
	C	Ser	Ser	Ser	Ser
	A	Tyr	Tyr	Stop	Stop
	G	Cys	Cys	Stop	Trp
C	U	Leu	Leu	Leu	Leu
	C	Pro	Pro	Pro	Pro
	A	His	His	Gln	Gln
	G	Arg	Arg	Arg	Arg
A	U	Ile	Ile	Ile	Met (start)
	C	Thr	Thr	Thr	Thr
	A	Asn	Asn	Lys	Lys
	G	Ser	Ser	Arg	Arg
G	U	Val	Val	Val	Val
	C	Ala	Ala	Ala	Ala
	A	Asp	Asp	Glu	Glu
	G	Gly	Gly	Gly	Gly

How was the genetic code solved? The first successful experiment was done by Marshall Nirenberg in 1961 (Nobel Prize, 1968). Nirenberg added a synthetic RNA, polyuridine (an RNA in which all the bases were uracil, U) to a cell-free protein-synthesizing system containing all the amino acids. He found a tremendous increase in the incorporation of phenylalanine in the resulting polypeptides. Since UUU is the only codon present in polyuridine, it must be a codon for phenylalanine. Similarly, polyadenosine led to the synthesis of polylysine, and polycytidine led to polyproline. Thus AAA ≡ Lys, and CCC ≡ Pro. Later, other synthetic polyribonucleotides with known repeating sequences were found to give other polypeptides with repeating amino acid sequences, and in this way, the complete code was unraveled.

EXAMPLE 18.3 A polyribonucleotide was prepared from the tetranucleotide UAUC. When it was subjected to peptide-synthesizing conditions, the polypeptide (Tyr—Leu—Ser—Ile)$_n$ was obtained. What are the codons for these four amino acids?

Solution The polyribonucleotide must have the sequence

$\underline{\text{U A U C}}\,\underline{\text{U A U C}}\,\underline{\text{U A U C}}\cdots$

If we divide the chain into codons, we get

$$\left.\begin{array}{l}\text{UAU—CUA—UCU—AUC}\cdots\\ \text{ Tyr — Leu — Ser — Ile }\cdots\end{array}\right\}\text{ this sequence repeats}$$

In this way, the meaning of four codons is disclosed.

PROBLEM 18.7 A polynucleotide made from the dinucleotide UA turned out to be (Tyr—Ile)$_n$. How does this outcome confirm the results in Example 18.3? What new information does this experiment give?

The genetic code is universal for all organisms on earth and has remained invariant through all the years of evolution. Consider what would happen if the "meaning" of a codon were changed. The result would be a change in the amino acid sequence of most proteins synthesized by that organism. Many of these changes would undoubtedly be disadvantageous. Hence there is a strong natural selection *against* changing the code. It was recently demonstrated, from a statistical analysis of *t*RNAs, that the genetic code cannot be older than 3.8 (±0.6) billion years and thus is not older than, but almost as old as, our planet.*

PROBLEM 18.8 Mutations (caused by radiation, cancer-producing agents, or other means) may replace one base with another or may add or delete a base. What would happen

---

* Manfred Eigen (Nobel Prize, 1967) and coworkers, *Science* **1989,** *244,* 673–679.

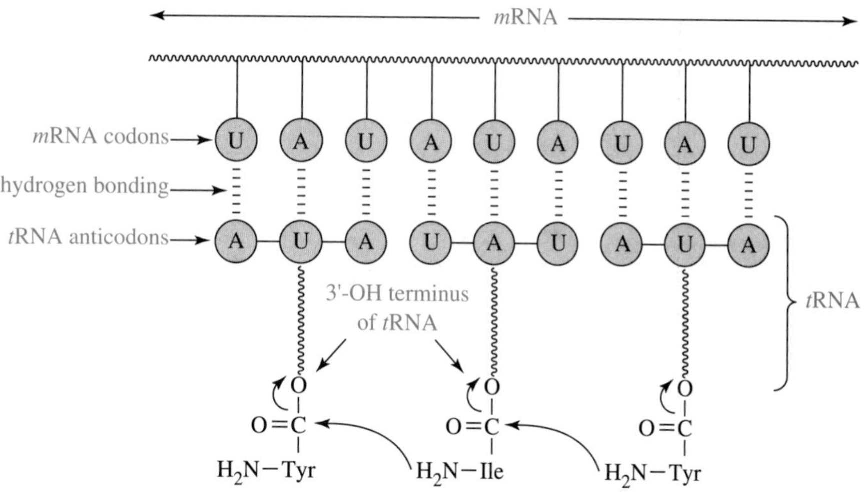

FIGURE 18.6 Schematic representation of protein biosynthesis. The amino group of one amino acid (Ile) displaces the carboxyl group of another amino acid (Tyr) from the terminal 3′-hydroxyl group of its *t*RNA (AUA). This continues down the *m*RNA template. The example would give the tripeptide unit Tyr–Ile–Tyr (see Problem 18.7).

to the protein produced if the sequence UUU were mutated to UCU? If UCU were mutated to UCC? What advantage is there to the genetic code in having redundant codons?

Proteins are biosynthesized using *m*RNA as a template (Figure 18.6). Amino acids, each attached to its own unique *t*RNA, are brought up to the *m*RNA, where the anticodon on the *t*RNA matches up, through hydrogen bonding, with the codon on the *m*RNA. Enzymes then link the amino acids together, detach them from their *t*RNAs, and detach the *t*RNAs from the *m*RNA so that the process can be repeated.

Coordinated reactions involving many types of molecules are required for protein biosynthesis. The molecules include *m*RNA, *t*RNA, scores of enzymes, the amino acids, phosphate, and many others. In spite of these requirements, this complex process happens with remarkable speed. It is estimated that a protein containing as many as 150 amino acid units can be assembled biosynthetically in less than a minute! When we compare this with solid-phase peptide synthesis (Sec. 17.13), efficient as it is for a laboratory procedure, we see that as chemists we still have a long way to go if we are to emulate nature.

18.13 Other Biologically Important Nucleotides

The nucleotide structure is a part not only of nucleic acids, but also of several other biologically active substances. Some of the more important of these species are described here.

Adenosine exists in several different phosphate forms. The 5′-*mono*phosphate, *di*phosphate, and *tri*phosphate, as well as the 3′,5′-cyclic monophosphate, are key intermediates in many biological processes.

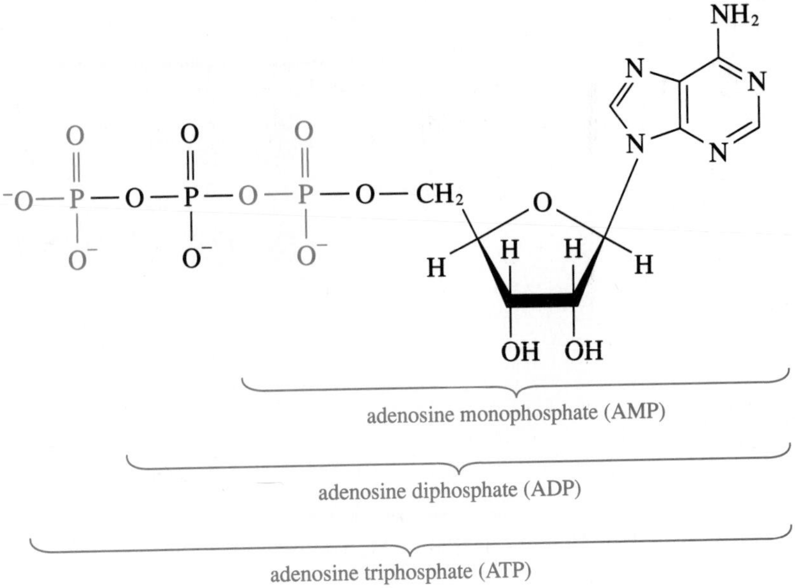

adenosine monophosphate (AMP)

adenosine diphosphate (ADP)

adenosine triphosphate (ATP)

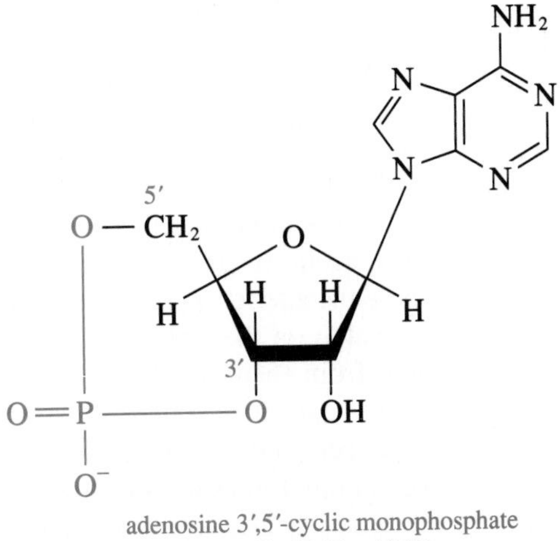

adenosine 3′,5′-cyclic monophosphate
(cyclic AMP; cAMP)

ATP contains two phosphoric anhydride bonds, and considerable energy is released when ATP is hydrolyzed to ADP and further to AMP. These reactions often provide energy for other biological reactions.

Cyclic AMP is a mediator of certain hormonal activity. When a hormone outside a cell interacts with a receptor site on the cell membrane, it may stimulate cAMP synthesis inside the cell. The cAMP in turn acts *within* the cell to regulate some biochemical process. In this way, a hormone need not penetrate a cell to exert its effect.

Four important coenzymes contain nucleotides as part of their structures. We have already mentioned **coenzyme A** (for its structure, see Figure 10.1), which contains ADP as part of its structure. It is a biological acyl-transfer agent and plays a key role in fat metabolism. **Nicotinamide adenine dinucleotide (NAD)** is a coenzyme that dehydrogenates alcohols to aldehydes or ketones, or the reverse process: reduces carbonyl groups to alcohols. It consists of two nucleotides linked by the 5′ hydroxyl group of each ribose unit.

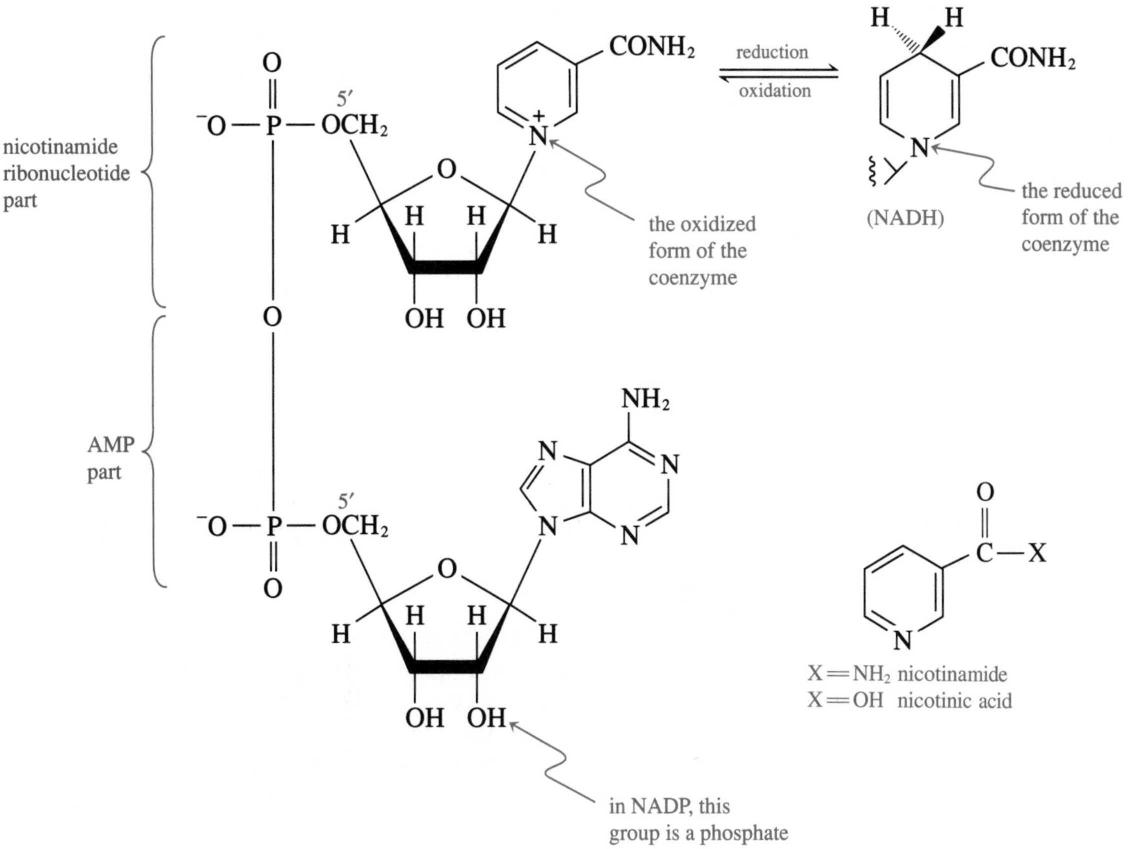

nicotinamide ribonucleotide part

AMP part

$CONH_2$ reduction / oxidation $CONH_2$

the oxidized form of the coenzyme

(NADH)

the reduced form of the coenzyme

in NADP, this group is a phosphate

$X{=}NH_2$ nicotinamide
$X{=}OH$ nicotinic acid

nicotinamide adenine dinucleotide (NAD)

When NADP (a phosphate ester of NAD) oxidizes an alcohol to a carbonyl compound, the pyridine ring in the nicotinamide part of the coenzyme is reduced to a dihydropyridine, giving NADPH. The reverse process occurs when NADPH reduces a carbonyl compound to an alcohol. Nicotinic acid is a B vitamin needed for synthesis of this coenzyme. Its deficiency causes the chronic disease pellagra.

Flavin adenine dinucleotide (FAD) is a yellow coenzyme involved in many biological oxidation-reduction reactions. It consists of a riboflavin part (vitamin B_2) connected to ADP. The reduced form has two hydrogens attached to the riboflavin part.

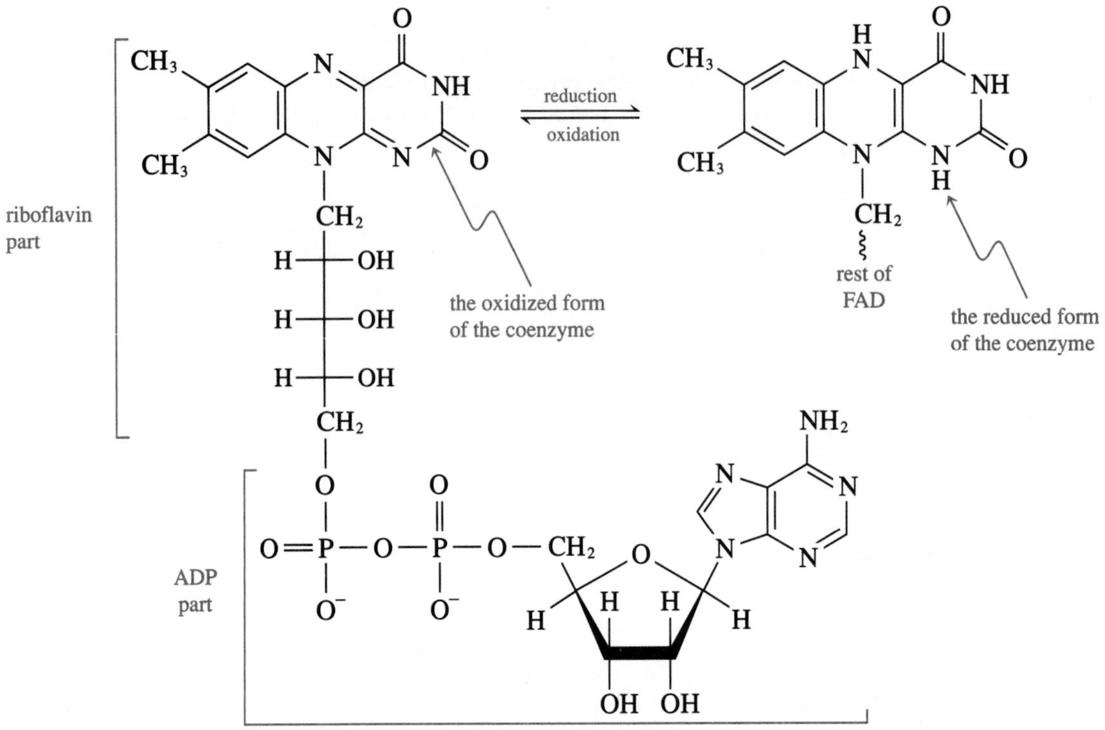

riboflavin part

ADP part

reduction / oxidation

the oxidized form of the coenzyme

the reduced form of the coenzyme

flavin adenine dinucleotide (FAD)

Vitamin B_{12} (cobalamine), which is essential for the maturation and development of red blood cells, is an incredibly complex molecule that includes a nucleotide as part of its structure (Figure 18.7). The related **coenzyme B_{12}** contains a second nucleotide unit. Both of these molecules have a central cobalt atom surrounded by a macrocyclic molecule containing four nitrogens, similar to a porphyrin (see "A Word About Porphyrins," p. 396). But the cobalt has two additional ligands attached to it, above and below the mean plane of the nitrogen-containing rings. One of these ligands is a ribonucleotide of the unusual base, **5,6-dimethylbenzimidazole.** The other ligand is a cyanide group in the vitamin and a 5-deoxyadenosyl group in the coenzyme. In each case, there is a direct carbon-cobalt bond. The reactions catalyzed by coenzyme B_{12} usually involve replacement of the Co—R group by a Co—H group.

Vitamin B_{12}, which is produced by certain microorganisms, cannot be synthesized by humans and must be ingested. Only minute amounts are required, but pernicious anemia can result from its deficiency.

Vitamin B_{12}, with its remarkable array of functionality and chirality, is one of the most complex molecules ever to have been created in an organic laboratory. Its synthesis was completed in 1973 by R. B. Woodward* and A. Eschenmoser and their students.

---

*Robert Burns Woodward (Harvard University) received the Nobel Prize in chemistry in 1965 for his many contributions to the "art of organic synthesis." He is regarded by many organic chemists as perhaps the greatest practitioner of this art.

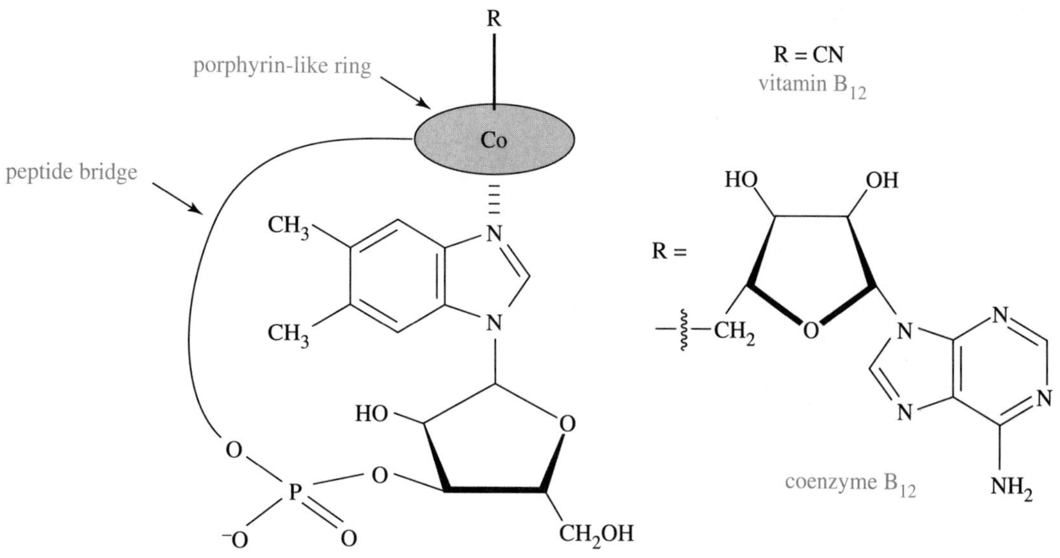

FIGURE 18.7 Schematic representation of vitamin B_{12} and coenzyme B_{12}.

A Word About . . .

35. Nucleic Acids and Viruses

Viral infections—from influenza and the common cold to the more serious herpes infections and AIDS (acquired immune deficiency syndrome)—account for about 60% of illnesses, contrasted with about 15% for bacterial infections. Since the 1940s, chemists have developed all sorts of highly effective antibiotics (sulfa drugs, penicillins, tetracyclines, and others) that are effective against bacterial infections, but progress with antiviral agents has been slower and more difficult. Why is this so, and what hope do we have for combating viral infections?

The problem is one of selectivity. Any drug must *selectively* kill pathogens in the presence of other living cells. Fortunately, there are sufficient biochemical differences between the metabolisms of bacterial and of mammalian cells to allow selectivity; thus, safe antibiotics have been developed. Viruses present a more difficult problem because, during their replicative cycle, they become physically and functionally incorporated into host cells, and one must find biochemical features that selectively attack the virus without damaging the host.

Viruses are exceedingly small; they consist of a protein coat surrounding an inner core of nucleic acid. The core carries all the genetic information for their reproduction. Broadly, there are two classes of viruses, in which the nucleic acid is either DNA (herpes viruses) or RNA (flu virus and HIV, the human immunodeficiency virus responsible for AIDS). When viruses infect a host cell, they first attach to the membrane of the host cell, then penetrate the membrane and shed their protein coat, and the viral nucleic acid enters the host cell nucleus, where it replicates. The new nucleic acid then leaves the nucleus and combines with structural proteins to form new viruses, which are then expelled from the cell. The normal metabolic machinery of the cell is essentially hijacked and forced into manufacturing the viral components instead of its own. The host cell becomes, in a sense, a virus factory.

Any one of these steps (from attachment to the membrane to final ejection) are targets for interference by an antiviral agent. One trick, for example, might be to design a molecule very much like the DNA nucleosides, but sufficiently different for it to put a roadblock in the DNA synthesis scheme. One example of a successful antiviral agent based on this idea is **acyclovir** [9-(2-hydroxyethoxymethyl) guanine (ACV), developed by Gertrude B. Elion* and coworkers at the Wellcome Research Laboratories (North Carolina). Notice that ACV mimics one of the natural nucleosides, 2′-deoxyguanosine, differing only (but importantly) in lacking part of the sugar portion of the molecule (shown in color).

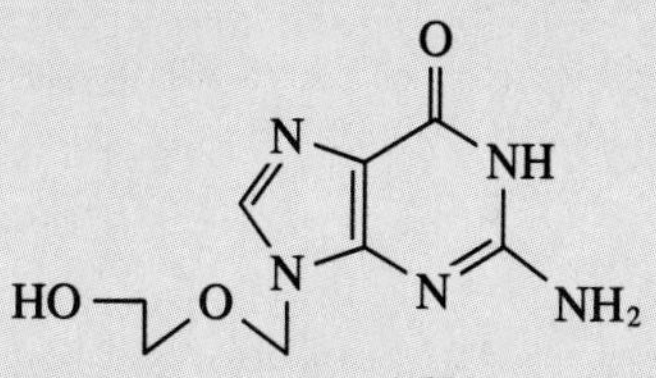

ACV

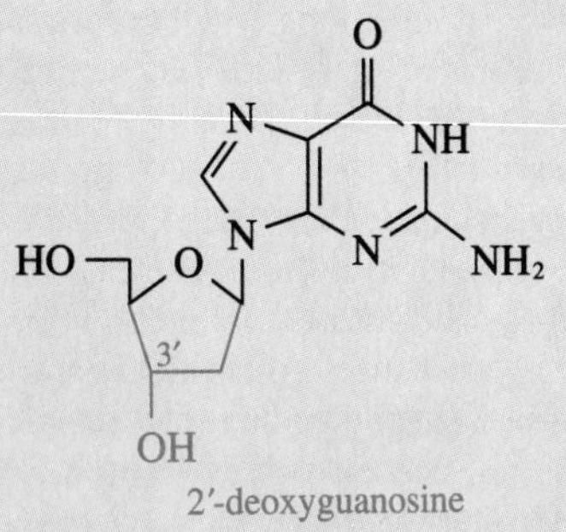

2′-deoxyguanosine

ACV is effective against herpes simplex virus (HSV). It has been used clinically since about the 1980s and has decreased suffering and saved lives. It is especially useful in treating genital herpes. It functions by selectively interfering with at least two enzyme-catalyzed processes essential for viral reproduction. Also, if it is incorporated into a DNA chain, it acts as a chain terminator because it lacks the 3′-hydroxyl group essential for attaching the next nucleotide unit.

AIDS, another virally transmitted disease, has evolved into an epidemic of worldwide proportions. At this writing, there are three FDA-approved drugs used to treat HIV-infected individuals. *Zidovudine* (or *AZT*, 3′-azido-3′-deoxythymine), an analog of the DNA nucleoside 2′-deoxythymidine, has been in use the longest.

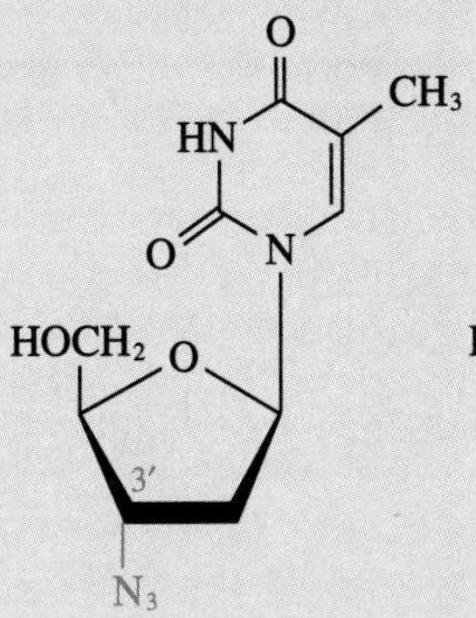

zidovudine (AZT)

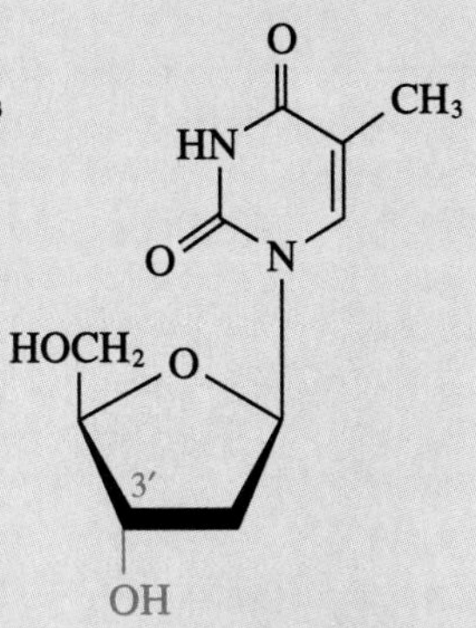

2′-deoxythymidine

Didanosine (*DDI*, 2′,3′-dideoxyinosine) and *zalcitabine* (*DDC*, 2′,3′-dideoxycytidine) are two other nucleoside analogs that are now used to treat the AIDS virus. Numerous other compounds are in clinical trials.

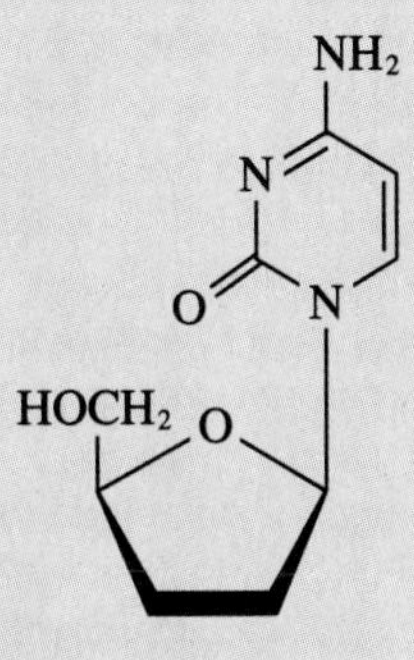

DDC

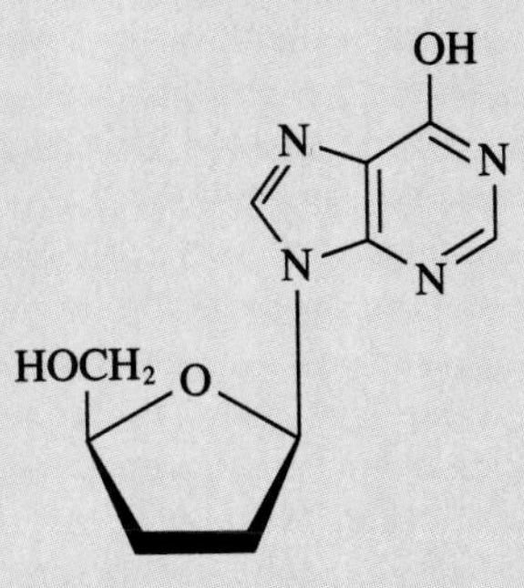

DDI

* Gertrude B. Elion, George Hitchings, and James Black shared the 1988 Nobel Prize in physiology and medicine.

AZT and other nucleoside analogs interfere with reverse transcriptase, the enzyme HIV uses to transcribe its RNA genome into a DNA copy. The enzyme is fooled into incorporating the drug instead of the natural nucleotide into the growing DNA chain, but the growth then stops because there is no OH group (an azido group or hydrogen atom instead) at the 3′-position. But AZT is hardly an ideal drug, due to its toxicity to bone marrow (where blood cells are produced). Also, the AIDS virus becomes resistant to these drugs, so there is still a great need for new methods of therapy.*

Many other approaches to the control of the HIV virus are also being researched. For example, CD4 is the protein on cell surfaces to which HIV binds when it infects a cell; molecules that combine CD4 with protein toxins might inhibit viral entry into cells. And vaccines are also being researched. One day, perhaps through application of methods such as those outlined here or new ones yet undiscovered, effective antiviral agents will be developed. It is certainly one of the major medical challenges of our times.

* For an overview of the present status and future prospects for HIV therapies, see Margaret I. Johnston and Daniel F. Hoth in *Science* **1993,** *260,* 1286–1293.

REACTION SUMMARY

1. Hydrolysis of Nucleic Acids (Sec. 18.2)

$$\text{DNA} \xrightarrow[\text{enzyme}]{H_2O} \text{nucleotides} \xrightarrow[HO^-]{H_2O} \text{nucleosides} \xrightarrow[H^+]{H_2O} \text{heterocyclic bases} + \text{2-deoxyribose}$$

The behavior of RNA is identical to DNA except ribose is produced instead of 2-deoxyribose.

2. Hydrolysis of Nucleosides (Sec. 18.5)

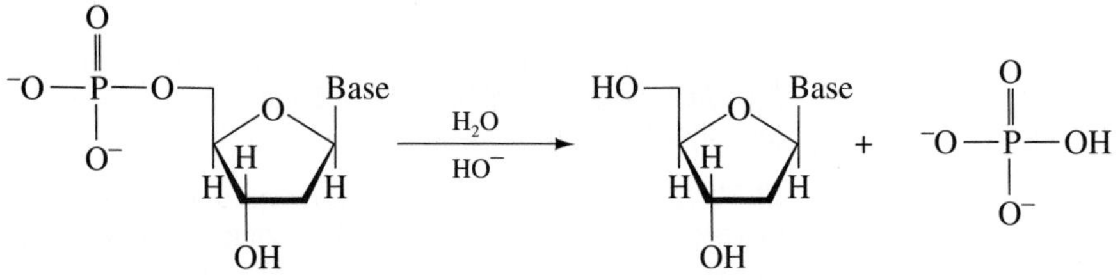

$$\text{nucleotide} \xrightarrow[HO^-]{H_2O} \text{nucleoside} + \text{inorganic phosphate}$$

nucleotide nucleoside inorganic phosphate

ADDITIONAL PROBLEMS

18.9 Write the structural formula for an example of each of the following:
a. a pyrimidine base **b.** a purine base **c.** a nucleoside **d.** a nucleotide

18.10. Examine the structures of adenine and guanine (Figure 18.1). Do you expect their rings to be planar or puckered? Explain. What about the pyrimidine bases, cytosine and thymine?

18.11. Draw the structure of each of the following nucleosides:
a. cytidine (from β-D-ribose and cytosine)
b. deoxyadenosine (from β-2-deoxy-D-ribose and adenine)
c. uridine (from β-D-ribose and uracil)
d. deoxyguanosine (from β-2-deoxy-D-ribose and guanine)

18.12. Write an equation for the complete hydrolysis of adenosine-5′-monophosphate (AMP) to its component parts.

18.13. Using Table 18.1 as a guide, write the structures of the following nucleotides:
a. guanosine 5′-monophosphate **b.** 2′-deoxythymidine 5′-monophosphate

18.14. Draw the structures of the following DNA-derived dinucleotides:
a. A—T **b.** G—T **c.** A—C

18.15. Draw the structures of the following RNA-derived dinucleotides:
a. A—U **b.** G—U **c.** A—C

18.16. Consider the DNA-derived tetranucleotide A—A—T—C. What products will be obtained when this tetranucleotide is hydrolyzed by each of the following?
a. base **b.** base, followed by acid

18.17. Draw the structures of the following RNA components:
a. UUU **b.** UAA **c.** ACA

18.18. Draw a structure showing the hydrogen bonding between uracil and adenine, and compare it with that for thymine and adenine (page 537).

18.19. A segment of DNA contains the following base sequence:

5′ A—A—G—C—T—G—T—A—C 3′

Draw the sequence of the complementary segment, and label its 3′ and 5′ ends.

18.20. For the DNA segment in Problem 18.19, write the *m*RNA complement, and label its 3′ and 5′ ends.

18.21. Consider the following *m*RNA sequence:

5′ A—G—C—U—G—C—U—C—A 3′

Draw the segment of DNA double helix from which this sequence was derived, using the schematic method at the right of Figure 18.4. Be sure to show the 5′ and 3′ ends of each strand.

18.22. Explain how the double-helical structure of DNA is consistent with Chargaff's analyses for the purine and pyrimidine content of DNA samples from various sources.

18.23. The codon CAC corresponds to the amino acid histidine (His). How will this codon appear in the DNA strand from which it was transcribed? in the complement of that strand? Be sure to label the 5′ and 3′ directions.

18.24. Consider Table 18.2. Will any changes occur in the resultant biosynthesized protein by a purine → purine mutation in the third base of a codon? in a pyrimidine → pyrimidine mutation in the third base of a codon? If so, describe the change.

18.25. From Table 18.2, are mutations in the first or second base of a codon more or less serious than mutations at the third base?

18.26. A *m*RNA strand has the sequence

$-^{5'}$CCAUGCAGCAUGCCAAACUAAUUAACUAGC$^{3'}-$

What peptide would be produced? (Don't forget the start and stop codons!)

18.27. What would happen if the first U in the sequence in Problem 18.26 were deleted?

18.28. What peptide would be synthesized from the following DNA sequence?

$^{5'}$TTACCGTCTGCTGCCCCCCAT$^{3'}$

18.29. What products would you expect to obtain from the complete hydrolysis of nicotine adenine dinucleotide (NAD)? See page 545 for its structure.

18.30. UDP-glucose is an activated form of glucose involved in the synthesis of glycogen. It is a nucleotide in which α-D-glucose is esterified at C-1 by the terminal phosphate of uridine diphosphate (UDP). From this description, draw the structure of UDP-glucose.

18.31. **Caffeine,** the alkaloid stimulant in coffee and tea, is a purine with the following formula:

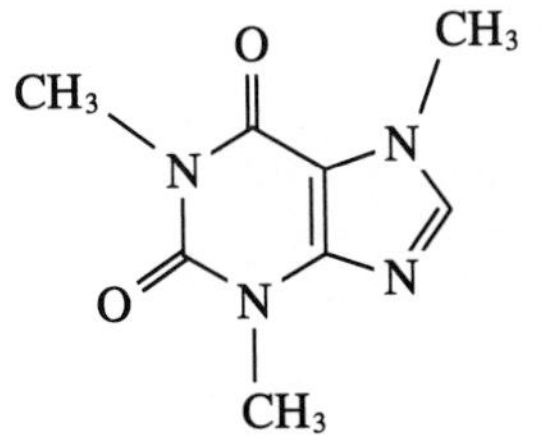

caffeine

Compare its formula with those of adenine and guanine. Do you expect caffeine to form N-glycosides with sugars such as 2-deoxy-D-ribose?

18.32. *5-Fluorouracil-2-deoxyriboside* (FUdR) is used in medicine as an antiviral and antitumor agent. From its name, draw its structure.

18.33. *Psicofuranine* is a nucleoside used in medicine as an antibiotic and antitumor agent. Its structure differs from that of adenosine only in having $—CH_2OH$ attached with α geometry at C-1′. Draw its structure.

Appendix A. Bond energies for the dissociation of selected bonds in the reaction $A—X \rightarrow A\cdot + X\cdot$ (in kcal/mol)

I. Single bonds	*Bond energies (kcal/mol)*								
A—X	*X = H*	*F*	*Cl*	*Br*	*I*	*OH*	*NH_2*	*CH_3*	*CN*
$CH_3—X$	105	108	84	70	57	92	85	90	122
$CH_3CH_2—X$	100	108	80	68	53	94	84	88	
$(CH_3)_2CH—X$	96	107	81	68	54	94	84	86	
$(CH_3)_3C—X$	96		82	68	51	93	82	84	
H—X	104	136	103	88	71	119	107	105	124
X—X	104	38	59	46	36			90	
Ph—X	111	126	96	81	65	111	102	101	
$CH_3C(O)—X$	86	119	81	67	50	106	96	81	
$H_2C{=}CH—X$	106								
$HC{\equiv}C—X$	132								

II. Multiple bonds	*Bond energies (kcal/mol)*
$H_2C{=}CH_2$	163
$HC{\equiv}CH$	230
$H_2C{=}NH$	154
$HC{\equiv}N$	224
$H_2C{=}O$	175
$C{\equiv}O$	257

Appendix B. Bond lengths of selected bonds (in angstroms, Å)

I. Single bonds				II. Double bonds	
Bond	Length (Å)	Bond	Length (Å)	Bond	Length (Å)
H—H	0.74	H—C=	1.08	C=C	1.33
H—F	0.92	H—Ph	1.08	C=O	1.21
H—Cl	1.27	H—C≡	1.06		
H—Br	1.41	C—C	1.54		
H—I	1.61	C—N	1.47		
H—OH	0.96	C—O	1.43		
H—NH_2	1.01	C—F	1.38		
H—CH_3	1.09	C—Cl	1.77		
F—F	1.42	C—Br	1.94		
Cl—Cl	1.98	C—I	2.21		
Br—Br	2.29				
I—I	2.66				

III. Triple Bonds	
Bond	Length (Å)
C≡C	1.20
C≡N	1.16
C≡O	1.13

Index